SCORPIONATES II:
CHELATING BORATE LIGANDS

Dedicated to

Swiatoslaw Trofimenko

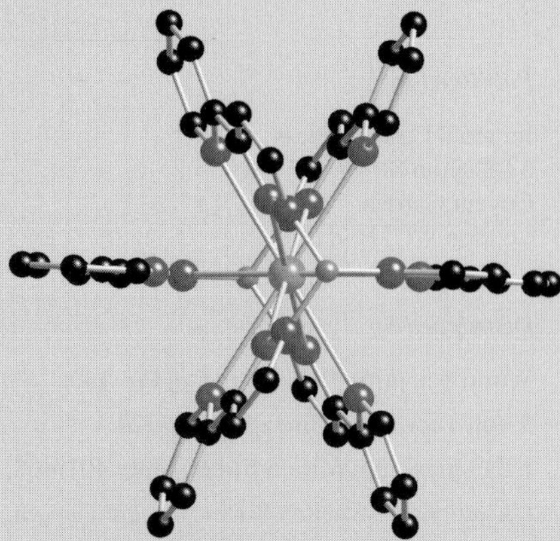

SCORPIONATES II: CHELATING BORATE LIGANDS

Dedicated to

Swiatoslaw Trofimenko

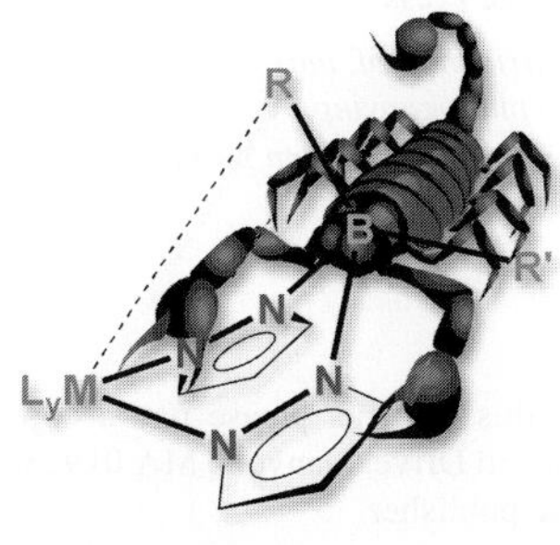

Claudio Pettinari
University of Camerino, Italy

Imperial College Press

Published by

Imperial College Press
57 Shelton Street
Covent Garden
London WC2H 9HE

Distributed by

World Scientific Publishing Co. Pte. Ltd.
5 Toh Tuck Link, Singapore 596224
USA office: 27 Warren Street, Suite 401-402, Hackensack, NJ 07601
UK office: 57 Shelton Street, Covent Garden, London WC2H 9HE

British Library Cataloguing-in-Publication Data
A catalogue record for this book is available from the British Library.

SCORPIONATES II: CHELATING BORATE LIGANDS
Dedicated to Swiatoslaw Trofimenko

ISBN-13 978-1-86094-876-3
ISBN-10 1-86094-876-6

Typeset by Stallion Press
Email: enquiries@stallionpress.com

Printed by FuIsland Offset Printing (S) Pte Ltd, Singapore

Dedicated to Jerry Trofimenko

Foreword

In 1999, Swiatoslaw ("Jerry") Trofimenko published *Scorpionates: The Coordination Chemistry of Polypyrazolyborate Ligands*, a book that comprehensively described an area of chemistry that had, as its genesis, his synthesis of simple pyrazolyboron compounds at DuPont in the 1960s. Shortly thereafter, Jerry realized that these molecules were effective ligands and thereby provided the inspiration for many (including myself) to utilize them in a manifold of ways. For example, polypyrazolylborate ligands have been widely used in bioinorganic chemistry, organometallic chemistry, and materials chemistry.

Not long after the publication of *Scorpionates*, Jerry realized that, because of the intense activity of this area, it would soon be out of date. As such, he planned writing a second book: *Scorpionates–2*. Unfortunately, Jerry's health was failing, and he realized that he would not be able to complete this task. Therefore, Jerry suggested to Claudio Pettinari, an aficionado and long-time practitioner of scorpionate chemistry, that he write it! And Claudio bravely accepted this challenge! By doing so, Claudio has not only provided a great service for coordination chemists, but has also provided us with a long lasting memory of Jerry who unfortunately died before this project came to fruition.

Claudio has done an excellent job and the book is jam-packed with information from more than 1300 papers covering 1999 to 2007. The book follows largely the same style as *Scorpionates*, but Claudio has extended the scope to include recent studies on scorpionate ligands that feature heterocycles other than pyrazoles. Jerry would have been proud of Claudio's achievement. My only question is has Claudio started thinking about a possible *Scorpionates-3*? If not, to whom will he pass on the torch?

Gerard Parkin
Columbia University, USA
August 2007

Jerry Trofimenko: A Testimonial

Jerry Trofimenko was an extraordinary chemist, full of enthusiasm and passion toward chemistry. His ligands, which he liked to call scorpionates, many of them made with his own hands, to be shared with dozens of colleagues around the world, constitute one of the most important ligand families in coordination and organometallic chemistry. Molecular inorganic chemistry would look very different today without the scorpionates, in particular without the tris(pyrazolyl)borates. In both his personal and professional life, Jerry was an example to all of us and we will always remember him with great affection.

Ernesto Carmona
Universidad de Sevilla, Spain

Thanks to Jerry Trofimenko, we do not have to read Biology journals to learn about Scorpions. The people friendly varieties popularized by Jerry are so hard to resist. They are everywhere. The spread of Scorpion(ate)s also benefited by his unique multi-gram production methods. We will miss him.

H. V. Rasika Dias
University of Texas at Arlington, USA

After joining at the late Nobumasa Kitajima's laboratory, I used hydrotris(pyrazolyl)borates, especially $Tp^{^iPr_2}$ and $Tp^{^tBu,^iPr}$ ligands with copper. These alkyl substituted ligands are called as "second generation." By using these ligands, we succeeded in preparations of good structural models in the active site of copper containing proteins such as oxy-hemocyanin, blue copper proteins, monooxygenase, etc. Recently, new scorpionates containing pyrazolyl arms with additional donor atoms have been expanded and these have some potential for new coordination chemistry. Therefore, the "scorpionates" have plunged into the "third generation."

Kiyoshi Fujisawa
University of Tsukuba, Japan

The amazing thing about Jerry was despite the fact that his professional career frequently drew him away from the laboratory, he never stopped promoting poly(pyrazolyl)borate and related chemistry. His introduction of "second generation" ligands gave new life to the field 20 years after he invented it. He would help anyone out that he could with ideas, ides that were always very good, or with actual compounds from his chemical archives. It was amazing how well he knew this entire field of this chemistry.

Dan Reger
University of South Carolina, USA

As is common in our profession, I first learned of Swiatoslaw "Jerry" Trofimenko through his scientific publications. Indeed, his research into the ligands he termed scorpionates served as inspiration for one of my initial scientific projects, on related tris(thioether)borate ligands. In the intervening years I had an occasional opportunity to correspond with Jerry. Then, on February 28, 1997 I met Jerry for the very first time during a seminar visit to the University of Delaware. Jerry was a visiting scholar in Klaus Theopold's laboratory having recently retired from DuPont. The first encounter did not disappoint! As was common for those who met Jerry during his Delaware days, I was presented with a gift — two vials containing metal scorpionate complexes that would be "perfect" for chemistry relevant to work I described during my seminar. He was a modern day Johnny Appleseed determined to spread scorpionates across the worldwide chemical community. When I received an offer to move to the University of Delaware, I was so very excited to become one of Jerry's colleagues. With some frequency Jerry would attend our group meetings, providing additional guidance and support to my students. Over past decade, we shared numerous conversations — about science, culture, politics, and family. Each topic brought a special sense of energy and excitement in Jerry's eye — none more so than we spoke of his family. Jerry was a proud husband, father, and grandfather.

The last few years, Jerry taught us something even more special and important. He taught by example how to live with dignity and courage. It was difficult to see his body deteriorate from the cancer that ultimately claimed his life. His mind never did. Despite the great pain, he never once complained. He took each day as a gift, continued to live his life focusing on his professional and personal pursuits.

Charles G. Riordan
University of Delaware, USA

I had the good fortune of meeting Jerry Trofimenko for the first time at the Xth ICOMC meeting in Toronto in 1981. By that time we have used his "scorpionate" ligands for a while, but to meet the "man" and "talk shop" with him was a real thrill, especially since Jerry had this insatiable thirst for chemistry and for scorpionate chemistry in particular. This chance encounter quickly blossomed into a friendship, kept up mostly by correspondence, letters first and then numerous email messages; Jerry was a very quick correspondent. Jerry visited Edmonton in 1989, as an External Examiner of the PhD thesis for one of Bill Graham's graduate students, and I had the unique pleasure of visiting him at his famous labs at DuPont in 1993. The first hand experience of seeing Jerry's labs and witnessing the drawers full of his various scorpionate ligands was indeed a sight not to be forgotten. The lab tour was followed by a memorable day visiting DuPont and surroundings, a stop at his charming house all the while Jerry regaling me with stories on one subject or another. Of course a visit to Jerry's labs was never complete without a "shopping list" of scorpionate ligands and, over the years, I was very fortunate to benefit from his generosity. On one occasion the research sample arrived in a somewhat unconventional way, a simple envelope addressed to me. Imagine my surprise when I opened the envelope and, instead of a note from Jerry, I found 'white powders' sealed in small plastic bags! Jerry was an unabashed champion and promoter of his scorpionate ligands and was always urging me to try out his most recent inventions. Of course, I could not keep up with Jerry's prolific output but over the years, his scorpionate ligands proved to be very kind to us in the fields of lanthanide and actinide chemistry, and continue to do so. One notable success, that pleased Jerry also, was the use of the Tp^{py} ligand which gave beautiful 12-coordinate icosahedral U(III) and Ln(III) complexes and a cherished joint publication with Jerry, Jon McCleverty, Michael Ward, and Arnie Rheingold.

I am very fortunate to have met, interacted with and befriended Jerry Trofimenko. His clever invention, the polypyrazolyl borate ligands have made a major impact on my scientific career and his friendship has enriched my personal life. Jerry will be greatly missed, but his legacy will live forever in the many old, and new disciples of the scorpionates.

Josef Takats
University of Alberta, Canada

Dr. Swiatoslaw 'Jerry' Trofimenko and I first met around 1996. Little did I know at the time, that a casual lunch among inorganic chemists, arranged by our mutual friend Arnie Rheingold, would lead to a collaboration that was to last for a decade. Back then, Jerry was about to retire from his long-held position at DuPont, but he was clearly not ready to give up chemistry. Indeed, he was looking for a "scientific address" (his words), to continue his work in scorpionate chemistry and to keep engaged with the chemistry community. It was one of those win–win opportunities that come along every once in a while; as a coordination chemist, I was of course aware of Jerry's work, and my students had been using his tris(pyrazolyl)borate ligands for years, but here was the chance to gain the world's authority on these ligands as a colleague and resource! On the spot, I offered Jerry an office and space in my laboratory. From then on, until his passing in 2007, he was a colleague, friend, and valued member of my research group and the Department of Chemistry and Biochemistry at the University of Delaware.

Jerry Trofimenko was a passionate synthetic chemist, and his book *Scorpionates* is a monument to his scientific love and life's crowning achievement. The structural diversity so readily apparent in this second edition is the result of many hours spent in the lab, and of sharing the products and his enthusiasm with all who would express an interest. Many a visitor left Jerry's office with a sample of the latest "designer ligand," and the ubiquity of his molecules is due in part to Jerry's generosity. His collaborators live and work all over the world. Aside from chemistry, Jerry was a kind, accomplished, and wholly unpretentious man. Among the seven languages he spoke, we shared some jokes and poetry in German. At social gatherings, which he and his wife Martha attended together, he might spontaneously take to the piano and play a few songs. He never complained, and came to the lab almost to the very end. I am glad we met.

Klaus H. Theopold
University of Delaware, USA

When I met Jerry for the first time, I had known of him already for a generation, having started my own career in chemistry with boron–nitrogen compounds in 1964. He visited me in Freiburg in the 1990s, having been transferred to a Polish location by Dupont. I do not have to repeat his curriculum vitae or his achievements here. But I have to mention that my own achievements in biomimetic zinc chemistry would not have been possible without his ligands, the pyrazolylborates, which he taught us to prepare and of which he gave us early samples.

It is a shame that a man of so many talents and of so much enthusiasm never really made it into the academic world. That is where he belonged, and not in a business and production oriented environment. He never complained to me about his, which raises my regard for him even further.

Fortunately our community became aware of the value of his contributions early enough. So in these last 10 years he could enjoy a little bit the recognition and the honors which he deserved, while he was already fighting his losing battle with cancer. There should have been more recognition, but that will be compensated by the long time during which chemists will read his name in textbooks and in citations.

I count myself happy to have been a friend of Swiatoslaw Trofimenko.

Heinrich Vahrenkamp
Universität Freiburg, Germany

Acknowledgments

Many thanks to my colleagues at UNICAM University: Augusto Cingolani, Professor of Chemistry, for his trust in my capacities; to Prof. Fabio Marchetti, for reading my book and offering valuable advice and for help with proofreading; to Dr. Corrado Di Nicola and Dr. Riccardo Pettinari for their support in checking relevant literature; to Martha and Zoya Trofimenko for helping me with English and for the book revision; to Prof. Gerard Parkin for his continuous support in many aspects; and to Prof. Ernesto Carmona, Rasika Dias, Kiyoshi Fujisawa, Daniel Rabinovich, Dan Reger, Charles G. Riordan, Josef Takats, Klaus H. Theopold, Heinrich Vahrenkamp, and Mike Ward for some color pictures and for advice on their subjects of specialization. I sincerely appreciate the secretarial assistance of Wanda Tan in the preparation of the manuscript. Special thanks go to my wife, Loredana, for continual encouragement and moral support, and to my son, Lorenzo, for the moments of intense happiness.

Contents

Chapter 1

Introduction

1.1. General Considerations

In 1966 Swiatoslaw Trofimenko, while at the Central Research Department of the DuPont Company, began working on boron–pyrazole chemistry. He reported a new class of ligands, the poly(pyrazolyl)borates, of general structure $[RR'B(pz^x)_2]^-$ (Fig. 1.1) (pz^x = pyrazol-1-yl; R, R′ = H, alkyl, aryl or pz^x) which he later named as scorpionates.

These molecules, combining some features of the cyclopentadienyl and β-diketonate ligands, are extremely versatile because the number of pyrazolyl groups and the substituents thereon or at the B center can be easily modified in order to obtain ligands with different steric and electronic profiles. They have been extensively used to prepare complexes of main and transition group metals and are particularly useful for the synthesis of monomeric derivatives in which the coordination sphere about the metal can be carefully controlled. Since their discovery,[1] a very large number of publications[2–5] has appeared describing also the possible application of the poly(pyrazolyl)borates, or Trofimenko's ligands, in an extraordinarily wide range of chemistry, from modeling the active site of metalloenzymes, through analytical chemistry and organic synthesis, to catalysis and material science.

The poly(pyrazolyl)borates are often defined as spectator ligands, generally not interfering with the reaction scenarios occurring at the

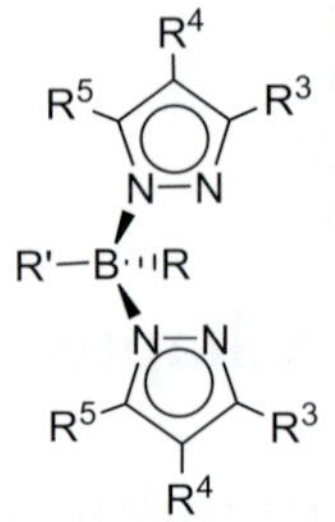

Fig. 1.1. General structure of poly(pyrazolyl)borates.

metal centers. However, it has been demonstrated that several different coordination modes of poly(pyrazolyl)borate ligands are possible, being responsible for the diverse chemical behavior of their compound and a number of examples are known in which they may have a noninnocent participation through hapticity changes. In fact, the term scorpionate has been used to describe the interchange between bidentate and tridentate coordination modes of these ligands. Like the pincers of a scorpion (Fig. 1.2), these versatile tripodal ligands introduced by Trofimenko (Fig. 1.3) bind a metal with nitrogen heteroatoms from two pyrazole rings attached to a central boron atom. The third pyrazole ring (or the R group), attached to the boron, rotates forward like a scorpion's tail to "sting" the metal.

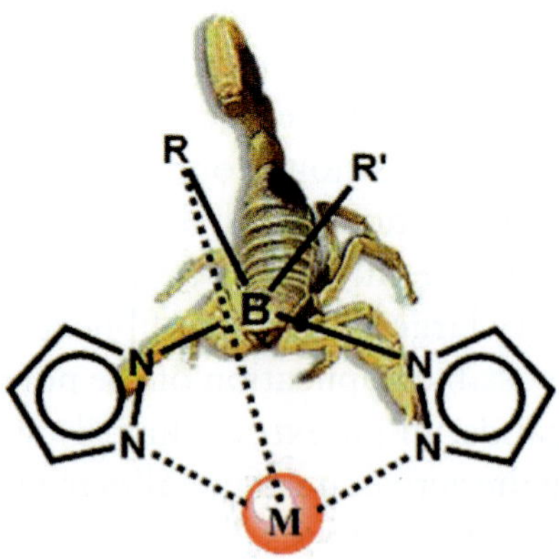

Fig. 1.2. The "scorpion" image used to visualize the poly(pyrazolyl)borates binding.

Fig. 1.3. Trofimenko and Pettinari at Camerino University (September 2004).

The development of the "second generation" of scorpionates,[6,7] bearing bulky substituents (R) such as ^{t}Bu, Ph or Cum on the 3-position of the pyrazoles and also of the "nonpyrazolyl scorpionate" ligands,[8] containing a boron core but a different donor atom from N, resulted in large increase of the articles (>2800) on the chemistry of these ligands, over 200 different scorpionates being known and complexes with *ca.* 70 elements of the periodic table being reported.

In 2003, Trofimenko and his "scorpionate" ligands were guests of honor (Fig. 1.4) at an ACS symposium to celebrate 35 years of chemistry accomplished with the pyrazolylborate and related tripodal ligands.

In 2005, in a mail directed to me, Trofimenko said:

> ... when I wrote the book "Scorpionates" I had in mind to follow, after some years, with another book "Scorpionates-2", which would continue where the first book stopped, i.e. it would contain materials from references beyond the book (which has 1568 references). It would also be an all-inclusive reference work, following the organization of the original book. And so I continued to collect references beyond the book's range, and have now about 500 of them, including

Fig. 1.4. The ACS symposium held in new orleans (March 2003).

> all papers cited in those references. I am sure that I did not catch a number of the newer ones, but I did what I could....
>
> Unfortunately, my health is not the best these days, and I came to the conclusion that planning to write the second book is for me unrealistic...Would you consider writing something like the "Scorpionates-2" book or, perhaps, a large article (since 500 is too few references for a book) in the style of the original book?[9]

I have accepted his proposal and this book would absolve this very difficult task by providing a "full" coverage of the scorpionate chemistry from 1999 to date, so that it could be used as reference source in addition to the first book on "Scorpionates", and also could provide information on the areas more recently investigated and developed (as that of mercaptoimidazolylborates). After a careful analysis of the literature data, I have found more than 1300 papers that reported on scorpionates in the last 8 years and for this reason a book seems the more valid and useful instrument to describe advances in the area. In memory of Trofimenko, I wanted to use the scheme adopted in the first edition, so that the first chapter covers the general features of the ligands, their coordination modes, general and new procedures adopted for their synthesis, the abbreviation system, the reviews published on these molecules describing synthesis, uses and applications of scorpionate derivatives, as well as a list of the new poly(pyrazolyl)borate ligands (also third generation), their synthesis and related metal derivatives. In the second chapter, I will describe the first generation homoscorpionates, i.e. complexes of Tp and Tp* grouped according to metal, whereas in the third chapter I will extend my discussion to more complicated homoscorpionate (second generation) ligands, grouped by ligand.

The fourth chapter deals with heteroscorpionate chemistry, whereas Chaps. 5, 6 and 7 have been dedicated to the chemistry of nonpyrazole based ligands, i.e. scorpionates based on thioimidazolones, thioethers, and diaryl- and dialkyl-phosphines as donors, respectively.

Specific applications of scorpionate ligands are dealt with in Chap. 8 which describes catalytic, enzyme modeling, C–H bond activation and metal extraction studies. Chapter 9 will briefly indicate some emerging areas and possible further developments and perspectives.

1.2. Scorpionate Ligands: Abbreviation System

Following the system proposed by Curtis *et al.*,[10,11] the abbreviation Tp is used for the simplest member of this family, i.e. the hydrotris(pyrazol-1-yl)borate (also indicated as $[HB(pz)_3]^-$) and Tp* for hydrotris(3,5-dimethylpyrazol-1-yl)borate (also indicated as $[HB(3,5\text{-}Me_2pz)_3]^-$), the two most frequently used ligands. Tp^R denotes a substituted tris(pyrazolyl-1-yl)borate. The following rules have been generally employed:

(a) any nonhydrogen substituent in the 3-position of the pz ring is denoted by a superscript. Hydrotris(3-methylpyrazol-1-yl)borate is denoted as Tp^{Me} and hydrotris(3-*t*-butyl-pyrazol-1-yl)borate is written as Tp^{tBu}. When rearrangement of the 3-substituted ligand has taken place and Tp^R transformed in $[HB(3\text{-}Rpz)_2(5\text{-}Rpz)]^-$, the ligand is denoted as Tp^{R*}. Analogously, if a double rearrangement occurs yielding a $[HB(3\text{-}Rpz)(5\text{-}Rpz)_2]^-$, the ligand is denoted with a double asterisk: Tp^{R**}, i.e. the $[HB(3\text{-}Mspz)(5\text{-}Mspz)_2]^-$ is abbreviated Tp^{Ms**};

(b) when there are four identical pyrazolyl groups bound to boron, the ligand will be denoted as pz^0Tp^R, i.e. $(3\text{-Mepz})_4B$ is also indicated pz^0Tp^{Me}. The not substituted tetrakis ligand is denoted pzTp;

(c) boron substituents are written preceding "Tp": for instance, phenyltris(pyrazol-1-yl)borate is PhTp;

(d) the 5-substituent follows the 3-substituent as a superscript, separated by a comma; for instance, $HB(3\text{-Ph-5-Mepz})_3$ is denoted as $Tp^{Ph,Me}$. When the 3 and the 5 substituents are identical, the superscript R-substituent is followed by 2: for instance hydrotris(3,5-diisopropylpyrazol-1-yl)borate is Tp^{iPr_2}. In the case of $HB(3,5\text{-}Me_2pz)_3$ the systematic abbreviation would be Tp^{Me_2} although Tp* continues to be the most used notation;

(e) when the ligand exhibits substituents in the 4-position a 4R superscript is present. $HB(4\text{-}CF_3pz)_3$ is Tp^{4CF_3}, $HB(3\text{-Me-4-Brpz})_3$ is $Tp^{Me,4Br}$. The only exception is when the ligand has both 3 and 5 position already indicated, so that $HB(3,4,5\text{-}Me_3pz)$ ligand is Tp^{*Me};

(f) poly(indazolyl)borates are represented as benzopyrazolylborates, Tp^{Bo}, and the fusion of the benzo ring to pz (3,4- or 4,5) is denoted by a superscript of 3 or 4 preceding "Bo";

(g) heteroscorpionate ligands will be abbreviated as "Bp". In the case of the C- and B-substituted ligands the same rules above indicated must be used. Ph_2Bp, means diphenylbis(pyrazol-1-yl)borate, whereas Bp^{tBu} is bis(3-*tert*-butylpyrazol-1-yl)borate;

(h) a general homoscorpionate ligand with unspecified substituents will be denoted as Tp^x, and a general pyrazolyl group will be pz^x; Bp^x and pz^0Tp^x are general bis- and tetrakis-pyrazolylborate, respectively;

(i) bis- and tris(thioimidazolyl)borates are generally denoted as Bm^x and Tm^x, respectively. Bm^R and Tm^R denote substituted ligands at the 1-positions of the imidazole rings. A methyl substituent in the 1-position can be omitted, so Tm denotes the hydrotris(methimazolyl)borato ligand. In the case of the C- and B-substituted ligands the same rules adopted for Tp ligands must be used, so $PhTm^{Ph}$ indicates the ligand phenyltris(1-phenyl-2-thioxoimidazolyl)borate;

(l) hydrotris(thioxotriazolyl)borate and dihydrobis(thioxotriazoly)borate will be indicated in this book as Tr^{x} and Br^{x}, respectively, in accordance with the author's use; hydrotris (mercaptothiazolyl)borate and hydrotris(mercaptobenzothiazolyl)borate will be indicated as Tz and Tbz, respectively; whereas $HB(tz)_3$ and $H_2B(tz)_2$ will be employed for tris and bis(triazol-yl)borate, respectively.

1.3. Scorpionate (poly(pyrazolyl)borate) Features

Two families of scorpionate ligands (pyrazole-based) may be distinguished:

(a) homoscorpionates, in which the pseudo-axial R group in Fig. 1.1 is another pyrazolyl group (pz^{x}) identical to the two bridging pz^{x} groups. These ligands typically coordinate the metal as tripodal N_3-donors, exhibiting a C_{3v} symmetry. If both R and R′ are pz^{x}, we have still a homoscorpionate;

(b) heteroscorpionates, where the coordinating pseudo-axial R group is anything but pz^{x}. Heteroscorpionates also include ligands where R is another pyrazolyl group (pz^{y}) different from pz^{x}. They can act in either bidentate or tridentate manner, depending on the type of the R and R′ substituents on boron.

Most of poly(pyrazolyl)borate complexes $RR'B(pz^{R})_2M(L)_y$ have a well-defined feature: a six-membered ring $B(\mu\text{-}pz^{R})_2M$, generally in the boat form, with the pseudo-equatorial R′ pointing away from the metal roughly along the B–M axis, and the pseudo-axial R group being directed toward the metal (Fig. 1.5), to interact with it, or simply screen it from other ligands. The R substituent protrudes in space past the metal, enveloping it, and forming a protective pocket of varying size and shape.

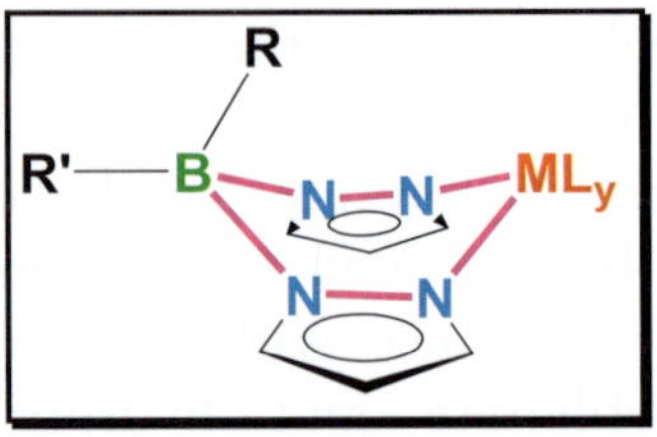

Fig. 1.5. The six-membered ring in poly(pyrazolyl)borate complexes.

The steric effects of the Tp^x ligand, relevant to understanding the metal complexes structure and properties, can be evaluated and quantified by the ligand cone angle (Θ), i.e. the "cone" swept out by the ligand at the metal center. However, this parameter, as also the wedge angle, depends not only on the ligand substituents but also on the length of the N–M bond.

All the Tp^x ligands have a cone angle larger than 180°. The ligands of small cone angle are characterized by a strong tendency to form $[M(Tp^x)_2]$ complexes with divalent first row transition metals, and the inability to form stable $[MX(Tp^x)]$ species (X = halide or R). Ancillary ligands can easily coordinate the metal as the Tp^x cone angle becomes smaller and the wedge angle larger.

Cone angles for several Tp^x ligands (averaged) have been calculated using X-ray crystallography data of selected complexes (a compilation based on first row transition metals),[12,13] connecting the center of the metal atom to the outermost point of the R-group in the 3-position and taking into account its van der Waals radii for all hydrogen atoms.[14] However, the choice of thallium complexes $[Tl(Tp^x)]$, the structures of which have been established by X-ray crystallography,[15–18] as "standard" systems for calculating the cone angles seemed to be the best choice. The values found can be used to establish a relative steric hierarchy for the various Tp^x ligands (selected relevant values are reported in Table 1.1).[19]

A progressive increase in the steric hindrance of Tp^x ligands influences the depth of the hydrophobic pocket around the metal as shown in Fig. 1.6.

Table 1.1. Cone and Wedge angles of the common Tp^x ligands.

Complex	Cone angle	Wedge angle
Tp	183	70
Tp^{Cpr}	223	68
Tp^{Cbu}	234	51
Tp^{Br_3}	234	60
$Tp^{4Bo,3Me}$	235	68
$Tp^{(CF_3)_2}$	237	49
Tp*	239	67
$Tp^{iPr,4Br}$	243	28
$Tp^{tBu,Me}$	243	31
Tp^{tBu}	251	29
Tp^{Cpe}	253	46
$Tp^{Cy,4Br}$	273	46
$Tp^{3Bo,7tBu}$	277	33
Tp^{Cy}	281	53

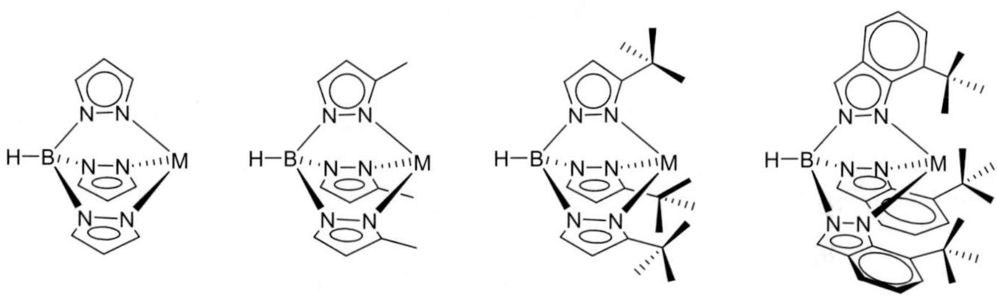

Fig. 1.6. The depth of the hydrophobic pocket around the metal increases with increasing in steric hindrance of Tp^x.

Some ligands exhibit unique features: for example Tp^{Ms}, having a phenyl group essentially orthogonal to the pyrazolyl plane and a minimal amount of rotational freedom, presents the smallest wedge angle between the Tp^x ligands. This creates a differently shaped pocket around the coordinated metal, compared with hitherto known Tp^{Ar} ligands (Fig. 1.7).[20]

On the other hand, the tethering of the 3-phenyl group leads to an increase or decrease of the cone angle of the ligand, depending on

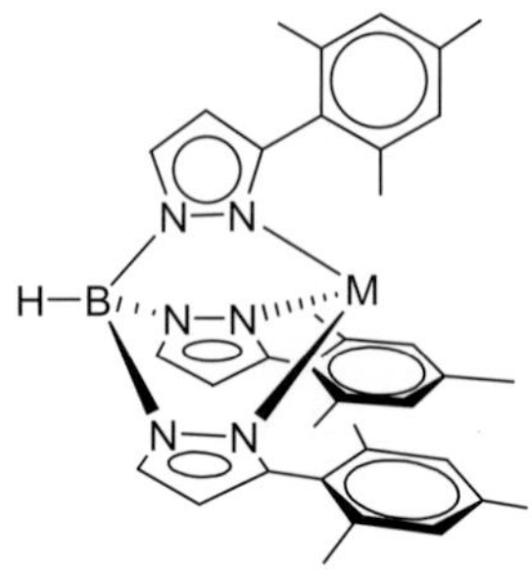

Fig. 1.7. Tp^{Ms} creates a shaped pocket around the metal.

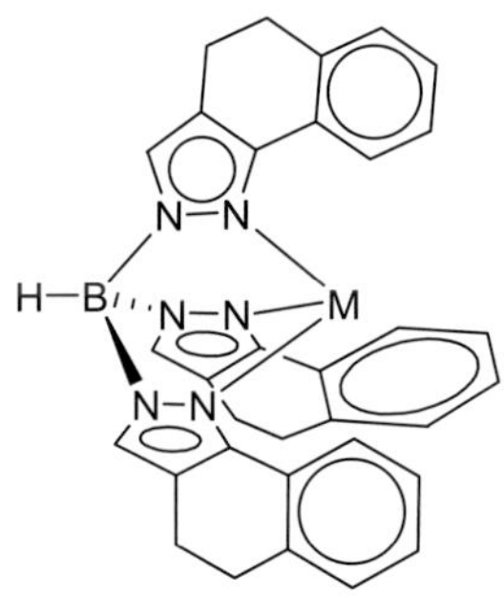

Fig. 1.8. The tethering of the 3-phenyl group with the CH_2CH_2 chain leads to a significant increase of the cone angle.

the length of the tether, and to an increase of the wedge angle, as compared with an untethered 3-phenyl group. In the case of the short tether, $-CH_2-$, the wedge angle is 82.4° close to that of the parent ligand Tp. With the longer zig-zagging tether, $-CH_2CH_2-$ (Fig. 1.8), the cone angle is very large, especially when a 5-Me group is present on the pyrazolyl ring, approaching that of the 3-^{t}Bu group, the wedge angle still being relatively open (~44°). The coordination chemistry of these ligands reflects such spatial restrictions. All the tethered ligands display significant tendency to form octahedral $[M(Tp^x)_2]$ complexes, while the formation and stability of half-sandwich complexes is inverse to the tightness of the tether.[21]

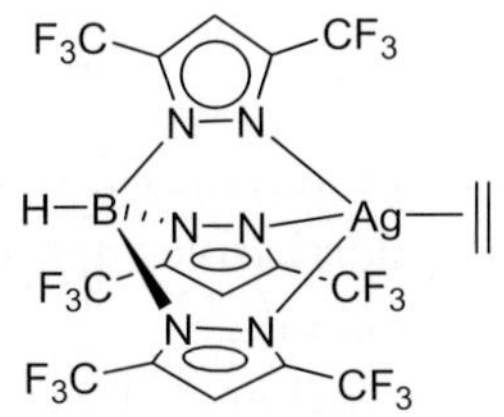

Fig. 1.9. $Tp^{(CF_3)_2}$ is a protective "Teflon coated" container for the $Ag(C_2H_4)$ fragment.

The highly fluorinated $Tp^{(CF_3)_2}$ ligand (Fig. 1.9) essentially serves as a protective "Teflon coated" container for M(I)–L (L = CO, C_2H_4) fragments.[22,23]

Some Bp and Tp ligands with expanded denticity via potential oxygen donors on the 3-phenyl substituent were able to coordinate through the *o*-methoxy substituent. These ligands may be particularly useful in lanthanide and actinide coordination chemistry.[24]

The ligand hydrotris[3-(2′-thienyl)pyrazol-l-yl)]borate (= Tp^{Tn}) is reported to have the second-lowest steric hindrance among the known poly(pyrazolyl)borates, also if *a priori* tris(pyrazolyl)borates with the 3-(2′-thienyl) substituent were expected to fit between those with 3-phenyl and those with 3-methyl substituents in terms of steric hindrance. Tp^{Tn} forms octahedral $[M(Tp^{Tn})_2]$ complexes with first-row transition-metal ions, but fails to yield stable $[MX(Tp^{Tn})]$ species, except with Zn(II).[12]

Comparison of the structural, physical, and spectroscopic properties of similar compounds with homologous ligands allowed to assign the relative electron-donating or -withdrawing capabilities of Tp^x ligands. Tp and Tp* ligands have been compared with their formally iso-electronic analogous Cp and Cp*. In particular $M(Cp^x)(CO)_3$ and $M(Tp^x)(CO)_3$ radicals and anions, where M is Cr, Mo, and W, have been employed to derive a series of electrodonating ability.[25–27] Stereoelectronic effect differences have been inferred from IR υ(CO) values and oxidation potentials for the carbonyl anions.[28] Tp^x ligands seems stronger electron donors and the following trend in ligand electron-releasing capability: $Tp^* > Tp \approx Cp^* > Cp$ has been proposed.

However, conflicting reports can be found in the literature on the electron-donating properties of Tp and Tp* in comparison to Cp and Cp*. An investigation on C–H activation reactions by [Ir(Tp*)] complexes suggest that [Ir(Tp*)] derivatives are less electron-rich than the corresponding [Ir(Cp*)] complexes.[29]

A report by Tellers and coworkers[30] has summarized data useful in comparing the electron density at Tp^x- and Cp^x-bearing metal centers. A consistent trend across the periodic table in the relative electron-donating abilities of these two important ligand classes is clearly lacking. Instead, the electron donor ability varies with the identity of the metal, its oxidation state, and the other ligands of the complex. These differences are undoubtedly the result of differences in the bonding nature of Tp^x and Cp^x: Tp^x is a weak-field ligand possessing relatively hard nitrogen σ donors, while Cp^x is relatively soft and capable of π donation. Tp^x also strictly enforces an octahedral geometry about the metal center, in contrast to Cp^x.

Kitajima and Tolman[31] found that the relative electron-donating or electron-releasing properties of the more hindered Tp^x ligands can be gained by comparing υ(CO) data for sets of metal–carbonyl complexes that differ only in their pyrazolyl ring substituents as $[Cu(Tp^x)(CO)]$,[32–35] $[Rh(\eta^2\text{-}Tp^x)(CO)_2]$,[21] and $[Mo(Tp^x)(CO)_2(NO)]$.[7,36,37] Higher values for the carbonyl stretching frequencies reflect lower electron density at the metal center and decreased electron donation by the Tp^x ligand for compounds within each set. The following trend of electron-donating ability of Tp^x ligands has been derived:

$$\begin{aligned} Tp^{R_2}(R = \text{alkyl}) &> Tp^{tBu} \approx Tp^{iPr} \approx Tp^{Me} \approx Tp^{Ms} > Tp > Tp^{Ph_2} \\ &\approx Tp^{Ph} > Tp^{CF_3,Tn} > Tp^{iPr,4Br} > Tp^{CF_3,Me} \end{aligned}$$

in which the order is generally expected on the basis of the electronic effects due to the substituents.

Database analysis and molecular mechanics were used to determine the conformational flexibility of tridentate scorpionate ligands. It has been reported that tris(pyrazolyl)borate ligands act like molecular vises, opening their tripodal structure for larger metals and closing around smaller metal ions. The Tp^x ligand conformation flexibility has been compared with that of the isosteric and isoelectronic neutral tris(pyrazolyl)methanes. Tris(pyrazolyl)methanes and tris(pyrazolyl)borates have similar conformational flexibilities. It has been found that placing sterically hindered groups on the central carbon or boron has only a minor effect on the geometry of the tris(pyrazolyl)methanes and tris(pyrazolyl)borates. However, it does influence the flexibility of the ligands (Table 1.2), particularly when they have to open far from their ideal geometry.[38]

Some ligands, such as 1,5-cyclooctanediylbis(pyrazol-1-yl)borate (Fig. 1.10), are very suitable for the study of C–H–M interaction, due

Table 1.2. Selected parameters (A, deg) defining the flexibility of all RTp ligands in a data set of $RB(pz)_3$ complexes with any group at positions 3, 4 and 5 of the pyrazolyl ligands, including hydrogen (RTp^x) or only hydrogen (RTp) (adapted from Ref. 38).

Data	Shortest values in RTp^x [39]	Longest value in RTp^x [40]	Average value in RTp^x
M–N	1.89	2.80	2.22
Bite size	2.73	2.97	2.90
N–M–N	94.3	66.5	82.2
	Shortest values in RTp[41]	Longest value in RTp[42]	Average value in RTp
M–N	1.78	2.93	2.22
Bite size	2.73	3.17	2.95
N–M–N	100	67.3	84.0

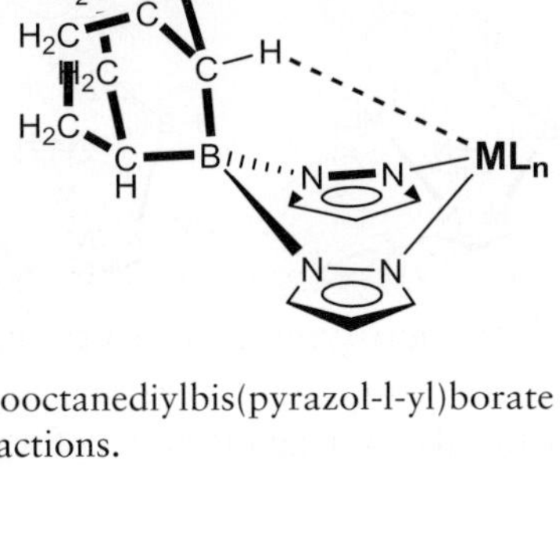

Fig. 1.10. The 1,5-cyclooctanediylbis(pyrazol-l-yl)borate ligand gives complexes showing C–H···M interactions.

to the presence of a single bridgehead hydrogen well-directed at the coordinated metal.[43]

1.4. Scorpionate Coordination Modes

Tris(pyrazolyl)borates, Tp^x, generally coordinate as tridentate κ^3-N,N',N''-ligands through three nitrogen atoms of the pyrazole rings (Fig. 1.11(a)), providing effective steric shielding of the metal center. Besides the very common κ^3-N,N',N''-coordination mode, the tridentate κ^3-N,N',B–H type (Fig. 1.11(b)),[18] the bidentate κ^2-N,N' (Fig. 1.11(c))[44] and κ^2-$N,B-H$ coordination (Fig. 1.11(d))[45,46] have been also reported. In contrast to the relatively easy formation of

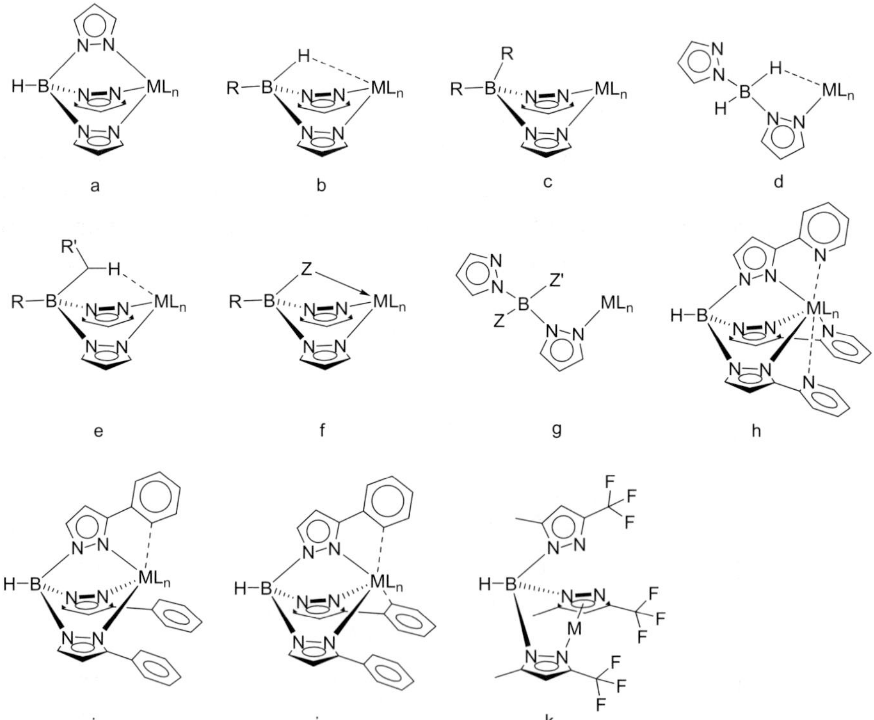

Fig. 1.11. Selected coordination modes of scorpionates.

agostic B–H···M bonds with Bp^x ligands, the formation of agostic B–C–H···M (Fig. 1.11(e)) bonds occurs only in limited cases. Ligand of type $R(Z)Bp^x$ are generally tridentate, due to the presence of a heteroatom Z which can coordinate to the metal (Fig. 1.11(f)). Lower denticity, such as κ^1-*N*, has been also reported (Fig. 1.11(g)).[46–48] The first ionic complex containing a κ^0-Tp^x ligand was reported by Carmona and coworkers.[49] The stepwise change in the denticity of poly(pyrazolyl)borate ligands from κ^3 to κ^0 may have important implications in catalytic uses of Tp metal complexes. Expansion of Tp^x denticity beyond κ^3 can occur by way of the 3-R substituent containing additional donor atoms, as in the hexadentate Tp^{py} (Fig. 1.11(h)).[50,51] On the other hand, tetradenticity has been reported in Tp^x ligands where R does not contain donor atoms, either by way of agostic bonding,[52] or through cyclometalation taking place at one of the aliphatic R groups per ligand (Fig. 1.11(i)).[53] Tp^{Ph} can also coordinate in a κ^5-fashion (Fig. 1.11(j)).[54] A π interaction between a poly(pyrazolyl)borate ligand and a metal center (η^5-pz) has been reported in the complex $[\{(KTp^{CF_3,Me})_2(CuCO_3)\}_2]$ (Fig. 1.11(k))[55] in which one of the coordination hemispheres of the potassium ion is occupied by two carbonate oxygen atoms, one N atom of the $Tp^{CF_3,Me}$ ligand, and three fluorine atoms. On the opposite side, the potassium ion is located directly above the center of a pyrazole ring, thus allowing a comparison with the η^5-cyclopentadienyl coordination in polymeric K(Cp).[56]

1.5. Reviews

Trofimenko reviewed his work on poly(pyrazolyl)borate chemistry and pyrazole-derived ligands in 1971[57] and 1972.[58] In the 1972 contribution Tp^x ligands were contained as a separate category. In 1986[59] Trofimenko reported a summary on the coordination chemistry of poly(pyrazolyl)borates. Its review in 1993[2] covers the 1984–1993 period and contains the systematic abbreviation system that we employed in this book and which could be adapted to most types of substitution on the Tp ligand, greatly facilitating writing chemical formulas for Tp^x complexes. Shaver also commented the Tp^x

coordination chemistry.[60] McCleverty, who carved out large specialty research areas involving Tp* complexes of low-valent Mo and W, presented an overview of his work in 1983.[61]

In 1988, Niedenzu reviewed the pyrazaboles,[62] whereas in 1992, Canty and coworkers described the Pd- and Pt–Tp^x coordination chemistry.[63] In 1994, Kitajima and Moro-oka presented many copper–dioxygen complexes of Tp^x,[64] whereas the lanthanides and actinides complexes were exhaustively investigated and described by Santos and Marques.[65] Parkin, author of a review of Tp-based models for carbonic anhydrase[66] and of a discussion on the effect of Tp^x-ligation on Grignard reagents,[67] reported also metal hydroxides, hydrides, and organometallics derived from hindered poly(pyrazolyl)borate ligands.[68] The organometallic and bioinorganic chemistry of hindered Tp^x ligands, developed by Kitajima and Tolman, was reviewed in 1995.[31] A year later Etienne presented a review on the coordination chemistry of Tp ligands with V, Nb, and Ta,[69] whereas Reger reported Ga and In derivatives of Bp^x and Tp^x.[70] In 1997[19] and 1998,[71] Janiak reviewed Tp^x–Tl derivatives. In 1998 McCleverty and Ward described the use of scorpionate ligands to form a variety of bridged polynuclear complexes of Mo.[72] Some Tp-based Mo and W pterine enzyme models were discussed in a review of Young and Wedd.[73] Cobalt scorpionate complexes were reviewed by Theopold *et al.*[74]

The book *Scorpionates: The Coordination Chemistry of Polypyrazolylborate Ligands*, organized according to the ligand type and to the coordinated metal, is the reference work for the papers published on poly(pyrazolyl)borates until April 1999.[75] A review on this book by Templeton appeared in *Journal of American Chemical Society* in 2000.[76] More recently, Trofimenko described also the most important features of poly(pyrazolyl)borates, from their synthesis and naming to their main applications.[77]

The scorpionate ligands and their father were guests of honor at a symposium to celebrate 35 years of chemistry with Tp^x and related ligands.[78] *Polyhedron* also dedicated a special issue to this topic, the first paper being presented by Trofimenko on development of Tp^x ligand system from its inception.[79]

A general review on organometallic complexes of Tp^x spanning from nontransition metals to copper, silver, and gold species has been written by Sadimenko.[80] A chapter dealing with the coordination chemistry of poly(pyrazolyl)borate and scorpionate ligands has been inserted in both editions of *Comprehensive Coordination Chemistry*.[81,82] An overview of progress reached with scorpionate ligands was given by Pettinari in 2004.[83] Gold, silver, and copper Tp^x complexes, catalysts for carbene and insertion reactions have been described by Pérez.[84,85] Non-Cp ancillaries in organogroup III metal chemistry including Tp^x have been reviewed by Piers and Emslie.[86] Studies on transition-metal complexes supported by Tp^x are reviewed, with emphasis on dioxygen complexes, by Akita *et al.*[87] who indicated in a subsequent paper the perspectives opened by Tp^x in Co–O_2 chemistry.[88] A review on copper–dioxygen complexes also focuses on structure and spectroscopic properties of $Cu(Tp^x)/O_2$ species.[89] In a subsequent review Akita and Hikichi examined not only the inorganic but also the organometallic chemistry of oxo, hydroxo, peroxo, alkylperoxo, and hydroperoxo species of Mn, Co, Fe, Ni, and Pd.[90]

Coordinatively unsaturated organometallic systems based on Tp^x ligands, in particular $[M(Tp^x)R]$ and $[(Tp^x)MM'L]$ complexes ($M = Fe, Co, Ni$; $M' = Ru, Co$), have been fully analyzed by Akita.[91]

Fe/Cu metalloenzymes, also based on Tp and their NO_x redox chemistry, is described by Wasser *et al.*[92] Information on zinc poly(pyrazolyl)borate chemistry related to zinc enzymes is provided by Vahrenkamp in 1999.[93] Bioinorganic zinc chemistry explored through synthetic analogs of zinc enzymes containing tripodal Tp^x ligands is well condensed in a feature article[94] of Parkin that in 2001 in a chapter for *Metal Ions in Biological Systems* reported a detailed investigation on synthetic analogs that mimic function and structure of active sites of zinc enzymes.[95] The same author summarized in 2004 all synthetic analogs relevant to the structure and function of zinc enzymes.[96] In the same year Edelmann[97] pointed out new developments in the coordination chemistry of Tp^x, showing that these ligands can be more versatile than reported by several authors. More recently some $[Cu^{II}(Tp^x)]$ complexes have been reported as models for dioxygen activation in dopamine β-monooxygenase and peptidylglycine α-hydroxylating monooxygenase.[98] A personal perspective describing the use of $[Cu^{II}(Tp^x)]$ systems to understand copper protein active sites has been presented by Tolman.[99] Copper and silver derivatives of scorpionates were also reviewed by Pettinari *et al.* in 2004.[100]

An overview of fluorinated tris(pyrazolyl)borate ligands and their use in the synthesis of silver adducts containing group 14 ligands has been presented by Dias and Fianchini.[101]

Carmona along with other researchers reported rhodium and iridium tris(pyrazoyl)borate chemistry giving great emphasis to C–H activation.[102] A perspective study on the generation and reactivity of sterically hindered iridium carbene containing Tp^x ligands based on a presentation given at Dalton Discussion was published by the Carmona's group in 2003.[103] Iridium carboxycarbene complexes were fully presented in 2005.[104] Ru complexes containing Tp^x have been reviewed by Kirchner in 1999,[105] in a paper referred to as being in print in the Trofimenko *Scorpionates* book. The structural features, acidity, and chemical properties of dihydrogen/dihydride complexes of group VIII metals were described by Jia and Lau.[106]

In a review on non-Cp type homogeneous catalytic systems for olefin polymerization, complexes based on Tp^x–Sc,–Hf,–Ti,–Zr,–Y have been also described.[107]

The lanthanides chemistry using Tp^x ligands has been reviewed by Takats and coworkers.[108] Uranium complexes containing multidentate ligands, including also scorpionates family have been reported by Sessler *et al.*[109]

A general review on rhenium complexes containing poly(pyrazolyl)borates appeared in 1998.[110] In 2004, an overview of the progress involving group VII elements and actinide was presented by Santos and coworkers,[111] who very recently described also rhenium and technetium complexes anchored by scorpionates prepared for radiopharmaceutical application.[112] Keane and Harman described exhaustively the coordination chemistry of a new generation of π-basic dearomatization agents {M(Tp)(L)(π-acid)}.[113] Jones *et al.*[114] reported a review on the role of unstable alkane transition-metal complexes, including also Tp^x derivatives, in the processes of catalytic C–H activation, whereas Norris and Templeton more specifically described the hydrocarbon C–H activation with $[Pt(Tp^x)]$ complexes.[115]

In a review on the π-bonding and the lone pair effect in multiple bonds between heavier main group elements, Power described the structure of $[Ga(Tp^{tBu2})S]$.[116]

$[M(Tp)(CN)_x]$ complexes were extensively discussed in a report on the design of single chain magnets through cyanide bearing six-coordinate complexes.[117]

Bp^{py} and Tp^{py}, multinucleating ligands, and their coordination and supramolecular chemistry have been reported by Ward *et al.*[118]

A detailed discussion on spin crossover in poly(pyrazolyl)borate iron and cobalt complexes has been performed by Long *et al.*[119]

Tris(pyrazolyl)borates have been reported also in two chapters on B–H[120] and X–H[121] activation (X = Si, Ge, Sn) in the book *Metal Dihydrogen and σ-Bond Complexes*, edited by Kubas.

Metric parameters for the $[X]^+[Mo(Tp^x)(CO)_3]$ and $[Mo(Tp^x)(allyl)(CO)_3]$ complexes have been reviewed by Wołowiec and coworkers.[122]

An overview of the chemistry of the divalent lanthanide hydride $[Yb(\mu\text{-}H)(Tp^{tBu,Me})]_2$ is reported by Ferrence and Takats.[123]

Templeton reviewed a number of transition-metal η^2-vinyl complexes also containing Tp^x ligands,[124] whereas Hall reported DFT calculations suitable for designing and testing alternative ligand schemes for transition iridium organometallic complexes, one chapter of the review being dedicated to Tp^x ligands.[125] Pettinari contributed also some chapters dedicated to Tp^x–organotin chemistry in two reviews on organotin compounds.[126,127]

A brief description of the third generation of Tp^x is given by Fujisawa and Okamoto.[128]

1.6. Synthesis of Scorpionate Ligands

Poly(pyrazolyl)borate ligands can be prepared, through a more general reaction, by heating BH_4^- in molten Hpz^x (Scheme 1.1). A large

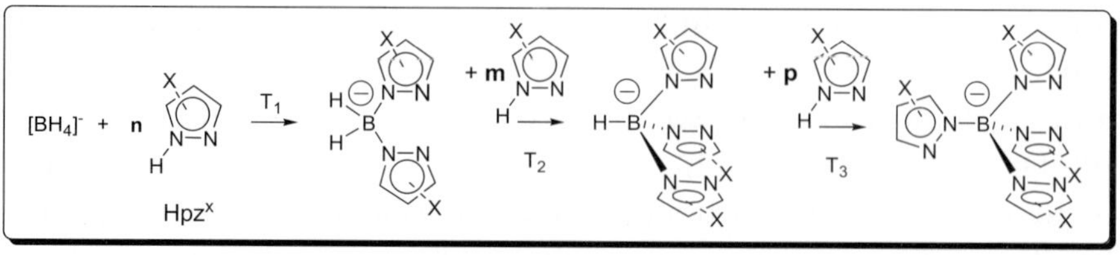

Ligand	Conditions	$n+m+p$	T
Bp^x	Reaction of substituted pz^x and MBH_4 in refluxed anhydrous DMF (M = Li, Na or K). Recrystallization from aromatic high-boiling solvents.	2.3	120–130°C
Tp^x	Hpz^x liquid: Neat-reaction with distillation of the excess pyrazole to prevent $pzTp^x$ formation. Required emanating hydrogen measurement.	4	170–190°C
Tp^{3R}	Hpz^{3R}: reflux, with rapid stirring in 3 or 4-methylanisole.	3.5	Boiling T
$Tp^{3R,5R}$	$Hpz^{3R,5R}$: Neat reaction with sublimation of the excess pyrazole. No $pzTp^x$ formation.	6–8	180–220°C
$pzTp^x$	Limited to 5-unsusbstituted Hpz^x: No solvent required, distillation or sublimation of Hpz^x. Possible decomposition.	6	>180°C

Scheme 1.1. Synthesis of poly(pyrazol-1-yl)borates.

variety of 1-H pyrazoles may be employed to synthesize the ligands by this route, with the exception of those containing functionalities incompatible with the borohydride ion. With asymmetric pyrazoles, such as 3-R-pyrazoles, this reaction proceeds with boron being bonded to the least hindered nitrogen atom. The only exceptions are indazole (benzopyrazole) and indazoles with substituents at positions other than 7, which bond through the more hindered nitrogen atom.[129,130]

The reaction of MBH_4 with Hpz^x can be used, through careful temperature control, to yield bis-, tris-, and in the case of 5-unsubstituted pyrazoles, tetrakis(pyrazolyl)borates. Synthesis of the parent ligands Bp, Tp, and pzTp has been detailed by Trofimenko.[131]

Boron halides, hydrides, arylboronic acids and corresponding esters can be employed as boron sources, when specific R on the boron atoms are requested. For example, the boron-substituted RTp^x ($RB(pz^x)_3$) may be synthesized according to the following reaction (1.1) by using pyrazolate ion and excess pyrazole:

$$RBX_2 + 3[pz^x]^- \rightarrow RB(pz^x)_3 + 2X^- \tag{1.1}$$

Janiak *et al.* found that dibromo(methyl)borane $MeBBr_2$ reacts with Hpz^x at room temperature in the presence of NEt_3 and Tl(OEt) to form $[Tl(MeTp^x)]$ in high yield.[132] For the synthesis of $R_2B(pz^x)$ species, sufficient pyrazolate ion is required to be present to convert $[R_3B(pz^x)]$ quickly to $[R_2B(pz^x)_2]$, otherwise the $R_2B(pz^x)$ species will dimerize to the stable pyrazabole $R_2B(\mu\text{-}pz^x)_2BR_2$.

1.7. Regiochemistry

In the synthesis of Tp^x ligands using 3(5)-monosubstituted pyrazoles or pyrazoles whose anions are not of C_{2v} symmetry, the following products can be formed: Tp^{3R}, Tp^{5R}, and $[HB(pz^{3R})_n(pz^{5R})_{3\text{-}n}]$. Usually, larger substituents end up in the pyrazolyl ring 3-position, relatively distant from the B–N bond. This tendency is pronounced when the steric differences between R(3) and R(5) are large. This regioselectivity is found in the reaction of 3(5)-Mepz with $[BH_4]^-$, leading to Bp^{Me},[133] and to Tp^{Me},[134] as well as in the regiospecific synthesis of Tp^{Ph},[6] and Tp^{tBu}.[7] On the other hand, a mixture of Tp^R and Tp^{5R} isomers can be obtained when the steric differences between the pyrazolyl rings substituent in the 3-position and in the 5-position are less substantial. Reaction of $Hpz^{iPr,Me}$ with $[BH_4]^-$ resulted in the formation of $Tp^{iPr,Me}$ and its isomer $[HB(pz^{iPr,Me})_2(pz^{Me,iPr})]$ (= $Tp^{iPr,Me*}$; an asterisk (∗) indicates an isomeric species) in an approximate 4:1 ratio.[135] The regioselectivity can be explained with the fact that boron–nitrogen bond formation, involving a concerted loss of hydrogen, proceeds through a less sterically encumbered transition state, when bonding occurs to the less hindered N(1) rather than to N(2). The regiochemistry of homoscorpionate ligands prepared from indazoles can be summarized as follows:

1. Indazole itself and indazoles containing alkyl or aryl substituents in positions 3, 4, 5, and 6 form ligands where boron is bonded to the more hindered N-1 which represent a special class of homoscorpionate ligands containing a protective pocket.
2. Indazoles containing both 3- and 7-substituents (including a fused 6,7-benzo ring) do not yield regiochemically pure ligands, although bonding to the less hindered N-2 is still preferred.

The electronic effects can also play an important role: the syntheses of $Tp^{CF_3,Me}$,[136] and $Tp^{CF_3,Tn}$,[137] in which electronically quite different substituents are present, are highly regioselective with the larger CF_3 group residing exclusively in the 3-position. The preference of an electron-withdrawing group for the 3-position in Tp^x ligands is likely due to the inductive electron-withdrawing effect of the CF_3 group, which makes the distal nitrogen more basic, favoring tautomer with the CF_3 group in the 3-position.[138] The relative influence of steric and electronic effect on the course of ligand synthesis is evident from the preparation of Tp^{Ms} (Hpz^{Ms} = 3-mesitylpyrazole)[20] in which the asymmetric isomer [HB(3-Rpz)$_2$(5-Rpz)] (= Tp^{Ms^*})[139] was the major product. A similar situation occurs, for instance, in the synthesis of Tp^{Np}[140] and Tp^{Ant}.[15]

Sometimes a regiochemically pure ligand Tp^R undergoes rearrangement to Tp^{R^*} during the course of complex formation. Tp^{iPr} and $Tp^{iPr,4Br}$ having large substituent in the 3-position which prevent [M(Tp^R)$_2$] formation, yields octahedral cobalt(II) complexes containing, upon a borotropic migration of the original ligands, Tp^{iPr^*} and $Tp^{iPr,4Br^*}$, respectively.[36] [Ir(Tp^{Me})(COD)] has been found to rearrange in solution first to [Ir(Tp^{Me^*})(COD)] and then, on heating, to [Ir{HB(pz^{5Me})$_2$(pz^{3Me})}(COD)].[141]

1.8. New Ligands

1.8.1. Tp^{CHPh_2}

The ligand hydrotris[3-(diphenylmethyl)pyrazol-1-yl]borate, Tp^{CHPh_2}, has been prepared by Trofimenko and coworkers[12] and its coordination chemistry compared with that of Tp^{iPr}. The complexes $[MX(Tp^{CHPh_2})]$ (M = Co, Ni, Zn; X = Cl, NCO, NCS), $[Pd(Tp^{CHPh_2})(\eta^3\text{-methallyl})]$, $[Co(Tp^{CHPh_2})(acac)]$, and $[Co(Tp^{CHPh_2})(Tp^x)]$ have been synthesized and characterized. $[Tl(Tp^{CHPh_2})]$, $[CoCl(Tp^{CHPh_2})]$, $[Co(Tp^{CHPh_2})(NCS)(DMF)]$, $[Ni(Tp^{CHPh_2})(NCS)(DMF)_2]$, $[Co(Tp^{CHPh_2})(acac)]$, $[Co(Tp^{CHPh_2})(Ph_2Bp)]$, $[Co(Tp^{CHPh_2})(Bp^{Ph})]$, $[Co(Tp^{CHPh_2})(Tp)]$, and $[Ni(Tp^{CHPh_2})]_2[C_2O_4](H_2O)_2$ were also structurally characterized.[142]

1.8.2. Tp^{cpd}

Trofimenko and coworkers[143] also reported the ligand hydrotris[3-(carboxypyrrolidido)pyrazol-1-yl]borate Tp^{cpd} (the Tl salt has been investigated by X-ray single-crystal study), and its coordination chemistry toward La, Nd, and Sm salts. $[Sm(Tp^{cpd})_2]PF_6$, structurally characterized, contains a lanthanide ion in a 10-coordinate environment consisting of one N_3O_3 hexadentate Tp^{cpd} ligand (Fig. 1.12), and one N_2O_2 tetradentate Tp^{cpd} ligand.[143]

1.8.3. Tp^{4tBu}, $Tp^{Tol,4tBu}$, and $Tp^{iPr,4tBu}$

Homoscorpionate ligands containing a *tert*-butyl group as substituent in the 4-position (Tp^{4tBu}), ($Tp^{Tol,4tBu}$), and ($Tp^{iPr,4tBu}$) were synthesized with the aim to investigate the restriction of rotation of the 3-R group through "remote control". While the 4-*tert*-butyl group does not alter the coordination chemistry of Tp^{4tBu} and $Tp^{Tol,4tBu}$, the coordination chemistry of $Tp^{iPr,4tBu}$ is different with respect to that of Tp^{iPr}, $Tp^{iPr,4tBu}$ acting as a "tetrahedral enforcer" (Fig. 1.13). $[Co(Tp^{4tBu})(Tp^{Np})]$, $[Tl(Tp^{Tol,4tBu})]$,

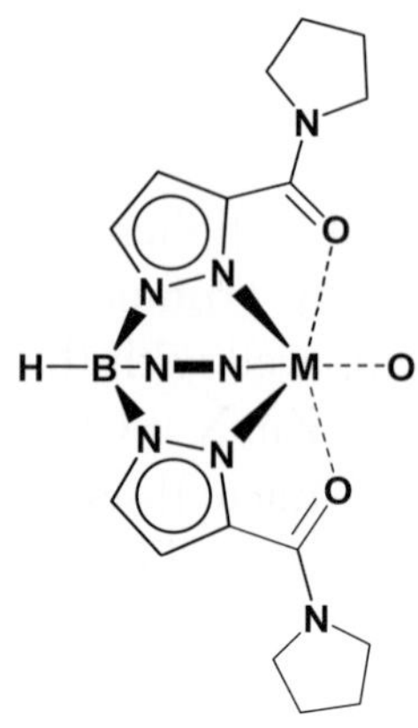

Fig. 1.12. The hexadentate N_3O_3 environment provided by the Tp^{cpd} ligand.

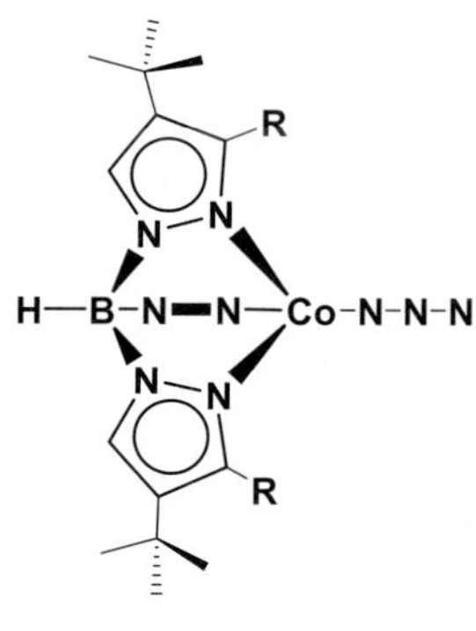

Fig. 1.13. $[Co(Tp^{iPr,4tBu})(N_3)]$.

$[Rh(Tp^{Tol,4tBu})(CO)_2]$, $[CoCl(Tp^{iPr,4tBu})]$, $[Co(Tp^{iPr,4tBu})(N_3)]$, and $[Co(Tp^{iPr,4tBu})(NCS)]$ have been structurally characterized.[144]

1.8.4. $Tp^{(2,4(OMe)2Ph)}$ and $Bp^{(2,4(OMe)2Ph)}$

To expand the scorpionate ligand denticity beyond the number of pz rings, Trofimenko and coworkers[24] reported two ligands, hydrotris[3-(2,4-dimethoxyphenyl)pyrazol-1-yl]borate $Tp^{(2,4(OMe)_2Ph)}$ and dihydrobis [3-(2,4-dimethoxyphenyl)pyrazol-1-yl]borate $Bp^{(2,4(OMe)_2Ph)}$, able to coordinate through both pyrazolyl nitrogens and *o*-methoxy

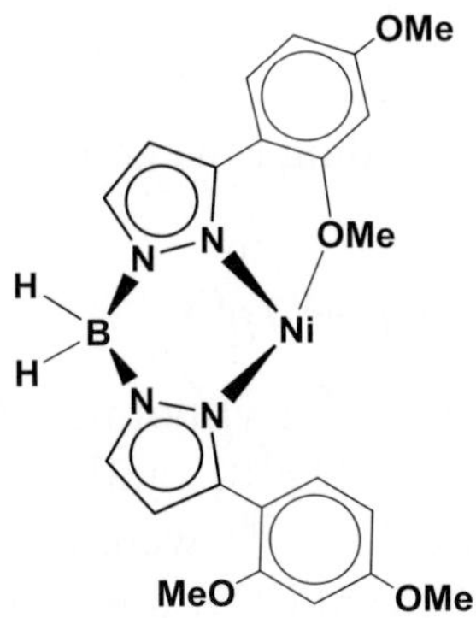

Fig. 1.14. $[Ni(Bp^{(2,4(OMe)2Ph)})_2]$ (only one scorpionate ligand is shown).

groups on the phenyl rings and giving a secondary chelation occurring through a six-membered ring. $[Ni(Bp^{(2,4(OMe)_2Ph)})_2]$ (Fig. 1.14) and $[Tl(Tp^{(2,4(OMe)_2Ph)})]$ have been also structurally characterized.

1.8.5. *Tp^{Me,mt3} and Tp^{Me,mt4}*

In a subsequent paper Trofimenko and others[145] reported the synthesis of the ligands hydrotris(3-methyl-4,5-propylenepyrazol-1-yl)borate $Tp^{Me,mt3}$ and hydrotris(3-methyl-4,5-butylenepyrazol-1-yl)borate $Tp^{Me,mt4}$, and the structures of their representative cobalt and rhodium complexes, $[Co(Tp^{Me,3mt})(Tp^{Np})]$ and $[Rh(\kappa^2\text{-}Tp^{Me,4mt})(COD)]$.

1.8.6. *Tp^{Fu,Me}*

The new ligand hydrotris(3-(2′-furyl)-5-methylpyrazolyl)borate $Tp^{Fu,Me}$ reacts with zinc salts forming complexes $[ZnX(Tp^{Fu,Me})]$ (X = Cl, Br, I, NCS, CH_3COO, CF_3COO). When zinc(II) perchlorate was employed, however, $[Zn(Tp^{Fu,Me})_2]$ afforded preferably. Careful control of the reaction conditions allows the formation of the "enzyme model" $[Zn–OH(Tp^{Fu,Me})]$, which mimics carbonic anhydrase by inserting $CO_2$and CS_2 in methanol, producing [Zn–OCOOMe($Tp^{Fu,Me}$)] and [Zn–SCSOMe($Tp^{Fu,Me}$)] able to deprotonate trifluoroacetamide, yielding thus [Zn–NHCOCF$_3$($Tp^{Fu,Me}$)].[146]

1.8.7. *Tp^{Cy}, Tp^{Cy,4Br}, and pz^{0}Tp^{Cy}*

Hydrotris(3-cyclohexylpyrazol-1-yl)borate Tp^{Cy}, tetrakis(3-cyclohexylpyrazol-1-yl)borate pz^0Tp^{Cy}, and hydrotris(3-cyclohexyl-4-bromopyrazol-1-yl)borate $Tp^{Cy,4Br}$ were prepared as Tl(I) salts and converted to tetrahedral [MX(Tp^x)] and octahedral [M(Tp^x)(Tp^y)] complexes, as well as to the dinuclear [Ni(Tp^{Cy})]$_2$(CO_3) and the mononuclear [Ni($Tp^{Cy,4Br}$)($pz^{Cy,4Br}$)$_3$$H_2$] (which may be regarded as the free acid of the novel dianionic ligand [Ni($Tp^{Cy,4Br}$)-($pz^{Cy,4Br}$)$_3$]$^{2-}$. [CoCl(Tp^{Cy})], [CoCl($Tp^{Cy,4Br}$)], [Co($Tp^{Cy,4Br}$)-(NCS)], [Ni(Tp^{Cy})]$_2$(CO_3), [Ni($Tp^{Cy,4Br}$)($pz^{Cy,4Br}$)$_3$](H)$_2$, and [Mo(Tp^{Cy})(η^3-methallyl)(CO)$_2$] were also structurally characterized. The paramagnetic heteroleptic complexes [Co(Tp^{Cy})(Tp)], [Co(Tp^{Cy})(Tp*)], [Co($Tp^{Cy,4Br}$)(Tp)], and [Co($Tp^{Cy,4Br}$)(Tp*)] were investigated by NMR. The homoleptic derivatives [Co(Tp^{Cy})$_2$] and [Co($Tp^{Cy,4Br}$)$_2$] rearrange thermally to [Co(Tp^{Cy^*})$_2$] and [Co($Tp^{Cy,4Br^*}$)$_2$], respectively.[147]

1.8.8. *Tp^{Bn,Me} and Tp^{Bn,4Ph}*

In order to explore how the coordination chemistry of the Tp^{Ph} ligand changes with the introduction of a methylene spacer between the pyrazolyl 3-C and a phenyl group, Trofimenko developed the synthesis of the ligands hydrotris(3-benzyl-5-methylpyrazol-1-yl)borate $Tp^{Bn,Me}$ and hydrotris(3-benzyl-4-phenylpyrazol-1-yl)borate $Tp^{Bn,4Ph}$. The dynamic behavior of a number of metal complexes containing these ligands was investigated by NMR. Structures of [Tl($Tp^{Bn,Me}$)], [Tl($Tp^{Bn,4Ph}$)], [Co($Tp^{Bn,Me}$)(Tp^{Np})], [Mo($Tp^{Bn,Me}$)(CO)$_2$NO], [Co($Tp^{Bn,4Ph}$)(Tp)], and [Mo($Tp^{Bn,Me}$)(η^3-methallyl)(CO)$_2$] were determined by X-ray crystallography. While in $Tp^{Bn,Me}$ the benzyl group is freely rotating and provides less steric hindrance to the coordinated metal than a neopentyl group, steric hindrance is increased in the $Tp^{Bn,4Ph}$, where the rotation of the benzyl substituent is restricted by the 4-phenyl substituent.[148]

1.8.9. *$Tp^{(7azain)}$*

The ligand $Tp^{(7azain)}$ has been obtained from the reaction of KBH_4 with 7-azaindole. It has been employed in the synthesis of $[Zn(Cl)(Tp^{(7azain)})]$ and $[Cu(Tp^{(7azain)})(PPh_3)]$. $K(Tp^{(7azain)})$ has a dimeric structure linked together by two K^+ ions, $[Zn(Cl)(Tp^{(7azain)})]$ a symmetric tripodal structure with all three 7-azaindolyl groups being coordinated to the Zn^{II} center, whereas $[Cu(Tp^{(7azain)})(PPh_3)]$ has an asymmetric structure with two of the 7-azaindolyl groups being coordinated to the Cu^I center and the third 7-azaindolyl group uncoordinated.[149] The same ligand interacts with $[RuCl(Cp^*)]_4$ yielding $[Ru(\kappa^3\text{-}H,N,N-Tp^{(7azain)})(Cp^*)]$ which upon interaction with CO forms $[Ru(\kappa^2\text{-}H,N\text{-}Tp^{(7azain)})(Cp^*)(CO)]$. Both Ru species have been structurally characterized.[150]

1.8.10. *$Tp^{(3,5\text{-}(CH_2)_3)}$ and $Tp^{(3,4\text{-}(CH_2)_6)}$*

The structure of two thallium scorpionates derived from 3(5),4-trimethylene- and 3(5),4-hexamethylenepyrazole, $Tp^{(3,5\text{-}(CH_2)_3)}$ and $Tp^{(3,4\text{-}(CH_2)_6)}$, respectively, has been determined by multinuclear NMR, both in solution and in the solid state. As for the other 3-substituted pyrazolyl derivatives, these compounds can exist, in principle, in four isomeric forms (Fig. 1.15) depending on the substitution pattern. The isomeric distribution has been found the same in both phases suggesting a kinetic origin.[151]

1.8.11. *Tp^{Br_3}*

The synthesis of hydrotris(3,4,5-tribromopyrazol-1-yl)borate, (Tp^{Br_3}), described also in the first edition, has been reported by Trofimenko and coworkers in detail in 2002.[152] $[Mo(Tp^{Br_3})(\eta^3\text{-methallyl})(CO)_2]$, $[Pd(Tp^{Br_3})(\eta^3\text{-methallyl})]$, and $[Rh(Tp^{Br_3})(CO)_2]$ have been structurally characterized and compared with [Mo(Tp*)-

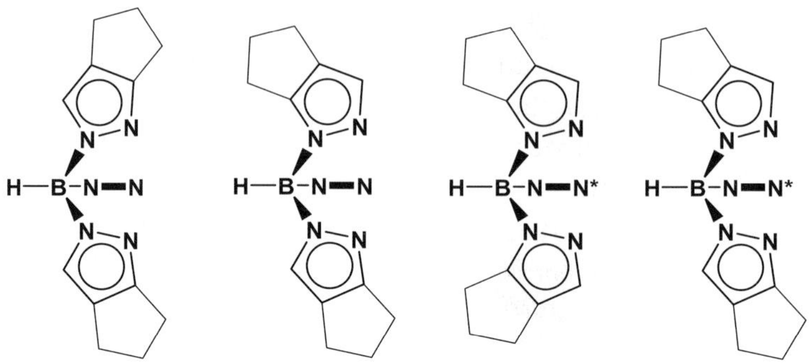

Fig. 1.15. Possible isomeric forms for $Tp^{(3,5-(CH_2)_3)}$ (the asterisk indicates a 3-substituted pyrazolyl ring).

(η^3-methallyl)$(CO)_2$], and [Pd(Tp*)(η^3-methallyl)]. Tp^{Br_3} has a unique feature among scorpionates of containing no C–H bonds, thus making its organometallic derivatives more suitable for spectroscopic (IR, NMR) studies with respect to other Tp^x ligands.

The complex [Cu(Tp^{Br_3})(NCMe)] reported by Pérez's group has been demonstrated an excellent catalyst for the regioselective carbene-transfer reaction to tertiary C–H bonds of hydrocarbons.[153] The choice of the scorpionate was dictated by the hypothesis that the existence of electron-withdrawing groups on ligands in the coordination sphere could favor insertion reaction, due to the enhancement in the electrophilicity of the metal–carbene intermediate. [Cu(Tp^{Br_3})(NCMe)] catalyzes the insertion of a nitrene group into the carbon–hydrogen bond of cyclohexane and benzene, as well as into the primary C–H bonds of the methyl groups of toluene and mesitylene, in moderate to high yield, respectively.[154] It catalyzes the transfer of the nitrene unit and its subsequent insertion into the sp^3 C–H bonds of alkyl aromatic and cyclic ethers or the sp^2 C–H bonds of benzene,[155] and catalyzes the addition of the :$CHCO_2Et$ unit (derived from EDA) to benzene forming a cycloheptatriene ring,[156] in analogy with the Büchner reaction. The same copper complex, dissolved in the perfluoropolyether FOMBLIN® has been also employed as a catalyst for the styrene cyclopropanation reaction with ethyl diazoacetate.[157] [Ag(Tp^{Br_3})]$_2$·CH_3COCH_3 and [Ag(Tp^{Br_3})(THF)] also catalyze the insertion of the :$CHCO_2Et$ group from EDA into the saturated C–H bonds of several C5, C6, and C8 linear and branched alkanes.[158]

$[Ir(C_2H_4)_2(Tp^{Br_3})]$ photochemically activates the C–H bonds of *n*-pentane. The process consists of two steps, the first one being the formation of the isolable alkyl-hydride intermediate $[IrH(C_2H_4)(C_5H_{11})(Tp^{Br_3})]$, the second a β-hydride elimination generating a terminal olefin complex $[IrH_2(C_5H_{10})(Tp^{Br_3})]$. This reaction can be extended to diethyl ether, the reaction proceeding by the activation of one of the β-CH bonds with formation of the alkyl-hydride $[IrH(C_2H_4)(CH_2CH_2OCH_2CH_3)(Tp^{Br_3})]$ and then yielding $[IrH_2(CH_2{=}CHOCH_2CH_3)(Tp^{Br_3})]$. These processes compete with the intramolecular C–H activation of one of the coordinated ethylene ligand by the Ir center in $[Ir(C_2H_4)_2(Tp^{Br_3})]$ affording the hydride-vinyl complex $[Ir(CH{=}CH_2)(C_2H_4)(Tp^{Br_3})]$. When the same reactions was carried out at high-temperature, the hydride-α,ω-butenyl complex $[Ir(H)(CH_2CH_2CH{=}CH_2)(Tp^{Br_3})]$ and the hydride-crotyl derivative $[IrH(C_4H_7)(Tp^{Br_3})]$ have been obtained.[159]

1.8.12. Bp^{Br_3}

The ligand dihydrobis(3,4,5-tribromopyrazol-1-yl)borate, (Bp^{Br_3}), and its derivative $[Mo(Bp^{Br_3})(\eta^3\text{-methallyl})(CO)_2]$ have been prepared and characterized by Trofimenko and coworkers who compared the structure of the Mo complex with that of $[Mo(Bp^*)(\eta^3\text{-methallyl})(CO)_2]$.[152]

1.8.13. $Tp^{Br,Ph,Br}$, $Tp^{Br,p\text{-}Tol,Br}$, $Tp^{Br,p\text{-}ClPh,Br}$, $Tp^{p\text{-}ClPh,4Br}$, *and* $Tp^{Ph,Me,Br}$

The new ligands $Tp^{Br,Ph,Br}$, $Tp^{Br,p\text{-}Tol,Br}$, and $Tp^{Br,p\text{-}ClPh,Br}$, first examples of an "atypical" B–N bond to the most sterically hindered pyrazole nitrogen, exhibit all aryl substituents in the 5-position of the ligand, forming a protective pocket around the B-H bond. These complexes display a rather high B–H stretch frequency in the IR region. On the other hand $Tp^{p\text{-}ClPh,4Br}$ and $Tp^{Ph,Me,Br}$, containing no outer bromine substituents, show normal B–N bonding to the least-hindered

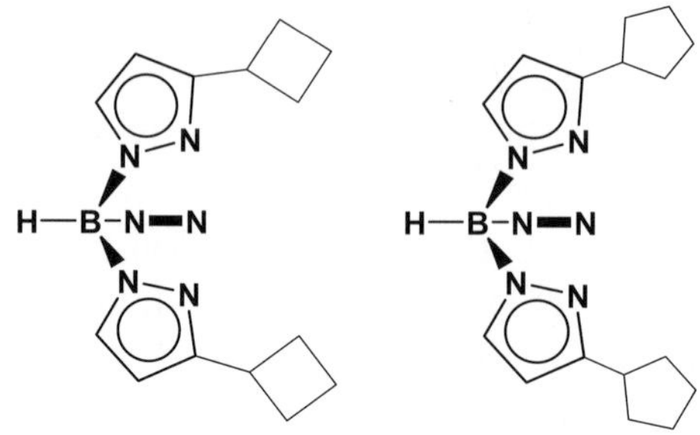

Fig. 1.16. Tp^{Cpe} (left) and Tp^{Cbu} (right).

nitrogen. These ligands, isolated as their Tl salts, have been employed to form [Cu(NCMe)(Tp^x)], also tested as catalysts in carbene and nitrene transfer reactions.[160]

1.8.14. *Tp^{Cpe} and Tp^{Cbu}*

The Tl derivatives of the new ligands Tp^{Cpe} and Tp^{Cbu} (Fig. 1.16) have been prepared and structurally characterized. The cone and wedge angles of these two compounds have been compared with those of [Tl(Tp^{Cpr})] and [Tl(Tp^{Cy})] and it has been reported that the cone angle of [Tl(Tp^{Cbu})] is relatively similar to [Tl(Tp^{Cpr})], whereas the cone angle of [Tl(Tp^{Cpe})] is roughly midway between those of [Tl(Tp^{Cpr})] and [Tl(Tp^{Cy})].[161]

1.8.15. *$Tp^{\alpha Nt}$ and $Tp^{\beta Nt}$*

The homoscorpionate ligands $Tp^{\alpha Nt}$ and $Tp^{\beta Nt}$ (Fig. 1.17) were synthesized by using 3-(α-naphthyl)pyrazole and 3-(β-naphthyl)pyrazole, respectively, bonded either through the 1-position $Tp^{\alpha Nt}$ or through the 2-position $Tp^{\beta Nt}$. $Tp^{\beta Nt}$ resembles other Tp^{Ar} ligands, where Ar is a *para*-substituted phenyl group, although the 2-naphthyl substituent provided a substantially deeper pocket around the coordinated metal while retaining rotational freedom. By contrast, rotation of the 1-naphthyl substituent in $Tp^{\alpha Nt}$ was restricted, which led to complexes of lower symmetry than those of the $Tp^{\beta Nt}$ ligand and, at times, as in the case of [Co($Tp^{\alpha Nt}$)(Tp*)], to a structure devoid of any symmetry

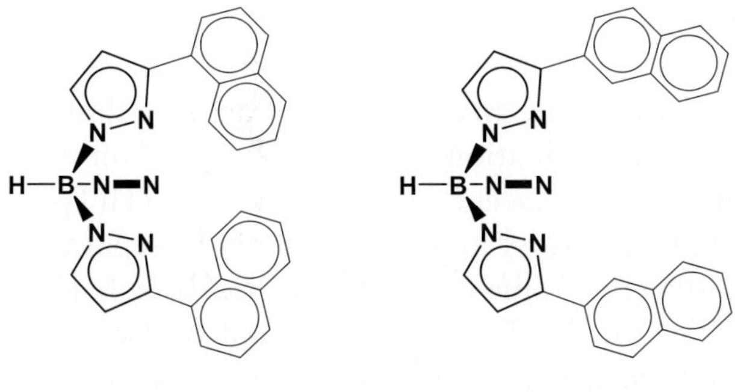

Fig. 1.17. $Tp^{\alpha Nt}$ (left) and $Tp^{\beta Nt}$ (right).

elements. $Tp^{\alpha Nt}$ could find some use in the preparation of asymmetric catalysts.[162]

1.8.16. $Tp^{Me_2Bn,Me}$

[K($Tp^{Me_2Bn,Me}$)] and [Tl($Tp^{Me_2Bn,Me}$)] have been reported by Biagini *et al.* (Fig. 1.18). The reaction of $Tp^{Me_2Bn,Me}$ with $TiCl_4$ gave, after rearrangement of the ligand (a double 1,2-borotropic shift), [$TiCl_3${HB($pz^{Me_2Bn,Me}$)(pz^{Me,Me_2Bn})$_2$}].[163]

1.8.17. Tp^{4py}, Bp^{4py}, pz^0Tp^{4py}, Bp^{3py}, Tp^{3py}, and pz^0Tp^{3py}

The hydro-tris[3-(4-pyridyl)pyrazol-1-yl]borate, (Tp^{4py}), has been isolated as its Tl(I) complex which shows an infinite one-dimensional polymeric chain, in which the Tl(I) center is connected to a pendant

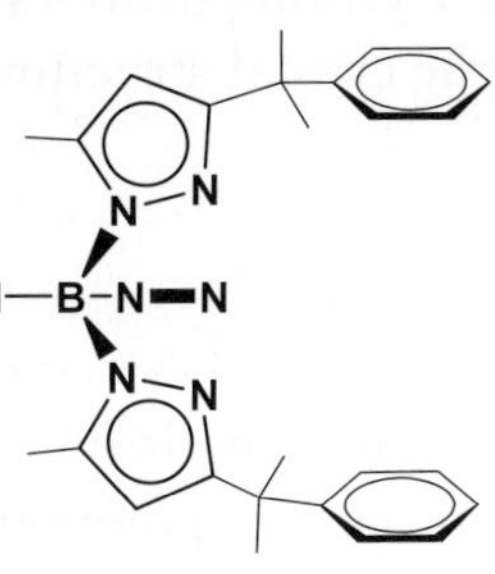

Fig. 1.18. $Tp^{Me_2Bn,Me}$.

4-pyridyl residue from an adjacent complex unit to build up the chain. The Bp^{4py} and its Tl derivative has been also described.[164] $[Tl(Bp^{3py})]$, $[Tl(Bp^{4py})]$, and $[Tl(Tp^{4py})]$ are one-dimensional polymeric chains via coordination of one of their pendant pyridyl units to the Tl(I) center of an adjacent complex fragment; $[Tl(Tp^{3py})]$ contains the three pendant pyridyl units coordinated to separate Tl(I) neighbors and yielding two-dimensional polymeric sheet. In $[Tl(pz^0Tp^{3py})]$, and $[Tl(pz^0Tp^{4py})]$ the Tl(I) is coordinated by two or three of the four pyrazolyl arms, respectively; bridging interactions of pendant 4-pyridyl groups with adjacent Tl(I) centers result in a two-dimensional sheet. In $[Ag(pz^0Tp^{4py})]$ the silver ion is coordinated by two pyrazolyl rings, and two bridging pyridyl ligands from other complex units, resulting in a one-dimensional chain consisting of pairs of cross-linked zigzag chains. $[Cu(Tp^{4py})]$ and $[Cd(OAc)(Tp^{3py})]$ are dimers, stabilized by π-π stacking interactions between sections of the two ligands. $[Ni(Tp^{3py})_2]$ is monomeric, with an octahedral coordination geometry. $[Re(CO)_3(Tp^{4py})]$ is also monomeric.[165]

1.8.18. Tp^{Bupy}

The potassium salt of the hexadentate tris[3-{(4-*t*butyl)-pyrid-2-yl}-pyrazol-1-yl]hydroborate (Tp^{BuPy}) has been described together with the crystal structure of $[La(NO_3)_2(Tp^{BuPy})]$.[166]

1.8.19. $Bp^{(COC)py}$

The potassium salt of the tetradentate bis[3-(2-pyridyl)-5-(methoxymethyl)pyrazol-1-yl]-dihydroborate ($Bp^{(COC)Py}$) has been described together with the crystal structures of $[La(X)(Bp^{(COC)Py})_2]$ ($X = NO_3$ or OTf).[166]

1.8.20. Bp^{2pyr}

Bp^{2pyr}, in which each pyrazolyl ring is functionalized with a pyrazin-2-yl group at the C2 position, is a potentially chelating tetradentate ligand with two externally directed N atoms available for additional metal–ion binding. The crystal structure of [Tl(Bp^{2pyr})] indicates it as a simple mononuclear complex with the Tl ion coordinated in the N_4 binding pocket of the ligand and the externally directed N atoms involved only in intermolecular N···H–C hydrogen bonding interactions. [Pb(Bp^{2pyr})$_2$]. Et_2O contains a Pb center nine-coordinate by two tetradentate chelating ligands and the ninth donor being a pyrazinyl N4 atom from an adjacent molecule.[167]

1.8.21. $Tp^{(4py),Me}$, $Tp^{(4\text{-}(2\text{-}Mepy)),Me}$, $Tp^{(CONHPh),Me}$, and $Tp^{(CONHC(Me_3)),Me}$

Four new Tp^x ligands, bearing pyridyl and carboxamide substituents at the 3-positions of the pyrazole rings, were prepared by Vahrenkamp group (Fig. 1.19).[168] Structure determination of their K salts was also carried out (Fig. 1.20). Their coordinative properties were explored by preparing [Zn(X)(Tp^x)] (X = Cl, Br, I, NO_3, OAc, phenolate, thiophenolate, and diorganophosphate) and the cationic complexes $[Zn(Tp^x)(L)]^+$ (L = MeOH or Hpz). Polar Tp^x ligands favor coordination numbers higher than four for zinc, either by inducing bidentate coordination of the coligands X and L, using the carboxamide oxygen atoms for coordination, or by linking two [Zn(X)(Tp^x)] units through

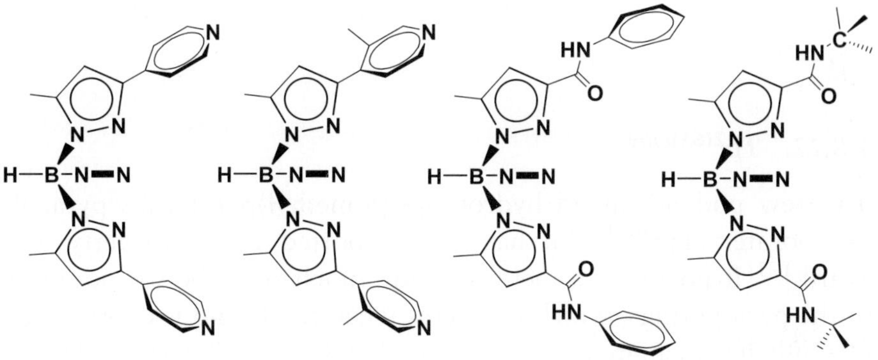

Fig. 1.19. The four new ligands recently reported by Vahrenkamp group.[168] From left to right, respectively: $Tp^{(4py),Me}$, $Tp^{(4\text{-}(2\text{-}Mepy)),Me}$, $Tp^{(CONHPh),Me}$, and $Tp^{(CONHC(Me3)),Me}$.

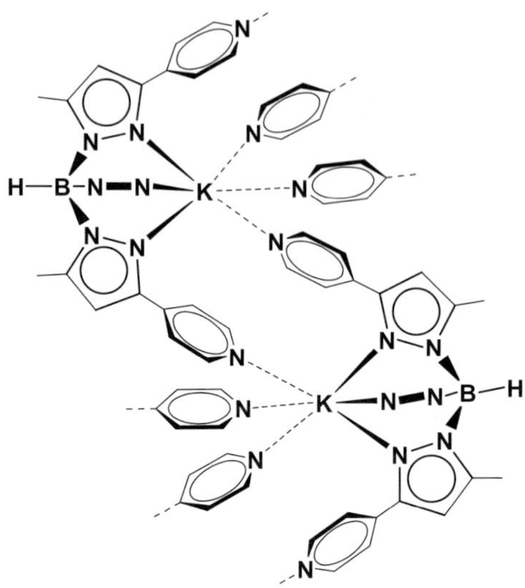

Fig. 1.20. Coordination of K^+ ion in $K(Tp^{(4py),Me})$.

the pyridyl nitrogen atoms. Coordination dimers and polymers were reported.[168] The same ligands were employed for the synthesis of mono and dinuclear Zn–hydroxo species. The $Zn–OH_2$, Zn–OH, Zn–OH–Zn, and $Zn–O_2H_3–Zn$ coordination motifs were found.[169] The same authors investigated the reactivity of selected species toward hydrolyzable substrates. $[ZnOH(Tp^{4\text{-}py,Me})]$ inserted CO_2 and CS_2 yielding $[Zn(OCOOMe)(Tp^{4\text{-}py,Me})]$ and $[Zn(SCSOMe)Tp^{4\text{-}py,Me}]$, respectively.[170]

1.8.22. $Tp^{2(6Me)py}$

The new podand ligand hydrotris[3-(6-methyl)pyridin-2-ylpyrazol-1-yl]borate, $(Tp^{2(6Me)py})$, has been reported by McCleverty and Ward.[171] It possesses a methyl group attached to the C6 positions of the pyridyl rings, which interferes when the ligand is coordinated in a fully hexadentate manner. In fact, $[M(Tp^{2(6Me)py})(NO_3)_2(H_2O)]$ (M = Eu, Tb or Gd) (Fig. 1.21) showed partial dissociation of the podand occurring to relieve potential steric problems: either one or two of the pyridyl groups are not coordinated, such that $Tp^{2(6Me)py}$

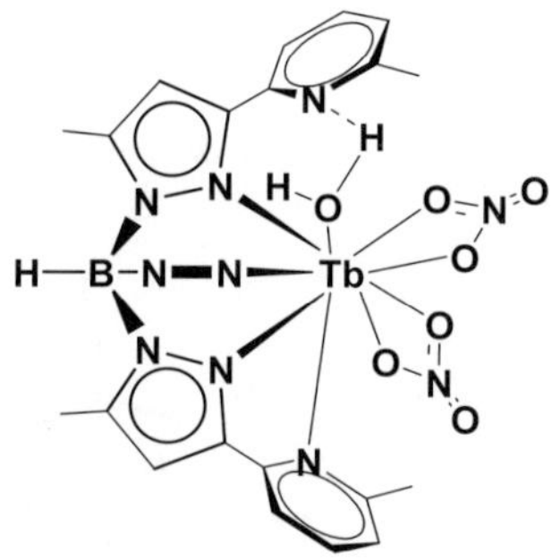

Fig. 1.21. The structure of $[Tb(Tp^{2(6Me)py})(NO_3)_2(H_2O)]$.

is penta- or tetra-dentate. Both structural forms are present in single crystals of the Gd and Eu complexes, suggesting facile interconversion between them in solution. Luminescence and NMR studies for these complexes have been reported and compared with those of $[M(Tp^{2py})(NO_3)_2]$.[171]

1.8.23. $Tp^{Ad,iPr}$

$Tp^{Ad,iPr}$, the most hindered hydrotris(pyrazolyl)borate ligand, has been reported by Fujisawa *et al.*[172] $[CuCl(Tp^{Ad,iPr})]$ has been also prepared and characterized by X-ray diffraction.[172]

1.8.24. Bp^{4CN} *and* $Bp^{Ph,4CN}$

$[Co(Bp^{4CN})(Tp^{Np})]$, $[Co(Bp^{4CN})(Tp^{Cy})]$, and $[Co(Bp^{4CN})_2(DMF)_2]$ contain the heretoscorpionate Bp^{4CN}, a new ligand able to lead oligomeric or polymeric structures. It has been shown that when sterically hindered Tp^x or DMF solvent is employed, the CN group in Bp^{4CN} cannot coordinate to the metal ion of a neighboring molecule. $[Co(Bp^{4CN})(Tp^{Np})]$ contains a five-coordinate Co center both in the solid and solution state.[173]

The synthesis of $Bp^{Ph,4CN}$ has been reported by Eichhorn and coworkers.[174] The monomeric homoleptic square-planar copper complex $[Cu(Bp^{Ph,4CN})_2]$ has been prepared by the reaction of

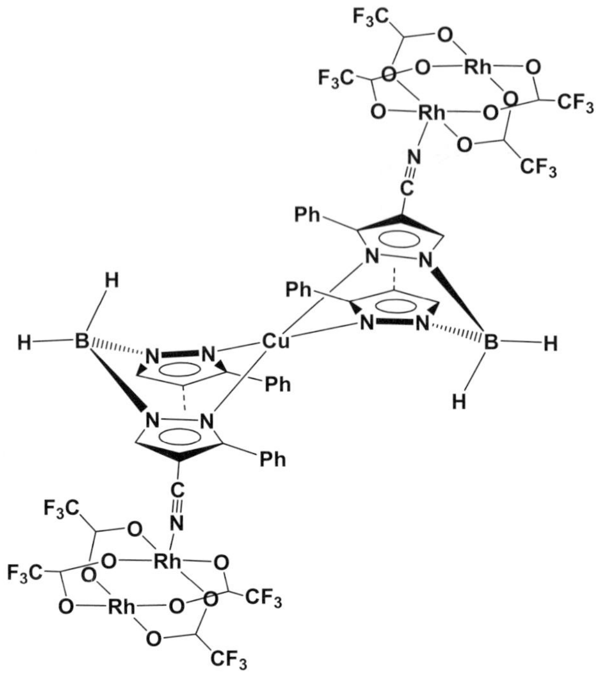

Fig. 1.22. Structure hypothesized for the coordination polymer formed by $[Cu(Bp^{Ph,4CN})_2]$ with $Rh_2(CF_3COO)_4$.

$Bp^{Ph,4CN}$ with $Cu(NO_3)_2{\cdot}2.5H_2O$. Protection of a susbtituent in the 3-position inhibits polymerization reactions,[174] however, when $[Cu(Bp^{Ph,4CN})_2]$ reacts with $Rh_2(CF_3COO)_4$, a coordination polymer forms for which a structure has been proposed consisting of $Rh_2(CF_3COO)_4$ moieties linked to $[Cu(Bp^{Ph,4CN})_2]$ through the CN groups (Fig. 1.22). This compound has been characterized by IR, EPR, and ESI MS spectroscopy.[175]

1.8.25. $Tp^{4Bo,6\text{-}COOEt}$ *and* $Tp^{4Bo,6\text{-}CH_2SEt}$

Two new tripodal ligands, able to anchor complexes onto surfaces, have been synthesized by Rapenne's group.[176] They integrate ester or thioether functions at the 6-position of the indazoles. Potassium

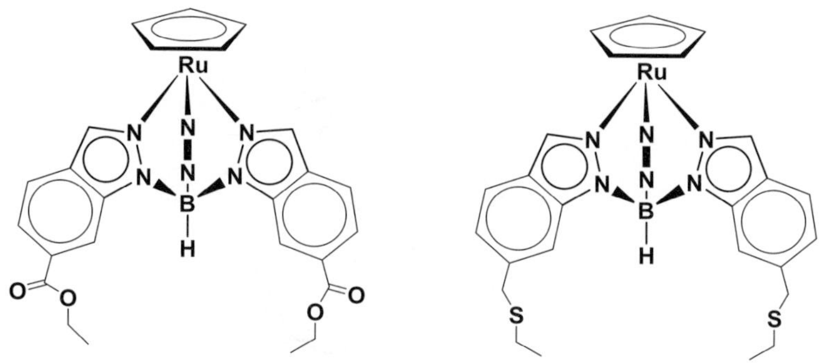

Fig. 1.23. The two piano-stool-shaped [Ru(Cp)-($Tp^{4Bo,6\text{-}COOEt}$)] and [Ru(Cp)-($Tp^{4Bo,6\text{-}CH_2SEt}$)].

hydrotris[6-(ethoxycarbonyl)indazolyl]borate K($Tp^{4Bo,6\text{-}COOEt}$) and potassium hydrotris{6-[(ethylsulfanyl)methyl]indazolyl}borate K-($Tp^{4Bo,6\text{-}CH_2SEt}$) react with $[Ru(Cp)(MeCN)_3]PF_6$ yielding two piano-stool-shaped complexes that were characterized by X-ray diffraction (Fig. 1.23). Comparison with the synthesized unfunctionalized analog showed that the three 6-substituted functions do not interfere with the coordination site and are particularly well oriented for surface deposition.[176]

1.8.26. *pzTp^{camph}, Tp$^{CH_2N(Me)EtPh}$, (BBN)Bp$^{+/-camph}$, and (BBN)Bp$^{+menthone}$*

The sodium salt of the optically active tetrakis(4,5,6,7-tetrahydro-7,8,8-trimethyl-2*H*-4,7 methanoindazolyl)borate, Na(pz-Tp^{camph}) was prepared by the reaction of 4,5,6,7-tetrahydro-7, 8,8-trimethyl-2*H*-4,7-methanoindazole with $NaBH_4$. Analogously the sodium salt of hydrotris((−)-3(5)-methyl-1-phenylethylaminomethylpyrazolyl)borate, Na($Tp^{CH_2N(Me)EtPh}$) (Fig. 1.24) was obtained by the reaction of (−)-3(5)-methyl-1-phenylethylaminomethylpyrazole with $NaBH_4$. These ligands were used in metal-catalyzed (Cu and Rh) enantioselective cyclopropanation reactions of styrene and several diazoacetates.[177]

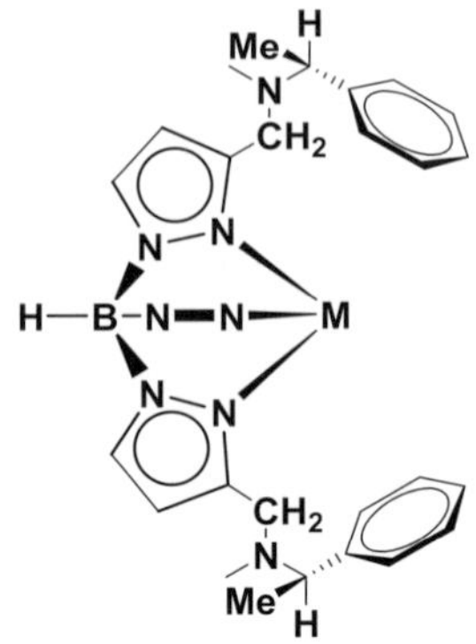

Fig. 1.24. $Tp^{CH_2N(Me)EtPh}$.

Chisholm *et al.*[178] reported the synthesis and the structure of chiral C2 and C1 symmetric M{(cyclooctane-1,5-diyl)bis[(4*S*,7*R*)-7,8,8-trimethyl-4,5,6,7-tetrahydro-4, 7-methano-2H-indazol-2-yl]borato-κN^1,κN^1} M($BBNBp^{+camph}$) (M = K or Tl), K(cyclooctane-1, 5-diyl)[(4*S*,7*R*)-7,8,8-trimethyl-4,5,6,7-tetrahydro-4,7-methano-2H-indazol-2-yl][(4*S*,7*R*)-7,8,8-trimethyl-4,5,6,7-tetrahydro-4,7-methano-2H-indazol-2-yl]borato-κN^1, κN^1} · 3THF, K($BBNBp^{+/-camph}$), and K{(cyclooctane-1,5-diyl)bis[(4R,7R)-7-isopropyl-4-methyl-4,5, 6,7-tetrahydro-2H-indazol-2-yl]borato-κN^1,κN^1} · 4THF, K(BBN-$Bp^{+menthone}$).[178]

1.8.27. Tp^{4Bz}

Hydrotris[3-(4-benzonitrile)-pyrazol-1-yl]borate Tp^{4Bz}, reported by Batten *et al.*[179] and structurally characterized as potassium salt, reacts with MX_2 salts yielding octahedral monomeric $[M(Tp^{4Bz})_2]$ (solv) species (M = Mn, Co, Ni, and Cd). The nitrile groups do not coordinate.[179]

1.8.28. $Tp^{CO_2Et,Me}$

$Tp^{CO_2Et,Me}$, a hydrogen-bond accepting ligand, was synthesized by Carrano and coworkers who also reported stable penta- or hexa-coordinate metal aquo complexes $[M(Tp^{CO_2Et,Me})(H_2O)_x]$ ($x = 2$ or

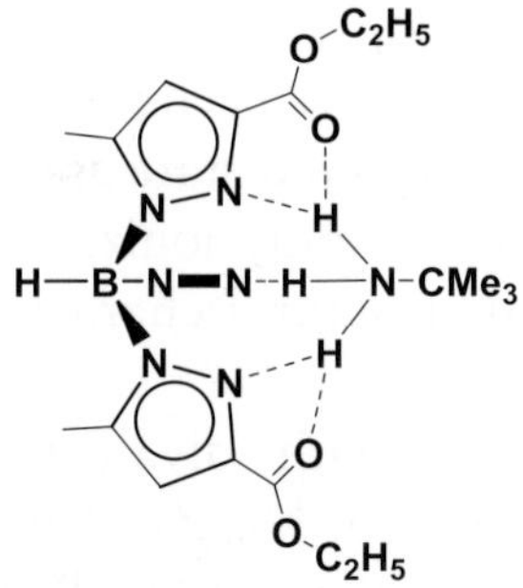

Fig. 1.25. $[Tp^{CO_2Et,Me}][NH_3R]$, a rare example of host–guest complex.

3) (M = Ni, Co, Mn, and Cu) which are unprecedented for the alkyl substituted Tp analogs.[180]

$[Tp^{CO_2Et,Me}][NH_3R]$, a host–guest complex, as revealed by an X-ray crystal structure of the *tert*-butylammonium analog (Fig. 1.25) has been isolated during the transamidation reactions to convert 3-ester-substituted $Tp^{CO_2Et,Me}$ into 3-amido-substituted ligands.[181]

The water molecules in $[M(Tp^{CO_2Et,Me})(H_2O)_3]$ (M = Ni, Co, Mn, or Cu) have been displaced by neutral bidentate ligands L giving $[M(Tp^{CO_2Et,Me})(L)H_2O]$.[182]

$[Zn(NO_3)(Tp^{CO_2Et,Me})]$, $[Zn(OAc)(Tp^{CO_2Et,Me})(H_2O)]$, $[\{Zn(Tp^{CO_2Et,Me})\}_2(OH)](ClO_4)$, and $[Zn(Tp^{CO_2Et,Me})(H_2O)_3](ClO_4)$ have been isolated and structurally characterized. These derivatives are stabilized by internal H-bonding between the water or hydroxide and the ester carbonyls of the donor. $[\{Zn(Tp^{CO_2Et,Me})\}_2(OH)](ClO_4)$ appears to catalyze self-transesterification reaction in alcoholic solvents. The expansion of the coordination sphere from four to five in $[Zn(OAc)(Tp^{CO_2Et,Me})(H_2O)]$ or the dimerization in $[\{Zn(Tp^{CO_2Et,Me})\}_2(OH)](ClO_4)$ suggests that a more sterically restrictive ester substituted ligand is needed to enforce the desired mononuclear four-coordinate pseudo-tetrahedral geometry characteristic of zinc metalloenzyme active sites.[183]

1.8.29. *Bp*^{CO_2Et,Me}

The ligand $Bp^{CO_2Et,Me}$, prepared as potassium salt in kerosene solution using ethyl 3-methylpyrazole-5-carboxylate and KBH_4, reacts with CuCl and PPh_3 yielding [Cu($Bp^{CO_2Et,Me}$)(PPh_3)]. $Bp^{CO_2Et,Me}$ and [Cu($Bp^{CO_2Et,Me}$)(PPh_3)] have been fully characterized by elemental analyses and FT-IR in the solid state and by NMR (^1H, ^{13}C, and ^{31}P) spectroscopy and electrospray ionization mass spectrometry in solution. A single-crystal structural characterization is reported for [Cu($Bp^{CO_2Et,Me}$)(PPh_3)].[184]

$Bp^{CO_2Et,Me}$ reacts also with divalent metals giving [M($Bp^{CO_2Et,Me}$)$_2$] (M = Mn, Fe, Co, Ni, Zn, Cu, Pb, and Cd). All complexes have been fully characterized by elemental analyses and FT–IR in the solid state and by NMR spectroscopy and electrospray ionization mass spectrometry in solution. In [Cu($Bp^{CO_2Et,Me}$)$_2$] and [Zn($Bp^{CO_2Et,Me}$)$_2$] the metals are four-coordinated and only bound to the nitrogen atoms of $Bp^{CO_2Et,Me}$ [185]

1.8.30. $Tp^{C(Me_2)CH_2OMe}$

A potential hemilabile tris(pyrazolyl)borate ligand bearing ether appendages (Fig. 1.26), $Tp^{C(Me_2)CH_2OMe}$, is reported by Chisholm's group.[186] Structural characterization of its Li^+, Na^+, K^+, Tl^+, and Ca^{2+} complexes evidenced its η^3, η^5, η^6, and μ-binding modes.

1.8.31. $PhTp^{tBu}$

[$PhTp^{tBu}$]H is one of the rare [Tp^x]H structurally characterized: its pyrazolyl groups are not symmetrical disposed with C_{3v} symmetry.

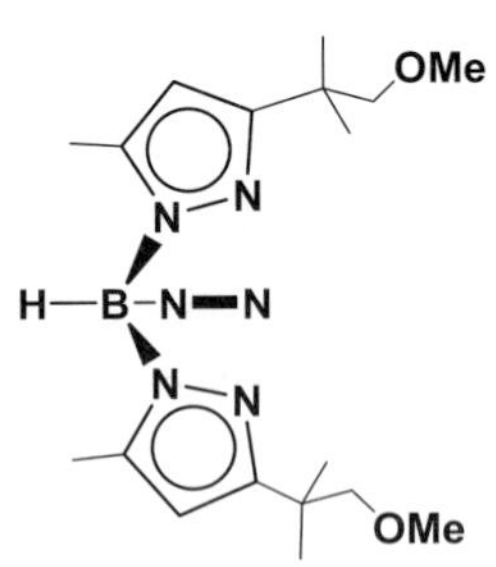

Fig. 1.26. The hemilabile ligand $Tp^{C(Me_2)CH_2OMe}$.

Tl(PhTp^{tBu}) shows an unprecedented structure in which one of *tert*-butylpyrazolyl groups is rotated by *ca.* 90° and the Tl interacts with the nitrogen attached directly to the boron via a π-orbital component of the aromatic π-system of the pyrazolyl nucleus. [Tl(PhTp^{tBu})] is stereochemically nonrigid on the NMR spectroscopic timescale in solution at room temperature.[187]

1.8.32. *^{t}BuTp^R (R = H, Me, ^{i}Pr, ^{t}Bu, Ph) and ^{t}BuBp**

The lithium and thallium salts of ^{t}BuTp^R (R = H, Me, ^{i}Pr, ^{t}Bu, Ph) have been prepared and characterized. It is noteworthy that the reaction between ^{t}BuBH$_3$Li0 · 5Et$_2$O and excess Hpz* afforded ^{t}BuBp* due to steric congestion at B (Scheme 1.2). The solid-state ^{7}Li and ^{11}B NMR spectra of [Li(^{t}BuTp^{iPr})] suggest that this salt exists as a mixture of axial and equatorial isomers. The partial hydrolysis of [Li(^{t}BuTp^{tBu})] afforded the dimeric [Li{^{t}BuB(pz^{tBu})$_2$-

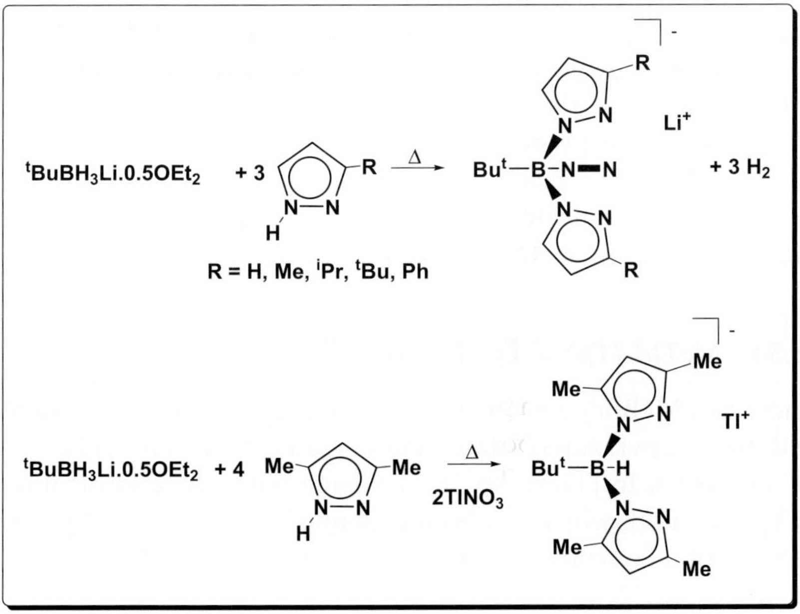

Scheme 1.2. Synthesis of ^{t}BuTp^R and ^{t}BuBp*.

(μ-OH)}]$_2$ containing two Li cations encapsulated by the heteroscorpionate [{^{t}BuB(pz^{tBu})$_2$(μ-OH)}] ligands. The homoleptic, paramagnetic low-spin iron(III) complexes [Fe(^{t}BuTp)$_2$]PF$_6$ and [Fe(^{t}BuTp^{Me})$_2$]PF$_6$ have been also reported and characterized.[188]

When FeCl$_2$(THF)$_{1.5}$ reacts with the sterically hindered thallium salt of (^{t}BuTp^{tBu}), the complex *trans*-FeCl$_2$(Hpz^{tBu})$_4$ is formed, the only isolable product appearing to be the first example of a *trans*-dichloro pseudo-octahedral iron(II) complex bearing four neutral pyrazole ligands.[189]

1.8.33. *FlTp and FlTp^{tBu}*

Two hybrid Tp^x ligands, FlTp and FlTp^{tBu} have been synthesized as Li salts and characterized by X-ray. The reaction of [Li(FlTp)] and ZnCl$_2$ afforded [(THF)$_3$Li(FlTp)ZnCl$_2$] in which a ZnCl$_2$ moiety is chelated by two pyrazolyl ligands while the third pz ring coordinates to a Li(THF)$_3$ fragment. [(THF)$_3$Li(FlTp)ZnCl$_2$] gradually decomposes into the mononuclear species [Zn(FlTp)$_2$]. In the compounds decribed by Wagner and coworkers,[190] the hybrid ligands FlTp and FlTp^{tBu} adopt a conformation with only two pyrazolyl rings bonded to the central metal, the third pyrazolyl acting as dangling substituent. The fluorenyl substituent of [Li(FlTp)] may be deprotonated with KH.[190] Degradation reactions of these scorpionates were observed in the presence of transition-metal salts MX$_2$ to give transition-metal adducts containing neutral pyrazoles.[191]

1.8.34. *MeTp^x (Tp^x = Tp, Tp* or Tp^{Me})*

Reaction of MeBBr$_2$ with pyrazoles in the presence of TlOEt and NEt$_3$ yields the tris(pyrazolyl)borate thallium salts [Tl(MeTp^x)] (Tp^x = Tp, Tp* or Tp^{Me}). In [Tl(MeTp*)] a bridging bidentate coordination of MeTp* is found, which has been explained as sterically enforced, on the basis of a comparison between the structure of [Tl(MeTp*)] and [Tl(MeTp^{Me})].[132]

1.8.35. *Fc(Tp^{Ph})*

A series of ferrocene-based mono and bifunctional tris(pyrazolyl)-borates bearing either methyl or phenyl substituents at the 3-position has been prepared via a *trans*-amination reaction. The lithium salts have been converted in the analogous Tl species and [Tl(Fc(Tp^{Ph}))] has been also structurally characterized. These compounds feature monomeric units with tridentate Fc(Tp^{Ph}) in striking contrast to the sterically less congested [Tl(Fc(Tp))] that forms polymeric rods in the solid state. In the same paper the authors reported the structure of Tl(Tp) for purpose of comparison.[192]

1.8.36. *CymTp, Cym′Tp, and Cym′Tp^R Tl (R = $CH_2(C_6H_{11})$)*

Reaction of $CymBBr_2$ and $Cym'BBr_2$ with Hpz in the presence of triethylamine gave new cymanthrene based ligands as their Tl salts. The structure of [Tl(CymTp)], and [Tl(Cym′Tp)] is similar to that of [Tl(FcTp)]. [Tl(CymTp)] features a polymeric structure with bridging Tp fragments whereas [Tl(Cym′Tp)] shows an oligomeric structure. [Tl(Cym′Tp^R)] (R = $CH_2(C_6H_{11})$), ligand of high solubility in non-polar solvents, has been also reported.[193]

The di- and trinuclear complexes [$Mn(CO)_3$(CymTp)] (Fig. 1.27), [$Mn(CO)_3$(CymBp(OH))], and [Zn$(CymTp)_2$] have been also reported.[194] The structure of [$Mn(CO)_3$(CymTp)], exhibiting a cyclopentadienyl ring together with a scorpionate moiety in the same molecule, indicates that both ligand types (Cp and Tp) exert a very similar electronic influence on the respective $Mn(CO)_3$ fragment. [$Mn(CO)_3$(CymBp(OH))] is obtained when the reaction between

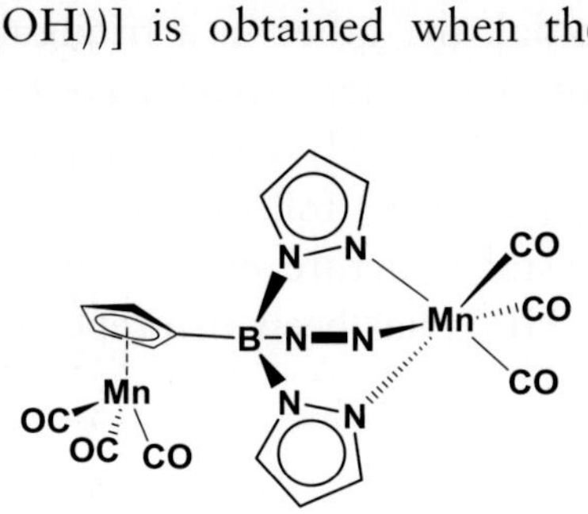

Fig. 1.27. The dinuclear [$Mn(CO)_3$(CymTp)].

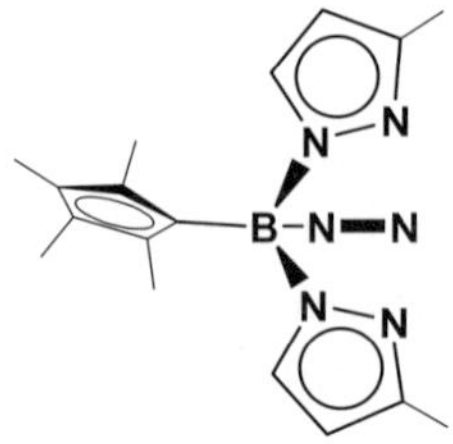

Fig. 1.28. The dianionic ligand $(C_5Me_4Tp^{Me})^{2-}$.

[Tl(CymTp)] and $Mn(CO)_5Br$ is performed without exclusion of air and moisture.[194]

Wagner and coworkers[195] reported also some rare examples of complexes featuring a chelating Cp/scorpionate hybrid ligand obtained by photolytic decarbonilation of K[(CymTp)] which leads to the constrained geometry complex $K[(OC)_2Mn(C_5H_4–B-((\mu\text{-pz})(pz)_2)]$.

1.8.37. $C_5Me_4Tp^{Me}$

Marques and coworkers[196] reported the new dianionic ligand $(C_5Me_4Tp^{Me})^{2-}$ (Fig. 1.28) and characterized it as lithium and potassium salt. In the solid state $[Li_2(C_5Me_4Tp^{Me})(THF)_3]$ is monomeric, one lithium being tetracoordinate by two nitrogen atoms of the Tp^x ligand and two oxygen atoms from THF, the second lithium being η^5-coordinated to the C_5Me_4 ring, one N from Tp^x and one O from the third THF molecule. $[K_2(C_5Me_4Tp^{Me})(THF)_2]_\infty$ consists of infinite puckered chains, one K atom being described as distorted square-pyramidal, the other in a triangular-pyramidal coordination environment. These derivatives have been prepared by the reaction of $B(NMe_2)_2(C_5Me_4H)$ with 2 equiv. of Hpz^{Me} and 1 equiv. of $M(pz^{Me})$ (M = Li, K) followed by the deprotonation with ^{n}BuLi or K. The crystal structure of the intermediate monoanion $[Li\{C_5Me_4HTp^{Me}\}(THF)_2]$ has been also reported.[196]

1.8.38. *IpcTp*

Na[(Ipc)Tp], prepared by a new general synthetic method (Scheme 1.3), has been used to synthesize $[Mn\{(Ipc)Tp\}(CO)_3]$.[197]

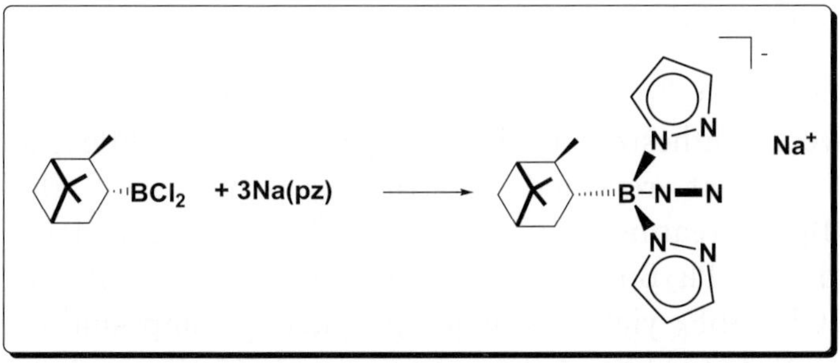

Scheme 1.3. Synthesis of Na[(Ipc)Tp].

1.8.39. *MeTp*^{CF_3}

The thallium derivative of MeTp^{CF_3} has been synthesized via a two-step process using the corresponding pyrazole, Li[$MeBH_3$], and thallium(I) acetate. Reaction of [Tl(MeTp^{CF_3})] with CuBr in the presence of ethylene gave [Cu(MeTp^{CF_3})(C_2H_4)] which reacts with [$(Bn)_2$ATI]SnCl to yield [(MeTp^{CF_3})Cu]←Sn(Cl)[$(Bn)_2$ATI], containing an unsupported Cu(I)–Sn(II) bond.[198] [Ag(MeTp^{CF_3})(C_2H_4)] has been also prepared and characterized.[199]

1.8.40. *PhTp*^{CF_3}

[Ag(PhTp^{CF_3})(C_2H_4)] obtained from the lithium poly(pyrazolyl)-borate salt, AgOTf and ethylene, has a planar three-coordinate silver center; whereas the ethylene-free [Ag(PhTp^{CF_3})]$_n$ adopts a helical structure with a hexagonal pore.[199]

1.8.41. *MeTp*$^{(C_2F_5)_3}$

The new B-methylated ligand [MeTp$^{(C_2F_5)_3}$], reported by Rasika Dias and Wang,[200] enables the isolation of the lithium adduct [Li(MeTp$^{(C_2F_5)_3}$)] exhibiting a *fac*-N_3F_3 coordination, and the formation of two stable silver species [Ag(MeTp$^{(C_2F_5)_3}$)CO] and [Ag(MeTp$^{(C_2F_5)_3}$)(C_2H_4)].[200]

1.8.42. $Ph_2Bp^{R,R'}$

The ligands $Ph_2Bp^{R,R'}$ (R = Ph, R′ = H, and R = R′ = CH_3) were synthesized by heating $pz^{R,R'}$ and $NaBPh_4$, a method similar to that reported for the preparation of the parent $Na(Ph_2Bp)$. Treatment of $Na(Ph_2Bp^{Ph})$ with $[Cu(MeCN)_4]SbF_6$ afforded $[Cu(Ph_2Bp^{Ph})(Hpz^{Ph})]$. Reaction of $Na(Ph_2Bp^*)$ with $[Cu(MeCN)]SbF_6$ yields as major product, a compound containing two new N-donor ligands bound to a single Cu(I) ion that adopts a linear coordination geometry (Scheme 1.4). In the new ligand (pz^{mod}), one pyrazolyl ring is missing from each boron center. The second product contains both pz^{mod} and a bridging Ph_2Bp^* ligand coordinated also through the aromatic Ph rings.[201]

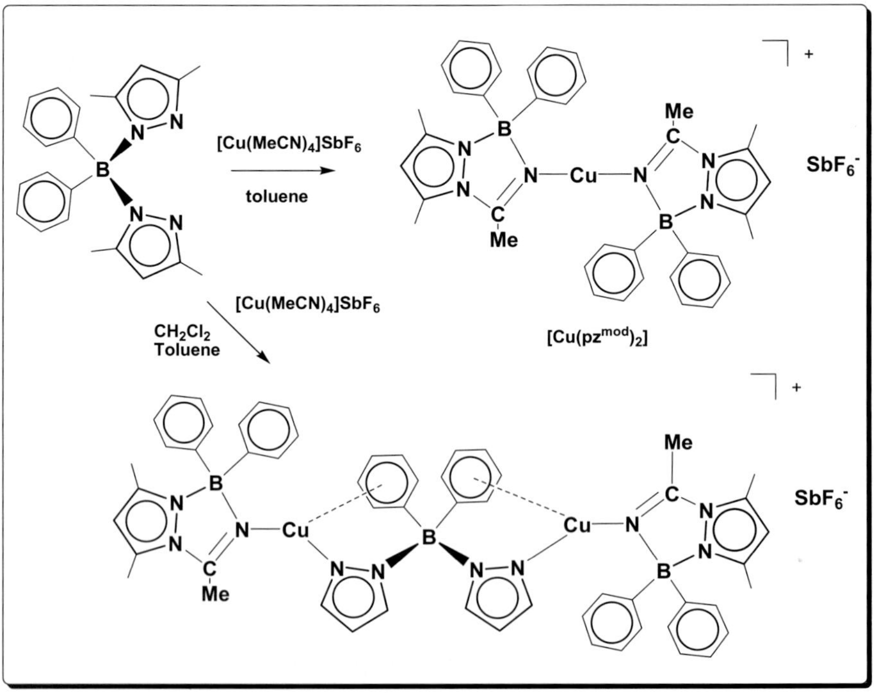

Scheme 1.4. Reaction of Ph_2Bp^* with $[Cu(MeCN)_4]SbF_6$.

1.8.43. *MeTp^{Ms} and MeTp^{tBu}*

MeTp^{Ms} has been synthesized by Dias and Wang as Tl salt.[202] The copper complex [CuCl(MeTp^{Ms})] has been employed to polymerize cleanly the aniline dimer, *N*-(4-aminophenyl)aniline under mild conditions to obtain polyaniline.[203] [Ag(CO)(MeTp^{Ms})], [Ag(C_2H_4)(MeTp^{Ms})] and [Tl(MeTp^{tBu})] have been also described and characterized.[204]

1.8.44. *"Third generation" scorpionates*

Daniel Reger introduce the term "third generation" poly(pyrazolyl)borate ligands to designate ligands that are specifically functionalized at the noncoordinating "back" position (Fig. 1.29).

1.8.44.1. *(p-IC$_6$H$_4$)Tp^R (R = H or Me)*

(*p*-IC_6H_4)Tp^x (Tp^x = Tp or Tp^{Me}), prepared according to Scheme 1.5, reacts with $FeBr_2$ giving the low-spin Fe[(*p*-IC_6H_4)Tp]$_2$

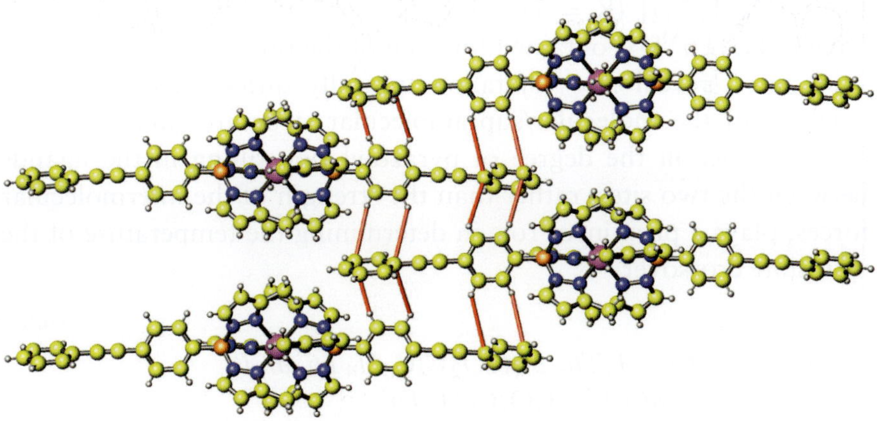

Fe[(*p*-$C_6H_5C{\equiv}CC_6H_4$)B(3-Mepz)$_3$]$_2$

Fig. 1.29. "Third generation" scorpionates.

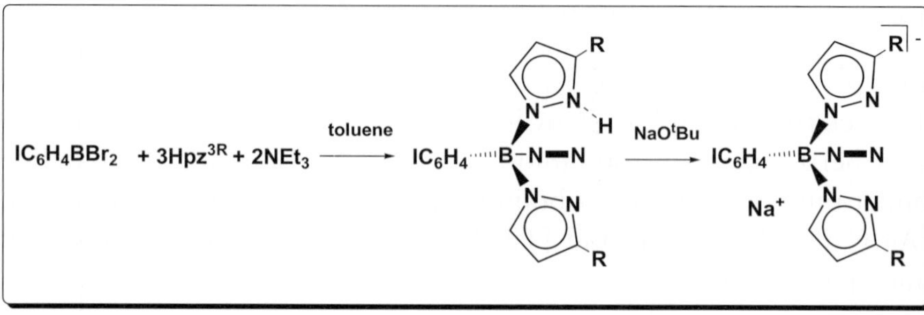

Scheme 1.5. Synthesis of [$(p\text{-}IC_6H_4)Tp^*$.

and the high-spin Fe[$(p\text{-}IC_6H_4)Tp^{Me}]_2$. The latter, can exist as two polymorphs.[205]

1.8.44.2. *(p-PhC$_2$C$_6$H$_4$)Tp^{Me}, (p-Me$_3$SiC$_2$C$_6$H$_4$)Tp^R, and (p-HC$_2$C$_6$H$_4$)Tp^R(R = H or Me)*

Sonogashira coupling reactions (Scheme 1.6) of terminal alkynes with [Fe{$(p\text{-}IC_6H_4)Tp^{Me}\}_2$] yielded [Fe{$(p\text{-}PhC_2C_6H_4)Tp^{Me}\}_2$], [Fe{$(p\text{-}Me_3SiC_2C_6H_4)Tp^R\}_2$] (R = H, Me), and [Fe{$(p\text{-}HC_2C_6H_4)Tp^R\}_2$] (R = H, Me). The crystal structures of [Fe{$(p\text{-}PhC_2C_6H_4)Tp^{Me}\}_2$] obtained for each of the three distinct electronic spin states, reveal two crystallographically different iron(II) sites. Analysis of the molecular/supramolecular structures indicates that the difference in the degree of pyrazolyl ring tilting in the ligands between the two sites, rather than the strength of the intermolecular forces, plays a prominent role in determining the temperature of the spin-state crossover.[206]

1.8.44.3. *p-BrC$_6$H$_4$Tp, p-(HO$_2$C)C$_6$H$_4$Tp, and p-(^{t}BuO-Phe-CO)C$_6$H$_4$Tp*

Kuchta and coworkers[207] reported the synthesis of p-BrC$_6$H$_4$Tp and [PtMe$_3${p-(HO$_2$C)C$_6$H$_4$Tp}] from [PtMe$_3${p-BrC$_6$H$_4$Tp}]. [PtMe$_3${p-(HO$_2$C)C$_6$H$_4$Tp}] can be readily coupled to biomolecules such as amino acids, as *L*-phenylalanine-*tert*-butylester, to provide species

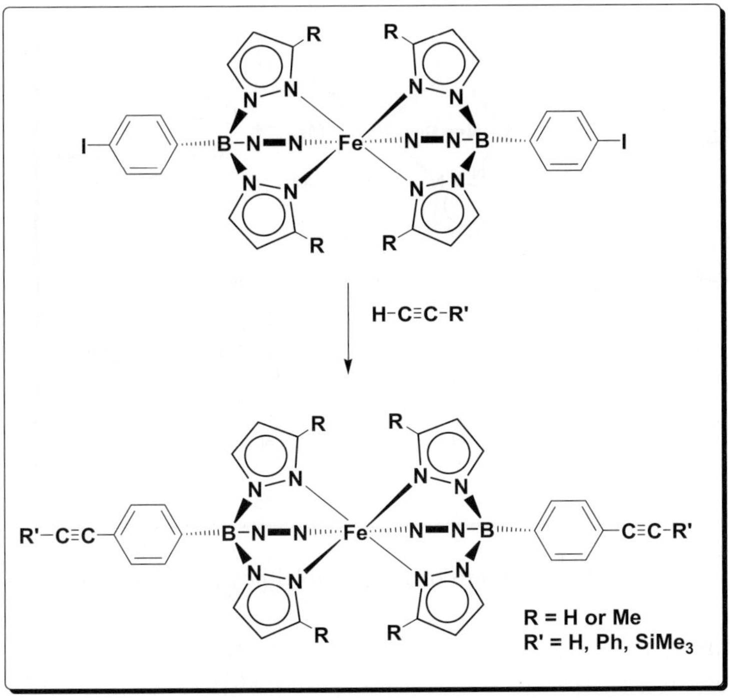

Scheme 1.6. Sonogashira coupling reactions for the synthesis of new scorpionates.

as [$PtMe_3${p-(^{t}BuO-Phe-CO)C_6H_4Tp}] (Scheme 1.7) which has been structurally characterized in the solid state by X-ray diffraction. This compound constitutes the first example of a tris(pyrazolyl)borate bioconjugate.[207]

1.8.44.4. *(pz)$_3$CCH$_2$OCH$_2$–C$_2$–C$_6$H$_4$Tp and {3,5-[(pz)$_3$CCH$_2$OCH$_2$]$_2$C$_6$H$_3$C$_2$}C$_6$H$_4$Tp*

The reaction of $HC_2CH_2OCH_2C(pz)_3$ with [Fe{(*p*-IC_6H_4)Tp}$_2$] yields the bitopic, metalloligand Fe[$(pz)_3CCH_2OCH_2$–C_2–C_6H_4-Tp]$_2$, structurally characterized. The analogous reaction with 3,5-[$(pz)_3CCH_2OCH_2$]$_2C_6H_3(C_2H)$ yields the tetratopic metalloligand Fe[{3,5-[$(pz)_3CCH_2OCH_2$]$_2C_6H_3C_2$}C_6H_4Tp]$_2$ (Fig. 1.30).[208]

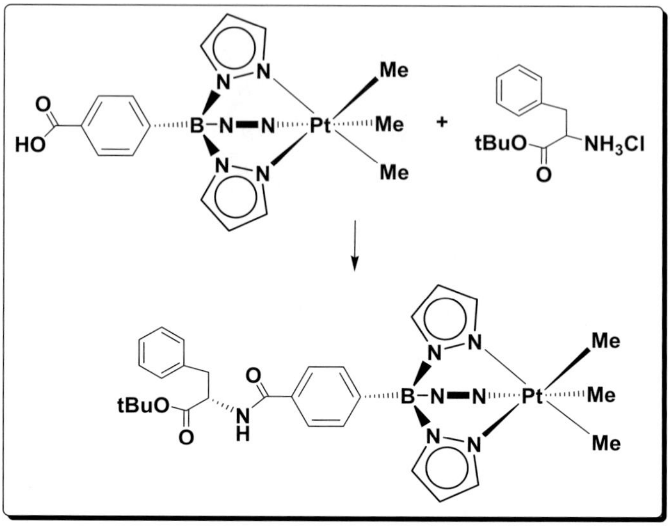

Scheme 1.7. Synthesis of tris(pyrazolyl)borate bioconjugate.

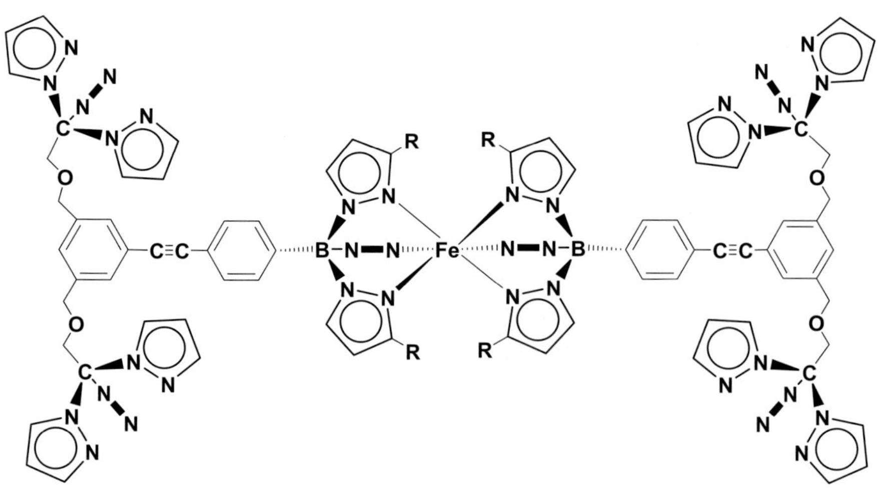

Fig. 1.30. The tetratopic metalloligand Fe[{3,5-[(pz)$_3$CCH$_2$OCH$_2$]$_2$C$_6$H$_3$C$_2$}C$_6$-H$_4$Tp]$_2$.

1.8.45. *Me_2NTp*

Bieller *et al.*[209] reported the syntheses and structures of the M(Me_2NTp) (M = Na, K), consisting of polymeric chains containing subunits where Me_2NTp acts as tridentate ligand toward Na^+ and K^+.

1.8.46. *Fc(Me)Bp, Fc_2Bp, and 1,1′-fc[(Me)Bp]$_2$*

Reaction of $FcB(Me)NMe_2$, Fc_2BNMe_2, and 1,1′-fc[$B(Me)NMe_2$]$_2$ with 1:1 mixtures of pyrazole and K(pz) in refluxing THF gave K[Fc(Me)Bp], K[Fc_2Bp], and K_2{1,1′-fc[(Me)Bp]$_2$}. K[Fc(Me)Bp] and K[Fc_2Bp] are centrosymmetric dimers with short K$\cdots$Cp contacts suggesting an η^5-coordination mode of the potassium ion. The crystal lattice of the ditopic ligand K_2{1,1′-fc[(Me)Bp]$_2$} consists of coordination polymer strands featuring essentially the same structural motif that has been observed for the monotopic derivatives. These scorpionate ligands are promising building blocks for the preparation of ferrocene-containing multiple-decker sandwich complexes.[210]

1.8.47. *1,1′-fc(BMe_2pz)$_2$*

Li^+, Na^+, K^+, Rb^+, and Cs^+ salts of the ditopic $\{1,1'\text{-fc}(BMe_2pz)_2\}^{2-}$ are multiple-decker sandwich complexes with the M^+ ions bound to the π faces of the ferrocene cyclopentadienyl rings in an η^5 manner. The lithium complex reveals discrete trimetallic entities with each lithium ion being coordinated by only one cyclopentadienyl ring. The sodium salt consists of polyanionic zig-zag chains where each Na^+ ion bridges the Cp of two ferrocene moieties. $[\text{–CpR–Fe–CpR–M}^+\text{–CpR–Fe–CpR–M}^+\text{–}]_\infty$ (R = $[\text{–BMe}_2\text{pz}]^-$, M = K, Rb, and Cs) exist as linear columns in the solid state.[211] In [Li_2{1,1′-fc(BMe_2pz)$_2$}] the lithium ions are coordinated by two solvent molecules (Et_2O), the pyrazolyl side-arm and the cyclopentadienyl ring. [Li_2{1,1′-fc(BMe_2pz)$_2$}] can be regarded as a trinuclear segment of the polymeric structure $[\text{–}(C_5H_4R)Li(C_5H_4R)Fe\text{–}]_\infty$ (R = BMe_2pz).[212]

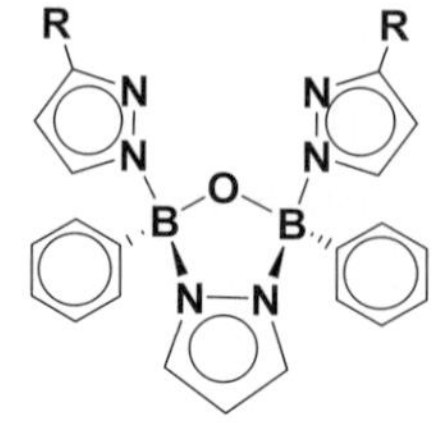

Fig. 1.31. [pz^R(Ph)B(μ-pz)(μ-O)B(Ph)pz^R)].

1.8.48. *[pz(Ph)B(μ-pz)(μ-O)B(Ph)pz)] and [pz^{Ph}(Ph)B(μ-pz)(μ-O)B(Ph)pz^{Ph})]*

Trans-coordinating tridentate poly(pyrazolyl)borate ligands have been reported by Wagner and coworkers who describe the synthesis of [pz(Ph)B(μ-pz)(μ-O)B(Ph)pz)] (Fig. 1.31) and of [pz^{Ph}(Ph)B-(μ-pz)(μ-O)B(Ph)pz^{Ph})] as well as the synthesis and the structural properties of [$FeCl_2${pz(Ph)B(μ-pz)(μ-O)B(Ph)pz)}], [CuCl{pz(Ph)B(μ-pz)(μ-O)B(Ph)pz)}], and [FeCl{pz(Ph)B(μ-pz)(μ-O)B(Ph)pz)}(py)].[213]

1.8.49. *[1,n-(^{t}BuBp)$_2$C$_6$H$_4$], [1,4-(Tp)$_2$C$_6$H$_4$], and [(Bp)$_2$(C$_6$H$_4$)$_2$]*

The Wagner research group reported the synthesis and crystal structure analysis of the lithium salt of the ditopic *p*-phenylene-bridged bis(pyrazol-1-yl)borate [Li_2{p-C_6H_4-(^{t}BuBp)$_2$}]. The dinuclear complex [Mn_2(THF)$_2$(μ-Cl)$_2${*p*-C_6H_4-(^{t}BuBp)$_2$}] has been obtained through a salt metathesis. Reaction of [Li_2{*p*-C_6H_4-(^{t}BuBp)$_2$}] with 2 equiv. of [Ti(NMe_2)$_3$Cl] gave the dinuclear titanium compound [Ti_2(NMe_2)$_6${*p*-C_6H_4-(^{t}BuBp)$_2$}], whereas reaction with [Ti(NMe_2)$_2$$Cl_2$] and water afforded [[$Ti_2$($NMe_2$)$_2$$Cl_2$]$_2$(μ-O){*p*-$C_6H_4$-(^{t}BuBp)$_2$}].[214]

[K_2{1,4-(^{t}BuBp)$_2$$C_6H_4$}], [(Li,K)$_2${1,3-(^{t}BuBp)$_2$$C_6H_4$}], and [$K_2$-{1,4-(Tp)$_2$$C_6H_4$}] have been prepared by using 1,4- and 1,3-diborylated benzene derivatives. [K_2{1,3-(^{t}BuBp)$_2$$C_6H_4$}] possesses a polymeric structure in the solid state. The unsymmetrically substituted hydrolysis product $[Li_2\{1,3\text{-}(^tBuBp)_2(^tBuB(OH)pz)C_6H_4\}]$ contains discrete dimers formed by four lithium cations encapsulated by the organic ligands. The 9,10-dihydro-9,10-diboraanthracene derivative $[K_2\{(Bp)_2(C_6H_4)_2\}]$ (Fig. 1.32) bears two pyrazolyl substituents at each of its boron atoms and stands out due to its stereorigidity.[215]

$[Tl_2\{1,3\text{-}C_6H_4(^tBuBp)_2\}]$, $[(PPh_3)Cu\{1,3\text{-}C_6H_4(^tBuBp)_2\}Cu(PPh_3)]$, and $[Ag(PPh_3)\{1,3\text{-}C_6H_4(^tBuBp)_2\}Ag(PPh_3)]$ containing the ditopic 1,3-phenylene-bridged heteroscorpionate (Fig. 1.33) have been prepared and investigated by ^1H, ^{13}C, ^{11}B, and ^{31}P NMR spectroscopy.[216]

1.8.50. *{CH₂=CHCH₂Bp(CH₂PPh₂)}, {CH₂=CHCH₂Tp}, and their carbosilane dendrimer*

Oro and coworkers described the synthesis of a novel hybrid pyrazolate/phosphine anionic ligand $\{CH_2{=}CHCH_2Bp(CH_2PPh_2)\}$

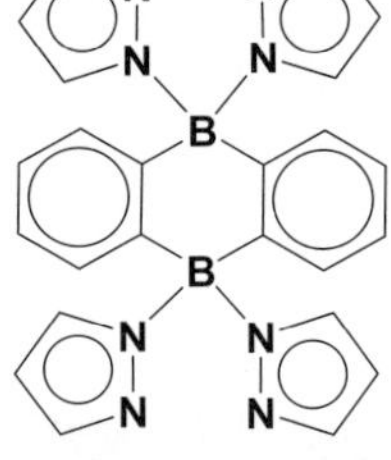

Fig. 1.32. The 9,10-dihydro-9,10-diboraanthracene $[K_2\{(Bp)_2(C_6H_4)_2\}]$.

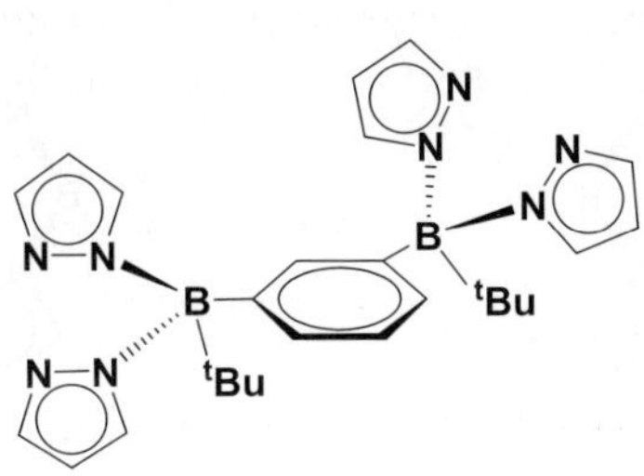

Fig. 1.33. The ditopic 1,3-phenylene-bridged $1,3\text{-}C_6H_4(^tBuBp)_2$.

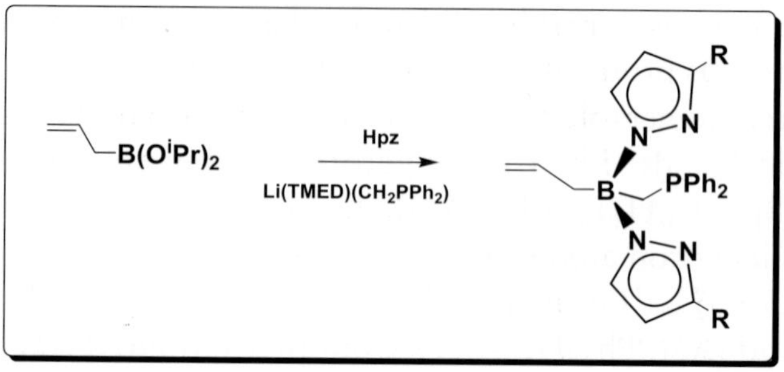

Scheme 1.8. Formation of a hybrid pyrazolate/phosphine ligand.

(Scheme 1.8). Coordination of this ligand in a *fac* tridentate fashion occurs in the complexes [M(COD){$CH_2{=}CHCH_2Bp(CH_2PPh_2)$}] (M = Rh, Ir). The ligand has been linked to the periphery of a carbosilane dendrimer, resulting in the polyanionic dendrimer [Li(TMED)]$_4$[Si{$(CH_2)_3SiMe_2(CH_2)_3Bp(CH_2PPh_2)$}$_4$] (Fig. 1.34), which gave the corresponding metallodendrimer with four rhodium atoms.[217]

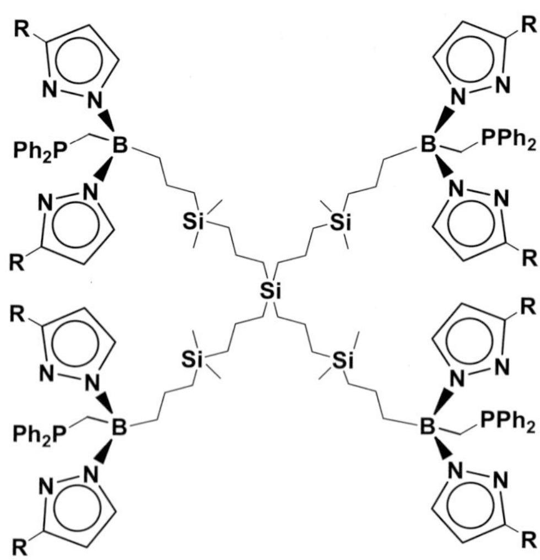

Fig. 1.34. The polyanionic dendrimer able to form a metallodendrimer with four Rh(I) olefin moieties.

A modified tris(pyrazolyl)borate ligand has been prepared analogously by reaction of $CH_2{=}CHCH_2B(O^iPr)_2$ with potassium pyrazolate and pyrazole to give $K[CH_2{=}CHCH_2Tp]$ which acts as a bi or tridentate ligand in its mononuclear complexes $[CH_2{=}CHCH_2\text{-}TpM(LL)]$ (M = Rh, LL = NBD; TFB; (CO)(PPh_3); M = Ir, LL = COD). The borate-containing dendrimers $Si[(CH_2)_3SiMe_2\text{-}(CH_2)_3B(O^iPr)_2]_4$, $Si[(CH_2)_3SiMe\{(CH_2)_3SiMe_2(CH_2)_3B(O^iPr)_2\}_2]_4$, and $Si[(CH_2)_3SiMe\{(CH_2)_3SiMe[(CH_2)_3SiMe_2(CH_2)_3B(O^iPr)_2]_2\}_2]_4$ react with Kpz and Hpz yielding the corresponding polyanionic dendrimers which contain 4, 8, and 16 Tp groups symmetrically located around the dendritic peripheries. These unusual polyanionic dendrimers are excellent scaffolds to support metal centers, as shown by their reactions with $[\{Rh(\mu\text{-}Cl)(NBD)\}_2]$.[218]

1.8.51. $HB(tz^{tBu,Me})_3$

3-*tert*-butyl-5-methyl-1,2,4-triazole reacted with KBH_4 to give tris(3-*tert*-butyl-5-methyl-1,2,4-triazolyl)borate, ($HB(tz^{tBu,Me})_3$). $HB(tz^{tBu,Me})_3$ gave the four-coordinate $[CoCl(HB(tz^{tBu,Me})_3)]$ and the five-coordinate $[CoNO_3(HB(tz^{tBu,Me})_3)]$ and $[Zn(OAc)(HB(tz^{tBu,Me})_3)]$ complexes. $K(HB(tz^{tBu,Me})_3)$, more water soluble than $K(Tp^{tBu,Me})$, can lead to functional models for enzyme active sites in an aqueous environment and yields water-soluble analogs of Tp catalysts.[219]

1.8.52. $H_2B(tz^*)_2$

$K[H_2B(tz^*)_2]$ (tz^* = 3,5-dimethyl-1,2,4-triazolyl) has been prepared and its reaction with copper and zinc salt investigated. The complexes $[Cu\{H_2B(tz^*)_2\}_2(H_2O)]$ and $[Zn\{H_2B(tz^*)_2\}_2(H_2O)]$ have been investigated by single-crystal X-ray, both being five-coordinated with four nitrogen atoms from triazolyl and one oxygen atom from the water molecule. In both compounds, the hydrogen atoms of H_2O are connected by hydrogen bonding with N atoms of two adjacent complex molecules to form 2-D planes.[220]

1.8.53. *{$H_2B(tz^{NO_2})_2$}*

{$H_2B(tz^{NO_2})_2$} has been prepared in DMAC and characterized as potassium salt. [MCl{$H_2B(tz^{NO_2})_2$}(H_2O)$_2$] (M = Zn or Cd) have been obtained by metathesis of K{$H_2B(tz^{NO_2})_2$} with $ZnCl_2$ and $CdCl_2$, respectively. The complexes likely contain a metal core in which the ligand is coordinated in the κ^2-N,N'- or κ^4-N,N',O,O'-fashion. A single-crystal structural characterization is reported for K{$H_2B(tz^{NO_2})_2$}, polymeric showing K $\cdots$ N and K $\cdots$ O interactions.[221]

[Cu{$H_2B(tz^{NO_2})_2$}(PR_3)$_2$], [Cu{$H_2B(tz^{NO_2})_2$}(dppe)] and [Cu{H_2-B(tz^{NO_2})$_2$}(PR_3)] have been synthesized from the reaction of CuCl with K{$H_2B(tz^{NO_2})_2$}, and mono or bidentate tertiary phosphines. Selected complexes have also been tested against a panel of several human tumor cell lines.[222]

[Ag{$H_2B(tz^{NO_2})_2$}(PR_3)$_x$] have been also synthesized and characterized. [Ag{$H_2B(tz^{NO_2})_2$}(P(m-Tol)$_3$)$_2$], in which the borate ligand acts as monodentate donor, has been also structurally characterized.[223]

1.8.54. *mt^{R}Bp (R = ^{t}Bu or ^{i}Pr)*

The anionic bis(pyrazolyl)(thioimidazolyl)borate ligands mt^{R}Bp (R = ^{t}Bu or ^{i}Pr) were prepared by reaction of K(Tp) with the corresponding thioimidazoles (Hmt^R) in the melt at 150°C. They were used in the synthesis of zinc complexes such as [ZnCl(mt^{tBu}Bp)] and [Zn(SC_6H_4-p-Cl)(mt^{iPr}Bp)], structurally characterized.[224] The same ligands react with zinc nitrate or zinc chloride and several thiolates, yielding mt^{R}Bp zinc–thiolate complexes. Structure determinations have been carried out, that evidenced tetrahedral ZnN_2S_2 coordination. Upon reaction with MeI, [ZnI(L)] (L = H_2B(pz)(mt^{Bu}), H_2B(mt^{iPr})$_2$) is formed.[225]

1.8.55. *{HB(pz^{Ph,Me})(mt$^{(2,6\text{-}Me_2Ph)}$)$_2$}*

{HB(pz^{Ph,Me})(mt$^{(2,6\text{-}Me_2Ph)}$)$_2$} (Fig. 1.35) reacts with Zn(ClO_4)$_2$ giving [Zn($OClO_3$){HB(pz^{Ph,Me})(mt$^{(2,6\text{-}Me_2Ph)}$)$_2$}], which partly

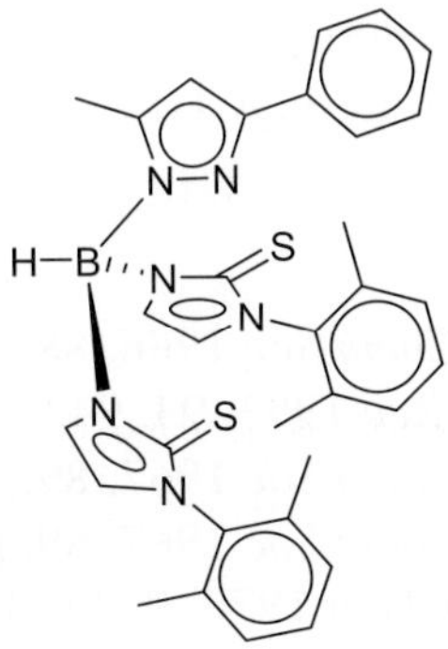

Fig. 1.35. $\{HB(pz^{Ph,Me})(mt^{(2,6\text{-}Me_2Ph)})_2\}$.

decomposes in CH_2Cl_2 upon attempts to replace the perchlorate ion by *p*-nitrophenolate. Two equivalents of the released $mt^{(2,6\text{-}Me_2Ph)}$ react with CH_2Cl_2 to form a dithioacetal able to coordinate two zinc moiety.[226]

1.8.56. *[MeB(Im^{N-Me})$_2$(pz^x)]*

The monoanionic tripodal ligands $[MeB(Im^{N-Me})_2(pz^x)]$ (Im^{N-Me} = 1-methylimidazol-2-yl, pz^x = pz^{Me}, pz^{Ph} or pz^{Me3}) have been synthesized and their rectivity toward Ni tested. It has been reported that the coordination mode (κ^3 vs κ^2) to the metal centers depends on the steric congestion around the boron center. The replacement of the B-bonded pz^{Me3} by acetate anion (Fig. 1.36) occurred

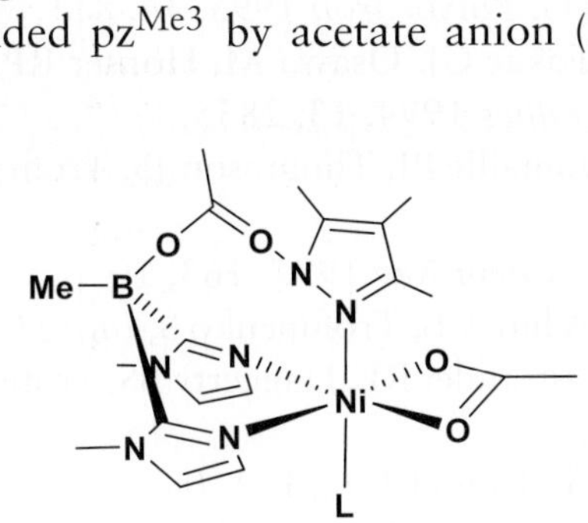

Fig. 1.36. The Ni complex containing the $[MeB(Im^{N-Me})_2(OAc)]$ ligand obtained upon replacement of pz^{Me3} by the acetate anion.

during the metathesis reaction of $[MeB(Im^{N-Me})_2(pz^{Me3})]$ with $Ni(OAc)_2 \cdot 4H_2O$ in $MeOH/CH_2Cl_2$.[227]

References

1. Trofimenko S, *J Am Chem Soc* 1966, **88**, 1842.
2. Trofimenko S, *Chem Rev* 1993, **93**, 943.
3. Trofimenko S, *J Am Chem Soc* 1967, **89**, 3904.
4. Trofimenko S, *J Am Chem Soc* 1967, **89**, 6288.
5. Trofimenko S, *Inorg Chem* 1970, **9**, 2493.
6. Calabrese JC, Trofimenko S, Thompson JS, *J Chem Soc Chem Commun* 1986, 1122.
7. Trofimenko S, Calabrese JC, Thompson JS, *Inorg Chem* 1987, **26**, 1507.
8. Garner M, Reglinski J, Cassidy I, Spicer MD, Kennedy AR, *Chem Commun* 1996, 1975.
9. Pettinari C, *Angew Chem Int Ed* 2007, **46**, 5652.
10. Curtis MD, Shiu K-B, Butler WM, *Organometallics* 1983, **2**, 1475.
11. Curtis MD, Shiu K-B, Butler WM, *J Am Chem Soc* 1986, **108**, 1550.
12. Calabrese JC, Domaille PJ, Trofimenko S, Long GJ, *Inorg Chem* 1991, **30**, 2795.
13. Rheingold AL, Haggerty BS, Trofimenko S, *J Chem Soc Chem Commun* 1994, 1973.
14. Rheingold AL, Liablc-Sands LM, Trofimenko S, *Chem Commun* 1997, 1691.
15. Han R, Parkin G, Trofimenko S, *Polyhedron* 1995, **14**, 387.
16. Yoon K, Parkin G, *Polyhedron* 1995, **14**, 811.
17. Le Cloux DD, Tokar GJ, Osawa M, Houser RP, Keyes MC, Tolman WB, *Organometallics* 1994, **13**, 2855.
18. Calabrese JC, Domaille PJ, Thompson JS, Trofimenko S, *Inorg Chem* 1990, **29**, 4429.
19. Janiak C, *Coord Chem Rev* 1997, **163**, 107.
20. Rheingold AL, White CB, Trofimenko S, *Inorg Chem* 1993, **32**, 3471.
21. Rheingold AL, Ostrander RL, Haggerty BS, Trofimenko S, *Inorg Chem* 1994, **33**, 3666.
22. Dias HVR, Jin W, Kim H-J, Lu H-L *Inorg Chem* 1996, **35**, 2317.
23. Dias HVR, Jin W, Wang Z, *Inorg Chem* 1997, **36**, 6205.
24. Rheingold AL, Haggerty BS, Liable Sands LM, Trofimenko S, *Inorg Chem* 1999, **38**, 6306.

25. McNeil JH, Watkins WC, Baird MC, Preston KF, *Organometallics* 1992, **11**, 2761.
26. Curtis MD, Shiu KB, Butler WM, Huffman JC, *J Am Chem Soc* 1986, **108**, 3335.
27. Shiu KB, Lee LY, *J Organomet Chem* 1988, **348**, 357.
28. Skagestad V, Tilset M, *J Am Chem Soc* 1993, **115**, 5077.
29. Tellers DM, Bergman RG, *J Am Chem Soc* 2000, **122**, 954.
30. Tellers DM, Skoog SJ, Bergman RG, Gunnoe TB, Harman WD, *Organometallics* 2000, **19**, 2428.
31. Kitajima N, Tolman WB, *Progress Inorg Chem* 1995, **43**, 419.
32. Mealli C, Arcus CS, Wilkinson JL, Marks TJ, Ibers JA, *J Am Chem Soc* 1976, **98**, 711.
33. Kitajima N, Koda T, Hashimoto S, Kitagawa T, Moro-oka Y, *J Am Chem Soc* 1991, **113**, 5664.
34. Ruggiero CE, Carrier SM, Antholine WE, Whittaker JW, Cramer CJ, Tolman WB, *J Am Chem Soc* 1993, **115**, 11285.
35. Bruce MI, Ostazewski APP, *J Chem Soc Dalton Trans* 1973, 2433.
36. Trofimenko S, Calabrese JC, Domaille PJ, Thompson JS, *Inorg Chem* 1989, **28**, 1091.
37. Reynolds SJ, Smith CF, Jones CJ, McCleverty JA, *Inorg Synth* 1985, **23**, 4.
38. De Bari H, Zimmer M, *Inorg Chem* 2004, **43**, 3344.
39. Sohrin Y, Kokusen H, Matsui M, *Inorg Chem* 1995, **34**, 3928.
40. Apostolidis C, Rebizant J, Kanellakopulos B, vonAmmon R, Dornberger E, Müller J, Powietzka B, Nuber B, *Polyhedron* 1997, **16**, 1057.
41. Han R, Parkin G, *Inorg Chem* 1992, **31**, 983.
42. Weis K, Vahrenkamp H, *Inorg Chem* 1997, **36**, 5589.
43. Trofimenko S, Calabrese JC, Thompson JS, *Inorg Chem* 1992, **31**, 974.
44. Chauby V, Serra Le Berre C, Kalck P, Daran J-C, Commenges G, *Inorg Chem* 1996, **35**, 6354.
45. Dias HVR, Lu H-L, *Inorg Chem* 2000, **39**, 2246.
46. Malbosc F, Kalck P, Daran J-C, Etienne M, *J Chem Soc Dalton Trans* 1999, 271.
47. Bucher UE, Currao A, Nesper R, Rüegger H, Venanzi LM, Younger E, *Inorg Chem* 1995, **34**, 66.
48. Gutiérrez E, Hudson SA, Monge A, Nicasio MC, Paneque M, Ruiz C, *J Organomet Chem* 1998, **551**, 215.

49. Paneque M, Sirol S, Trujillo M, Gutiérrez-Puebla E, Monge MA, Carmona E, *Angew Chem Int Ed* 2000, **39**, 218.
50. Amoroso AJ, Thompson AM, Jones PL, McCleverty JA, Ward MD, *J Chem Soc Chem Commun* 1994, 2751.
51. Amoroso AJ, Jeffery JC, Jones PL, McCleverty JA, Rees L, Rheingold AL, Sun Y, Takats J, Trofimenko S, Ward MD, Yap GPA, *J Chem Soc Chem Commun* 1995, 1881.
52. Keyes MC, Young Jr, VG, Tolman WB, *Organometallics* 1996, **15**, 4133.
53. Takahashi Y, Hikichi S, Akita M, Moro-oka Y, *Organometallics* 1999, **18**, 2571.
54. Slugovc C, Mereiter K, Trofimenko S, Carmona E, *Chem Commun* 2000, 121.
55. Hu Z, Gorun SM, *Inorg Chem* 2001, **40**, 667.
56. Dinnebier RE, Behrens U, Olbrich F, *Organometallics* 1997, **16**, 3855.
57. Trofimenko S, *Acc Chem Res* 1971, **4**, 17.
58. Trofimenko S, *Chem Rev* 1972, **72**, 497.
59. Trofimenko S, *Progress Inorg Chem* 1986, **34**, 115.
60. Shaver A, *J Organomet Chem Libr* 1977, **3**, 157.
61. McCleverty JA, *Chem Soc Rev* 1983, **12**, 331.
62. Niedenzu K, Advances in Boron and the Boranes, Molecular Structure and Energetics Vol. 5, Liebman JF, Greenberg JA, Williams RE, (eds.), VXH Publishers, p. 357, 1988.
63. Byers PK, Canty AJ, Honeyman RT, *Adv Organomet Chem* 1992, **34**, 1.
64. Kitajima N, Moro-oka Y, *Chem Rev* 1994, **94**, 737.
65. Santos I, Marques N, *New J Chem* 1995, **19**, 551.
66. Han R, Looney A, McNeill K, Parkin G, Rheingold AL, Haggerty BS, *J Inorg Biochem* 1993, **49**, 105.
67. Parkin G, *Handbook of Grignard Reagents*, Silverman GS, Rakita PE, (eds.), Marcel Dekker, Inc. p. 291, 1996.
68. Parkin G, *Adv Inorg Chem* 1995, **42**, 291.
69. Etienne M, *Coord Chem Rev* 1996, **156**, 201.
70. Reger DL, *Coord Chem Rev* 1996, **147**, 571.
71. Janiak C, *Main Group Met Chem* 1998, **21**, 33.
72. McCleverty JA, Ward MD, *Acc Chem Res* 1998, **31**, 842.
73. Young CG, Wedd AG, *Chem Commun* 1997, 1251.

74. Theopold KH, Reinaud OM, Doren D, Konecny R, *Stud Surf Sci Catal* (3rd World Congress on Oxidation Catalysis) 1997, **110**, 1081.
75. Trofimenko S, *Scorpionates, The coordination Chemistry of Polypyrazolylborate Ligands*, Imperial College Press, 1999.
76. Templeton JL, *J Am Chem Soc* 2000, **122**, 5670.
77. Trofimenko S, *J Chem Ed* 2005, **82**, 1715.
78. Ritter SK, *Chem Eng News* 2003, **81**, 40.
79. Trofimenko S, *Polyhedron* 2004, **23**, 197.
80. Sadimenko AP, *Adv Heter Chem* 2001, **81**, 167.
81. Shaver A, *Compr Coord Chem* 1987, **2**, Ch. 13.6, 245.
82. Pettinari C, Santini C, *Compr Coord Chem II* 2004, **1**, 159
83. Pettinari C, *La Chimica e l'Industria* 2004, **86**, 94.
84. Díaz-Requejo MM, Pérez PJ, *J Organomet Chem* 2005, **690**, 5441.
85. Díaz-Requejo MM, Belderrain TR, Nicasio MC, Pérez PJ, *Dalton Trans* 2006, 5559.
86. Piers WE, Emslie DJH, *Coord Chem Rev* 2002, **233–234**, 131.
87. Akita M, Hikichi S, Moro-oka Y, *J Synth Org Chem Jpn* 1999, **57**, 619.
88. Hikichi S, Akita M, Moro-oka Y, *Coord Chem Rev* 2000, **198**, 61.
89. Mirica LM, Ottenwaelder X, Stack TDP, *Chem Rev* 2004, **104**, 1013.
90. Akita M, Hikichi S, *Bull Chem Soc Jpn* 2002, **75**, 1657.
91. Akita M *J Organomet Chem* 2004, **689**, 4540.
92. Wasser IM, de Vries S, P. Moënne-Loccoz, Schröder I, Karlin KD, *Chem Rev* 2002, **102**, 1201.
93. Vahrenkamp H, *Acc Chem Res* 1999, **32**, 589.
94. Parkin G, *Chem Commun* 2000, 1971.
95. Parkin G, *Metal Ions in Biological Systems*, Vol. 38, Marcel Dekker Inc., p. 411, 2001.
96. Parkin G, *Chem Rev* 2004, **104**, 699.
97. Edelmann FT, *Angew Chem Int Ed* 2001, **40**, 1656.
98. Gherman BF, Heppner DE, Tolman WB, Cramer CJ, *J Biol Inorg Chem* 2006, **11**, 197.
99. Tolman WB, *J Biol Inorg Chem* 2006, **11**, 261.
100. Pettinari C, Cingolani A, Gioia Lobbia G, Marchetti F, Martini D, Pellei M, Pettinari R, Santini C, *Polyhedron* 2004, **23**, 451.
101. Dias HVR, Fianchini M, *Comm Inorg Chem* 2007, **28**, 73.
102. Slugovc C, Padilla-Martínez I, Sirol S, Carmona E, *Coord Chem Rev* 2001, **213**, 129.

103. Carmona E, Paneque M, Poveda ML, *Dalton Trans* 2003, 4022.
104. Carmona E, Paneque M, Santos LL, Salazar V, *Coord Chem Rev* 2005, **249**, 1729.
105. Slugovc C, Schmid R, Kirchner K, *Coord Chem Rev* 1999, **185–186**, 109.
106. Jia G, Lau C-P, *Coord Chem Rev* 1999, **190–192**, 83.
107. Park S, Han Y, Kim SK, Lee J, Kim HK, Do Y, *J Organomet Chem* 2004, **689**, 4263.
108. Marques N, Sella A, Takats J, *Chem Rev* 2002, **102**, 2137.
109. Sessler JL, Melfi PJ, Dan Pantos G, *Coord Chem Rev* 2006, **250**, 816.
110. Paulo A, Correia JD, Santos I, *Trends Inorg Chem* 1998, **5**, 57.
111. Paulo A, Correja JDG, Campello MPC, Santos I, *Polyhedron* 2004, **23**, 331.
112. Santos I, Paulo A, Correia JD, *Top Curr Chem* 2005 **252**, 45.
113. Keane JM, Harman WD, *Organometallics* 2005, **24**, 1786.
114. Jones WD, Vetter AJ, Wick DD, Northcutt TO, *ACS Symp Series* 2004, **885**, 56.
115. Norris CM, Templeton JL, *ACS Symp Ser* 2004, **885**, 303.
116. Power PP, *Chem Rev* 1999, **99**, 3463.
117. Lescouëzec R, Toma LM, Vaissermann J, Verdaguer M, Delgado FS, Ruiz-Pérez C, Lloret F, Julve M, *Coord Chem Rev* 2005, **249**, 2691.
118. Ward MD, McCleverty JA, Jeffery JC, *Coord Chem Rev* 2001, **222**, 251.
119. Long GJ, Grandjean F, Reger DL, *Top Curr Chem* 2004, **233**, 91.
120. Coordination and activation of B–H and other X–H and X–Y bonds, in *Metal Dihydrogen and σ-Bond Complexes*, Kubas (ed.), SpringerLink, pp. 417–439, Chap. 13, 2002.
121. Coordination and activation of Si–H, Ge–H and Sn–H bonds in *Metal Dihydrogen and σ-Bond Complexes*, Kubas (ed.), SpringerLink, pp. 327–364, Chap. 11, 2002.
122. Kisała J, Białońska A, Ciunik Z, Kurek S, Wołowiec S, *Polyhedron* 2006, **25**, 3222.
123. Ferrence GM, Takats J, *J Organomet Chem* 2002, **647**, 84.
124. Fronhapfel DS, Templeton JL, *Coord Chem Rev* 2000, **206–207**, 199.
125. Webster CE, Hall MB, *Coord Chem Rev* 2003, **238–239**, 315.
126. Pettinari C, *Main Group Met Chem* 1999, **22**, 661.
127. Pettinari C, *Recent Res Devel Organomet Chem* 2001, **4**, 71.
128. Fujisawa K, Okamoto K, *Kagaku* 2007, **62**, 64.

129. Rheingold AL, Yap GPA, Trofimenko S, *Inorg Chem* 1995, **34**, 759.
130. Rheingold AL, Haggerty BS, Yap GPA, Trofimenko S, *Inorg Chem* 1997, **36**, 5097.
131. Trofimenko S, *Inorg Synth* 1970, **12**, 99.
132. Janiak C, Braun L, Girgsdies F, *J Chem Soc Dalton Trans* 1999, 3133.
133. McCurdy Jr, WH, *Inorg Chem* 1975, **14**, 2292.
134. Niedenzu K, Niedenzu PM, Warner KR, *Inorg Chem* 1985, **24**, 1604.
135. Cano M, Heras JV, Jones CJ, McCleverty JA, Trofimenko S, *Polyhedron* 1990, **9**, 619.
136. Ghosh CK, Hoyano JK, Krentz R, Graham WAG, *J Am Chem Soc* 1989, **111**, 5480.
137. Han R, Ghosh P, Desrosiers PJ, Trofimenko S, Parkin G, *J Chem Soc Dalton Trans* 1997, 3713.
138. Lopez C, Claramunt RM, Trofimenko S, Elguero J, *Can J Chem* 1993, **71**, 678.
139. Niedenzu K, Trofimenko S, *Topics Current Chem* 1986, **131**, 1.
140. Calabrese JC, Trofimenko S, *Inorg Chem* 1992, **31**, 4810.
141. Albinati A, Bovens M, Ruegger H, Venanzi LM, *Inorg Chem* 1997, **36**, 5991.
142. Rheingold AL, Liable-Sands LM, Golen JA, Yap GPA, Trofimenko S, *Dalton Trans* 2004, 598.
143. Rheingold AL, Incarvito CD, Trofimenko S, *J Chem Soc Dalton Trans* 2000, 1233.
144. Rheingold AL, Liable-Sands LM, Golan JA, Trofimenko S, *Eur J Inorg Chem* 2003, 2767.
145. Rheingold AL, Liable-Sands LM, Trofimenko S, *Inorg Chem* 2000, **39**, 1333.
146. Maldonado Calvo JA, Vahrenkamp H, *Inorg Chim Acta* 2006, **359**, 4079.
147. Trofimenko S, Rheingold AL, Liable-Sands LM, *Inorg Chem* 2002, **41**, 1889.
148. Rheingold AL, Zakharov LN, Trofimenko S, *Inorg Chem* 2003, **42**, 827.
149. Song D, Jia WL, Wu G, Wang S, *Dalton Trans* 2005, 433.
150. Saito T, Kuwata S, Ikariya T, *Chem Lett* 2006, **35**, 1224.
151. Claramunt RM, Santa María D, Elguero J, Trofimenko S, *Polyhedron* 2004, **23**, 2985.

152. Rheingold AL, Liable-Sands LM, Incarvito CL, Trofimenko S, *J Chem Soc Dalton Trans* 2002, 2297.
153. Caballero A, Díaz-Requejo MM, Belderraín TR, Nicasio MC, Trofimenko S, Pérez PJ, *J Am Chem Soc* 2003, **125**, 1446.
154. Díaz-Requejo MM, Belderraín TR, Nicasio MC, Trofimenko S, Pérez PJ, *J Am Chem Soc* 2003, **125**, 12078.
155. Fructos MR, Trofimenko S, Díaz-Requejo MM, Pérez PJ, *J Am Chem Soc* 2006, **128**, 11784.
156. Morilla ME, Díaz-Requejo MM, Belderrain TR, Nicasio MC, Trofimenko S, Pérez PJ, *Organometallics* 2004, **23**, 293.
157. Urbano J, Izarra R, Gómez-Ariza JL, Trofimenko S, Diaz-Requejo MM, Peréz PJ, *Chem Commun* 2006, 1000.
158. Urbano J, Belderraín TR, Nicasio MC, Trofimenko S, Díaz-Requejo MM, Pérez PJ, *Organometallics* 2005, **24**, 1528.
159. Rodríguez P, Díaz-Requejo MM, Belderraín TR, Trofimenko S, Nicasio MC, Pérez PJ, *Organometallics* 2004, **23**, 2162.
160. Yap GPA, Jove F, Urbano J, Alvarez E, Trofimenko S, Díaz-Requejo MM, Pérez PJ, *Inorg Chem* 2007, **46**, 780.
161. Rheingold AL, Yap GPA, Zakharov LN, Trofimenko S, *Eur J Inorg Chem* 2002, 2335.
162. Rheingold AL, Liable-Sands LM, Trofimenko S, *Inorg Chem* 2001, **40**, 6509.
163. Biagini P, Calderazzo F, Marchetti F, Romano AM, Spera S, *J Organomet Chem* 2006, **691**, 4172.
164. Davies GM, Jeffery JC, Ward MD, *New J Chem* 2003, **27**, 1550.
165. Adams H, Batten SR, Davies GM, Duriska MB, Jeffery JC, Jensen P, Lu J, Motson GR, Coles SJ, Hursthouse MB, Ward MD, *Dalton Trans* 2005, 1910.
166. Bell ZR, Motson GR, Jeffery JC, McCleverty JA, Ward MD, *Polyhedron* 2001, **20**, 2045.
167. Mann KLV, Jeffery JC, McCleverty JA, Ward MD, *Polyhedron* 1999, **18**, 721.
168. Pérez Olmo C, Böhmerle K, Steinfeld G, Vahrenkamp H, *Eur J Inorg Chem* 2006, 3869.
169. Ibrahim MM, Pérez Olmo C, Tekeste T, Seebacher J, He G, Maldonado Calvo JA, Böhmerle K, Steinfeld G, Brombacher H, Vahrenkamp H, *Inorg Chem* 2006, **45**, 7493.

170. Pérez Olmo C, Böhmerle K, Vahrenkamp H, *Inorg Chim Acta* 2007, **360**, 1510.
171. Reeves ZR, Mann KLV, Jeffery JC, McCleverty JA, Ward MD, Barigelletti F, Armaroli N, *J Chem Soc Dalton Trans* 1999, 349.
172. Fujisawa K, Tada N, Ishikawa Y, Higashimura H, Miyashita Y, Okamoto KI, *Inorg Chem Commun* 2004, 7, 209.
173. Rheingold AL, Incarvito CD, Trofimenko S, *Inorg Chem* 2000, **39**, 5569.
174. Siemer CJ, Goswami N, Kahol PK, Van Stipdonk MJ, Eichhorn DM, *Inorg Chem* 2001, **40**, 4081.
175. Siemer CJ, VanStipdonk MJ, Kahol PK, Eichhorn DM, *Polyhedron* 2004, **23**, 235.
176. Carella A, Vives G, Cox T, Jaud J, Rapenne G, Launay J-P, *Eur J Inorg Chem* 2006, 980.
177. Singh UP, Babbar P, Hassler B, Nishiyama H, Brunner H, *J Mol Catal A Chem* 2002, **185**, 33.
178. Chisholm MH, Iyer SS, Streib WE, *New J Chem* 2000, **24**, 393.
179. Batten SR, Duriska MB, Jensen P, Lu J, *Aust J Chem* 2007, **60**, 72.
180. Hammes BS, Carrano MW, Carrano CJ, *J Chem Soc Dalton Trans* 2001, 1448.
181. Hammes BS, Luo X, Carrano MW, Carrano CJ, *Angew Chem* 2002, **114**, 3393.
182. Hammes BS, Luo X, Chohan BS, Carrano MW, Carrano CJ, *J Chem Soc Dalton Trans* 2002, 3374.
183. Hammes BS, Luo X, Carrano MW, Carrano CJ, *Inorg Chim Acta* 2002, **341**, 33.
184. Alidori S, Pellei M, Pettinari C, Santini C, Skelton BW, White AH, *Inorg Chem Commun* 2004, 7, 1075.
185. Bandoli G, Dolmella A, Gioia Lobbia G, Papini G, Pellei M, Santini C, *Inorg Chim Acta* 2006, **359**, 4036.
186. Chisholm MH, Gallucci JC, Yaman G, *Chem Commun* 2006, 1872.
187. Kisko JL, Hascall T, Kimblin C, Parkin G, *J Chem Soc Dalton Trans* 1999, 1929.
188. Graziani O, Hamon P, Thépot J-Y, Toupet L, Szilágyi PA, Molnár G, Bousseksou A, Tilset M, Hamon J-R, *Inorg Chem* 2006, **45**, 5661.
189. Graziani G, Toupet L, Hamon J-R, Tilset M, *Inorg Chim Acta* 2002, **341**, 127.

190. Bieller S, Bolte M, Lerner H-W, Wagner M, *J Organomet Chem* 2005, **690**, 1935.
191. Bieller S, Haghiri A, Bolte M, Bats JW, Wagner M, Lerner H-W, *Inorg Chim Acta* 2006, **359**, 1559.
192. Herdtweck E, Peters F, Scherer W, Wagner M, *Polyhedron* 1998, **17**, 1149.
193. Guo S-L, Bats JW, Bolte M, Wagner M, *J Chem Soc Dalton Trans* 2001, 3572.
194. Ilkhechi AH, Guo S-L, Bolte M, Wagner M, *Dalton Trans* 2003, 2303.
195. Kunz K, Vitze H, Bolte M, Lerner H-W, Wagner M, *Organometallics* 2007, **26**, 4463.
196. Roitershtein D, Domingos Â, Marques N, *Organometallics* 2004, **23**, 3483.
197. Bailey PJ, Pinho P, Parsons S, *Inorg Chem* 2003, **42**, 8872.
198. Dias HVR, Wang X, Diyabalanage HVK, *Inorg Chem* 2005, **44**, 7322.
199. Dias HVR, Wu J, Wang X, Rangan K, *Inorg Chem* 2007, **46**, 1960.
200. Dias HVR, Wang X, *Dalton Trans* 2005, 2985.
201. Schneider JL, Young Jr, VG, Tolman WB, *Inorg Chem* 2001, **40**, 165.
202. Dias HVR, Wang X, *Polyhedron* 2004, **23**, 2533.
203. Dias HVR, Wang X, Rajapakse RMG, Elsenbaumer RL, *Chem Commun* 2006, 976.
204. Dias HVR, Fianchini M, *Angew Chem Int Ed* 2007, **46**, 2188.
205. Reger DL, Gardinier JR, Smith MD, Shahin AM, Long GJ, Rebbouh L, Grandjean F, *Inorg Chem* 2005, **44**, 1852.
206. Reger DL, Gardinier JR, Gemmill WR, Smith MD, Shahin AM, Long GJ, Rebbouh L, Grandjean F, *J Am Chem Soc* 2005, **127**, 2303.
207. Kuchta MC, Gemel C, Metzler-Nolte N, *J Organomet Chem* 2007, **692**, 1310.
208. Reger DL, Gardinier JR, Bakbak S, Semeniuc RF, Bunz UHF, Smith MD, *New J Chem* 2005, **29**, 1035.
209. Bieller S, Bolte M, Lerner H-W, Wagner M, *Z Anorg Allg Chem* 2006, **632**, 319.
210. Ilkhechi AH, Bolte M, Lerner H-W, Wagner M, *J Organomet Chem* 2005, **690**, 1971.
211. Ilkhechi AH, Mercero JM, Silanes I, Bolte M, Scheibitz M, Lerner H-W, Ugalde JM, Wagner M, *J Am Chem Soc* 2005, **127**, 10656.

212. Ilkechi AH, Scheibitz M, Bolte M, Lerner H-W, Wagner M, *Polyhedron* 2004, **23**, 2597.
213. Bieller S, Bolte M, Lerner H-W, Wagner M, *Chem Eur J* 2006, **12**, 4735.
214. Bieller S, Bolte M, Lerner H-W, Wagner M, *Inorg Chem* 2005, **44**, 9489.
215. Bieller S, Zhang F, Bolte M, Bats JW, Lerner H-W, Wagner M, *Organometallics* 2004, **23**, 2107.
216. Zhang F, Bolte M, Lerner H-W, Wagner M, *Organometallics* 2004, **23**, 5075.
217. Casado MA, Hack V, Camerano JA, Ciriano MA, Tejel C, Oro LA, *Inorg Chem* 2005, **44**, 9122.
218. Camerano JA, Casado MA, Ciriano MA, Oro LA, *Dalton Trans* 2006, 5287.
219. Jernigan FE, Sieracki NA, Taylor MT, Jenkins AS, Engel SE, Rowe BW, Jové FA, Yap GPA, Papish ET, Ferrence GM, *Inorg Chem* 2007, **46**, 360.
220. Shen W-Z, Kang F, Sun Y-J, Cheng P, Yan S-P, Liao D-Z, Jiang Z-H, *Inorg Chem Commun* 2003, **6**, 408.
221. Pellei M, Benetollo F, Gioia Lobbia G, Alidori S, Santini C, *Inorg Chem* 2005, **44**, 846.
222. Marzano C, Pellei M, Alidori S, Brossa A, Gioia Lobbia G, Tisato F, Santini C, *J Inorg Biochem* 2006, **100**, 299.
223. Santini C, Pellei M, Alidori S, Gioia Lobbia G, Benetollo F, *Inorg Chim Acta* 2007, **360**, 2121.
224. Benkmil B, Ji M, Vanhrenkamp H, *Inorg Chem* 2004, **43**, 8212.
225. Ji M, Benkmil B, Vahrenkamp H, *Inorg Chem* 2005, **44**, 3518.
226. He G, Vahrenkamp H, *Z Anorg Allg Chem* 2007, **633**, 51.
227. Fujita K, Hikichi S, Akita M, Moro-oka Y, *J Chem Soc Dalton Trans* 2000, 117.

Chapter 2

Homoscorpionates — First Generation

2.1. General Considerations

This chapter summarizes all metal complexes, reported from 1999 to date, containing the three ligands Tp, pzTp, and Tp*.

2.2. Group 1: Li, Na, and K

2.2.1. *Li*

The compound [Li(Tp*)(Hpz*)]·Hpz* is the unique reported lithium compound, structurally characterized, containing a homoscorpionate ligand. It shows a LiN_4 distorted tetrahedral geometry with a tridentate Tp* (Fig. 2.1). The crystal packing is dominated by hydrophobic methyl–methyl interactions.[1]

2.2.2. *Na*

The structure of Na(Tp) is found to be an infinite ribbon containing monodentate, bridging and pendant pyrazolyl units (Fig. 2.2). The coordination sphere of the sodium center in Na(Tp) is completed by two water molecules.[2]

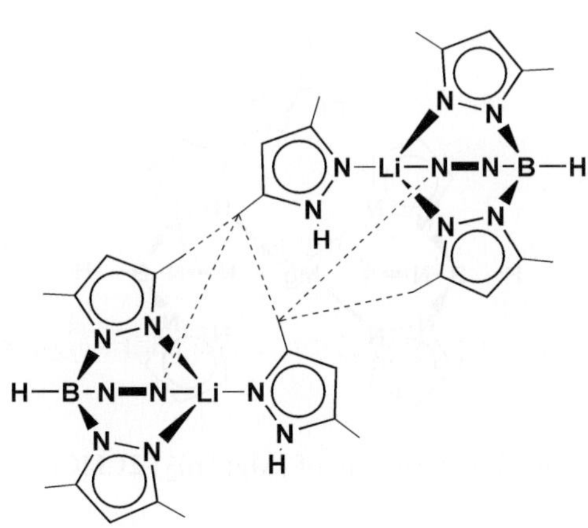

Fig. 2.1. The hydrophobic Me–Me interactions in [Li(Tp*)(Hpz*)]·Hpz*.

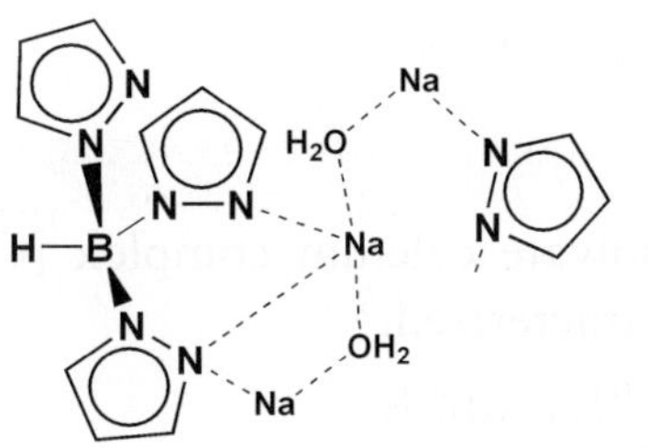

Fig. 2.2. Structure of Na(Tp).

2.2.3. *K*

The electronic structure of K(Tp*) has been investigated using gas-phase photoelectron spectroscopy and DFT calculations, also reported for K(Tp).[3]

2.3. Group 2: Mg and Ca

2.3.1. *Mg*

The chloroform disolvate $[Mg(Tp)_2]\cdot 2CHCl_3$ has been structurally characterized and its geometry is very similar to that found for the unsolvated $[Mg(Tp)]_2$.[4] The inclusion of chloroform

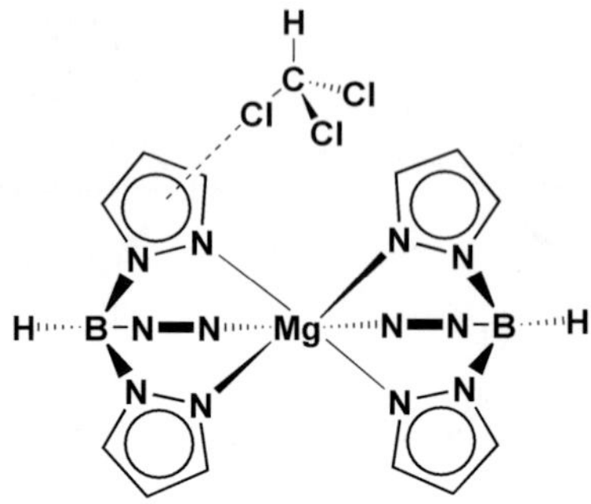

Fig. 2.3. Structure of $[Mg(Tp)_2]\cdot 2CHCl_3$.

molecules on three-fold rotation axes plays a role in stabilizing the three-dimensional supramolecular architecture through facial pz $\cdots$ Cl $\cdots$ pz interactions (Fig. 2.3).[5]

2.3.2. *Ca*

The chloroform disolvate calcium complex $[Ca(Tp)_2]\cdot 2CHCl_3$ has been structurally characterized.[5]

2.4. Group 3: Sc and Y (lanthanides and actinides are listed separately)

2.4.1. *Sc*

Tp* reacts with $ScCl_3(THF)_3$ yielding $[ScCl_2(Tp^*)(THF)]$. Its reaction with alkyl lithium reagents LiR (R = Me, CH_2SiMe_3, $CH(SiMe_3)_2$) gave in most of the cases lithium salts of Tp*. $[Sc(Tp^*)(CH_2SiMe_3)_2(THF)]$, fluxional at room temperature and structurally characterized, has been synthesized via alkane elimination.[6]

2.4.2. *Y*

$[YCl_2(Tp^*)(L)]$ (L = bipy, phen) and $[YCl_3(Tp^*)][Na(L)_3]$ (L = dmp) have been prepared and characterized by spectroscopy.

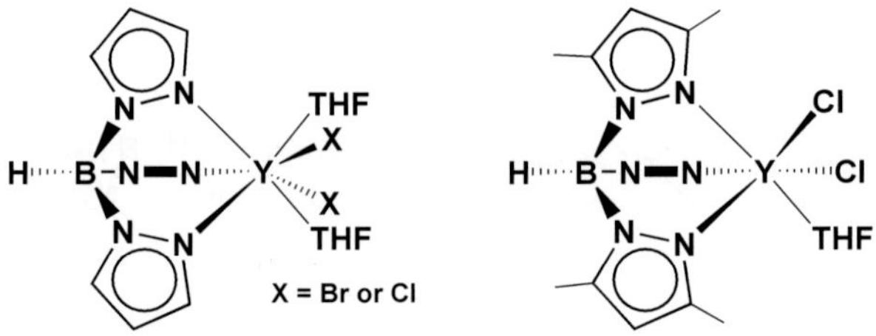

Fig. 2.4. The $[YX_2(Tp^x)(THF)_n]$ complexes ($n = 1$ or 2) described in Ref. 8.

$[YCl_2(Tp^*)(bipy)]$ is an entry for the synthesis of complexes containing bipy in a radical anionic form. Reaction of $[YCl_2(Tp^*)(bipy)]$ with an excess of sodium amalgam gives $[Y(Tp^*)_2(bipy)(THF)_2]$.[7]

$[YX_2(Tp^x)(solvent)_n]$ ($Tp^x = Tp$ or Tp^*; X = Cl, Br, or I, solvent = THF or pz*), $[YCl(Tp^*)(Bp^*)]$, and $[YBr(Tp)(Bp^{Ph})]$ have been prepared and characterized. Steric, dynamic, and coordination behavior are heavily influenced by the choice of the scorpionate ligand substituents. The THF molecule in the complexes undergoes a dynamic solvation/desolvation equilibrium (Fig. 2.4). $[YCl_2(Tp^*)(Hpz^*)]$, distorted capped octahedral, has also been structurally characterized.[8]

2.5. Group 4: Ti, Zr, and Hf

2.5.1. *Ti*

$[Ti(Tp)_2]$, a stable hard donor complex, has been demonstrated to possess a controlled reaction chemistry with simple chalcogen oxidants. $[Ti(Tp)_2]$, $[Ti(Tp)_2]PF_6$, $[Ti{=}Se(\kappa^2\text{-}Tp)(\kappa^3\text{-}Tp)]$, $[Ti(\kappa^2\text{-}Tp)(\kappa^3\text{-}Tp)(\kappa^2\text{-}S_2)]$, and $[Ti(\kappa^2\text{-}Tp)(\kappa^3\text{-}Tp)]_2(\mu\text{-}O)$ (Fig. 2.5) have been prepared and characterized by X-ray.[9]

The electronic features and photochemistry of $[TiCl_3(Tp)]$ and $[TiCl_3(Tp^*)]$ in THF have been investigated, and it has been reported that reactive decay of the excited states produced either $Tp^{\blacksquare}$ or $Tp^{\blacksquare *}$, respectively, and metal center Ti(III) radicals via homolytic cleavage of the Tp → Ti (Tp* → Ti) bond. $[TiCl_2(Tp)(THF)]$ and $[TiCl_2(Tp^*)(THF)]$ were prepared by stoichiometric reduction of

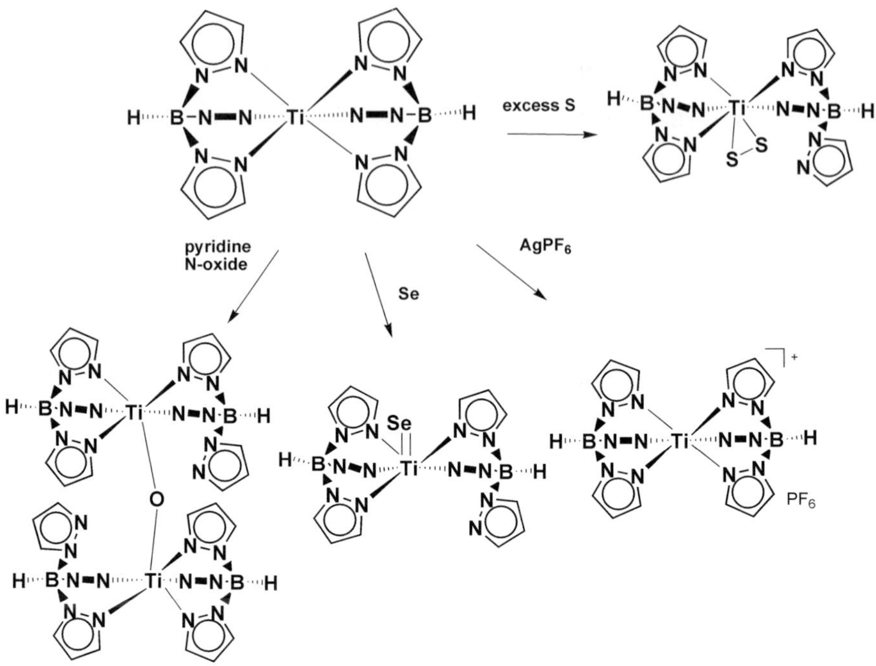

Fig. 2.5. Reactivity of $[Ti(Tp)_2]$ toward chalcogen oxidants.

$[TiCl_3(Tp)]$ and $[TiCl_3(Tp^*)]$ with Li_3N. In $[TiCl_2(Tp^x)(THF)]$ the THF ligand can be replaced by the stable nitroxyl radical TEMPO to form $[TiCl_2(Tp)(TEMPO)]$ and $[TiCl_2(Tp^*)(TEMPO)]$.[10]

$[TiCl_2(Tp^*)(LL)]$ (LL = 2-oxy-6-methylpyridine, 2,4-dimethyl-6-oxypyrimidine), rigid in solution, have been prepared by the reaction of $[TiCl_3(Tp^*)]$, the structure of which has been determined by X-ray single-crystal study, with the lithium salt or the protic form of the corresponding hydroxyl-pyridine or -pyrimidine (Fig. 2.6).[11]

$[Ti(OR)_nCl_{3-n}(Tp)]$ complexes have been prepared and characterized, and their activity toward olefin polymerization is tested. The first studies of the chain transfer processes of this family of catalysts were reported, and the results were discussed in terms of the ligand Tp^x steric hindrance.[12] The molecular structure of $[TiCl_2(O^tBu)(Tp^*)]$ has also been determined.[13]

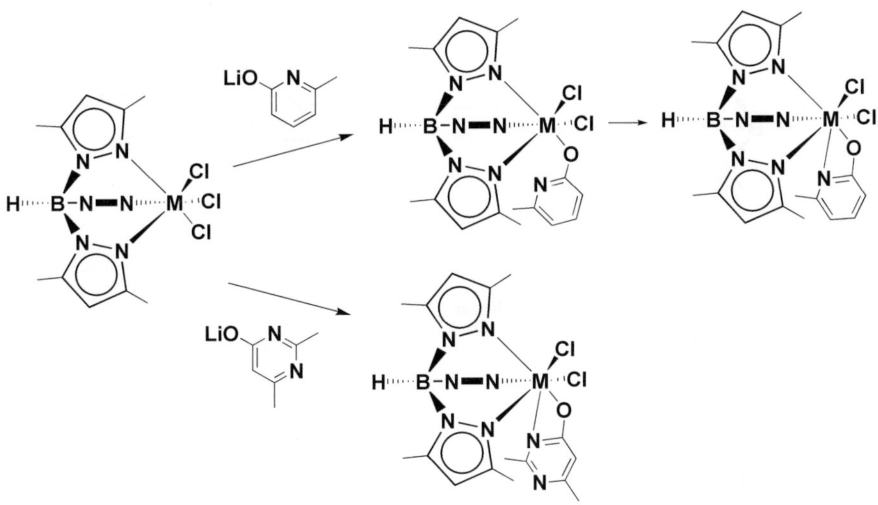

Fig. 2.6. The reaction of [$TiCl_3$(Tp*)] with oxypyridine and oxypyrimidine.

Titanium amido complexes [Ti(Tp)(NMe_2)$_3$] and [Ti(Tp*)-(NMe_2)$_3$] were synthesized by the reactions of [TiCl(NMe_2)$_3$] with the corresponding Tp^x salts. These complexes exhibit octahedral coordination geometry around the metal center. DFT calculations at the B3PW91 level were also performed.[14]

2.5.2. *Zr*

Zirconium amido complexes [Zr(Tp)(NMe_2)$_3$] and [Zr(Tp*)-(NMe_2)$_3$] were prepared and characterized. The dichloromethane solvate form of [Zr(Tp*)(NMe_2)$_3$] shows an octahedral geometry.[14]

Jordan and Lee described the synthesis of [Zr(Tp*)(CH_2Ph)$_3$] and the unusual reactivity of the cationic complex [Zr(Tp*)(CH_2Ph)$_2$][B-(C_6F_5)$_4$] (Fig. 2.7). This compound rearranges to [Zr{($PhCH_2$)(H)-B(μ-pz*)$_2$}(η^2-pz*)(CH_2Ph)][B(C_6F_5)$_4$] at 0°C in dichloromethane and reacts with 2 equiv. of 2-butyne giving [Zr(CH_2-Ph)(Tp*)(CMe=CMeCMe=CMeCH_2Ph)][B(C_6F_5)$_4$] quantitatively by double insertion into a Zr-CH_2Ph bond. [Zr{($PhCH_2$)(H)B(μ-pz*)$_2$}(η^2-pz*)(CH_2Ph)][B(C_6F_5)$_4$] reacts with 2 equiv. of 2-butyne

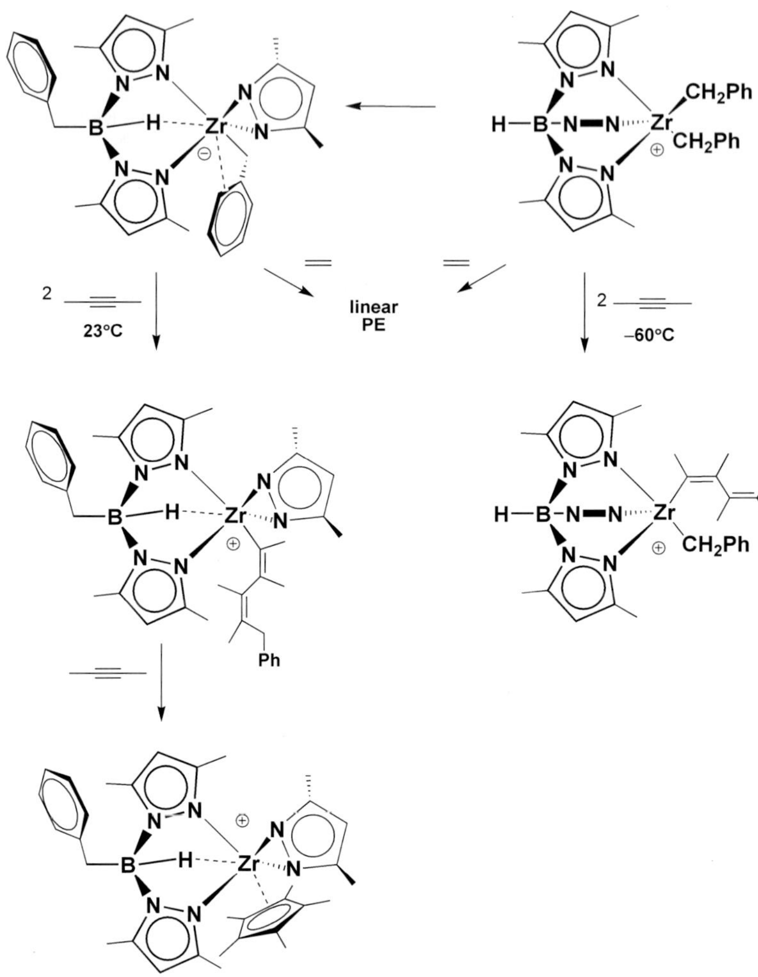

Fig. 2.7. The unusual reactivity of the cationic complex [Zr(Tp*)$(CH_2Ph)_2$]-[$B(C_6F_5)_4$].

to form [Zr{$(PhCH_2)$(H)B(μ-pz*)$_2$}(η^2-pz*)(CMe=CMeCMe=CMe-CH_2Ph)][$B(C_6F_5)_4$] and with 3 equiv. of 2-butyne also giving *cis*-β-methylstyrene and [Zr{$(PhCH_2)$(H)B(μ-pz*)$_2$}(Cp*)(η^2-pz*)][$B(C_6$-$F_5)_4$] that was derivatized to [Zr{$(PhCH_2)$(H)B(μ-pz*)$_2$}(Cp*)(η^2-pz*)]Cl by reacting with [^{n}Bu$_3$($PhCH_2$)N]Cl.[15]

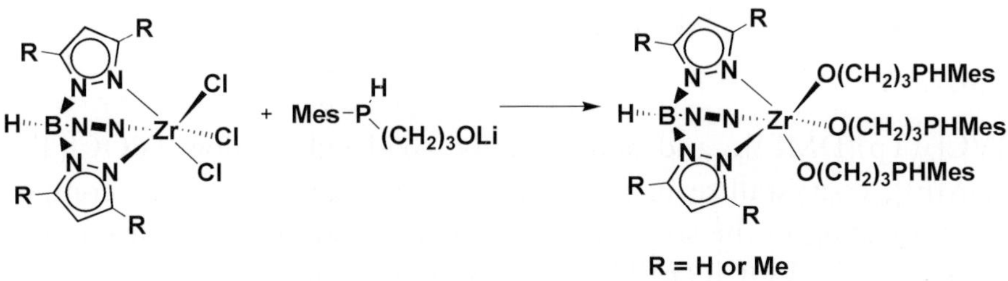

Fig. 2.8. Synthesis of $[Zr\{O(CH_2)_3PHMes\}_3(Tp^x)]$.

The fluxional complex $[ZrCl_2(Tp^*)(LL)]$ (LL = 4,6-dimethyl-2-thiolatepyrimidine) has been prepared by the reaction of $[ZrCl_3(Tp^*)]$ with the lithium salt or the protic form of the thiolato pyrimidine.[11]

The derivatives $[Zr\{O(CH_2)_3PHMes\}_3(Tp^x)]$ [Tp^x = Tp or Tp*] (Fig. 2.8) have been prepared and spectroscopically investigated by the Hey-Hawkins group.[16]

The analogous $[Zr(OCH_2PPh_2)_3(Tp)]$ reacts with $[Mo(CO)_3(NCEt)_3]$ to give the heterodinuclear complex $[Zr(\mu\text{-}OCH_2PPh_2)_3(Tp)Mo(CO)_3]$.[17]

The mixed-ligand complex $[Zr(acac)_3(Tp^*)]$ has been prepared and characterized by IR, NMR, and EPR methods. Its capability to undergo oxidation and reduction reactions was also examined by electrochemical methods.[18]

Patiño *et al.*[19] reported FT-Raman and IR spectra of $[ZrCl_3(Tp^*)]$ and $[Zr(OSiPh_3)Cl_2(Tp^*)]$. B–H stretching vibrations in these complexes are shifted to higher wavenumbers with respect to starting potassium salt K(Tp*).

Hill and Smith reported the synthesis of $[ZrCl(\eta\text{-}C_8H_8)(\kappa^2\text{-}Tp^*)]$ in which the metal is electronically unsaturated (16 electrons), despite the capability of Tp* to increase the metal coordination number.[20]

2.5.3. *Hf*

FT-Raman and IR spectra of $[HfCl_3(Tp^*)]$ and $[Hf(OSiPh_3)Cl_2(Tp^*)]$, which show the B–H stretching vibrations shifted to higher wavenumbers with respect to starting potassium salt K(Tp*), were reported.[19]

2.6. Group 5: V, Nb, and Ta

2.6.1. *V*

$[VCl_2(Tp)(DMF)]$, and its partially oxidized analog [VOCl(Tp)(DMF)], cocrystallize in an approximately 1:1 ratio, both components existing in the solid state in a pseudo-octahedral coordination geometry.[21] [VOCl(Tp)(Hpz)], also pseudo-octrahedral, has been structurally characterized.[22]

[VO(acac)(Tp)], indicated as a potential antidiabetic, was prepared by reaction of Tp with $[VO(acac)_2]$ in MeOH. Structural and spectroscopic properties, as also air and solution stabilities, have also been reported.[23]

[VO(acac)(Tp*)] and $[VO(acac)(Tp^*)]\cdot CH_3CN$ were prepared and characterized by IR, UV–Vis, and NMR techniques.[24] [VO(acac)(Tp*)] and $[VO(acac)_2(Tp^*)]$ have been characterized by IR, NMR, and EPR methods. Thermoanalytical studies on [VO(acac)Tp*] and $[VO(acac)_2Tp^*]$ were performed by TG and DTG methods to study the influence of the chelating ligand L on the thermal stability of these metal acetylacetonate complexes.[18]

The reaction of $[VO(Ph\text{-}acac)_2]$ with Tp and Tp* gave $[VO(Ph\text{-}acac)(Tp)]\cdot 2H_2O$ and $[VO(Ph\text{-}acac)(Tp^*)]\cdot 2H_2O$, respectively. The oxovanadium complex $[V_2O_4(Tp)]_2\cdot 2CH_3CN$ was obtained by the reaction of $VOSO_4\cdot 3H_2O$ with Tp in MeOH.[25]

[VO(Tp)(SCN)(pzH)] and $[VO(SCN)(Tp^*)(Hpz^*)](SCNH)_2$ were prepared by one-pot reaction of $VOSO_4\cdot nH_2O$, KSCN with K(Tp) or Na(Tp*), Hpz or Hpz*. They have been characterized by IR, UV–vis, fluorescence spectra and X-ray diffraction. The electronic structures and the bonding characters of both complexes were investigated by *ab initio* calculations.[26]

Ligand displacement of $[VOCl_2(MeCN)_2(H_2O)]$ by Tp* afforded [V(O)(Cl)(Tp*)(MeCN)]. Hydrolysis of this chloro-complex yielded a dinuclear bis(μ-hydroxo) complex, $[V(O)(\kappa^3\text{-}Tp^*)(\mu\text{-}OH)_2V(O)(\kappa^2\text{-}Tp^*)]$, which consists of *syn*-arranged V=O fragments, having different coordination geometries of the vanadium centers. Intramolecular hydrogen-bonding interaction between one of the two hydroxo

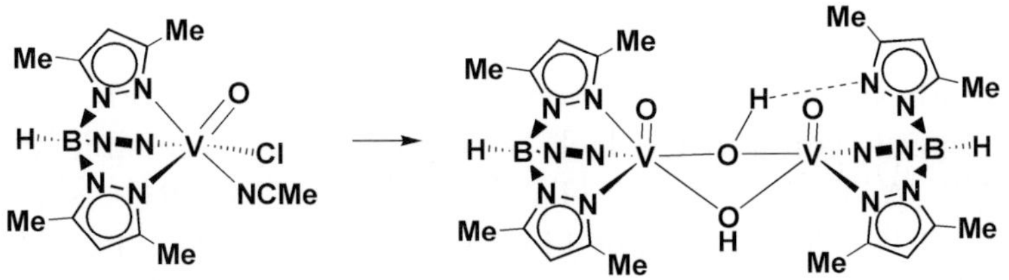

Fig. 2.9. Hydrolysis of $[V(O)(Cl)(Tp^*)(MeCN)]$ gave $[V(O)(\kappa^3\text{-}Tp^*)(\mu\text{-}OH)_2V(O)\text{-}(\kappa^2\text{-}Tp^*)]$.

groups and the noncoordinated pyrazolyl nitrogen atom in κ^2-Tp* was observed (Fig. 2.9).[27]

Reaction of mer-$VCl_3(THF)_3$ with Tp* and $[NEt_4]CN$ in MeCN affords $[NEt_4][V^{III}(Tp^*)(CN)_3]\cdot H_2O$; aerobic oxidation affords $[NEt_4][V^{IV}(O)(Tp^*)(CN)_2]$, which upon interaction with $Mn(OTf)_2$ and bipy gave $\{[V(O)(Tp^*)(CN)_2]_2[Mn^{II}(bipy)_2]_2[OTf]_2\}\cdot 2MeCN$. Magnetic measurements indicate that these complexes exhibit $S = 1$, 1/2, and 4 spin ground states, respectively.[28]

Synthesis, structure,[29] and electronic and magnetic properties[30] of $[VCp^R(Tp)]$ and $[VCp^R(Tp)]^+$ are reported. Oxidation of [VCp(Tp)] in MeCN solution yields $[VCp(Tp)(MeCN)]^+$, whereas similar reaction in CH_2Cl_2 solution yields $[VCp(Tp)]^+$, which reacts with σ-donor ligands L (L = CN^tBu and PMe_3), to form $[VCp(Tp)L]^+$. The structures of $[VCp(Tp)(MeCN)]^+$ and $[VCp(Tp)(PMe_3)]^+$ show that considerable distortion of the sandwich moiety has occurred to accommodate coordination of the extra ligand. The d^2 complex $[VCp(Tp)]^+$ displays magnetic behavior consistent with an orbitally nondegenerate ground state.

Reaction of the edge-bridged double cubane cluster $[V_2Fe_6(Tp)_2\text{-}S_8(PEt_3)_4]$ with hydrosulfide affords the clusters $[V_2Fe_6(Tp)_2S_9\text{-}(SH)_2]^{3-,4-}$.[31]

$[VO(O_2)(Tp)(Hpz)]$ and $[VO(O_2)(pzTp)(Hpz)]$ were synthesized and characterized by elemental analysis, IR, UV–vis, and NMR spectra. $[VO(O_2)(Tp)(Hpz)]$ investigated by X-ray diffraction is significant for haloperoxidase enzymes.[32]

2.6.2. *Nb*

The synthesis of niobium complexes of Tp and other scorpionate ligands suffered the lack of simple, easily accessible starting material. A series of [Nb(Tp*)(CO)(PhC≡CMe)(RC≡N)] complexes (R = Me, Et, $PhCH_2$, 4-$MeOC_6H_5$, Ph, 4-$CF_3C_6H_5$) has been obtained by CO displacement from [Nb(Tp*)$(CO)_2$-(PhC≡CMe)]. PhC≡N is thermally displaced by PMe_2Ph or PhC≡CMe to give [Nb(Tp*)(CO)(PhC≡CMe)(PMe_2Ph)] and [Nb(Tp*)(CO)$(PhC≡CMe)_2$], respectively (Fig. 2.10). Protonation of [Nb(Tp*)(CO)(PhC≡CMe)(RC≡N)] with HBF_4 induces nitrile/ alkyne coupling to give a niobacycle [Nb(Tp*)(CO)(η^2-CPh=NH)(PhC≡CMe)][BF_4], whereas upon protonation of [Nb(Tp*)(CO)$(PhC≡CMe)_2$] the niobacycle [NbF(Tp*){C(Ph)C(Me)-C(CMe=CP hH)O}], structurally characterized, is afforded. The intermediate [Nb(Tp*)(CO)(η^2CPh=NH)(PhC≡CMe)][BF_4] has been characterized only by NMR. Protonation of [Nb(Tp*)-(CO)(PhC≡CMe)(PPh_2C≡CPh)] yields the stable [Nb(Tp*)(CO)-{$Ph_2P(H)PC$≡CPh}(PhC≡CMe)][BF_4].[33]

The complex [NbCl(^{i}Pr)(Tp*)(PhC≡CMe)] (Fig. 2.11) exhibits a β-agostic interaction in the structure. The conformation of the alkyl group is such that the agostic methyl group lies in the Cα–Nb–Cl plane, and the nonagostic one, in a wedge formed by two pyrazole rings. NMR studies suggested that the restricted rotation about the Nb–C bond allows an equilibrium between the agostic species and a minor α-agostic rotamer. Analogous behavior has been found in

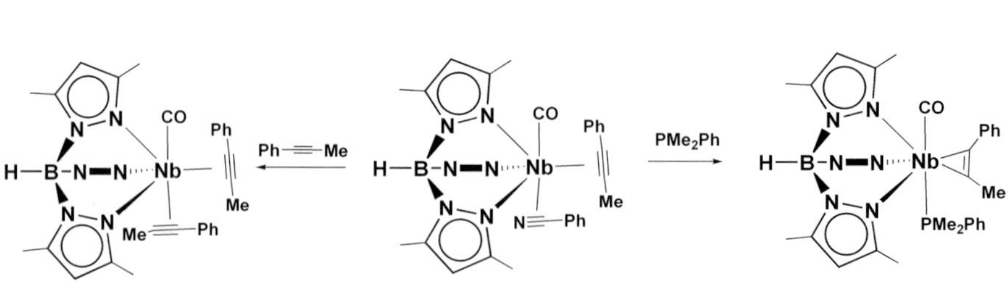

Fig. 2.10. Reactivity of [Nb(Tp*)(CO)(PhC≡CMe)(PhC≡N)] (the alkyne in [Nb(Tp*)(CO)(PhC≡CMe)(PMe_2Ph)] is not a conventional two-electron donor; the metallacyclopropane form has been adopted in this drawing and in the following ones).

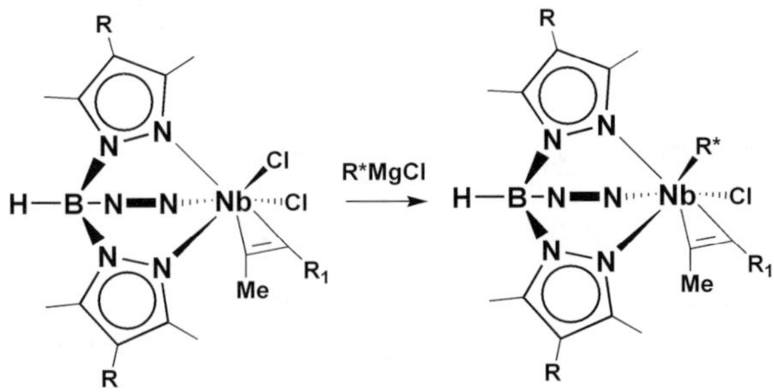

Fig. 2.11. Synthesis of the secondary alkyl chloro complexes [Nb(Cl)(R*)(Tp^x)-($R_1C{\equiv}CMe$)].

the related species [NbCl(^{i}Pr)(Tp^x)($R_1C{\equiv}CMe$)] (Tp^x = Tp*, R_1 = Me; Tp^x = Tp^{*Cl}, R_1 = Ph). [NbCl(*sec*-Bu)(Tp*)(PhC≡CMe)], structurally characterized, exists as two diastereomers, one of which has a β-agostic methyl group with the ethyl group located in the wedge formed by two pyrazole rings, the other having a β-agostic methylene group in the solid state, and the methyl group sits in the wedge. Calculations have been performed on [NbCl(CH_2Me)(Tp*)(HC≡CMe)], an α-agostic species.[34]

Pyridine silylimido species have been employed for the synthesis of [$NbCl_2$(Tp*)($NSiMe_3$)] through a standard procedure.[35]

Cationic methyl complexes [NbMe(L)(Tp*)(MeC≡CMe)][BAr_F] (L = OEt_2, PMe_2Ph, PEt_3) models of the active species in alkene polymerization, have been also described. The diethyl ether adduct exhibits Tp* playing its characteristic directing role, one ethyl group of the OEt_2 ligand occupying the wedge between two of the three pyrazole rings.[36]

Reduction of the niobium(IV) complexes [$NbCl_3$(Tp^x)] (Tp^x = Tp* or Tp^{*Cl}) with zinc in the presence of a terminal alkyne HC≡CR gives [$NbCl_2$(Tp^x)(HC≡CR)] (R = H, Ph, C_6H_4-4-Me, CH_2Ph, CF_3). This is a new strategy different from the synthesis reported above only with internal alkynes. [$NbCl_2$(Tp*)(HC≡CR)] (R = Ph, C_6H_4-4-Me, CH_2Ph, CF_3) (Fig. 2.12) have been structurally characterized. In the crystal structure of [$NbCl_2$(Tp*)(HC≡CCF_3)] one independent

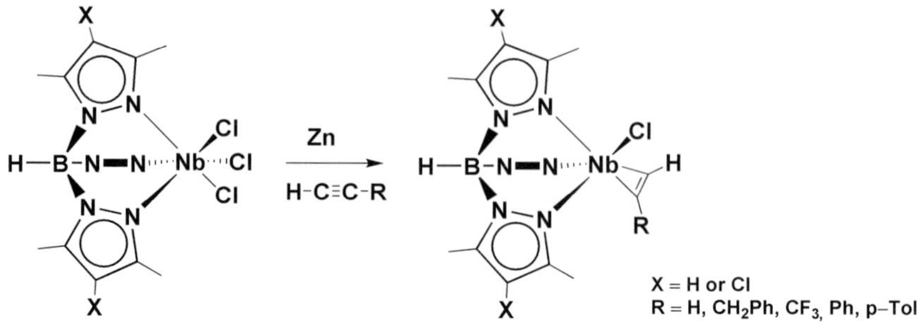

Fig. 2.12. Reduction of [$NbCl_3(Tp^x)$] with zinc in the presence of HC≡CR gives [$NbCl_2(Tp^x)$(HC≡CR)].

molecule having the CF_3 group proximal to Tp* exhibits H···F bridges between the alkyne CF_3 group and the methyl group at the 3-position of the Tp* ligand. Oxo- and imido-niobium complexes can be prepared analogously. R–C–H agostic ethyl complexes [NbCl(Tp*)-(μ-H-$CHCH_3$)(HC≡CR)] (R = Ph, C_6H_4-4-Me, CH_2Ph) have been synthesized and their thermolysis investigated.[37]

[$Nb(Me)_2$(Tp*)($CH_3C{\equiv}CCH_3$)] and [NbCl(Me)(Tp*)($CH_3C{\equiv}CCH_3$)] obtained from the corresponding dichloride, have been studied by X-ray diffraction. No α-agostic interaction seems present in these compounds, as pointed by NMR and IR studies. On this basis Etienne and coworkers conclude that α-agostic structures with this kind of compounds are predominantly the result of steric interactions between the alkyl moiety and the bulky Tp* ligand.[38]

Cycloalkyl niobium derivatives [NbX(R)(Tp*)(MeC≡CMe)] (X = Cl, R = *c*-C_3H_5; X = Br, R = *c*-C_3H_5, X = Cl, R = *c*-C_5H_9, X = Cl, R = *c*-C_6H_{11}) have been synthesized from the reaction of [$NbCl_2$(Tp*)(MeC≡CMe)] with the corresponding Grignard reagents. [NbCl(*c*-C_3H_5)(Tp*)(MeC≡CMe)] exhibits no sign of C–H agostic interaction either in the solid state (X-ray) or in solution. The NMR spectra of [NbCl(*c*-C_5H_9)(Tp*)(MeC≡CMe)] and [NbCl(*c*-C_6H_{11})(Tp*)(MeC≡CMe)] are temperature-dependent as a consequence of an equilibrium between a major α-agostic species and a minor nonagostic one. These species have been investigated also by hybrid QM/MM calculations.[39] However, structural, spectroscopic, and theoretical evidences indicate that an unusual α-C–C agostic interaction is present in [NbCl(*c*-C_3H_5)(Tp*)(MeC≡CMe). The steric demands of the pendant methyl groups in the Tp* ligand preferentially destabilize one or both of the C–H agostic alternatives.[40]

The last member of the family [NbCl(Tp*)(*c*-C_nH_{2n-1})] ($n = 3$–6) has been also recently reported: [NbCl(*c*-C_4H_7)(Tp*)(MeC≡CMe)] is similar to cyclopentyl and cyclohexyl analogues, and very different from the cyclopropyl derivatives. The cyclobutyl complex exhibits an α-C–H agostic interaction in the dominant isomer, with no evidence for the α-C–C agostic character found for the smaller ring that seems to be unique for the cyclopropyl complex.[41]

The substitution reactions of [$NbCl_2$(BnSC≡CSBn)(Tp*)], structurally characterized, have been recently investigated.[42]

2.6.3. *Ta*

Only few tantalum complexes are known. [$TaCl_2$(Tp*)(PhC≡CCH_3)] has been prepared from the reaction between K(Tp*) and [$TaCl_3$(PhC≡CCH_3)]. The reaction of [$TaCl_2$(Tp*)(PhC≡CCH_3)] with MeLi gives [$TaMe_2$(Tp*)(PhC≡CCH_3)], whereas the reaction with MgEtX yields the α-agostic ethyl complex [TaCl(μ-H-$CHCH_3$)(Tp*)(PhC≡CCH_3)] (Fig. 2.13) that has been characterized crystallographically. This compound reacts with MeLi forming in solution the mixed [Ta(Tp*)(Me)(μ-H-$CHCH_3$)(PhC≡CCH_3)].[43]

Reaction of Tp* with [$TaCl_2$(tht)]$_2$(μ-Cl)$_2$(μ-tht) gives the unsymmetrical [(TaTp*)(μ-Cl)$_2$(μ-tht)($TaCl_3$)], structurally characterized as

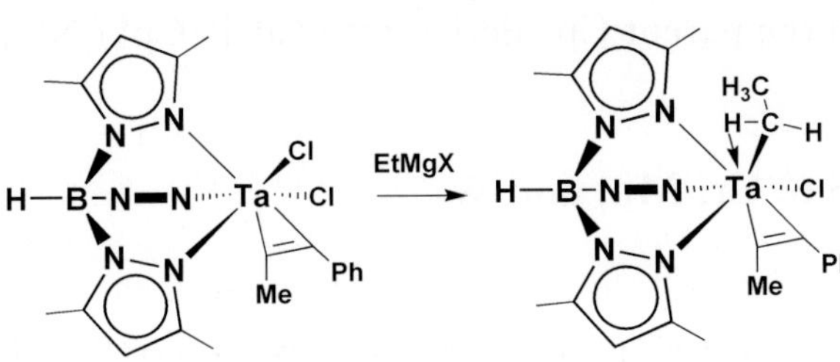

Fig. 2.13. Synthesis of the α-agostic ethyl complex [TaCl(Tp*)(μ-H-$CHCH_3$)-(PhC≡CCH_3)].

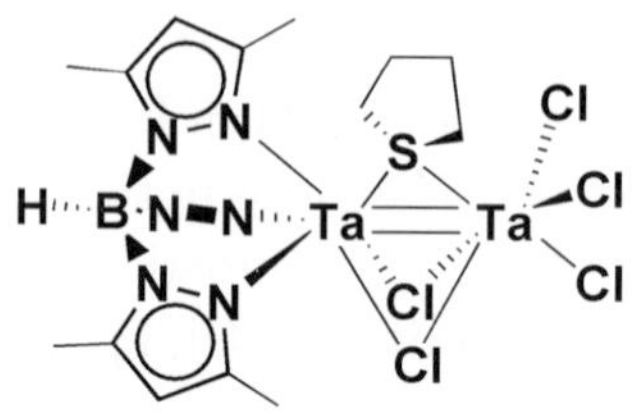

Fig. 2.14. The unsymmetrical [(TaTp*)(μ-Cl)$_2$(μ-tht)(TaCl$_3$)].

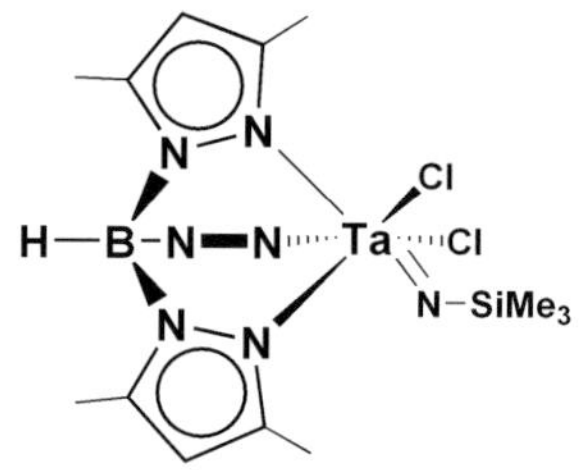

Fig. 2.15. [TaCl$_2$(Tp*)(NSiMe$_3$)].

a diamagnetic doubly bonded Ta=Ta complex (Fig. 2.14). The X-ray structure of [TaCl$_2$(Tp*)(PhC≡CMe)], containing the (4e)-alkyne in the symmetry plane of the molecule, has also been reported.[44] Pyridine silylimido complexes are good starting materials for the synthesis of [TaCl$_2$(Tp*)(NSiMe$_3$)] (Fig. 2.15) in which the tantalum center is octahedrally coordinated by *fac*-binding nitrogen atoms of the Tp* ligand, two chlorine atoms, and one nitrogen atom of the silylimido ligand. The structure and reactivity of this complex have been compared with those of the parent Cp* derivative [TaCl$_2$(Cp*)(NSiMe$_3$)].[35]

2.7. Group 6: Cr, Mo, and W

2.7.1. *Cr*

[Cr(Cp*)(Tp)] has been prepared by the reaction of [CrCl(Cp*)]$_2$ and Tp in THF. This complex has a high-spin ($S = 2$) electronic configuration from 5 to 300 K. The crystal structure reveals bond lengths

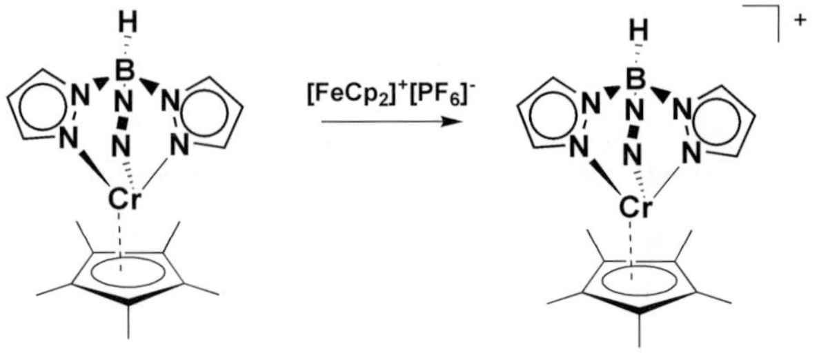

Fig. 2.16. Oxidation of [Cr(Cp*)(Tp)] with [FeCp$_2^*$]$^+$salts.

typical for a high-spin configuration with a pronounced Jahn–Teller distortion.[45]

Reaction of $CrCl_3(THF)_3$ with K(Tp) in THF leads to the formation of the complex K[$CrCl_3$(Tp)], which, upon salt metathesis, yields [Ph_4P][$CrCl_3$(Tp)] in which Cr shows a distorted octahedral geometry with the Tp bonded as N_3-donor tripod ligand. The latter species, in the presence of MAO, leads to the formation of an active catalyst for the polymerization of ethylene.[46]

[Cr(Cp*)(Tp)]$^{n+}$ complexes have been synthesized by reaction of a {Cr(Cp*)} precursor with Tp. Oxidation of [Cr(Cp*)(Tp)] with [Fe(Cp*)$_2$]$^+$ salts provided easy access to the corresponding Cr(III) species (Fig. 2.16). This complex shows reversible Cr(III)/Cr(II) redox couples.[29]

Synthesis, structure,[29] electronic and magnetic properties of [Cr-(Cp^R)(Tp)] and [Cr(Cp^R)(Tp)]$^+$ (R = H, Me) have been reported.[30] The trivalent homoleptic [Cr(Tp)$_2$]$^+$, [Cr(Tp*)$_2$]$^+$, and [Cr(pzTp)$_2$]$^+$ have also been characterized by electrochemical and spectroscopic studies.[47]

Long and Harris reported the complex [Cr(CN)$_3$(Tp)]$^-$ which enables isolation of two metastable isomers of a face-centered cubic cluster [(Tp)$_8$(H_2O)$_6$Cu$_6$Cr$_8$(CN)$_{24}$]$^{4+}$.[48]

2.7.2. *Mo*

[Et_4N][Mo(Tp*)(CO)$_3$] reacts with S_8 giving a mixture of [Et_4N]-[Mo(Tp*)S(S_4)] and [Et_4N][MoO(Tp*)(S_4)], whereas only the sulfido

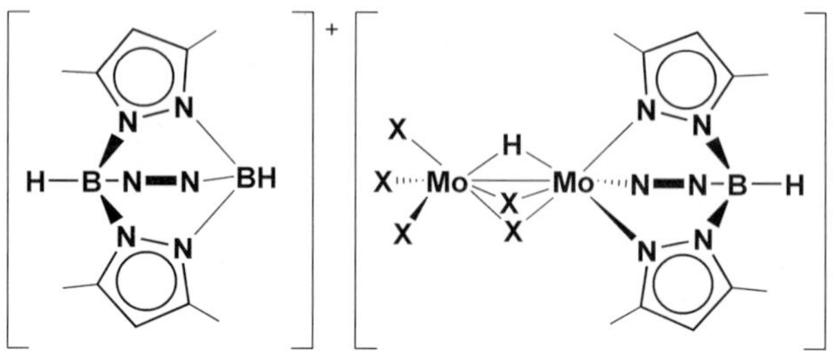

Fig. 2.17. $[HB(pz^*)_3BH][X_3Mo(\mu\text{-}X)_2(\mu\text{-}H)Mo(Tp^*)]$.

$[Et_4N][Mo(Tp)S(S_4)]$ has been obtained by using the unsubstituted Tp. $[Et_4N][Mo(Tp^*)S(S_4)]$ and $[Et_4N][Mo(Tp)S(S_4)]$ containing octahedral anions have been structurally characterized.[49] Reaction of $[Mo\{\kappa^3\text{-}Tp)(\equiv CC_6H_4Me\text{-}4)(CO)_2]$ with S_8 in THF at reflux temperature affords $[Mo(O)\{\kappa^3\text{-}Tp\}(\kappa^3\text{-}S_2CC_6H_4Me\text{-}4)]$, a rare example of a mononuclear tridentate dithiocarboxylato complex.[50]

$Mo_2(OAc)_4$ and Tp* in glyme at room temperature, upon addition of $(CH_3)_3SiCl$ and $(CH_3)_3SiBr$ gave $[HB(pz^*)_3BH][Cl_3Mo(\mu\text{-}Cl)_2(\mu\text{-}H)Mo(Tp^*)]$ and $[HB(pz^*)_3BH][Br_3Mo(\mu\text{-}Br)_2(\mu\text{-}H)Mo(Tp^*)]$, respectively (Fig. 2.17). $[HB(pz^*)_3BH][MoBr_4(Hpz^*)_2]$ was also reported. The anions of the Tp* complexes consist of two octahedra sharing a common triangular face, the unsymmetrical metal centers being chelate by a tridentate Tp* ligand. The formation of $[HB(pz^*)_3BH]^+$ is due to fast B–N bond cleavage of the Tp* ligand,[51] whereas reaction of $Mo_2(Ac)_4$ with Tp* in diglyme at room temperature followed by addition of excess Me_3SiBr yielded $[HTp^*]^+[Br_3Mo(\mu\text{-}Br)_3Mo(Tp^*)]^-$, $[Mo_4(\mu_3\text{-}O)_2(\mu_2\text{-}O)(\mu_2\text{-}OH)_2(O)_4(Hpz^*)_6](Br)_4(Hpz^*)_2$, and $[MoBr_2(Tp^*)(Hpz^*)]$ that have been structurally characterized. $[HTp^*]^+[Br_3Mo(\mu\text{-}Br)_3MoTp^*]^-$ consists of two octahedra sharing a common triangular face so that the three Br atoms bridge the Mo atoms.[52]

$[Mo(Tp)(NO)(L)(\eta^2\text{-}L')]$ (L = Meim or NH_3, and L′ = alkene, alkyne, ketone, polyaromatic hydrocarbon, or aromatic heterocycle) were prepared and their properties discussed.[53]

Reaction of $[Et_4N][Mo(Tp)(CO)_3]$ with red Se gave the distorted octahedral $[Et_4N][MoTp(CO)_2(\eta^2\text{-}Se_2)]$ that by reaction with

$MeOSO_2CF_3$ in MeCN, was converted into the selenido-bridged dimolybdenum complex [{Mo(Tp)$(CO)_2$}$_2$(μ-Se)].[54]

The oxygen atom transfer reactivity of [Mo(O)Cl_2(Tp*)] with PMe_3, PEt_3, and $PPhMe_2$ has been described by Nemykin,[55] the reactions proceeding through the diamagnetic phosphoryl intermediate [Mo(O)Cl(Tp*)(OPR_3)] (OPR_3 = $OPMe_3$, $OPEt_3$, $OPPhMe_2$). The structures of [Mo(O)Cl(Tp*)($OPMe_3$)] and [Mo(O)Cl(Tp*)($OPPhMe_2$)] have also been reported. These compounds have been converted to solvent-coordinated species, [Mo(O)Cl(Tp*)(solv)] (solv = MeCN or DMF). The stable [$MoOCl_2$(Tp)] has also been oxidized under mass spectrometric conditions (Fig. 2.18). The products distribution has been followed by isotope labeling experiments and energy-dependent electrospray mass spectrometry. Oxygen atom transfer and loss of a chlorine atom from the resulting species have been found.[56,57]

The six-coordinate [MoO(Tp*)(*p*-O-C_6H_4-OC_2H_5)$_2$]$^+$ has been prepared by [MoO(Tp*)(*p*-O-C_6H_4-OC_2H_5)$_2$]. The deep blue color of the Mo complex in acetonitrile is bleached by addition of tertiary

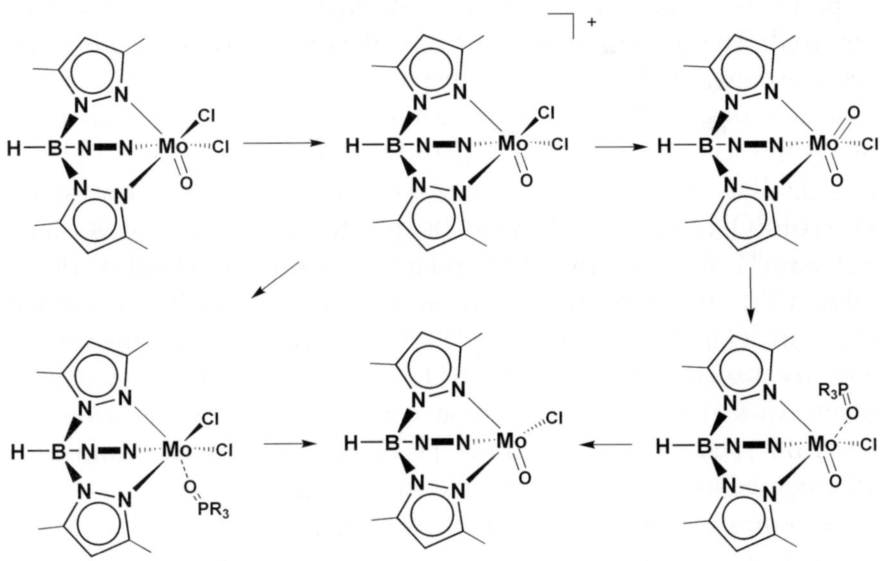

Fig. 2.18. The oxidation products of [$MoOCl_2$(Tp)] observed under mass spectrometric conditions.

phosphines such as PMe_3, PEt_3, PPh_3, and $PEtPh_2$, and the product of the PPh_3 reaction has been investigated by mass spectrometry. The isotopic distribution agrees well with that expected for the $[Mo(p\text{-}O\text{-}C_6H_4\text{-}OC_2H_5)_2(Tp^*)]^+$ ion, consistent with the loss of an O atom from $[MoO(Tp^*)(p\text{-}O\text{-}C_6H_4\text{-}OC_2H_5)_2]^+$.[58]

$[MoO_2(OPh)(Tp^*)]$ reacts with PEt_3 in toluene giving $[MoO(OPh)(Tp^*)]_2(\mu\text{-}O)$, while hydrolysis of $[Mo(O)Cl_2(Tp^*)]$ yields $[MoO(Cl)(Tp^*)]_2(\mu\text{-}O)$, which exhibits a corner shared bioctahedral structure with a single bridging oxo ligand linking $[MoO(X)(Tp^*)]$.[59]

The amido nitrosyl molybdenum complex $[MoCl(Tp^*)(NO)(NH_2)]$ reacts with alcohols, yielding the bisalkoxo complexes $[Mo(OR)_2(Tp^*)(NO)]$ (R = Me, Et, ^{n}Pr, ^{n}Bu), analyzed by IR and ^{1}H-NMR spectroscopy.[60] Topaloğlu and McCleverty also reported the synthesis and characterization of amido- and amido(monoalkylamido)-Mo(Tp*)(NO) species[61] and of $[MoCl(Tp^*)(NO)(OC_6H_4PPh_2\text{-}p)]$.[62]

Reaction of $[Mo(O)Cl_2(Tp^*)]$ with *m*-fluoroaniline in toluene afforded the pair of geometric isomers of $[Mo(O)Cl(Tp^*)](\mu\text{-}O)[Mo(Cl)(Tp^*)(\equiv NC_6H_4F)]$. The reaction with *m*-iodoaniline yielded the compound $[Mo(O)Cl(Tp^*)](\mu\text{-}O)[Mo(Cl)(Tp^*)(\equiv NC_6H_4I)]$. Both complexes were characterized by IR and ^{1}H-NMR spectroscopy, FAB mass spectrometry, and X-ray crystallography.[63] Reaction of $[Mo(O)Cl_2(Tp^*)]$ with *p*-methoxy and *p*-nitroaniline in the presence of Et_3N gave $[Mo(O)Cl(Tp^*)](\mu\text{-}O)[Mo(Cl)(Tp^*)(NC_6H_4R)]$ (R = OMe, NO_2). The single-crystal X-ray study of $[Mo(O)Cl(Tp^*)](\mu\text{-}O)[Mo(Cl)(Tp^*)(NC_6H_4OMe)]$ was also reported,[64] whereas $[Mo(O)Cl_2(Tp^*)]$ with $H_2NC_6H_4R\text{-}4$ (R = OEt; OPr) in refluxing toluene in the presence of Et_3N yielded $[Mo(O)Cl(Tp^*)](\mu\text{-}O)[MoCl(Tp^*)(NC_6H_4OR\text{-}4)]$. From the analogous reaction between $[Mo(O)Cl_2(Tp^*)]$ and $C_6H_5NH_2$, $[\{Mo(O)Cl(Tp^*)\}_2(\mu\text{-}O)]$ is obtained. These compounds were characterized by mass spectrometry, IR and ^{1}H NMR spectroscopy, and also investigated by cyclic voltammetric studies.[65]

Reaction of $[MoCl_2(Tp^*)(NO)]$ with substituted pyridine (py-R) gave $[MoCl(Tp^*)(NO)(py\text{-}R)]$ (R = 4-(1-butyl)pentyl, 4-Ph, 3-Ph, 4-benzoyl, 4-acetyl, 3-acetyl, 4-cyano, 3-cyano, 4-Cl, 3-Cl, or py-R =

isoquinoline). [MoCl(Tp*)(NO)(4-CNpy)] and [MoCl(Tp*)(NO)(3-benzoylpy)], distorted octahedral, have been structurally characterized. These 17-electron complexes (formally Mo(I)) have been investigated by UV/Vis/NIR spectrochemistry. They undergo a chemically reversible one-electron reduction to the 18-electron monoanion at potentials that are sensitive to the nature of the pyridyl substituent R.[66] A UV/Vis/NIR spectroelectrochemical study on a series of dinuclear [{Mo(O)Cl(Tp*)}$_2$(μ-L$_2$)] (L$_2$ = O$_2$-donor bridging ligand: 1,4-[O(C$_6$H$_4$)$_n$O]$^{2-}$ (n = 1–4), 1,3-[O(C$_6$H$_4$)$_n$O]$^{2-}$, 1,4-[OC$_6$H$_4$XC$_6$H$_4$O]$^{2-}$(X = CH$_2$, S, or SO$_2$), and mononuclear complexes [{Mo(O)Cl(Tp*)}(O-C$_6$H$_4$R)] (R = H or OMe) has also been reported; most of the species described undergo two one-electron oxidations and two one-electron reductions. The crystal structures of the distorted octahedral [{Mo(O)Cl(Tp*)}(O-C$_6$H$_4$SC$_6$H$_4$OH)] and [{Mo(O)Cl(Tp*)}$_2$(μ-1,4-[OC$_6$H$_4$SC$_6$H$_4$O])] have been determined, and molecular orbital calculations have been performed.[67]

The trinuclear complexes [Mo(NO)Tp*{OC$_6$H$_4$EpyMo(NO)Tp*-Cl}$_2$] (E = nothing, CH=CH and CH$_2$CH$_2$; py = C$_5$H$_4$N or C$_5$H$_3$N), [{(NO)(Tp*)ClMo(OC$_6$H$_4$)}$_2$pyMoCl(Tp*)(NO)], and [Cl-Mo(NO)(Tp*{OC$_6$H$_3$Me[CH=CHpyMo(NO)Tp*Cl]$_2$}] have also been described. Reduction of these species by cobaltocene in tetrahydrofuran/dichloromethane mixtures affords complexes having three 17-valence electron centers with one unpaired electron per metal center.[68]

[Fe(Cp){C$_5$H$_4$QNHMX(NO)(Tp*)}] (Q = nothing, M = Mo, X = Cl, Br, I; M = W, X = Cl; Q = C$_6$H$_4$, M = Mo, X = Cl, Br, I; M = W, X = Cl; Q = CH=CHC$_6$H$_4$ or N=NC$_6$H$_4$, M = Mo, X = Cl), containing 16-valence electron metal nitrosyl centers, [Fe(Cp){C$_5$H$_4$QpyMoCl(NO)(Tp*)}] (py = 4-pyridyl; Q = CH=CH, CH=CHCO, N=CH, and C$_6$H$_4$CH=CH), [Fe(C$_5$Me$_4$H){C$_5$H$_4$CH=CHpyMoCl(NO)(Tp*)}], and [Fe(C$_5$Me$_4$H)(C$_5$Me$_4$QpyMoCl(NO)(Tp*)] (Q = CH=CH or CH=N), some of which contain 17-valence electron molybdenum nitrosyl centers, have been synthesized and characterized electrochemically and spectroelectrochemically.[69]

$[Mo^{V}(O)Cl(Tp^{*})]^{+}$, coordinated to naphtholate ligands, and dinuclear complexes based on bridging ligands containing two naphtholate or naphtholate and one phenolate terminus, have been prepared and characterized by electrochemistry and UV/Vis/NIR spectroelectrochemistry.[70]

The reaction between $[Mo(O)Cl_2(Tp^{*})]$ and 4-bromoaniline in refluxing toluene gave geometric isomers of $[Mo(O)Cl(Tp^{*})](\mu\text{-}O)[Mo(Cl)(\equiv NC_6H_4Br)Tp^{*}]$, both structurally characterized, whereas a similar reaction between $[Mo(O)Cl_2(Tp^{*})]$ and 4-chloroaniline yielded only one product, $[MoTp^{*}(O)Cl](\mu\text{-}O)[MoTp^{*}(Cl)(\equiv NC_6H_4Cl)]$.[71]

$[MoXTp^{*}(NO)(ZC_6H_4CN\text{-}p)]$ (X = Cl, Z = O and NH; X = I, Z = CN) were synthesized and characterized by ^{1}H-NMR, IR, and FAB mass spectroscopy.[72]

$[Mo(O)Cl_2(Tp^{*})]$ reacts with 1,3-, 1,5-, 1,6-, 2,6-, and 2,7-dihydroxynaphthalene, yielding dinuclear complexes $[\{Mo^{V}(O)Cl(Tp^{*})\}_2(\mu\text{-}C_{10}H_6O_2)]$ which have been investigated by UV–vis/NIR spectroscopy and variable-temperature magnetic susceptibility studies.[73]

Dinuclear paramagnetic $[\{MoCl(Tp^{*})(NO)\}_2(\mu\text{-}L)]$ (L = 4,4′-bis[2-(4-pyridyl)ethen-1-yl)]-biphenyl; 4,4′-bis[2-(4-pyridyl)ethen-1-yl)]-terphenyl; 4,4′-bis[2-(4-pyridyl)ethen-1-yl)]-benzophenone; 4,4′-bis[2-(4-pyridyl)ethen-1-yl)]-benzyl; 6,6′-bis[2-(4-pyridyl)ethen-1-yl)]-2,2′-bipyridine), and $[\{Mo(O)Cl(Tp^{*})\}_2(\mu\text{-}L)]$ ($H_2L = HOC_6H_4OC(S)OC_6H_4OH$; $HOC_6H_4OS(O)OC_6H_4OH$; $HOC_6H_4OC(O)C_6H_4C(O)(S)OC_6H_4OH$) have been synthesized and characterized. The EPR investigations indicated these species as weakly coupled dinuclear compounds.[74]

Syntheses, electrochemical and spectroscopic (UV, Vis, NIR, and EPR) properties of $[\{MoCl(Tp^{*})(NO)\}_2(\mu\text{-}L)]$ containing two-redox active $Mo(Tp^{*})$ units linked by a conjugated bridging ligand L have been reported by McCleverty and others. Some of these complexes exhibit intense, redox-switchable absorbance in the near-IR region.[75]

Ward *et al.*[76] described some dinuclear complexes containing both ferrocenyl and oxomolybdenum(IV) moiety $[Mo(O)Cl(Tp^{*})]$, linked by conjugate bridges (Fig. 2.19).

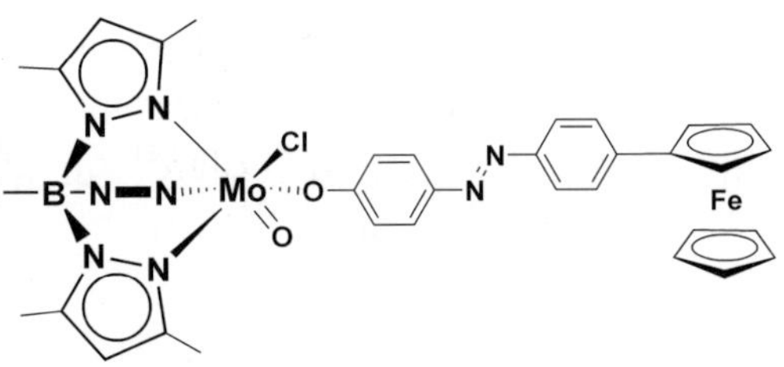

Fig. 2.19. A [Mo(O)Cl(Tp*)] and a ferrocenyl moiety linked by a conjugate bridge.

Some dinuclear complexes have been prepared in which two {Mo(O)Cl(Tp*)} fragments are attached to either end of a bis-*p*-phenolate bridging ligand $[(4,4'\text{-}OC_6H_4)\text{–}X\text{–}(4,4'\text{-}C_6H_4O)]^{2-}$ (X = CH=CH, CH=CH–CH=CH, CH=CH–CH=CH–CH=CH, 2,5-thiophenediyl, 2,5:2′,5′-bithiophenediyl, 2,5:2′,5′:2′,5′-terthiophenediyl, C≡C, N=N, C(O), CH=CH(1,4-C_6H_4)CH=CH]. Electrochemical, UV/Vis/NIR spectroelectrochemical, and magnetic measurements have also been reported.[77]

[MoCl(NO)(3-PhCH=CHCOpy)(Tp*)] and [MoCl(NO)(3,5-(MeCO)$_2$py}(Tp*)], octahedral species, have been prepared and characterized. They underwent reduction and oxidation, giving anionic and cationic species, respectively.[78]

[Mo(O)(biph)(Tp*)] has been prepared and investigated by electrochemical and UV/Vis/NIR spectroelectrochemical methods.[79]

Reaction of [$MoO_2Cl_2(THF)_2$] (X = Cl or Br) with Tp yielded [MoO_2Cl(Tp)].[80] The reaction of [MoO_2Cl(Tp*)] with pyridine-based ligands L (L = py, 4,4′-bipy, 4-Phpy, bpe) in toluene in the presence of PPh_3 gave mononuclear [Mo(O)Cl(Tp*)(L)] (L = py, 4,4,-bipy, 4-Phpy) and binuclear [{Mo(O)Cl(Tp*)}$_2$(μ-L)] (L = 4,4′-bipy, bpe). The octahedral [{Mo(O)Cl(Tp*)}$_2$(μ-(4,4′-bipy))] and [{Mo(Tp*)(O)Cl(L)}$_2$(μ-O)] have been structurally characterized.[81]

Treatment of allenic electrophiles with [Mo(CO)$_3$(py)$_3$] and Tp or treatment of an enyne with [MoH(Tp)(CO)$_3$], followed by decarbonylation, cleanly produced stable, new η^3-dienyl complexes such as [Mo(Tp)(CO)$_2$(η^3-CH_2CHC=C(CH_3)$_2$)] (Fig. 2.20).[82]

[{Mo(CO)(Tp^x)}(RC≡CMe)(η^1/η^2-O=CHR′)][BAr$_F$] (Tp^x = Tp or Tp*) have been used to investigate σ/π aldehyde binding.[83]

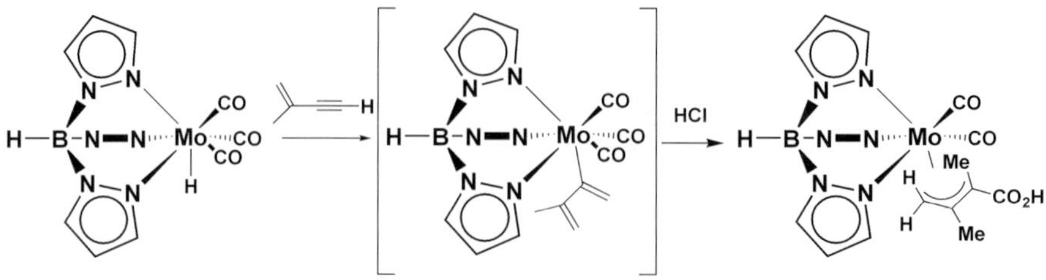

Fig. 2.20. Synthesis of the η^3-dienyl complexes $[Mo(Tp)(CO)_2(\eta^3\text{-}CH_2CHC{=}C(CH_3)_2)$.

Reactions of $[Mo(O)Cl_2(Tp^*)]$ with dilithium dichalcogenolate carboranes $[(THF)_3LiE_2C_2B_{10}H_{10}(THF)]_2$ [E = S, Se, Te] afforded $[Mo(O)(Tp^*)E_2C_2B_{10}H_{10}]$. The structure of the Se derivative was determined by single-crystal X-ray study. The sulphur complex $[Mo(O)(Tp^*)S_2C_2B_{10}H_{10}]$ converts in the air into a $[Hpz^*][Mo(O)$-$(Tp^*)(\mu^2\text{-}O)_2Mo(O)S_2C_2B_{10}H_{10}]$, also structurally characterized.[84]

$[Mo(Tp)(CO)_2]$-mediated cycloadditions of 3-substituted pyranyl and pyridinyl π-complexes represent the first enantiocontrolled [5+3] cycloadditions described to date and provide a new and efficient synthetic approach to oxa- and aza[3.3.1]bicyclics of high enantiomeric purity.[85] An "organometallic chiron" strategy employing single enantiomers of $[Mo(Tp)(CO)_2(\eta^3\text{-pyridinyl})]$ has been reported for the synthesis of an optically active tropane alkaloid.[86]

[5 + 2] cycloaddition of the pyridinyl π-complex in Fig. 2.21 to methyleneoxindole gave the spirooxindole complex in high enantiomeric purity. This complex has been converted to pyrrolidine.[87]

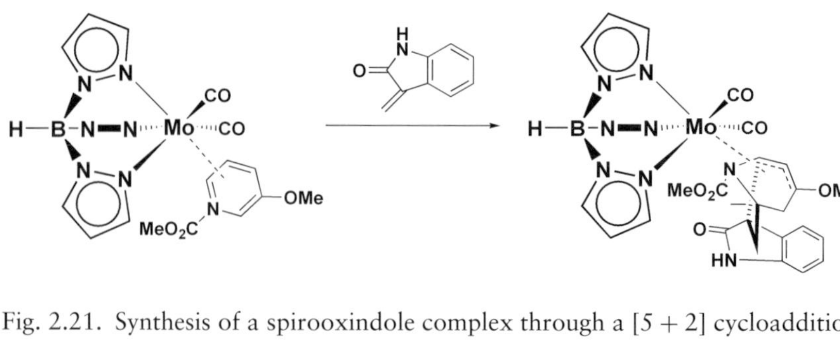

Fig. 2.21. Synthesis of a spirooxindole complex through a [5 + 2] cycloaddition.

[Mo(Tp)$(CO)_2$(5-alkenyl-η-2,3,4-pyranyl)diene] acts as an excellent chiral scaffold for the efficient regio- and enantiocontrolled synthesis of highly functionalized 1-oxadecaline derivatives through a transition metal-mediated Diels–Alder reaction.[88]

Cationic [MoTp$(CO)_2$(dihydropyridine)] complexes formed from 3-oxo-3,6-dihydro-2*H*-pyridine-1-carboxylic acid benzyl ester are scaffolds from which either 2,3,6- or 2,5,6-trisubstituted-1,2,3,6-tetrahydropyridines can be prepared in a regio- and stereodefined way.[89]

[Mo(Tp)$(CO)_2$(5-oxopyranyl)] and [Mo(Tp)$(CO)_2$(5-oxopyridinyl)] were prepared and used for the synthesis of substituted oxa- and aza[3.2.1], and [4.3.1] bycyclics via an unprecedented Mo-mediated 1,5-"Michael-type reaction".[90]

The syntheses and molecular structures of [Mo(Tp*)(NO)(bdt)], [Mo(Tp*)(NO)(tdt)], and [Mo(Tp*)(NO)($bdtCl_2$)] have been reported by Enemark *et al.*[91] These derivatives are the first structurally characterized {MoNO} compounds containing a $[Mo(Tp^*)(NO)]^{2+}$ fragment coordinated to an ene-1,2-dithiolate. The properties of these species have been investigated also by cyclic voltammetry, IR, NMR, electronic absorption, and He I photoelectron (PES) spectroscopies. [MoO(Tp*)($bdtCl_2$)] has also been prepared and characterized. This compound exhibits a distorted pseudo-octahedral coordination geometry with the Mo atom linked by a terminal oxo atom, two sulfur donor atoms from $bdtCl_2$ and Tp* in the κ^3-form. The coordination environment is similar to that in [MoO(Tp*)(bdt)]. The similarity of IR, EPR, and electronic absorption spectroscopic results suggests that the electron-withdrawing nature of chlorine does not influence significantly the electronic structure of the Mo(V) center, whereas redox potentials and gas-phase ionization energies are sensitive to remote substituent effects.[92] Kirk compares the molecular and electronic structures of the first- and second-generation models of the SO active site [MoO(Tp*)(bdt)] and [MoO(L′S)(bdt)] (L′SH = (2-dimethylethane-thiol)bis(3,5-dimethylpyrazolyl)methane). The electron paramagnetic resonance (EPR) parameters for both of these models are compared with those measured for the Mo(V) form of both low and high pH forms of SO in order to determine what insight these structural models can offer about the electronic structure of the active site of SO.[93]

The compounds [MoO(Tp*)(qdt)] and [MoO(Tp*)(tdt)] have been investigated by cyclic voltammetry and photoelectron, magnetic circular dichroism, and electronic absorption spectroscopies, and the experimental data have been interpreted in the context of *ab initio* molecular orbital calculations on a variety of dithiolate dianion ligands.[94] X-ray crystallography and resonance Raman spectroscopy have been used to characterize [MoO(Tp*)(qdt)] which is an important benchmark oxomolybdenum mono-dithiolene model system relevant to various pyranopterin Mo enzyme active sites, including sulfite oxidase.[95] [MoO(Tp*)(bdt)] has been investigated by gas-phase photoelectron spectroscopy and DFT in order to investigate the interactions between the sulfur π-orbitals of arene dithiolates and high-valent transition metals as minimum molecular models of the active site features of pyranopterin Mo–W enzymes.[96]

Temperature-dependent measurements of potential and electron-transfer rate constant are reported for the electrochemical reduction of a series of [MoO(Tp*)(X,Y)], where X,Y is a series of bidentate 1,2-disubstituted aliphatic or aromatic ligands in which oxygen donors are replaced sequentially by sulfur.[97] Solution redox potentials and heterogeneous electron transfer rate constants for [(Tp*)Mo(E)(SS)] [E = O, NO; SS = dithiolate] are also reported.[98]

A series of mononuclear octahedral thiomolybdenyl derivatives [MoX_2(Tp*)S] (X, X_2 = anion, dianion) has been synthesized and characterized by Young and coworkers.[99] These complexes provide insights into the electronic nature and chemistry of the catalytically and biologically important thiomolybdenyl units.[99]

Reaction of [$MoCl_2$(Tp*)S] with a variety of phenols and thiols yielded [MoX_2(Tp*)S] (X = 2-(ethylthio)phenolate, 2-(*n*-propyl)phenolate, phenolate; X_2 = benzene-1,2-dithiolate, 4-methylbenzene-1,2-dithiolate, benzene-1,2-diolate) characterized by microanalysis, mass spectrometry, IR, EPR, UV–visible spectroscopy, and X-ray crystallography, all derivatives containing a tridentate facial Tp*.[100] The electron paramagnetic resonance spin Hamiltonian parameters of these complexes together with their molybdenyl analogs have been calculated using density functional theory.[101]

[MoOCl(Tp*)(*p*-OC_6H_4X)] and [MoO(Tp*)(*p*-$OC_6H_4X)_2$] (X = OEt, OMe, Et, Me, H, F, Cl, Br, I, and CN) were investigated by electrochemical techniques and gas-phase photoelectron spectroscopy to study the effect of the remote substituent (X) on electron-transfer reactions at the oxomolybdenum core. The first ionization energies for [MoO(Tp*)(*p*-$OC_6H_4X)_2$] (X = OEt, OMe, H, F, and CN) were determined by photoelectron spectroscopy.[102]

Wadepohl and coworkers investigated the hydroboration reactions of Fischer carbyne metal complexes [Mo(≡CR)(Tp*)$(CO)_2$] and prepared [Mo(Tp*)$(CO)_2(\eta^2$-B(R′)CH_2R}] (R′ = Et, R = C_6H_4Me-4). A β-agostic C-H· · · M interaction is present between the metal and the boryl ligand (Fig. 2.22).[103]

The first air-stable anionic carbene derivatives with a metal-centered negative charge [Mo=C$(CN)_2$(Tp*)$(CO)_2]^-$ (Fig. 2.23) and [Mo=C(CN)(R)(Tp*)$(CO)_2]^-$ (R = substituted alkyl) have been prepared by the addition of cyanide anion to the corresponding chlorocarbine complexes. [Mo=C$(CN)_2$(Tp*)$(CO)_2]^-$ and [Mo=C(CN)(R)(Tp*)$(CO)_2]^-$ are stereochemically rigid. The cyanoalkylcarbene could be doubly alkylated at the CN nitrogen to give the cationic dialkylaminoalkyne complex [Mo{η^2(C,C)Me_2N-C≡C-R}(Tp*)$(CO)_2]^+$. [Mo=C(CN)C$(CN)_2$Me(Tp*)$(CO)_2]^-$ eliminates [MeC$(CN)_2]^-$ in the presence of MeI and gives [Mo(≡C-CN)-(Tp*)$(CO)_2$], as a first example of the cyanocarbyne complex.[104] [$(C_2H_5)_4$N][Mo=C$(CN)_2$(Tp*)$(CO)_2$] has also been structurally characterized.[105]

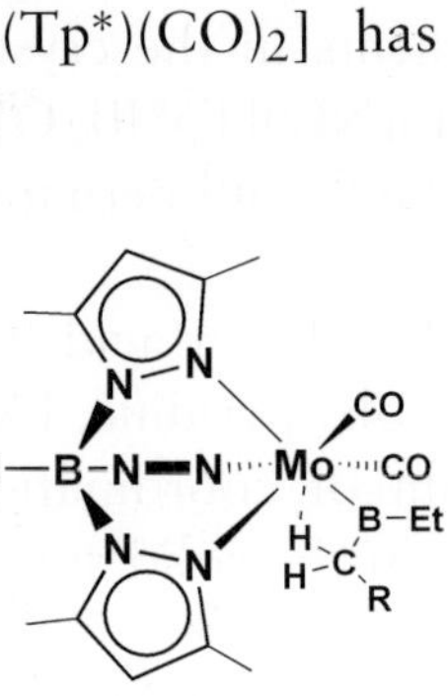

Fig. 2.22. The β-agostic interaction in [Mo(Tp*)$(CO)_2${η^2-B(Et)CH_2R}].

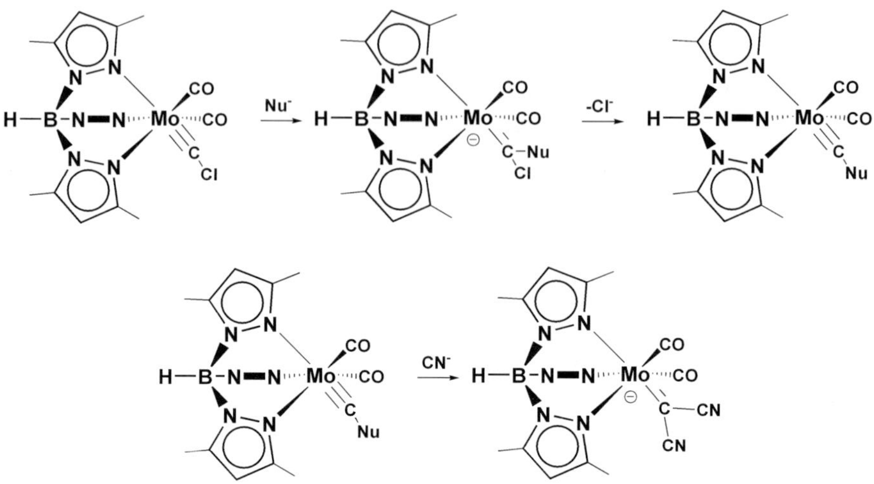

Fig. 2.23. Syntheses proposed in Ref. 104 for the preparation of $[Mo{\equiv}C(Nu)(Tp^*)(CO)_2]^-$ and of the first stable anionic carbene derivative $[Mo{=}C(CN)_2(Tp^*)(CO)_2]^-$.

[X{Tp*}(NO)MoOMo(NO){Tp*}Y] (X = Y = OH, OMe, OEt, OCOMe, or OCOPh; X = I, Y = OH, OMe, NHMe, or NHEt) have been synthesized and characterized spectroscopically. Cyclic voltammetry demonstrated that, when X = Y = OH, OCOPh, or OMe, and X = I, Y = OH or NHMe, this kind of complex undergoes a one-electron reduction and a one-electron oxidation. A single-crystal X-ray diffraction study evidenced the presence of four different components in the crystal: [I{Tp*}(NO)MoOMo(NO){Tp*}(NHMe)], $[\{Mo(NO)[Tp^*]I\}_2O]$, $[I\{HB(pz^*)_2(pz^{*,I})\}(NO)MoOMo(NO)\{Tp^*\}(NHMe)]$, and perhaps $[I\{HB(pz^*)_2(pz^{*,I})\}(NO)MoOMo(NO)\{Tp^*\}I]$.[106]

The arene complexes $[Mo(Tp)(\pi\text{-acid})(L)(\eta^2\text{-arene})]$ (π-acid = CO or NO, L = 1-alkylimidazole, pyridine, PMe_3, arene prochiral) exist as a dynamic equilibrium of coordination diastereomers in solution. On the other hand, in the solid state only one diastereomer is present.[107]

The fragment [Mo(Tp)(NO)(Meim)] also forms stable η^2 complexes with acetone, methyl formate, and DMF. [Mo(Tp)(NO)-

$(PMe_3)(\eta^2\text{-DMF})]$ is one of the rare examples of η^2-amide complexes.[108]

Reduction of $[MoBr_2(NO)Tp]$ with sodium amalgam in the presence of 1.5 equiv. of Meim and an excess of cyclohexene produces [Mo(Tp)(NO)(Meim)(cyclohexene)]. The reduction of $[MoBr_2$-(NO)(Tp)] in the presence of Meim with a series of aromatic molecules such as naphthalene, furan, and thiophene yielded stable Mo^0 complexes also (Fig. 2.24).[109]

The fragment [Mo(Tp)(NO)(Meim)] was also chosen as π-base because of its exceptionally low reduction potential. [Mo(Tp)(NO)-(MeIm)(η^2-cyclohexadiene)] is prepared by the reduction of [Mo(Br)-Tp(NO)(Meim)] (THF, sodium dispersion) in the presence of 1,3-cyclohexadiene. When a solution of [Mo(Tp)(NO)(MeIm)(η^2-cyclohexadiene)] is treated with ethyl vinyl ketone in CH_2Cl_2, no reaction occurs after three days (25°C). However, when the experiment is repeated at −40°C with the addition of 25 mol% of $BF_3{\cdot}OEt_2$, a molybdenum-bound 1-bicyclo[2.2.2]oct-5-en-2-ylpropan-1-one complex can be isolated (Fig. 2.25).[110]

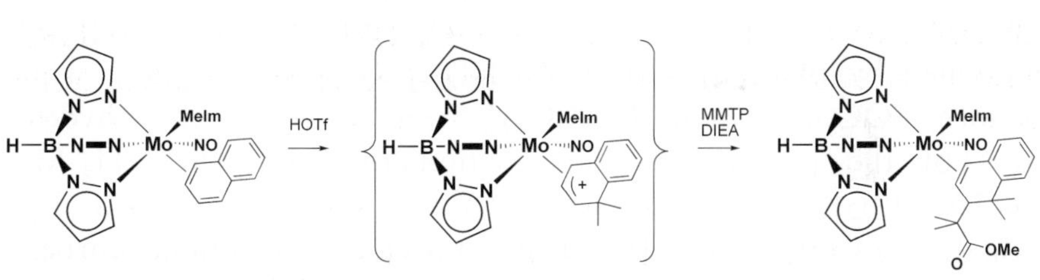

Fig. 2.24. Tandem addition sequence to [Mo(Tp)(NO)(MeIm)(η^2-naphthalene)].

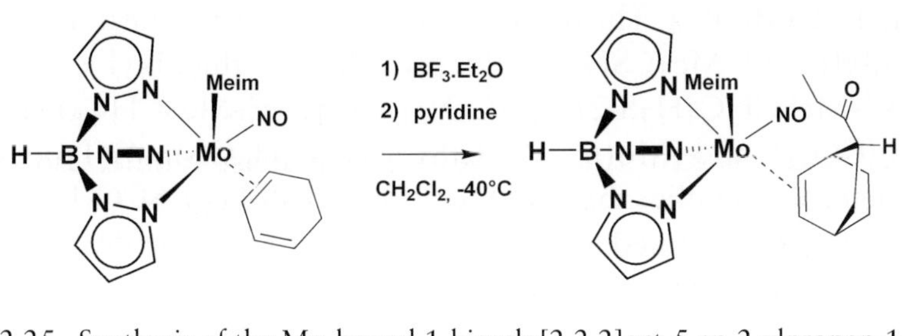

Fig. 2.25. Synthesis of the Mo-bound 1-bicyclo[2.2.2]oct-5-en-2-ylpropan-1-one.

Tp has been employed also for the stabilization of the π-allylmolybdenum(II) complex [Mo(Tp)$(CO)_2$(η^3-CH_2CH=$CHCO_2$-Me)].[111]

Reaction of [Mo$(CO)_3$(MeCN)Cl(η^2$SCNMe_2$)] with Tp gave [Mo(η^2$SCNMe_2$)$(CO)_2$(Tp)] which may be obtained also from the reaction of (i) K[Mo$(CO)_3$(Tp)] with Me_2NC(=S)Cl; (ii) [Mo$(CO)_3$-(η^6-C_7H_8)] with Me_2NC(=S)Cl and Tp; and (iii) Tp with [Mo-(η^2$SCNMe_2$)Cl$(CO)_2${PPh_3}$_2$]. Sequential treatment of either [Mo$(CO)_3$$(NCMe)_3$] or [Mo$(CO)_3$(η^6-$C_7H_8$)] with ClC(=S)OR (R = C_6H_4Me-4, C_6H_5) and Tp provides [Mo(η^2-SCOR)$(CO)_2$(Tp)] (R = C_6H_4Me-4, C_6H_5). Addition of [$Fe_2(CO)_9$] to [Mo(≡CN^iPr_2)-$(CO)_2$(Tp)] provided the thermally unstable complex [MoFe(μ-CN^iPr_2)$(CO)_5${Tp}], which reacts with sulfur to form [MoFe(μ-SCN^iPr_2)$(CO)_5${Tp}].[112]

[Mo(≡CC_6H_4Me-4)(Tp)$(CO)_2$], [Mo(≡CC_6H_4OMe-4)(Tp)-$(CO)_2$], and [Mo(≡C_4H_3S-2)(Tp)$(CO)_2$] react with methylthiirane to provide primarily [Mo(η^2-SCR)(Tp)$(CO)_2$] in addition to small amounts of the corresponding dithiocarboxylates [Mo(η^2-S_2CR)(Tp)$(CO)_2$]. The mixed selenothiocarboxylates [Mo(η^2-SSeCR)(Tp)$(CO)_2$] (R = C_6H_4OMe-4, C_4H_3S-2) are obtained by treating [Mo(η^2-SCR)(Tp)$(CO)_2$] with Li_2Se_2, while [Mo(η^2-SOCC_6H_4Mc-4)(Tp)$(CO)_2$] results from the hydrolysis of the product of the reaction of [Mo(η^2-SCC_6H_4-Me-4)(Tp)$(CO)_2$] with [Me_2SSMe]BF_4. Alkylation of [Mo(η^2-SCC_6H_4-Me-4)(Tp)$(CO)_2$] with [Et_3O]BF_4 provides the thiolatocarbene salt [Mo(η^2-EtSCC_6H_4Me-4)(Tp)$(CO)_2$], while the analogs [Mo(η^2-MeSCC_6H_4Me-4)(Tp)$(CO)_2$]BF_4 and [Mo(η^2-MeSCC_4H_3S-2)(Tp)$(CO)_2$]BF_4 result from the reactions of [Mo(≡CR)(Tp)$(CO)_2$] with [Me_2SSMe]BF_4. The reactions of these salts with the nucleophiles Li[Et_3BH] and Me_3CSH finally provide the thioalkyl complexes [Mo(η^2-MeSCHC_4H_3S-2)(Tp)$(CO)_2$], [Mo(η^2-MeSCHC_6H_4OMe-4)(Tp)$(CO)_2$], and [Mo(η^2-MeSC(SCMe_3)C_6H_4OMe-4)(Tp)$(CO)_2$]. The heterobimetallic thioaroyl complex [MoFe(μ-SCC_6H_4Me-4)-(Tp)$(CO)_5$] is obtained from the reaction of [Mo(η^2-SCC_6H_4Me-4)(Tp)$(CO)_2$] with [$Fe_2(CO)_9$].[113]

$[MoI(Tp^*)(CO)_3]$ with AgS_2CNEt_2 affords $[Mo(Tp^*)(S_2CNEt_2)(CO)_2]$ which reacts with propylene sulfide giving $[Mo(Tp^*)S(S_2CNEt_2)]$. Interaction of $[Mo(Tp^*)S(S_2CNEt_2)]$ with dimethyl acetylenedicarboxylate produced $[Mo(Tp^*)\{S_2C_2(CO_2Me)_2\}(SCNEt_2\text{-}\kappa^2 S,C)]$, a seven-coordinate complex exhibiting distorted pentagonal bipyramidal coordination with $C1$ symmetry, the thiocarboxamide unit being orientated along one of the W–S(dithiolene) bonds with its NEt_2 group projecting away from the Tp* ligand.[114]

$[Mo(Tp^*)(CO)_3(\eta^3\text{-allyl})]$ requires higher potentials for a generally irreversible oxidation to proceed. The product of oxidation of $[Mo(Tp^*)(CO)_3I]$ with oxygen was a tetranuclear linear molybdenum(V) complex with terminal metal ions coordinated to Tp* ligands and inner molybdenum(V) ions surrounded exclusively by oxygen donors.[115]

Reduction of $[M(Tp^*)(CO)_2(\eta\text{-RC}{\equiv}\text{CR}')]X$ (M = Mo, X = PF_6, R = R′ = Ph, C_6H_4OMe-4, or Me; R = Ph, R′ = H; M = W, X = BF_4, R = R′ = Ph or Me; R = Ph, R′ = H) with $[Co(\eta\text{-}C_5H_5)_2]$ gave paramagnetic $[M(Tp^*)(CO)_2(\eta\text{-RC}{\equiv}\text{CR}')]$, characterized by IR and ESR spectroscopy. X-ray structural studies on the redox pair $[Mo(Tp^*)(CO)_2(\eta\text{-PhC}{\equiv}\text{CPh})]$ and $[Mo(Tp^*)(CO)_2(\eta\text{-PhC}{\equiv}\text{CPh})][PF_6]$ have been carried out. Reduction of $[Mo(Tp^*)(CO)(NCMe)(\eta\text{-MeC}{\equiv}\text{CMe})][PF_6]$ with $[Co(\eta\text{-}C_5H_5)_2]$ in CH_2Cl_2 gives $[MoCl(Tp^*)(CO)(\eta\text{-MeC}{\equiv}\text{CMe})]$, via nitrile substitution in $[Mo(Tp^*)(CO)(NCMe)(\eta\text{-MeC}{\equiv}\text{CMe})]$, whereas a similar reaction with $[Mo(Tp^*)(CO)\{P(OCH_2)_3CEt\}(\eta\text{-MeC}{\equiv}\text{CMe})]^+$ gives the phosphite-containing radicals $[Mo(Tp^*)(CO)\{P(OCH_2)_3CEt\}(\eta\text{-MeC}{\equiv}\text{CMe})(Tp^*)]$. ESR spectroscopic studies and DFT calculations on $[Mo(Tp^*)(CO)L(\eta\text{-MeC}{\equiv}\text{CMe})]$ {L = CO or $P(OCH_2)_3CEt$} have also been performed.[116] The d^4 halide complexes $[MX(Tp^*)(CO)(\eta\text{-RC}{\equiv}\text{CR})]$ {X = F, Cl, Br, or I; R = Me or Ph; M = Mo or W} undergo one-electron oxidation to the d^3 monocations $[MX(Tp^*)(CO)(\eta\text{-RC}{\equiv}\text{CR})(Tp^*)]^+$ (isolable when M = W, R = Me).[117]

The stable cyanide-bridged linkage isomers, $[(o\text{-}O_2C_6Cl_4)L(OC)_2Ru(\mu\text{-XY})M(CO)(Tp^*)PhC{\equiv}CPh]$ (M = Mo or W, L = PPh_3 or $P(OPh)_3$, XY = CN or NC) have been prepared and characterized by IR spectroscopy and cyclic voltammetry, and [(*o*-$O_2C_6Cl_4$)(Ph_3P)$(OC)_2$Ru(μ-NC)Mo(Tp*)(CO)(PhC≡CPh)] by single-crystal X-ray study. [(*o*-$O_2C_6Cl_4$)L$(OC)_2$Ru(μ-XY)M(Tp*)-(CO)(PhC≡CPh)] are also firstly oxidized at the catecholate ligand; whereas a second oxidation, and a one-electron reduction, is based on the {M(Tp*)(CO)(PhC≡CPh)} fragment. Chemical oxidation of the same species with $AgBF_4$, gave the paramagnetic monocations and [(*o*-$O_2C_6Cl_4$)L$(OC)_2$Ru(μ-XY)M(Tp*)(CO)(PhC≡CPh)]$^+$.[118]

[M(Tp*)$(CO)_2${η^2-(BnS)C≡C(S)}] (Mo, W) has been obtained by the reductive removal of one benzyl group from [M(Tp)$(CO)_2${η^2-(BnS)C≡C(SBn)}](PF_6). [M(Tp*)$(CO)_2${η^2-(BnS)C≡C(S)}] has been employed as monodentate ligand in homoleptic pentanuclear complexes with nickel(II) and palladium(II).[119] Substitution reactions of [Mo(Tp*)(BnSC≡CSBn)$(CO)_2$]PF_6 have also been investigated and the structural characterization of [Mo(Tp*)(BnSC≡CSBn)(CO)-(MeCN)]PF_6, [Mo(Tp*)(BnSC≡CSBn)(CO)Cl], and [Mo(Tp*)(Bn-SC≡CSBn)Cl_2] [Mo(Tp*)(BnSC≡CSBn)ClF] reported.[42]

[MoOL_2(Tp)] (L = OPh, SPh, or Cl) and related nitrosyl complexes, as also the dinuclear B–B linked complex reported in Fig. 2.26, have been investigated by electron magnetic resonance techniques, electron nuclear double resonance, and hyperfine sublevel correlation spectroscopy: ^{11}B hyperfine and quadrupolar couplings have been measured.[120]

[MoCl_2(Tp*)(NO)] reacts with an equimolar quantity of the zinc[121] and copper[122] Schiff base complexes yielding heterobinuclear complexes in which the redox fragment [MoClTp*(NO)]$^+$ is attached to the functionalized Schiff base derived by

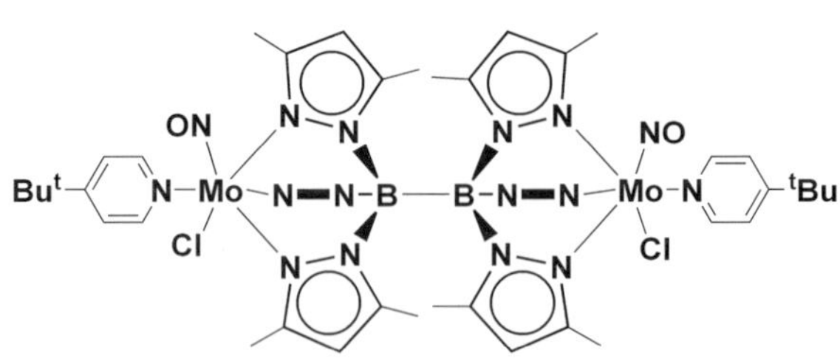

Fig. 2.26. The dinuclear [(4-^{t}Bupy)(NO)(Cl)Mo(pz)$_3$BB(pz)$_3$Mo(Cl)(NO)(4-^{t}Bupy)].

condensation of 2,5-dihydroxybenzaldehyde and salicylaldehyde with α, ω-diammines. These complexes were characterized by IR, UV, and ^{1}H NMR spectroscopy.

The mixed-ligand complex [MoO_2(Tp*)(acac)] has been prepared and characterized by IR, NMR, and EPR methods.[18]

Reis has reported dipole moments, linear, and nonlinear polarizabilities of [Mo(NO)(Tp*)Cl{OC_6H_4[CH= CHC_6H_4N-$(CH_3)_2$-4]-4}] calculated at the SCF and second-order Møller–Plesset correlation (MP2) level.[123]

Reaction of [Mo(Cp)(Ind)$(NCMe)_2$][BF_4]$_2$ with pzTp produced [Mo(Cp)(Ind)(κ^2-pzTp)]BF_4 (Fig. 2.27).[124]

Walker and Ogura[125] reported a useful method for the construction of large phorphirin assembly by the attachment of bulky {Mo^V=O(Tp*)} moiety onto the periphery that are mutually *trans* (Fig. 2.28). These structures are of interest to bio-inspired chemistry and material science.[125]

Reaction of [$MoCl_2(NNPh)_2$(dme)] with Tp* afforded the bis(isodiazene) complex [MoCl(Tp*)$(NNPh_2)_2$], octahedral, having a facially coordinated Tp* ligand and linear $NNPh_2$ ligands.[126]

The reaction between [$MoCl_2${$(CH_2)_n$(2-$C_6H_4N)_2$}(dme)] (n = 1; 2) and Tp* in refluxing THF gives [MoCl(Tp*){$(CH_2)_n$(2-C_6-$H_4N)_2$}]. The reaction between Tp* and [$MoCl_2(NAr)_2$(dme)] or [$MoCl_2$(NAr)(N^tBu)(dme)] (Ar = 2,6-$^i Pr_2C_6H_3$) yields [Mo(Cl)-(Tp*)$(NAr)_2$] and [Mo(Cl)(Tp*)(NAr)(N^tBu)], respectively, exhibiting *fac* coordination for the tridentate Tp*.[127] [Mo(O)Cl_2(N-Ms)(dme)] reacts with Tp* giving [Mo(O)Cl(Tp*)(N-Ms)].[128] The chloride substitution reaction of [$MoCl_2$(N-Ms)$_2$(dme)] with Tp* in

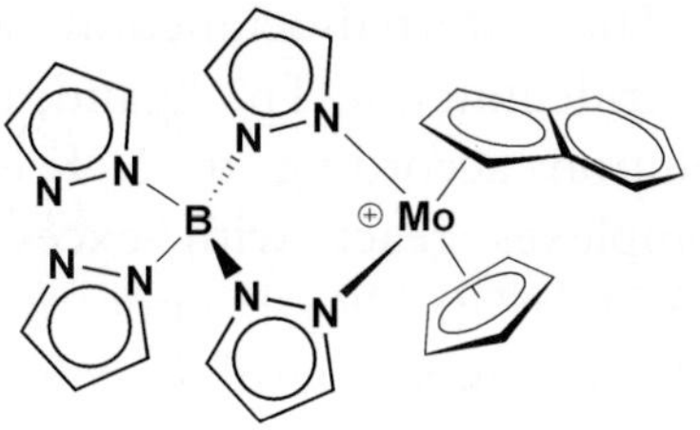

Fig. 2.27. [Mo(Cp)(Ind)(κ^2-pzTp)]BF_4.

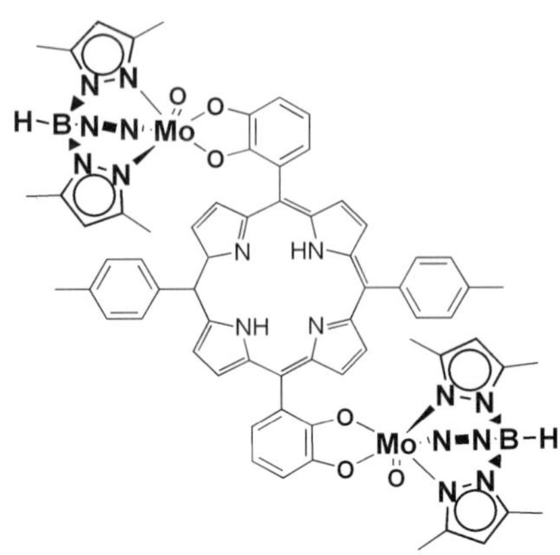

Fig. 2.28. A bio-inspired phorphirin molybdenum scorpionate derivative.

THF resulted in [MoCl(Tp*)(N-Ms)$_2$], which upon reaction with MeMgCl in THF gave [Mo(Me)(Tp*)(N-Ms)$_2$]. [Mo(OH)(Tp*)(N-Ms)$_2$] was produced in the presence of wet methanol under aerobic conditions. These complexes were characterized by ^{1}H NMR and IR spectroscopy, and the chloride and hydroxo species by X-ray single-crystal analyses.[129]

Raman spectroscopic investigation of the complex [MoO$_2$-Me(Tp*)] is described. A characteristic shift of the metal oxygen stretching vibrations is observed, which can be used as an indicator for the donor ability of the ligands Tp*.[130]

The reaction of [MoO(Tp*)(S$_2$PR$_2$)] with propylene sulfide produces the distorted octahedral [MoO(Tp*)S{SP(S)R$_2$} characterized by microanalysis, mass spectrometry, cyclic voltammetry, spectroscopy (IR, NMR, UV–vis, and X-ray absorption), and X-ray crystallography. The distorted-octahedral isopropyl and phenyl derivatives feature a tridentate *fac*-Tp* ligand; the solid-state structures persist in solution according to S K-edge X-ray absorption data. The complexes react with excess cyanide to form thiocyanate and [MoO(Tp*)(S$_2$PR$_2$)], under anaerobic conditions, or [MoO$_2$(Tp*)(S$_2$PR$_2$)], under aerobic conditions. [MoO(Tp*)-S{SP(S)R$_2$}] reacts with PPh$_3$ and SPPh$_3$ giving MoO(Tp*)-(S$_2$PR$_2$)]; whereas [MoO$_2$(TP*)(S$_2$PR$_2$)] reacts with cobaltocene or

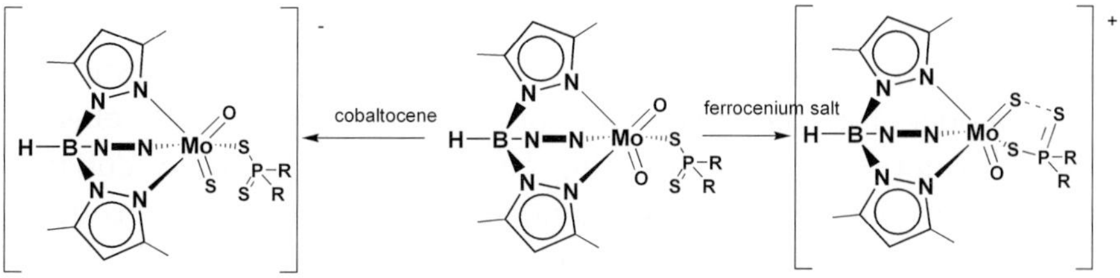

Fig. 2.29. Reactivity of $[MoO_2(Tp^*)(S_2PR_2)]$.

hydrosulfide ion producing $[MoO(Tp^*)S(S_2PR_2)]^-$ (Fig. 2.29), and with ferrocenium salts yielding $[MoO(Tp^*)(S_3PR_2)]^+$.[131]

Bruce *et al.* reported the synthesis of $[Mo(\equiv CC\equiv CSiMe_3)(Tp^x)(CO)_2]$ (Tp^x = Tp or Tp*). $[Mo(\equiv CC\equiv CSiMe_3)Tp(CO)_2]$ reacts with $AuCl(PPh_3)/K_2CO_3$ in MeOH affording $[Mo\{\equiv CC\equiv CAu(PPh_3)\}(Tp)(CO)_2]$. The latter, upon interaction with $Co_3(\mu_3\text{-}CBr)(\mu\text{-dppm})(CO)_7$, gives $[\{(Tp)(OC)_2Mo\}\equiv CC\equiv CC\equiv\{Co_3(\mu\text{-dppm})(CO)_7\}]$. Reaction of $[Mo(\equiv CBr)(CO)_2(Tp^*)]$ with $Co_3\{\mu^3\text{-}C(C\equiv C)_2Au(PPh_3)\}(\mu\text{-dppm})(CO)_7$ yields $[\{Tp^*(OC)_2Mo\}\equiv C(C\equiv C)_2C\equiv\{Co_3(\mu\text{-dppm})(CO)_7\}]$. The C_3 complexes $[Tp^x(OC)_2Mo\equiv CC\equiv CRu(dppe)Cp^*]$ were obtained from the interaction of $[Mo(\equiv CC\equiv CSiMe_3)(CO)_2(Tp^x)]$ with $[RuCl(dppe)Cp^*]$ via KF-induced metalla-desilylation reactions. Reactions between $[Mo(\equiv CBr)(CO)_2Tp^*]$ and $[Ru\{(C\equiv C)_nAu(PPh_3)\}(dppe)Cp^*]$ (n = 2, 3) afforded $[\{Tp^*(OC)_2Mo\}\equiv C(C\equiv C)_n\{Ru(dppe)Cp^*\}]$.[132]

A new procedure for the preparation of molybdenum single cubanes is introduced by the reaction of $[Mo(Tp)S(S_4)]^-$ [49] with $FeCl_2$/NaSEt to afford $[(Tp)MoFe_3S_4Cl_3]^-$. $[(Tp)MoFe_3S_4L_3]^-$ (L = EtS, p-MeC_6H_4S) have also been prepared. Reduction of $[(Tp)MoFe_3S_4Cl_3]^-$ with borohydride gives $[(Tp)MoFe_3S_4Cl_3]^{2-}$. Trigonal structures are found for $(Bu_4N)[(Tp)MoFe_3S_4Cl_3]$, $(Bu_4N)[(Tp)MoFe_3S_4EtS_3]$, and $(Bu_4N)_2[(Tp)MoFe_3S_4Cl_3]\cdot MeCN$. The availability of both $[(Tp)MoFe_3S_4Cl_3]^-$ and $[(Tp)MoFe_3S_4Cl_3]^{2-}$ allows the first comparison of structures and ^{57}Fe isomer shifts of $[MoFe_3S_4]^{3+,2+}$ in a constant ligand environment. Reaction of $[(Tp)MoFe_3S_4Cl_3]^-$ with excess Li_2S in acetonitrile affords the double cubane $\{[(Tp)MoFe_3S_4Cl_2]_2(\mu_2\text{-}S)\}^{2-}$.[133] The double cubane $[(Tp)_2Mo_2Fe_6S_8(PEt_3)_4]$ system has been investigated as a model for the reactivity of the nitrogenase P^N cluster. Four distinct processes for the conversion of double cubanes to PN-type clusters are documented, affording the products $[(Tp)_2Mo_2Fe_6S_9(SR)_2]^{3-}$, $[(Tp)_2Mo_2Fe_6S_8(OMe)_3]^{3-}$, and $[(Tp)_2Mo_2Fe_6S_7(OMe)_4]^{2-}$. The reverse transformation of a PN-type cluster to an edge-bridged double cubane has been performed through the reaction of $[(Tp)_2Mo_2Fe_6S_8(OMe)_3]^{3-}$ with Me_3SiX that yields $[(Tp)_2Mo_2Fe_6S_8X_4]^{2-}$ (X = Cl, Br).[134] Reaction of $[(Tp)_2Mo_2Fe_6S_8(PEt_3)_4]$ with hydrosulfide affords the clusters $[(Tp)_2Mo_2Fe_6S_9(SH)_2]$,$^{3-,4-}$ established as the first structural (topological) analogs of the P^N cluster of nitrogenase. Utilizing SeH^- as a surrogate reactant for SH^- in the system, a series of selenide clusters $[(Tp)_2Mo_2Fe_6S_8SeL_2]^{3-}$ (L^- = SH^-, SeH^-, EtS^-, CN^-) was prepared and characterized by ^1H NMR spectra and X-ray. The isostructural clusters $\{[(Tp)_2Mo_2Fe_6S_9](\mu_2\text{-}S)\}_2^5$ and $\{[(Tp)_2Mo_2Fe_6S_8Se](\mu_2\text{-}Se)\}_2^5$ were also reported.[31]

$[(Tp)_2Mo_2Fe_6S_8L_4]^z$ (z =4-, 3-, 2-, 1-; L = F^-, Cl^-, N_3^-, PhS^-) contains the core unit $M_2Fe_6(\mu_3{-}S)_6(\mu_4{-}S)_2$ in which two $MoFe_3S_4$ cubanes are coupled by two $Fe(\mu_4\text{-}S)$ interactions to form a centrosymmetric edge-bridged double cubanc cluster. Some of these clusters are synthetic precursors to $[(Tp)_2Mo_2Fe_6S_9L_2]^{3-}$, which possess the same core topology as the P^N cluster of nitrogenase. $[(Tp)_2Mo_2Fe_6S_8Cl_4]^{4-,3-,2-}$ were isolated and structurally characterized. Under oxidizing conditions, $[(Tp)_2Mo_2Fe_6S_8Cl_4]^{2-}$cleaves with the formation of a single-cubane $[(Tp)MoFe_3S_4Cl_3]^{1-}$. The synthesis and structure of $[(Tp)_2Mo_2Fe_6S_9F_2{\cdot}H_2O]^{3-}$, analog of the P^N cluster of nitrogenase, have been described.[135]

Reaction of the single-cubane $[(Tp)MoFe_3S_4(PEt_3)_3]^+$ and of the edge-bridged double-cubane $[(Tp)_2Mo_2Fe_6S_8(PEt_3)_4]$ with cyanide affords $[(Tp)MoFe_3S_4(CN)_3]^{2-}$ and $[(Tp)_2Mo_2Fe_6S_8(CN)_4]^{4-}$, respectively. Reduction of $[(Tp)MoFe_3S_4(CN)_3]^{2-}$ with $KC_{14}H_{10}$ yields $[(Tp)MoFe_3S_4(CN)_3]^{3-}$.[136]

The reaction of $[Mo_3S_4(H_2O)_9]^{4+}$ with Tp produced $[Mo_3S_4(Tp)_3]X{\cdot}nH_2O$ (n = 1, 2, or 4; X = Cl, OTf, PF_6,

BF_4). An X-ray structure analysis of $[Mo_3S_4(Tp)_3]Cl{\cdot}4H_2O$ revealed that each molybdenum atom is bonded to the Tp ligand. $[Mo_3OS_3(Tp)_3]PF_6{\cdot}H_2O$ has been obtained via the reaction of $[Mo_3OS_3(H_2O)_9]^{4+}$ with Tp in the presence of NH_4PF_6. The reaction of $[Mo_3S_4(Tp)_3]X{\cdot}nH_2O$ with Fe in CH_2Cl_2 gave $[Mo_3FeS_4X(Tp)_3]$ (X = Cl, Br). X-ray structure analysis of the chloride species has revealed the existence of a cubane-type core Mo_3FeS_4.[137]

Sengar and Basu have reported and characterized some dioxo-Mo^{VI} complexes of formula $[MoO_2(Tp^*)(p\text{-}SC_6H_4Dn)]$ (Dn = dendritic unit).[138]

2.7.3. *W*

The scorpionate tungsten chemistry continues to be very rich and limited to the homoscorpionate ligands. Analogous Mo and W complexes are often reported in the same article; tungsten complexes are also described in the previous section, and references to Mo derivatives can also be found in this chapter.

Chiral oxo-alkyne tungsten(IV) complexes $[W(O)(I)(Tp^*)(RC{\equiv}CR')]$ (R = R′ = H; R = H, R′ = Me; R = R′ = Me; R = Me, R′ = Ph) have been prepared by the reaction of $[W(O)(I)(Tp^*)(CO)]$ with Me_3NO in the presence of an excess alkyne. $[W(O)(I)(Tp^*)(MeC{\equiv}CPh)]$, structurally characterized, was isolated as a 2:1 mixture of alkyne rotamers. Different synthetic routes to $[W(O)(I)(Tp^*)(HC{\equiv}CMe)]$ have also been reported. When ^{n}BuLi was added to $[W(O)(I)(Tp^*)(HC{\equiv}CMe)]$, deprotonation of the terminal acetylene site was found, and electrophilic addition yielded the vinylidene species $[W(=C=CMe_2)(O)(I)(Tp^*)]$ and $[W\{=C=C(Me)(H)\}O)(I)(Tp^*)]$.[139]

The air-stable anionic stereochemically rigid carbene complex $[W{=}C(CN)_2(Tp^*(CO)_2)]^-$ has been prepared by the addition of CN^- to the corresponding chlorocarbine derivative.[104]

$[W{\equiv}C\text{-}OAr(Tp^*)(CO)_2]$ (Ar = Ph, $p\text{-}C_6H_4Me$, $p\text{-}C_6H_4OMe$) described in the first edition[140] have been converted into $[W(\eta^3\text{-}(C_\alpha,C_\beta,C_\gamma){=}CPh\text{-}CH{=}CH(O(p\text{-}C_6H_4R)))(Tp^*)(CO)_2][X]$ (R = H, X = BF_4; R = Me, X = BF_4; R = OMe, X = BF_4; R =

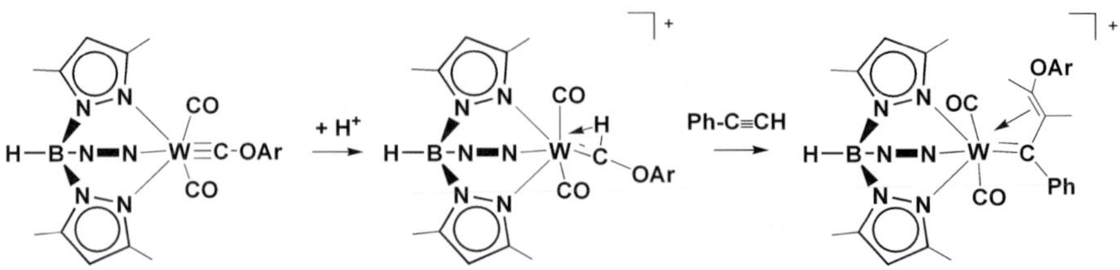

Fig. 2.30. Proton addition and phenylacetylene insertion in [W≡C-OAr(Tp*)(CO)$_2$].

H, X = BAr$_F$) upon proton addition and phenylacetylene insertion (Fig. 2.30).[141] The intermediate agostic carbene complex [W=C(H)O(*p*-C$_6$H$_4$OMe)(Tp*)(CO)$_2$][BAr$_F$] was characterized by IR, ^{1}H NMR, and ^{13}C NMR spectroscopy. Addition of a base to the cationic vinylcarbene complex [W(η^3-(C$_\alpha$,C$_\beta$,C$_\gamma$)=CPh-CH=CH(O(*p*-C$_6$H$_4$R)))Tp*(CO)$_2$][X] yielded prevalently the metallafuran complex [W(κ^2-(C$_\alpha$,O)=CPh-CH=CH-O)(Tp*)(CO)$_2$]. [W≡C-CPPh=CH(OTol)(Tp*)(CO)$_2$] was also found in the reaction mixture of [W(η^3-(C$_\alpha$,C$_\beta$,C$_\gamma$)=CPh-CH=CH(O(*p*-C$_6$H$_4$Me)))(Tp*)(CO)$_2$][BF$_4$] with the base. Addition of HBF$_4$ to metallacyclopropene complex resulted in complete conversion of the latter compound to the metallafuran complex at room temperaturc, whereas when acid was added to the metallacyclopropene complex at low temperature [W(η^3-(C$_\alpha$,C$_\beta$,C$_\gamma$)=CPh-CH=CH(OH))(Tp*)(CO)$_2$][BF$_4$] was formed, identified by ^{1}H NMR.[141] Addition of Li[Et$_3$BH] to [W(η^3-(C$_\alpha$,C$_\beta$,C$_\gamma$)=CPh-CH=CH(O(*p*-C$_6$H$_4$R)))(Tp*)(CO)$_2$][X] produces double hydride addition and loss of aryloxide at C$_\gamma$ of [W(η^3-(C$_\alpha$,C$_\beta$,C$_\gamma$)=CPh-CH=CH(O(*p*-C$_6$H$_4$R)))(Tp*)(CO)$_2$][X], to form the η^2-vinyl complex [W(η^2-(C$_\alpha$,C$_\beta$,)CPh=CHCH$_3$)(Tp*)(CO)$_2$].[142]

Hydride addition to an *E*/*Z* mixture of the coordinated imine in [W(Tp*)(CO)(PhC≡CMe)(NH=CMeEt)][BAr$_F$] yields an amido complex with high diastereoselectivity through *E*/*Z* imine isomer reactivity differences.[143] The amido species [W(Tp*)(CO)(PhC≡CMe)(NHCHRR′)] (R/R′ = Me/Me; Et/Et) were synthesized according to literature methods.[144] [W(Tp*)-(CO)(PhC≡CMe)(NHCHMeEt)] has been obtained in a 1:1

diastereomer ratio. Their oxidation produces imine complexes, [W(Tp*)(CO)(PhC≡CMe)(NH=CR′)][BAr_F]. In particular [W(Tp*)(CO)(PhC≡CMe)(NH=CMeEt)][BAr_F] exists as a 75:25 ratio mixture of *E* and *Z* isomers (NOESY characterization). The corresponding azavinylidene complexes were also prepared by deprotonation of imine complexes with KH in THF.

By using $LiHBEt_3$ as the nucleophile, hydride was added to the terminal alkyne carbon of [W(Tp*)$(CO)_2$(η^2-RCH_2C≡CH)][OTf] (R = H, Pr^n, Ph), [W(Tp*)$(CO)_2$(η^2-C(CH_2R)=CH_2)] being formed (Fig. 2.31). Deprotonation of η^2-vinyl (or 1-metallacyclopropene) with ^{n}BuLi or ^{t}BuLi removes a β′-hydrogen forming [Li][W(Tp*)$(CO)_2$(η^2-CH_2=C=CHR)]. Finally, quenching of the allene intermediates for R = H, ^{n}Pr with MeI results in methylation giving [W(Tp*)$(CO)_2$((η^2-C(CHRMe)=CH_2)]. [W(Tp*)$(CO)_2$((η^2-C(CH(CHOH)Ph)Pr)=CH_2)] has also been prepared by the addition of benzaldehyde to the anionic η^2-propylallene intermediate.[145]

The chiral [W(O)(Tp*)(H_2O)(RC≡CR)][OTf] (R = H, Me) has been prepared by halide abstraction from [W(O)(Tp*)(I)(RC≡CR)]. [W(O)(Tp*)(OTf)(HC≡CH)] has also been isolated. Both the species undergo deprotonation and isomerization when exposed to Al_2O_3, giving [W$(O)_2$(Tp*)(CH=CH_2)] and [W$(O)_2$(Tp*){C(Me)=C(H)(Me)}]. The presumed intermediates, the neutral oxo-hydroxo compounds [W(O)(OH)(Tp*)(RC≡CR)] (R = H, Me), can be accessed by deprotonation of [W(O)(Tp*)(H_2O)(RC≡CR)][OTf] with NaOH. [W(O)(OH)(Tp*)(HC≡CH)]

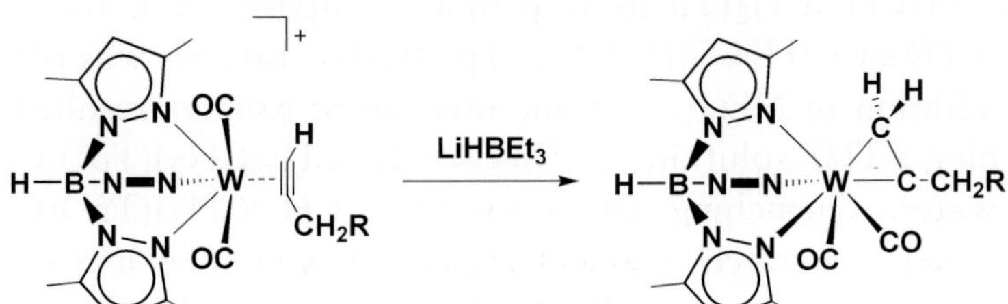

Fig. 2.31. Synthesis of [W(Tp*)$(CO)_2$(η^2-C(CH_2R)=CH_2)].

converts to $[W(O)_2(Tp^*)(CH{=}CH_2)]$ upon heating. X-ray structural data have provided solid-state structures of $[W(O)(Tp^*)(H_2O)(MeC{\equiv}CMe)][OTf]$ and $[W(O)_2(Tp^*)(CH{=}CH_2)]$.[146]

The X-ray structures of $[W(Tp^*)(I)(CO)\{MeC{\equiv}CCHMe(CH_2)_4I\}]$, $[W(Tp^*)(I)(CO)\{Me(PhCH_2)HCC{\equiv}CCHMeCH(OH)Ph\}]$, and $[W(Tp^*)(CO)(NHCHMePh)\{MeC{\equiv}CMe\}]$ have also been reported.[147]

The electrophilic reactivity of $[W{=}CH_2(Tp^*)(CO)(PhC{\equiv}CMe)][PF_6]$ was tested with both neutral nucleophiles and anionic nucleophiles to give $[W\text{-}CH_2\text{-}Nu(Tp^*)(CO)(PhC{\equiv}CMe)]^{+,0}$. The phthalimide and diethyl malonate derivatives were structurally characterized. Reaction of $[W{=}CH_2(Tp^*)(CO)(PhC{\equiv}CMe)][PF_6]$ with cyclohexene sulfide produces a sulfur atom transfer to form a π-bound thioformaldehyde complex. The structure of $[W{=}CH_2(Tp^*)(CO)(PhC{\equiv}CMe)][BAr_F]$ has also been reported.[148]

$[W{\equiv}CPMe_2Ph(Tp^*)(CO)_2]^+$ adds $PhCH_2MgCl$ and $CH_2{=}CHCH_2MgCl$ at C_α yielding the zwitterionic intermediates $[W{=}C(CH_2R)(PMe_2Ph)(Tp^*)(CO)_2]$ (R = Ph, $CH{=}CH_2$), which upon PMe_2Ph dissociation give neutral $[W{\equiv}CCH_2R(Tp^*)(CO)_2]$. Deprotonation at C_β provides the reactive vinylidene anions $[W{=}C{=}C(H)(R)(Tp^*)(CO)_2]^-$. $[W{=}C{=}C(H)(Ph)(Tp^*)(CO)_2]^-$ adds electrophiles at C_β: $[W{=}CCHCH{=}CH_2(Tp^*)(CO)_2]^-$ reacts with MeI at C_β to generate $[W{\equiv}CCH(CH_3)(CH{=}CH_2)(Tp^*)(CO)_2]$, which upon treatment with ^{n}BuLi forms $[W{=}C{=}C(CH_3)CH{=}CH_2(Tp^*)(CO)_2]^-$. Its reaction with MeI at C_β gives $[W{\equiv}CC\text{-}(CH_3)_2CH{=}CH_2(Tp^*)(CO)_2]$, although some methylation occurs at C_δ to generate the conjugated carbine $[W{\equiv}CC(CH_3){=}C(H)(CH_2CH_3)(Tp^*)(CO)_2]$. Isomerization of $[W{\equiv}CCH_2CH{=}CH_2(Tp^*)(CO)_2]$ to a 4:1 mixture of *E* and *Z* vinyl carbyne $[W{\equiv}CC(H){=}C(H)(CH_3)(Tp^*)(CO)_2]$ isomers was observed upon addition of NEt_3; the same mixture of isomers resulted from quenching a THF solution of $[W{=}C{=}CH\text{-}(CH{=}CH_2)(Tp^*)(CO)_2]^-$ with water. Quenching $[W{=}C{=}C(CH_3)(CH{=}CH_2)(Tp^*)(CO)_2]^-$ with water generated $[WCC(CH_3){=}C(H)(CH_3)(Tp^*)(CO)_2]$, the vinyl carbyne isomer of $[W{\equiv}CC(H)(CH_3)CH{=}CH_2(Tp^*)(CO)_2]$.

Addition of $[W{\equiv}CPMe_2Ph(Tp^*)(CO)_2]^+$ to the parent vinylvinylidene anion occurs at C_α and forms a reactive zwitterionic phosphonium carbene intermediate that loses PMe_2Ph to form the dinuclear $[(Tp^*)(CO)_2W{\equiv}CCH_2CH{=}CHC{\equiv}W(CO)_2(Tp^*)]$, which undergoes deprotonation, forming $[(Tp^*)(CO)_2W{\equiv}CC(H)C(H)C(H)C{\equiv}W(CO)_2(Tp^*)]^-$.[149]

Reaction of $[W(O)(I)(Tp^*)(CO)]$ with MeLi followed by trapping of the resultant anion by $ClSiPh_2Me$ gave a single isomer of $[W(O)(I)(=C(Me)OSiPh_2Me)(Tp^*)]$. The acyl intermediate $[W(O)(I)(C(O)Me)(Tp^*)][Li]$ was detected *in situ* by NMR (Fig. 2.32). Protonation of $[W(O)(I)(C(O)Me)(Tp^*)][Li]$ with benzoic acid yielded the neutral hydroxy carbene complex $[W(O)(I)(=C(Me)OH)(Tp^*)]$, whereas reaction of $[W(O)(I)(Tp^*)(CO)]$ with PhLi yielded $[W(O)(I)(C(O)Ph)(Tp^*)][Li]$, which is trapped with $ClSiPh_2Me$, giving $[W(O)(I)(=C(Ph)OSiPh_2Me)(Tp^*)]$. $[W(O)(OTf)(=C(Me)\text{-}OSiPh_2Me)(Tp^*)]$ has also been reported. Reaction of $LiCuMe_2$ with $[W(O)(I)(Tp^*)(CO)]$ produced $[W(O)(Me)(Tp^*)(CO)]$, which upon protonation in dichloromethane in the presence of MeCN yields methane and $[W(O)(Tp^*)(CO)(NCMe)]^+$, while upon protonation in MeCN produces $[W(O)(Tp^*)(NCMe)(=C(Me)OH)]^+$.[150]

Reaction of $[W(O)(I)(Tp^*)(CO)]$ with Ag(OTf) gave $[W(O)\text{-}(OTf)(Tp^*)(CO)]$ that upon dissolution in MeCN generated $[W(O)(Tp^*)(NCMe)(CO)][OTf]$ in which the triflate can be substituted by $[BAr_F]$. Photolysis of $[W(O)(I)(Tp^*)(CO)]$ in MeCN promoted loss of carbon monoxide and gave $[W(O)(I)(Tp^*)(NCMe)]$ (Fig. 2.33). Chromatography of $[W(O)(I)(Tp^*)(NCMe)]$ on alumina

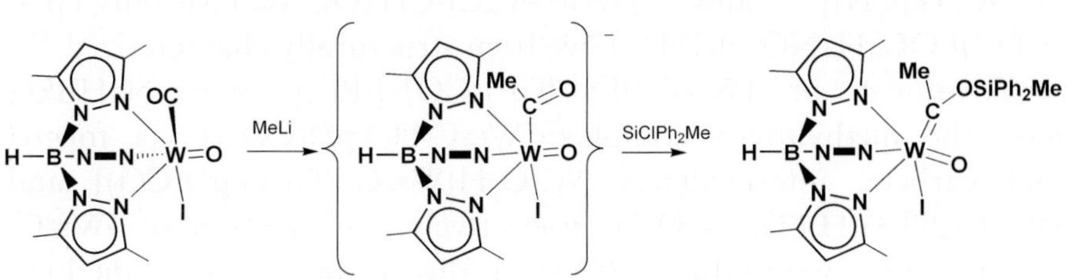

Fig. 2.32. The reaction of $[W(O)(I)(Tp^*)(CO)]$ with MeLi and with the trapping $SiClPh_2Me$.

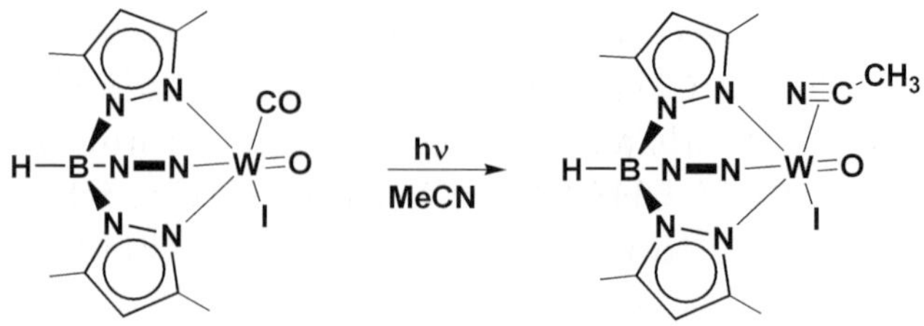

Fig. 2.33. Photolysis of [W(O)(I)(Tp*)(CO)] in MeCN yields [W(O)(I)(Tp*)-(NCMe)].

led to oxidation and formation of [W(O)(I)(Tp*)(NC(O)Me)]. The dimer [(Tp*)(O)(I)W(NC(Me)C(Me)N)W(I)(O)(Tp*)] results from the reductive coupling of two nitrile ligands. Methylation of η^2-nitrile [W(O)(I)(Tp*)(NCMe)] by MeOTf generates the η^2-iminoacyl species [W(O)(I)(Tp*)(MeNCMe)][OTf].[151]

[W(η^2-(C,C)-O=CC-PR'_2Ph)(Tp*)(CO)(*p*-OC_6H_4R)] (R = NO_2, R′ = Me; R = NO_2, R′ = Ph; R = CN, R′ = Me; R = CN, R′ = Ph; R = Cl, R′ = Ph) has been prepared from the reaction of [W≡CPR_3(Tp*)$(CO)_2$][PF_6] with aryloxide nucleophiles. Also, [W≡CO(*p*-$C_6H_4NO_2$)(Tp*)(CO)], [W≡CO(*p*-C_6H_4CN)(Tp*)(CO)], and [W≡CO(*p*-C_6H_4Cl)(Tp*)$(CO)_2$] have been reported. The ketenyl complexes [W(η^2-(C,C)-O=CC-PMe_2Ph)(Tp*)(CO)(*p*-$OC_6H_4NO_2$)] and [W(η^2-(C,C)-O=CC-PPh_3)(Tp*)(CO)(*p*-$OC_6H_4NO_2$)] were methylated at the ketenyl oxygen forming cationic alkyne complexes [W(η^2-(C,C)-MeOC≡C-PR'_2Ph)(Tp*)(CO)(*p*-$OC_6H_4NO_2$)][OTf] (R′ = Me or Ph). The “pseudo-octahedral” [W(η^2-(C,C)-O=CC-PMe_2Ph)-(Tp*)(CO)(*p*-$OC_6H_4NO_2$)], [W(η^2-(C,C)-O=CC-PPh_3)(Tp*)(CO)-(*p*-OC_6H_4CN)], and [W(η^2-(C,C)-CH_3OC≡C-PMe_2Ph)(Tp*)-(CO)(*p*-$OC_6H_4NO_2$)][OTf] have been structurally characterized.[152]

Treatment of [W≡C-PPh_3(Tp*)$(CO)_2$][PF_6] with $NaHBEt_3$ gave the methylidyne complex [W≡C-H(Tp*)$(CO)_2$] via formyl and carbene intermediates [(C(O)H)W≡C-PPh_3(Tp*)(CO)] and [W=C(PPh_3)(H)(Tp*)$(CO)_2$], respectively. Protonation of [W≡C-H(Tp*)$(CO)_2$] with $HBF_4\cdot Et_2O$ yields the cationic α-agostic methylidene complex [W=CH_2(Tp*)$(CO)_2$][BF_4]. [W≡C-H(Tp*)$(CO)_2$] can be deprotonated with alkyllithium yielding [W≡C-Li(Tp*)$(CO)_2$]

that adds electrophiles at the carbide carbon to generate [W≡C-R(Tp*)(CO)$_2$] (R = CH_3, $SiMe_3$, I, C(OH)Ph_2, CH(OH)Ph, and C(O)Ph). The p*K*a of [W≡C-H(Tp*)(CO)$_2$] was also determined. Addition of excess $NaHBEt_3$ to [W≡C-H(Tp*)(CO)$_2$] generates the anionic methylidene complex [Na][W=CH_2(Tp*)(CO)$_2$], which upon the addition of PhSSPh gives [W(η^2-CH_2-SPh)(Tp*)(CO)$_2$].[153] Protonation of [Na][W=CH_2(Tp*)(CO)$_2$] in the presence of trapping phosphine ligands such as PMe_3, PMe_2Ph, $PMePh_2$ yielded seven-coordinate complexes of formula [W(PR_3)Me(Tp*)(CO)$_2$] which undergo equilibrium with their CO-insertion products [W(PR_3)(η^2-C(O)Me(Tp*)(CO)$_2$]. ^{13}C spin saturation transfer NMR experiments have been carried out in order to probe the stereoselectivity of CO insertion in the W–Me bonds. The seven-coordinated capped octahedron [WMe(Tp*)(CO)$_2$($PMePh_2$)], [W(η^2-C(O)Me)(Tp*)(CO)$_2$(PMe_3)], and two isomeric forms of [W(η^2-C(O)Me)(Tp*)(CO)$_2$(PR_3)], have been structurally characterized. Trapping of the protonated carbene derivative with phenylacetylene gave the η^1-acyl compound [W(η^1-C(O)Me)-(Tp*)(CO)(PhC≡CH)].[154]

[W≡C-OMe(Tp*)(CO)$_2$], reported by Templeton *et al.*[155] undergoes reversible protonation at the carbyne carbon in the presence of [H(Et_2O)$_2$][BAr_F] to yield the α-agostic carbene complex [Tp*(CO)$_2$W=C(H)OMe][BAr_F].

Nitrogen ligands, coordinated to [W(Tp*)(CO)$_2$]$^+$ and [W(Tp*)(CO)(PhC≡CMe)]$^+$, accessible from free nitriles and amines, including nitrile, azavinylidene, imido, imine, amido, amine, and nitrido, can be interconverted by addition of nucleophiles or electrophiles as well as by hydride abstraction or proton removal.[156] [W(Tp*)(CO)(PhC≡CMe)(NH=CMeEt)][BAr_F] was treated with $NaHBEt_3$, giving [W(Tp*)(CO)(PhC≡CMe)(NH_2CHMeEt)][BAr_F] which has been structurally characterized.[157] Chirality at tungsten in the complexes[157] was assigned by considering Tp* as an η^3-ligand and using the Baird/Sloan modification of the Cahn–Ingold–Prelog priority rules.

Reaction of [W(Tp*)(CO)$_2$(PhC≡CMe)][OTf] with excess pyrrolidine yielded the chiral [W(Tp*)(CO)(PhC≡CMe)(NCH_2-

$CH_2CH_2CH_2)$] that upon oxidation with I_2 formed the η^1-imine [W(Tp*)(CO)(PhC≡CMe)($\overline{N{=}CHCH_2CH_2CH_2}$)][$BAr_F$]. [W(Tp*)(CO)(PhC≡CMe)($\overline{N{=}CMeCH_2CH_2CH_2}$)][$BAr_F$] is synthesized by 2-methyl-1-pyrrolidine addition to the labile [W(Tp*)$(CO)_2$-(PhC≡CMe)][OTf]. Only one pair of enantiomers ($S_W R_C/R_W S_C$) of [W(Tp*)(CO)(PhC≡CMe)($NCHMeCH_2CH_2CH_2$)] is detected in the reaction of [W(Tp*)$(CO)_2$(PhC≡CMe)][OTf] with deprotonated 2-methylpyrrolidine. Protonation of [W(Tp*)(CO)-(PhC≡CMe)($NCH_2CH_2CH_2CH_2$)] using [$H(OEt_2)_2$][BAr_F] yields [W(Tp*)(CO)(PhC≡CMe)($NHCH_2CH_2CH_2CH_2$)][BAr_F].[158]

[W(Tp^x)(CO)(RC≡CMe)(η^1/η^2-O=CHR′)][BAr_F] (Tp^x = Tp or Tp*) exhibits different σ/π aldehyde coordination depending on the aldehyde and on the ancillary ligand Tp^x.[83]

Irradiation of [W(H)Tp*$(CO)_3$] in the presence of alkynes yields η^2-vinyl, η^2-acyl, metallafuran, and carbyne complexes likely through *cis*-insertion of alkynes into a photogenerated [WH(Tp*)$(CO)_2$], while *cis* 2,1-insertion initially leads to η^2-vinyl and η^3-allyl, and subsequent *cis*-1,2-insertion produces η^2-acyl, metallafuran, and carbyne species. [W(η^2(*C*,*O*)-C(Bu^n)CHC(=O)-CH=CHBu^n)(Tp*)$(CO)_2$] has been characterized by X-ray. Support for the η^1-vinyl intermediates is derived from the η^2-acyl complex [W(η^2(*C*,*O*)-C(=O)CH=CHC$(CH_3)_3$)Tp*$(CO)_2$]. Irradiation of [W(H)Tp*$(CO)_3$] with $Me_3SiC{\equiv}CCH_3$ yields two isomers of [W(η^3-syn-$CH_2CHCHSiMe_3$)(Tp*)$(CO)_2$]. When this allyl complex was heated, rearrangement to a carbyne species occurred (Fig. 2.34).[159]

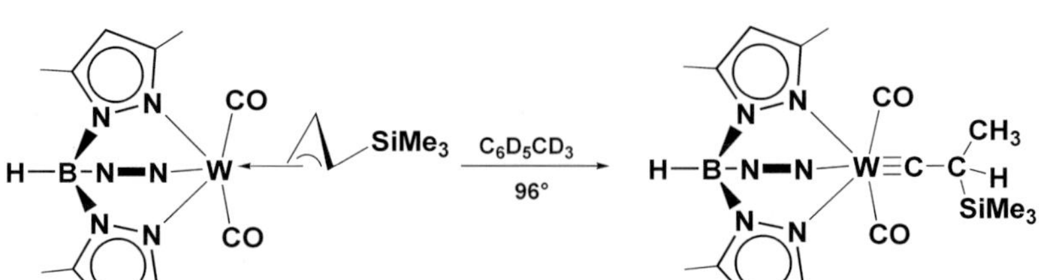

Fig. 2.34. Synthesis of a carbine species from [W(η^3-syn-$CH_2CHCHSiMe_3$)(Tp*)-$(CO)_2$].

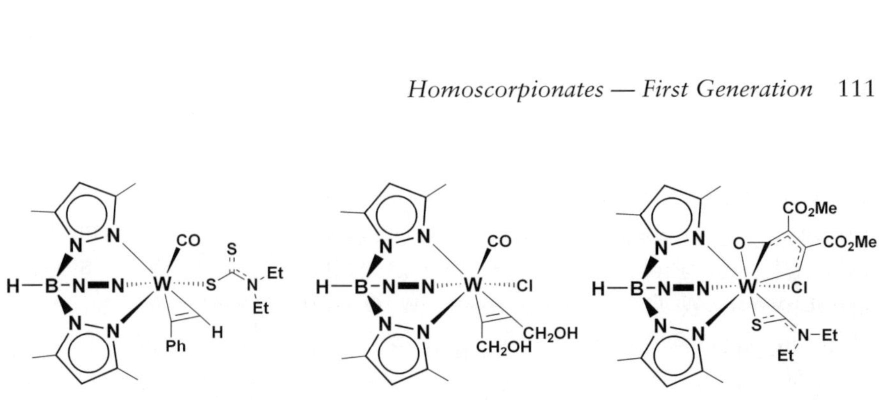

Fig. 2.35. Products of the reaction of $[W(Tp^*)(S_2CNEt_2)(CO)_2]$ with PhC≡CMe, $OHCH_2C{\equiv}CCH_2OH$, and DMAD.

$[W(Tp^*)(S_2CNEt_2)(CO)_2]$ reacts with PhC≡CMe, $OHCH_2C{\equiv}CCH_2OH$, and DMAD giving $[W(Tp^*)(S_2CNEt_2\text{-}\kappa S)(\eta^2\text{-}PhC{\equiv}CH)(CO)]$ (**I**), $[WCl(Tp^*)(\eta^2\text{-}HOCH_2C{\equiv}CCH_2OH)(CO)]$ (**II**), and $[W(Tp^*)\{OCCCO_2Me)=C(CO_2Me)S\text{-}\kappa^3 O, C, S)\}(SCNEt_2\text{-}\kappa^2 C, S)]$ (**III**) (Fig. 2.35), respectively. The octahedral $[W(Tp^*)(S_2CNEt_2\text{-}\kappa S)(\eta^2\text{-}PhC{\equiv}CH)(CO)]$, $[WCl(Tp^*)(\eta^2\text{-}HOCH_2C{\equiv}CCH_2OH)(CO)]$, and $[W(Tp^*)\{OCC(CO_2Me)=C(CO_2Me)S\text{-}\kappa^3 O, C, S)\}(SCNEt_2\text{-}\kappa^2 C, S)]$ exhibit a tridentate *fac*-Tp* ligand.[160]

$[W(O)(Tp^*)(S_2CNEt_2)]$, distorted octahedral with axial oxo, equatorial dithiocarbamate and facial tridentate Tp* was synthesized by O-atom transfer from pyridine N-oxide to $[W(Tp^*)(S_2CNEt_2)(CO)_2]$.[161] The synthesis and characterization of $[W(Tp^*)Cl_3]$ have also been reported. Interaction of $[W(Tp^*)S(S_2CNEt_2)]$ with DMAD produced the seven-coordinate $[W(Tp^*)\{S_2C_2(CO_2Me)_2\}(SCNEt_2\text{-}\kappa^2 S, C)]$ exhibiting a distorted pentagonal bipyramidal coordination.[114] $[WX(Tp^*)S_2]$ reacts with alkynes producing [WX(Tp*)(dithiolene)]. The synthesis and characterization of $[W(OPh)(Tp^*)\{S_2C_2(CO_2Me)_2\}]$ and $[W(SePh)(Tp^*)\{S_2C_2(Ph)(2\text{-quinoxalinyl})\}]$ have been reported, the former being a distorted octahedral complex with a tridentate Tp*.[162]

$[WO_2(Tp^*)X]$ reacts with B_2S_3 or P_4S_{10} giving the oxothio- and bis(thio)-tungsten(VI) complexes [WO(Tp*)SX] (X = Cl) and $[WS_2(Tp^*)X]$ (X = Cl, S_2PPh_2). The reaction of $[WS_2(Tp^*)Cl]$ with (i) PPh_3 in pyridine and (ii) dimethyl sulfoxide affords [WO(Tp*)SCl], which undergoes metathesis with alkali metal salts

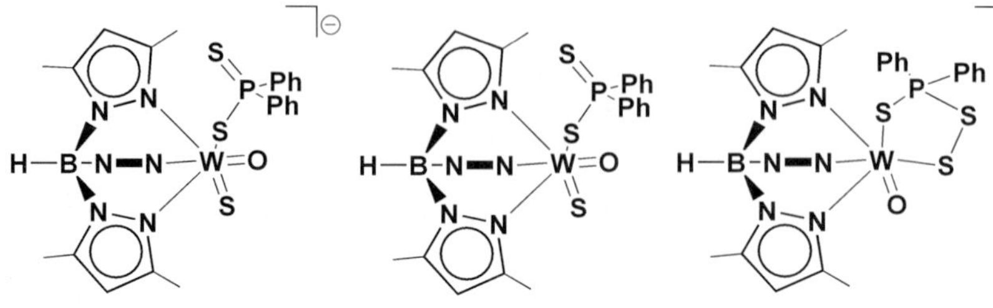

IV V VI

Fig. 2.36. Redox interplay behavior of oxo-thio-tungsten-Tp* species.

yielding [WO(Tp*)SX] and [WS_2(Tp*)X] [X = OPh, SPh, SePh, (−)-mentholate]. These diamagnetic complexes exhibit NMR spectra indicative of C_1, [WO(Tp*)SX], or C_s symmetry, [WS_2(Tp*)X].[163]

Synthesis and characterization of [W(O)(Tp*)S(S_2PR_2-*S*)] (**IV**) (Fig. 2.36) and [W(O)(Tp*)S(pyS-*S*)], (R = OEt, Ph; pyS = pyridine-2-thiolate) have been reported by Young *et al.*[164] The 1,2-dichloroethane hemisolvate of [W(O)(Tp*)S(S_2PPh_2-*S*)] exhibits a distorted octahedral structure in which Tp* acts as a facial tridentate ligand. X-ray absorption and extended X-ray absorption fine structure results support an oxo-thio formulation for [W(O)(Tp*)S(pyS-*S*)]. These complexes are reduced to the corresponding oxo-thio-W(V) anions, [W(O)(Tp*)S(S_2PR_2-*S*)]$^-$ (**V**) (Fig. 2.36) and [W(O)(Tp*)S(pyS-*S*)]$^-$, which exhibit highly anisotropic EPR spectra. They have been oxidized to form the EPR-active (dithio)oxo-W(V) cations, [W(O)(Tp*)(S_3PR_2-*S*, *S*′)]$^+$ (**VI**) (Fig. 2.36) and [W(O)(Tp*)(pyS_2-*N*, *S*)]$^+$ (pyS_2 = pyridine-2-dithio). The former, distorted octahedral, has been isolated as BF_4 salt. The structures and redox behavior of these complexes are compared with the related molybdenum species [Mo(O)(Tp*)S(S_2PR_2-*S*)] and [Mo(O)(Tp*)(pyS_2-*N*, *S*)].[165]

Thio-W^{VI} complexes such as [WS_2Cl(Tp*)] fail to react with cyanide. [WO(Tp*)S(S_2PPh_2-*S*)] and [W(Tp*)S_2(S_2PPh_2-*S*)] undergo S atom transfer to tertiary phosphines to produce [WO(Tp*)(S_2PPh_2-*S*, *S*′)] and [W(Tp*)S(S_2PPh_2-*S*, *S*′)], respectively. The X-ray crystal structures of [W(Tp*)S_2(S_2PPh_2)]·0.5CH_2Cl_2 and

$[W(Tp^*)S(S_2PPh_2)]$ have been determined, both complexes showing six-coordinate, distorted octahedral tungsten centers ligated by tridentate *fac*-Tp*. $[W(E)(Tp^*)(S_2PPh_2)]$ (E = O, S) reacts with S-donors yielding $[W(E)(Tp^*)S(S_2PPh_2)]$, whereas reaction with O-donors produces mixtures of $[WO(Tp^*)E(S_2PPh_2)]$ and $[W(E)(Tp^*)S\{SP(O)Ph_2\text{-}S\}]$.[166] The dinuclear $[(Tp^*)SW(\mu\text{-}S)_2WS(S_2PPh_2)]$ has been prepared from the reaction of $[WO_2Cl(Tp^*)]$ with PPh_3 in refluxing py upon addition of $NH_4S_2PPh_2$. This compound has been structurally characterized.[167]

$[WOX(Tp^*)(CO)]$ (X = Cl, Br, I) can be synthesized by the reaction of O_2 or pyridine N-oxide (X = $S_2P(OPr^i)_2$, S_2PPh_2) with $[WX(Tp^*)(CO)_2]$. Analogous species $[WSX(Tp^*)(CO)]$ result from the reactions of $[WX(\kappa^2\text{-MeCN})(Tp^*)(CO)]$ with propylene sulfide. $[WO(Tp^*)(S_2PPh_2)(CO)]$ and $[WS(Tp^*)(S_2\text{-}PPh_2)(CO)]\cdot 0.5CHCl_3$ are distorted octahedral with a *fac* tridentate Tp* ligand and mutually *cis*, monodentate chalcogenido, carbonyl, and dithiophosphinato ligands. In refluxing toluene, $[WOI(Tp^*)(CO)]$ converts into purple, mixed-valence $[(Tp^*)W^{III}I(CO)(\mu\text{-}O)W^{V}OI(Tp^*)]$ which contains a nearly linear μ-oxo bridge connecting distorted octahedral tungsten centers (Fig. 2.37).[168]

Paramagnetic $[W(Tp^*)S\{S_2C_2(CO_2Me)_2\}]$ is formed from the reactions of DMAD with the sulfur-rich complexes $NEt_4[Tp^*WS_3]$; the analogous $[WO(Tp^*)\{S_2C_2(CO_2Me)_2\}]$ results from the hydrolysis of the sulfido compound. The "organoscorpionate" $[W\{S_2C_2(CO_2Me)_2\}\{SC_2(CO_2Me)_2\text{-}Tp^*\}]$ has been isolated from the reactions of $NEt_4[Tp^*WS_3]$ with excess DMAD. $[WO(Tp^*)\{S_2C_2\text{-}(CO_2Me)_2\}]$ and $[W\{S_2C_2(CO_2Me)_2\}\{SC_2(CO_2Me)_2\text{-}Tp^*\}]$ have

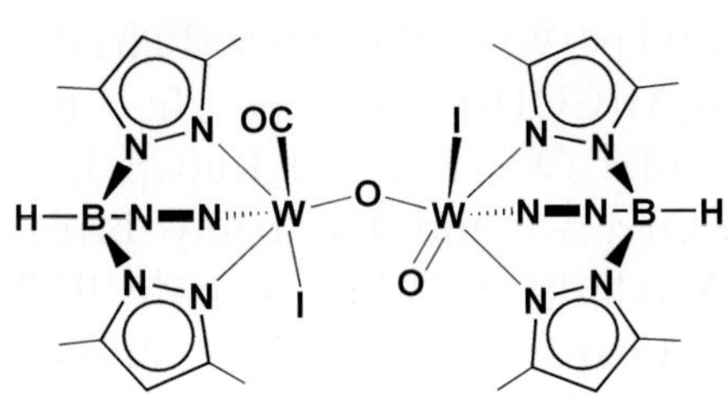

Fig. 2.37. The mixed-valence compound $[(Tp^*)W^{III}I(CO)(\mu\text{-}O)W^{V}OI(Tp^*)]$.

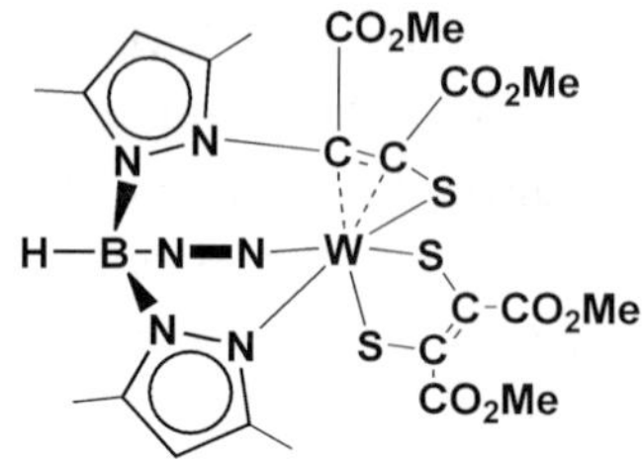

Fig. 2.38. The pentadentate "organoscorpionate" in $[W\{S_2C_2(CO_2Me)_2\}\{SC_2\text{-}(CO_2Me)_2\text{–}Tp^*\}]$.

been characterized by X-ray crystallographic techniques. The latter contains a novel pentadentate "organoscorpionate" ligand, coordinated through two pyrazolyl N atoms and a tridentate *S, C, C*-unit appended to N-β of the third (uncoordinated) pyrazolyl group (Fig. 2.38). $[WX\{S_2C_2(CO_2Me)_2\}(Tp^*)]$ (X = S, O) were also reported.[169]

$[Et_4N][W(Tp^*)(CO)_3]$ reacted with S_8 in DMF yielding $[Et_4N]\text{-}[W(Tp^*)S_3]$. The reaction with $[Et_4N][W(Tp)(CO)_3]$ yields both $[Et_4N][W(Tp)S_3]$ and the by-product $[Et_4N][\{WO(S_2)_2\}_2(\mu\text{-}S)]$. $[Et_4N][W(Tp^*)S_3]$ reacts with $[PtCl_2(COD)]$ forming $[Et_4N][W(Tp^*)\text{-}S(\mu\text{–}S)_2PtCl_2]$. $[Et_4N][W(Tp^*)S_3]$, $[Et_4N][W(Tp)S_3]$, and $[Et_4N]\text{-}[W(Tp^*)S(\mu\text{–}S)_2PtCl_2]$, all containing octahedral anions, have been structurally characterized.[49]

Reaction of $[Et_4N][W(Tp^*)S_3]$ with $[M(PPh_3)_4]$ (M = Pt or Pd), $[(PtMe_3)(\mu_3\text{-}I)_4]$, $[M(COD)(PPh_3)_2]PF_6$ (M = Ir, Rh), $[Rh(COD)(dppe)][PF_6]$, $[IrCp^*(MeCN)_3][PF_6]$, $[Ru(Cp^*)\text{-}(MeCN)_3][PF_6]$, and $[M(CO)_3(MeCN)_3]$ (M = Mo, W) in MeCN or MeCN/THF at room temperature afforded the doubly bridged — $[Et_4N][(Tp^*)W(=S)(\mu\text{–}S)_2M(PPh_3)]$ (M = Pt or Pd), $[(Tp^*)W(=S)(\mu\text{–}S)_2M(COD)]$ (M = Ir, Rh), $[(Tp^*)W(=S)\text{-}(\mu\text{–}S)_2Rh(dppe)]$, $[(Tp^*)W(=S)(\mu\text{–}S)_2Ru(Cp^*)]$, and $[Et_4N][(Tp^*)\text{-}W(=S)(\mu\text{–}S)_2W(CO)_3]$ — and the triply bridged complexes — $[(Tp^*)W(\mu\text{–}S)_3PtMe_3]$, $[(Tp^*)W(\mu\text{–}S)_3IrCp^*][PF_6]$, and $[Et_4N]\text{-}[Tp^*W(\mu\text{–}S)_3Mo(CO)_3]$. The X-ray crystal structures of $[Et_4N][(Tp^*)W(=S)(\mu\text{–}S)_2M(PPh_3)]$(M = Pt or Pd), $[(Tp^*)W(\mu\text{–}S)_3\text{-}PtMe_3]$, $[(Tp^*)W(=S)(\mu\text{–}S)_2Rh(COD)]$, $[(Tp^*)W(\mu\text{–}S)_3IrCp^*][PF_6]$,

$[Et_4N][(Tp^*)W(\mu{-}S)_3Mo(CO)_3]$, $[(Tp^*)W(=S)(\mu{-}S)_2Ru(Cp^*)]$, and $[Et_4N][(Tp^*)W(=S)(\mu{-}S)_2W(CO)_3]$ have been reported.[170]

The *in situ*-generated synthetic synthons $[W(Tp^*)(CO)_3(NCMe)]X$ ($X = BF_4$ or OTf) and $[W(Tp^*)(NCEt)(CO)_3][OTf]$ have been prepared and characterized by IR and NMR studies. The cationic moiety of these complexes exhibits a carbonyl-capped octahedral structure with C_s symmetry and undergo slow nitrile ligand exchange,[171] whereas sequential reaction of *fac*-$[W(CO)_3(NCMe)_3]$ with $Me_2NC(=S)Cl$ and K(Tp) gave $[W(Tp)(\eta^2\text{-}SCNMe_2)(CO)_2]$; replacing $Me_2NC(=S)Cl$ with $ClC(=S)OC_6H_4\text{-}Me\text{-}4$ led to $[WCl(Tp)(CO)_3]$.[112]

The $[W(Tp^*)(CO)_2(\eta^2\text{-}C_2S_2)]_2$ synthon is able to form the isostructural dithiolene complexes $[\{W(Tp^*)(CO)_2(\eta^2\text{-}C_2S_2)\}_2M]$ (M = Ni, Pd, Pt).[172] A facile access to η^2-*C*, *C'*-acetylenedithiolate complexes and a full characterization of the heterobimetallic complex $[\{Tp^*W(CO)_2\}(C_2S_2)\{(\eta^5\text{-}C_5H_5)Ru(PPh_3)\}]$ has also been reported by Seidel *et al.*[173]

$[W(Tp^*)(CO)_2(BnSC{\equiv}CS)]$ has been obtained by the reductive removal of the benzyl group at $[W(Tp^*)(CO)_2(BnSC{\equiv}CSBn)]PF_6$, and has been employed as a ligand in a combined organometallic/Werner-type heterobimetallic complex with Cu(I).[174] $[M(Tp^*)(CO)_2\{\eta^2\text{-}(BnS)C{\equiv}C(S)\}]$ (Mo, W) has been obtained by the reductive removal of one benzyl group in $[M(Tp^*)(CO)_2\{\eta^2\text{-}(BnS)C{\equiv}C(SBn)\}](PF_6)$. The by-product, $[W(CO)(Tp^*)\{C(O)Bn\}\{\eta^2\text{-}(BnS)C{\equiv}C(SBn)\}]$, is also reported. $[M(Tp^*)(CO)_2\{\eta^2\text{-}(BnS)C{\equiv}C(S)\}]$ have been employed as monodentate ligands in homoleptic pentanuclear complexes with nickel(II) and palladium(II).[119]

X-ray structural and EPR spectroscopic studies of the redox-related pairs $[WX(CO)(Tp^*)(MeC{\equiv}CMe)]^z$ (X = F, Cl, Br, and I; $z = 0$ and +1) have been reported by the Connelly group.[175] The d_4 halide complexes $[MX(Tp^*)(CO)(RC{\equiv}CR)]$ {X = F, Cl, Br, or I; R = Me or Ph; M = Mo or W} undergo one-electron oxidation to the d^3 $[MX(Tp^*)(CO)(RC{\equiv}CR)]^+$ (isolable when M = W, R = Me). X-ray structural studies on $[WX(CO)(MeC{\equiv}CMe)]^z$ (X = Cl and Br, $z = 0$ or +1), the ESR spectra of the cations $[WX(Tp^*)(CO)(RC{\equiv}CR)]^+$ (X = F, Cl, Br or I; R = Me or Ph), and DFT calculations on

$[WX(Tp^*)(CO)(MeC{\equiv}CMe)]^z$ (X = F, Cl, Br, and I; $z = 0$ and $+1$) have also been reported.[117]

The d^2/d^3 redox pair $[WF_2(Tp^*)(PhC{\equiv}CPh)]^z$ ($z = +1$ or 0), has been obtained by a series of stepwise reactions involving sequential one-electron oxidation followed by ligand substitution. These octahedral compounds have been investigated by CV, EPR, and X-ray diffraction studies.[176]

$[M{\equiv}C\text{-}As{=}C(NMe_2)_2(Tp^*)(CO)_2]$ (M = W or Mo) has been prepared by condensation of $[M{\equiv}CCl(Tp^*)(CO)_2]$ with $Me_3SiAs{=}C(NMe_2)_2$. Methylation of the distorted octahedral $[M{\equiv}C\text{-}As{=}C(NMe_2)_2(Tp^*)(CO)_2]$ at the arsenic atom yields $[M{\equiv}C\text{-}As(Me)C(NMe_2)_2(Tp^*)(CO)_2](OTf)$.[177]

$[W({\equiv}CC(CH_3)_3)Cl(Tp)(NHR)]$ (R = C_6H_5, 4-Br-C_6H_4, 3,5-$(CF_3)_2C_6H_3$, H) has been obtained from the reaction of $[W({\equiv}CC(CH_3)_3)Cl_2(Tp)]$ with M(NHR) (M = Li, K). When an excess of Li(NHPh) was employed, $[W({\equiv}CC(CH_3)_3)Cl(Tp)(NHPh)_2]$ was formed. The presence of substoichiometric quantities of HCl or H_2O yielded tautomerization of $[W({\equiv}CC(CH_3)_3)Cl(Tp)(NHPh)]$ (Fig. 2.39) to the corresponding imido alkylidene. $[WO(CHC(CH_3)_3)Cl(Tp)]$ and $[WO({\equiv}CC(CH_3)_3)Cl(Tp)]$ were prepared in good yield when the reaction was carried out in the presence of excess H_2O. $[W({\equiv}CC(CH_3)_3)(Tp^*)(NHPh)Cl]$ was obtained from the reaction of $(CH_3)_3SiN(H)Ph$ with $[W({\equiv}CC(CH_3)_3)(Cl)_3(dme)]$ followed by addition of K(Tp*). $[W({\equiv}CC(CH_3)_3)Cl(Tp)(NHPh)_2]$ and $[W({\equiv}CC(CH_3)_3)Cl(Tp^*)(NHPh)]$ were structurally characterized,

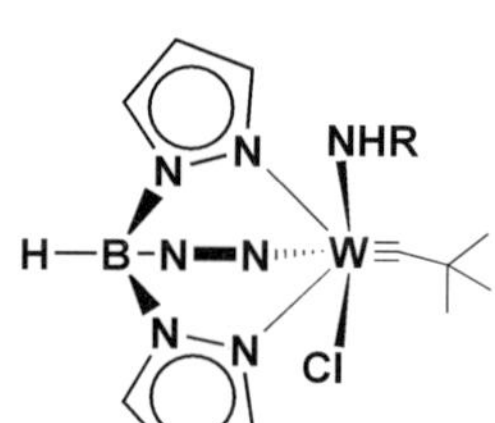

Fig. 2.39. $[W({\equiv}CC(CH_3)_3)Cl(Tp)(NHR)]$.

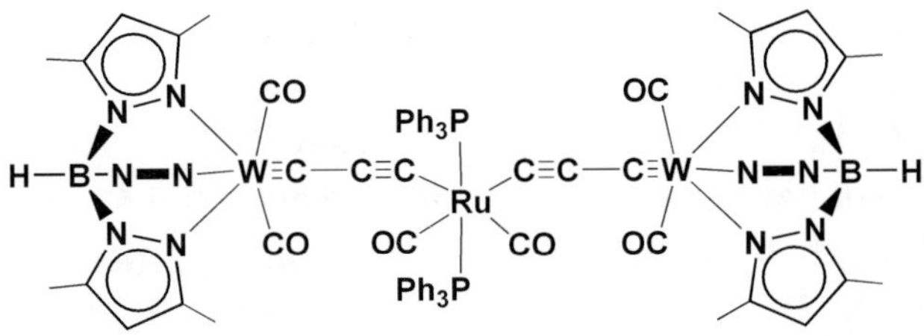

Fig. 2.40. Structure of the heteronuclear complex $[(Ph_3P)_2(CO)_2Ru\{C_3W(CO)_2(Tp^*)\}_2]$.

the amido groups being *syn* with respect to the neopentylidyne ligands likely due to the steric hindrance of the Tp^x donors.[178]

The reaction of $[(Tp^x)(CO)_2WC_6W(CO)_2(Tp^x)]$ (Tp^x = Tp or Tp*) with $[Ru(CO)_2(PPh_3)_3]$ generates $[(Ph_3P)_2(CO)_2Ru\{C_3W(CO)_2(Tp^x)\}_2]$ (Fig. 2.40) via $[(Ph_3P)_2(CO)_2Ru\{\eta^2\text{-}(Tp^x)(CO)_2WC_6W(CO)_2(Tp^x)\}]$, an intermediate coupled with $[(Ph_3P)_2(CO)_2Ru\{C_3W(CO)_2(Tp^x)\}\{HgC_3W(CO)_2(Tp^x)\}]$ also found in the reaction of $[Ru(CO)_2(PPh_3)_3]$ with $[Hg\{C_3W(CO)_2(Tp^x)\}_2]$.[179]

F^--mediated desilylation of $[W(C{\equiv}C{-}C{\equiv}CSiMe_3)(Tp)(CO)_2]$ in the presence of $HgCl_2$ gave $[Hg\{C{\equiv}C{-}C{\equiv}W(Tp)(CO)_2\}_2]$ structurally characterized as a DMSO hexasolvate.[180] The fluoride-mediated protodesilylation of $[W({\equiv}CC{\equiv}CSiMe_3)(Tp)(CO)_2]$ in the presence of Vaska's complex gave the tricarbido trimetallic complex $[IrH\{(C{\equiv}CC{\equiv}W(CO)_2(Tp)\}_2(CO)(PPh_3)_2]$, which exhibits a linear nine-membered WC_3IrC_3W chain containing pseudo-octahedral W and Ir centers.[181] Bruce reported the synthesis of $[W({\equiv}CC{\equiv}CSiMe_3)(Tp)(CO)_2]$ and its reaction with $AuCl(PPh_3)/K_2CO_3$ in MeOH which affords $[W\{{\equiv}CC{\equiv}CAu(PPh_3)\}(Tp)(CO)_2]$. This gold complex reacts with $[Co_3(\mu_3\text{-}CBr)(\mu\text{-}dppm)(CO)_7]$ to give $[\{Tp(CO)_2W\}{\equiv}CC{\equiv}CC{\equiv}\{Co_3(\mu\text{-}dppm)(CO)_7\}]$. $[\{Tp(CO)_2W\}{\equiv}CC{\equiv}C\{Ru(Cp^*)(dppe)\}]$ has also been reported.[132]

$[W({\equiv}CR)(Tp)(CO)_2]$ (R = $C_6H_3Me_2$-2,6) reacts with thionyl chloride yielding $[W({\equiv}CR)Cl_2(Tp)]$, which was also obtained from the reaction of $[W({\equiv}CR)(Tp)(CO)_2]$ with $[Fe(\eta\text{-}C_5H_5)_2]Cl$. The crystal structures of $[W({\equiv}CC_6H_2Me_3\text{-}2,4,6)(Tp)CO)_2]$ and $[W({\equiv}CR)Cl_2(Tp)]$ (R = $C_6H_3Me_2$-2,6) have also been reported.[182]

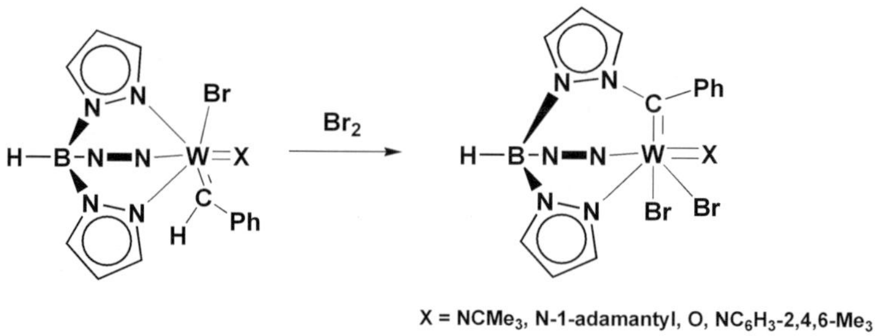

Fig. 2.41. Oxidation of [W(CHPh)(X)Br(Tp)] with Br_2 produces the insertion of the benzylidyne unit into the W–N bond of the W(Tp) cage.

[W(≡CPh)(Tp)$(CO)_2$] reacts with Br_2 affording [W(≡CPh)-Br_2(Tp)] that interacts with amines or water, H_2X (X = $NCMe_3$, N-1-adamantyl, NC_6H_2-2,4,6-$(CH_3)_3$, O), forming [W(CHPh)(X)Br(Tp)]. The oxidation of [W(CHPh)(X)Br(Tp)] with Br_2 was also investigated. In all cases the insertion of the benzylidyne unit into the W–N bond of the W(Tp) cage affords complexes [W{C(Ph)-$C_3H_3N_2$-BH$(pz)_2$}(X)(Br_2)] (Fig. 2.41).[183]

Reaction of [M≡C–P=C$(NMe_2)_2$(Tp*)$(CO)_2$] (M = W, Mo) with AuCOCl yielded the trinuclear complex [M≡C–P$(AuCl)_2$C$(NMe_2)_2$(Tp*)$(CO)_2$]. The same reaction carried out with [M≡C-As=C$(NMe_2)_2$(Tp*)$(CO)_2$] yields [ClAuC$(NMe_2)_2$] and [{M≡C-As(Tp*)$(CO)_2$}$_3$]. All compounds have been characterized by IR and NMR, [M≡C-P$(AuCl)_2$C$(NMe_2)_2$(Tp*)$(CO)_2$] (M = Mo, W), and by X-ray.[184]

Wadepohl *et al.*[103] also investigated the hydroboration reactions of Fischer carbyne metal complexes [W(≡CR)(Tp*)$(CO)_2$]. The boryl-metal complexes [W(Tp*)$(CO)_2${η^2-B(R′)CH_2R}] (R′ = Et or Ph, R = C_6H_4Me-4 or Me) has been prepared. A β-agostic C-H· · · M interaction is present between the metal and the boryl ligand.[103]

The bimetallic complexes [L_nM≡C–C≡C-RuH-(CO)(Hpz*)-$(PPh_3)_2$] and [L_nM≡C–C≡C-RuH-(CO)$(PPh_3)_3$] (L_nM = W$(CO)_2$-Tp*, W$(CO)_2$Tp) were prepared and reacted with PPh_3 to form [L_nM≡C–C≡C-RuH(CO)(L)$(PPh_3)_2$] (L = CO, Hpz*, 2,4,6-Me_3-CNC_6H_2). [(Tp*)$(CO)_2$W≡C–C≡C-RuH-(CO)(Hpz*)$(PPh_3)_2$] has pseudo-octahedral geometries at both the W and Ru centers.[185]

[W≡C–C≡C–R(Tp^x)$(CO)_2$] (Tp^x = Tp or Tp*, R = $SiMe_3$ or Ph) have been prepared through a stepwise fashion by using [$W(CO)_6$], Li[C≡C–R], $(CF_3CO)_2O$, and M(Tp^x) (M = Na or K). The donating ability of the tripodal Tp^x ligands has been investigated by IR and ^{13}C-NMR spectroscopy.[186]

Mononuclear [W(O)Cl(OAr)(Tp*)] and dinuclear [{W(O)C-(Tp*)}$_2$(μ-bis(phenolate))] complexes have been synthesized characterized by electrochemistry, EPR, magnetic susceptibility, UV/Vis/NIR spectroelectrochemistry, and X-ray with the aim to evaluate the electronic and magnetic interaction between the redox-active, paramagnetic metal centers.[187]

[W(O)(Tp*)(biph)] have been structurally characterized and investigated by electrochemical and UV/Vis/NIR spectroelectrochemical methods.[79]

[W(O_2)(Me)(Tp*)] has been investigated by Raman spectroscopy. A characteristic shift of the metal oxygen stretching vibrations is observed, which can be used as an indicator for the donor ability of the ligands Tp*.

[W(Tp)(π-acid)(L)(η^2-arene)] (π-acid = CO or NO, L = 1-alkylimidazole, pyridine, PMe_3, arene prochiral) exists as a dynamic equilibrium of coordination diastereomers in solution but only one diastereomer is present in the solid state.[107] Harman *et al.* have found that the bound arene in [W(Tp)(NO)(PMe_3)(η^2-benzene)] is activated toward Diels–Alder cycloaddition and substitution of the arene with DMF, yielding a complex containing a η^2-coordinated amide.[188]

[W(Tp)(NO)(PMe_3)(L)] (L = 2-methylfuran or 2,5-dimethylfuran) react with various dipolarophiles (*N*-methylmaleimide, *N*-phenylmaleimide, acrylonitrile) giving 7-oxanorbornene complexes through a two-step reaction sequence (Fig. 2.42). Attempts to obtain the intact 7-oxanorbornene were unsuccessful. The crystal structures of [W(Tp)(NO)(PMe_3)((8,9-η^2)-1,4,7-trimethyl-10-oxa-4-azatricyclo-[5.2.1.0^{2,6}]dec-8-ene-3,5-dione)] and [W-(Tp)(NO)(PMe_3)((8,9-η^2)-1-methyl-4-phenyl-10-oxa-4-azatricyclo-[5.2.1.0^{2,6}]dec-8-ene-3,5-dione)] support the stereochemical assignments of the cycloadducts.[189]

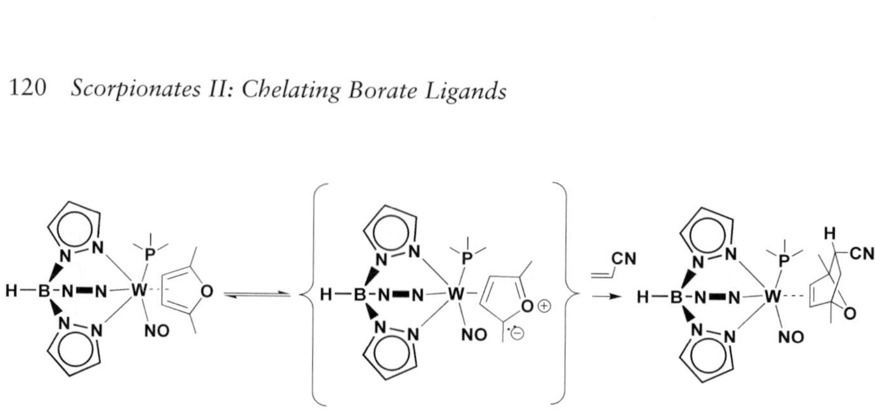

Fig. 2.42. Synthesis of [W(Tp)(NO)(PMe_3)(L)] (L = 7-oxanorbornene).

Upon interaction of the π-basic metal fragment {W(Tp)(NO)(L′)} with L (L′ = MeIm or PMe_3; L = benzene, acetaldehyde, acetone, ethyl acetate, methyl formate, and *N,N*-dimethylformamide) stable [W(Tp)(NO)(PMe_3)(η^2-L)] complexes formed. When the tungsten fragment is combined with a carboxylic acid, oxidative addition yields the seven-coordinate hydride complex [W(Tp)(NO)(H)(PMe_3)C(=O)R] (Fig. 2.43).[108]

[W(Tp)(NO)(PMe_3)(η^2-benzene)] reacted with an excess of phenol to yield the two steroisomers of [W(Tp)(NO)(PMe_3)(η^2-2H-phenol)] (Fig. 2.44), which in the presence of base and electrophilic species such as benzaldehyde, alkyl iodides, and Michael acceptors, is able to form new C–C bonds. Methyl and ethyl iodide react at C2 to form 2-alkyl-2H-phenol complexes, whereas the Michael acceptors react at C4 to give 4-alkyl-4H-phenol complexes. The crystal and molecular structures of the 2-ethyl-2H-phenol and of the phenyl *o*-quinone methide complexes have been reported.[190]

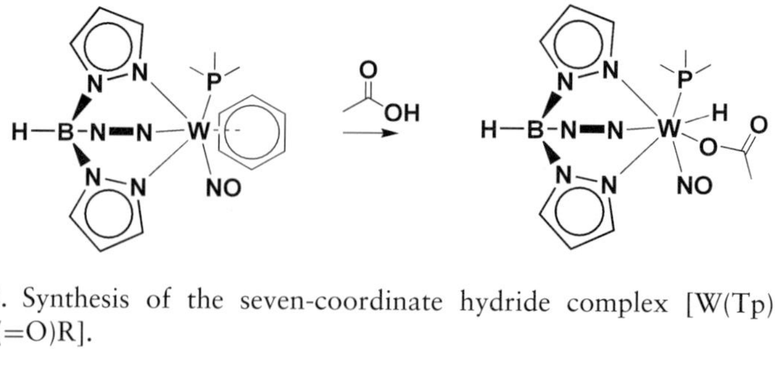

Fig. 2.43. Synthesis of the seven-coordinate hydride complex [W(Tp)(NO)(H)-(PMe_3)C(=O)R].

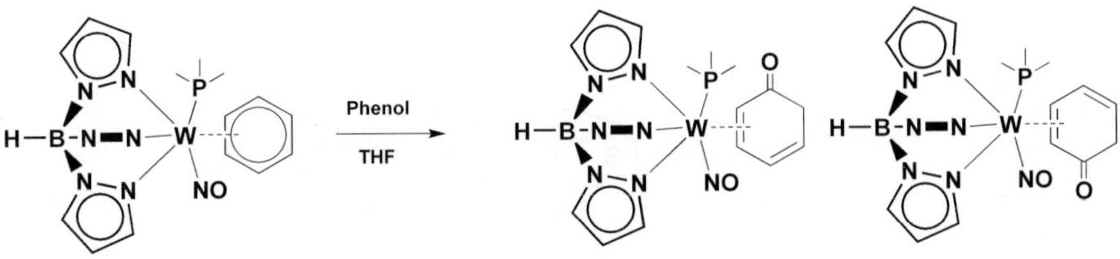

Fig. 2.44. The reaction of [W(Tp)(NO)(PMe$_3$)(η^2-benzene)] with phenol gives both stereoisomers of [W(Tp)(NO)(PMe$_3$)(η^2-2,4-cyclohexadienone)].

[W(Tp)(NO)(PMe$_3$)(η^2-thiophene)] has been prepared by the reduction of [W(Br)(Tp)(NO)(PMe$_3$)] in the presence of thiophene. Hydrogenation of the uncoordinated double bond in [W(Tp)(NO)(PMe$_3$)(η^2-thiophene)] has been performed and sulfur–carbon, sulfur–oxygen, and carbon–carbon bond-forming reactions were also investigated.[191] Interaction of [M(Tp)(NO)(L′)(η^2-L)] (M = W, L = nitrile, ketone, or alkene, L′ = PMe$_3$, Meim) with electrophiles such as alkyl halides and Brønsted acids has been investigated, and some nitrilium complexes, for example, [W(Tp)(NO)(PMe$_3$)(η^2-acetonitrilium)](OTf), spectroscopically characterized. In the same paper, an interesting example of a η^2-acetonium complex [Mo(Tp)(NO)(Meim)(η^2-methylacetonium)]OTf, is reported.[192] Harman *et al.* reported [W(Tp)(PMe$_3$)(NO)(2-(dimethylamino)pyrimidine)] existing in equilibrium with a η^2-bound isomer, an efficient diene for Diels–Alder cycloaddition reactions.[193] A large-scale synthesis of [W(Tp)(NO)(PMe$_3$)(η^2-arene)] (arene = benzene, anisole, dimethoxybenzene, trifluorotoluene, and naphthalene) as well as of their precursors [W(Tp)(NO)(CO)$_2$], [W(Tp)(NO)Br$_2$], and [W(Tp)(NO)(PMe$_3$)Br] have also been proposed.[194]

DFT methods have been employed to characterize bonding and charge distribution in [W(Tp)NO(L)], and also in the analogous Re and Mo species. Structural details illustrate the ability of W, Re, and Mo species to dearomatize complexed benzene, which is extensive for all but the most weakly bound species with L = NCMe and CO. Re and W dearomatizing agents have been indicated as economic alternatives to osmium dearomatizing agents.[195]

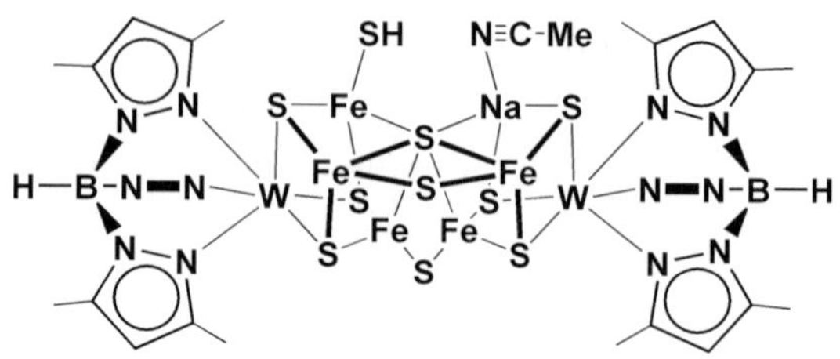

Fig. 2.45. $[(Tp^*)_2W_2Fe_5S_9Na(SH)(MeCN)]^{3-}$, a cluster showing a core topology similar to the P^N cluster of nitrogenase.

The cubane cluster $[(Tp^*)WFe_3S_4Cl_3]^-$ has been synthesized and converted to $[(Tp^*)WFe_3S_4(SEt)_3]^-$ with NaSEt. Reaction of $[W(Tp^*)S_3]^{2-}$ with $FeCl_2$/NaSEt forms $[(Tp^*)WFe_2S_3Cl_2(SEt)]^-$ with a cuboidal $[WFe_2(\mu^2{-}S)_2(\mu^3\text{-}S)(\mu^2\text{-}SR)]^{2+}$ core. The edge-bridged double $[(Tp^*)_2W_2Fe_6S_8(PEt_3)_4]$ has been prepared by reaction of $[(Tp^*)WFe_3S_4Cl_3]^-$ with excess Et_3P, whereas the same tungsten complex reacts with a system composed of excess Et_3P, BH_4^-, and HS^-, yielding a mixture of products, from which $[(Tp^*)_2W_2Fe_5S_9Na(SH)(MeCN)]^{3-}$ formed (Fig. 2.45). This cluster, as the closely related $[(Tp)_2Mo_2Fe_6S_9(SH)_2]^{3-}$, exhibiting a core topology similar to the P^N cluster of nitrogenase, has been structurally characterized as Et_4N^+ salt.[196]

Enemark *et al.*[98] reported solution redox potentials and heterogeneous electron transfer rate constants for [W(E)(Tp*)(SS)] (E = O, NO; SS = dithiolate). Cyclic voltammograms reveal a one-electron quasi-reversible oxidation process for [WO(Tp*)(tdt)], which has the most negative electrochemical potentials among a large series of Mo and W investigated.[98]

2.8. Group 7: Mn and Re

2.8.1. *Mn*

The Mn(II) complexes, $[Mn(Tp)_2]$ and $[Mn(Tp^*)_2]$, have been prepared by the reaction of $Mn(OAc)_2$ with the corresponding scorpionates. These monomeric neutral complexes were characterized by

IR and X-ray single-crystal studies. They possess similar coordination modes around the Mn centers.[197] The divalent and trivalent homoleptic $[Mn(Tp)_2]^{+/0}$, $[Mn(Tp^*)_2]^{+/0}$, and $[Mn(pzTp)_2]^{+/0}$ have been characterized by electrochemical, magnetic, and spectroscopic techniques.[47]

$[Mn(Tp)_2SbF_6]$, $[Mn(Tp^*)_2SbF_6]$, and $[Mn(pzTp)_2SbF_6]$ have been synthesized by the oxidation of the corresponding $[Mn(Tp^x)_2]$ (Tp^x = Tp, Tp*, or pzTp) compounds with $NOSbF_6$. The Mn(III) complexes are low-spin in solution and in the solid state. X-ray crystallography confirms their uncommon low-spin character.[198]

2.8.2. *Re*

The chemistry of [Re(O)(Tp)X(Y)] (X = alkyl, aryl, halide, alkoxide, Y = OTf), exhaustively investigated mainly by Mayer and Brown[199,200] and by Santos *et al.*,[201,202] and widely reported in the first edition,[203] continued to attract much attention.

Reaction of [Re(O)(OH)Cl(Tp*)] with concentrated aqueous HF yielded [Re(O)(F)Cl(Tp*)]. The analogous [Re(O)(F)X(Tp*)] (X = I, OTf, F) have also been prepared, but no analogous species formed with the unsubstituted Tp. When $[Re(O)I_2(Tp)]$ reacts under reflux with an excess of NaF in MeCN $\{[\kappa^2\text{-H(F)Bpz}_2]Re(O)\}_2(\mu\text{-pz})_2(\mu\text{-O})$ is the unique product identified. The difference in the reactivity of Tp- and Tp*-containing rhenium-oxo species has been ascribed to the greater steric protection of the boron caused by the methyls in the Tp* ligand.[204]

[Re(O)X(Cl)(Tp*)] (X = Cl, I) was obtained from the reaction of [Re(O)OH(Cl)(Tp*)] with HX. $[Re(O)(Cl)_2(Tp^*)]$ reacts with 1 or 2 equiv. of $LiPh/ZnCl_2$ or Et_2Zn giving the corresponding oxo-aryl and oxo-alkyl complexes [Re(O)Ph(Cl)(Tp*)], $[Re(O)(Ph)_2(Tp^*)]$, and [Re(O)Et(Cl)(Tp*)]. The alkoxide complexes [Re(O)(OR)(X)(Tp*)] can be prepared from the reaction of $[Re(O)(Cl)_2(Tp^*)]$ with ROH or catechol. Halide metathesis with AgOTf or alkoxide metathesis with Me_3SiOTf yielded triflate derivatives [Re(O)X(OTf)(Tp*)] (X = halide, H, Et, Ph, OEt, OPh), whereas hydride complexes [(Tp*)Re(O)H(X)] (X = Cl or H) have been produced from the corresponding alkoxide complexes with BH_3·THF. The oxidation of [Re(O)Ph(OTf)(Tp*)] with Me_2SO gives a phenoxide complex, whereas oxidation of [Re(O)Ph(OTf)(Tp*)] with pyridine N-oxide produces acetaldehyde. [Re(O)H(OTf)(Tp*)] is oxidized by the same reagents to [Re(O)$_3$(Tp*)] (Fig. 2.46). Tp* imparts greater stability to the rhenium(V) derivatives, allowing their preparation to be relatively simple. It is worthy to note that the stability and the crowding of Tp* inhibit reactions with oxygen atom donors, so that heating is often required and the resulting oxidations are more difficult than for analogous Tp compounds.[205]

The electrochemical properties of a number of isostructural rhenium(V) oxo, rhenium(V) imido, osmium(VI) nitrido, of formula [M(E)(X)(Y)(Tp^x)] (Tp^x = Tp or Tp*, E = O, N-tolyl, N; X, Y = hydrocarbyl, halide, triflate), have been described. The reactivity of these complexes as inner-sphere oxidants does not correlate with their peak reduction potentials, whereas the ease of the oxidation of these compounds well parallels their reactivity as oxidants.[206]

[Re(O)$_3$(Tp*)] has been shown able to catalyze O atom transfer from epoxides *cis*-stilbene oxide and styrene oxide to PPh_3.

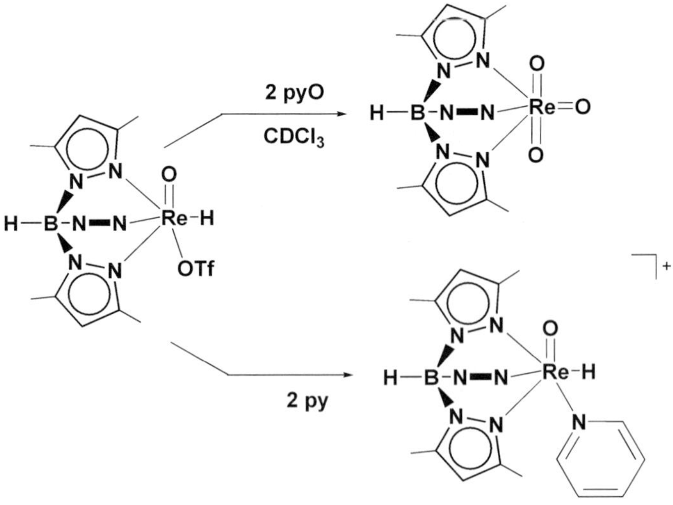

Fig. 2.46. The reaction of [Re(O)H(OTf)(Tp*)] with pyO and excess py.

Reduction of $[Re(O)_3(Tp^*)]$ with PPh_3 in THF is fast above room temperature to form likely the highly reactive species as $[Re(O)_2(Tp^*)]$. Spectral data and chromatographic separation, however, lead only to $[Re(O)(OH)_2(Tp^*)]$, that upon exposure to ethanol, forms $[Re(O)(OEt)(OH)(Tp^*)]$.[207]

Reductive cyclocondensation of alkane-1,2-dithiols or alkane-1-thio-1,2-diols with $[Re(O)_3(Tp^*)]$ and PPh_3 has been investigated and $[Re(O)(Tp^*)(SCH_2CH_2S)]$ has also been structurally characterized. The 1,2-diolato complexes are less thermally stable than the sulfur analogs, stable in hydrocarbon solvents for more than one week at 120°C. DFT calculations have been performed on selected derivatives.[208]

$[Re(O)_2(Tp^*)]$ generated *in situ* from $[Re(O)_3(Tp^*)]$ and PPh_3, reacts stereospecifically with epoxides and episulfides. When *cis*- and *trans*-cyclooctene oxides are used, a ring-expanded Re(V) diolate is formed. However, when an episulfide such as ethylene sulfide (Fig. 2.47) or *cis*- or *trans*-cyclooctene sulfide is employed, a dithiolate species is obtained.[209]

Analogous complexes containing Tp and $HC(pz)_3$ ligands, in detail $[ReOCl_2(Tp)]$, and $[ReOCl_2(HCpz_3)]^+$, $[ReCl_2(Tp)(OPPh_3)]$ and $[ReCl_2(HCpz_3)(OPPh_3)]Cl$, have been prepared and characterized by X-ray crystallography. Bimolecular reduction of $[ReOCl_2(Tp)]$ and of its analogue by triarylphosphines indicates a pronounced charge effect, the cationic species being reduced by PPh_3 about 1000 times faster than its neutral. The effect of charge on ground and transition states in atom transfer reactions as also ligand substitution rates have also been described.[210]

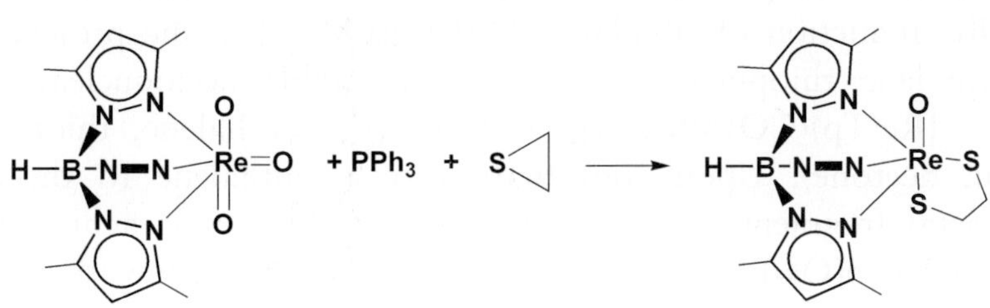

Fig. 2.47. The reaction between $[Re(O)_3(Tp^*)]$, PPh_3, and ethylene sulfide yields a dithiolate complex.

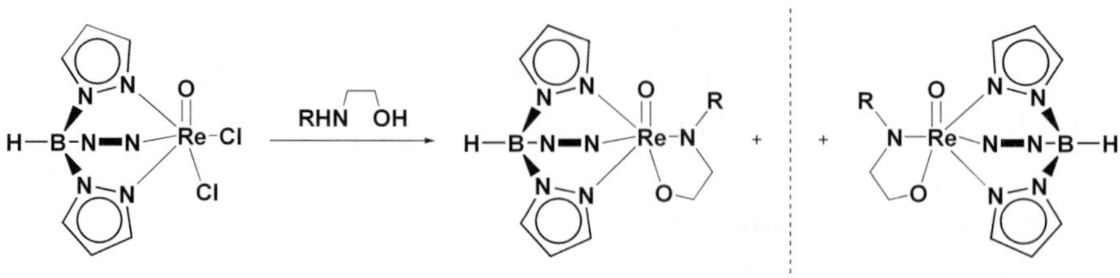

Fig. 2.48. Synthesis of enantiomeric oxorhenium complexes by using achiral amino alcohols.

[$ReOCl_2$(Tp)] has been employed for the synthesis of chiral oxorhenium(V) complexes: the two diastereomers of [ReO(η-O(CH_3)CH_2CH_2O-O,O)(Tp)], [ReO(η^2-N(CH_3)CH_2CH_2O-*N,O*)-(Tp)] and [ReO(η^2-N(^{t}Bu)CH_2CH_2O-*N,O*)(Tp)] enantiomers (Fig. 2.48), [ReO(η^2-N(CH_3)CH(CH_3)CH(Ph)O-*N,O*)(Tp)], and [ReO(η^2-N(CH_2)$_3$CHCO_2-*N,O*)(Tp)]. These species, investigated by IR and vibrational circular dichroism (VCD) spectra are promising candidates for the observation of parity violation (symmetry breaking due to the weak nuclear force). The absolute configuration determination of all compounds, together with their conformational analysis, has been carried out.[211]

[Re(V)(Cl)$_2$(O)(Tp)] reacts with PMe_3 yielding [Re^{III}(Cl)$_2$-Tp(O=PMe_3)], labile, which, upon reacting with excess L (L = PMe_3, NH_3, P(OMe)$_3$, ^{t}BuNC, py) yields [Re(Cl)$_2$(Tp)(L)]. Reduction of Re^{III} to Re^I in [Re(Cl)$_2$(Tp)(L)] with Na/Hg under CO in the presence of cyclohexene gave [Re(Tp)(CO)(PMe_3)(η^2-cyclohexene)], that upon interaction with Ag(OTf) forms the one-electron oxidation product [Re(Tp)(CO)(PMe_3)(η^2-cyclohexene)]OTf. Under reflux, this compound dissociates to [Re(Tp)(OTf)(CO)(PMe_3)]. Finally, reduction of [Re(Tp)(OTf)(CO)(PMe_3)] in the presence of naphthalene, thiophene, and furan yields stable mononuclear complexes [Re(Tp)(CO)(PMe_3)(η^2-L′)] (L′ = naphthalene, thiophene, furan, acetone). Upon coordination of naphthalene to the chiral metal fragment, excellent stereo-differentiation is achieved.[212] [Re^{III}(Cl)$_2$Tp(O=PMe_3)], in refluxing dme, dissociates the labile phosphine oxide, also in refluxing dme. When this reaction sequence is performed in the presence of excess bipy and 1 equiv. of Tl(OTf),

[Re(Cl)(Tp)(bipy)][OTf] is formed. The Tp and bipy ligand set are able to stabilize rhenium in four oxidation states {Re^{IV}–Re^{I}}, and this stability has been ascribed to the ability of the nitrogen heterocyclic ligands to act as both weak π-donors and π-acceptors depending on the demand of the metal. [Re^{II}(Cl)(Tp)(bipy)] and [Re^{II}(Tp)(bipy)(η^2-cyclopentene)] have also been reported.[213]

Cycloreversion of 4-methoxystyrene from [Re(O)(OCH_2CHPhO)-(Tp*)] has been investigated and it has been proposed that the reaction proceeds via a significantly asynchronous but concerted transition state. Computation modeling studies as also a Hammett study on cycloreversions of substituted styrene from a series of different [Re(O)(Tp*)(diolato)] complexes have also been reported.[214]

[ReO($OSiMe_3$)(η^2-pzTp)$(L)_2$]X (X = Cl, L = 4-(NMe_2)C_5H_4N, Meim, 1/2 dmpe, py; X = I, L = py) have been prepared by the reaction of [ReO_2(η^2-pzTp)$(L)_2$]X with $SiXMe_3$. All species are cationic in solution and rearrange in the case of neutral monodentate ligand to [ReOX($OSiMe_3$){η^2-pzTp}(L)].[215]

[ReO(Tp)(η^2-N-X)] have been synthesized from the reaction of [$ReOCl_2$(Tp)] with chiral bidentate ligands in which η^2-N-X = alcoholates or amidates derived from (1*S*,2*R*)-ephedrine, (1*S*,2*S*)-diphenylethylenediamine, and L-proline. These chiral-at-Re complexes have been fully characterized by ^{1}H NMR, IR, circular dichroism, polarimetry, and X-ray crystallography, and have been found to be stable and resistant to oxo-transfer when subjected to harsh conditions.[216]

The reaction of [$Re(O)_3$(Tp)] and [$Re(O)_3$(PhTp)] with diphenyl ketene, has led to the isolation of six novel [3 + 2] cycloaddition products. These air-stable solids, as a result of [3 + 2] addition of the O=Re=O motif across the ketene C=C double bond (Fig. 2.49), have been characterized by single-crystal X-ray diffraction and ^{13}C NMR and IR spectroscopies.[217]

Tp* (and also Tp^{Ph}) decomposes during the reaction with [$ReOCl_3(PPh_3)_2$] and $[NEt_4]_2[Re(CO)_3Br_3]$, respectively. The neutral pyrazoles, derived from B–N bond breaking, form complexes with the rhenium(V) oxo and the rhenium(V) tricarbonyl cores.[218]

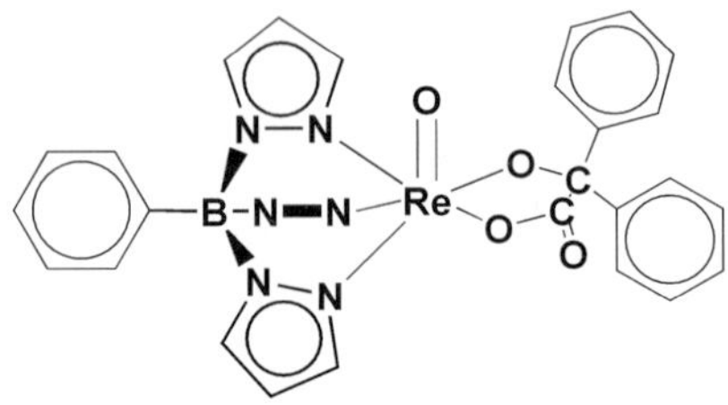

Fig. 2.49. The [3 + 2] cycloaddition product formed from the reaction of $[Re(O)_3(PhTp)]$ with diphenyl ketene.

$[Re(O)_3(Tp)]$ reacts with the Lewis acid $[B(C_6F_5)_3]$ giving $[Re(O)_2(Tp)\{OB(C_6F_5)_3\}]$.[219]

Synthesis and spectroscopic characterization of $[Re(Tp)(CO)PMe_3(L)]$ (L = cyclohexene, cyclopentene, naphthalene, phenanthrene, thiophene, 2-methylthiophene, furan, or acetone) have been reported. The sterochemical and the stability of the aromatic molecules bound to the $[Re(Tp)(CO)(PMe_3)]$ synthons have also been evidenced through the solid-state single crystal study of $[Re(Tp)(CO)(PMe_3)(cyclohexene)]$. $[Re(OTf)(Tp)(CO)(PMe_3)]$ has been prepared and its crystal structure reported.[220] $[Re(Tp)(\pi\text{-acid})(L)(\eta^2\text{-arene})]$ complexes (π-acid = CO or NO, L = 1-alkylimidazole, py, PMe_3, arene prochiral) exist as a dynamic equilibrium of coordination diastereomers in solution.[107]

$[Re(Tp)(CO)(L)(\eta^2\text{-naphthalene})]$ undergoes tandem electrophile/nucleophile addition reactions with a high degree of regiocontrol depending on the auxiliary ligand, L. When L = PMe_3, from the reaction of $[Re(Tp)(CO)(PMe_3)(\eta^2\text{-naphthalene})]$ with triflic acid and a silyl ketene acetal, a 1,4-addition product is formed, whereas with $[Re(Tp)(CO)(L)(\eta^2\text{-naphthalene})]$ (L = py, N,N-dimethylaminopyridine, Meim, or NH_3) a 1,2-addition product is afforded.[221]

The asymmetric π-basic moiety $[Re(Tp)(CO)(Meim)]$ gives thermally stable complexes with ethyl acetate, acetic anhydride, N-methylsuccinimmide, N-acetylpyrrole, and N-methylmaleimide. X-ray and NMR data for $[Re(Tp)(CO)(Meim)(\eta^2\text{-N-methylsuccinimide})]$ have also been reported.[222]

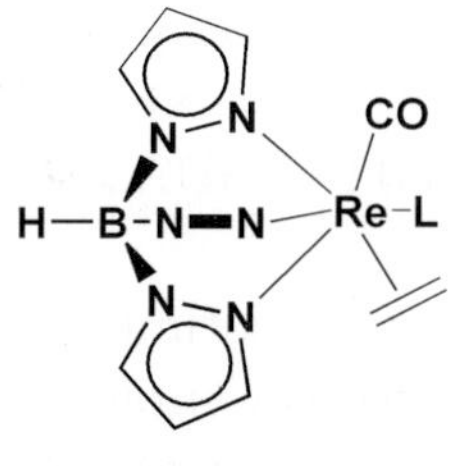

Fig. 2.50. [Re(Tp)(CO)(L)(η^2-ethylene)].

[Re(Tp)(CO)(L)(η^2-ethylene)] (L = ^{t}BuNC, PMe_3, py, Meim, NH_3) (Fig. 2.50) have been prepared by the reaction of [Re(Tp)(CO)(L)(η^2-cyclohexene)] with ethylene. These complexes have been used to determine the rates of the propeller-like rotation of ethylene about the ethylene–rhenium bond using spin-saturation transfer experiments at low temperatures.[223]

Rhenium asymmetric π-basic dearomatization agents [Re(Tp)(CO)(L)(L′)] (L = ^{t}BuNC, py, PMe_3, Meim, or NH_3; L′ = di-hapto coordinated ligand) can also be prepared through alternative routes as that in Fig. 2.51. By changing the ligands L and L′, it is possible to tune steric and electronic properties of the complexes and then stability and selectivity of the η^2-aromatic systems. [Re(Tp)CO)(L)(L′)] allows the modulation of diastereomeric selectivity, rate of dissociation, and air stability. These systems are less expensive, soluble in a wider range of solvents, and more conducive to common chromatographic purification techniques than $Os(NH_3)_5$ moieties.[224]

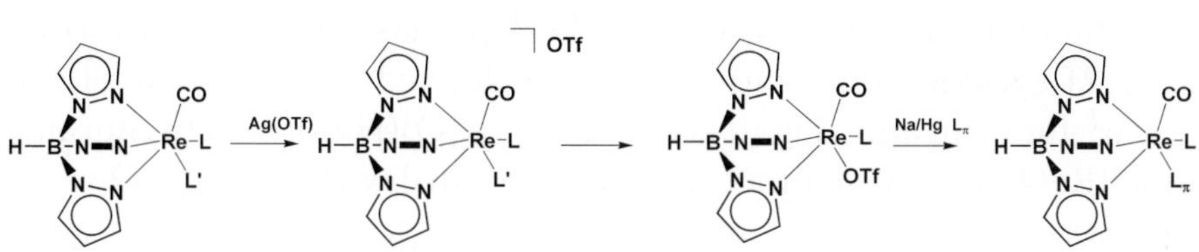

Fig. 2.51. Alternative route to the synthesis of dearomatization agents [Re(Tp)(CO)-(L)(L′)].

The bonding and charge distribution in the complexes of benzene with [Re(Tp)(CO)(L)] (L = NH_3, Him, py, phosphine, CO, MeNC) have been calculated by LSDA and B3LYP density functional methods.[195]

^1H NMR spin-saturation experiments have been carried out to investigate the mechanism for the interconversion of facial diastereomers of a variety of [Re(Tp)(CO)L(η^2-D)] (L = ^{t}BuNC, py, PMe_3, Meim, D = benzene, anisole, naphthalene, 1-methylpirrole, furan, or thiophene), and the rates and free energies of activation were calculated. The two fragments [Re(Tp)(CO)(PMe_3)] and $[Re(Cp)(NO)(PPh_3)]^+$ have been compared in order to understand the role of the ancillary ligand.[225]

(*R*)-α-pinene has been employed to resolve [Re(Tp)(CO)-(Meim)(η^2-benzene)]. Each enantiomer of the [Re(Tp)(CO)(Meim)] system can be obtained due to the differing rates of pinene substitution for the two diastereomers of [Re(Tp)(CO)(Meim)(η^2-(*R*)-α-pinene)]. Racemization studies have been performed for both the rhenium benzene and pinene systems. These rhenium complexes, enantiomerically enriched, can serve as convenient precursors to other resolved aromatic complexes.[226]

[Re(Tp)(CO)(L)(η^2-furan)] (L = PMe_3, ^{t}BuNC) undergoes dipolar cycloaddition with TCNE (= tetracyanoethylene) yielding 7-oxabicycloheptene derivatives. Analogously, [Re(Tp)(CO)-(L)(η^2-2-methylfuran)] gives [Re(Tp)(CO)(^{t}BuNC)(methyloxabicycloheptene)]. When [Re(Tp)(CO)(Meim)(η^2-furan)] was employed, TCNE acts as oxidizing toward the complex but cycloadditions can be achieved by DMAD as the electrophile. Three different products are formed: one in which DMAD undergoes a cycloaddition with the carbonyl ylide form of the complex, and two that are coordination diastereomers where DMAD has undergone a formal [2 + 2] cycloaddition with the uncoordinated double bond of the 4,5-η^2-furan ligand (Fig. 2.52). [Re(Tp)(CO)(Him)(2-methylfuran)] and [Re(Tp)(CO)(Meim)(2,5-dimethylfuran)] yielded only [2 + 2] products.[227]

In [Re(Tp)(CO)(Meim)(η^2-L)] (L = 1-methylpyrrole, 2-methylpyrrole, and 2,5-dimethylpyrrole), a η^2-1*H*- or a η^2-3*H*-pyrrole

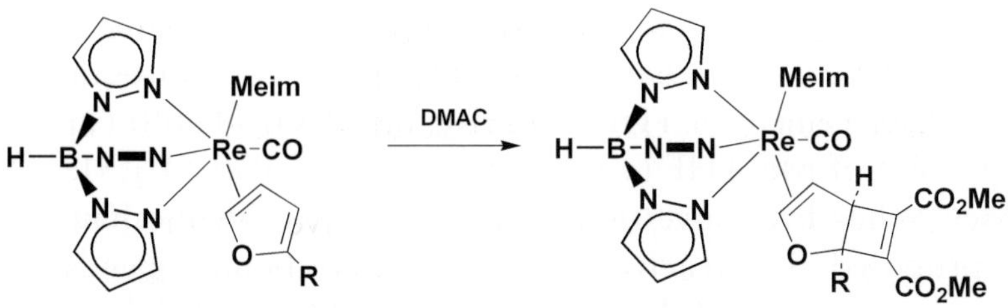

Fig. 2.52. [2 + 2] Cycloaddition of DMAD to the uncoordinated double bond of the 4,5-η^2-furan.

complex is present depending on the nature of the pyrrole. In contrast, when the pyrrole is used, oxidative addition occurs across the N–H bond, yielding either a *N*-pyrrolyl or *N*-pyrrolyl hydride. Protonation or methylation of the η^2-pyrrole complexes and the reduction of the resulting pyrrolium species have also been described.[228]

[Re(Tp)(π-acid)(L)(L_π)] (π-acid = CO or NO^+, L = Meim, Buim, py or PMe_3, L_π = arene), significantly more basic than the analogous pentaammineosmium(II) arene complexes, upon protonation with moderate acids gave stable η^2 and η^3 arenium complexes (Fig. 2.53). [Re(Tp)(CO)(Meim)(5,6-η^2-2H-anisolium)](OTf) has been structurally characterized.[229]

[Re(Tp)(CO)(^{t}BuNC)(2,3-η^2-5-R-furan)] (R = H, Me), obtained as a mixture of coordination diastereomers, reacts with aldehydes RC(=O)H (R = Me, Ph, 3-furyl) in the presence of $BF_3{\cdot}OEt_2$ to yield dihydrofuran complexes. Dihapto-coordinated *trans*-2-alkyl-3-acyl-2,3-dihydrofuran, *trans*-2,3-dialkyl-2,3-dihydrofuran, or 4a,7a-dihydro-4H-furo[2,3-d][1,3]dioxine have been obtained depending

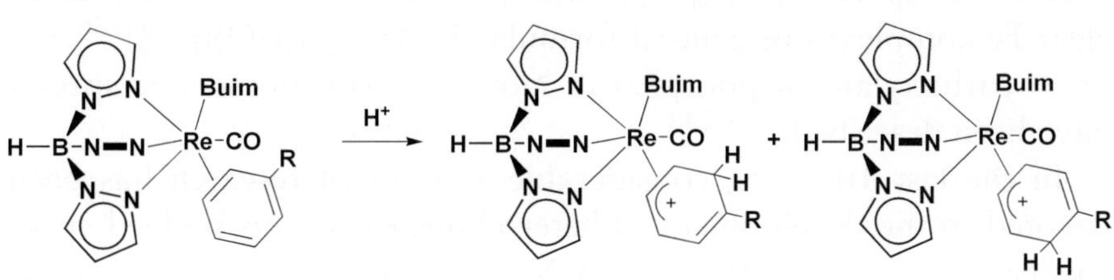

Fig. 2.53. Protonation of [Re(Tp)(π-acid)(L)(L_π)] yields stable η^2 and η^3 arenium complexes.

on the reaction conditions. These species, isolated as mixtures of coordination diastereomers, have been separated by chromatography.[230]

The displacement of THF from [Re(Tp^x)(CO)$_2$(THF)] (Tp^x = Tp or Tp*) and of Me$_n$THF ($n = 1, 2$) from [Re(Tp^x)(CO)$_2$(Me$_n$THF)] by MeCN has been investigated and a dissociative mechanism has been proposed. The reactivity of the Re-THF bond depends on the electronic properties of the ancillary ligands (Tp, Tp*).[231]

[Re(Tp)(NTol)X$_2$] reacts with RMgX, RLi, and organozinc reagents, yielding [Re(Tp)(NTol)(R)X] (R = Ph, Me, Et, ^{i}Pr, ^{n}Bu; X = Cl or I; Tol = *p*-tolyl) or [Re(Tp)(NTol)(R)$_2$] (R = Ph, Me). Iodide or triflate metathesis with Ag(OTf) produces [Re(Tp)(NTol)X(OTf)] (X = Ph, Et, Cl OTf). [Re(Tp)(NTol)(Et)I] reacts with excess Ag(OTf) generating the ethylene hydride cation [Re(Tp)(NTol)(η^2-C$_2$H$_4$)(H)(OTf)], that upon rearrangement yields [Re(Tp)(OTf)(NTol)(Et)], whereas reaction of [Re(Tp)(NTol)(Ph)I] with AgPF$_6$ gave [{Re(Tp)(NTol)(Ph)I}$_2$Ag][PF$_6$] containing two Re-I-Ag linkages. Excess py displaces OTf in [Re(Tp)(NTol)Ph(OTf)] to give [Re(Tp)(NTol)Ph(py)](OTf). Triflate substitution is slow in [Re(Tp*)(O)Ph(OTf)]. [Re(Tp)(NTol)(Ph)$_2$], [Re(Tp)(NTol)(Me)$_2$], [Re(Tp)(OTf)$_2$(NTol)] [{Re(Tp)(NTol)(Ph)I}$_2$Ag][PF$_6$], and [Re(κ^2-Tp)(NTol)Ph(phen)][OTf] have been structurally characterized.[232]

2.9. Group 8: Fe, Ru, and Os

2.9.1. *Fe*

The early work on iron scorpionates is limited to homoleptic octahedral species [Fe(Tp^x)$_2$] and [Fe(Tp^x)$_2$]X[233–235] and dinuclear Fe complexes of general formula [Fe(Tp^x)]$_2$(μ-O)(μ-Z)[236–238] (Z = carboxylato or phosphato). Not many organometallic species have been described.[239–243]

In the last 10 years, considerable amount of research has been devoted to the development of heterodinuclear cyano-bridged cubic cluster based on the [Fe^{III}(Tp)(CN)$_3$]$^-$ synthon which directs the formation of cyanide-bridged compounds with interesting structures and magnetic properties.

The mononuclear PPh_4-*fac*-$[Fe^{III}(Tp)(CN)_3]_3 \cdot H_2O$ and the tetranuclear *fac*-$\{[Fe^{III}(Tp)(CN)_2(\mu\text{-}CN)]_3Fe^{III}(H_2O)_3\} \cdot 6H_2O$ have been synthesized and characterized by X-ray diffraction analysis by Julve *et al.*[244] PPh_4-*fac*-$[Fe^{III}(Tp)(CN)_3]_3 \cdot H_2O$ is a low-spin iron(III) compound with three cyanide ligands in *fac* arrangement and a tridentate Tp ligand building a distorted octahedral environment around the iron atom. *fac*-$\{[Fe^{III}(Tp)(CN)_2(\mu\text{-}CN)]_3Fe^{III}(H_2O)_3\} \cdot 6H_2O$ is the first example of a molecular species containing three peripheral low-spin iron(III) ions linked to a central high-spin iron(III) cation by single cyanide bridges. Variable-temperature magnetic susceptibility measurements of *fac*-$\{[Fe^{III}(Tp)(CN)_2(\mu\text{-}CN)]_3Fe^{III}(H_2O)_3\} \cdot 6H_2O$ (Fig. 2.54) reveal the occurrence of a significant ferromagnetic coupling between the three peripheral low-spin iron(III) centers, and the central high-spin iron(III) ion leading to a low-lying nonet spin state.

Mononuclear $(LH_2)[Fe(Tp)(CN)_3]_2 \cdot 2H_2O \cdot 2MeOH$ (L = C-meso-3,14-dimethyl-2,6,13, 17-tetraazatricyclo[16.4.0.0]docosane), and two cyano-bridged bimetallic complexes $\{[Fe(Tp)(CN)_3]_2\text{-}[Mn(MeOH)_4] \cdot 2MeOH\}$ and $[Mn_2Fe_2(Tp)_2(CN)_6(4,4'\text{-bipy})_2]\text{-}(ClO_4)_2 \cdot 4MeCN$ have also been prepared and their structures and magnetic properties were studied. $(LH_2)[Fe(Tp)\text{-}(CN)_3]_2 \cdot 2H_2O \cdot 2MeOH$ exhibits a hydrogen-bonded dimeric structure of two $[Fe(Tp)(CN)_3]^-$ anions and two water molecules. The reaction of 4,4′-bipy with $\{[Fe(Tp)(CN)_3]_2[Mn(MeOH)_4]\} \cdot 2MeOH$

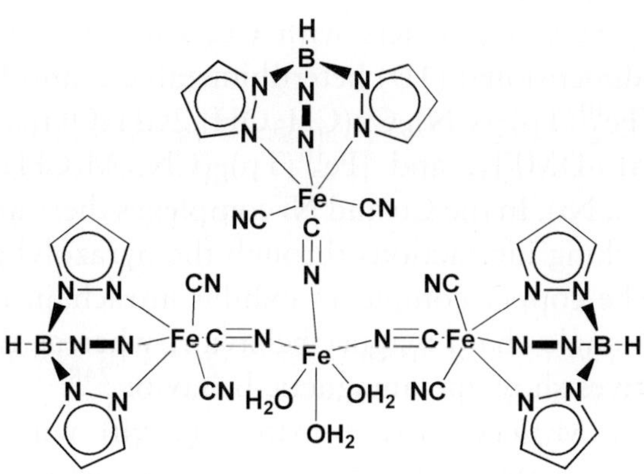

Fig. 2.54. *fac*-$\{[Fe^{III}(Tp)(CN)_2(\mu\text{-}CN)]_3Fe^{III}(H_2O)_3\} \cdot 6H_2O$.

gave $[Mn_2Fe_2(Tp)_2(CN)_6(4,4'\text{-bipy})_2](ClO_4)_2 \cdot 4MeCN$, which exhibits a discrete tetranuclear square Fe_2Mn_2 core. The manganese ions are surrounded by two 4,4′-bipy and bridged by two $[Fe(Tp)(CN)_3]^-$ moieties through cyanide-bridging ligands to form a cyclic tetranuclear structure. The magnetic properties of $\{[Fe(Tp)(CN)_3]_2[Mn(MeOH)_4]\} \cdot 2MeOH$ and $[Mn_2Fe_2(Tp)_2(CN)_6(4,4'\text{-bipy})_2](ClO_4)_2 \cdot 4MeCN$ are found to be dominated by a weak antiferromagnetic interaction of low-spin Fe^{III} and high-spin Mn^{II}.[245]

Reaction of $[NEt_4][Fe^{III}(Tp^*)(CN)_3]$ with $Mn(OTf)_2$ in DMF affords rectangular clusters $[(Tp^*)Fe^{III}(CN)_2(\mu\text{-CN})Mn^{II}(DMF)_4]_2[OTf]_2 \cdot 2DMF$, while tosylate salts afford one-dimensional networks $[Mn^{II}(DMF)_2(\mu\text{-OTs})(\mu\text{-NC})_2(NC)Fe^{III}(Tp^*)]_n$ containing embedded $[(Tp^*)_2Fe^{III}_2Mn^{II}_2(CN)_6]^{2+}$ clusters via *in situ* trapping; additional magnetic studies of these complexes indicate that the $[Fe^{III}(Tp^*)(CN)_3]^-$ centers are highly anisotropic and are antiferromagnetically coupled to adjacent Mn^{II} centers.[246] $[Fe^{III}(Tp)(CN)_3]^-$ has also been employed as a building block for the synthesis of the trinuclear $[Fe_2(Tp)_2(CN)_6Ni(cyclam)] \cdot 2H_2O$, and of the one-dimensional polymer, $[Fe^{III}(Tp)(CN)_3Cu(dien)]ClO_4 \cdot H_2O$. In the latter complex, a zigzag chain is formed by combining $[Fe(Tp)(CN)_3]^-$ and $[Cu(dien)]^{2+}$ fragments alternately through cyano bridges. The magnetic properties of these complexes have also been investigated.[247]

$(Bu_4N)[(Tp)Fe(CN)_3]$ reacts with Cu, Co, and Ni cations, giving four one-dimensional (1D) heterobimetallic cyano-bridged chain complexes $[Fe^{III}_2(Tp)_2(CN)_6Cu(CH_3OH) \cdot 2CH_3OH]_n$, $[(Tp)_2Fe^{III}_2(CN)_6Cu(DMF) \cdot DMF]_n$, and $[Fe^{III}_2(Tp)_2(CN)_6M(CH_3OH)_2 \cdot 2CH_3OH]_n$ ($M = Co$, Ni). In the Co and Ni complexes there are weak interchain π–π stacking interactions through the pyrazolyl groups of the Tp ligands. The copper complexes exhibit intrachain ferromagnetic coupling and single-chain magnet behavior, whereas the nickel and cobalt derivatives show metamagnetic behavior.[248]

Reaction of $M(OTf)_2$ (M = Mn, Co, Ni) with $[NEt_4][Fe^{III}(Tp^*)(CN)_3]$ in DMF affords three isostructural rectangular clusters $\{[Fe^{III}(Tp^*)(CN)_3M^{II}(DMF)_4]_2[OTf]_2\} \cdot 2DMF$. Susceptibility measurements suggest that $\{[Fe^{III}(Tp^*)(CN)_3M^{II}(DMF)_4]_2[OTf]_2\}\cdot$ 2DMF (M = Co, Ni) exhibits incipient single-molecule magnetic behavior below 2 K.[249]

Using $[Fe(Tp)(CN)_3]^-$, a 1D chain polymer containing Ru_2(II,III) paramagnetic dinuclear groups, $[Ru_2(O_2C^tBu)_4][Fe(Tp)(CN)_3]$ has also been prepared and characterized. Magnetic measurements indicate antiferromagnetic interaction between Ru_2 units and low-spin Fe^{III} ions.[250] The $[Fe(Tp)(CN)_3]^-$ synthon has been employed in the synthesis of $[(Tp)_8(H_2O)_6Cu^{II}_6Fe^{III}_8(CN)_{24}]^{4+}$, a face-centered-cubic cluster exhibiting SMM-type behavior,[251] and also as a building block toward a $[Cu(bipy)_2]^{2+}$ core giving $[Fe(Tp)(CN)_3][Cu(bipy)_2]ClO_4\cdot CH_3OH$ which exhibits ferromagnetic coupling between Cu and Fe.[252]

Heterobimetallic hexanuclear $[\{Fe(Tp)(CN)_3\}_4\{M(MeCN)(H_2O)_2\}_2]\cdot 10H_2O\cdot 2MeCN$ (M = Ni, Co, Mn) have been synthesized in the H_2O-MeCN solution. The Ni complex is a canted antiferromagnet that undergoes a field-induced spin-flop-like transition. The Co and Mn species show antiferromagnetic intracluster coupling.[253]

Two cyano-bridged heterotrinuclear complexes $[Fe^{III}_2(Tp)_2(CN)_6Mn^{II}(C_2H_5OH)_4]\cdot .2C_2H_5OH$ and $[Fe^{III}_2(Tp)_2(CN)_6Ni^{II}(en)_2]$ have been prepared and structurally characterized. A weak antiferromagnetic interaction between the Mn^{II} and Fe^{III} ions has been found in the former complex.[254] The one-dimensional (1D) cyano-bridged heterobimetallic chain, $[Fe^{III}_2(Tp)_2(CN)_6Cu(CH_3OH).2CH_3OH]_n$, exhibiting intrachain ferromagnetic coupling and superparamagnetic behavior, has been reported by You *et al.*[255] (Fig. 2.55).

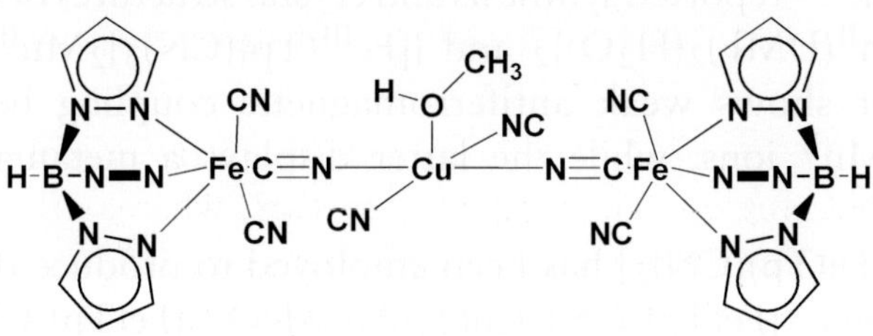

Fig. 2.55. The trimetallic unit of the one-dimensional $[Fe^{III}_2(Tp)_2(CN)_6Cu(CH_3OH)\cdot 2CH_3OH]_n$.

$[M\{HC(pz^*)_3\}(H_2O)_3]^{2+}$ (M = Ni, Co, Fe) reacts with $[Bu_4N][Fe^{III}(Tp)(CN)_3]$ yielding the pentanuclear cyano-bridged clusters, $[Fe^{III}_3(Tp)_3(CN)_9M_2\{HC(pz^*)_3\}_2]ClO_4 \cdot 15H_2O$ (M = Ni, Co) and $[Fe^{III}_3(Tp)_3(CN)_9Fe^{II}_2\{HC(pz^*)_3\}_2]BF_4 \cdot 15H_2O$. Single-crystal X-ray analyses have been carried out for all these derivatives. Antiferromagnetic interaction was observed in the latter complex.[256]

The assembling of $[Mn(5\text{-MeOsalen})(H_2O)]^+$ and $[Fe(Tp)(CN)_3]^-$ affords the one-dimensional zig-zag chain $[Fe(Tp)(CN)_3Mn(5\text{-MeOsalen}) \cdot 2CH_3OH]_n$ in which a ferromagnetic Fe(III)–Mn(III) couplings and $D < 0$ anisotropy on Mn(III) is found.[257]

Treatment of $[NEt_4][Fe^{III}(pzTp)(CN)_3]$ with $Ni(OTf)_2$ in DMF, followed by excess 2,2,2-tris(pyrazolyl)ethanol (L) in CH_2Cl_2, affords $[Fe^{III}(pzTp)(CN)_3]_4[Ni^{II}L]_4 \cdot [OTf]_4 \cdot 10DMF \cdot Et_2O$ that crystallizes as a cationic octanuclear $Fe^{III}_4Ni^{II}_4$ in which the metals reside in alternate corners of a distorted molecular box and that exhibits slow relaxation of the magnetization.[258] Reaction of $[NEt_4][Fe^{III}(pzTp)(CN)_3]$ with $Ni^{II}(OTf)_2$ and 1,5,8,12-tetraazadodecane (L) affords $\{[Fe^{III}(pzTp)(CN)_3]_2[Ni^{II}L]\} \cdot 1/2MeOH$, while the same reaction carried out in the presence of bipy yields $\{[Fe^{III}(pzTp)(CN)_3]_2[Ni^{II}(bipy)_2]\} \cdot 2H_2O$. Magnetic measurements have been reported for all complexes.[259] Li *et al.*[260] also reported $[Fe^{III}(pzTp)(CN)_3]_4[Ni^{II}(L)]_4[OTf]_4 \cdot solvent$ (L = bifunctional 1-S(acetyl)-tris(pyrazolyl)alkanes $(pz)_3C(CH_2)_nSAc$ (n = 6, 10), which exhibit slow relaxation of the magnetization.

Reaction of $[Cu(Me_3tacn)(H_2O)_2](ClO_4)_2$ with $(Bu_4N)[Fe(Tp)(CN)_3]$ in a mixture of ethanol and acetonitrile yields $[Cu_3(Me_3tacn)_3Fe_2(Tp)_2(CN)_6](ClO_4)_4 \cdot 2H_2O$.[261]

Gao *et al.*[262] reported synthesis and crystal structures of $\{[Fe^{III}(Tp)(CN)_3]_2Mn^{II}(DMF)_2(H_2O)\}_2$ and $\{[Fe^{III}(Tp)(CN)_3]_2Mn^{II}(DMF)_2\}_n$. The former shows weak antiferromagnetic coupling between the Fe^{III} and Mn^{II} ions, while the latter displays a metamagnetic-like behavior.[262]

$(Bu_4N)[Fe(Tp)(CN)_3]$ has been employed to produce the tetranuclear clusters, $[Fe(Tp)(CN)_3Cu(Tp)]_2 \cdot 2H_2O$, $[Fe(Tp)(CN)_3Cu(bpca)]_2 \cdot 4H_2O$, $[Fe(Tp)(CN)_3Ni(tren)]_2(ClO_4)_2 \cdot 2H_2O$, and $[Fe(Tp)(CN)_3Ni(bipy)_2]_2[(Tp)Fe(CN)_3]_2 \cdot 6H_2O$. $[Fe(Tp)(CN)_3Cu(Tp)]_2 \cdot$

$2H_2O$ and $[Fe(Tp)(CN)_3Ni(tren)]_2(ClO_4)_2 \cdot 2H_2O$ possess similar square structures and exhibit intermolecular π–π stacking interactions through pyrazolyl groups of Tp ligands which lead to 1D chain structures. $[Fe(Tp)(CN)_3Ni(bipy)_2]_2[Fe(Tp)(CN)_3]_2 \cdot 6H_2O$ shows a 3D network structure due to the coexistence of π–π stacking effects and hydrogen-bonding interactions.[263]

Electronic structure and intramolecular exchange constants have been calculated (spin-polarized DFT calculations) for three cyanide-bridged molecular magnets, $[Fe^{III}(Tp^*)(CN)_3M^{II}(DMF)_4]_2$-$(OTf)_2 \cdot 2DMF$ (M^{II}=Mn, Co, Ni).[264]

The reaction between $K[Tp(Fe)(CN)_3]$ and $M(ClO_4)_2 \cdot 6H_2O$ (M = Co or Ni) in MeCN/THF gives $[(H_2O)_{12}Co_6Fe_8(Tp)_8$-$(CN)_{24}](ClO_4)_4 \cdot 12THF \cdot 7H_2O$ and $[(H_2O)_{12}Ni_6Fe_8(Tp)_8(CN)_{24}]$-$(ClO_4)_4 \cdot 12THF \cdot 7H_2O$.[265]

The octanuclear $[(Tp)Fe_4Ni_4(CN)_{12}] \cdot H_2O \cdot 24MeCN$, obtained by the reaction of $[Tp(Fe)(CN)_3]^-$ with $[Ni(NO_3)Tp]$, is a single molecule magnet with ground-spin state, $S = 6$.[266]

The synthon $[(pzTp)Fe(CN)_3]$ reacts with $[Cu(Me_3tacn)(H_2O)_2]$-$(ClO_4)_2$ affording the pentanuclear cyano-bridged cluster $[(pzTp)_2$-$(Me_3tacn)_3Cu_3Fe_2(CN)_6](ClO_4)_4 \cdot 4H_2O$, whereas its reaction with $[Ni(phen)(CH_3OH)_4](ClO_4)_2$ or $Zn(OAc)_2 \cdot 2H_2O$ affords the molecular box $[(pzTp)_4(phen)_4Ni_4Fe_4(CH_3OH)_4(CN)_{12}](ClO_4)_4 \cdot 4H_2O$ and the rectangular cluster $[(pzTp)_2Zn_2Fe_2(OAc)_2(H_2O)_2(CN)_6]$.[267]

$[(pzTp)Fe(CN)_3]_4[Ni(L)]_4[OTf]_4$ (L = 1-S-(acetyl)tris(pyrazolyl)-decane) has been also prepared and covalently linked onto the electrodes yielding a dominant conduction pathway.[268]

The reaction of $FeCl_3 \cdot 6H_2O$, K(Tp), and KSCN gives both $[Fe(Tp)_2][Fe(Tp)(NCS)_3]$ and $[Fe(Tp)_2]_3[Fe(NCS)_6]$. The bond lengths and angles indicate that these complexes are formed by $[Fe(Tp)_2]^+$ cations where Fe^{III} ions are in a typical low-spin state, and counter-ions $[Fe(Tp)(NCS)_3]^-$ and $[Fe(NCS)_6]^{3-}$ where Fe^{III} ions are in a high-spin state.[269] $[Fe(Tp)_2]_3[Fe(NCS)_6]$ has also been structurally characterized: the anion and the cation are stacked into a three-dimensional supramolecular aggregate with weak $C-H \cdots \pi$ interactions.[270]

$[Fe(Tp)_2][FeCl_4]\cdot MeCN$, a compound obtained during the attempted synthesis of polynuclear iron complexes, exhibits weak C–H$\cdots$Cl/N interaction between the cation $[Fe(Tp)_2]$ and the anion $[FeCl_4]^-$.[271]

Electrochemical, magnetic, and spectroscopic properties are reported for divalent and trivalent homoleptic $[Fe(Tp)_2]^{+/0}$, $[Fe(Tp^*)_2]^{+/0}$, and $[Fe(pzTp)_2]^{+/0}$.[272] Ligand field strengths for these metal complexes increase as Tp* < Tp < pzTp, which reflects the importance of steric rather than electronic effects on spectroscopic properties. Metal-centered redox potentials become more negative as pzTp < Tp < Tp*, which follows the electron-donating ability of the ligands.[47]

$[Fe(\kappa^3\text{-}Tp)(CO)(PMe_3)(COMe)]$ is prepared by one-step reaction of *cis,trans*-$[Fe(CO)_2(PMe_3)_2MeI]$ with K(Tp). Visible light irradiation of this complex in toluene for 1 h afforded the octahedral $[Fe(Tp)(CO)(PMe_3)Me]$. Steric and electronic properties are compared with those of related cyclopentadienyl complexes, and suggests that relative donating abilities follow the trend Cp, Cp*, C_5Ph_5 > Tp, and structural data indicate that Tp is the most sterically demanding ligand.[273]

Synthesis and structure of $[Fe(Cp^R)(Tp)]$ have been reported,[29] and the electronic and magnetic properties have been studied and compared to their homoleptic analogs $[Fe(Cp^R)_2]$ and $[Fe(Tp)_2]$. $[Fe(Cp^*)(Tp)]$ displays a spin equilibrium ($S = 0 \Leftrightarrow S = 2$) in solution, investigated by variable temperature NMR spectroscopy.[30]

A difluorocarbene complex (Fig. 2.56) is implicated in the formation of the ferraoxetene $[Fe\{\kappa^2\text{-}C(N^iPr_2)OCF_2\}(Tp)(CO)]$ obtained from the reaction of $[Fe\{\eta^2\text{-}OCN^iPr_2\}(CF_3)(CO)_2(PPh_3)]$ with K(Tp). $[Fe\{\kappa^2\text{-}C(N^iPr_2)OCF_2\}(Tp)(CO)]$ undergoes an unusual acid hydrolysis with HBF_4(aq)/CO to provide the isonitrile salt $[Fe(CN^iPr)(Tp)(CO)_2]PF_6$.[274]

Synthesis, structure, and physical properties of $[Fe_2O(O_2CCH_2\text{-}C_{10}H_7)_2(Tp)_2]$ and $[Fe_2O((O_2CCH_2CH_2)_2\text{-}C_{10}H_6)(Tp)_2]$ have been reported. These complexes exhibit moderately strong intramolecular antiferromagnetic exchange between the high-spin ferric ions.[275]

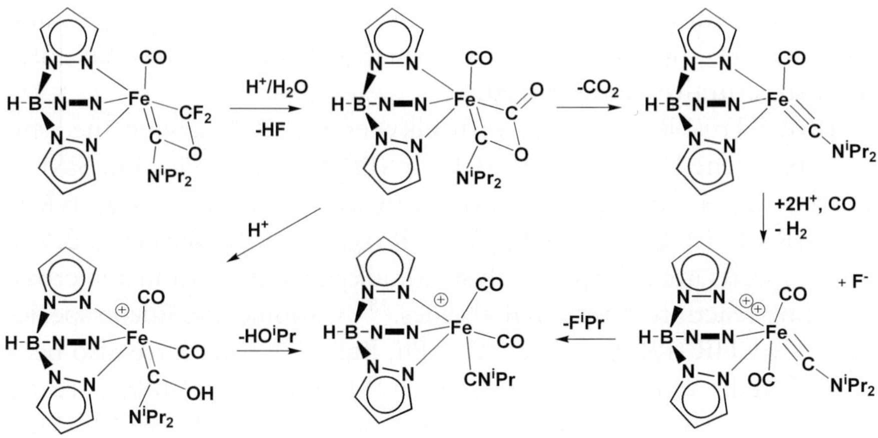

Fig. 2.56. Mechanism proposed for the ferraoxetene/isonitrile interconversion.

[Fe(Cat)(Tp*)(Hpz*)] [Cat = catecholate = $C_6H_4O_2$, 3,5-$^tBu_2C_6H_2O_2$, $C_6Cl_4O_2$) have been prepared and their reactivity toward O_2 investigated. Single-crystal X-ray study of [Fe(Cat)-(Tp*)(Hpz*)] (Cat = $C_6H_4O_2$, 3,5-$^tBu_2C_6H_2O_2$) has also been performed.[276]

2.9.2. *Ru*

Ruthenium–poly(pyrazolyl)borate complexes, [RuX(Tp)(arene)],[277–279] [RuX(Tp^x)(diolefin)],[280,281] [RuX(Tp^x)$(CO)_n$],[282–284] and [RuX(L)H_2(Tp^x)][285,286] have been widely reported in the first edition. Ruthenium, however, continued from 1999 to date to exhibit a rich and varied homoscorpionate chemistry.

The octahedral [RuCl(Tp)$(DMSO)_2$] has been prepared and treated with an excess of HC≡CR (R = Ph, C_6H_9, COOMe, *n*-Hex) and allyl alcohol to form stable (allyloxy)carbene complexes [Ru(=C(CH_2R)OCH_2CH=CH_2)Cl(Tp)], key intermediates on the pathway to β, γ-unsaturated ketones. When R = C_7H_{15}, [Ru(η^3-H_2C=$CHCH_2COC_7H_{15}$)Cl(Tp)] has been isolated. [RuCl(Tp)(COD)] has been used for the synthesis of the corresponding (allyloxy)carbene when Me_3SiC≡$CSiMe_3$ was employed.

The octahedral $[Ru(=C(CH_2Hex)OCH_2CH=CH_2)Cl(Tp)]$ and $[Ru(=C(CH_2SiMe_3)CH_2\text{-}CH=CH_2)Cl(Tp)]\cdot 1/2C_6H_5CH_3$ have also been structurally characterized.[287]

$[RuCl(Tp)(PR_3)_2]$ reacts with alkynes $HC\equiv CR'$ giving the neutral vinylidene $[Ru=C=CHR'(Cl)(Tp)(PR_3)_2]$ ($R' = Ph$, ${}^tBu$, $SiMe_3$, H) that upon further interaction with LDA and diphenylacetylene form $[Ru(PhC=C(Ph)C\equiv CR')(Tp)(PR_3)_2]$ ($R' = Ph$, Bu^t) (Fig. 2.57). These species are efficient catalysts for alkyne dimerization and cross-coupling reaction of terminal alkynes.[288] Cationic vinylidene species $[Ru=C=CHR'(PR_3)_2(Tp)]^+$ ($R' = Ph$, Bu^t, COOMe) have also been synthesized. When $[Ru=C=CHCOOMe(PR_3)_2(Tp)][BPh_4]$ reacts with $HC\equiv CCOOMe$ the *E*-stereoisomer of $[Ru(=C=CCOOMe)\text{-}CH=CHCOOMe(Tp)(PEt_3)_2][BPh_4]$ is formed. The σ-η^1-butadienyl complex $[Ru(CH=C(COOMe)CH=CHCOOMe(Tp)(PEt_3)_2]$ has been synthesized by the reaction of the vinylvinylidene complex with $NaBH_4$.[288]

The reaction of $[RuCl(Tp)(COD)]$ with PR_3 ($PR_3 = PPh_2{}^iPr$, P^iPr_3, PPh_3) and propargylic alcohols, $HC\equiv CCPh_2OH$, $HC\equiv CCFc_2OH$, and $HC\equiv C(Ph)MeOH$, has been investigated. When $PR_3 = PPh_2{}^iPr$, P^iPr_3, and $HC\equiv CCPh_2OH$ were employed, the 3-hydroxyvinylidene $[Ru(=C=CHC(Ph)_2OH)Cl(Tp)(PPh_2{}^iPr)]$ and $[Ru(=C=CHC(Ph_2)OH)Cl(Tp)P^iPr_3]$ were obtained. With $PR_3 = PPh_2{}^iPr$ and $HC\equiv CCFc_2OH$ as well as with $PR_3 = PPh_3$ and $HC\equiv CCPh_2OH$, dehydration takes place affording the allenylidene complexes $[Ru(=C=C=CFc_2)Cl(Tp)(PPh_2{}^iPr)]$ and $[Ru(=C=C=CPh_2)Cl(Tp)(PPh_3)]$. Similarly, with $PPh_2{}^iPr$ and $HC\equiv CC(Ph)MeOH$, rapid elimination of water results in the formation of the vinylvinylidene complex $[Ru(=C=CHC(Ph)=CH_2)Cl(Tp)(PPh_2{}^iPr)]$. $[RuCl(Tp)(PR_3)]$ reacts with $HC\equiv C(CH_2)_nOH$ ($n = 2, 3, 4, 5$) yielding six- and seven-membered cyclic oxycarbene complexes $[Ru(=C_4H_6O)Cl(Tp)(PR_3)]$, $[Ru(=C_5H_8O)Cl(Tp)(PR_3)]$, and $[Ru(=C_6H_{10}O)Cl(Tp)(PR_3)]$. On the other hand, by using 1-ethynylcyclohexanol, $[Ru(PPh_2{}^iPr)\text{-}(=C=HC_6H_9)Cl(Tp)]$ is formed. The vinylcarbyne complexes $[RuCl(Tp)(PPh_2{}^iPr)(\equiv C\text{-}CH=Ph_2)]^+$, $[Ru(\equiv C\text{-}CH=CFc_2)Cl(Tp)\text{-}$

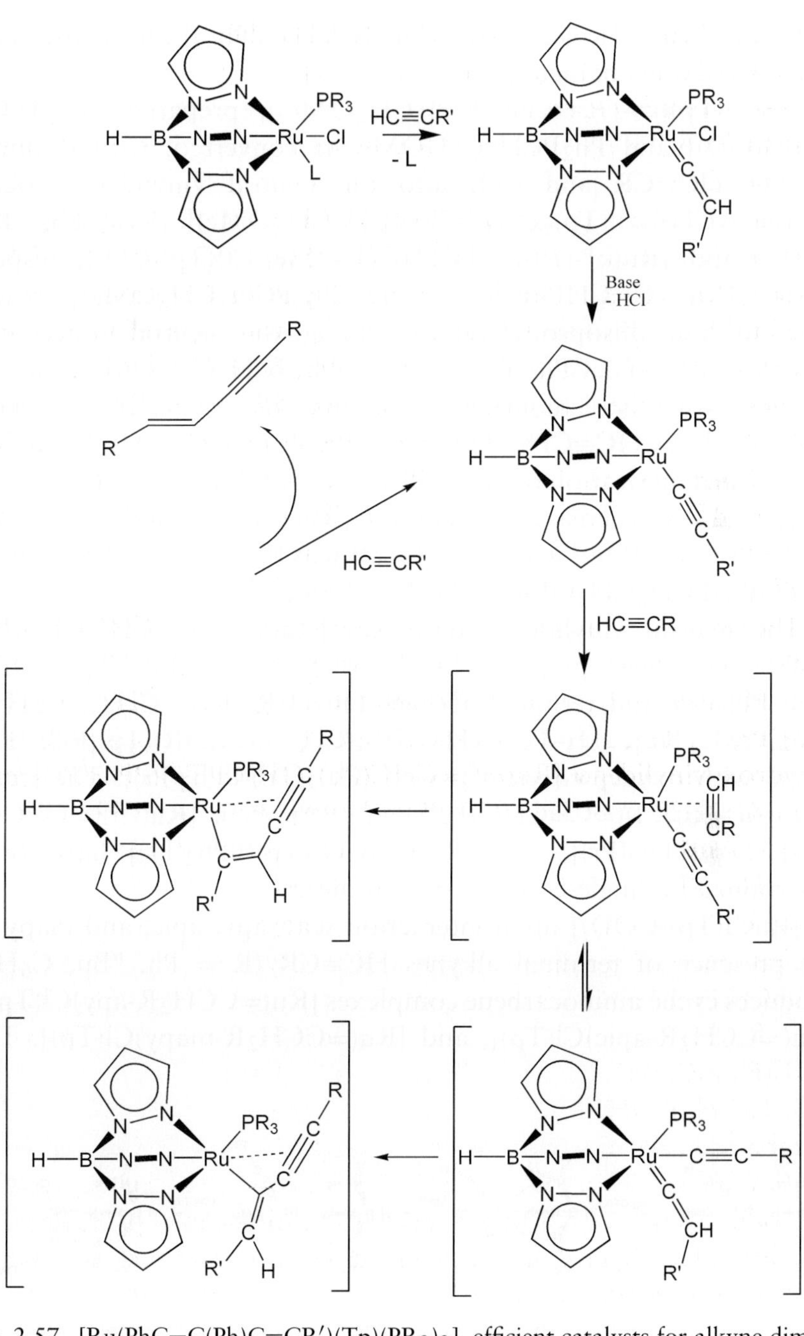

Fig. 2.57. [Ru(PhC=C(Ph)C≡CR′)(Tp)(PR_3)$_2$], efficient catalysts for alkyne dimerization and cross-coupling reaction of terminal alkynes.

$(PPh_2{}^iPr)]^+$, and $[Ru(\equiv C\text{-}CH{=}CPh_2)Cl(Tp)(PPh_3)]^+$ have also been synthesized and characterized.[289]

$[RuCl(Tp)(\kappa^2\text{-}(P, O)\text{-}Ph_2PCH_2CH_2OMe)]$, prepared by [RuCl-(Tp)(COD)] and $Ph_2PCH_2CH_2OMe$, is converted, with terminal alkynes HC≡CR and CO, into the neutral vinylidene species $[Ru(=C=CHR)Cl(Tp)(\kappa^1\text{-}(P)\text{-}Ph_2PCH_2CH_2OMe)]$ (R = Ph, ^{n}Bu, C_6H_9) and $[Ru(\kappa^1\text{-}(P)\text{-}Ph_2PCH_2CH_2OMe)(Cl)(Tp)(CO)]$, respectively. $[Ru(=C=CHPh)Cl(Tp)(\kappa^1\text{-}(P)\text{-}Ph_2PCH_2CH_2OMe)]$ reacts with lithium diisopropylamide yielding the neutral α-acetylide complex $[Ru(-C\equiv CPh)(Tp)(\kappa^2\text{-}(P, O)\text{-}Ph_2PCH_2CH_2OMe)]$ which couples with stoichiometric amounts of HC≡CPh to give $[Ru(=C(Ph)=CHC\equiv CPh)(Tp)(\kappa^1\text{-}(P)\text{-}Ph_2PCH_2CH_2OMe)]$ and catalyzes the dimerization of HC≡CR (R = Ph, $SiMe_3$, ^{n}Bu, ^{t}Bu, C_6H_9) to give enynes, whereas the reaction of $[Ru(=C=CHPh)Cl(Tp)(\kappa^1(P)\text{-}Ph_2PCH_2CH_2OMe)]$ with Ag(OTf) affords $[Ru(=C=CHPh)(Tp)(\kappa^2\text{-}(P,O)\text{-}Ph_2PCH_2CH_2OMe)]OTf$ (Fig. 2.58).[290]

The neutral vinylidene complexes $[Ru(Cl)(=C=CHR)(Tp)(PPh_2{}^iPr)]$ have been prepared by the reaction of [RuCl(Tp)(COD)] with $PPh_2{}^iPr$ and terminal alkynes HC≡CR (R = C_6H_5, C_4H_3S, C_6H_4OMe, Fc, C_6H_4–Fc, C_6H_9). $[Ru(Cl)(=C=CHR)(Tp)(PPh_2{}^iPr)]$ reacts with oxygen to yield $[Ru(Cl)(CO)(Tp)(PPh_2{}^iPr)]$. The structures of $[Ru(Cl)(=C=CHC_4H_3S)(Tp)(PPh_2{}^iPr)]$, $[Ru(Cl)(=C=CH\text{-}C_6H_9)(Tp)(PPh_2{}^iPr)]$, and $[Ru(Cl)(CO)(Tp)(PPh_2{}^iPr)]$ have been determined by single-crystal X-ray studies.[291]

[RuCl(Tp)(COD)] upon interaction with apy, apic, and mapy in the presence of terminal alkynes HC≡CR (R = Ph, ^{n}Bu, C_6H_9) produces cyclic aminocarbene complexes $[Ru(=CCH_2R\text{-}apy)Cl(Tp)]$, $[Ru(=CCH_2R\text{-}apic)Cl(Tp)]$, and $[Ru(=CCH_2R\text{-}mapy)Cl(Tp)]$. The

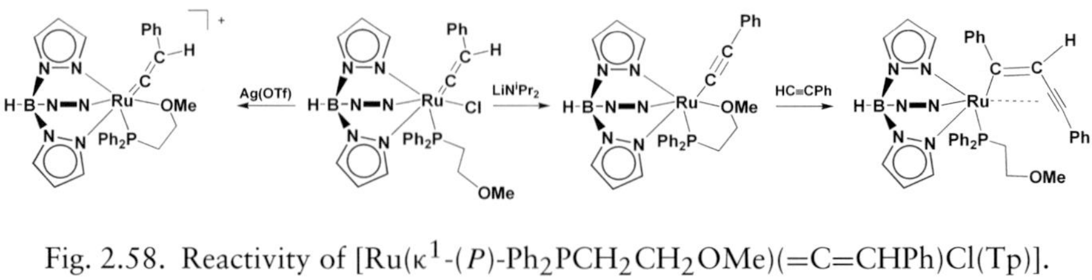

Fig. 2.58. Reactivity of $[Ru(\kappa^1\text{-}(P)\text{-}Ph_2PCH_2CH_2OMe)(=C=CHPh)Cl(Tp)]$.

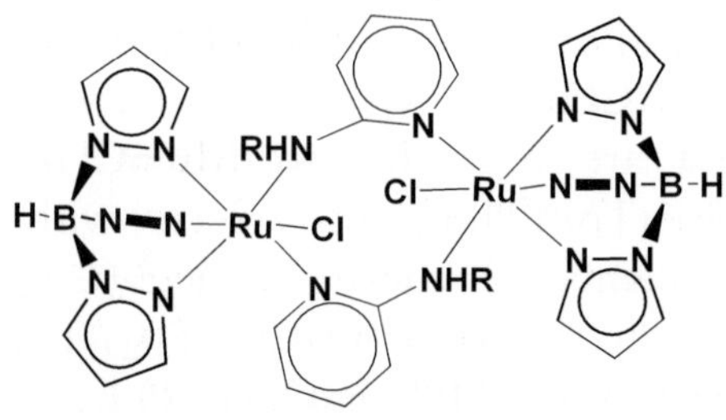

Fig. 2.59. The dimeric $[Ru(\mu\text{-apy})Cl(Tp)]_2$.

dimeric $[Ru(\mu\text{-apy})Cl(Tp)]_2$ (Fig. 2.59) and $[Ru(\mu\text{-apic})Cl(Tp)]_2$ were also isolated. $[Ru(=CCH_2C_6H_9\text{-apic})Cl(Tp)]$ and $[Ru(\mu\text{-apic})Cl(Tp)]_2$ were characterized by single-crystal X-ray study.[292]

Specific pyridine-based aldimines and aminals act as chelating ligands toward the {Ru(Tp)(L)} moieties (L = MeCN, PMe_3, $SbPh_3$) giving κ^2-N,N′-coordinated cyclic imine complexes which can rearrange into cyclic aminocarbene complexes. [RuCl(Tp)(COD)] also reacts readily with the imines py-N=CHR (R = Ph, p-$MeOC_6H_4$, Np) at high temperature yielding the aminocarbene complexes [RuCl(Tp){=C(R)NH-py}].[293]

$[Ru(\eta^2\text{-}O_2CCHPh_2)(Tp)(PPh_3)]$ has been prepared and allowed to react with diphenylcyclopropene yielding the metallacycle $[Ru(\kappa^2\text{-}(C,O)\text{-}C(=CHCHPh_2))OC(CHPh_2)=O(Tp)(PPh_3)]$ (Fig. 2.60). Small quantities of the carbene species $[Ru=CHCH=C(Ph)_2(\eta^1\text{-}O_2CCHPh_2)(Tp)(PPh_3)]$ have also been detected. $[Ru(\eta^2\text{-}O_2CCHPh_2)(Tp)(PPh_3)]$ also reacts with phenyldiazomethane giving

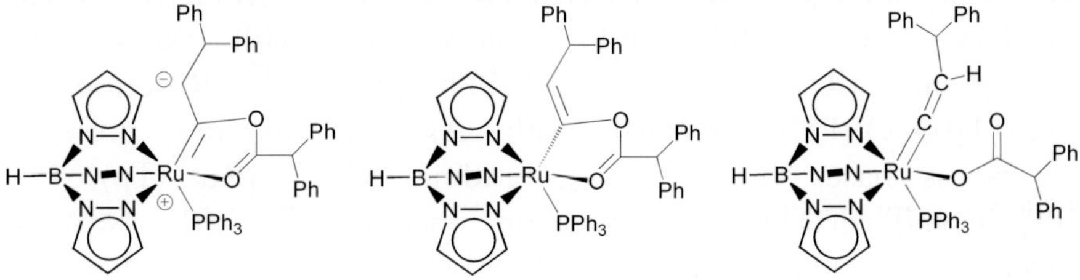

Fig. 2.60. Tautomeric forms of $[Ru(\kappa^2\text{-}(C,O)\text{-}C(=CHCHPh_2))OC(CHPh_2)=O(Tp)\text{-}(PPh_3)]$.

[Ru=C=CHPh(Tp)(PPh$_3$)(η^1-O$_2$CCHPh$_2$)]. The latter compound and [Ru=CHCH=C(CH$_3$)$_2$(η^1-O$_2$CCHPh$_2$)(Tp)(PCy$_3$)] release free HO$_2$CCHPh$_2$ and are converted to [Ru=CHPh(Cl)(Tp)(PPh$_3$)]. [Ru(η^2-O$_2$CCHPh$_2$)(Tp)(PPh$_3$)] also reacts with phenylacetylene yielding the five-membered chelate [Ru(κ^2-(*C*,*O*)-C(=CHPh)OC-(CHPh$_2$)=O)(Tp)(PPh$_3$)]. [Ru(η^2-O$_2$CCHPh$_2$)(Tp)(PPh$_3$)], [Ru(κ^2-(*C*,*O*)-C(=CHCHPh$_2$)OC(CHPh$_2$)=O(Tp)(PPh$_3$)], and [Ru(κ^2-(*C*,-*O*)-C(=CHPh)OC(CHPh$_2$)=O)(Tp)(PPh$_3$)] have been structurally characterized. [Ru=CHPh(η^1-O$_2$CCHPh$_2$)(Tp)(PPh$_3$)] has been reported as an active initiator for the ring-opening metathesis polymerization of norbornene.[294]

Reaction of [RuCl(PPh$_3$)$_2$(Tp)] with 4-ethynyl toluene yielded [RuCl(=CHC$_6$H$_4$Me-4)(PPh$_3$)(Tp)] which upon reaction with [EtNH$_2$][S$_2$CNEt$_2$] yielded the metallacycle [RuC{(=CHC$_6$H$_4$Me-4)SC(NEt$_2$)S}(PPh$_3$)(Tp)], whereas reaction of [RuCl(PPh$_3$)$_2$(Tp)] with diphenylpropynol and AgPF$_6$ yielded [Ru(=C=C=CPh$_2$)-(PPh$_3$)(Tp)]PF$_6$ (Fig. 2.61). The same reaction carried out in the absence of silver salts gave the neutral complex [RuCl(=C=C=CPh$_2$)-(PPh$_3$)(Tp)]. The reaction of both complexes with Na(S$_2$CNMe$_2$) afforded [Ru{C(=C=CPh$_2$)SC(NMe$_2$)S}(PPh$_3$)(Tp)].[295]

[RuCl(Tp)(PPh$_2$NHR)$_2$] reacts with terminal alkynes HC≡CR′ (R′ = Ph, *p*-C$_6$H$_4$Me, ^{n}Bu) and propargylic alcohols to give the azaphosphacarbene complexes [Ru(κ^2-(*C*,*P*)=C(CH$_2$R′)N(R)PPh$_2$)-(κ^1(*P*)-PPh$_2$NHR)(Tp)]$^+$ and [Ru(κ^2-(*C*,*P*)=C(CH=CPh$_2$)N(Pr^n)-PPh$_2$)(κ^1-(*P*)-PPh$_2$NH^{n}Pr)(Tp)]$^+$. The intermediate [Ru(PPh$_2$-NHR)$_2$(=C=C=CPh$_2$)(Tp)]$^+$ has also been isolated.[296]

[Ru(Tp)(OTf)(CO)(PPh$_3$)] reacts with LiNHPh affording the amido complex [Ru(Tp)(NHPh)(CO)(PPh$_3$)] which exhibits a pyramidal amido moiety, and with excess aniline giving [Ru(Tp)(NH$_2$Ph)(CO)(PPh$_3$)](OTf).[297] Reaction of [Ru^{II}(Tp)(CO)(PPh$_3$)(NHPh)] with Ag(OTf) in the presence of excess bases (Et$_3$N or 2,6-lutidine) yields [Ru^{IV}(Tp)(CO)-(PPh$_3$)(NHPh)](OTf)$_2$. Counter-ion methathesis with NaBAr$_F$ produces [Ru^{IV}(Tp)(CO)(PPh$_3$)(NHPh)](BAr$_F$)$_2$. Reversible deprotonation of [Ru^{IV}(Tp)(CO)(PPh$_3$)(NHPh)](OTf)$_2$ with NaN(SiMe$_3$)$_2$

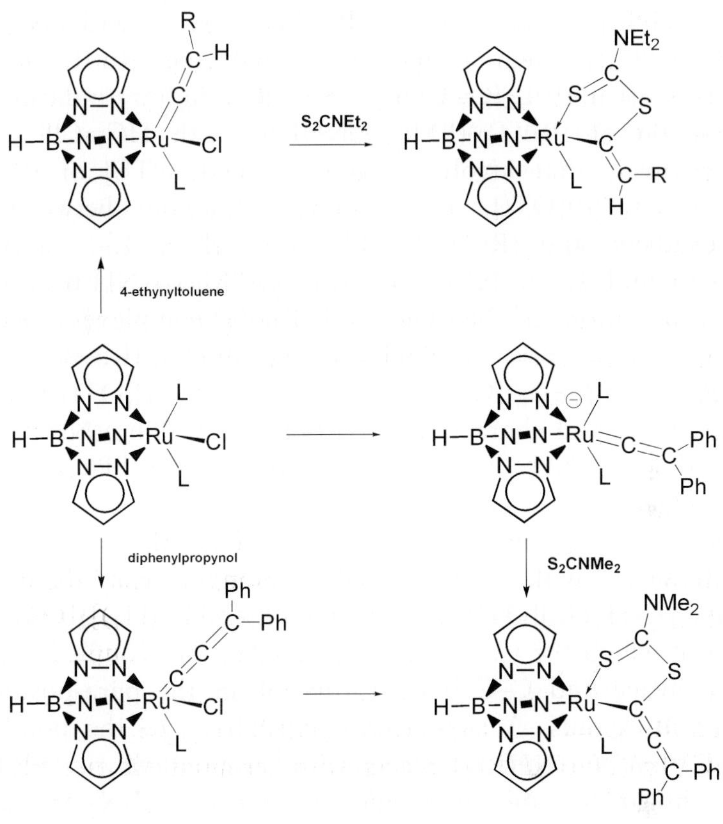

Fig. 2.61. Reactivity of $[RuCl(L)_2(Tp)]$ (L = PPh_3) toward 4-ethynyltoluene and diphenylpropynol.

and HBF_4 at -78°C generates a thermally unstable d^4 imido complex $[Ru^{IV}(Tp)(CO)(PPh_3)(NHPh)](OTf)$.[298]

$[Ru(Tp)(L)(L')(NH_2R)][OTf]$ ($L = L' = PMe_3$, $P(OMe)_3$ or L = CO and $L' = PPh_3$; R = H or ^{t}Bu) have been synthesized and characterized, their deprotonation yielding $[Ru(Tp)(L)(L')(NH_2)]$ and $[Ru(Tp)(PMe_3)_2(NH^tBu)]$, which reacts with HC≡CPh forming ruthenium-amine/phenylacetylide ion pairs. Heating a benzene solution of the $[Ru(Tp)(PMe_3)_2(NH_2^tBu)][PhC_2]$ ion pair, results in the formation of the Ru(II) phenylacetylide complex $[Ru(Tp)(PMe_3)_2(C{\equiv}CPh)]$. The phenyl amido complexes [Ru-

(Tp)L_2(NHPh)] (L = PMe_3 or $P(OMe)_3$) react with 10 equiv. of HC≡CPh at high temperature producing Ru(II) acetylide complexes [Ru(Tp)L_2(C≡CPh)]. Kinetic studies indicate that the reaction of [Ru(Tp)$(PMe_3)_2$(NHPh)] with HC≡CPh occurs via a pathway that involves [Ru(Tp)$(PMe_3)_2$(OTf)] or [Ru(Tp)-$(PMe_3)_2$(NH_2Ph)][OTf] as catalysts. Reaction between 1,4-cyclohexadiene and [Ru(Tp)(L)(L′)(NH_2)] (L = L′ = PMe_3 or L = CO and L′ = PPh_3) or [Ru(Tp)$(PMe_3)_2$(NH^tBu)] results in the formation of benzene and Ru–H complexes. [Ru(Tp)-$(PMe_3)_2$(H)], [Ru=C=C(H)Ph(Tp)$(PMe_3)_2$][OTf], [Ru=C(CH_2Ph)-{N(H)Ph}(Tp)$(PMe_3)_2$][OTf], and [Ru(Tp)$(PMe_3)_3$][OTf] have also been prepared and characterized. [Ru(Tp)(CO)(PPh_3)(NH_3)][OTf], [Ru(Tp)$(PMe_3)_2$(NH_3)][OTf], and [Ru(Tp)(CO)(PPh_3)(C≡CPh)] have been structurally characterized.[299]

From the reaction of [Ru(Tp)L_2(NHPh)] (L = CO, PMe_3 or $P(OMe)_3$) with AgOTf, both mononuclear diamagnetic [Ru(Tp)L_2(NH_2Ph)][OTf] and [Ru(Tp)L_2(NHC_6H_4)]$_2$[OTf]$_2$ are formed in which the two ruthenium moieties are coupled via C–H bond cleavage and C–C bond formation at the para-position of the anilido ligands. A significant contribution to the bonding is derived from a resonance structure corresponding to the Ru(II) centers linked by a diimine ligand. A single-crystal X-ray study of [Ru(Tp){$P(OMe)_3$}$_2$NH(C_6H_4)]$_2$[OTf]$_2$ has also been carried out. The coupled products are formed via initial single-electron oxidation followed by C–C bond formation. The existence of two geometrical isomers around the rigid HN–C_6H_4–C_6H_4–NH bridges for the aryl-coupled complexes has been suggested by VT NMR spectra.[300]

[Ru(Tp)(L)(L′)(NH_2)] (L/L′ = neutral two-electron donor ligands such as PMe_3, $P(OMe)_3$, PPh_3, CO) and [Ru(Tp)(PPh_3)$(NH_2)_2$]Li have been reported and their reactivity investigated. Also, [Ru-(Tp)(L)(L′)(NH_2)] deprotonates HC≡CPh at room temperature giving [Ru(Tp)(L)(L′)(NH_3)][PhC_2] ion pairs.[301] [Ru(Tp)L_2(NH_2-Ph)][OTf] and [Ru(Tp)L_2(NHPh)] (L = PMe_3 or $P(OMe)_3$), including solid-state X-ray diffraction studies of [Ru(Tp)-$(PMe_3)_2$(NH_2Ph)][OTf], [Ru(Tp)$(P(OMe)_3)_2$(NH_2Ph)][OTf], and [Ru(Tp)$(P(OMe)_3)_2$(NHPh)], have been reported by the same authors who also evaluated the pK_a's of some [Ru(Tp)] complexes and the impact of the filled d_π-manifold on barrier to the rotation of the Ru-NHPh moieties.[302]

[Ru(Tp)(L)(L′)(ArN_2)](BF_4)$_2$ (Ar = C_6H_5, 4-$CH_3C_6H_4$; L = $P(OEt)_3$ or $PPh(OEt)_2$, L′ = PPh_3; L = L′ = $P(OEt)_3$) have been synthesized by the reaction of [Ru(Tp)(η^2-H_2)-(L)(L′)]$^+$ with aryldiazonium cations. Hydrazine complexes [Ru(Tp)($RNHNH_2$)-(L)(L′)]BPh_4 (R = H, CH_3, C_6H_5, 4-$NO_2C_6H_4$; L = $P(OEt)_3$ or $PPh(OEt)_2$, L′ = PPh_3; L = L′ = $P(OEt)_3$) were also obtained by the reaction of the [Ru(Tp)(η^2-H_2)LL′]$^+$ cation with an excess of hydrazine. The X-ray crystal structure determination of [Ru(Tp)(CH_3NHNH_2){$P(OEt)_3$}(PPh_3)]BPh_4 is reported. [Ru(Tp)-(ArN=NH)(L)(L′)]BPh_4 derivatives (Ar = C_6H_5, 4-$CH_3C_6H_4$) can be prepared following different procedures such as reaction of [RuH(Tp)(L)(L′)] with aryldiazonium cations in CH_2Cl_2, reaction of aryldiazenido [Ru(Tp)(L)(L′)(ArN_2)](BF_4)$_2$ with $LiBHEt_3$ in CH_2Cl_2, and oxidation of [Ru(Tp)($ArNHNH_2$)(L)(L′)]BPh_4 complexes with $Pb(OAc)_4$ in CH_2Cl_2 at −30°C.[303]

Reacting [Ru(=C=CHR)(Cl)(Tp)(PPh_3)] with excess $HBF_4 \cdot Et_2O$ yields [Ru(≡CCH_2R)(Cl)(Tp)(PPh_3)][BF_4] (R = ^{t}Bu, R = ^{n}Bu, R = Ph). [Ru(=C=CHPh)(Tp)(OTf)(PPh_3)], is formed upon reaction of [Ru(=C=CHR)(Cl)(Tp)(PPh_3)] with MeOTf (Fig. 2.62).[304]

[RuCl(Tp)(dippe)] and [RuCl(Tp)(PEt_3)$_2$] react with HN=CPh_2 and $NaBPh_4$ yielding [Ru(Tp)(HN=CPh_2)(dippe)]BPh_4 and [Ru(Tp)-(HN=CPh_2)(PEt_3)$_2$]BPh_4 (Fig. 2.63), respectively. The former compound has also been structurally characterized. The reaction of

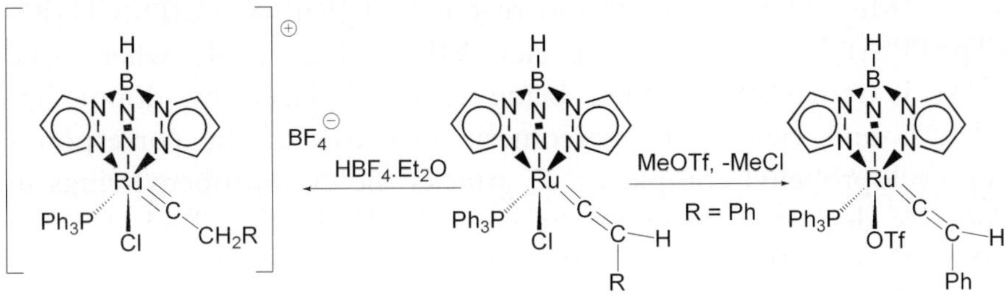

Fig. 2.62. Reaction of [Ru(=C=CHR)(Cl)(Tp)(PPh_3)] with HBF_4 and MeOTf.

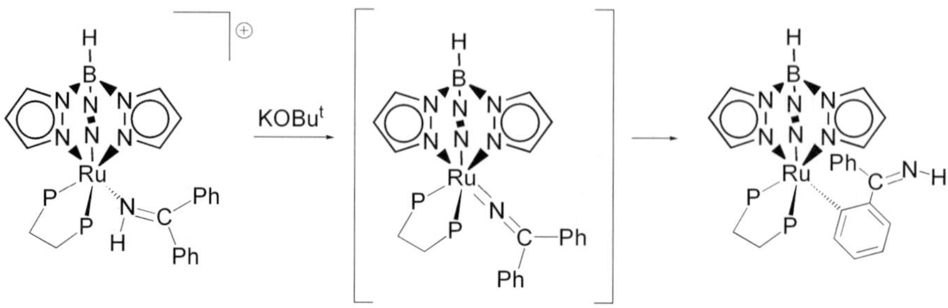

Fig. 2.63. Synthesis of [Ru(Tp)(σ-η^1-$C_6H_4C(Ph)$=NH)(dippe)] (the isopropyl groups of dippe are omitted for clarity).

both imine species with K(OBu^t) produces [Ru(Tp)(σ-η^1-C_6H_4-C(Ph)=NH)(dippe)] and [Ru(Tp)(σ-η^1-C_6H_5)$(PEt_3)_2$], respectively. The reaction of [RuCl(Tp)(PMe^{i}Pr$_2$)(MeCN)] with HN=CPh_2 yields [RuCl(Tp)(PMe^{i}Pr$_2$)(HN=CPh_2)] that is able to coordinate a second molecule of imine upon chloride dissociation forming the cationic [Ru(Tp)(PMe^{i}Pr$_2$)(HN=CPh_2)$_2$]$^+$.[305]

[Ru=C=C=CPhR(Tp)L_2]SbF_6 (L_2 = 2PPh_3 or dppf, R = Ph or ferrocenyl) have been synthesized and their spectroscopic and electrochemical features are reported. [RuCl(Tp)(dppf)] and [Ru=C=C=CPh_2(Tp)(dppf)]SbF_6 were also characterized by X-ray crystallography.[306]

[$RuCl_2(Hpz)_2(PPh_3)_2$] reacts with Na(Tp) and H–C≡C–Ph yielding [Ru(C≡C–Ph)(Tp)$(PPh_3)_2$], structurally characterized. Electrophilic addition of organic halides to the latter species gave a number of new cationic vinylidene complexes [Ru{=C=C(Ph)-CH_2R}(Tp)$(PPh_3)_2$]$^+$ (R = CN, HC=CH_2; CH=C$(CH_3)_2$, Ph, C(O)OMe). The deprotonation reaction of [Ru{=C=C(Ph)CH_2R}-(Tp)$(PPh_3)_2$]$^+$ yields the species **VII** in Fig. 2.64, where one PPh_3 is remarkably labile, being easily replaced by donor ligands L and yielding diastereomeric mixtures of the corresponding cyclopropenyl complexes. I_2 attacks the cyclopropenyl rings in the CN derivative forming the species **VIII** in Fig. 2.64. Finally, [Ru{=C=C(Ph)CH_2R}(Tp)$(PPh_3)_2$]$^+$ reacts with ^{n}BuNC in the presence of MeOH giving [Ru{C(OMe)=C(Ph)CH_2CN}(Tp)(PPh_3)-(^{n}BuNC)].[307]

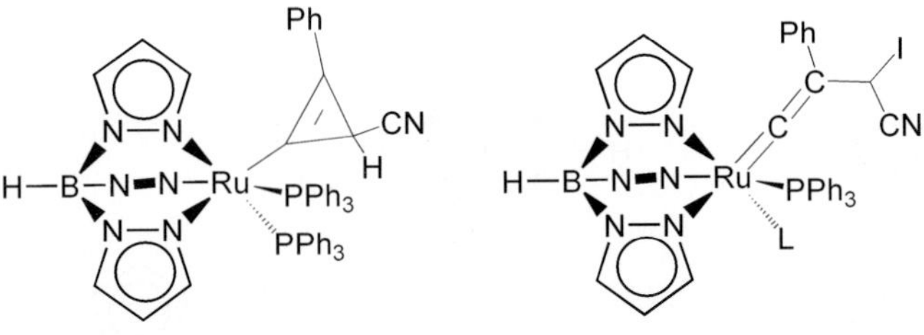

VII VIII

Fig. 2.64. I_2 attacks the cyclopropenyl complex VII, yielding the vinylidene species VIII.

Nelson and Wilson have synthesized, characterized (IR, ^1H, ^{13}C, ^{31}P NMR, and X-ray), and also investigated the reactivity of the complexes [Ru(Tp)(DPVP)(MeCN)Cl], [Ru(Tp)$(DPVP)_2$Cl], [Ru(Tp)(DPVP)(DMPP)[4+2]Cl], [Ru(Tp)$(DPVP)_2$(MeCN)]PF_6, [Ru(Tp)$(DMPP)_2$[4+2]Cl], and [Ru(Tp)(dppe)(Me_2CO)]PF_6 (Fig. 2.65).[308]

The unusual [Ru(κ^3-(*C*, *C*)-1-(2,4,6-Me_3Ph)-3-(4,6-Me_2Ph-2-methylidene)-4,5-dihydroimidazol-2-ylidene)(Cl)(κ^3-(*N*, *N*, *N*)-{(Cl)-Tp})], structurally characterized, has been obtained from the reaction of [Ru(*SPY*-5-34)-dichloro-(κ^2-(*C*,*O*)-2-formylbenzylidene)-(H_2IMes)] (H_2IMes=1,3-bis(2,4,6-Me_3Ph)-4,5-dihydroimidazol-2-ylidene) with K(Tp). The double C–H activation of a methyl group of the H_2IMes resulted in the formation of a chelating N-heterocyclic biscarbene ligand (Fig. 2.66) and release of the former 2-formylbenzylidene as 2-methylbenzaldehyde. The reaction of K(Tp) with (*SPY*-5-34)-dichloro(κ^2(*C*,*O*)-2-ethoxycarbonylbenzylidene)-(H_2IMes)Ru, gave the expected product [Ru(H_2IMes)(2-ethoxycarbonylbenzylidene)Cl(κ^3-Tp)], also characterized by single-crystal X-ray structure determination. Studies performed on the relative activities of both species in model ring opening metathesis polymerizations exhibit a pronounced thermal latency.[309]

[RuX(Tp)(COD)] complexes (X = Br, I, CN) have been prepared by the reaction of [RuCl(Tp)(COD)] with KX, whereas $[Ru(Tp)(COD)(py)]^+$, $[Ru(Tp)(COD)(DMSO)]^+$, and $[Ru(Tp)(COD)(MeCN)]^+$ are formed when [RuCl(Tp)(COD)] reacts with an equimolar quantity of Ag(OTf) in the presence of

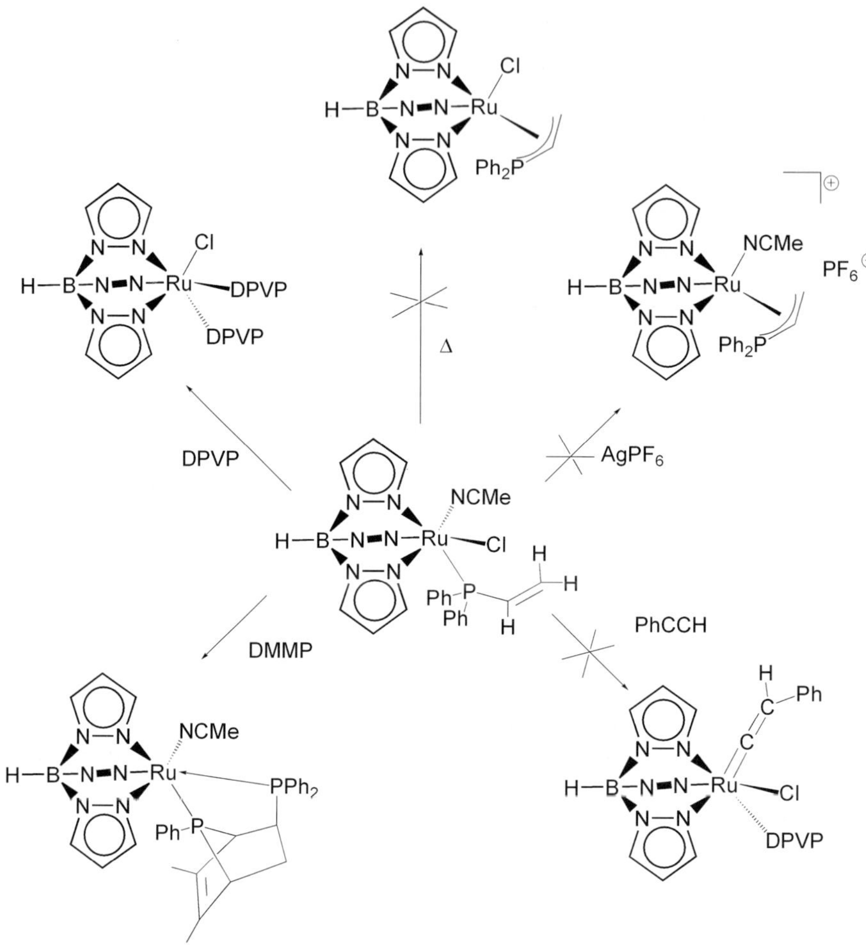

Fig. 2.65. Reactivity of [Ru(Tp)(DPVP)(MeCN)Cl].

a corresponding monodentate ligand. All compounds have been spectroscopically and structurally characterized, the ^{15}N NMR spectroscopic data revealing a linear correlation of ^{15}N chemical shifts with Ru–N (*trans* to X(L)) bond distances.[310]

K(quin) is able to substitute Tp in [Ru(Tp)(COD)Cl] yielding [Ru(COD)(quin)$_2$]. Treatment of [Ru(quin)Cl(η^6-*p*-MeC$_6$H$_4{}^i$Pr)] with K(Tp) gave [Ru(quin)(κ^1-*N*-Tp)(η^6-*p*-MeC$_6$H$_4{}^i$Pr)] containing

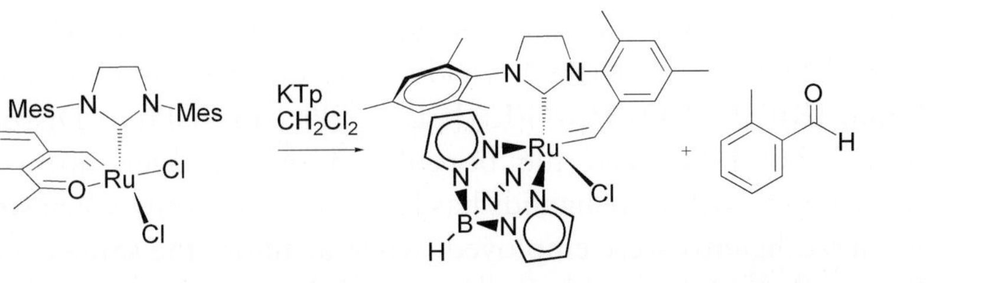

Fig. 2.66. Formation of the N-heterocyclic biscarbene upon double C–H activation of a methyl group of the H_2IMes.

a κ^1 coordinated Tp ligand (Fig. 2.67). The same reaction, carried out in the presence of Ag(OTf), yields [Ru(η^6-p-$MeC_6H_4{}^iPr$)(quin)(κ^1-Hpz)]OTf.[311]

[RuCl(Tp)(COD)] reacts with $RC_6H_4SSC_6H_4R$ (R = H, Me) in DMF yielding [Ru(Tp)(κ^2-(*S*, *S*)-$SC_6H_3RSC_6H_4R$)CO], an unusual product which involves decomposition of the solvent DMF and acts as a HCl scavenger.[312] The complexes [RuCl(Tp^x)$(PhCN)_2$] (Tp^x = Tp, pzTp) have been prepared from *trans*-[$RuCl_2(PhCN)_4$] and K(Tp^x). Its reaction with K(Tp*) gave [Ru(pzTp)(pz*)$(Hpz^*)_2$]. Upon protonation with $HBF_4(Et_2O)$, [Ru(pzTp)(pz*)$(Hpz^*)_2$] was converted to the fairly stable [Ru(pzTp)$(Hpz^*)_3$]BF_4 which has been employed as a halogeno-anion receptor. [Ru(pzTp)(pz*)$(Hpz^*)_2$], [Ru(pzTp)$(Hpz^*)_3$]$BF_4\cdot CHCl_3$, [Ru(pzTp)$(Hpz^*)_2(OH_2)$]$O_3SC_6H_4$-$CH_3\cdot CH_3OH$, and [RuCl(pzTp)$(Hpz^*)_2$]$\cdot CHCl_3$ were structurally characterized.[313] [RuCl(Tp)(DMF)(tri-(*N*-pirrolyl)phosphine)] has been structurally characterized, the Ru atom being in an octahedral coordination, by a tridentate Tp, a monodentate O-donor

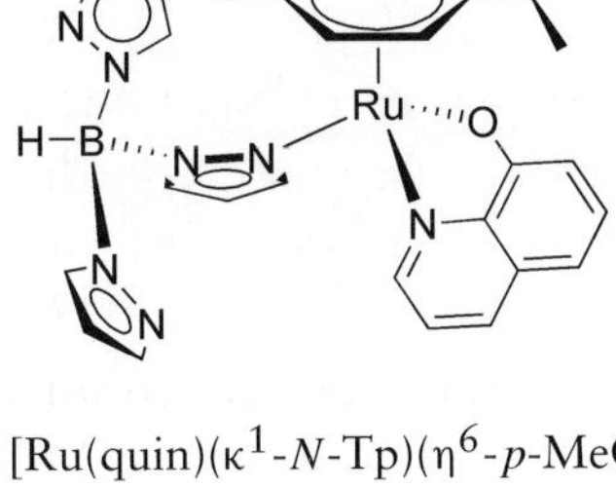

Fig. 2.67. [Ru(quin)(κ^1-*N*-Tp)(η^6-*p*-$MeC_6H_4{}^iPr$)].

DMF, and a monodentate phosphine.[314] [Ru(CO)(L)$_3$HCl] reacts with K(Tp*) and K(Bp*) yielding, upon breaking of the B–N bond, [RuHCl(CO)(Hpz*)(L)$_2$] (L = PPh$_3$ or AsPh$_3$). Dissociation of the bond between the boron and the nitrogen atom of the poly(pyrazolyl)borate ligands has been observed, only when sterically hindered ligands were employed, whereas under the same conditions Bp and Bp^{4Br} form stable Ru^{II} species.[315]

The interaction of [Ru(OTf)(Tp)(PMe$_3$)$_2$] with CsOH·H$_2$O gave [Ru(OH)(Tp)(PMe$_3$)$_2$] that, upon heating in C$_6$D$_6$, yielded [Ru(OD)(Tp)(PMe$_3$)$_2$] through H/D exchange at the hydroxide ligand.[316] When [Ru(OTf)(Tp)(PMe$_3$)$_2$], H$_2$O, and C$_6$D$_6$ were stirred at 100°C the catalytic H/D exchange produces C$_6$D$_x$H$_y$. A mechanism for this H/D exchange (Fig. 2.68) has been hypothesized, which proceeds through PMe$_3$ ligand dissociation and reversible addition of a benzene C–D bond across the Ru–OH bond to likely form the not-identified [Ru(Tp)(PMe$_3$)$_2$(OHD)] species.[316] At high temperatures (90°C–130°C), [RuX(Tp)(PMe$_3$)$_2$] (X = OH, OPh, Me, Ph, or NHPh) undergoes regioselective hydrogen–deuterium exchange with deuterated arenes. For X = OH, OPh, Me, Ph, or NHPh, isotopic exchange occurs at the Tp 4-positions with only minimal deuterium

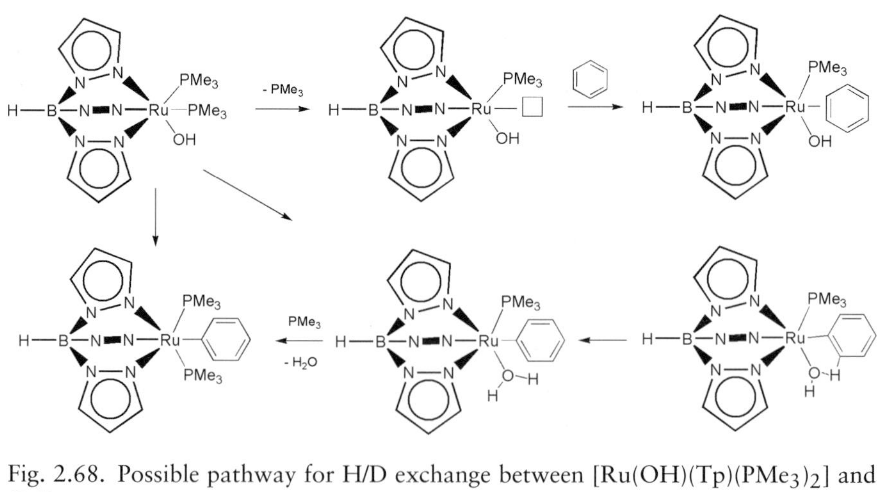

Fig. 2.68. Possible pathway for H/D exchange between [Ru(OH)(Tp)(PMe$_3$)$_2$] and C$_6$D$_6$.

incorporation at the Tp 3- or 5-positions, whereas in the case of [RuPh(Tp)(PMe_3)(NCMe)], the H/D exchange occurs at 60°C at all three Tp positions and the phenyl ring. Finally, [RuCl(Tp)(PMe_3)$_2$], [Ru(OTf)(Tp)(PMe_3)$_2$], and [Ru(Tp)SH(PMe_3)$_2$] do not undergo H/D exchange in C_6D_6 even after long reaction times at very high temperatures.[317]

[RuH(Tp)(PPh_3)(CH_3CN)] is able to catalyze the H/D exchange between CH_4 and some deuterated organic solvents (benzene-d_6, tetrahydrofuran-d_8, diethyl ether-d_{10}, and dioxane-d_8). Theoretical study indicates that the σ-complexes [Ru(Tp)(PPh_3)(η^2-HR)] are active species in the exchange processes, where the reversible transformations of [RuH(Tp)(PPh_3)(η^2-HR)] to [RuH(Tp)(PPh_3)-(η^2-H_2)R] is the most important step. The seven-coordinate [Ru(R)(H)(H)(Tp)(PPh_3)], derived from the oxidative addition of H-R to the metal center, is described as the transition state. The exchange processes involve transformations of the (η^2-H-R) species to the (η^2-H_2) species followed by H–H rotation in the latter. [RuH(Tp)(PPh_3)(CH_3CN)] also catalyzes H/D exchange between H_2 and the deuterated solvents.[318]

[Ru(OR)(Tp)(PMe_3)$_2$] (R = H or Ph) reacts with excess phenylacetylene giving [Ru(C≡CPh)(Tp)(PMe_3)$_2$], the reaction likely proceeding through a pathway that involves [Ru(OTf)(Tp)(PMe_3)$_2$] as catalyst. From the interaction of [Ru(OH)(Tp)(PMe_3)$_2$] with 1,4-cyclohexadiene, [Ru(H)(Tp)(PMe_3)$_2$] and benzene have been obtained (Fig. 2.69). The paramagnetic [Ru(OH)(Tp)(PMe_3)$_2$][OTf] is produced upon single-electron oxidation of [Ru(OH)(Tp)(PMe_3)$_2$] with Ag(OTf).[319]

[Ru(Tp)(CO)(NCMe)(Ph)], upon interaction with carbodiimides, gives amidinate complexes [Ru(Tp)(CO){*N*, *N*-(Ph)NC(Ph)N(Ph)}], whereas its interaction with *N*-methylacetamide yields [Ru(Tp)-(CO){(*N*,*O*-OC(Me)N(Me)}]. [Ru(Tp)(CO)(NCMe)(Ph)] also reacts with C≡N(^{t}Bu) forming [Ru(Tp)(Ph)(CO){C≡N(^{t}Bu)}], that upon heating in the presence of PMe_3 is in equilibrium with [Ru(Tp)-(CO){C(Ph)=N^{t}Bu}(PMe_3)]. On the other hand, the reaction of [Ru(Tp)(Ph)(NCMe)CO] with aromatic aldehydes at high temperatures results in C–H activation and decarbonylation of the aldehyde

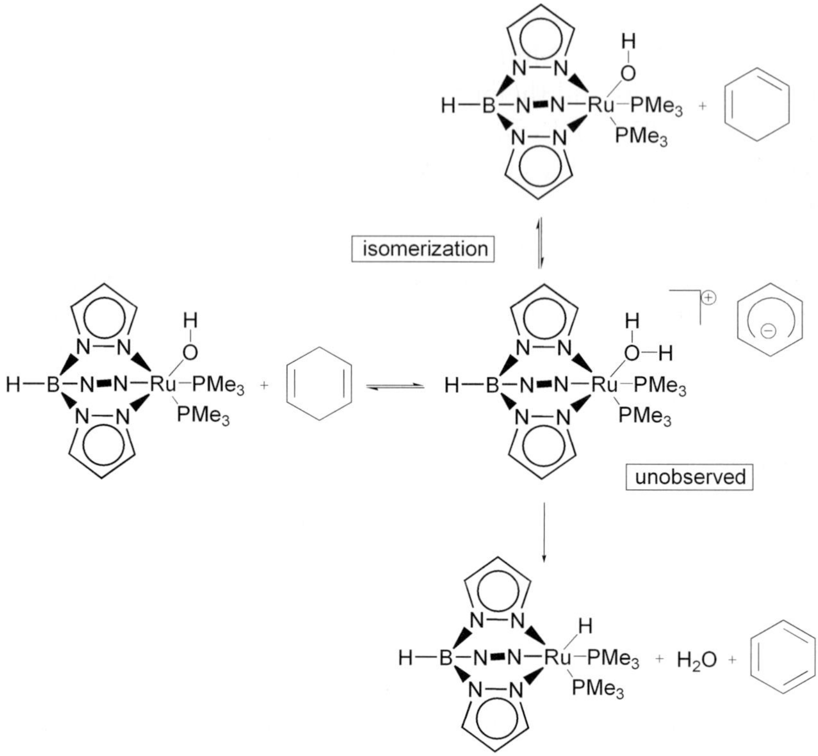

Fig. 2.69. The reaction of $[Ru(OH)(Tp)(PMe_3)_2]$ with 1,4-cyclohexadiene.

to form $[Ru(Ar)(Tp)(CO)_2]$ (Ar = phenyl or *p*-tolyl) and free benzene. DFT calculations have been reported for investigating coordination and insertion reactions of ethylene, methyleneimine, formaldehyde, HN=C=NH, and C≡NH into the Ru–Ph bond of the model fragment {Ru(Tab)(Ph)(CO)} (Tab = tris(azo)borate).[320]

Albertin *et al.*[321] reported that $[RuCl(Tp)L(PPh_3)]$ (L = $P(OEt)_3$, $PPh(OEt)_2$), and $[RuCl(Tp)\{P(OEt)_3\}_2]$ can be prepared by reacting $[RuCl(Tp)(PPh_3)_2]$ with an excess of phosphite. These complexes react with $NaBH_4$ in ethanol giving $[RuH(Tp)L(PPh_3)]$ and $[RuH(Tp)\{P(OEt)_3\}_2]$ that, upon protonation with Brønsted acids, led to thermally unstable dihydrogen species, $[Ru(\eta^2\text{-}H_2)(Tp)L(PPh_3)]^+$ and $[Ru(\eta^2\text{-}H_2)(Tp)\{P(OEt)_3\}_2]^+$. $[Ru(Tp)(H_2O)L(PPh_3)]BPh_4$,

[Ru(Tp)(CO)L(PPh_3)]BPh_4, and [Ru(Tp)(CH_3CN)L(PPh_3)]BPh_4 (L = $P(OEt)_3$) were prepared upon substitution of H_2 by the P-donor. Vinylidene [Ru(Tp){=C=C(H)R}L(PPh_3)]BPh_4 (R = Ph, ^{t}Bu) and allenylidene [Ru(Tp)(=C=C=CR^1R^2)L(PPh_3)]BPh_4 derivatives ($R^1 = R^2$ = Ph; R^1 = Ph; R^2 = Me) were also prepared by reacting dihydrogen complexes with the corresponding alkynes. [Ru(C≡CR)(Tp)L(PPh_3)] were prepared by the deprotonation of the vinylidene [Ru(Tp){=C=C(H)R}L(PPh_3)]BPh_4 with NEt_3. [RuCl(Tp)L(PPh_3)] reacts with $SnCl_2{\cdot}2H_2O$ yielding [Ru($SnCl_3$)-(Tp)L(PPh_3)].[321]

High-pressure NMR monitoring of the catalytic hydrogenation of CO_2 to formic acid, by employing the solvento metal hydride species [RuH(Tp)(PPh_3)(CH_3CN)] as catalyst, shows that CO_2 is readily inserted into Ru-H to form the metal formate [Ru(η^1-OCHO)(Tp)(PPh_3)(MeCN)]·H_2O, in which the formate ligand is intermolecularly hydrogen-bonded to a water molecule. A second formato species [Ru(η^1-OCHO)(Tp)(PPh_3)(H_2O)], containing a coordinated H_2O, exists in solution. Both species are not within the major catalytic cycle of the reaction. The key species in the cycle is [RuH(Tp)(PPh_3)(H_2O)], which could be generated by a ligand displacement reaction of [RuH(Tp)(PPh_3)(MeCN)] with H_2O. [RuH(Tp)(PPh_3)(H_2O)] is able to transfer a proton and a hydride simultaneously to CO_2 to yield formic acid in a concerted manner, itself being converted to a transient hydroxo species, which then associates a H_2 molecule. [RuH(Tp)(PPh_3)(H_2O)] is regenerated via σ-metathesis between the hydroxo and η^2-H_2 ligands. Theoretical calculations have been carried out to study the structural and energetic aspects of species involved in this catalytic cycle (Fig. 2.70).[322,323]

A promoting effect of alcohols in the catalytic hydrogenation of CO_2 to formic acid catalyzed by [Ru(Tp)$(PPh_3)_2$(H_2O)]BF_4 has been observed by Yin *et al.*[324] NMR monitoring indicates two formate species as observable intermediates: [Ru(Tp)$(PPh_3)_2$(η^1-OCHO)]·MeOH and [Ru(Tp)(PPh_3)(MeOH)(η^1-OCHO)]. It has been proposed that the transient hydride intermediate [Ru(Tp)(PPh_3)(ROH)(H)] is able to transfer a hydride and a proton simultaneously to an approaching CO_2 molecule.

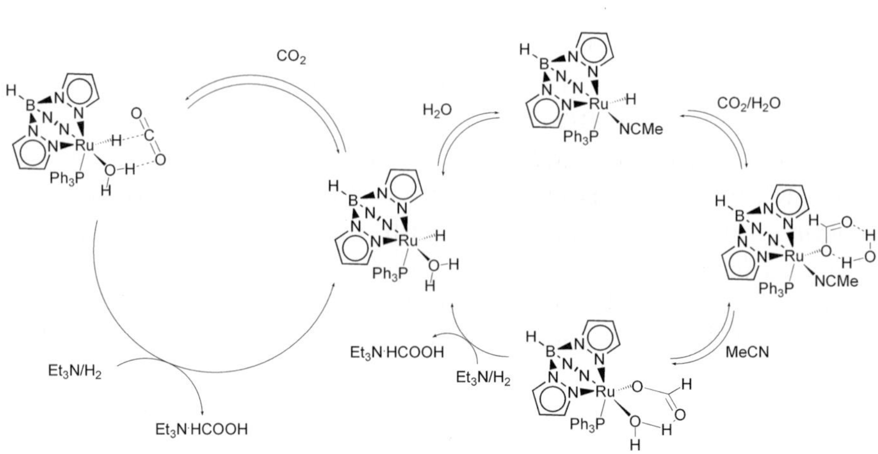

Fig. 2.70. Catalytic cycle proposed for the hydrogenation of CO_2 to HCOOH by employing [RuH(Tp)(PPh_3)(MeCN)] as catalyst.

The hydride complex [Ru(H)(Tp)(MeCN)(PPh_3)] reacts with $HSiR_3$ yielding [Ru(Tp)(PPh_3)(H)(η^2-$HSiR_3$)] ($R_3 = Et_3$, $(EtO)_3$, Ph_3, HEt_2, HPh_2, H_2Ph) in which the two hydrogen atoms are equivalents due to rapid fluxionality between the groups bonded to Ru. Molecular orbital calculations have been performed in order to assess stereochemistry and interconversion processes. [Ru(Tp)(PPh_3)(H)(η^2-$HSiR_3$)] reacts reversibly with pressurized H_2, MeCN, and PPh_3 yielding [Ru(Tp)(H_2)(H)(PPh_3)], [Ru(Tp)(MeCN)(H)(PPh_3)], and [Ru(Tp)$(PPh_3)_2$(H)], respectively.[325]

[Ru(Tp)(Ph)(NCMe)CO] reacts with the electron-rich olefins, ethyl vinyl sulfide and 2,3-dihydrofuran, yielding [Ru(Tp)(CO)(μ-SEt)]$_2$ and [Ru(Tp)(CO)(NCMe)(C≡CCH_2CH_2OH)] through a transformation that involves stoichiometric C–S and C–H/C–O bond cleavage, respectively.[326] [Ru(Tp)(Me)(CO)(NCMe)] reacts with pyrrole forming [Ru(Tp)(CO){κ^2-*N*,*N*-(H)N=C(Me)(NC_4H_3)}]. Mechanistic studies indicate that the most likely reaction pathway involves metal-mediated N–H activation of pyrrole to form [Ru(Tp)(CO)(*N*-pyrrolyl)(NCMe)], followed by C–C bond formation and proton transfer (Fig. 2.71).[327]

The monoacetylide complex [RuCl(C≡CR)(Tp)(NO)] (R = Ph, *p*-$CH_3C_6H_4$, ^{t}Bu, CH_2CH_2OH, CH_2OH, C$(Me)_2$OH, C$(Ph)_2$OH)

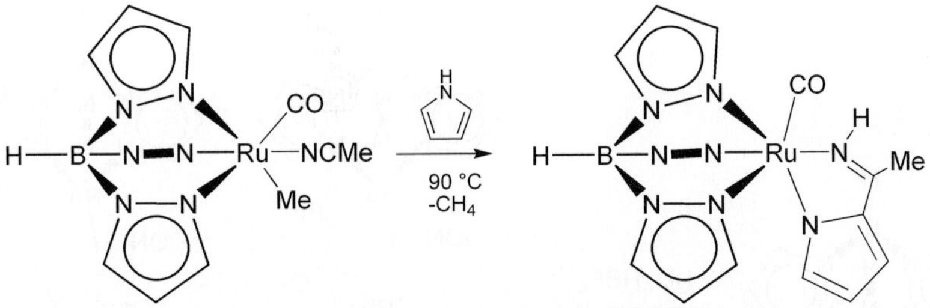

Fig. 2.71. The reaction of [Ru(Tp)(Me)(CO)(NCMe)] with pyrrole yields [Ru(Tp)(CO){κ^2-*N*,*N*-(H)N=C(Me)(NC_4H_3)}].

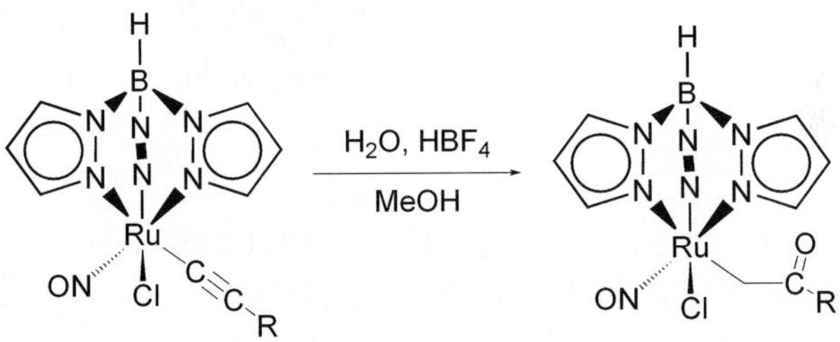

Fig. 2.72. [RuCl(C≡CR)(Tp)(NO)] reacts with HBF_4 yielding [RuCl(CH_2C(O)R)-(Tp)(NO)].

interacts with $HBF_4 \cdot Et_2O$ in methanol giving [RuCl(CH_2-C(O)R)(Tp)(NO)] (R = Ph, *p*-$CH_3C_6H_4$, ^{t}Bu, CH_2CH_2OH, CH_2OH) from the reaction of η^2-coordinated of 1-alkyne species with adventitious water (Fig. 2.72). [RuCl(C(O)CH=CR'_2)(Tp)(NO)] (R′ = Me, Ph) has been obtained by hydration via an allenylidene species.[328]

[Ru(C≡CPh)$_2$(Tp)(NO)] reacts with 1 equiv. of H_2O in the presence of $HBF_4 \cdot Et_2O$ in distilled MeOH giving the metallacycle complex [Ru{CH=C(Ph)C(O)CH(Ph)}(Tp)(NO)] and the bis(ketonyl) complex [Ru(CH_2C(O)Ph)$_2$(Tp)(NO)]. The former exists as a mixture of two diastereomers with a varying abundance ratio. The *p*-Tol derivative [Ru{CH= C(C_6H_4Me)-C(O)CH(C_6H_4Me)}(Tp)(NO)] was also isolated by similar methods. [Ru(C(O)CH_2Ph)(CH_2C(O)Ph)(Tp)(NO)] and the regioisomer of [Ru(CH_2C(O)Ph)$_2$(Tp)(NO)] were obtained in THF.

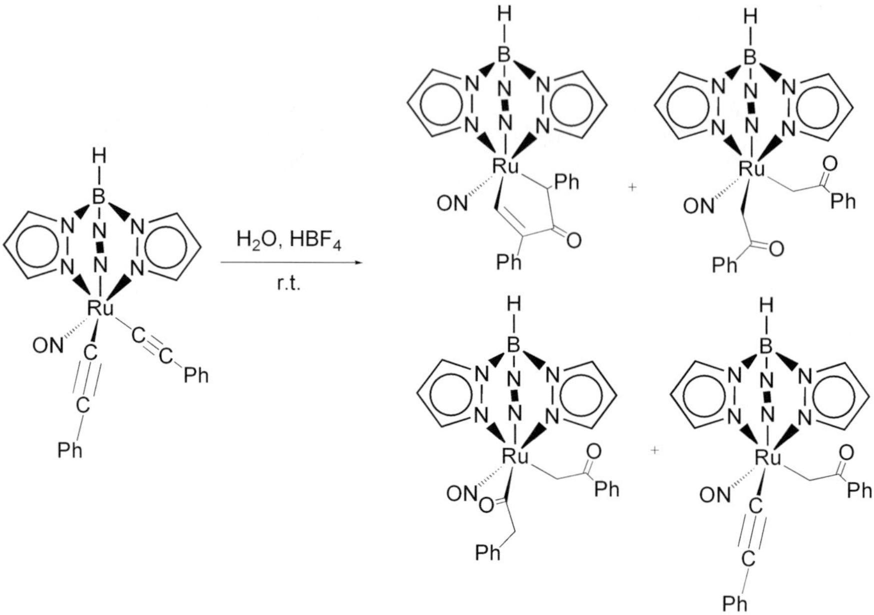

Fig. 2.73. The four products isolated from the reaction of $[Ru(C{\equiv}CPh)_2(Tp)(NO)]$ with H_2O in the presence of $HBF_4{\cdot}Et_2O$ in MeOH or THF as solvent.

The $[Ru(C{\equiv}CPh)_2(Tp)(NO)]/HBF_4{\cdot}Et_2O/H_2O$ reaction system in MeOH at 0°C also produces $[Ru(C{\equiv}CPh)(CH_2C(O)Ph)(Tp)(NO)]$ (Fig. 2.73).[329]

The reaction of $[RuCl(C{\equiv}CPh)(Tp)(NO)]$ with PPh_3 in the presence of $HBF_4{\cdot}Et_2O$ gives (*E*)-$[RuCl\{CH{=}C(PPh_3)Ph\}(Tp)(NO)]BF_4$, whereas $[RuCl\{C{\equiv}CC(R)_2OH\}(Tp)(NO)]$ (R = Me, Ph, H) upon similar treatments with PPh_3 produces the γ-phosphonio-alkynyl $[RuCl\{C{\equiv}CC(Me)_2PPh_3\}(Tp)(NO)]BF_4$, the α-phosphonio-allenyl $[RuCl\{C(PPh_3){=}C{=}CPh_2\}(Tp)(NO)]BF_4$, and the γ-hydroxy-β-phosphonio-alkenyl (*E*)-$[RuCl\{CH{=}C(PPh_3)CH_2OH\}(Tp)(NO)]$-$BF_4$, respectively. From $[Ru(C{\equiv}CPh)_2(Tp)(NO)]$, (*E, E*)-$[Ru\{CH{=}C(PPh_3)Ph\}_2(Tp)(NO)](BF_4)_2$ was produced at room temperature. When the same reaction was carried out at 0°C, the alkynyl β-phosphonio-alkenyl complex (*E*)-$[Ru(C{\equiv}CPh)\{CH{=}C(PPh_3)Ph\}(Tp)(NO)]BF_4$ has been isolated (Fig. 2.74) that,

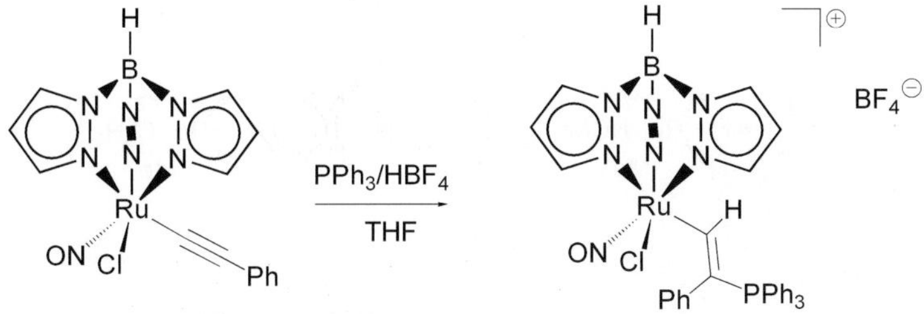

Fig. 2.74. Synthesis of the alkynyl β-phosphonio-alkenyl complex (*E*)-$[Ru(C{\equiv}CPh)\{CH{=}C(PPh_3)Ph\}(Tp)(NO)]BF_4$.

upon hydration in the presence of $HBF_4{\cdot}Et_2O$, forms (*E*)-$[Ru\{CH_2C(O)Ph\}\{CH{=}C(PPh_3)Ph\}(Tp)(NO)]BF_4$.[330]

$[Ru(C{\equiv}CSiMe_3)Cl(Tp)(NO)]$ reacts with HBF_4 or HCl in MeOH yielding a mixture of $[Ru(C{\equiv}CH)(Cl)(Tp)(NO)]$ and $[Ru\{C(O)CH_3\}(Cl)(Tp)(NO)]$, whereas the reaction of $[Ru(C{\equiv}CSiMe_3)(Cl)(Tp)(NO)]$ with HCl in CH_2Cl_2 gave the η^1-α-chlorovinyl $[Ru\{C(Cl){=}CH_2\}(Cl)(Tp)(NO)]$. $[Ru(C{\equiv}CSiMe_3)_2(Tp)(NO)]$ shows the same reactivity, its proton-assisted hydration affording $[Ru\{C(O)CH_3\}_2(Tp)(NO)]$ while the HCl-treatment led to the formation of $[Ru\{C(Cl){=}CH_2\}_2(Tp)(NO)]$.[331]

$[Ru(C{\equiv}NR)(Ph)(Tp)(PMe_3)]$ (R = ^{t}Bu, CH_2Ph, CH_2CH_2Ph) complexes have been investigated by DFT studies in order to probe the energetics of the catalytic intra- and intermolecular hydroarylation of isonitriles.[332]

[Ru(Me)(Tp)(CO)(NCMe)] also promotes C–H bond activation at the 2-position of furan and thiophene yielding methane and [Ru(Ar)(Tp)(CO)(NCMe)] (Ar = 2-furyl or 2-thienyl). [Ru(2-thienyl)(Tp)(CO)(NCMe)] and $[Ru(\mu\text{-}C,S\text{-thienyl})(Tp)(CO)]_2$ were characterized structurally. [Ru(2-furyl)(Tp)(CO)(NCMe)] also catalyzes the formation of 2-ethylfuran from ethylene and furan. DFT calculations have been performed on the C–H activation of furan by {Ru(Me)(Tab)(CO)} (Tab = tris(azo)borate).[333]

[Ru(Me)(Tp)(CO)(NCMe)] reacts with arenes to initiate C–H activation and loss of methane via initial arene/acetonitrile ligand exchange. This reaction produces the active catalyst [Ru(Ph)-(Tp)(CO)(NCMe)] (Fig. 2.75). Subsequent C–H activation of another

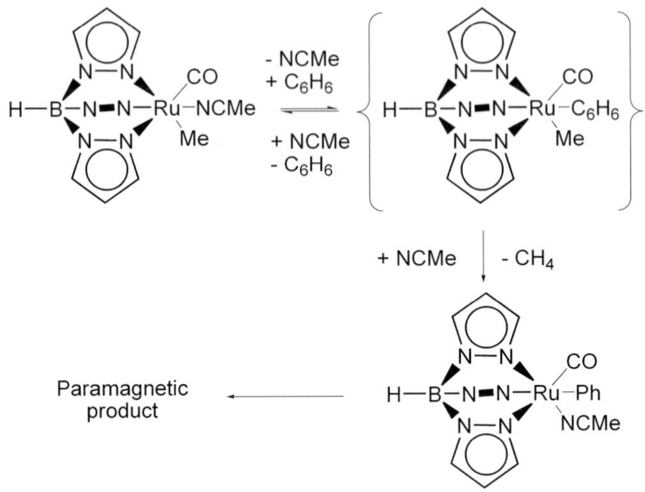

Fig. 2.75. Formation of the active catalyst [Ru(Ph)(Tp)(CO)(NCMe)].

equiv. of arene allows elimination of alkylbenzene and re-formation of the catalyst.[334]

[Ru(Tp)(Ph)(CO)(NCMe)] is able to catalyze the hydroarylation reactions of olefins. [Ru(Ph)(Tp)(CO)(η^2-olefin)] has been hypothesized as the species that precedes the insertion step, in which the presence of the strong π-acid CO likely results in a preferred olefin orientation, the C=C bond being parallel to the Ru–phenyl bond.[335]

The impact of ancillary ligand nature on the C–H activation bonds, including catalytic hydroarylation and hydrovinylation/ oligomerization of ethylene, has been investigated by using complexes of the type [RuR(Tp)(L)(NCMe)] (R = Ph or Me, L = CO or PMe_3). It has been reported that [RuPh(Tp)(PMe_3)(NCMe)] and [RuPh(Tp)(CO)(NCMe)] get separately converted to [Ru(Tp)(L)(η^3-C_4H_7)] (L = CO, PMe_3) via initial Ru-mediated ethylene C–H activation. Heating mesitylene solutions of [Ru(Tp)(L)(η^3-C_4H_7)] under ethylene pressure results in the catalytic formation of butenes and hexenes.[336]

Methylation of [Ru(Tp)$(CO)_2$(THF)][PF_6] gave [Ru(Tp)$(CO)_2$-(Me)], that upon reaction with Me_3NO in MeCN produces

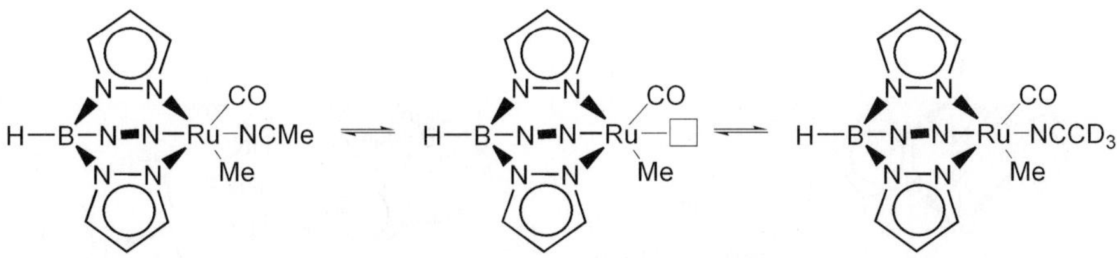

Fig. 2.76. Exchange of MeCN with CD_3CN in [Ru(Me)(Tp)(CO)(NCMe)].

[Ru(Me)(Tp)(CO)(NCMe)], in which the bound acetonitrile undergoes exchange with CD_3CN yielding [Ru(Tp)(Me)(CO)($NCCD_3$)], the rates of the exchange reactions being independent of CD_3CN concentration (Fig. 2.76).[337]

Lau *et al.*[338] described a catalytic H/D exchange between organic compounds and D_2O by using [Ru(H)(Tp)(PPh_3)(MeCN)]. It has been found that [Ru(H)(Tp)(PPh_3)(MeCN)] reacts with H_2O giving the acetamido species [Ru(Tp)(H_2O)(PPh_3)(NHC(O)CH_3)]. [Ru(H)(Tp)(PPh_3)(MeCN)] has also been found to promote the catalytic hydration of nitriles, whereas the chloro analog [RuCl(Tp)(PPh_3)(MeCN)] does not show activity.[339]

Single-electron oxidation of [Ru^{II}(Tp)(L)(L′)(R)] (L = CO, L′ = NCMe, R = Me or CH_2CH_2Ph; L = L′ = PMe_3, R = CH_3) with Ag(OTf) gives [Ru(Tp)(L)(L′)(OTf)] through alkyl elimination reactions. The formation of ethylbenzene and styrene in the oxidation of [Ru(Tp)(CH_2CH_2Ph)(CO)(NCMe)] is also consistent with the generation of alkyl radicals.[340]

The air-sensitive [Ru(Cl)(Tp)(PR_3)(DMF)] (PR_3 = PPh_3, $PPh_2{}^iPr$, P^iPr_3, and PCy_3) have been prepared by refluxing a DMF solution of [RuCl(Tp)(COD)] and PR_3. [Ru(Cl)(Tp)(PR_3)(DMF)], on exposure to air in the presence of CCl_4, is converted into [$RuCl_2$(Tp)(PR_3)] that reacts with $NaBH_4$ affording [Ru(Tp)(PR_3)(H)(η^2-H_2)]. Reduction of [$RuCl_2$(Tp)(PR_3)] with Zn in the presence of MeCN and py affords the diamagnetic $[Ru(Tp)(PR_3)(MeCN)_2]^+$ and [RuCl-(Tp)(PR_3)(py)].[341]

[RuCl(Tp)(L)] reacts with $NaPF_6$ yielding the enantiomerically and diastereomerically pure dinitrogen-bridged complexes

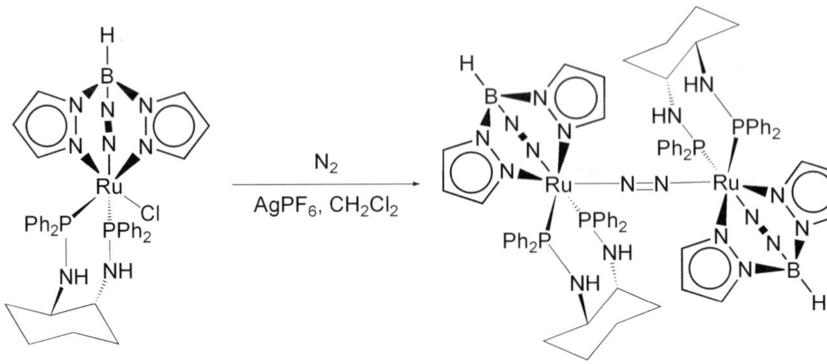

Fig. 2.77. Synthesis of [{Ru(Tp)(1,2-bis(diphenylphosphinoamino)cyclohexane)}$_2$(μ-N_2)][PF_6]$_2$.

[{Ru(Tp)(L)}$_2$(μ-N_2)][PF_6]$_2$ (L = *R*, *R*- or *S*, *S*-1,2-bis(diphenylphosphinoamino)cyclohexane) (Fig. 2.77). [{Ru(Tp)(L)}$_2$(μ-N_2)][PF_6]$_2$ with neat methacrolein furnishes the labile species [Ru(Tp)-(methacrolein)(L)][PF_6].[342]

[Ru(CH=CHCH$_2$PPh$_3$)X(CO)(PPh$_3$)$_2$]$^+$ (X = Cl, Br) reacts with K(Tp) and NaBPh$_4$ giving [Ru(CH=CHCH$_2$PPh$_3$)(Tp)(CO)-(PPh$_3$)$_2$]BPh$_4$, whereas [Ru(CH=CHCH(OEt)$_2$)Cl(CO)(PPh$_3$)$_2$] upon reaction with K(Tp) gives [Ru(CH=CHCH=CHPh)(Tp)(CO)-(PPh$_3$)]. Reaction between [Ru(CH=CHCH$_2$PPh$_3$)(Tp)(CO)(PPh$_3$)$_2$]-BPh$_4$, NaN(SiMe$_3$)$_2$, and benzaldehyde produces [Ru(CH=CHCH=CHPh)(Tp)(CO)(PPh$_3$)]. Analogous Wittig reaction of [Ru(CH=CHCH$_2$PPh$_3$)(Tp)(CO)(PPh$_3$)$_2$]BPh$_4$ with NaN(SiMe$_3$)$_2$ and terephthaldicarboxaldehyde yields the bimetallic complex [Ru-(Tp)(CO)(PPh$_3$)]$_2$(μ-CH=CH-CH=CH-C$_6$H$_4$–CH=CH-CH=CH)] (Fig. 2.78).[343]

The η^2-dihydrogen derivatives [Ru(Tp)(L$_2$)(H$_2$)]$^+$ (L$_2$ = dppm, dppp, or (PPh$_3$)$_2$) react with O_2 affording the paramagnetic Ru^{III}superoxo complexes [Ru^{III}(Tp)(L$_2$)(O$_2$)]$^+$, which do not exhibit antiferromagnetic coupling between the Ru^{III} ion (d^5, S = 1/2) and the coordinated superoxide radical (S = 1/2). In THF, the superoxo moiety of the complex readily abstracts a hydrogen atom from the solvent generating the hydroperoxo (-OOH$^-$) group, which then

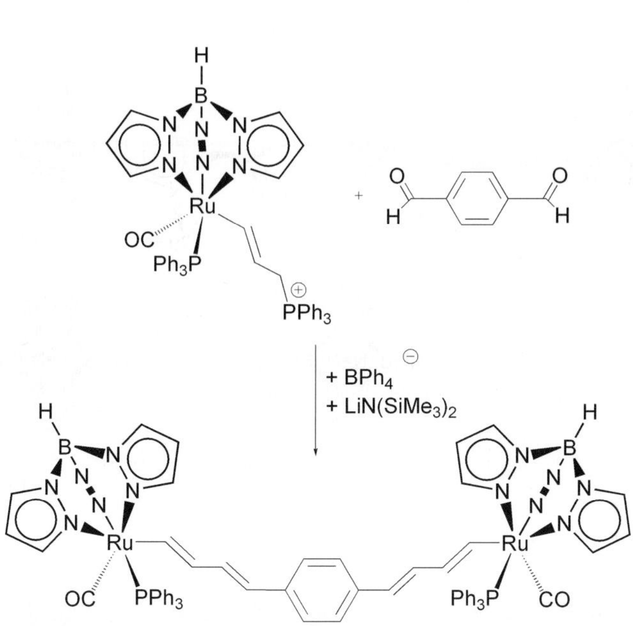

Fig. 2.78. Synthesis of the bimetallic complex $[Ru(Tp)(CO)(PPh_3)]_2(\mu\text{-CH=CH-CH=CH-}C_6H_4\text{-CH=CH-CH=CH})]$.

changes into the hydroxo ligand by transferring an oxygen atom to a phosphine ligand (Fig. 2.79). DFT calculation at the B3LYP level on the model complex $[Ru(Tp)(PH_3)_2(O_2)]^+$ shows that the $[Ru^{III}(O_2^-)]$ superoxo structure with a triplet state is more stable than the $[Ru^{IV}(O_2^{2-})]$ structure with a singlet state.[344]

Reactions of $[Ru_2(RCOO)_4Cl]$ (R = Me, Ph) with K(Tp) afforded the oxo-bridged diruthenium(III) complexes, $[Ru_2(\mu\text{-O})(\mu\text{-RCOO})_2(Tp)_2]$ (R = Me, Ph) that were converted into $[Ru_2(\mu\text{-OH})(\mu\text{-RCOO})_2(Tp)_2](PF_6)$ and in the oxo-bridged mixed-valence diruthenium(III,II) $[Ru_2(\mu\text{-O})(\mu\text{-RCOO})_2(Tp)_2](PF_6)$ by the reaction with $(NH_4)_2Ce(NO_3)_6$. $[Ru_2(\mu\text{-O})(\mu\text{-MeCOO})_2(Tp)_2]$ consists of a $\{Ru_2(\mu\text{-oxo})(\mu\text{-carboxylato})_2\}$ core with two terminal face-capping Tp ligands. When $[Ru_2(\mu\text{-O})(\mu\text{-PhCOO})_2(Tp)_2]$ was reduced with sodium amalgam, $[Ru(PhCOO)(Tp)(CH_3CN)_2]$ was obtained, which upon exposure to air or dioxygen regenerates $[Ru_2(\mu\text{-O})(\mu\text{-PhCOO})_2(Tp)_2]$. This work demonstrates the versatile redox states of diruthenium centers terminally capped by Tp ligands.[345]

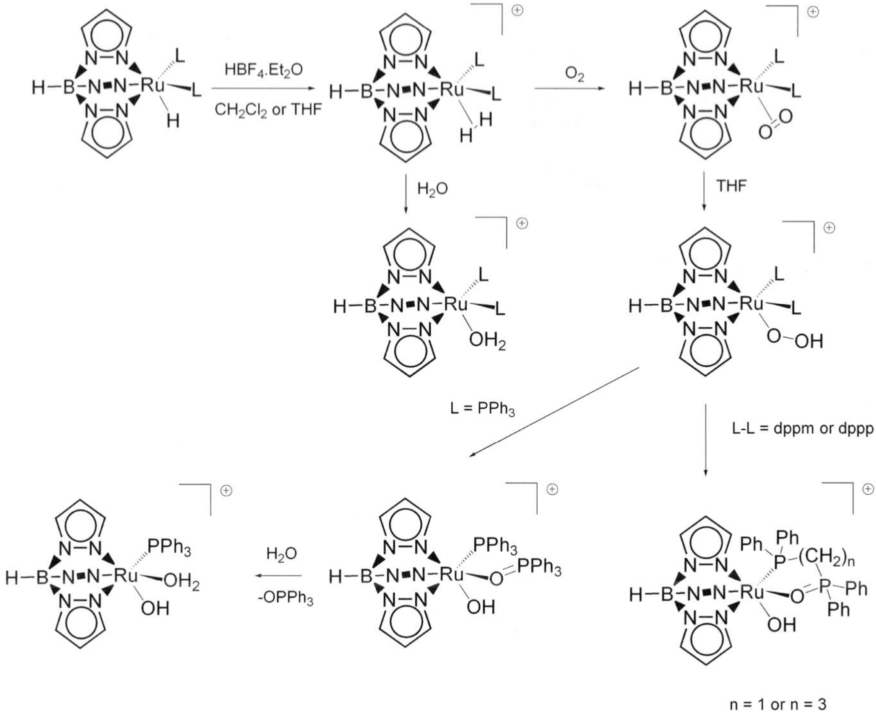

Fig. 2.79. Synthesis and reactivity of $[Ru(Tp)(L_2)(H_2)]^+$.

Synthesis, characterization, and reactivity of Ru(II) complexes of the type *cis*-$[Ru(\kappa^2\text{-pzTp})(L)_2]^{n+}$, (L = α-diimine, e.g. bipy or phen) have been reported. These complexes exhibit strong visible absorption assigned as a $\pi^*(L) \leftarrow d\pi(Ru)$ metal-to-ligand charge-transfer (MLCT) transition characteristic of the *cis*-L_2Ru^{2+} kernel. Electronic spectra and cyclic voltammetry measurements indicate that the relative π-acceptor abilities of the coordinated pzTp are between those of two pyrazolyl and two pyridines in the order: $(1H\text{-pyrazolyl})_2 \sim \text{pzTp} < (\text{pyridine})_2$. Uncoordinated pz groups of *cis*-$[(bipy)_2Ru(\kappa^2\text{-pzTp})]$ can coordinate a second Ru moiety yielding a sterically hindered, localized-valence $[(cis,cis\text{-}(bipy)_2Ru^{II}(\mu\text{-pzTp})Ru^{II}(bipy)_2]^{3+}$ dimer which is photoreactive and undergoes an asymmetrical photocleavage in MeCN, yielding *cis*-$[(bipy)_2Ru^{III}(\kappa^2\text{-pzTp})]^{2+}$and *cis*-$[(bipy)_2Ru^{II}(MeCN)_2]^{2+}$.[346]

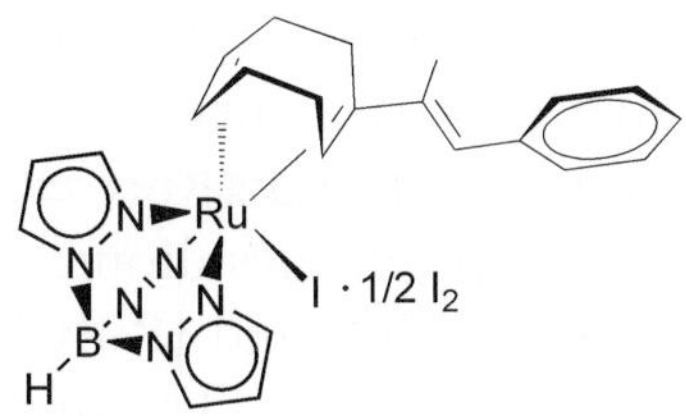

Fig. 2.80. $[Ru(Tp)I(C_{16}H_{17}I)]{\cdot}0.5I_2$.

Synthesis and electrochemistry of $[Ru(\kappa^3\text{-}Tp^x)_2]$ complexes have also been reported and the relative contribution (ΔE(L)) of the corresponding Tp^x to the $E^{-1/2}$ values were evaluated.[347]

$[Fc(CH{=}CH)_3\text{-}RuCl(CO)(PPh_3)_2]$ reacts with K(Tp) forming the six-coordinate heterobimetallic complex $[Fc(CH{=}CH)_3Ru(Tp)\text{-}(CO)(PPh_3)]$.[348]

$[Ru(Tp)I(C_{16}H_{17}I)]{\cdot}0.5I_2$ (Fig. 2.80) has been obtained as a cocrystal of an iodine-bearing neutral ruthenium complex and diiodine. The centrosymmetric diiodine molecule links a pair of Ru complexes via halogen bonds to the Ru-bound iodide anions, thus forming a linear I···I–I···I chain.[349]

The diastereomerically pure [RuX(Tp)(*R*, *R*-dippach)] (X = Cl, H) as well as [RuX(Tp)(dippae)] (X = Cl, H) react with H_2 and $NaBAr_F$ in fluorobenzene yielding the stable $[Ru(H_2)(Tp)(R,R\text{-}dippach)][BAr_F]$ and $[Ru(H_2)(Tp)(dippae)][BAr_F]$ which react readily with N_2 giving $[Ru(Tp)(N_2)(R,R\text{-}dippach)][BAr_F]$ and $[Ru(Tp)(N_2)(dippae)][BAr_F]$. Proton transfer processes have been investigated by NMR study. $[Ru(Tp)(H_2)(R,R\text{-}dippachH)]^{2+}$ and $[Ru(Tp)(H_2)(dippaeH)]^{2+}$ result from the protonation of $[Ru(Tp)(H_2)(R,R\text{-}dippach)][BAr_F]$ and $[Ru(Tp)(H_2)(dippae)][BAr_F]$ at one of the NH groups of the respective phosphinoamine ligands. These species undergo slow tautomerization to their monohydride isomers $[RuH(Tp)(R,R\text{-}dippachH_2)]^{2+}$ and $[RuH(Tp)(dippaeH_2)]^{2+}$.[350]

$[RuCl(CH{=}CH\text{-}CH{=}CH\text{-}C_6H_4\text{-}R\text{-}p)(CO)(PPh_3)_2]$ reacted with K(Tp) to give $[Ru(CH{=}CH\text{-}CH{=}CH\text{-}C_6H_4\text{-}R\text{-}p)(Tp)(CO)(PPh_3)]$, the structure of which has been confirmed by X-ray diffraction. The

NLO properties (hyper-Rayleigh scattering measurements) reveal a Tp ligand role on the hyperpolarizability.[351]

$[Ru\{C{\equiv}CPh_2(OH)\}_2(Tp)NO]$ and $[Ru(C{\equiv}CC_6H_4Me)\{C{\equiv}CPh_2(OH)\}_2(Tp)NO]$ react with *p*-toluenesulfonic acid, giving $[Ru\{C(=C=CPh_2)C(O)C(=CPh_2)\}(Tp)NO]$ and $[Ru\{C(=C=CPh_2)C(O)CH(C_6H_4Me_2)\}(Tp)NO]$, respectively. Treatment of these four-membered metallacycles with HCl gave the ring-opened HCl adducts $[RuCl\{C(=C=CPh_2)C(O)CH=CPh\}(Tp)NO]$ and $[RuCl\{C(=C=CPh_2)C(O)CH_2(C_6H_4Me)\}(Tp)NO]$.[352]

Reaction of $[RuX(Tp)(CO)_2]$ (X = Br, I) with Me_3NO in MeCN afforded [RuX(Tp)(CO)(NCMe)] (X = Br, I). Their reactions with either neutral isocyanides or anionic dialkyldithiocarbamates gave [RuX(Tp)(CO)(CNR)] (X = Br, R = $PhCH_2$; X = Br, R = ^{t}Bu; X = I, R = $PhCH_2$; X = I, R = ^{t}Bu) and $[Ru(Tp)(CO)(\eta^2{-}S_2CNR_2)]$ (R = Me, Et), respectively. From the reaction of [RuX(Tp)(CO)(NCMe)X] with RSH–Et_3N, (*cis*)-$[Ru_2(Tp)_2(CO)_2(\mu\text{-}X)(\mu\text{-}SR)]$ (X = I, R = ^{i}Pr; X = Br, R = ^{t}Bu; X = I, R = ^{t}Bu), (*trans,anti*-1)-$[Ru_2(Tp)_2(CO)_2(\mu\text{-}S^iPr)_2]$, (*cis, syn*)-$[Ru_2(Tp)_2(CO)_2(\mu\text{-}S^iPr)_2]$, (*trans,anti*-1)-$[Ru_2(Tp)_2(CO)_2(\mu\text{-}S^tBu)_2]$, and (*cis,anti*-2)-$[Ru_2(Tp)_2(CO)_2(\mu\text{-}S^iPr)(\mu\text{-}S^tBu)]$ formed. The first diruthenium sulfenate, (*cis*)-$[Ru_2(Tp)_2(CO)_2(\mu\text{-}I)(\mu\text{-}S(O)^iPr)]$ has also been reported.[353]

Reaction of $[RuNO_2(bipy)(MeCN)_3]PF_6$ with Tp gave $[RuNO_2(Tp)(bipy)]$. Its reaction with HPF_6 afforded $[Ru(Tp)(bipy)(NO)][PF_6]_2$. These derivatives, prepared by using ^{15}N-labeled nitro or nitrosyl groups, have been characterized by ^{1}H and ^{15}N NMR and IR spectroscopy, cyclic voltammetry, and single-crystal structure determination for $[Ru(Tp)(bipy)NO][PF_6]_2$.[354]

A nanostructured molecular film containing (μ-hydroxo)bis(μ-carboxylato) diruthenium(III) units, $[Ru_2^{III}(\mu\text{-}OH)(\mu\text{-}CH_3COO)_2(Tp)_2]^+$, has been prepared by an *in situ* conversion of its μ-oxo precursor, $[Ru_2^{III}(\mu\text{-}O)(\mu\text{-}CH_3COO)_2(Tp)_2]$ in a Nafion membrane matrix.[355]

Kirchner *et al.*[356] have compared the isoelectronic $[Ru(Tp)(MeCN)_3]PF_6$ and $[Ru(Cp)(MeCN)_3]PF_6$. The acetonitrile exchange kinetics support the fact that the Tp ligand strongly prefers a six-coordinate environment around the Ru metal, promoting substitution reactions following a dissociative pathway. The large cone angle of Tp has been indicated responsible for this preference. The approximately octahedral $[Ru(Tp)(MeCN)_3]PF_6$ has been investigated by single-crystal X-ray study.[356]

2.9.3. *Os*

The Os chemistry has early been exhaustively investigated by Mayer and Crevier[357,358] and by Meyer *et al.*[359] who have described a number of osmium-nitrido compounds and also mixed valence species as $[(Tp)Os^{III}Cl_2(NN)Cl_2Os^{II}(Tp)]$.[360] Relevant osmium Cp,[361] hydrido and dihydrogen,[362] carbonyl,[363,364] and heterometallic complexes[365] were also described in the first edition. In the last 8 years Mayer *et al.*[366] investigated mainly the redox activity of osmium-nitrido compounds. Atom transfer to the terminal nitrido ligand in $[Os(N)Cl_2(Tp)]$ has been employed to prepare a homologous series of chalconitrosyl complexes $[OsCl_2(Tp)(NO)]$, $[OsCl_2(Tp)(NS)]$, and $[OsCl_2(Tp)(NSe)]$. The selenium derivative has also been structurally characterized. Seleno- and sulfur-nitrosil compounds seem to have similar bonding and reactivity.[366]

$[Os^{IV}XCl_2(Tp)]$ have been prepared by the reaction of $[Os^{IV}(NPPh_3)Cl_2(Tp)]$ with protic acids HX (HOTf, HCl, HBr, CF_3COOH, and CH_3COOH). Upon addition of small amounts of H_2O or HCl to $[Os^{IV}(OTf)Cl_2(Tp)]$, $[Os^{IV}Cl_3(Tp)]$ is formed (Fig. 2.81(a)). The Os^{IV} species demonstrate a strong oxidant power. $[Os^{IV}(OTf)Cl_2(Tp)]$ reacts with PPh_3 or py yielding $[Os^{III}Cl_2(Tp)(L)]$ (L = PPh_3, py). Os^{III} species $[Cp_2^*Fe][Os(OTf)(Cl_2)(Tp)]$ can be prepared by reduction with decamethylferrocene (Cp_2^*Fe). They then undergo substitution forming $[Os^{III}Cl_2(Tp)(L)]$ (L = MeCN, C_6H_5CN, PPh_3, py, imH, and NH_3). Oxidation of $[Os^{III}Cl_2(Tp)(L)]$ with $[NO]BF_4$ in CH_2Cl_2 yielded $[OsCl_2(Tp)(MeCN)]BF_4$ (Fig. 2.81(b)).[367]

When the inert (due to the partial Os–N π-bonding and the oxidizing nature of the Os(IV) center) $[OsCl_2(Tp)(NHPh)]$ is

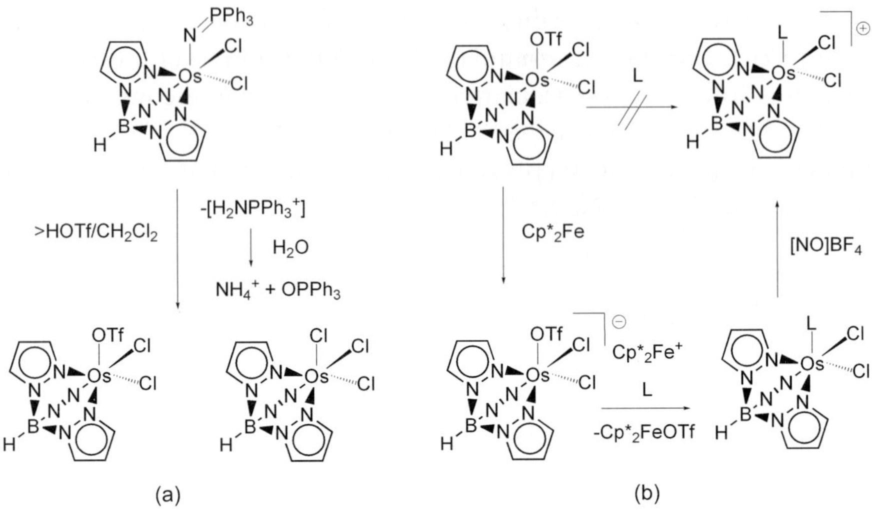

Fig. 2.81. Reactivity of $[Os^{IV}(NPPh_3)Cl_2(Tp)]$ (a) and $[Os^{IV}(OTf)Cl_2(Tp)]$ (b).

protonated by the reaction with triflic acid, $[OsCl_2(Tp)(NH_2Ph)]OTf$ is formed. This complex is extremely acidic and can be deprotonated by stoichiometric addition of weak bases such as Cl^- or H_2O. Upon heating, MeOTf removes the chloride from $[OsCl_2(Tp)(NHPh)]$ to give MeCl and $[Os(OTf)_2(Tp)(NHPh)]$, whereas $[Os\{NPh(MgBr)\}Cl_2(Tp)]$, obtained by the reaction of PhMgBr with $[Os(Cl)_2(Tp)N]$, is extremely basic and can be protonated by weak acids to generate $[OsCl_2(Tp)(NHPh)]$.[368]

$[Os(Cl)_2(Tp)N]$ obtained from $K[OsO_3(N)]$ and KTp in aqueous ethanolic HCl reacts with PhMgCl and analogous reagents with the transfer of a phenyl group to the nitrido ligand. The Os(IV) metalla-analido species have been described, which are readily protonated, giving the analido derivative $[OsCl_2(Tp)(NHPh)]$. $[Os(Cl)_2(Tp)N]$ reacts with the alkyl and arylboranes transferring one organic group to nitrogen and forming isolable borylamido complexes $[Os(Ph)(Tp)(N)(BPh_2)]Cl_2$, upon insertion of a nitrido ligand into a boron–carbon bond. $[Os(Ph)(Tp)(N)(BPh_2)]Cl_2$ undergoes hydrolysis generating $[OsCl_2(Tp)(NHPh)]$.[369]

The electrochemical reduction of $[Os(Cl)_2(Tp)N]$ in MeCN 3.5M in HPF_6 results in the formation of $[Os(Cl)_2(Tp)NH]$, that possess a distorted octahedral arrangement of ligands around the Os center and is not found in aqueous solution as a transient because it is unstable with regard to disproportionation. Addition of water to a MeCN solution of $[Os^{V}Cl_2(Tp)(NH)]$ led to the formation of $[Os^{III}Cl_2(Tp)(NH_3)]$ and $[Os^{VI}Cl_2(Tp)(N)]$ in a 1:2 ratio.[370]

The electrochemical properties of a number of isostructural osmium(VI)-nitrido complexes, $[Os(E)(X)(Y)(Tp^{x})]$ (Tp^{x} = Tp or Tp*, E = O, NTol, N; X, Y = hydrocarbyl, halide triflate) have been described. It has been reported that the reactivity of these complexes as inner-sphere oxidants does not correlate with their peak reduction potentials, whereas the ease of oxidation of these compounds well parallels their reactivity as oxidants.[206]

The pseudo-octahedral $[OsCl_2(Tp)(NH_2Ph)]$ is formed by heating $[Cp^{*}_2Fe][Os^{III}(OTf)Cl_2(Tp)]$ in the presence of aniline or by reduction and protonation of the osmium(IV) anilido complex $[OsCl_2(Tp)(NHPh)]$. $[OsCl_2(Tp)(NH_2Ph)]$, oxidized by O_2, yielded $[OsCl_2(Tp)(NHPh)]$, through a reaction which proceeds readily in the presence of catalytic base by a stepwise mechanism of pre-equilibrium proton transfer (PT) forming $[Os^{III}(Tp)NHPhCl_2]^-$, which then reacts with O_2 by outersphere electron transfer (ET). $[OsCl_2(Tp)(NH_2Ph)]$ is likely air-stable in the presence of acid, because this PT–ET mechanism in this case is prevented (Fig. 2.82).[371]

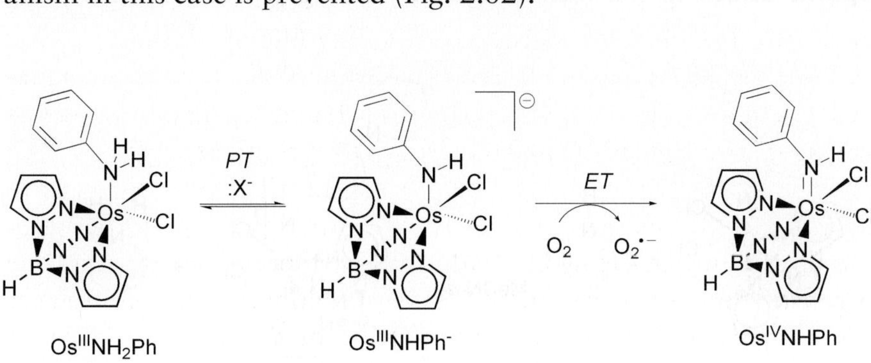

Fig. 2.82. Synthesis of $[OsCl_2(Tp)(NHPh)]$ through a PT–ET mechanism.

When amine nucleophiles such as piperidine and pirrolidine were added to the MeCN solutions of [$OsCl_2$(Tp)(NHPh)], [$OsCl_2$(Tp){NH-*p*-$C_6H_4(N(CH_2)_5)$}] and [$OsCl_2$(Tp){NH-*p*-$C_6H_4(N(CH_2)_4)$}], respectively, were formed, that, under anaerobic conditions, yielded [$OsCl_2$(Tp)(NH_2Ph)] upon oxidative removal of two H^+ and two e^-. The oxidation can be accomplished by [$OsCl_2$(Tp)(NHPh)] yielding [$OsCl_2$(Tp)(NH_2Ph)], or by O_2. *Ab initio* calculations on [$OsCl_2$(Tp)(NHPh)] show that the LUMO has significant density at the ortho and para positions of the anilido ligand. [$OsCl_2$(Tp){NH-*p*-$C_6H_4(N(CH_2)_5)$}] can be quantitatively formed also from piperidine and [$OsCl_2$(Tp){NH-*p*-C_6H_4Cl}] (Fig. 2.83).[372]

[$Os(Cl)_2$(Tp)N] reacts with PAr_3 affording [$OsCl_2$(Tp)($NPAr_3$)] (Ar = Tol, Ph, *p*-$CF_3C_6H_4$). These compounds upon protonation at the phosphiniminato nitrogen gave [$Os^{IV}Cl_2$(Tp)($NHPAr_3$)]OTf. Dichloromethane solutions of [$OsCl_2$(Tp)($NPAr_3$)] react with excess H_2O over 1 week forming the disproportionation products [$Os(Cl)_2$(Tp)N] and [$Os^{III}Cl_2$(Tp)($NHPAr_3$)]. X-ray structures of the phenyl derivatives have been reported.[373]

[$Os(OAc)_2$(Tp)(N)] is formed upon reaction of [$Os(Cl)_2$(Tp)N] with excess Ag(OAc) and its reaction with protic acids HX gives analogous [$Os(X)_2$(Tp)N] (X = Tfa, TCA, TBA, Br, Ox, or ONO_2). These compounds have been investigated by cyclic voltammetry, ^{15}N NMR, IR, and optical spectra. [$Os(X)_2$(Tp)N] reacts with PPh_3 by nucleophilic attack at the nitride ligand, yielding [$Os(X)_2$(Tp)($NPPh_3$)],

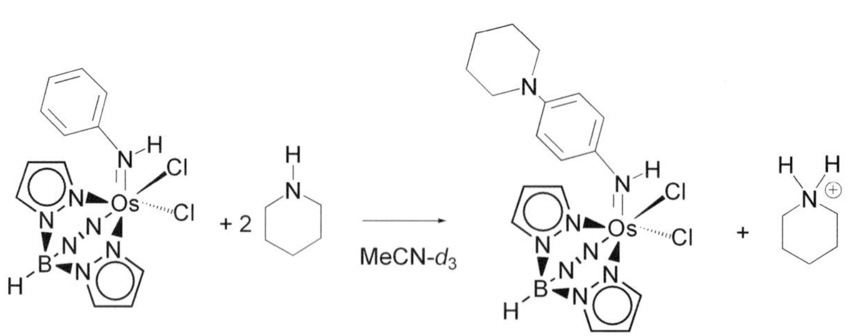

Fig. 2.83. Synthesis of [$OsCl_2$(Tp){NH-*p*-$C_6H_4(N(CH_2)_5)$}].

and with triphenylboron, BPh_3, with the formation of borylanilido complexes $[OsX_2(Tp)\{N(Ph)BPh_2\}]$.[374]

Stirring a benzene solution of $[Os(Cl)_2(Tp)N]$ under NO atmosphere results in the formation of the nitrosyl complex $[Os(Cl)_2$-$(Tp)(NO)]$ in accordance with the mechanism proposed in Fig. 2.84, confirmed by an ^{15}N-labeling experiment.[375]

$[Os(Cl)_2(Tp)(NO)]$ has also been obtained from the reaction of $[Os(Cl)_2(Tp)N]$ with aqueous $(^nBu_4N)(OH)$. This reaction is different with respect to that of $[Os(OAc)_2(Tp)(N)]$ with NaOH in Me_2CO/H_2O which gave $[Os(Tp)(OH)_2(N)]$, structurally characterized. $[Os(OAc)_2(Tp)(N)]$ with $Na^{18}OH$ in $H_2{}^{18}O/CD_3CN$ yields the doubly ^{18}O-labeled $[Os(Tp)(^{18}OH)_2(N)]$ and unlabeled acetate. The reaction of $[Os(OAc)_2(Tp)(N)]$ with the softer nucleophile thiosulfate affords the thionitrosyl complex $[Os(OAc)_2(Tp)(NS)]$.[376]

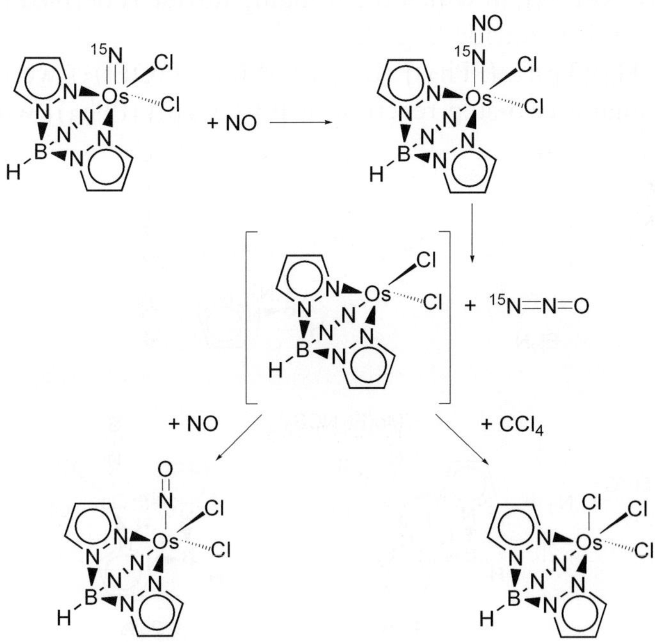

Fig. 2.84. Mechanism proposed for the formation of the nitrosyl complex $[Os(Cl)_2$-$(Tp)(NO)]$.

The reversible $4e^-/3H^+$ interconversion between NH_3 and N^{3-} in the coordination spheres of OsTp complexes has been described by Meyer *et al.*[377] Electrochemical or chemical reduction of the nitrido complex $[Os^{VI}(Cl)_2(Tp)N]$ gives Os^{II}-ammine species that after air-oxidation transforms into $[Os^{III}(Cl)_2(Tp)(NH_3)]$, structurally characterized. The latter species can be reoxidized electrochemically to the starting nitrido complex by a stepwise mechanism involving the loss of both electrons and protons and sequential Os(III → IV) and Os(IV → V) oxidation.

$[Os(Cl)_2(Tp)N]$ reacts with CS_2 and N_3^- in acetone yielding the corresponding thionitrosyl complex $[Os(Cl)_2(Tp)NS]$, SCN, and N_2. The ^{15}N labeled analogous is also reported. The structure of $[Os^{II}(Cl)_2(Tp)(NS)]$ that co-crystallizes with 11% of $[Os^{IV}(Cl)_3)(Tp)]$ is described.[378] $[Os(Cl)_2(Tp)NS]$ is formed in the reaction of $Mo(Et_2NCS_2)_3N$ with the electrophilic $[Os(Cl)_2(Tp)N]$ also. The most abundant product from this reaction is, however, $[OsCl_2(Tp)(\mu\text{-}N)Mo(S_2CNEt_2)_3]$, in which the bridging nitride is derived primarily from the osmium nitride (Fig. 2.85).[379]

$[Os(SnH_3)(Tp)(L)(PPh_3)]$ (L = $P(OMe)_3$, $P(OEt)_3$) were synthesized through a two-step reaction of $[OsCl(Tp)L(PPh_3)]$ with $SnCl_2$

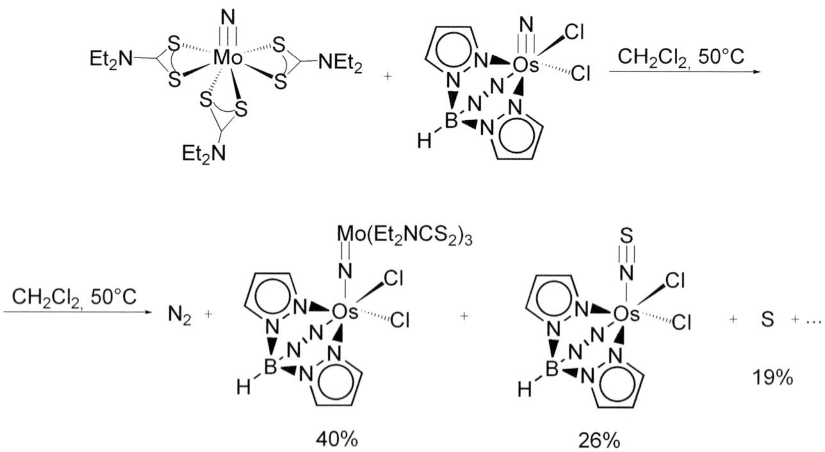

Fig. 2.85. Products isolated in the reaction of $Mo(Et_2NCS_2)_3N$ with the electrophilic $[Os(Cl)_2(Tp)N]$.

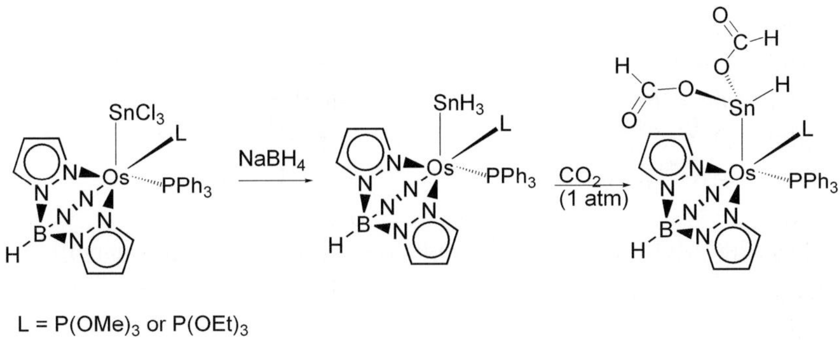

Fig. 2.86. The synthesis of $[Os(SnH_3)(Tp)(L)(PPh_3)]$ and its reaction with CO_2.

and $NaBH_4$ in ethanol. Both complexes were characterized spectroscopically, and $[Os(SnH_3)(Tp)\{P(OMe)_3\}(PPh_3)]$ also by X-ray crystal structure determination. Reaction of tin trihydride complexes with CO_2 yielded $[Os[SnH\{OC(H){=}O\}_2](Tp)L(PPh_3)]$ (Fig. 2.86).[380]

$[OsCl(Tp)L(PPh_3)]$ [L = $P(OMe)_3$, $P(OEt)_3$, and $PPh(OEt)_2$] were prepared by the reaction of $[OsCl(Tp)(PPh_3)_2]$ with an excess of phosphite. $[OsCl(Tp)L(PPh_3)]$ reacts with $NaBH_4$ in EtOH affording the hydride $[OsH(Tp)L(PPh_3)]$. The dihydrogen species $[Os(\eta^2\text{-}H_2)(Tp)L(PPh_3)]BPh_4$ were prepared by the protonation of the monohydrides with $HBF_4{\cdot}Et_2O$. Treatment of the hydride $[OsH(Tp)\{P(OEt)_3\}(PPh_3)]$ with $[4\text{-}CH_3C_6H_4N_2]BF_4$ gave the aryldiazene $[Os(4\text{-}CH_3C_6H_4N{=}NH)(Tp)\{P(OEt)_3\}(PPh_3)]BPh_4$. Instead, $[Os(4\text{-}CH_3C_6H_4N_2)(Tp)\{P(OEt)_3\}(PPh_3)](BF_4)_2$ was prepared by reacting the hydride $[OsH(Tp)\{P(OEt)_3\}(PPh_3)]$ first with MeOTf and then with $[4\text{-}CH_3C_6H_4N_2]BF_4$. Finally, the imine $[Os\{\eta^1\text{-}NH{=}C(H)Ar\}(Tp)\{P(OEt)_3\}(PPh_3)]BPh_4$ (Ar = C_6H_5, $4\text{-}CH_3C_6H_4$) were synthesized by reaction of $[OsH(Tp)\{P(OEt)_3\}(PPh_3)]$ with MeOTf and alkylazides.[381]

The polymer $[OsBr_2(COD)]_x$ reacts with K(Tp) in ethanol to afford the hydride [Os(H)(Tp)(COD)] and the bromide complex [Os(Br)(Tp)(COD)]. The latter reacts with sodium methoxide to afford [Os(OMe)(Tp)(COD)]. When [Os(H)(Tp)(COD)] reacts with MeOTf, loss of methane and formation of [Os(OTf)-(Tp)(COD)] are observed, that upon treatment with $MgMe_2$

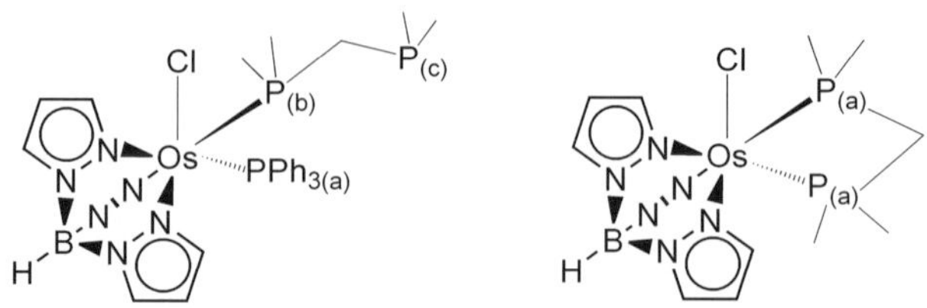

Fig. 2.87. [Os(Cl)(Tp)(κ^1-dmpm)(PPh_3)] and [Os(Cl)(Tp)(κ^2-dmpm)].

gave [Os(Me)(Tp)(COD)]. Finally, the addition of dmpm to [OsCl(Tp)$(PPh_3)_2$] yields a mixture of [Os(Cl)(Tp)(κ^1-dmpm)(PPh_3)] and [Os(Cl)(Tp)(κ^2-dmpm)] (Fig. 2.87); the latter reacts with MeLi to form [OsMe(Tp)(κ^2-dmpm)].[382]

[$OsH_3Cl(P^iPr_3)_2$] reacts with K(Tp) in THF to give [OsH_3(κ^2-Tp)$(P^iPr_3)_2$]. In toluene at 80°C, [OsH_3(κ^2-Tp)$(P^iPr_3)_2$] is transformed to [OsH_3(κ^3-Tp)$(P^iPr_3)_2$] in quantitative yield. Protonation of the latter with HBF_4 affords [Os(Tp)$(\eta^2\text{-}H_2)_2$(P^iPr_3)]BF_4, which in acetone releases the coordinated H_2 in a sequential manner. At room temperature, [Os(Tp)(η^2-H_2)(κ^1-$OCMe_2$)(P^iPr_3)]BF_4 is formed, while at 56°C the loss of both hydrogen molecules gives rise to the bis-solvento derivative [Os(Tp)$(\kappa^1\text{-}OCMe_2)_2$($P^iPr_3$)]$BF_4$.[383]

2.10. Group 9: Co, Rh, and Ir

2.10.1. *Co*

Cobalt complexes containing "first generation" homoscorpionates are not very common. Most of the early works described homoleptic $[Co(Tp^x)_2]^{n+}$[384,385] and heteroleptic $[Co(Tp^x)L]^{n+}$ (L = tacn,[386] dien,[386] SC_6F_5,[387] $n = 0$, 1) complexes.

[Co(Cp)(κ^3-Tp)], the first example of an organometallic cobalt complex which adopts a maximum spin configuration in preference to a potential low-spin alternative, has been prepared by the reduction of $[Co(Cp)(Tp)]^+I^-$ with $Co(Cp)_2$. [Co(Cp*)(Tp)], having a low-spin electronic configuration, has been oxidized to [Co(Cp*)(Tp)][PF_6] by reacting with [$FeCp_2$][PF_6].[388] Synthesis, structure,[29] and electronic and magnetic properties of [Co(Cp^R)(Tp)] and [Co(Cp^R)(Tp)]$^+$ have been reported.[30]

DFT studies have been used to predict geometries, energies, and spin states of a series of inorganic cobalt complexes containing Tp ligand. [Co(Tp)(CO)], a stable 16-electron compound with an unusual asymmetric geometry has been identified as a structural and electronic analog of an unstable intermediate in C–H bond activation. DFT studies on [(CoTp)$_2$O$_2$] provided a guide to the feasibility of proposed oxidation mechanisms.[389]

Electrochemical, magnetic, and spectroscopic properties are reported for the divalent and trivalent homoleptic derivatives [Co(Tp)$_2$]$^{+/0}$, [Co(Tp*)$_2$]$^{+/0}$, and [Co(pzTp)$_2$]$^{+/0}$.[47]

When the dinuclear [{M^{II}(Tp^x)}$_2$(μ-OH)$_2$] (Tp^x = Tp*, Tp^{*Br}, Tp^{*Me}, M = Ni or Co) reacts with 1 equiv. of H_2O_2, the corresponding [{M^{III}(Tp^x)}$_2$(μ-O)$_2$] was formed. The employment of a series of Tp$^{Me_2,4\text{-}X}$ (X = Me, H, Br) as a metal-supporting ligand makes it possible to isolate and structurally characterize the thermally unstable [{M^{III}(Tp^R)}$_2$(μ-O)$_2$]. Both [{M^{II}(Tp^R)}$_2$(μ-OH)$_2$] and [{M^{III}(Tp^R)}$_2$(μ-O)$_2$] contain five-coordinate metal centers with a slightly distorted square-pyramidal geometry. Thermal decomposition of the Ni^{III} and Co^{III} species results in the oxidation of the inner saturated hydrocarbyl substituents of the Tp^R ligand. Large k_H/k_D values obtained from the first-order decomposition rates of the Tp^{Me_3} and Tp$^{(CD_3)_2,Me}$ derivatives of Ni^{III} evidently indicate that the rate-determining step is a hydrogen abstraction from the primary C–H bond of the methyl substituents, mediated by the M^{III}_2(μ-O)$_2$ species. The Ni^{III} complex shows reactivity about 10^3 times greater than that of the cobalt analog. The oxidation ability of the M^{III}(μ-O)$_2$M^{III} core should be affected by the hindered Tp^R ligand system, which can stabilize the +2 oxidation state of the metal centers.[390]

The six-coordinated octahedral [Co(Tp*)(NCS)$_2$(Hpz*)]·H_2O·CH_3OH shows the hydrogen atoms of the water molecule connected by H-bond with S atoms of two adjacent molecules to form 1D chains.[391]

[Co(Tp*)(N_3)] and [Co(Tp*)(NCS)(MeOH)$_2$] were synthesized and characterized, the structure of the latter also being reported.

The dehydration kinetic measurements of HCO_3^-, catalyzed by these cobalt(II) complexes, were also investigated.[392]

2.10.2. *Rh*

Also, if the first Rh complex containing a homoscorpionate ligand has been prepared in 1969,[393] the Rh(I)Tp^x chemistry continues to attract many researchers, likely for its increasing application in C–H bond activation. [Rh(Tp^x)(diolefin)],[394,395] [Rh(Tp^x)(L)$_2$][396–399] (L = CO or PR$_3$), and Rh(III) alkyl complexes of general structure [Rh(Cl)-(CNCH$_2{}^t$Bu)(R)(Tp^x)][400,401] are the most important classes of compounds investigated until 1999.

[Rh(Tp*)(PPh$_3$)$_2$] and [Rh(Tp*)(P(4-C$_6$H$_4$F)$_3$)$_2$] react with HC≡CPh, 4-NO$_2$-C$_6$H$_4$CHO, and Ph$_3$SnH yielding [Rh(H)(C≡CPh)-(Tp*)(PR$_3$)], [Rh(H)(4-NO$_2$-C$_6$H$_4$CO)(Tp*)(PR$_3$)], and [Rh(H)-(Ph$_3$Sn)(Tp*)(PR$_3$)], respectively. [Rh(Tp*)(PPh$_3$)$_2$], structurally characterized, contains the ligand in the bidentate form, whereas Tp* acts as a tridentate in [Rh(H)(4-NO$_2$-C$_6$H$_4$CO)(Tp*)(PPh$_3$)], [Rh(H)(Ph$_3$Sn)(Tp*)(PPh$_3$)], and in [Rh(H)(C≡CPh)(Tp*)(P(4-C$_6$H$_4$-F)$_3$)] (Fig. 2.88).[402]

[Rh(Tp*)(PPh$_3$)$_2$] also reacts with dichloromethane giving [RhCl-(H)$_2$(PPh$_3$)$_2$(Hpz*)] and [RhCl$_2$(H)(PPh$_3$)$_2$(Hpz*)], whereas reaction with C$_6$F$_5$SH yields [(PPh$_3$)$_2$Rh(μ-SC$_6$F$_5$)$_2$Rh(SC$_6$F$_5$)(H)(PPh$_3$)-(pz*)]. Finally, the reaction of [Rh(Tp*)(PPh$_3$)$_2$] with HgCl$_2$ gave [{Rh(Cl)$_2$(PPh$_3$)$_2$}$_2$Hg].[403]

A study of the decomposition process of [Rh(Tp*)(PPh$_3$)$_2$] in 1,2-dichloroethane has been performed. [Rh(Tp*)(PPh$_3$)$_2$] undergoes facile, irreversible orthometalation of a PPh$_3$ ligand at room temperature in slightly polar solvents, such as THF and 1,2-dichloroethane, but not in PhCH$_3$. The structure of the orthometalated compounds has also been reported. [Rh(Tp*)(PPh$_3$)$_2$] shows catalytic activity for hydrothiolation and hydrophosphinylation in low-polarity solvents. [Rh(H)(COPh)(Tp*)(PPh$_3$)] and [Rh(H)(C≡CCO$_2$Et)(Tp*)(PPh$_3$)], formed by the oxidative addition of benzaldehyde and ethylpropiolate to [Rh(Tp*)(PPh$_3$)$_2$], respectively, upon loss of 1 equiv. of PPh$_3$ also yield the orthometalated compound (Fig. 2.89).[404]

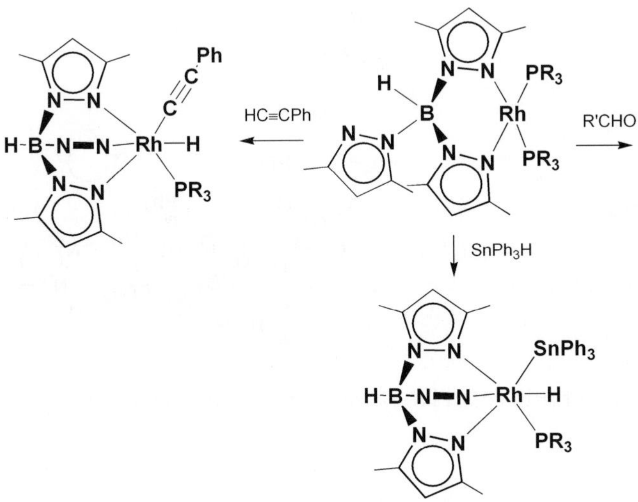

Fig. 2.88. The reaction of $[Rh(Tp^*)(PPh_3)_2]$ with HC≡CPh, R′CHO, and Ph_3SnH.

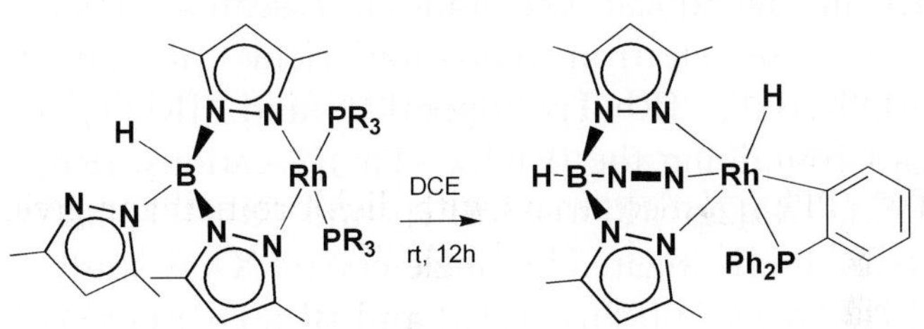

Fig. 2.89. Orthometalation of a PPh_3 in $[Rh(Tp^*)(PPh_3)_2]$ in DCE.

$[Rh(Tp^*)(PPh_3)_2]$ and $[Rh(Tp^*)(COD)]$ are active catalysts for alkyne hydrophosphinylation, although neither is as active as the Wilkinson's catalyst. $[Rh(PPh_3)_2(Tp^*)]$ is also able to effect hydrophosphinylation of a series of alkyne substrates with diphenylphosphine oxide. $[Rh(Tp^*)(PPh_3)_2]$ also reacts with diphenylphosphineoxide forming, upon decomposition, a product which involves exchange of the B–N and B–H bonds of the Tp* ligand with phosphine oxide (Fig. 2.90).[405]

$[Rh(Tp^*)(L)(CO)]$ (L = PPh_3, PCy_3, $P(NMe_2)_3$, $P(p\text{-Tol})_3$ or $P(m\text{-Tol})_3$), $[Rh(Tp^x)(PPh_3)_2]$ (Tp^x = Tp, Tp* or pzTp), and

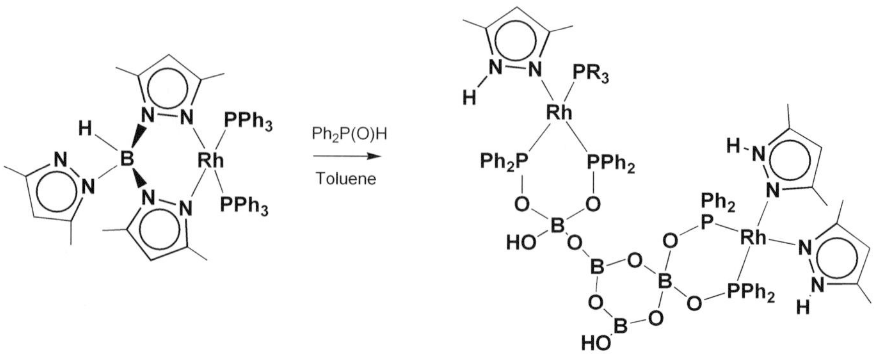

Fig. 2.90. The reaction of [Rh(Tp*)(PPh_3)$_2$] with diphenylphosphineoxide.

[Rh(Tp*)(dppe)], fluxional on the NMR timescale, adopt four-coordinate κ^2-structures, whereas [Rh(Tp*)(CO)(P(OPh)$_3$] has a distorted five-coordinate square-pyramidal structure with a long Rh–N contact in the apical site and an essential planar $RhCPN_2$ basal plane. One electron oxidation yields [Rh(Tp*)(CO)(L)]$^+$, [Rh(Tp^x)(PPh_3)$_2$]$^+$, [Rh(Tp*)(dppe)]$^+$, and [Rh(Tp*)(CO)(P(O-Ph)$_3$)]$^+$, all containing the [Rh^{II}(κ^3-Tp^x)]$^+$ cations. [Rh(Tp*)(CO)-(PPh_3)][PF_6], [Rh(Tp*)(PPh_3)$_2$][BF_4], and [RhH(Tp*)(PPh_3)(CO)]-[BF_4] have been investigated by single-crystal X-ray study. The structures of [Rh(Tp*)(CO)(PPh_3)][PF_6] and [Rh(Tp*)(PPh_3)$_2$][BF_4] and ESR data are consistent with a five-coordinate square-pyramidal geometry with the unpaired electron in a σ^* Rh–N$_{axial}$orbital.[406]

Reaction of MeI with the square-planar [Rh(κ^2-Tp*)(CO)(PMe_3)] at room temperature gives the pseudo-octahedral [Rh(Me)(κ^3-Tp*)(CO)(PMe_3)]I, fully characterized by spectroscopy and X-ray crystallography. A reaction mechanism in which nucleophilic attack by Rh on MeI occurs together with the coordination of the pendant pyrazolyl group has been proposed. In solution, [Rh(Me)(κ^3-Tp*)(CO)(PMe_3)]I is converted slowly into [Rh(COMe)-I(κ^3-Tp*)(PMe_3)]. Kinetic studies on the reactions of [Rh(κ^2-Tp*)(CO)(L)] (L = PMe_3, PMe_2Ph, $PMePh_2$, PPh_3, CO) with MeI support a S_N2 mechanism. [Rh(κ^2-Tp*)(CO)$_2$] reacts with MeI yielding [Rh(COMe)I(κ^3-Tp*)(CO)] (Fig. 2.91). The [Rh(κ^2-Bp*)(CO)(L)]

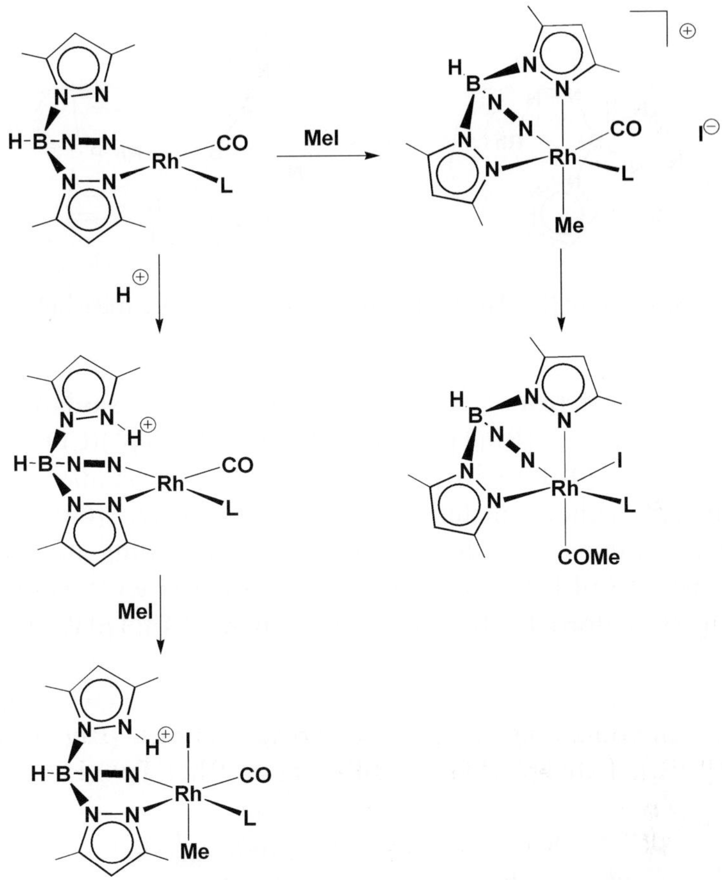

Fig. 2.91. Mechanism proposed for the reaction of the square-planar [Rh(κ^2-Tp*)(CO)(L)] with MeI.

complexes (L = CO, PPh_3) are much less reactive toward MeI than the Tp* analogs, indicating that the third pyrazolyl ring is the determinant for the accessibility of a κ^3 coordination mode.[407]

[Rh(Tp*)L] (L = tris(1-cyclohepta-2,4,6-trienyl)phosphane), in which Tp* is linked to rhodium through a single pyrazolyl group and a nonlinear B–H···Rh bridge, as established by X-ray single-crystal study, has been prepared and characterized by NMR spectroscopy (Fig. 2.92).[408]

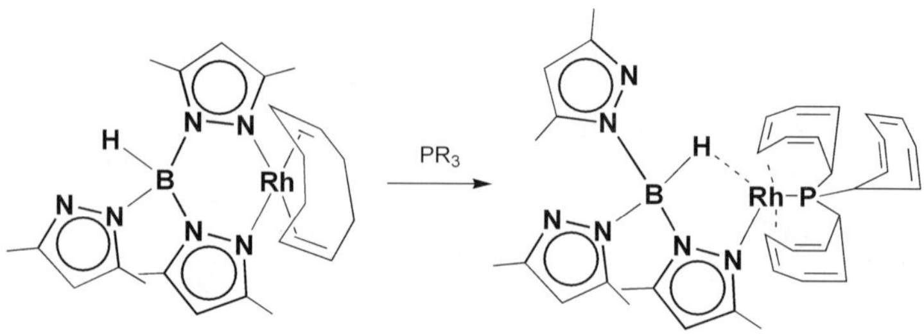

Fig. 2.92. Synthesis of [Rh(Tp*)L] (L = tris(1-cyclohepta-2,4,6-trienyl)phosphane).

The one-electron chemically reversible oxidation of [Rh(Tp*)(PPh$_3$)$_2$], [Rh(Tp*)(CO)(PPh$_3$)], [Rh(Tp*)(CO)(PCy$_3$)], and [Rh(Tp*)(CO)(P(OPh)$_3$)] (Fig. 2.93) has been studied by electrochemical methods including cyclic voltammetry (CV).[409] The oxidation lead to κ^3–Tp* bonding in the corresponding monocation.

[Rh(Tp*)(CO)L] (L = CO or PPh$_3$) also reacts with *o*-$C_6Cl_4O_2$-(3,4,5,6-tetrachloro-1,2-benzoquinone) giving [Rh(Tp*){C(O)OC$_6$-

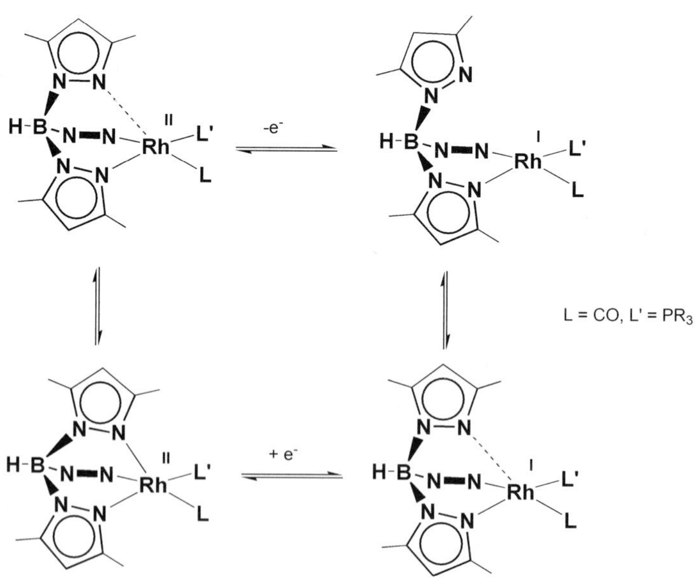

Fig. 2.93. One-electron reversible oxidation of [Rh(Tp*)(L)(L′)]$^+$.

Cl_4O}L]. X-ray structural study of [Rh(Tp*){C(O)OC_6Cl_4O}PPh_3] demonstrates CO insertion into one Rh–O bond of a rhodium–catecholate ring. Loss of the inserted CO on UV irradiation or thermolysis produced [Rh(Tp*)(*o*-$O_2C_6Cl_4$)L]. Thermal substitution of the CO ligand with L′ gave [Rh(Tp*)(*o*-$O_2C_6Cl_4$)L′] {L′ = $AsPh_3$, $P(OPh)_3$, py} that can be oxidized by $[NO]^+$ to the monocations $[Rh(Tp^*)(o\text{-}O_2C_6Cl_4)L']^+$.[410]

When 1 equiv. of PR_3 or $P(OR)_3$ was added to [Rh(κ^3-Tp*)$(CO)_2$], the monosubstituted species [Rh(Tp*)(CO)(PR_3)] and [Rh(Tp*)(CO)$P(OR)_3$] were formed. [Rh(Tp*)(CO)(PMe_3)] and [Rh(Tp*)(CO)($PMePh_2$)] have been found in the square-planar environment, whereas [Rh(Tp*)(CO)($P(OPh)_3$)] is consistent with an interaction between the dangling pyrazolyl group and the rhodium center. All the PR_3-containing species are square-planar both in solution and in the solid state (IR study), whereas with the exception of [Rh(Tp*)(CO){$P(OMe)_3$}], four-coordinate in solution, the phosphite species adopt a penta-coordinate geometry both in the solid and in the solution state. Exchange processes between dangling and coordinated pyrazolyl groups have been investigated by NMR.[411]

Hall and Webster compare predictions by DFT of the structure and relative energy of various isomers of the Tp^x ligand in [Rh(Tp^x)$(CO)_2$] (Tp^x = Tp* and Tp) (Fig. 2.94) and their IR and ^{11}B NMR spectra with the experimental data. The lowest energy structure of [Rh(Tp^x)$(CO)_2$] (Tp^x = Tp*, Tp^{Me} and Tp) is a nonclassical square-pyramidal (SPy) structure with a long metal apical ligand distance in rapid exchange with an equivalent SPy structure through a low-energy TBP transition state. A second higher energy minimum, a pseudosquare-planar complex with the third uncoordinated pz arm rotated approximately parallel with the metal ligand pseudoplane (SP1), is accessed by a second low-energy transition state. Finally, a third pseudosquare-planar minimum structure (SP2) is produced by a transition state which lengthens the Rh–$N_{(apical)}$ bond distance.[412]

Extensive DFT analyses of the structures and vibrational frequencies of [Rh(Tp*)$H_2(H_2)$], also through an inelastic neutron scattering study, have been performed by Eckert.[413] Geometry optimizations produced canted η^2-dihydrogen dihydride local minima of C_1

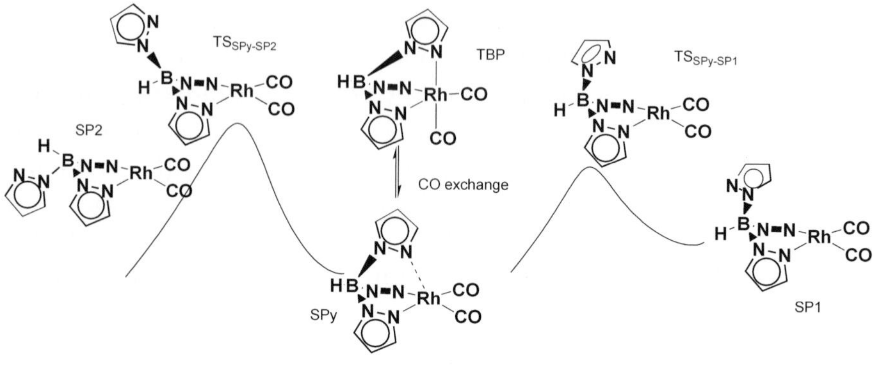

Fig. 2.94. Schematic representation of the potential energy surface for the interconversion of various [RhTp(CO)$_2$] structures.

symmetry. Optimization of the Rh structure generates a high-energy (>4 cal mol^{-1}) C_s transition-state [Rh^{III}(Tp)H$_4$] structure with a η^3–H_3^- ligand.

Reaction of [RhCl$_2$(Tp)(PPh$_3$)] with AgX (X = BF$_4$, NO$_3$, SbF$_6$) yielded heterotrinuclear derivatives [{Rh(Tp)(PPh$_3$)(μ-Cl)$_2$}$_2$Ag]X. All derivatives have been characterized by ^{1}H, ^{31}P and ^{13}C NMR spectroscopy, and FAB$^+$ mass spectrometry. [{Rh(Tp)(PPh$_3$)(μ-Cl)$_2$}$_2$Ag]BF$_4$ consists of two {Rh(Tp)(PPh$_3$)(μ-Cl)$_2$} units bonded to silver through two double μ-Cl bridges in an unusual distorted square-planar coordination environment (Fig. 2.95).[414]

[RhCl(Cp*)(Tp^x)] have been synthesized by the reaction of the appropriate [RhCl$_2$(Cp*)]$_2$ with K(Tp^x) (Tp^x = Tp or Tp*). These complexes have been spectroscopically and structurally characterized.

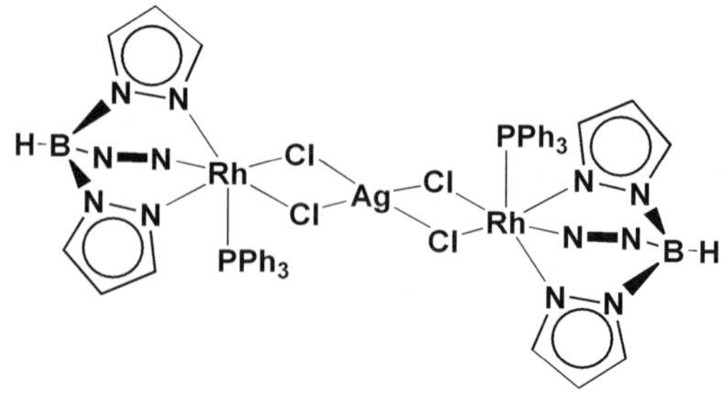

Fig. 2.95. The heterotrinuclear [{Rh(Tp)(PPh$_3$)(μ-Cl)$_2$}$_2$Ag]$^+$ cation.

Reaction between [RhCl(Cp*)(κ^2-Tp)] and $AgNO_3$ in MeCN gave the ionic [M(Cp*)(κ^3-Tp](NO_3), whereas by using [RhCl(Cp*)(κ^2-Tp*)] B–N bond cleavage occurs. B–N bond scission is also detected during the interaction of [$RhCl_2$(Cp*)] with K(Tp^x) when the reactions are carried out in air for extended periods of time, providing pyrazole adducts.[415]

Tp^x (Tp^x = Tp, Tp*, but also $Tp^{Me,4Me}$, $Tp^{Me_2,Cl}$, Tp^{iPr}, and Tp^{iPr4Br}) reacts with $RhCl_3 \cdot 3H_2O$ in MeOH giving [$RhCl_2$(Tp^x)(MeOH)]. Tp* interacts with [$RhCl_3$(MeCN)$_3$] in MeCN giving the octahedral [$RhCl_2$(Tp*)(pz*)]. Different species have been obtained by using $Tp^{CF_3,Me}$ and $Tp^{Me,Ph}$. The complexes [$RhCl_2$(Tp^x)(L)] (L = MeOH and MeCN) react with Cl^- in $CHCl_3$ forming the corresponding [$RhCl_3$(Tp^x)]$^-$ anions. The X-ray crystal structure of [PPh_4][$RhCl_3$(Tp*)] shows octahedral coordination at rhodium. The structure of this compound has been investigated by a molecular modeling study.[416]

[Rh(SPh)$_2$(Tp*)(MeCN)] readily undergoes substitution of MeCN by (Xyl)NC yielding [Rh(SPh)$_2$(Tp*)(XylNC)]. On the other hand, reaction of [Rh(SPh)$_2$(Tp*)(MeCN)] with terminal alkynes produces the rhodathiacyclobutene species [Rh(SPh)(Tp*){η^2-CH=CR(SPh)}] (R = aryl, alkyl) (Fig. 2.96). The molecular structures of [Rh(SPh)$_2$-(Tp*)(XylNC)] and [Rh(SPh)(Tp*){η^2-CH=CCH$_2$Ph(SPh)}] have been investigated by single-crystal X-ray diffraction. [Rh(SPh)$_2$(Tp*)-(MeCN)] and [Rh(Tp*)(COE)(MeCN)] catalyze regioselective addition of benzenethiol to terminal alkynes RC≡CH.[417]

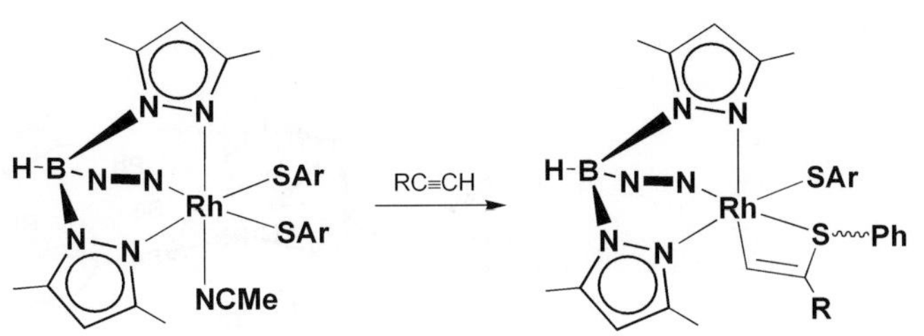

Fig. 2.96. The reaction of [Rh(SPh)$_2$(Tp*)(MeCN)] with terminal alkynes afforded rhodathiacyclobutene.

Reactions of [Rh(Tp*)(COE)(MeCN)] with 1 equiv. of PhEEPh (E = S, Se, Te) yield [Rh(Tp*)(EPh)$_2$(MeCN)], which with [(Ru(Cp*)$_4$(μ^3-Cl)$_4$] in THF at room temperature gives rise to the chalcogenolato-bridged dinuclear complexes [RhCl(Tp*)(μ-EPh)$_2$Ru-(Cp*)(MeCN)] (Fig. 2.97). [RhCl(Tp*)(μ-SPh)$_2$Ru(Cp*)(MeCN)] in solution was converted slowly into the sterically less-encumbered dinuclear complex [Rh(Cl)(Tp*)(SPh)(μ-η^1-S-η^6-Ph)Ru(Cp*)]. [Rh(Cl)-(Tp*)(μ-EPh)$_2$Ru(Cp*)(MeCN)], dissolved in THF and exposed to air, gives the peroxo complex [Rh(Cl)(Tp*)(μ-EPh)$_2$Ru(Cp*)-(η^{2}O$_2$)].[418]

[Rh(Tp*)(COE)(MeCN)] also reacts with 1 equiv. of the organic disulfides, TolSSTol (Tol = 4-MeC$_6$H$_4$), PySSPy (Py = 2-pyridyl),

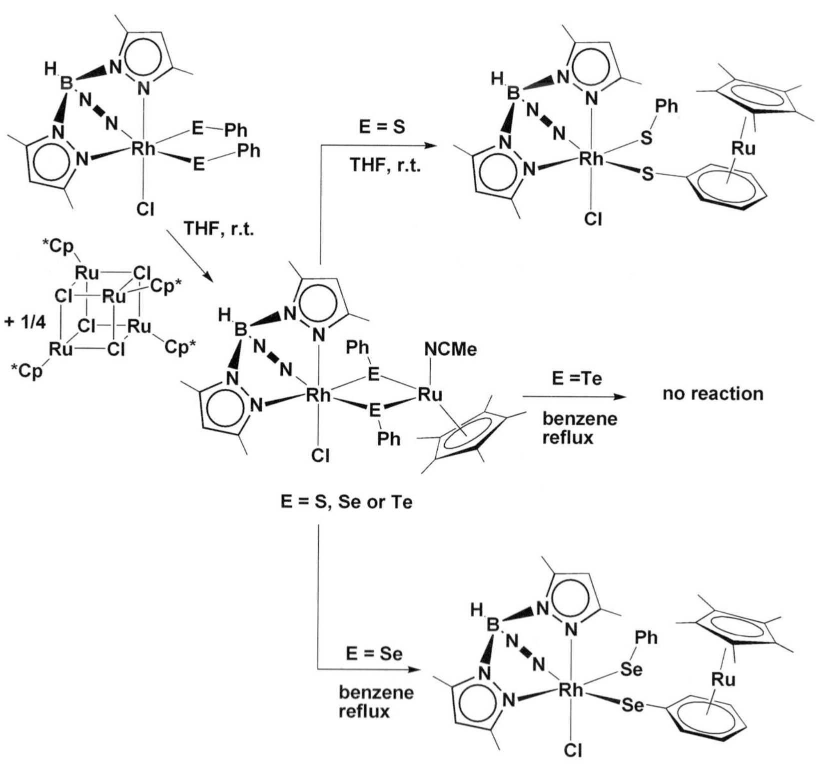

Fig. 2.97. Chalcogenolato-bridged dinuclear complexes [RhCl(Tp*)(μ-EPh)$_2$Ru-(Cp*)(MeCN)] and their reactivity.

and tetraethylthiuram disulfide yielding the mononuclear Rh^{III} complexes [Rh(Tp*)$(STol)_2$(MeCN)], [Rh(Tp*)(η^2-SPy)(η^1-SPy)], and [Rh(Tp*)(η^2-S_2CNEt_2)(η^1-S_2CNEt_2)], respectively, via the oxidative addition of the organic disulfides to the Rh^I center in [Rh(Tp*)(COE)(MeCN)]. In the case of [Rh(Tp)(COE)(MeCN)], the reaction with TolSSTol gave both the bis(thiolato) complex [Rh(Tp)$(STol)_2$(MeCN)] and the dinuclear [{Rh(Tp)(STol)}$_2$(μ-STol)$_2$]. [Rh(Tp*)$(SPh)_2$(MeCN)] and [Rh(Tp*)$(STol)_2$(MeCN)], treated with [{M(Cp*)Cl}$_2$(μ-Cl)$_2$] (M = Rh, Ir), gave the thiolato-bridged dinuclear complexes [RhCl(Tp*)(μ-SPh)$_2$M(Cp*)Cl]. The dirhodium complexes [RhCl(Tp)(μ-STol)$_2$Rh(Cp*)Cl] and [RhCl-(Tp*)(μ-SPh)$_2$Rh(Cp*)(MeCN)][PF_6] have also been synthesized and characterized.[419]

[Rh(Tp*)(η^4-1,5-COD)] photoisomerizes to [Rh(Tp*)(η^4-1,3-COD)] through an intramolecular (3,4) hydrogen shift of the coordinated 1,5-COD likely involving an allylrhodium(III) intermediate. The selective photolysis of complex [Rh(Tp*)(η^4-1,3-COD)] causes the dissociation of 1,3-COD forming the organometallic fragment {Rh(Tp*)} as its primary photoproduct, versatile entry point in the photochemical preparation of new hydrido Rh(III) complexes.[420] In detail, photolysis of [Rh(Tp*)(η^4-1,5-COD)] in benzene at 400 nm generates [Rh(Tp*)(η^4-1,3-COD)], which upon selected photolysis at 336 nm in the presence of $P(OMe)_3$, CH_3OH, and ^{t}Bu-acrylate can be used to produce [Rh(H)(C_6H_5)(Tp*){$P(OMe)_3$}], [Rh$(H)_2$(Tp*)(CO)], and [Rh(C_6H_5)(Tp*)($CH_2CH_2COO^t$Bu)], respectively.[421]

The photolysis of [Rh(Tp*)$(CO)_2$] at 308 nm in liquid Xe solution gives rise to two transient IR bands (1978 and 1996 cm^{-1}) that have been assigned to the CO stretching in [Rh(Tp*)$(CO)_2$]-Xe solvates, differing in the coordination geometry of the Tp* ligand. Some preliminary kinetic data for the reaction of [Rh(Tp*)$(CO)_2$] with cyclohexane in xenon evidenced the presence of the same transient IR bands (1978 and 1996 cm^{-1}) that decay with the formation of the Rh-alkyl hydride product [Rh(cy)(H)(Tp*)(CO)], absorbing at 2032 cm^{-1} (Fig. 2.98).[422]

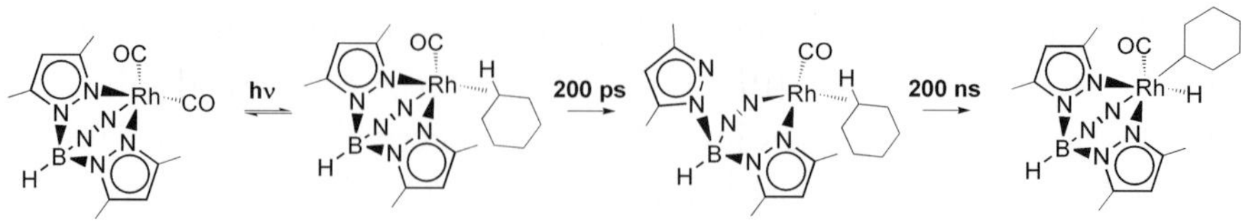

Fig. 2.98. The reaction of $[Rh(Tp^*)(CO)_2]$ with cyclohexane in xenon.

$[Rh(Tp^*)(C_2H_4)_2]$ easily decomposes in solution giving $[Rh(Tp^*)$-$(\eta^4\text{-}C_4H_6)]$, whereas in the solid state, it thermally decomposes to $[Rh(Tp^*)(\textit{syn}\text{-}C_3H_4Me)]$. $[Rh(Tp^x)(C_2H_4)_2]$ (Tp^x = Tp or Tp^*) reacts with PR_3 yielding $[Rh(Tp^x)(C_2H_4)(PR_3)]$, while the reaction of $[Rh(Tp)(C_2H_4)_2]$ with L′ = CO or CNR yields $[Rh(Tp)](\mu\text{-}L')_3$ (Fig. 2.99). The labile $[Rh(Tp^*)(C_2H_4)(PR_3)]$, upon reaction with

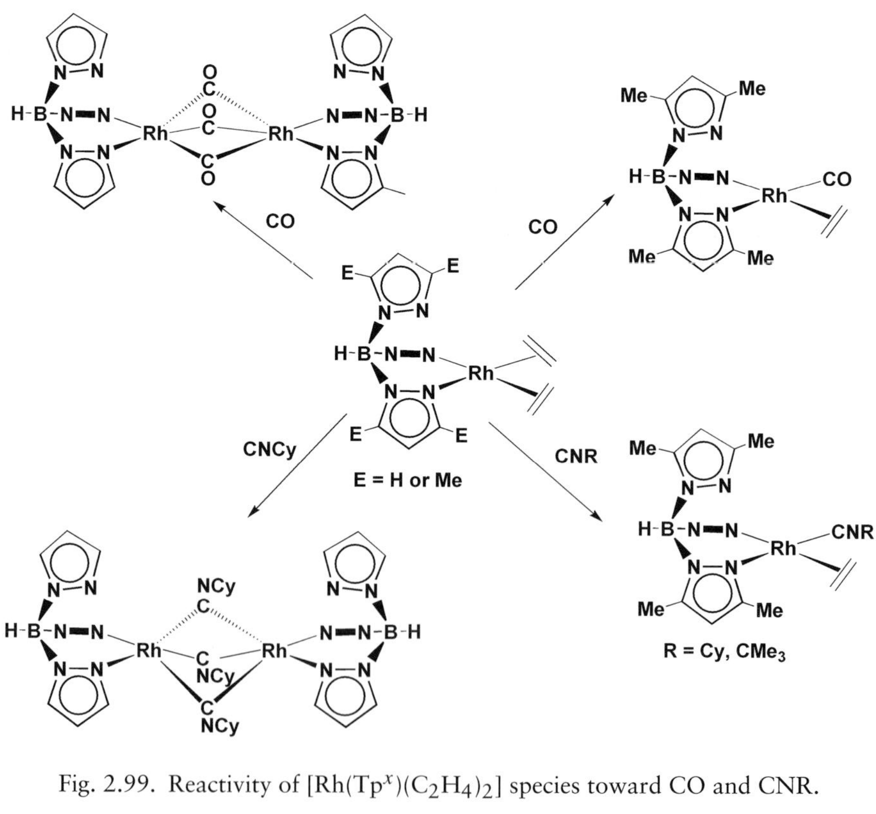

Fig. 2.99. Reactivity of $[Rh(Tp^x)(C_2H_4)_2]$ species toward CO and CNR.

O_2, forms the peroxo species [Rh(Tp*)(O_2)(PR_3)]. The dihydride compounds [Rh(Tp^x)H_2(PR_3)] have been obtained by the hydrogenation of the corresponding ethylene species. [Rh(Tp)](μ-CNcy)$_3$, [Rh(Tp*)(C_2H_4)(PEt_3)], and [Rh(Tp*)(μ-O_2)(PEt_3)] have been structurally characterized.[423]

[Rh(Tp*)(C_2H_4)$_2$] reacts with N-donors (L), such as MeCN or py, giving [Rh(C_2H_5)(Tp*)(CH=CH_2)(L)]. Upon heating at 60°C, [Rh(C_2H_5)(Tp*)(CH=CH_2)(MeCN)], converted to [Rh(Tp*)(C_2H_4)(MeCN)], is able to induce the activation of one of the C–H bonds in C_6H_6. The adducts [Rh(Tp*)(C_2H_4)(PR_3)] (R = Me or Et), also reported, are efficient reagents for the C–H activation of C_6H_6, py, or thiophene.[424] Four coordination modes of scorpionates in Rh complexes, i.e. κ^3-*N*,*N*′,*N*″-, κ^2-*N*,*N*′-, κ^1-*N*-, and κ^0-Tp^x have been observed by X-ray crystallography. The interconversion between the different forms and the denticity changes, PMe_3-induced, have been investigated by spectroscopy. Reaction of [Rh(Tp*)(C_2H_4)] with an equiv. of PMe_3 yields [Rh(κ^3-Tp*)(C_2H_4)(PMe_3)], that in excess of PMe_3 forms [Rh(κ^1-Tp*)(C_2H_4)(PMe_3)$_3$] structurally characterized. When PMe_3 was added to the dihydrido species [Rh(H)$_2$(κ^3-Tp*)(PMe_3)], analogous reaction occurred, [Rh(H)$_2$(κ^2-Tp*)(PMe_3)$_2$] being initially formed. Further addition of PMe_3 gives [Rh(H)$_2$(PMe_3)$_4$](κ^0-Tp*).[425] The step-by-step uncoordination of the pz^x rings of Tp^x, observed in Rh^I and Rh^{III} complexes, has been reported in detail in 2001.[426]

Oxidative addition of perfluorobenzyl iodide to [Rh(Tp)(C_2H_4)$_2$] or [Rh(Tp*)(CO)$_2$] afforded [Rh(Tp)I($CF_2C_6F_5$)(C_2H_4)] and [Rh-(Tp*)I($CF_2C_6F_5$)(CO)], respectively. Reactions of perfluoro-*n*-propyl iodide gave [Rh(Tp)I(C_3F_7)(CO)], [Rh(Tp*)I(C_3F_7)(C_2H_4)], [Rh(Tp)I(C_3F_7)(C_2H_4)], and [Rh(Tp*)(C_3F_7)(PMe_3)I]. Displacement of C_2H_4 from [Rh(Tp)I(C_3F_7)(C_2H_4)] by PMe_3 affords [Rh(Tp)(C_3F_7)(PMe_3)I] structurally characterized.[427]

The reaction of [Rh(Tp*)(COE)(MeCN)] with S_8 in MeCN yields the complex [Rh(Tp*)(S_5)(MeCN)], containing a chelating pentasulfido ligand, converted into [Rh(Tp*)(S_5)(CNXyl)] upon reaction with XylNC. [Rh(Tp*)(S_5)(CNXyl)] has also been structurally characterized.[428]

Jones and coworkers carried out some experiments in order to verify the involvement of alkane σ-complexes in oxidative addition/reductive elimination reaction of [Rh(R)H(Tp*)(L)] (L = $CNCH_2CMe_3$). They reported reductive elimination rates for the loss of alkane from [Rh(R)H(Tp*)(L)] (R = Me, Et, Pr, Bu, Pe, Hex) in deuterated benzene to generate [Rh(C_6H_5)D(Tp*)(L)]. The *iso*-propyl hydride complex [Rh(Tp*)($CHMe_2$)H(L)] has been found to rearrange to the *n*-propyl hydride complex [Rh(Tp*)($CH_2CH_2CH_3$)H(L)] through intramolecular reaction. The same behavior has been found for the *sec*-butyl complex. Analogous reactions have been investigated by preparing the corresponding deuteride complexes.[429] [Rh(CH_3)D(Tp*)(L)] rearranges to [Rh(CH_2D)H(Tp*)(L)] prior to the loss of CH_3D. Similarly, [Rh(CD_3)H(Tp*)(L)] rearranges to [Rh(CD_2H)D(Tp*)(L)] prior to the loss of CD_3H.[430]

The [Rh(Tp*)(CNneopentyl)] has also been shown to selectively activate the primary C–H terminus of acetonitrile, propionitrile, butyronitrile, and valeronitrile, the hydride formed being stable in C_6D_6.[431] The [Rh(Tp*)(CNneopentyl)] fragment, obtained photochemically from the carbodiimide complex, reacts with chloro-substituted alkanes to give primary C–H oxidative addition products. If the alkane has chlorine substitution β to the methyl group, then activation is followed by rapid β-elimination to give [Rh(II)(Cl)(Tp*)(L)] (L = neopentylisocyanide) and the corresponding olefin (Fig. 2.100). [Rh(Y)(X)(Tp*)(L)] (Y = H or Cl, X = 1-, 2-, or 3-chloropentyl) has also been spectroscopically characterized.[432]

2.10.3. *Ir*

The iridium homoscorpionate chemistry continues to be very rich and close to that of rhodium, several papers in the last period reporting analogous [M(Tp^x)(L)] (M = Rh or Ir) complexes. Hydride-,[433,434] phosphino-,[435,436] carbonyl-[437,438] and diene-Ir[439–442] complexes and their reactivity were widely reported in the first edition also.

[Ir(Tp*)(C_2H_4)$_2$] activates C–H bonds successfully in two molecules of benzene to generate a mixture of the bis(phenyl) complexes [Ir(C_6H_5)$_2$(Tp*)(N_2)] and [Ir(C_6H_5)$_2$(Tp*)]$_2$(μ-N_2). In the

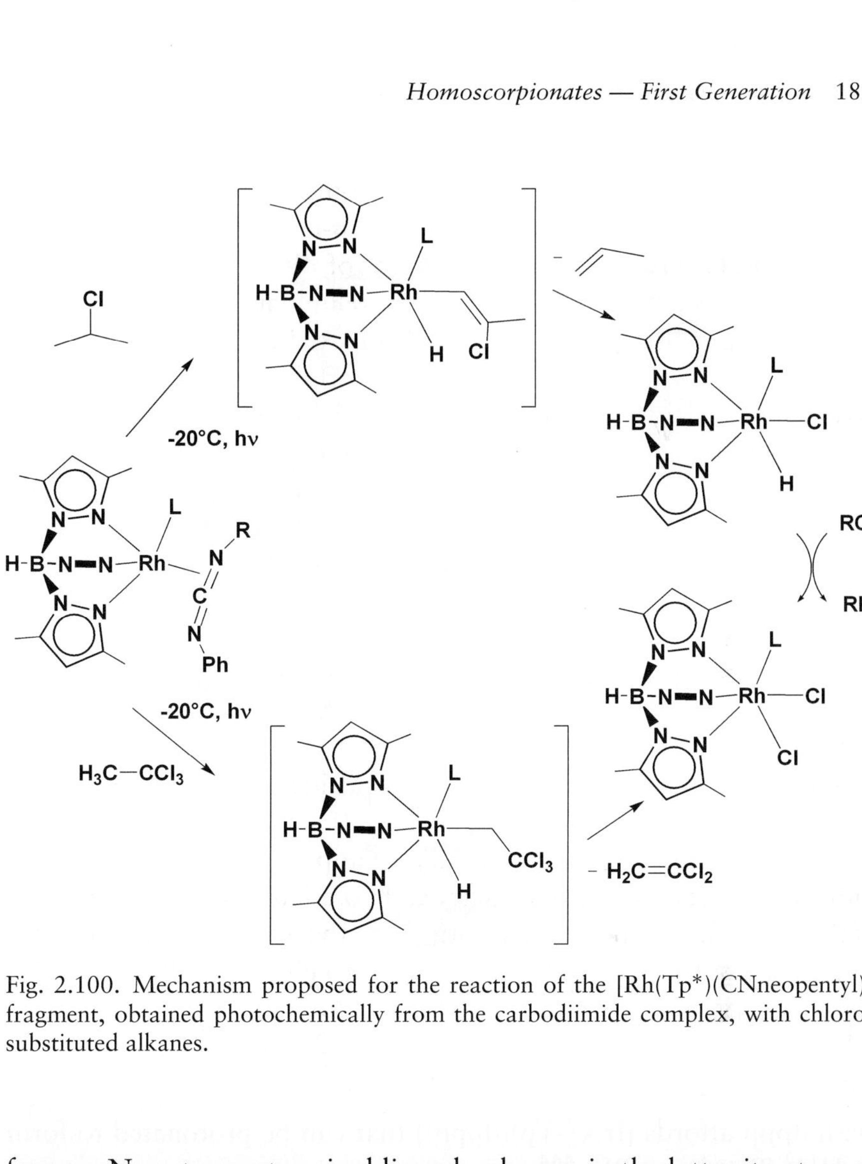

Fig. 2.100. Mechanism proposed for the reaction of the [Rh(Tp*)(CNneopentyl)] fragment, obtained photochemically from the carbodiimide complex, with chloro-substituted alkanes.

former, N_2 acts as a terminal ligand, whereas in the latter it acts as a bridge between two metal centers. [Ir(Tp*)(C_2H_4)$_2$] also activates the $–OCH_2–$ functionality of THF yielding a Fischer carbene compound [Ir=$\overline{C(CH_2)_3O}$(H)(^{n}Bu)(Tp*)] (Fig. 2.101).[443]

From the reaction of [Ir(Tp)(C_2H_4)$_2$] with dppe, complete displacement of Tp from the Ir coordination sphere and [Ir(dppe)$_2$](Tp) formation is observed, whereas the reaction of [Ir(Tp)(C_2H_4)$_2$] with dppm shows an intramolecular proton transfer from the methylene bridge to the iridium center yielding an iridium hydride species by oxidative C–H bond activation (Fig. 2.102). Finally, reaction

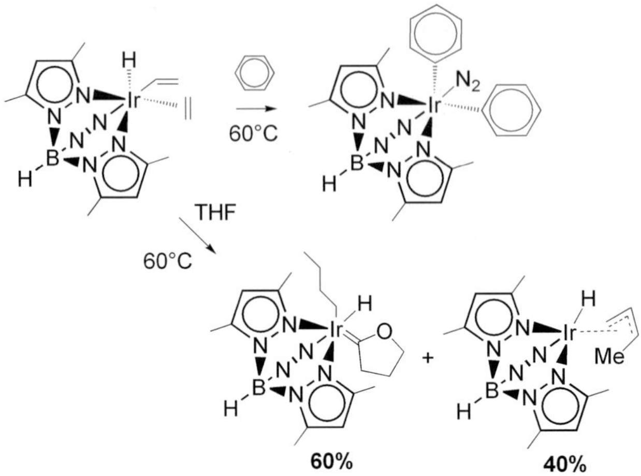

Fig. 2.101. CH activation through $[Ir(Tp^*)(H)(C_2H_3)(C_2H_4)_2]$.

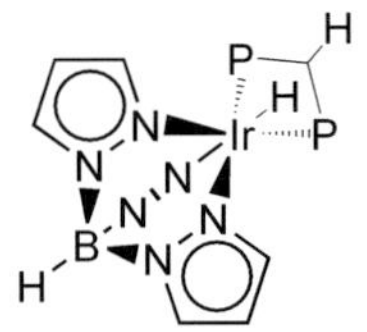

Fig. 2.102. The intramolecular proton transfer from the methylene bridge in a Ir(dppm)(Tp) adduct.

with dppp affords $[Ir(\kappa^2\text{-}Tp)(dppp)]$ that can be protonated to form $[IrH(\kappa^2\text{-}Tp)(dppp)]BF_4$.[444]

Heating a MeCN solution of $[Ir(Tp)(C_2H_4)_2]$ in the presence of DMAD leads to $[Ir\{CH_2CH_2C(CO_2Me){=}C(CO_2Me\}(Tp)(NCMe)]$. The reaction of $[Ir(Tp)(C_2H_4)_2]$ with DMAD in THF also yields $[Ir\{CH_2CH_2C(CO_2Me){=}C(CO_2Me\}(Tp)(THF)]$, whereas the same reaction in benzene gave $[Ir\{C_4(CO_2Me)_4\}(Tp)(CH_2{=}CH_2)]$. These results demonstrate that Tp^x ligands will provide an important entry into the novel alkyne-derived metallacycles, whose analogs have only a fleeting existence in chemistry involving softer ancillary ligands.[445]

2,5-dimethylthiophene (L) reacts with [Ir(Tp*)(C_2H_4)$_2$] yielding a mixture of [Ir(H)(Tp*)(*C*3-L)(*S*-L)] and [Ir{CH_2-(5-methy-3-vinylthiophene)}(H)(Tp*)], upon aromatic C–H bond activation. In excess of L [Ir{CH_2-(5-methyl-3-vinylthiophene)}(H)(Tp*)] yields [Ir{3-ethyl-5-methyl-2-((2,5-dimethylthiophen-3-yl)methyl)thiophen-yl}(H)(Tp*)] (Fig. 2.103).[446] It has been proposed that only sp^2 C–H activation is needed for the formation of [Ir(H)(Tp*)(*C*3-L)(*S*-L)], whereas the formation of [Ir{CH_2-(5-methy-3-vinylthiophene)}(H)(Tp*)] requires the activation of an sp^3 C–H bond and the formation of a new C–C bond (between vinyl and thienyl fragments).[447]

[Ir(C_6H_5)$_2$(Tp*)], generated *in situ*, induces both aromatic and aliphatic C–H bond activation reactions, along with C–C bond formation, when heated with benzene and 1,2-dimethoxyethane. The derivatives [Ir(=CH-OCH_2CH_2OMe)(H)(H)(Tp*)] and [Ir(=CH-OCH_2CH_2OMe)(H)(Ph)(Tp*)] were investigated by NMR

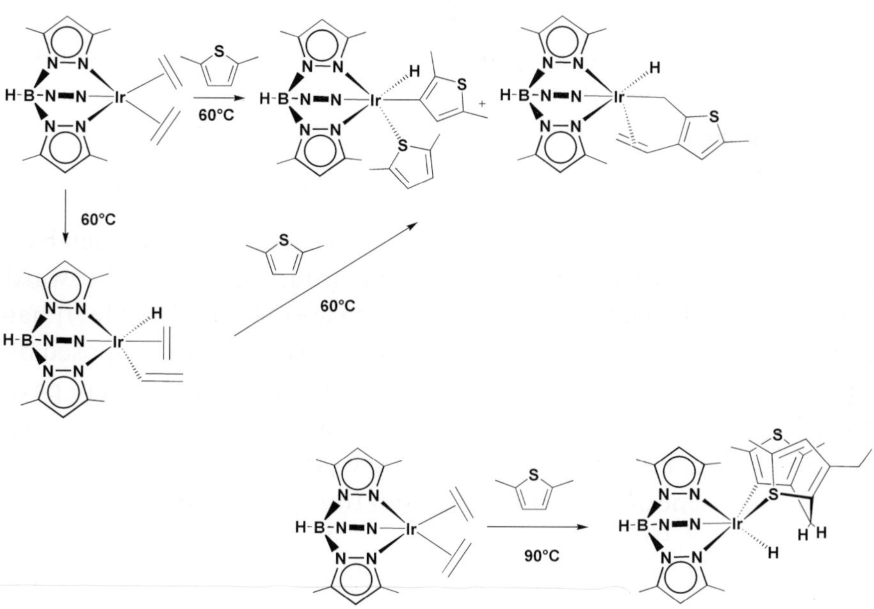

Fig. 2.103. The reaction of [Ir(Tp*)(C_2H_4)$_2$] with 2,5-dimethylthiophene.

spectroscopy.[448] Benzene solutions of [Ir(Tp*)(C_6H_5)$_2$] react with methyl ethers MeO^{t}Bu, MeO^{n}Bu, and MeOCH$_2$CH$_2$OMe to give [Ir(H)$_2$(=C(H)OR)(Tp*)] (R = ^{t}Bu, ^{n}Bu, CH$_2$CH$_2$OMe). Hydride carbenes of composition [Ir(H)(C_6H_5)(=C(H)OR)(Tp*)] are early intermediates of this reaction. Me–OR bond cleavage occurs in the reaction of [Ir(Tp*)(C_6H_5)$_2$] with MeO^{n}Bu, which allows the isolation of the hydride *n*-propyl carbonyl complex [Ir(H)(CH$_2$CH$_2$CH$_3$)-(Tp*)(CO)].[449]

[Ir(Tp)(PPh$_3$)(C_2H_4)] exists in solution as an equilibrium mixture of [Ir(Tp)(PPh$_3$)(C_2H_4)] and of [Ir(H)(*N*,*C*5,*N*-Tp)(PPh$_3$)$_2$]. Protonation of the unbound nitrogen of the pyrazole ring affords the cationic complex [Ir(H)(*N*,*C*5,*N*-HTp)(PPh$_3$)$_2$]BF$_4$. The structure of [Ir(Tp)(C_2H_4)$_2$] and [Ir(H)(*N*,*C*5,*N*-HTp)(PPh$_3$)$_2$]BF$_4$ has also been determined. [Ir(Tp)(PPh$_3$)$_2$] can be considered as an intermediate in the C–H bond activation. The key requirement for the thermal C–H bond activation reactions in Ir^I(Tp) complexes is the accessibility of a three-coordinate species containing a κ^2-Tp ligand.[450]

[IrH$_2$(Tp*)(PMe$_3$)$_2$] reacts with 2 equiv. of NBS yielding [IrBr$_2$-(Tp*)(PMe$_3$)$_2$] that can be selectively alkylated with MeLi to form [Ir(Me)Br(Tp*)(PMe$_3$)]. This compound reacts with Ag(OTf) giving [Ir(Me)(Tp*)(PMe$_3$)](OTf) that has been structurally characterized and employed to address the question of whether the Tp* system could be induced to undergo C–H activation. Its reactivity is reported in Fig. 2.104. Both ligand dissociation and C–H activation face high barriers as a result of the poorer electron donating ability of Tp* toward Ir.[451] In fact, the lack of reactivity of [Ir(Me)(OTf)(Cp*)(PMe$_3$)] has been ascribed to electronic effects rather than steric effects imposed by Tp*. Metathesis of the triflate with BAr$_F$ under N$_2$ atmosphere gave [Ir(Me)(Tp*)(N$_2$)(PMe$_3$)](BAr$_F$). This species, unreactive toward dative ligands, H$_2$, and hydrocarbons, is able to activate the C–H bonds of benzene and aldehydes, but not those of the saturated hydrocarbons (Fig. 2.105).[452]

Tellers and Bergman[453] reported the synthesis of [IrMe(Tp)-(PMe$_3$)(ClCH$_2$Cl)][BAr$_F$]·2CH$_2$Cl$_2$ which reacts with N$_2$ (50 atm)

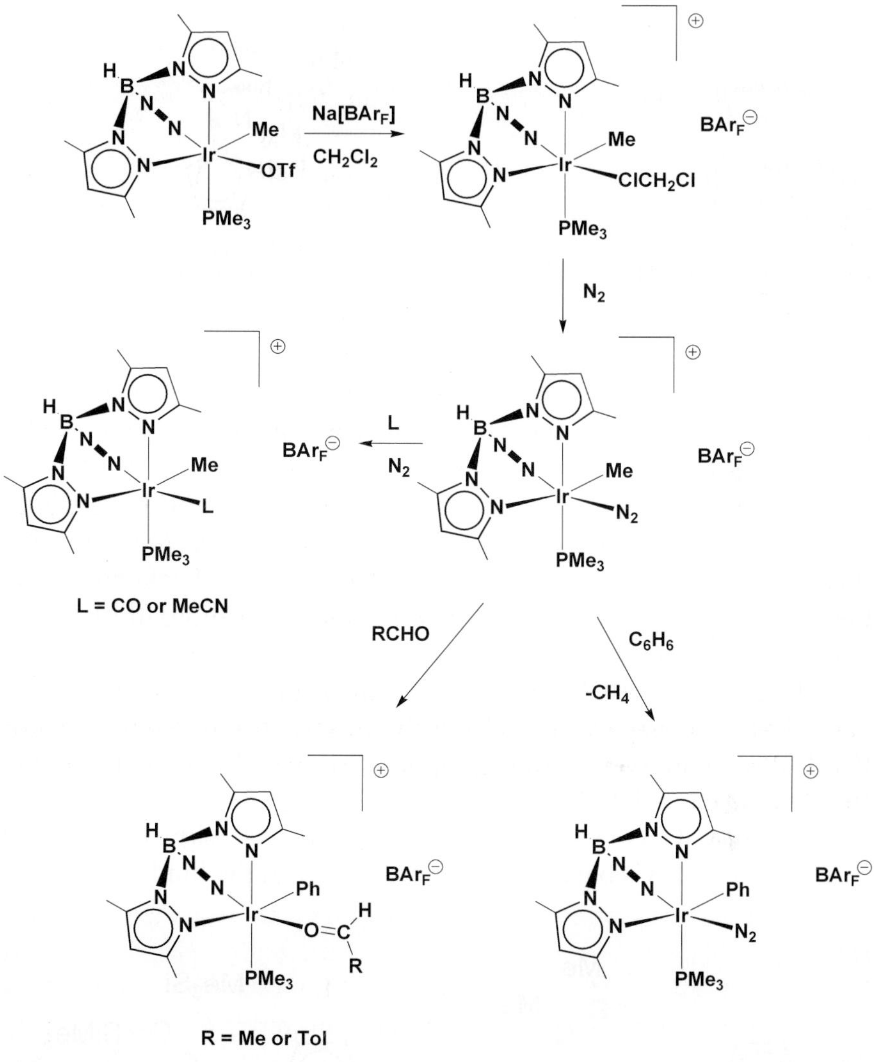

Fig. 2.104. The reactivity of [Ir(Me)(Tp*)(PMe_3)](OTf) toward CO, MeCN, RC(O)H, and benzene.

giving [IrMe(Tp)(PMe_3)(N_2)][BAr_F]. MeCN reacts as nucleophile toward [IrMe(Tp)(PMe_3)($ClCH_2Cl$)][BAr_F] yielding [IrMe(Tp)(PMe_3)(NCMe)][BAr_F]. Mechanistic studies on the ligand substitution processes have been proposed.[453]

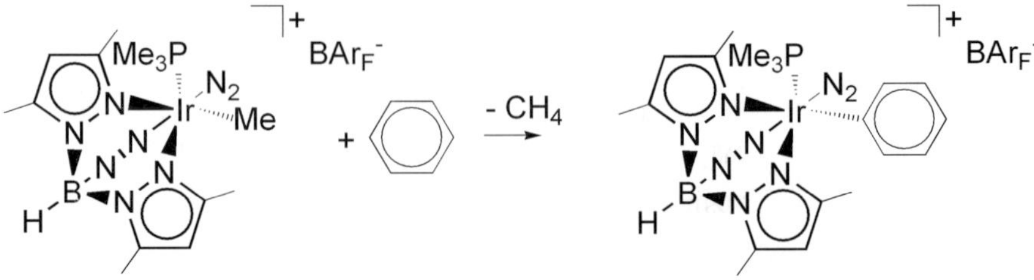

Fig. 2.105. $[Ir(Me)(Tp^*)(N_2)(PMe_3)](BAr_F)$ is able to activate the CH bond of benzene.

Reaction of $[Ir(Tp^*)(\eta^4\text{-}CH_2{=}C(Me)C(Me){=}CH_2)]$ with $Me_3SiC{\equiv}CSiMe_3$ afforded the vinylidene complex $[Ir(Tp^*)(\overline{C,C\text{-}CH_2\text{-}C(Me){=}C(Me)CH_2})({=}C{=}C(SiMe_3)_2)]$ (Fig. 2.106). When $[Ir(Ph)_2(Tp^*)(N_2)]$ reacted with $Me_3SiC{\equiv}CSiMe_3$, $[Ir({=}C{=}C(SiMe_3)_2)(Ph)_2(Tp^*)]$ was formed. Both species have been characterized by spectroscopy and $[Ir(C,C\text{-}CH_2C(Me){=}C(Me)CH_2)({=}C{=}C(SiMe_3)_2)(Tp^*)]$ also by X-ray diffraction. $[Ir(Ph)_2({=}C{=}C(SiMe_3)_2)(Ph)_2(Tp^*)]$ reacts with HBF_4, HCl, and CF_3COOH yielding $[Ir(Ph)({=}C(Me)Ph)(Tp^*)(Et_2O)]BF_4$, $[Ir(Ph)(Cl)(C(Me)Ph)(Tp^*)]$, and $[Ir(Ph)(\kappa^1\text{-}O_2CCF_3)({=}C(Me)Ph)(Tp^*)]$, respectively. The iridabenzene $[Ir(H)(CHC(Me)C(Me)CHC(Me))(Tp^*)]$ has also been reported and structurally characterized.[454]

$[Ir(Tp^*)(\eta^4\text{-}CH_2{=}C(Me)C(Me){=}CH_2)]$ reacts with a variety of soft and hard Lewis bases producing a change in the binding mode of

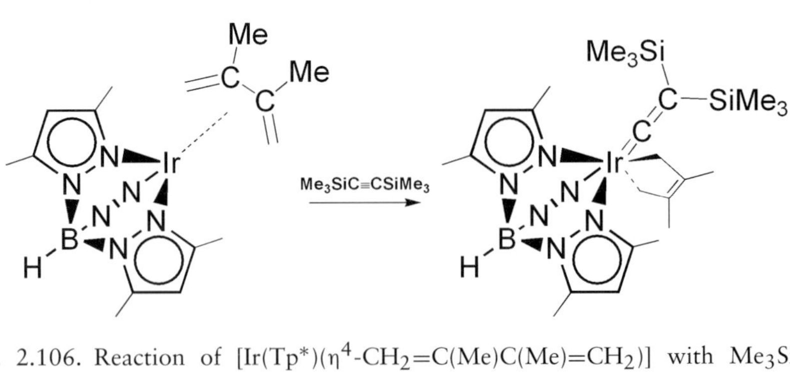

Fig. 2.106. Reaction of $[Ir(Tp^*)(\eta^4\text{-}CH_2{=}C(Me)C(Me){=}CH_2)]$ with $Me_3SiC{\equiv}CSiMe_3$.

the diene and affording [Ir(Tp*)(η^2:σ^2-$CH_2C(Me){=}C(Me)CH_2$)(L)] (L = CO, PMe_3, MeCN, py). The same reactivity has been displayed by [Ir(Tp*)(η^4-$CH_2{=}C(Me)CH{=}CH_2$)] and [Ir(Tp*)(η^4-$CH_2{=}CHCH{=}CH_2$)]. It has been proposed that metal–base Lewis acid centers of composition [Ir(R)(R′)(Tp*)] (R and R′ = alkyl groups) are of intermediate soft–hard nature.[455] Reaction of [Ir(Tp*)(η^4-$CH_2{=}C(Me)C(Me){=}CH_2$)] with PhC≡CPh, yielded, upon coupling of the diene and the alkyne, the compound **IX** in Fig. 2.107, which contains a bicyclic chelating ligand with alkyl and η^3-allyl functionalities and decomposes in C_6H_{12} yielding the iridafulvalene species **X**.[456]

MeCN reacts reversibly with **IX** giving rise to a mixture of the isomeric iridacyclopentenes [Ir($CH_2C(Me)(CMe{=}CH_2)C(Ph){=}C(Ph)$)(Tp*)(NCMe)] and [Ir($CH_2C(Me)$(*cis*-CPh=CHPh)$C(Me){=}CH$)(Tp*)(NCMe)].[456]

Unusual iridacycloheptatrienes, models for the proposed intermediates of the metal-catalyzed cyclotrimerization of alkynes, are produced under mild conditions. Most remarkably, some of these unsaturated iridacycles undergo a mild, regioselective oxidation which involves a distant C=C bond to yield unprecedented iridacyclohexadienes with a pendant, coordinated ketone functionality. [Ir(η^4-$CH_2{=}C(Me)C(Me){=}CH_2$)(Tp*)] reacts with 3 equiv. of DMAD at 60°C forming an iridacycloheptatriene complex upon coupling of three alkyne molecules. [Ir(Tp*)(C_2H_4)$_2$] and DMAD give

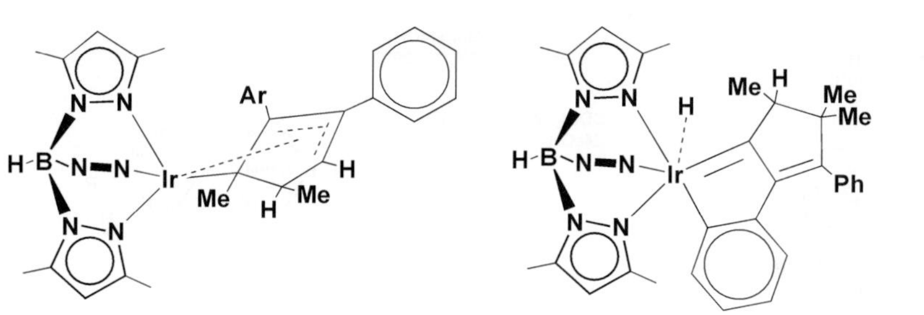

IX

X

Fig. 2.107. Compounds formed in the reaction of [Ir(Tp*)(η^4-$CH_2{=}C(Me)C(Me){=}CH_2$)] with PhC≡CPh.

rise to a compound containing a six-carbon organometallic ligand resulting from the coupling of two molecules of DMAD and one of ethylene followed by the isomerization of the metallacyclic moiety (stereospecific hydrogen shift) to give the observed hydrocarbyl chain. A benzannelated iridacycloheptatriene linkage formally results from the cycloaddition of one benzyne and two DMAD units when $[Ir(C_6H_5)_2(Tp^*)(N_2)]$ and DMAD reacted with cyclohexane.[457]

Reaction of $[Ir(Tp^*)(C_2H_4)(DMAD)]$ with CO and C_2H_4 gave $[Ir(Tp^*)(CO)(DMAD)]$ and $[Ir(Tp^*)$*cis*-$(C(R){=}C(R)H)(CH{=}CH_2)(NCMe)]$ (R = CO_2Me). $[Ir(Tp^*)(C_2H_4)(DMAD)]$ transforms in the bis(alkenyl) derivative $[Ir(Tp^*)($*cis*-$C(R){=}C(R)H)(CH{=}CH_2)(C_2H_4)]$ which slowly evolves to the alkenyl species $[Ir(Tp^*)(C_2H_4)($*cis*-$C(R){=}C(R)H)(CH_2CH_2CH{=}CH_2)]$.[458]

A new route to iridabenzenes involving the coupling of the primary alkene $MeCH{=}CH_2$ with the iridacycles in Fig. 2.108 have been reported by Paneque *et al.*[459] Allylic species, propene insertion, and molecular structures of two iridabenzene have also been reported.

Heating the anisoles 2,6-$Me_2C_6H_3OMe$ and 2,4,6-$Me_3C_6H_2OMe$ in the presence of the unsaturated species $[Ir(C_6H_5)_2(Tp^*)]$ gives the

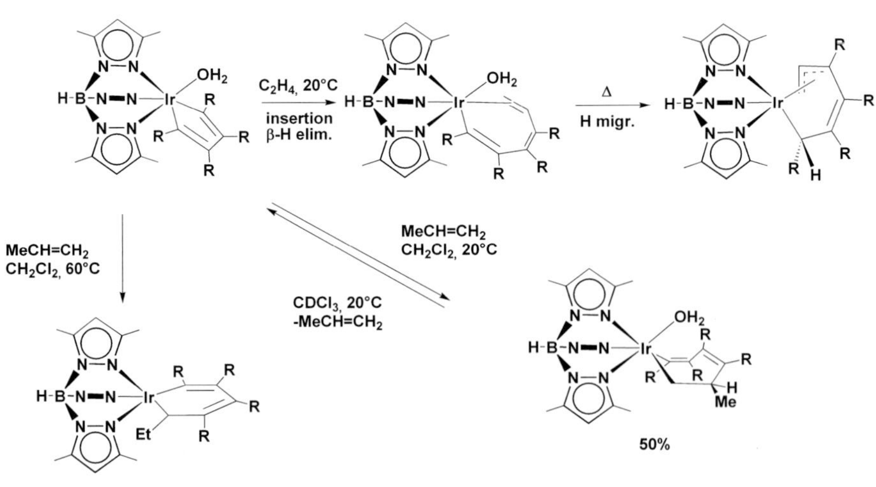

Fig. 2.108. A route to iridabenzenes involving the coupling of the primary alkene $MeCH{=}CH_2$ with iridacycles.

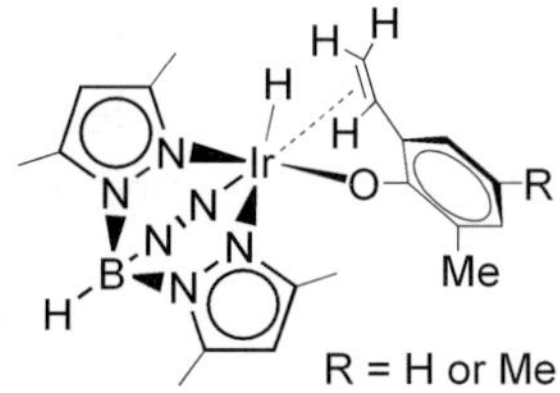

Fig. 2.109. [Ir(Tp*)H(anisolate)] derivatives.

compounds, shown in Fig. 2.109. Their treatment with $SiHEt_3$ gave $[Ir(Tp^*)(H)_3(SiEt_3)]$ and phenols 2-Me-4-R-6-CH_2CH_3-C_6H_3 (R = H or Me).[460]

[Ir(Tp*)(1-aryl-4,5-dimethylene-2-oxazolidinone)] complexes have been prepared and allowed to react with benzaldehyde or anisaldehyde yielding new complexes containing a bidentate ligand formed by coupling the two organic fragments (diene and aldehyde) (Fig. 2.110). [Ir(Tp*)(1-C_6H_4-4-Cl-4,5-dimethylene-2-oxazolidinones)] reacts with $Ph_2PC{\equiv}CPPh_2$ undergoing formal oxidative addition of the diene moiety to the metal center and coordination of the diphosphine in the κ^1 mode.[461]

The $\{Ir^{III}Tp^*\}$ fragment also allows the attainment of equilibrium mixture of hydride alkylidene **XI** and hydride alkene **XII** species shown in Fig. 2.111.[462]

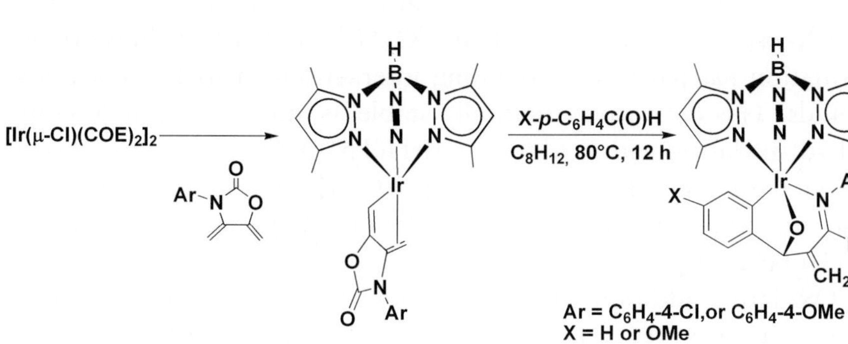

Fig. 2.110. Synthesis of [Ir(Tp*)(1-aryl-4,5-dimethylene-2-oxazolidinone)] and their reaction with aldehydes.

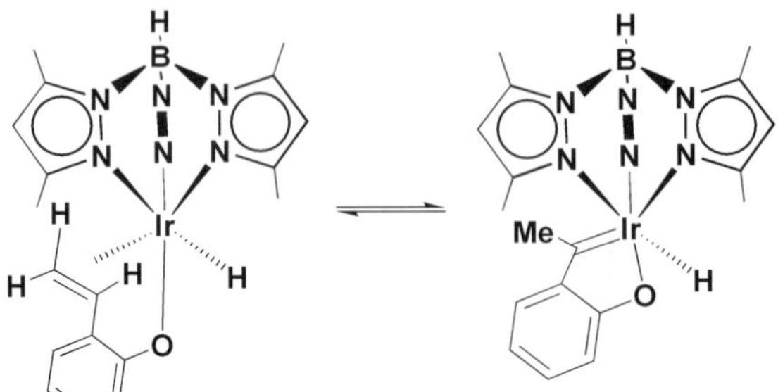

Fig. 2.111. The equilibrium between iridium hydride-alkylidene and -alkene isomers.

On the basis of NMR spectroscopy, the structure of [Ir(Tp)H_4] and [Ir(Tp*)H_4] has been hypothesized as supporting face-capped octahedral structure (C_{3v}) with each of the three Ir–H bonds *trans* to an Ir–N, and the fourth hydride capping the IrH_3 face. Density functional geometry optimizations and coupled cluster calculation on [Ir(Tp)H_4] indicated that a C_s edge-bridged octahedral tetrahedral structure and a C_1 η^2-dihydrogen, dihydride structure are local minima.[463]

The reaction of [Ir(Tp)$(CO)_2$] with water to give [Ir(CO_2H)H(Tp)(CO)], investigated by kinetic studies, is of second order. [Ir(Tp*)$(CO)_2$] also reacts more rapidly with adventitious water to give [Ir(CO_2H)H(Tp*)(CO)]. A proposed mechanism consistent with the relative reactivity of these species involves initial protonation of Ir(I) followed by nucleophilic attack on the carbonyl ligand. The X-ray crystal structure of [Ir(CO_2H)H(Tp*)(CO)] shows dimer formation *via* pairwise H-bonding interactions of hydroxycarbonyl ligands. This derivative, thermally stable, is amphoteric, undergoing dehydroxylation with acid to give $[IrH(Tp^*)(CO)_2]^+$ and decarboxylation with OH^- to give [IrH_2(Tp*)(CO)]. [Ir(CO_2H)H(Tp)(CO)] also undergoes thermal decarboxylation above *ca.* 50°C to give [IrH_2(Tp)(CO)].[464]

Oxidative addition of perfluorobenzyl iodide to [Ir(Tp^x)$(CO)_2$] (Tp^x = Tp or Tp*) afforded [Ir($CF_2C_6F_5$)(I)(Tp^x)(CO)]. [Ir(C_3F_7) (I)(Tp)(CO)] and [Ir(C_3F_7)(I)(Tp)(C_2H_4)] were obtained from the reaction of the corresponding starting reagent with perfluoro-*n*-propyl iodide. Attempted displacement of ethylene from $[Ir(C_3F_7)(I)(Tp)(C_2H_4)]$ by PMe_3 affords the salt $[Ir(C_3F_7)(Tp)(C_2H_4)(PMe_3)]I$. The molecular structure of $[Ir(CF_2C_6F_5)(I)(Tp)(CO)]$ has been determined.[427]

$[IrCl(Cp^*)(Tp^x)]$ species (Tp^x = Tp or Tp*) have been synthesized, spectroscopically and structurally characterized. Comparison has been made with analogous Rh species.[415]

Thompson *et al.*[465] reported syntheses and structures of [Ir(pzTp)($N,C^{2'}$-(2-para-tolylpyridine))$_2$] ($\equiv$[Ir(pzTp)(tpy)$_2$]) and [Ir(pzTp)($N,C^{2'}$-(4′,6′-difluorophenylpyridine))$_2$] ($\equiv$[Ir(pzTp)(dfppy)$_2$]). The Ir–N–N–B–N–N cycle of [Ir(pzTp)(tpy)$_2$] adopts a half-chair conformation, more flattened than the typical boat conformation. The poly(pyrazolyl)borate ancillary ligand affects the absorption and emission energies of these complexes by tuning the HOMO energies. The electron-withdrawing pzTp decreases the electron density on the iridium, stabilizing the metal-centered HOMOs.[465]

Carmona *et al.*[466] reported a series of cationic Ir complexes containing the neutral ligand HC(pz*)$_3$ and compared their reactivity and structure with those found for analogous Tp* derivatives. X-ray single-crystal studies carried out on (HC(pz*)$_3$) derivatives show a striking similarity between the structural parameters of Ir(HC(pz*)$_3$) and Ir(Tp*) species.[466]

Karam and coworkers[467] reported *in situ* regioselective hydrogenation of α,β-unsaturated aldehydes using [Ir(COD)Cl]$_2$ stabilized by Tp and Tp* (and also by Bp* and Tp^{3Me}). When using *trans*-cinnamaldehyde, selectivity toward the saturated aldehydes and the unsaturated alcohol depends on the reaction conditions and becomes more pronounced with the type of stabilizing ligands employed. When Tp is employed, the selectivity shifts to the saturated aldehydes, while with Tp* complexes the selectivity changes to the corresponding allylic alcohol.[467]

Two patents have been reported describing the synthesis of cyclometalated organometallic phosphorescent complexes {Ir(2,3-diarylpyrazine)}[468] and {Ir(2,3-diaryl-5,6,7,8-tetrahydroquinoxaline)}[469] containing pzTp, that are highly efficient light-emitting elements and light-emitting devices with increased recombination efficiency.

2.11. Group 10: Ni, Pd, and Pt

2.11.1. *Ni*

In the first edition of "Scorpionates," a very limited number of Ni compounds has been described. Some of them continued to attract attention, as for example the homoleptic derivatives [Ni(Tp)$_2$], [Ni(Tp*)$_2$], and [Ni(pzTp)$_2$],[470,471] for which in 2003 De Alwis and Schultz have reported electrochemical, magnetic, and spectroscopic properties,[47] whereas no other papers have been published on [Ni(Tp)(η^3-allyl)][472] and [NiX(R)(Tp^x)PR$_3$] complexes.[473]

A series of monomeric five-coordinate nickel-(*l*-cysteine)-Tp* complexes were reported by Desrochers *et al.*[474] [Ni(CysEt)(Tp*)] (Fig. 2.112), is a five-coordinate complex which crystallizes with a single methanol of solvation. [NiCys(Tp*)][K] exhibits magnetic and electronic features similar to [Ni(CysEt)(Tp*)]. These Ni(cysteine)(Tp*) derivatives undergo a rapid reaction with O$_2$ forming sulfinate species. The increased nucleophilicity of the nickel–sulfur center enhanced by electron donation from Tp* (compared with the analogous Tp^{PhMe} species) and supported by a trigonal bipyramidal geometry, is likely the explanation for the susceptibility of [NiCys(Tp*)]$^-$ to oxygen.

The crystal structure of the nonsolvated [Ni(Tp*)$_2$] consists of a highly distorted trigonal bipyramidal nickel center.[475] [NiX(Tp*)]

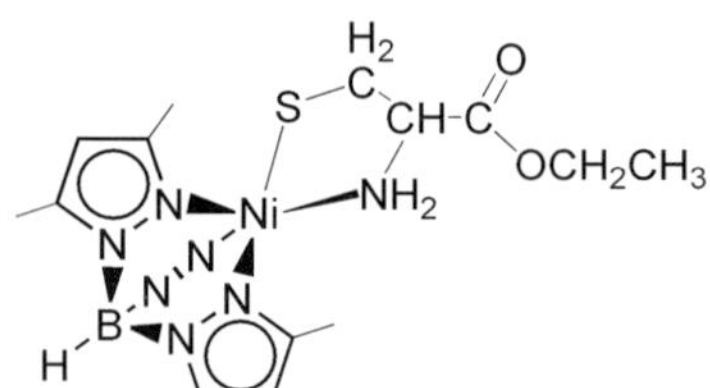

Fig. 2.112. [Ni(CysEt)(Tp*)].

(X = Cl, Br, I) have been characterized by electronic absorption spectroscopy in the visible and near-infrared (NIR) region and by high-frequency and field electron paramagnetic resonance (HFEPR) spectroscopy. The crystal structure of [NiBr(Tp*)], containing a four-coordinate nickel(II) ion (d^8) with approximate C_{3v} point group symmetry about the metal and a resulting $S = 1$ high-spin ground state, is reported. [NiX(Tp*)] complexes are "EPR silent" due to a sizable zero-field splitting.[476] [NiCl(Tp*)] has been tested toward ethylene polymerization.[477]

The high-spin [(Tp*)Ni-Ru(Cp)$(CO)_2$], prepared from the reaction of [NiBr(Tp*)] with K[RuCp$(CO)_2$] reacts with 2e-donors L (CO, CNCy, PPh_3) *via* spin crossover at Ni giving diamagnetic adducts [(Tp*)Ni-Ru(Cp)L(μ-CO)$_2$].[478]

When the dinuclear-bis(μ-hydroxo) derivatives [{Ni(Tp^x)}$_2$(μ-OH)] react with 1 equiv. of H_2O_2, the corresponding [{Ni^{III}(Tp^x)}$_2$-(μ-O)$_2$] (Tp^x = $Tp^{Me_2,X}$ = hydrotris(3,5-dimethyl-4-X-1-pyrazolyl)borate, X = Me, H, Br) have been obtained (Fig. 2.113). Both the initial μ-hydroxo and resulting μ-oxo derivatives exhibit a five coordinate metal center with a slightly distorted square-pyramidal geometry. Thermal decomposition of [{Ni^{III}(Tp^x)}$_2$(μ-O)$_2$] produces oxidation of the inner saturated hydrocarbyl susbtituents of the Tp^x ligand. The oxidation ability of the Ni^{III}(μ-O)Ni^{III} core is affected by the hindered Tp^x ligand system which can stabilize the +2 metal oxidation state.[390]

Some half-sandwich nickel(II) complexes have also been reported.[479] The nickel(II) ion in [Ni(Tp*)(pz*)$_2$(N_3)] is coordinated

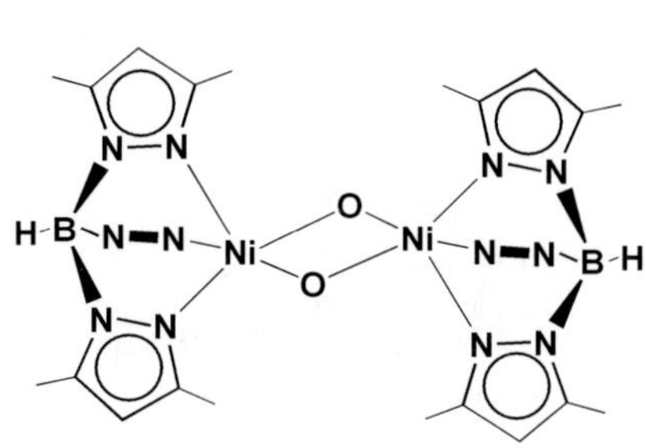

Fig. 2.113. Structure of the dinuclear [{Ni^{III}(Tp*)}$_2$(μ-O)$_2$].

by six nitrogen atoms to form a distorted N_6 octahedron. The structure of [Ni(Tp*)(pz*)(SCN)(H_2O)] shows two structurally independent units in which the nickel(II) ion is six-coordinated by five nitrogen atoms of Tp*, pz*, thiocyanate, and one oxygen atom of water molecule to form a N_5O octahedral environment. The coordination environment of [Ni(Tp*)(SCN)(CH_3OH)$_2$] is described as a N_4O_2 octahedron from three nitrogen atoms of Tp*, one nitrogen atom of thiocyanate and two oxygen atoms of coordinated methanol molecules.[479] In the half-sandwich compound [Ni(η^2-O_2N)(Tp*)(CH_3OH)], the nickel(II) center has been found in a N_3O_3 coordination environment.[480]

[Ni(Tp*)(BH_4)] and [Ni(Tp*)(BD_4)], stable toward air, boiling water, and high temperatures, have been characterized structurally and spectroscopically. [Ni(Tp*)(BH_4)] showed a six-coordinate geometry with the nickel(II) center being facially coordinated by three bridging hydrogen atoms from borohydride and a tridentate Tp* (Fig. 2.114). The IR data confirmed the presence of bridging hydrogen atoms, the υ(B–H)$_{terminal}$, and υ(B–H)$_{bridging}$ being shifted with respect to υ(B–D) of [Ni(Tp*)(BD_4)]. [Ni(Tp*)(BH_4)] and [Ni(Tp*)(BD_4)] readily reduce halocarbon substrates, leading to the complete series of [Ni(X)(Tp*)] complexes (X = Cl, Br, I). Nickel K-edge X-ray absorption spectroscopy (XAS) measurements were performed for [Ni(Tp*)(BH_4)], [Ni(Tp*)(BD_4)], and [Ni(Cl)(Tp*)]. While X-ray absorption near-edge spectroscopy indicates a similar six-coordinate geometries for [Ni(Tp*)(BH_4)] and [Ni(Tp*)(BD_4)], XAS indicates [Ni(Cl)(Tp*)] as pseudo-tetrahedral.[481]

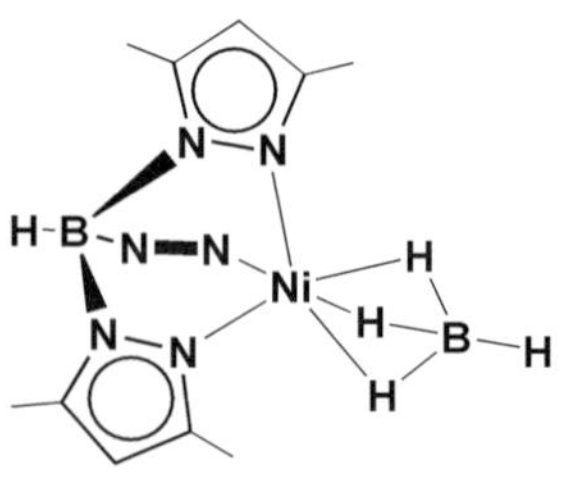

Fig. 2.114. The six-coordinate octahedral [Ni(Tp*)(BH_4)].

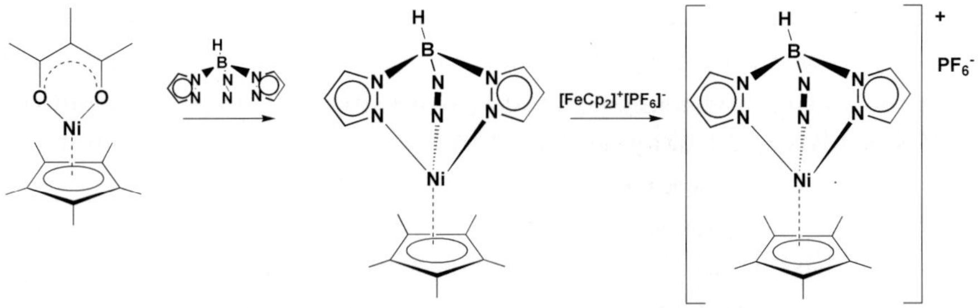

Fig. 2.115. Synthesis of $[Ni(Cp^R)(Tp)]^{n+}$.

Synthesis and structure of $[Ni(Cp^R)(Tp)]^{n+}$ (Fig. 2.115) is reported[29] and the electronic and magnetic properties have been studied and compared to their homoleptic analogs, $[Ni(Cp^R)_2]$ and $[Ni(Tp)_2]$. $[Ni(Cp^*)(Tp)]^+$ is a low-spin ($S = 1/2$) species. Orbital energies and compositions were calculated and compared for a series of homoleptic and mixed-sandwich complexes of Ni(II). The ability of Tp (vs Cp) to act as a δ-donor (with respect to the principal molecular axis) imparts significant ligand antibonding character to the δ-orbitals and results in decreased $\varepsilon_\pi - \varepsilon_\delta$ values compared to the metallocenes and an increased tendency toward high-spin complexes in the mixed-sandwich complexes.[30]

The preparation of several $[Ni(Tp^x)X]$ complexes characterized by different substituents on the pz^x rings is described in a patent[482] by Bianchini and coworkers, together with their use in the homo-, co-, and terpolymerization of α-olefins with aluminoxane co-catalysts.

The diamagnetic dicyanometalate(II) $[NEt_4][Ni^{II}(CN)_2(Tp^*)]\cdot MeCN\cdot 1/2Et_2O$ contains a pentacoordinate Ni from a tridentate Tp* and two CN^-.[483]

2.11.2. *Pd*

Until 1999 the palladium complexes containing homoscorpionates, described in literature, are exclusively Pd(II)-(η^3-allyl),[484,485] Pd(II) and Pd(IV)-alkyl,[486–488] Pd(II)-aryl,[486] Pd(phosphine),[489,490] and homoleptic $[Pd(Tp^x)_2]$ species.[491,492]

The sterically hindered [Pd(PPh_3)(Cl)(Tp)], metastable in solution in the presence of water, has been synthesized by the stepwise reaction of [$PdCl_2$($MeCN$)$_2$] with K(Tp), followed by PPh_3. A solution NMR study of [Pd(PPh_3)(Cl)(Tp)] indicates the ligand coordinate to palladium in a κ^2-fashion.[493]

Kläui *et al.*[494] prepared and investigated the palladium complexes [Pd(R)(Tp*)(PPh_3)] (R = Me, C(O)Me, *p*-Tol, C(O)*p*-Tol) nonrigid on the NMR scale at room temperature likely due to an intramolecular substitution of coordinated and noncoordinated pz* groups. [Pd(R)(Tp*)(PPh_3)] has been found in a square-planar coordination environment.[494] [Pd(*p*-Tol)(Tp*)(PPh_3)] exhibits catalytic activity in the copolymerization of CO and norbornadiene (Fig. 2.116). Upon CO insertion, [Pd{C(O)*p*-Tol}(Tp*)(PPh_3)] is formed which slowly inserts norbornadiene giving [Pd{$C_7H_8C(O)$*p*-Tol}(Tp*)(PPh_3)]. When CO and olefin were simultaneously added to [Pd(Tp*)(*p*-Tol)(PPh_3)], a polyketone was formed in few hours.[495]

The anionic derivative (K)[Pd(CH_2CMe_2-*o*-C_6H_4)(κ^2-Tp^x)], prepared *in situ* from [Pd(CH_2CMe_2-*o*-C_6H_4)(COD)] and K(Tp^x) (Tp^x = Tp or Tp*), reacts with 1 equiv. of *N*-methyl-*N*-nitroso-*p*-toluenesulfonammide yielding Pd^{IV}(alkyl) and Pd^{IV}(aryl) compounds [Pd^{IV}(NO_x)(CH_2CMe_2-*o*-C_6H_4)(κ^3-Tp^x)] (Tp^x = Tp or Tp*; $x = 1 - 3$). Single-crystal X-ray investigation of [Pd^{IV}(NO)(CH_2CMe_2-*o*-C_6H_4)(κ^3-Tp*)] evidenced the presence of strongly bent nitrosyl ligand (Fig. 2.117).[496]

The same metallacycle K[Pd(–CH_2CMe_2-*o*-C_6H_4)(κ^2-Tp)] reacts with CH_2Cl_2 giving the stable [Pd(–CH_2CMe_2-*o*-C_6H_4)(κ^3-Tp)-

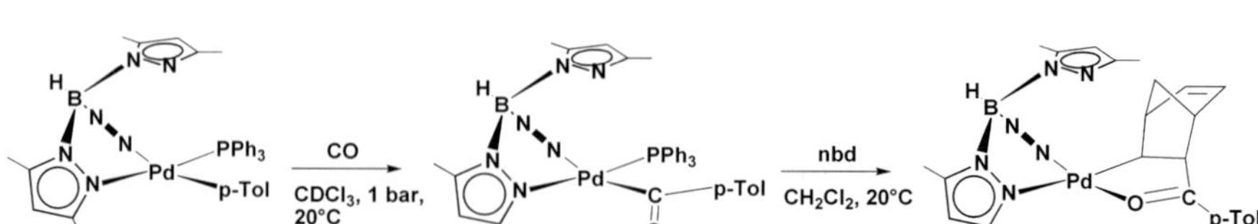

Fig. 2.116. The reaction of [Pd(*p*-Tol)(Tp*)(PPh_3)] with CO and NBD.

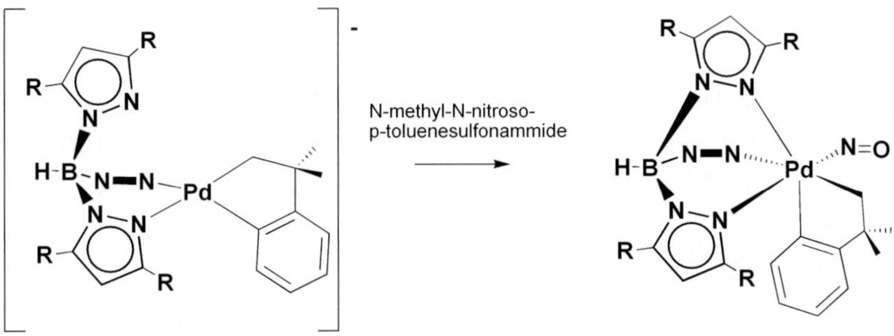

Fig. 2.117. Synthesis of $[\overline{Pd(NO)(CH_2CMe_2\text{-}o\text{-}C_6H_4})(\kappa^3\text{-}Tp^*)]$ containing a bent NO.

$(CH_2Cl)]$. The reaction with CH_2Br_2 and CH_2I_2 also yields six-membered metallacycles $[\overline{Pd(-CH_2CMe_2\text{-}o\text{-}C_6H_4(CH_2)})(\kappa^3\text{-}Tp)X]$ (X = Br or I), due to the insertion of CH_2 into the Pd–C_{aryl} bond. $[\overline{Pd(-CH_2CMe_2\text{-}o\text{-}C_6H_4(CH_2)})(Br)(\kappa^3\text{-}Tp)]$ has been structurally characterized (Fig. 2.118).[497] Furthermore, chemical oxidation of $(K)[\overline{Pd(-CH_2CMe_2\text{-}o\text{-}C_6H_4})(\kappa^2\text{-}Tp^x)]$, in the presence of additional ligands, yielded cationic Pd(IV) complexes (Fig. 2.119). An electrochemical study has been perfomed that supports the stability of the +4 oxidation state due to the electron-donor capability of the added co-ligand and to the steric features of the Tp^x ligand.[498]

When $[Pd(OH)(Tp^*)(py)]$ reacts with X-CH_2-Y (dicyanomethane, methyl cyanoacetate, or benzoylacetonitrile) at 0°C, the N/O-bound enolato complexes, $[Pd(X\text{-}CH\text{-}Y)(Tp^*)(py)]$, are obtained as kinetic products, which gradually get converted to the more stable C-bound enolato complexes $[Pd(CHXY)(Tp^*)(py)]$, upon warming to 50°C (Fig. 2.120). X-ray single-crystal studies of these species show the Pd center in a square-planar or square-pyramidal geometry. The N-bound enolato derivatives $[Pd(\text{-}X\text{-}CH\text{-}Y)(Tp^*)(py)]$ further reacted with dicyanomethane giving the 2-cyanoethenylamido complexes $[Pd\text{-}NH\text{-}C(CH_2CN){=}CCN\text{-}(CN)(Tp^*)(py)]$. It has been found that enolato complexes containing the bulkier Tp^{iPr_2} ligand yields only the N-bound isomer $[Pd(\text{-}X\text{-}CH\text{-}Y)(Tp^*)(py)]$, due to steric repulsion between the Tp^{iPr_2} ligand and the C-bound, secondary alkyl moiety.[499]

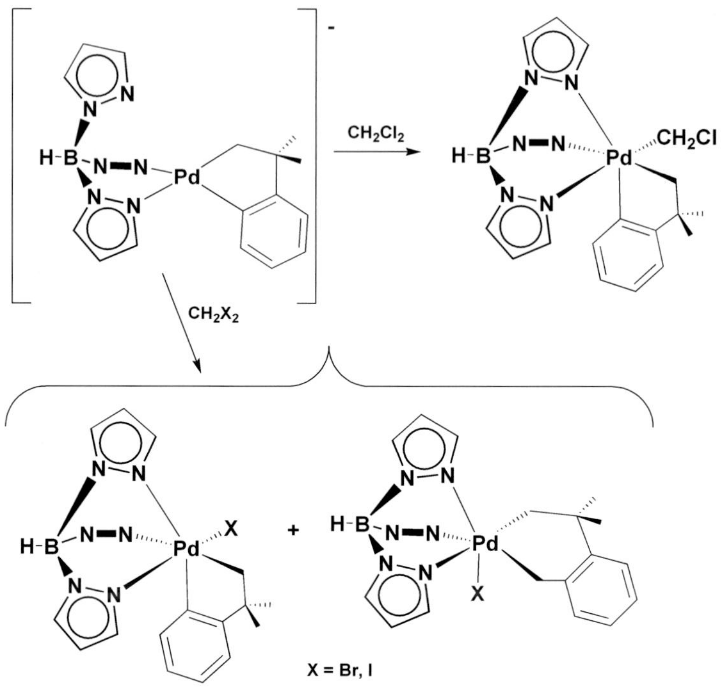

Fig. 2.118. Reaction of K[Pd(–CH_2CMe_2-*o*-C_6H_4)(κ^2-Tp)] with CH_2X_2.

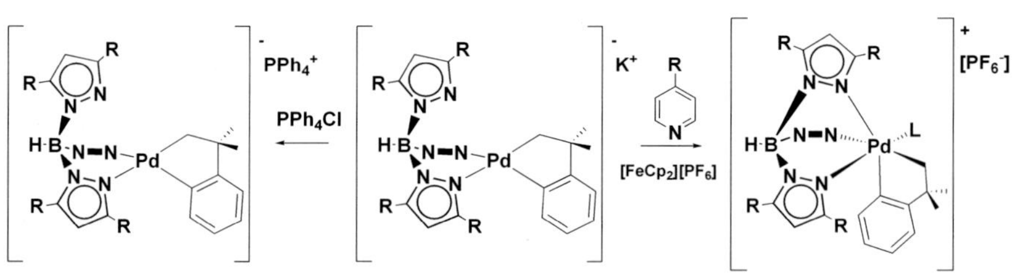

Fig. 2.119. The oxidation of (K)[$\overline{\text{Pd}(CH_2CMe_2\text{-}o\text{-}C_6}H_4$)($\kappa^2$-Tp^x)] in the presence of L.

[Pd(η^1,η^2-C_8H_{12}OMe)(Tp)]$^+$ was synthesized by the reaction of [Pd(η^1,η^2-C_8H_{12}OMe)Cl]$_2$ with Tp and characterized in solution by multinuclear and multidimensional low-temperature NMR spectroscopy.[500]

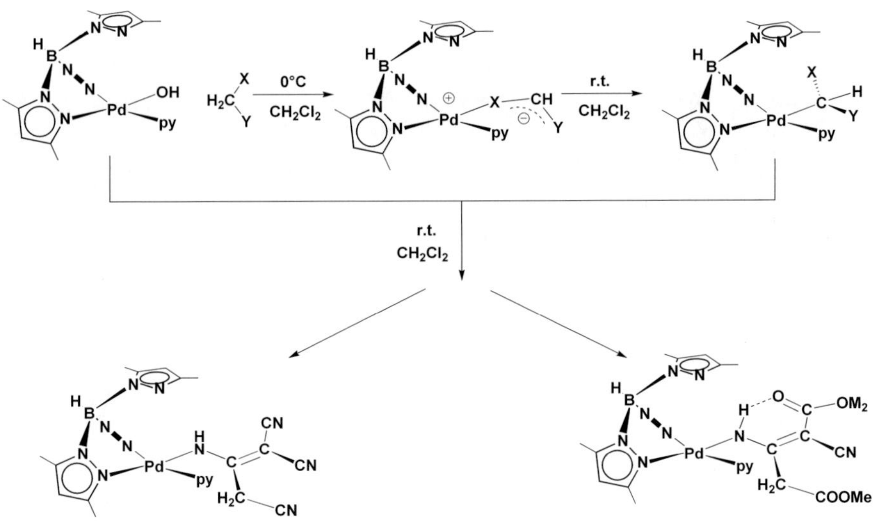

Fig. 2.120. Synthesis of the N-bound enolato derivatives [Pd(-X-CH-Y)(Tp*)(py)] and reaction with dicyanomethane.

2.11.3. *Pt*

Most of the articles published in the last 10 years on Pt(Tp^x) chemistry represents an evolution of the previous papers on $[Pt(Tp^x)(L)]^+$ (L = olefin, CO, CNR),[501–503] [PtR(Tp^x)],[504,505] and [PtR$_3$(Tp^x)].[506,507]

Upon interaction of K[PtMe$_2$(κ^2-Tp)] with SnMe$_2$Cl$_2$, [Pt(SnMe$_2$Cl)Me$_2$(κ^3-Tp)] is formed. The analogs [Pt(SnMe$_3$)Me$_2$(κ^3-Tp)], structurally characterized, exhibits a distorted octahedral geometry. The Pt–N distance *trans* to the SnMe$_3$ group is longer with respect to those *trans* to the Me groups, due to a greater *trans* influence of the trimethylstannyl group with respect to methyls.[508]

[PtMe$_2$H(Tp*)] formally inserts dioxygen into Pt–H to produce the hydroxoperoxo Pt(IV) species [PtMe$_2$(OOH)(Tp*)], which exhibits an octahedral structure with a κ^3-coordination of the Tp* ligand, confirmed by IR and single-crystal X-ray study. Thermolysis of [PtMe$_2$(OOH)(Tp*)] in benzene led predominantly to the hydroxo species [PtMe$_2$(OH)(Tp*)] (Fig. 2.121).[509]

[PtMe$_2$H(Tp*)] reacts with Et$_3$SiH forming [Pt(SiEt$_3$)(H)$_2$(Tp*)] which is the precursor to [PtH$_3$(Tp*)], a rare N-supported monomeric

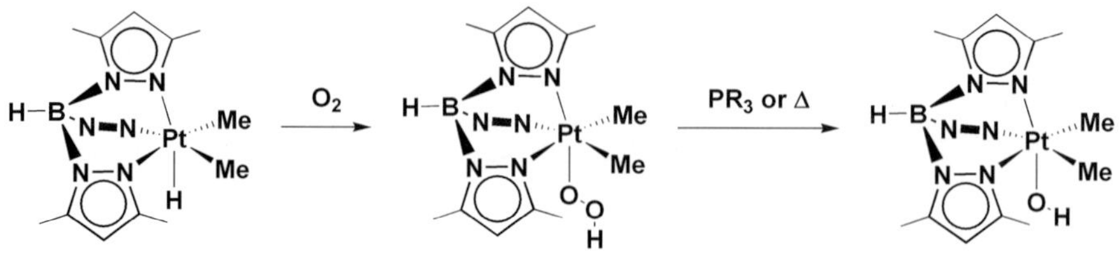

Fig. 2.121. Insertion of O_2 in [$PtMe_2H(Tp^*)$] and formation of [$PtMe_2(OH)(Tp^*)$].

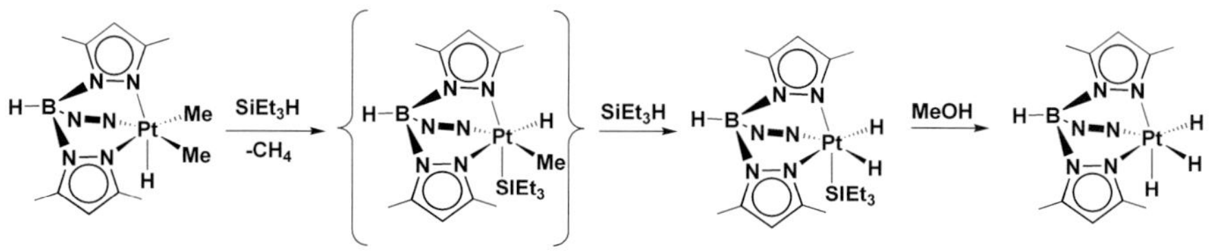

Fig. 2.122. Synthesis of [$Pt(SiEt_3)(H)_2(Tp^*)$] and [$PtH_3(Tp^*)$].

platinum trihydride complex characterized by IR and NMR studies (Fig. 2.122).[510]

[$PtMe(H)_2(Tp)$], air- and moisture-stable, characterized by NMR (^1H and ^{13}C) and FT-IR, has been prepared by the reaction of [PtMe(Tp)CO] with water. The reaction mechanism proposed consists of an initial nucleophilic attack by a water mechanism on the coordinated carbonyl in [PtMe(Tp)CO], followed by the release of CO_2 and protonation at the platinum center. [PtMe(Tp)CO] reacts with $HBF_4.OEt_2$ yielding [$Pt(\kappa^2$-TpH)(Me)(CO)][BF_4], structurally characterized.[511] [$PtMe(H)_2(Tp^*)$] can also be protonated at one of the pyrazolyl nitrogen atoms inducing methane elimination. The resulting platinum monohydride intermediate can be trapped by added ligands. In this manner, two chiral cationic complexes [$Pt(H)(L)(\kappa^2$-HTp*)][BAr_F]$_2$ (L = MeCN, CH_2=CH_2) have been prepared. When [$PtMe(H)_2(Tp^*)$] is protonated in the absence of trapping ligand, the hydride-bridged dinuclear complex [$Pt(\mu$-$H_2)(\kappa^2$-HTp*)]$_2$[BAr_F]$_2$ is formed, which is structurally identified (Fig. 2.123).[512]

[$PtMe(H)_2(Tp)$] is remarkably resistant to methane loss. When it is heated in methanol, methane or hydrogen loss is not observed.

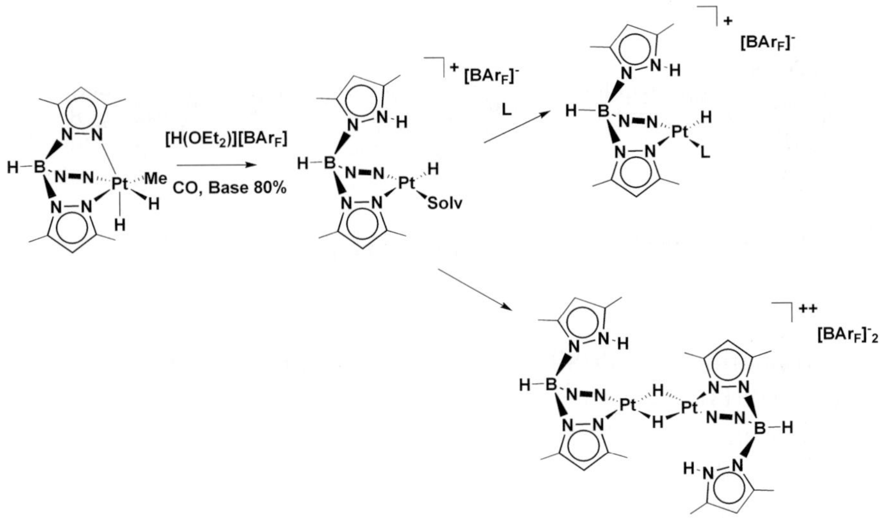

Fig. 2.123. Protonation of [PtMe(H)$_2$(Tp)] in the presence or absence of trapping ligand L.

When CD_3OD is used, reversible H/D scrambling of the hydrides and of the methyl hydrogens occurs. By studying the reactivity by DFT methods at the mPW1k/LANL2DZ+P//mPW1k/LANL2DZ level, it has been found that methane loss cannot occur due to the rigidity of the Tp ligand, not allowing the *trans* geometry required for the product of methane elimination, PtH(Tp) which has a very high energy. However, H/D scrambling of the methyl ligand is possible through the $\eta^{2\text{-CH}}$-CH_4 complex even though methane loss is not observed. In order to assess that the cause of the reactivity is the rigidity of the Tp systems, the model compound [PtMe(H)$_2$(NH$_3$)$_3$]$^+$ has also been investigated.[513] [PtMe(H)$_2$(Tp*)] also reacts with substoichiometric amounts of B(C$_6$F$_5$)$_3$ in aromatic solvents forming platinum(IV) aryl dihydride complexes via borane-induced reductive elimination of methane from [PtMe(H)$_2$(Tp*)] and subsequent activation of the aromatic solvent. [Pt(C$_6$H$_5$)H$_2$(Tp*)], [Pt(Tol)H$_2$(Tp*)], [Pt(3,4-Me$_2$C$_6$H$_3$)H$_2$(Tp*)], [Pt(3,5-Me$_2$C$_6$H$_3$)H$_2$(Tp*)], and [Pt(2,5-Me$_2$C$_6$H$_3$)H$_2$(Tp*)] have been described, and the octahedral [Pt(3,5-Me$_2$C$_6$H$_3$)H$_2$(Tp*)] also structurally characterized.[514]

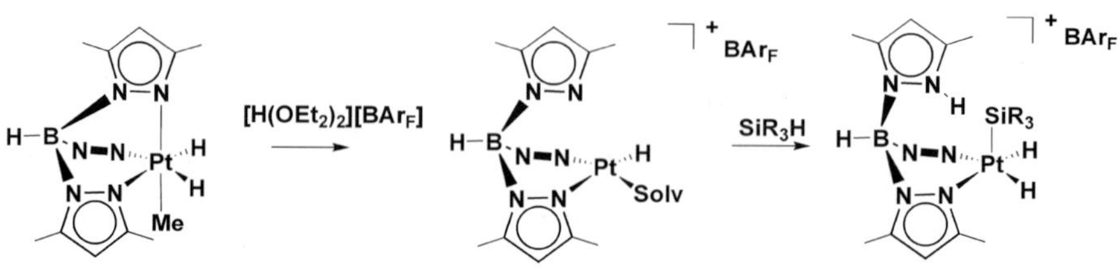

Fig. 2.124. Formation of $[Pt(H)_2(SiEt_3)(\kappa^2\text{-}HTp^*)][BAr_F]$.

Protonation of $[Pt(SiEt_3)(H)_2(\kappa^3\text{-}Tp^*)]$ with $[H(OEt_2)_2][BAr_F]$ yielded $[Pt(H)_2(SiEt_3)(\kappa^2\text{-}HTp^*)][BAr_F]$ which can also be obtained by protonation of $[PtMe(H)_2(\kappa^3\text{-}Tp^*)]$ and subsequent trapping with $SiEt_3H$ (Fig. 2.124). The chiral intermediate $[Pt(solv)(H)(\kappa^2\text{-}HTp^*)][BAr_F]$ has been trapped with SiR_3H ($R_3 = Et_3, Ph_3, Ph_2Et$). The triethylsilane complex $[Pt(H)_2(SiEt_3)(\kappa^2\text{-}HTp^*)][BAr_F]$ has been structurally characterized.[515]

Protonation of a pyrazole promotes reductive elimination of methane from $[PtMe(H)_2(Tp^x)]$ yielding an isolable cationic Pt(II) hydride complex after addition of a trapping ligand, whereas protonation of $[Pt(Ph)(H)_2(Tp^x)]$ with $[H(OEt_2)_2][BAr_F]$ yields the chiral compound $[Pt(H)(\kappa^2\text{-}HTp^x)(C_6H_6)][BAr_F]$ that, *via* an oxidative addition, forms the pentacoordinate $[Pt(H)(H)(C_6H_5)(\kappa^2\text{-}HTp^x)][BAr_F]$.[516] In fact, protonation of $[Pt(R')(H)(R)(Tp^*)]$ [$R' =$ aryl; $R = H$, Ph] with $[H(OEt_2)_2][BAr_F]$ results in the formation of cationic platinum(II) η^2-arene complexes $[Pt(\eta^2\text{-}HR')(R)(\kappa^2\text{-}HTp^*)][BAr_F]$ [$R = H; R' = MeC_6H_4, R = H; R' = 3{,}6\text{-}Me_2C_6H_3, R = H; R = R' = C_6H_5$]. Hydrogen exchange process between the bound benzene and the hydride ligand for $[Pt(\eta^2\text{-}HPh)(H)(\kappa^2\text{-}HTp^*)][BAr_F]$ indicates that it proceeds via arene C–H oxidative addition to give the five-coordinate platinum(IV) phenyl dihydride $[Pt(C_6H_5)(H)_2(\kappa^2\text{-}HTp^*)][BAr_F]$. The kinetic studies on these complexes provide information on the energetics of arene C–H bond activation.[517]

$[PtH_3(Tp^*)]$ has also been synthesized through the sequence of clean reactions reported in Fig. 2.125. The acid-assisted reductive elimination of hydrogen from $[PtH_3(Tp^*)]$ and of methane and hydrogen from $[PtMeH_2(Tp^*)]$ has been examined by the same authors,

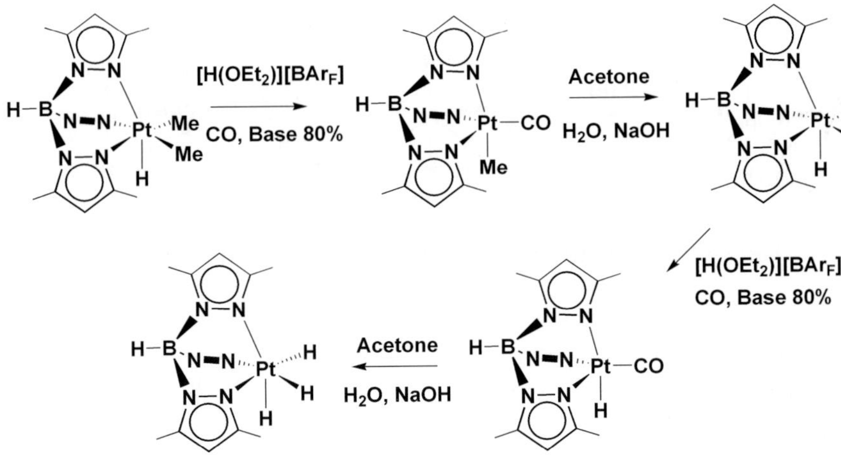

Fig. 2.125. Multistep synthesis of [PtH_3(Tp*)].

reporting the loss of H_2 from solutions containing platinum(IV) complexes of the type [Pt(R)$(H)_2$(Tp*)] (R = Me, H) upon protonation and interaction with CO. The formation of [Pt(R)(L)(κ^2-HTp*)][BAr_F] is described. The elimination reaction likely occurs from cationic six-coordinate [Pt(R)$(H)_2$(L)(κ^2-HTp*)][BAr_F] species. The geometry of [Pt(H)(CO)(Tp*)], structurally characterized, lies between square-planar and trigonal-bipyramidal coordination environment.[518]

Gentle heating of [$PtMe_2H$(Tp*)] in alkanes in the presence of $B(C_6F_5)_3$ results in C–H activation of alkanes R–H, to give [Pt(Me)(H)(R)(Tp*)] intermediates (Fig. 2.126). Further heating leads to the formation of [Pt(H)(Tp*)(η^2-olefin)] via methane elimination followed by β-hydride elimination. [Pt(H)(κ^2-HTp*)(η^2-cyclo-C_6H_{10})][BAr_F] has also been structurally characterized.[519]

Chiral [Pt(R)(L)(κ^2-HTp*)][BAr_F] (R = Me, L = MeCN, ^{t}BuNC, py, CO, $CH_2{=}CH_2$, PMe_2Ph; R = Ph, L = MeCN) have been prepared by the methods analogous to that above reported. Addition of 2 equiv. of PMe_2Ph as a trapping ligand produces *trans*-[Pt(Me)(κ^1-HTp*)$(PMe_2Ph)_2$][BAr_F], a square-planar complex containing a monodentate protonated tris(pyrazolyl)borate ligand (Fig. 2.127). Deprotonation of the pyrazolium ring gave the neutral platinum(II) complexes of the type [PtR(Tp*)(L)] (R = Me, L = MeCN, SMe_2, CO,

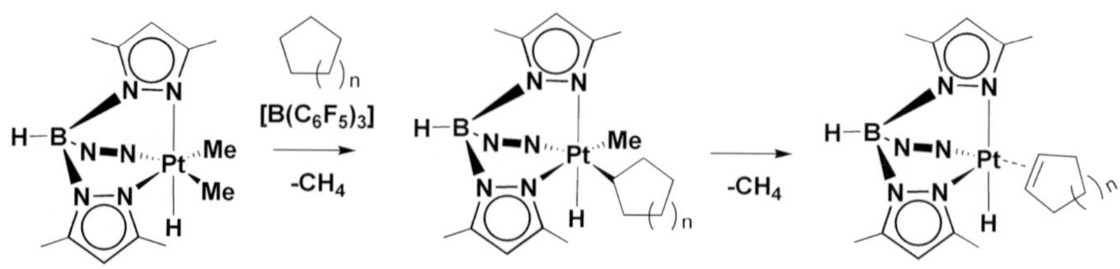

Fig. 2.126. C–H activation of cycloalkanes with $[PtMe_2H(Tp^*)]$.

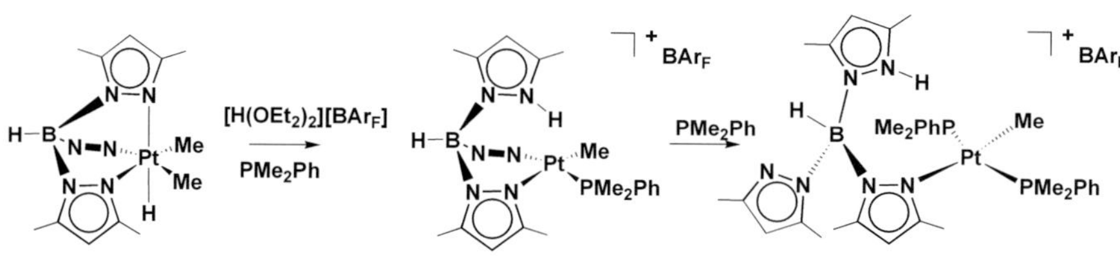

Fig. 2.127. Reaction of $[PtMe_2H(Tp^*)]$ with $[H(OEt_2)_2][BAr_F]$ and PMe_2Ph.

$CH_2{=}CH_2$; R = Ph, L = MeCN). These complexes exhibit square-planar coordination with a bidentate Tp* ligand, with the exception of $[PtMe(CH_2{=}CH_2)(Tp^*)]$ trigonal bipyramidal. $[PtMe(Tp^*)(CO)]$ is present in solution in both isomeric forms: the four-coordinate square-planar and five-coordinate trigonal-bipyramid species rapidly interconvert in solution. The structures of $[Pt(Me)(MeCN)(\kappa^2\text{-}HTp^*)][BAr_F]$, $[Pt(Me)(\kappa^2\text{-}HTp^*)(L)][BAr_F]$ (L = CO, $CH_2{=}CH_2$, PMe_2Ph), *trans*-$[Pt(Me)(\kappa^1\text{-}HTp^*)(PMe_2Ph)_2][BAr_F]$, and of the neutral square planar complex $[PtMe(Tp^*)(SMe_2)]$ have been determined by single-crystal X-ray study.[520]

Solution thermolysis of $[Pt^{IV}(Me)_2H(\kappa^3\text{-}Tp^*)]$ also induces reductive elimination of methane. C–D bond activation occurs after methane elimination in benzene-d_6, to yield $[Pt^{IV}(Me)(C_6D_5)D(\kappa^3\text{-}Tp^*)]$ (Fig. 2.128), that upon a second reductive elimination/oxidative addition reaction forms isotopically labeled methane and $[Pt^{IV}(C_6D_5)_2D(\kappa^3\text{-}Tp^*)]$. $[Pt(Me)(NCCD_3)(\kappa^2\text{-}Tp^*)]$ has been obtained upon elimination of methane from $[Pt^{IV}(Me)_2H(\kappa^3\text{-}Tp^*)]$. The σ-methane complex $[Pt^{II}(Me)(\kappa^2\text{-}Tp^*)(CH_4)]$ has been indicated

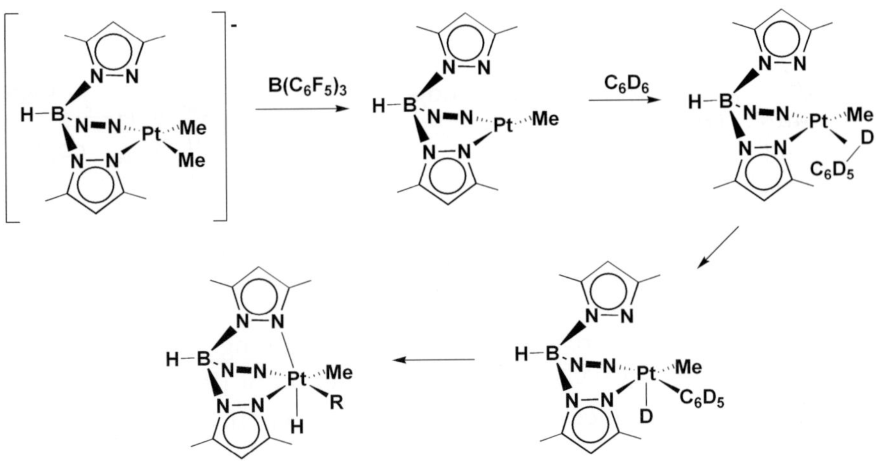

Fig. 2.128. C–D bond activation studies with $[Pt^{IV}(Me)_2H(\kappa^3\text{-Tp}^*)]$.

as intermediate in both the scrambling and reductive elimination processes. $[Pt(Ph)_2H(\kappa^3\text{-Tp}^*)]$ has also been structurally characterized.[521]

When nitrile ligands were added at low-temperature to $[Pt(R)(\kappa^2\text{-HTp}^*)(\eta^2\text{-}C_6H_6)][BAr_F]$ (R = H, Ph), $[Pt(C_6H_5)(H)(R)(NCR')(\kappa^2\text{-HTp}^*)][BAr_F]$ (R = Ph, H, R′ = Me, CMe_3, *p*-XC_6H_4) is formed, whereas at higher temperature irreversible loss of benzene occurred and $[Pt(R)(NCR')(\kappa^2\text{-HTp}^*)][BAr_F]$ afforded.[522]

[Pt(Me)(Tp)] is described as a new alkyne protecting group that can tolerate various reaction conditions and is particularly useful for electron-deficient acetylenes (Fig. 2.129). The protecting group can be used under a variety of reaction conditions, including basic and acidic media, environment for catalytic hydrogenation, and for chromate oxidation reaction. The alkyne product can be released from the protecting complex by CO under neutral conditions at room temperature.[523]

The structure of $[Pt(Me)_3(Tp^*)]$ was compared with that of the distorted octahedral Pt(IV) complex containing the $Tp^{(CF3)}$ ligand. The study supports a substantially weaker coordination for the fluorinated Tp ligand, the Pt–N being significantly longer and the Pt–Me slightly shorter in the fluorinated complex.[524]

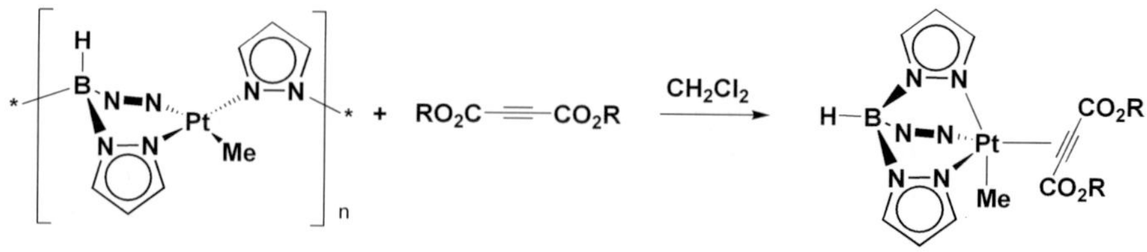

Fig. 2.129. [Pt(Me)(Tp)], a good alkyne protecting group.

[PtMe(Tp*)(CO)] reacts with MeLi forming the derivative [PtMe(κ^2-Tp*)(C(=O)Me)]Li, whereas reaction of [PtMe(Tp*)(CO)] with $NaHBEt_3$ produces [PtMe(C(=O)H(κ^2-Tp*)]. The reaction of [Pt(Me)(C(=O)R)(κ^2-Tp*)] with R′X yields [PtMe(R′)(C(=O)R)(κ^3-Tp*)] (R = Me, R′ = H; R = Ph, R′ = H; R = R′ = Me; R = Ph, R′ = Me; R = R′ = H) (Fig. 2.130). These species have been investigated by solution NMR.[525]

Experimental and theorethical results suggested that the proton exchange between water and Me group in [PtMe(Tp)(CO)] involves the formation and deprotonation of a σ-methane ligand.[526]

The *ortho*-metalated [Pt(CH_2CH_2-*o*-C_6H_4)(H)(Tp*)] has been prepared by heating [Pt(Ph)(η^2-CH_2=CH_2)(Tp*)] in benzene, phenyl migration to ethylene likely giving [PtCH_2CH_2Ph(Tp*)]. The *ortho*-metalated species afforded from gentle heating with the Lewis acid $B(C_6F_5)_3$ in ethylbenzene. Low-temperature protonation of [Pt(Ph)(η^2-CH_2=CH_2)(Tp*)] results in the formation of [Pt(C_6H_5)(η^2-CH_2=CH_2)(κ^{2}HTp*)][BAr_F]. Low-temperature proto nation of the *ortho*-metalated phenethyl hydrido species

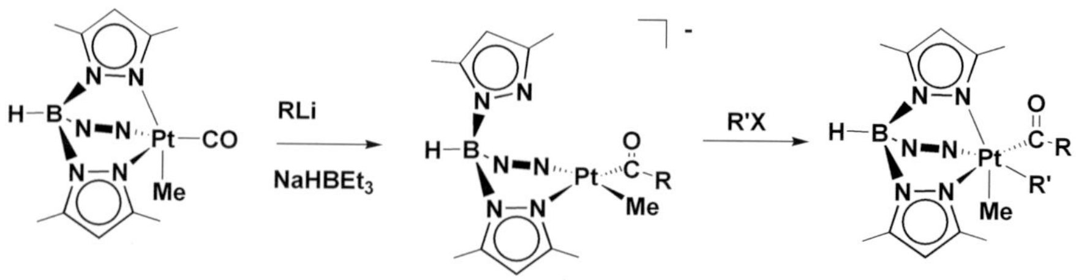

Fig. 2.130. Synthesis of [PtMe(R′)(C(=O)R)(κ^3-Tp*)].

followed by addition of MeCN upon reductive coupling of the Pt–H and the alkyl methylene group gave $[Pt(C_6H_4\text{-}2\text{-}CH_2CH_3)(\kappa^2\text{-}HTp^*)(NCMe)][BAr_F]$. The synthesis of the analogous metallacycles, $[Pt(CH_2CH(Me)\text{-}o\text{-}C_6H_4)(H)(Tp^*)]$ and $[Pt(CH_2CH_2\text{-}o\text{-}C_6H_3Et)(H)(Tp^*)]$ has been described. Heating $[Pt(Me)(H)_2(Tp^*)]$ with $B(C_6F_5)_3$ in either ethylbenzene or 2-propylbenzene gave $[Pt(Ar)(H)_2(Tp^*)]$. The 1,2-insertion products, $[Pt(CH(Me)CH_2\text{-}o\text{-}C_6H_4)(H)(Tp^*)]$ and $[Pt(CH_2CH(Me)\text{-}o\text{-}C_6H_4)(H)(Tp^*)]$, have been reported. The serendipitous synthesis of $[Pt(C_6H_5)(\kappa^2\text{-}HTp^*)(OH_2)][BF_4]$ is also described.[527]

The dimetallic platinum complex $[\{(C_6F_5)_2Pt\}_2(\kappa^2\text{-}pzTp)]^-$ has been characterized by NMR which shows the occurrence of a very fast intramolecular interconversion between the conformational isomers arising from the boat-like $M(N\text{–}N)_2B$ ring inversion (Fig. 2.131).[528]

$[Pt(\eta^1,\eta^2\text{-}C_8H_{12}OMe)Tp]^+$ and $[Pt(\eta^1,\eta^2\text{-}C_8H_{12}OMe)(Bp)]^+$ were synthesized by the reaction of the dimers $[Pt(\eta^1,\eta^2\text{-}C_8H_{12}OMe)Cl]_2$ with the corresponding poly(pyrazolyl)borate. These derivatives were characterized in solution by multinuclear and multidimensional low-temperature NMR spectroscopy. The five-coordinate $[Pt(\eta^1,\eta^2\text{-}C_8H_{12}OMe)Tp]^+$ was investigated by X-ray single-crystal studies and the relative cation–anion position (interionic structure) determined in solution at room and low temperature by ^{19}F and ^1H-HOESY NMR spectroscopy.[500]

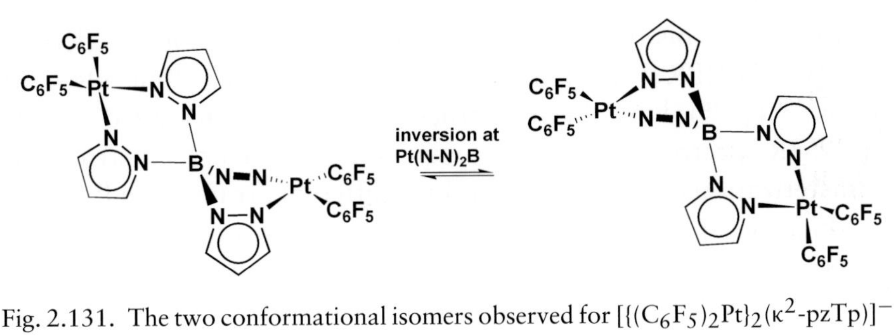

Fig. 2.131. The two conformational isomers observed for $[\{(C_6F_5)_2Pt\}_2(\kappa^2\text{-}pzTp)]^-$.

2.12. Group 11: Cu, Ag, and Au

2.12.1. *Cu*

Cu(I) and Cu(II) complexes containing homoscorpionates ligands have been widely reported due to their potential applications in catalysis and enzyme modeling. Homoleptic $[Cu(Tp^x)_2]$,[529,530] dimeric $[Cu(Tp^x)]_2$[531] and $[Cu(Tp^x)]_2O$,[532] $[Cu(Tp^x)(PR_3)]$[533,534] and $[Cu(Tp^x)(CO)]$[531] species have been characterized with a number of experimental techniques and theoretical calculations.

$[Cu(Tp)_2]$, synthesized according to literature,[529] has been structurally characterized, two independent, centrosymmetric half-molecules per asymmetric unit, both showing Jahn–Teller elongated six-coordinate geometries but different fluxionality regimes being present in the crystals. EPR and theoretical studies have been performed and the the d–d transition energies of the two independent molecules found have also been calculated by a time-dependent DF calculation.[535]

Reaction of $Cu(OAc)_2{\cdot}2H_2O$ with $K(Tp^x)$ (Tp^x = Tp or Tp*) and Hpz in methanol gave the dimeric complexes $[Cu(\mu\text{-pz})(\kappa^3\text{-}Tp^x)]_2$, structurally characterized (Fig. 2.132).[536]

It has been reported that it is possible to fix the preformed copper poly(pyrazolyl)borate complexes Cu(Tp*) and Cu(pzTp) on silica gel and use them under heterogeneous conditions as catalysts for the olefin cyclopropanation reaction. The catalytic activity is similar to that found in homogeneous conditions as a consequence of a ligand-to-support interaction that likely involve the hydroxyl groups of the silica gel surface and the borohydride B–H or the nitrogen atom of a pyrazolyl ring.[537]

Mono and bis-(trialkyl) and triarylphosphine copper derivatives $[Cu(Tp^x)PR_3]$ containing Tp, Tp*, or pzTp (but also Bp and $pz^{Me}Tp^{Me}$) have been synthesized and characterized by IR and NMR spectroscopies. Low-temperature single-crystal structural characterization indicated at least three different arrays: (a) a P_2CuN_2 coordination sphere, the scorpionate acting as a bidentate; (b) a $PCuN_3$ coordination sphere, the scorpionate acting as a tridentate; (c) a $PCuN_2$ coordination sphere, the scorpionate acting as

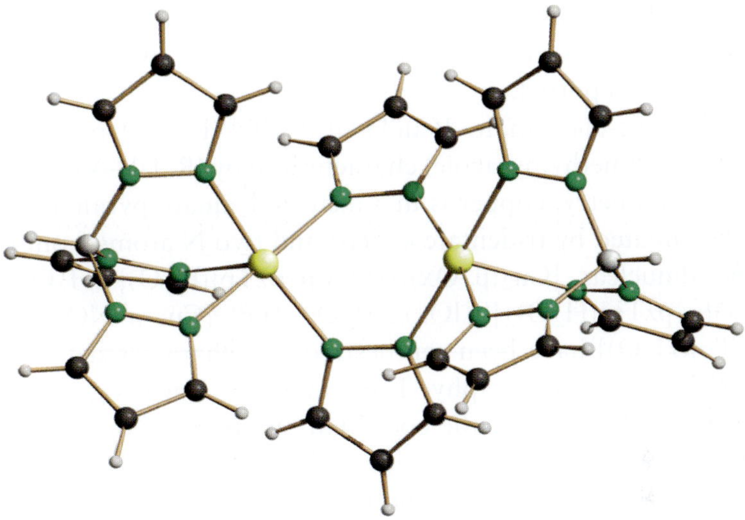

Fig. 2.132. The dimeric copper complex $[Cu(\mu\text{-pz})(\kappa^3\text{-Tp})]_2$.

a bidentate.[538] Volatility studies, electrospray mass spectra and IR in vapor phase have been carried out for a series of $[Cu(Tp^x)PR_3]$ (including Tp). XRD data of films obtained from MOCVD experiments on selected derivatives have also been reported.[539]

Zink *et al.* reported[540] the emission and the UV–vis absorption spectra of $[Cu(Tp)(AsPh_3)]$, $[Cu(Tp)(PPh_3)]$, and $[Cu(Tp)(NEt_3)]$. The spectra of the arsine complex contain some low-energy bands not present in the spectra of the amine and of the phosphine analogs. A lowest energy electronic transition has been assigned to ligand-to-ligand charge transfer with some contribution from the metal. Molecular orbital calculations PM3(tm) indicate HOMO to consist primarily of π orbitals on the Tp ligand and the LUMO to be primarily antibonding orbitals on the $AsPh_3$ ligand.

Copper(I) complexes $[Cu(L)(Tp^x)]$ (L = 4-(diphenylphosphane) benzoic acid, Tp^x = Tp, pzTp, or Tp*) have been synthesized and characterized by FT-IR in the solid state, and by NMR (^1H and $^{31}P\{^1H\}$) spectroscopy, ESI MS, and conductivity measurements in solution. The solution data are in accordance with partial dissociation of the sterically hindered complexes due to breaking of Cu–P and Cu–N bonds.[541,542] Chemiluminescence technique was used to evaluate the [Cu(L)(pzTp)] superoxide scavenging activity.[542] [Cu(pz^{0}Tp^{Me})(L)] has also been reported.

In the five-coordinate [Cu(Tp*)L]ClO_4 (L = 1-methylcarbaldimino-3,5-dimethylpyrazole) characterized by IR, UV–Vis, EPR, and cyclic voltammetry, copper is in a distorted square-pyramidal geometry coordinated by tridentate κ^3-Tp* and two N atoms from L.[543]

The dinuclear [Cu_2(μ-Ox)(NO_3){μ-pzTp}(H_2O)]·$2H_2O$, [Cu_2(μ-Ox){μ-pzTp}(H_2O)$_2$](ClO_4)·$2H_2O$, and [Cu_2(μ-NO_3)(NO_3)$_2${μ-pzTp}(H_2O)] have been synthesized and characterized by single-crystal X-ray crystallography. The oxalate derivatives generate one-dimensional (ID) chains, the coordinating atoms of the oxalate and pzTp occupying the basal positions of a CuN_2O_3 square-pyramidal coordination environment. In the 1D polymeric structure of the paramagnetic [Cu_2(μ-NO_3)(NO_3)$_2${μ-pzTp}(H_2O)], the copper(II) centers are linked through nitrate and pzTp, each copper atom being in a distorted octahedral CuN_2O_4 coordination geometry. Electronic spectra and VT magnetic susceptibility measurements in the range 20–300 K indicate that the oxalate complexes possess strong antiferromagnetic behavior. In these species, the pzTp acts as a bridge and appears apparently ineffective in mediating magnetic exchange interaction due to nonavailability of orbitals for exchange interactions.[544]

[Cu(Tp)]$_2$(μ-N_3)$_2$ has been synthesized and structurally characterized. The two Cu atoms have a distorted square-pyramidal coordination and are bridged by two end-on azide groups. Weak C–H···N contacts are present in the two-dimensional network.[545]

Aullón *et al.*[546] investigated, by DFT methods, the effect of the substituents of the tris(pyrazolyl)borate ligands on the structure of dinuclear [Cu_2(μ-O)$_2$(Tp^x)$_2$] complexes, analogs of the active site of oxyhemocyanin. They reported that the choice of peroxo or bisoxo form of the bridging oxygen is strongly influenced by the nature and position of the R substituents, because of variable substituent bridging oxygen interactions (R···O), as well as electronic effects. The electronic effects of the substituent at the 5-position are not significant. The energy profiles of compounds with several substitution patterns clearly indicate that, upon fluorination of position 3 of the pyrazolyl

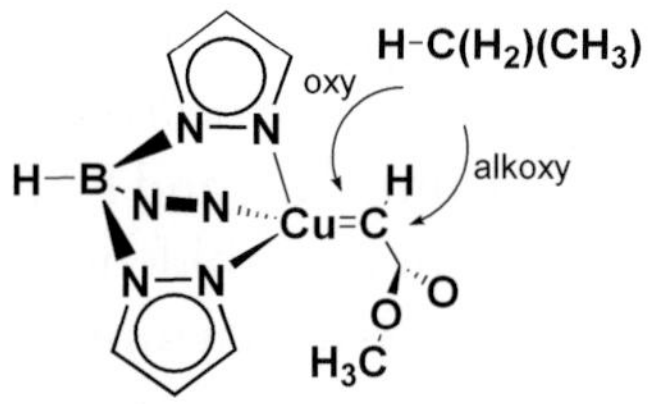

Fig. 2.133. The metallocarbene homoscorpionate [$Cu{=}C(H)(CO_2CH_3)(Tp)$], able to activate C–H bonds in alkanes.

rings (either directly or by attaching a fluoroalkyl group), it becomes harder to split the O–O bond, i.e. fluorination makes the bonded peroxo closer to free dioxygen. Quantum chemistry investigations on the cuprous tripodal species [Cu(Tp)] have also been performed. The structures of the complex, the coordination mode (η^2 side-on or η^1 end-on) of O_2 to Cu^I, and the charge transfer from the metal to dioxygen have been investigated together with the interaction with MeCN or CO. It has been reported that any factor favoring pyramidalization at copper favors charge transfer and thus coordination of the incoming O_2 moiety.[547]

Substituents' influences on the [$CuNO_3(Tp^x)$] structure have been described by Fujisawa *et al.*[548] which indicates the influence of both 5- and 3-positions in the structural parameters.[548]

A computational study of the activation of alkane (methane, ethane, propane, and butane) C–H bonds by the metallocarbene homoscorpionate [$Cu{=}C(H)(CO_2CH_3)(Tp)$] (Fig. 2.133) and [$Cu{=}C(H)(CO_2CH_3)(Tp^{Br3})$] has been performed with DFT Becke3LYP calculations. A low-barrier transition state where the key bond-breaking and bond-forming processes take place in a concerted way has been postulated. The transition state has several possible conformations.[549]

2.12.2. *Ag*

Until 1999, Ag(Tp^x) chemistry has been limited to phosphino and isonitrile complexes.[550–552]

The dinuclear $[Ag(Tp)]_2$, $[Ag(Tp^*)]_2$, $[Ag(pzTp)]_2$, (but also $[Ag(Tp^{Me})]_2$ and $[Ag(Tp^{4\text{-}Br})]_2$) have been synthesized and characterized by IR, far-IR, NMR (^1H and ^{13}C), and FAB MS spectroscopy, and their interaction toward unidentate or bidentate N- or S-donor ligands was investigated.[553] $[Ag(Tp^*)]_2$ is an asymmetric dimer, in which each Tp ligand bridges both metals by two pyrazolyl donors to one Ag(I) center and one to the other. Even if, at a first approximation, one Ag(I) can be considered four-coordinate and the other two-coordinate, two of the pyrazolyl donors around the "four-coordinate" metal centers are also involved in weak, asymmetric bridging interactions with the "two-coordinate" metal centers (Fig. 2.134).[554]

Synthesis and spectral data for $[Ag(Tp^x)(CNR)]$ and $[Ag(pz^xTp^x)(CNR)]$ (Tp^x = Tp^*, Tp^{Ph_2}; pz^xTp^x = pzTp or $pz^{Me}Tp^{Me}$; = ^tBu, Cy) have been reported. $[Ag(Tp^*)(CNR)]$ (R = ^tBu, Cy) are all unsolvated mononuclear arrays, with tridentate Tp and unidentate CNR making up the four-coordinate environment of the silver atom. [Ag(pzTp)(RCN)] reacts with imH yielding [Ag(pzTp)(imH)], whereas $[Ag(pz^{Me}Tp^{Me})(CNCy)]$ reacts with imH forming $[Ag(pz^{Me}Tp^{Me})(imH)(CNCy)]$.[555]

Silver(I) derivatives $[Ag(Tp^x)(P^iBu_3)]$ (Tp^x = Tp, Tp^* Tp^{Me}, Tp^{CF_3}, Tp^{4Br}), $[Ag(pz^xTp^x)(P^iBu_3)]$ (pz^0Tp^x = pzTp, pz^0Tp^{Me}) have been synthesized from the reaction of $[Ag(NO_3)(P^iBu_3)_2]$ with the appropriate potassium or sodium pyrazolylborates and characterized both

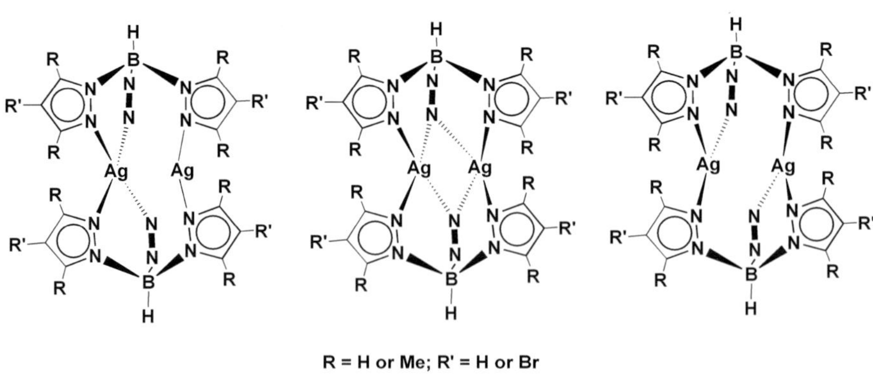

Fig. 2.134. Possible structures for the dinuclear $[Ag(Tp^x)]_2$.

in solution (^{1}H- and ^{31}P{^{1}H} NMR, ESI MS spectroscopy, conductivity) and in the solid-state (IR, single-crystal X-ray structure analysis). These complexes are air-stable, light-sensitive, and nonelectrolytes in CH_2Cl_2 and acetone in which they slowly decompose, yielding metallic silver and/or azolate species of formula $[Ag(pz^x)]_n$ upon breaking of the bridging B–N(azole) bond. The solid state structures of $[Ag(Tp)(P^iBu_3)]$, $[Ag(Tp^{Me})(P^iBu_3)]$, and $[Ag(Tp^{CF_3})(P^iBu_3)]$ show that the silver atom adopts a distorted tetrahedral coordination geometry. ^{31}P{^{1}H} NMR variable temperature NMR studies showed that the scorpionate ligand acts as a bidentate donor, whereas tridentate coordination is found for all tris(pyrazolyl)borate derivatives. ESI MS data suggest the existence in solution of species such as $[Ag(P^iBu_3)_2]^+$ upon dissociation of the Tp^x ligand, and also the formation of dimeric species of the form $[Ag_2(Tp^x)(P^iBu_3)_2]^+$.[556] $[Ag(pzTp)(PPh_3)_2]$ has also been described and structurally characterized.[557] Ag complexes containing Tp, Tp*, and pzTp ligands have also been investigated by ^{31}P cross-polarization magic-angle-spinning (CPMAS) NMR spectroscopy. In the same paper the structure of $[Ag(Tp^*)(PPh_3)]\cdot 2MeCN$ is described.[558]

Silver(I) derivatives containing bidentate tertiary phosphines and pzTp have been prepared from $AgNO_3$ and dppe, and dppf. Their solid state and solution properties have been investigated through analytical and spectroscopic measurements. The structures of [Ag(pzTp)(dppf)] have been determined by single-crystal X-ray studies, the tetrakis(pyrazolyl)borate being always bidentate with the silver(I) center in distorted tetrahedral geometries.[559]

A computational study of the activation of alkane (methane, ethane, propane, and butane) C–H bonds by silver carbene homoscorpionate $[Ag{=}C(H)(CO_2CH_3)(Tp)]$ and $[Ag{=}C(H)(CO_2CH_3)(Tp^{Br3})]$, analogous to that reported for copper,[549] has been performed with DFT Becke3LYP calculations.

2.12.3. *Au*

Gold-Tp^x chemistry[560,561] continues to be very limited.

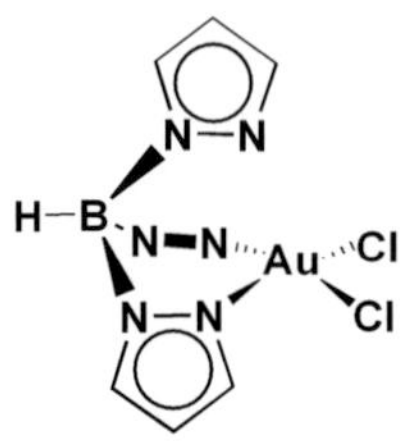

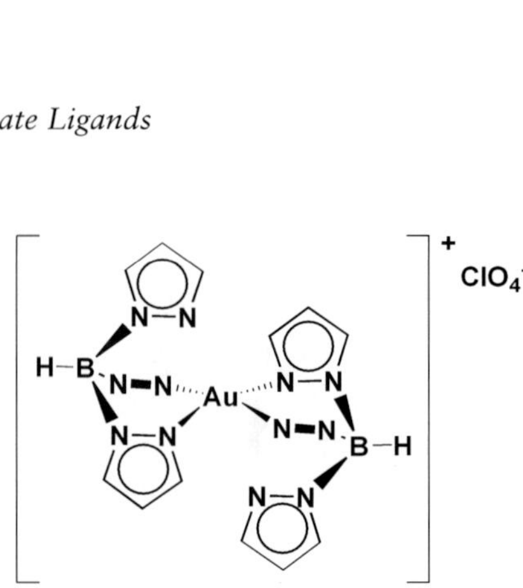

XIII XIV

Fig. 2.135. $[AuCl_2(\kappa^2\text{-}Tp)]$ (XIII) and $[Au(\kappa^2\text{-}Tp)_2]ClO_4$ (XIV).

Na(Tp) and $Na[AuCl_4]\cdot 2H_2O$ reacted in H_2O yielding $[AuCl_2(\kappa^2\text{-}Tp)]$ (**XIII**, Fig. 2.135). The same reaction carried out in the presence of $NaClO_4$ gave $[Au(\kappa^2\text{-}Tp)_2]ClO_4$ (**XIV**, Fig. 2.135). The reaction of Na(pzTp) with the cyclometalated gold derivatives $[AuRCl_2]$ and $NaClO_4$ yielded $[Au(R)(\kappa^2\text{-}pzTp)]ClO_4$ (R = κ^2-$(C,N)C_6H_4CH_2NMe_2$-2) (**XV**, Fig. 2.136), and in the absence of $NaClO_4$ $[Au(R)Cl(\kappa^2\text{-}pzTp)]$ (R = $C_6H_3(N{=}N{=}N\text{-}C_6H_4Me\text{-}4')$-2-Me-5) (**XVI**, Fig. 2.136). The square-planar $[AuCl_2(\kappa^2\text{-}Tp)]$ and $[Au(R)(\kappa^2\text{-}pzTp)]ClO_4.CHCl_3$ have been crystallographically characterized.[562]

Gold derivatives $[Au(Tp^x)(PR_3)]$ (Tp^x = Tp, or Tp*; R = Ph or ^{t}Bu) and $[Au(pzTp)(PR_3)_x]$ (x = 1 or 2, R = Ph or

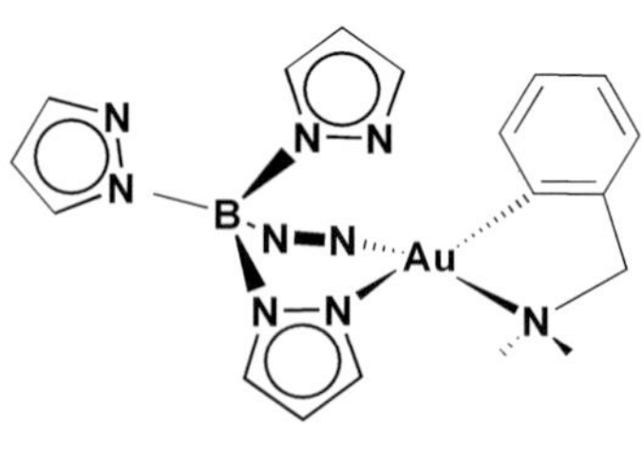

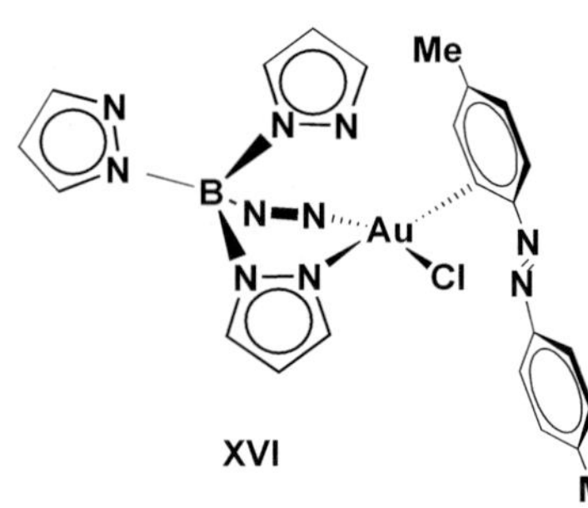

XV XVI

Fig. 2.136. The cyclometalated gold derivatives $[Au(R)(\kappa^2\text{-}pzTp)]ClO_4$ (XV) and $[Au(R)Cl(\kappa^2\text{-}pzTp)]$ (XVI).

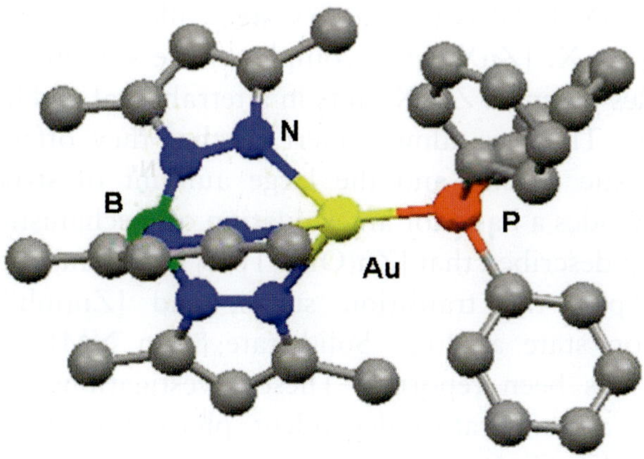

Fig. 2.137. The tetrahedral [Au(Tp*)(PPh$_3$)].

^{t}Bu) have been prepared and characterized both in solution (^{1}H- and ^{31}P{^{1}H}-NMR) and in the solid state (IR, single-crystal X-ray structure analysis, ^{31}P CPMAS). ^{31}P {^{1}H} NMR solution data suggest greater stability of the tetrakis(pyrazolyl)borate relative to those of tris(pyrazolyl)borate. All compounds are fluxional at room temperature. [Au(Tp*)(PPh$_3$)] (Fig. 2.137) has been compared with analogous coinage metal adducts [Cu(Tp*)(PPh$_3$)]·PPh$_3$ and [Ag(Tp*)(PPh$_3$)]·2MeCN. In [Au(Tp*)(PPh$_3$)], the gold atom adopts a distorted tetrahedral geometry. A number of Cu, Ag, and Au complexes containing scorpionate ligands have also been investigated by ^{31}P cross-polarization magic-angle-spinning (CPMAS) NMR spectroscopy.[558]

2.13. Group 12: Zn and Cd

2.13.1. *Zn*

Simple zinc(II)-homoscorpionate complexes have been long known.[563–565] Not much work has been performed in the last 10 years. An excellent review on the zinc pyrazolylborate chemistry related to zinc enzymes has been presented by Vahrenkamp[566] describing [ZnX(Tp*)] as a versatile system, allowing a wide variation of substituents X. [ZnX(Tp*)] complexes are suitable for studying the properties of single Zn–X units in a tetrahedral and hydrophobic environment. They are almost inert so that they offer themselves for mechanistic studies and the large amount of structural data acquired provides a basis for an evaluation of mechanistic pathways. Vahrenkamp described that [Zn(OH)(Tp*)] represents the transition, [Zn(X)(Y)(Tp*)] the transition states, and [Zn(inhibitor)(Tp*)] the transition state analogs. Solid-state ^{67}Zn NMR spectrum of $[Zn(Tp^*)_2]$ has been reported. These investigations have demonstrated that a temperature-dependent phase transition occurs at ca. 185 K for $[Zn(Tp^*)_2]$.[567]

2.13.2. *Cd*

Homoleptic $[Cd(Tp^x)_2]$,[568] tetrahedral $[CdX(Tp^x)]$,[568] and heteroleptic five-coordinate $[CdL(Tp^x)]$[569] have been described before 1999, the year in which Reger *et al.*[570] reported the syntheses and the solid-state structure of $[Cd(Tp^*)\{HC(pz^*)_3\}][BAr_F]$ (Fig. 2.138). This study allowed a comparison between the binding properties of tris(pyrazolyl)methanes and tris(pyrazolyl)borates.

In a "perspective study" in *Science*,[571] Parkin compared the first zinc–zinc bond reported by Carmona[572] with some $[M_2(Tp^*)_2]$ (M = Cd, Hg) complexes reported by Reger[573] and Gioia Lobbia *et al.*[574]

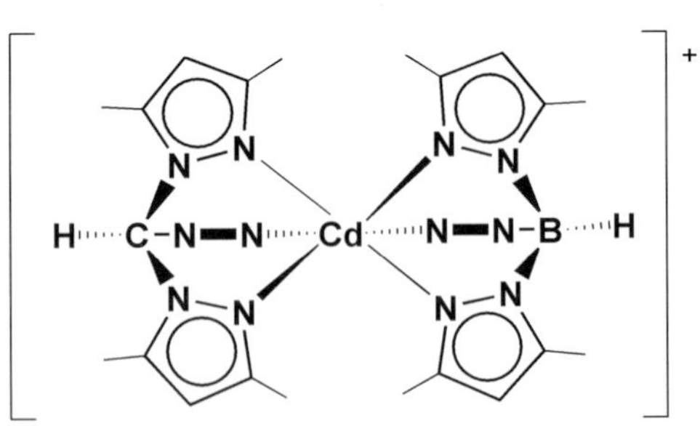

Fig. 2.138. The cation of $[Cd(Tp^*)\{HC(pz^*)_3\}][BAr_F]$.

2.14. Group 13: Ga, In, and Tl

2.14.1. *Ga*

Only one paper has been published in the last 10 years on Ga–Tp^x chemistry, previously investigated mainly by Parkin[575] and Reger.[576] Density functional studies have been carried out on the model gallium complexes [Ga(Tp)], [Ga(Tp)E] (E = O, S, Se or Te], and $[Ga_2(Tp)I_3]$ (Fig. 2.139) in order to understand the bonding in the analogous $[Ga(Tp^{^tBu_2})]$. Calculated and observed parameters are in good agreement. The gallium lone-pair in [Ga(Tp)] has a Ga–N antibonding character. The bonding[577] in [Ga(Tp)E] has been found to be almost entirely semipolar $[(GaTp)^+E^-]$, whereas the GaI_3 adduct can be described by the resonance structure $[Ga(Tp)]^+GaI_3^-$.

2.14.2. *In*

While numerous homoscorpionate complexes of indium have been previously reported,[578] in the last period only a limited number of papers appeared. The reaction between $[InCl_2(Tp^*)(NCMe)]$ and L_R^- (L_R^- = cyclopentadienyl)-tris(diorganylphosphi(ni)to-P)cobaltate(1-), $[\eta^5\text{-}C_5H_5Co\{PR_2(O)\}_3]^-$, R = OCH_3: $L_R = L_{OMe}$, R = Et: $L_R = L_{Et}$) yields $[In(Tp^*)(L_R)]Cl$. When $[In(L_{OMe})_2][PF_6]$ and $[In(Tp)_2][PF_6]$ were mixed together, heteroleptic $[In(Tp)(L_{OMe})]PF_6$ used formed (Fig. 2.140). $[In(Tp)(L_{OMe})][InCl_4]$ exhibits a *fac*-N_3O_3 coordination sphere. The hard indium(III) seems to prefer an antisymbiotic arrangement of ligands.[579]

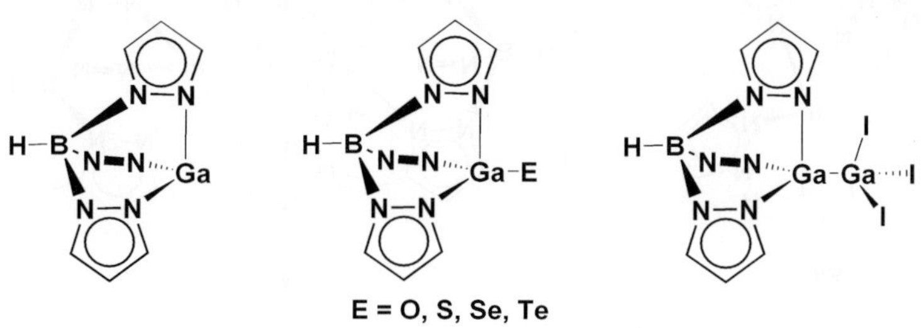

Fig. 2.139. Ga(Tp) derivatives.

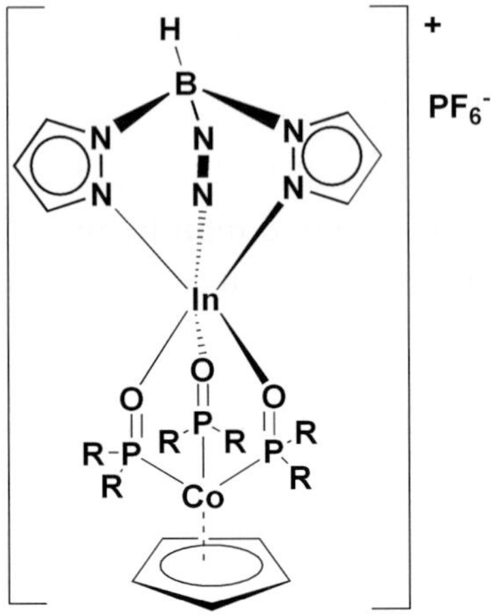

Fig. 2.140. The heteroleptic $[In(Tp)(L_{OMe})]PF_6$.

The structure in solution of $[In(Tp)(L_{OMe})][InCl_4]$ has been determined by 2D ^{1}H NOESY NMR spectroscopy.[580]

$[InI_2(Tp^*)(Hpz^*)]$ (**XVII**, Fig. 2.141) and $[(Tp^*)In(\mu\text{-}pz^*)_2(\mu\text{-}OH)In(Tp^*)]^+$ (**XVIII**, Fig. 2.141) have been prepared and characterized by single-crystal X-ray studies, the presence of coordinated pz* and Hpz* ligand being due to hydrolysis of the moisture-sensitive Tp* ligands.[581]

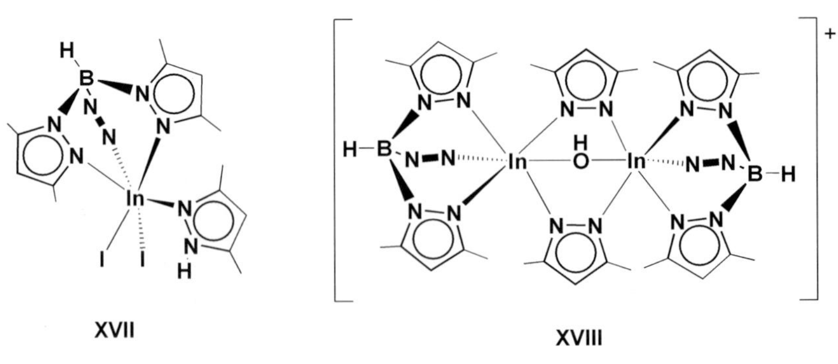

XVII

XVIII

Fig. 2.141. $[InI_2(Tp^*)(Hpz^*)]$ (XVII) and $[(Tp^*)In(\mu\text{-}pz^*)_2(\mu\text{-}OH)In(Tp^*)]^+$ (XVIII).

2.14.3. *Tl*

Janiak reported the structure of [Tl(Tp)], which exhibits pronounced $Tl\cdots\pi_{azolyl}$ interactions between neighboring molecules.[582]

2.15. Group 14: Ge, Sn, and Pb

2.15.1. *Ge*

Oxidation of the germanium(II) azide [$GeN_3(Tp^*)$] with HN_3 selectively yielded [$Ge(N_3)_3(Tp^*)$] (Fig. 2.142), the crystal structures of which have been determined and compared with those of other six-coordinate germanium(II) polyazides.[583] [$GeCl(Tp^x)$] derivatives were reported before 1999.[584,585]

2.15.2. *Sn*

[$SnR_nX_{4-n}(Tp^x)$][586–588] (n = 0–3; X = halide, R = alky or aryl group), previously reported, have been used as reagents for introducing a Tp^x ligand on group 4 and 5 metals and chromium. [$TiCl_3(Tp^*)$], [$ZrCl_3(Tp^*)$], and [$HfCl_3(Tp^*)$] have been obtained in good yields by the reaction of [$SnCl_{3-n}Bu_n(Tp^*)$] with the tetrahalide salt MCl_4 (M = Ti, Zr, or Hf). Analogously, the reaction of [$SnClBu_2(Tp^*)$] with a stoichiometric amount of [$NbCl_4(THF)_2$] and $TaCl_5$ gave [$NbCl_3(Tp^*)$] and [$TaCl_3(Tp^*)$][$TaCl_6$], respectively. This synthetic method has been applied to prepare vanadium [$VCl_2(Tp^*)(THF)$], [$VCl_2(Tp^*)(DMAP)$] (DMAP = dimethylaminopyridine), [$VCl_2(=NC_6H_3Me_2\text{-}2,6)(Tp^*)$], and chromium complexes [$CrCl_2(Tp^*)(L)$] (L = THF, H_2O or DMAP).[589]

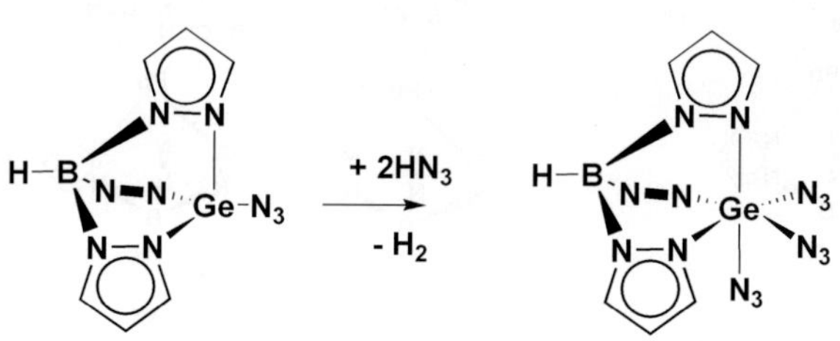

Fig. 2.142. Oxidation of [$GeN_3(Tp^*)$] with HN_3.

The reaction of transition metal-stannyltrichlorides ($MSnCl_3$, M = $Cp(CO)_2Fe$, $CH_3C_5H_4Mo(CO)_3$) with pzTp or Tp give the six-coordinate tin chelates [$MSnCl_2(Tp^x)$]. ($MSnCl_3$) reacts with Bp anomalously to give the unexpected trimetallic complexes Cl_2SnM_2, where M = $W(CO)_3Cp$, $Mo(CO)_3Cp$, $Fe(CO)_2Cp$.[590]

2.15.3. *Pb*

[Pb(μ-NCS)(Tp*)]$_2$ (**XIX**, Fig. 2.143) and [Pb(μ-NCS)(Tp)]$_n$ (**XX**, Fig. 2.143) have been synthesized and characterized. [Pb(μ-NCS)(Tp)]$_n$ has been obtained by the reaction of [PbCl(Tp)][591] with KSCN. [Pb(μ-NCS)(Tp*)]$_2$ has a dimeric structure with both μ-NCS-*S*,*N* and μ-NCS-*S*,*S* bridging SCN-ligands and [Pb(μ-NCS)(Tp)]$_n$ has an infinite 2D layered array in which three {Pb(Tp)}$^+$ groups sandwich a layer of SCN^- ligands.[592]

2.16. Group 15: Bi

2.16.1. *Bi*

The first poly(pyrazolyl)borate complex of Bi, [BiCl(Tp)$_2$(Hpz)] (Fig. 2.144), has been synthesized and characterized by Reglinski group.[593] The only other group-15 derivative to date reported is [$AsMe_2$(pzTp)].[585]

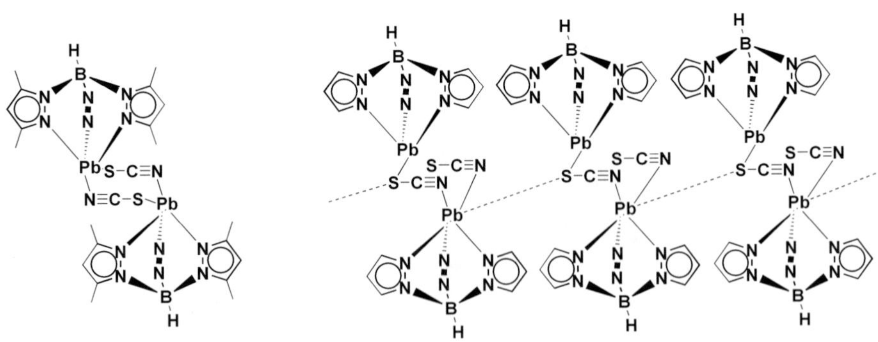

XIX XX

Fig. 2.143. [Pb(μ-NCS)(Tp*)]$_2$ (XIX) and [Pb(μ-NCS)(Tp)]$_n$ (XX).

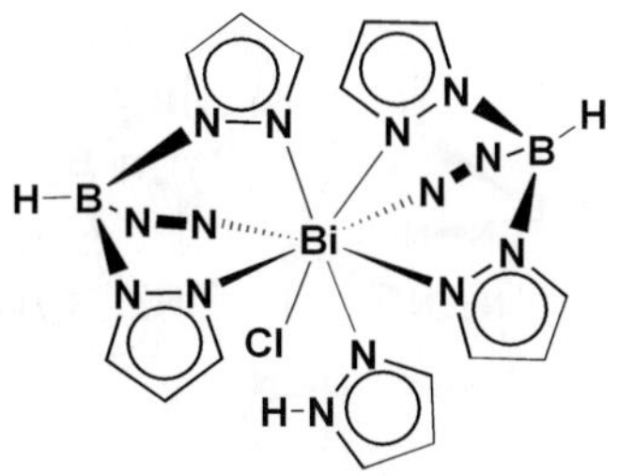

Fig. 2.144. [BiCl(Tp)$_2$(Hpz)], the first Bi(Tp^x) derivative.

2.17. Lanthanides

Lanthanides Tp^x chemistry is likely the field more developed in the last decade. A large number of derivatives has been synthesized in addition to the previously reported [Ln(Tp^x)$_3$],[594,595] [LnX(Tp^x)$_2$],[596,597] [LnX(Tp^x)$_2$(L)],[598,599] and [LnX$_3$(Tp^x)]$^-$[HL]$^+$.[598]

2.17.1. *La*

[La(Tp)$_3$] has been synthesized, structurally characterized, and its optical properties investigated.[600] [LaX$_3$(THF)$_n$] (X = Cl, I) reacts with 2 equiv. of K(Tp*) yielding [La(Tp*)$_2$X].[601] The formation of these species is always accompanied by the formation of [La(Tp*)$_2$(κ^2-pz*)] which arises from the decomposition of Tp*. The solid-state structures of [LaCl(Tp*)$_2$] and [La(Tp*)$_2$(κ^2-pz*)] (Fig. 2.145) were determined by single-crystal X-ray diffraction studies. Subsequent reactions of [LaX(Tp*)$_2$] with the alkali metal salts, MR (M = Li, R = CH$_2$SiMe$_3$, Ph, N(SiMe$_3$)$_2$; M = K, R = OAr), gave [M(Tp*)] as the major product. The mono-Tp bis(aryloxide) derivatives [La(OC$_6$H$_2$-2,6-^{t}Bu-4-Me)$_2$(Tp*)] were prepared by salt metathesis of [La(OC$_6$H$_2$-2,6-^{t}Bu-4-Me)$_3$] with K(Tp*).[601]

Complexes [LaCl$_2$(Tp*)(L)] (L = bipy or phen) have a seven-coordinate structure with capped octahedral geometry. Reaction of [LaCl$_2$(Tp*)(bipy)] with an excess of sodium amalgam gives [La(Tp*)$_2$(bipy)], which upon oxidation with iodine forms [La(Tp*)$_2$(bipy)]I.[602]

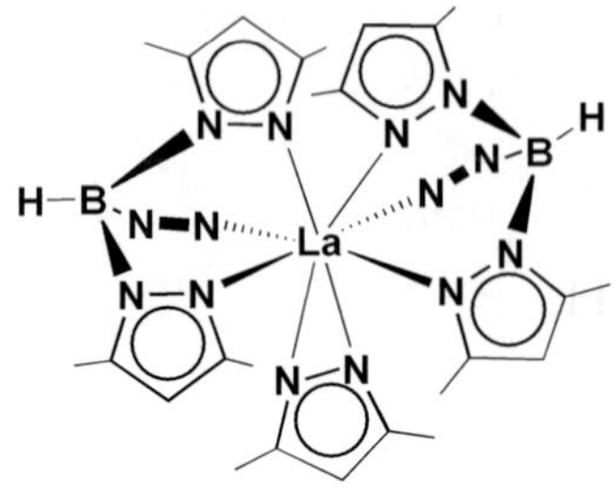

Fig. 2.145. $[La(Tp^*)_2(\kappa^2\text{-}pz^*)]$.

The insoluble dimer $[La(Tp^*)(\mu\text{-}BOp^{Me2})]_2$ has been obtained upon hydrolysis of $[La(Tp^*)_2X]$ (X = basic anionic ligand, BOp^{Me2} = $(HBO(pz^*)_2)^{2-}$). The formation of this compound is proposed to involve the intermediacy of $[Ln(OH)(Tp^*)_2]$, formed by protonolysis with adventitious water.[603]

2.17.2. *Ce*

Eight-coordinate fluxional species $[Ce(\beta\text{-dike})(Tp^*)_2]$ (β-dike = $(RC{=}O)_2CH$, R = Me, Ph, Bu^t) have been reported and their reactivity and spectroscopic features compared with those of Sm and Nd analogs.[604]

2.17.3. *Pr*

$[Pr(Tp)_3]$ has been reported, structurally characterized, and its absorption and luminescence spectra discussed. The global field experienced by Pr^{3+} and the individual ligand field strength due to the Tp ligand, as also nephelauxetic parameters were determined.[605]

2.17.4. *Nd*

$[NdX(Tp^*)_2]$ complexes have been prepared and their chemistry compared with that reported for the redox-active samarium species. $[NdCl(Tp^*)_2]$ is employed for the preparation of the seven-coordinate

[Nd(OAr)(Tp*)$_2$] (Ar = C_6H_4-4-X; X = H, ^{t}Bu, Br, Ph, OMe, F, Me) and [Nd(NPh$_2$)(Tp*)$_2$]. [NdCl(Tp*)$_2$] and [Nd(NPh$_2$)(Tp*)$_2$] are isostructural with the analogous samarium species. The same authors reported eight-coordinate fluxional species [Nd(β-dike)(Tp*)$_2$] (β-dike = (RC=O)$_2$CH, R = Me, Ph, Bu^t). The structure of [Nd{(Bu^{t}CO)$_2$CH}(Tp*)$_2$] has been determined. In an effort to prepare a neodymium hydride, the derivative [Nd(Tp*)$_2$(H$_2$BEt$_2$)] has been isolated serendipitously and structurally characterized.[604] [Nd(OC$_6$H$_2$-2,6-^{t}Bu-4-Me)$_2$(Tp*)] has been obtained in high yields by salt metathesis of [Nd(OC$_6$H$_2$-2,6-^{t}Bu-4-Me)$_3$] with 1 equiv. of K(Tp*).[601] The absorption spectrum of [Nd(Tp)$_3$] carried out at room and low temperature allows the insertion of the Tp ligand into empirical nephelauxetic and relativistic nephelauxetic series.[606] The thermal behavior of this compounds was also investigated through TG/DTG and DSC analyses.[607]

[NdX$_2$(Tp^x)(solvent)$_n$] (Tp^x = Tp or, Tp*; X = Cl, Br or I, solvent = THF or pz*) have been prepared and characterized. Steric, dynamic, and coordination behavior of these complexes are heavily influenced by the choice of the scorpionate ligand substituents. [NdI$_2$(Tp)(THF)$_2$] (XXI) and [NdI$_2$(Tp*)(THF)], distorted capped octahedral (XXII) (Fig. 2.146), have also been characterized structurally.[8]

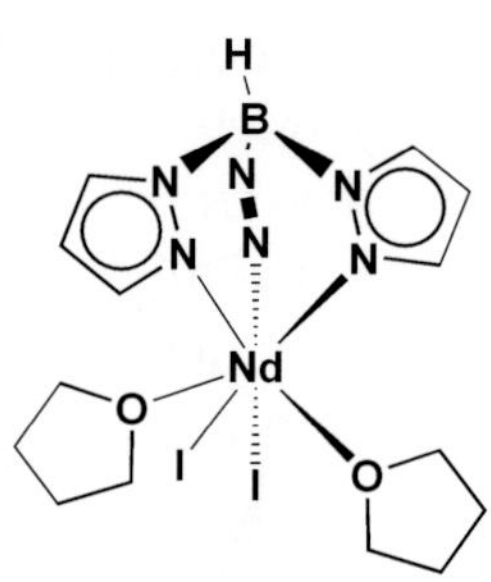

XXI

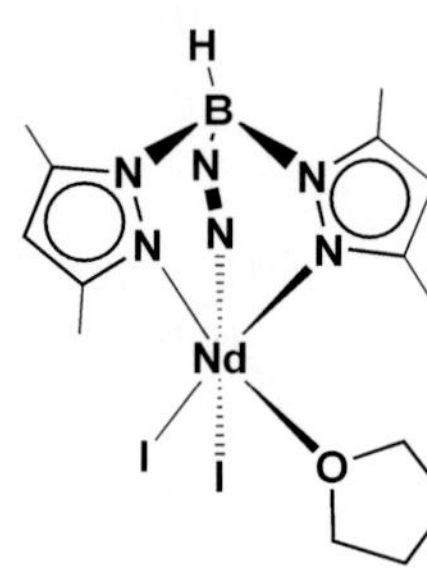

XXII

Fig. 2.146. [NdI$_2$(Tp)(THF)$_2$], (XXI) and [NdI$_2$(Tp*)(THF)], (XXII).

[(Tp)Nd(μ-OBz)$_4$Nd(Tp)] has been synthesized by the reaction of $NdCl_3$, K(Tp), and sodium benzoate, whereas the reaction of $NdCl_3$, K(Tp), sodium *p*-fluoro-benzoate, and Hpz resulted is the formation of [(Tp)(Hpz)Nd(μ-*p*-F-OBz)$_4$Nd(Hpz)(Tp)].[608]

2.17.5. *Sm*

Addition of Tl(Cp) to Sm(Tp*)$_2$ in THF resulted in the formation of [Sm(Tp*)$_2$(Cp)] that when sealed in a tube under vacuum and heated at 165°C overnight, transforms into [Sm(CpBp*)(κ^3-Tp*)] (Fig. 2.147), the structure of which has been established by single-crystal X-ray diffraction. A byproduct has also been identified in the residue as [Sm(Tp*)$_2$(κ^2-pz*)].[609]

Crystal field strengths, nephelauxetic effects, and experimentally based molecular orbital schemes (in the f range) of low-symmetric [Sm(η^5-Cp)(η^3-Tp*)(η^2-Tp*)], have been determined through absorption and luminescence spectra, carried out at room and low temperatures.[610]

[Sm(κ^2-S_2CNR_2)(Tp*)$_2$] (R = Et, Me) have been isolated from the reaction of ($R_2NC(S)S$)$_2$ with 2 equiv. of [Sm(Tp*)$_2$]. [Sm(Tp*)$_2$(κ^2-Y)] (Y = 2-SC_5H_4N, 2-SeC_5H_4N) have been prepared by reductive cleavage of 2,2′-dipyridyl disulfide or 2,2′-dipyridyl diselenide. [Sm(Tp*)$_2$(2-OC_5H_4N)] has also been reported. These compounds are all eight-coordinate with dodecahedral geometry.[611]

[Sm(Tp)$_3$] has been prepared and structurally characterized. In these complexes the Sm^{3+} central ion is coordinated by nine N

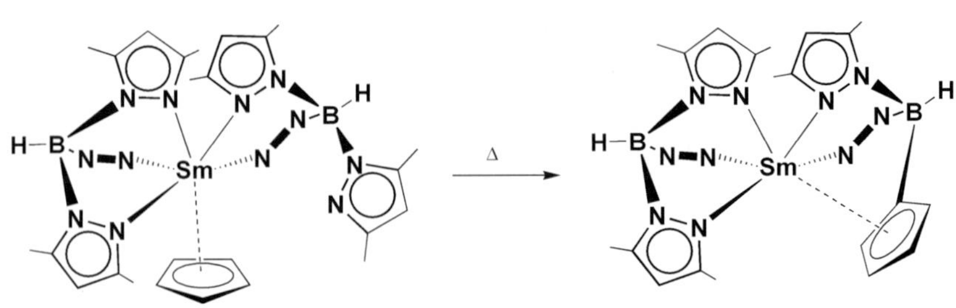

Fig. 2.147. The transformation of [Sm(Tp*)$_2$(Cp)] in [Sm(CpBp*)(κ^3-Tp*)].

atoms in the shape of a tricapped trigonal prism. [$Sm(Tp)_3$] has also been inserted into empirical nephelauxetic and relativistic nephelauxetic series.[612] Complexes containing Sm, coordinated by one, two, or three Tp^x ligands are described in a patent as potential light-emitting devices. Synthetic procedures are also reported and claimed.[613]

Addition of K(R) (R = $CH_2C_6H_4$-*o*-NMe_2, C_6H_4-*o*-CH_2NMe_2, $CH_2C_6H_5$) to a solution of [$SmCl(Tp^*)_2$] gave the insoluble [$Sm(Tp^*)_2$]. The same reaction carried out in the presence of protic substrates such as HOPh, HCp, HC≡CPh, HOC_6H_2-2,4,6-$^t Bu_3$, $HNPh_2$, and Hpz* produces [$Sm(Tp^*)_2(Y)$] (Y = OPh, C≡CPh, Cp, OC_6H_2-2,4,6-$^t Bu_3$, NPh_2, and pz*). Some of these species have been structurally characterized.[614]

Reaction of SmI_2 with 2 equiv. of Na(Tp*) gives the highly insoluble homoleptic [$Sm(Tp^*)_2$] (**XXIII**). [$SmX(Tp^*)_2$] species (X = F, Cl, Br, I, BPh_4) (**XXIV**) (Fig. 2.148) were prepared either via oxidation of [$Sm(Tp^*)_2$] or salt metathesis from SmX_3 (X = Cl, Br, I). The solid-state structures of these compounds were determined by single-crystal X-ray diffraction, the homoleptic bis-Tp* complex being six-coordinate with trigonal antiprismatic geometries. On the other hand, [$SmX(Tp^*)_2$] complexes exhibit structures varying from seven-coordinate molecular to six-coordinate ionic forms.[615]

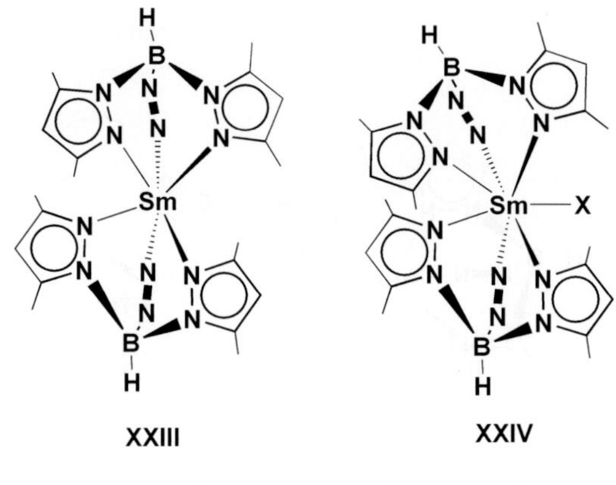

Fig. 2.148. [$Sm(Tp^*)_2$] (XXIII) and [$SmX(Tp^*)_2$] (XXIV).

The stable [Sm(β-ketoenolato){η^3-pzTp}$_2$] (β-ketoenolato = acac, salicylaldehydate, tetramethylheptane-3,5-dionate, 3-methyl-2,4-pentanedionate) has been prepared and characterized by ^{1}H NMR spectra which shows novel stereochemically nonrigid temperature-dependent motions of pzTp. [Sm(salicylaldehydate){η^3-pzTp}$_2$] has also been structurally characterized.[616]

[Sm(η^2-O_2)(Tp)$_2$] has been prepared by the reactions of [Sm(Tp)$_2$] with dioxygen in toluene at −75°C. The reaction of [Sm(Tp*)$_2$] with pyridine N-oxide yielded [Sm_2(μ-η^2:η^2-O_2)(Tp*)$_2$]·2THF. Upon addition of hexane to the mother liquid, the seven-coordinated dimeric molecule with a pseudo-octahedral coordination geometry [Sm(μ_3-O)(BOH)(Tp*)(pz*)$_2$]$_2$·2THF was formed.[617]

Reaction of [Sm(Tp*)$_2$] with Hg(C≡CPh)$_2$ in THF at ambient temperature afforded the monomeric, base-free lanthanide alkynide complex [Sm(C≡CPh)(Tp*)$_2$]. Thermolysis of the latter species in benzene afforded [Sm{HB(pz*)$_2$(C≡CPh)}(pz*)(Tp*)] upon unusual ligand rearrangement on the Sm(III) center (Fig. 2.149).[618]

While reactions of $SmCl_3 \cdot 6H_2O$ with K(Tp) afford [SmCl(Tp)$_2$-(Hpz)], similar reactions with K(pzTp) afforded anhydrous [HpzTp]. The homoleptic eight-coordinate complex [Sm(pzTp)$_3$] (Fig. 2.150), upon reaction with twofold moles of K(Tp) yielded a mixture of three major species [Sm(pzTp)$_n$(Tp)$_{3-n}$] (n = 2, 1, 0). [Sm(pzTp)$_2$(acac)], treated with twofold moles of K(Tp) gets converted quantitatively to [Sm(Tp)$_2$(acac)]. The Sm^{III} ion seems to prefer coordination of Tp

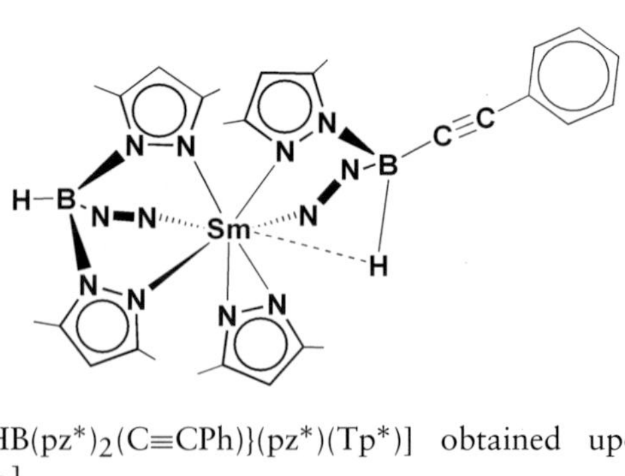

Fig. 2.149. [Sm{HB(pz*)$_2$(C≡CPh)}(pz*)(Tp*)] obtained upon thermolysis of [Sm(C≡CPh)(Tp*)$_2$].

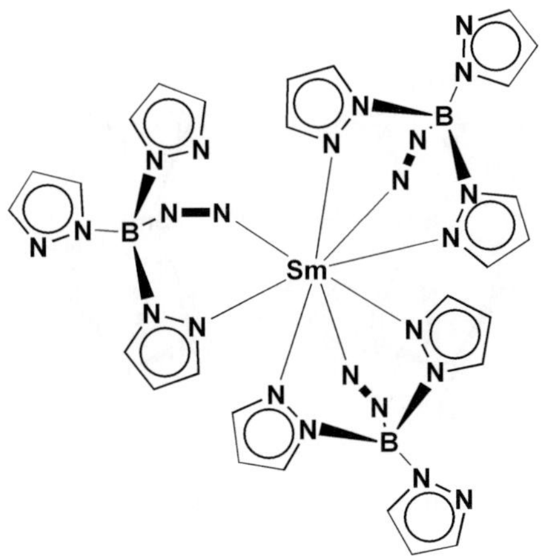

Fig. 2.150. [Sm(pzTp)$_3$].

ligand to that of pzTp, indicating less σ-donating electronic character of the latter. The complexes [Sm(pzTp)$_2$(L–L)] (L–L = β-ketoenolato) show fast intramolecular exchange of their coordinated and uncoordinated pyrazolyl groups.[619]

Reaction of [Sm(Tp*)$_2$Cl] with NaOR (R = Ph, Ph-Bu^t) yields [Sm(Tp*)$_2$OR]. Complexes [Sm(Tp*)$_2$ER] and [Sm(Tp^{*Et})$_2$ER] (E = S, R = Ph; E = S, R = Ph-4-Me; E = S, R = CH$_2$Ph; E = Se, R = Ph; E = Se, R = Tol; E = Se, R = CH$_2$Ph; E = Te, R = Ph) prepared by reductive cleavage of the dichalogenides, form an isoleptic series of seven-coordinate complexes with terminal chalcogenolate ligands. Little correlation has been found between the bond angle and bond length. In [Sm(Tp*)$_2$SePh] and [Sm(Tp*)$_2$SeTol] one of the Tp* ligands undergoes a major distortion arising from twisting around a B–N bond to give an effective eight-coordinate complex with a π-stacking of the Ph group with a pz ring.[620]

The reaction of [Sm(Tp*)$_2$] with Mn$_2$CO$_{10}$ upon reductive cleavage of the Mn–Mn bond gave [Sm(Tp*)$_2$][Mn(CO)$_5$], characterized by X-ray diffraction. [(Tp*)$_2$Sm(μ-O$_2$CH)Sm(Tp*)$_2$]Mn(CO)$_5$ (Fig. 2.151) has been isolated and characterized crystallographically.[621]

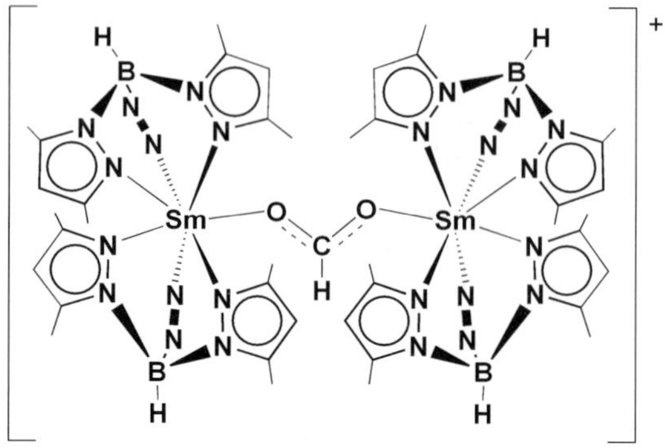

Fig. 2.151. The cation of $[(Tp^*)_2Sm(\mu\text{-}O_2CH)Sm(Tp^*)_2]Mn(CO)_5$.

$[Sm(Tp^*)_2]$ reacts with dichalcogenides giving isoleptic complexes $[Sm(Tp^*)_2ER]$ (E = O, S, Se, Te; R = phenyl). In the selenolate complexes an unusual distortion of the Tp* is found. $[Sm(Tp^*)_2(S_2CNR_2)]$ and $[Sm(Tp^*)(S\text{-}2\text{-}py)]$ are fluxional. $[Sm(Tp^*)_2]$ reacts with $Mn_2(CO)_{10}$ to give $[Sm(Tp^*)_2]Mn(CO)_5$ and small amounts of $[\{Sm(Tp^*)_2\}_2(\mu\text{-}O_2CH)]Mn(CO)_5$. An analogous reaction with $Re_2(CO)_{10}$ yielded $[Sm(Tp^*)_2]_2Re(CO)_5$ and $[Sm(Tp^*)_2]_2Re_4(CO)_{17}$.[622]

$[(acac)_2Co(Ox)Sm(Tp)_2]$ has been prepared and its structural and spectroscopic properties compared with those of analogous {{Cr(Ox)Ln} systems.[623] Near-infrared circular dichroism (NIR CD) spectra in the 4f–4f transitions has been investigated in dichloromethane and compared with those of the analogous Yb and Dy species. The solution NIR magnetic circular dichroism (MCD) spectra were also studied in conjunction with the NIR CD spectra. The compounds have also been structurally characterized.[624] Analogous La, Nd, Ho, ER, and Tm species were reported in the same work.

The diamagnetic tropolonato $[Sm(Trop)(Tp)_2]$ and the paramagnetic semiquinone $[Sm(DTBSQ)(Tp)_2]$ complexes have been reported by Gatteschi and coworkers.[625] The $[Sm(Trop)(Tp)_2]$ derivative was used as a reference for the qualitative determination of crystal-field effects in the exchange-coupled semiquinone derivatives.

The reaction of nitric oxide with $[Sm(Tp^*)_2]$ gave the seven-coordinate $[Sm(Tp^*)_2(NO_2)]$.[626]

Four binuclear samarium complexes $[Sm(\mu\text{-}p\text{-}XOBz)_2(Tp)]_2$ have been prepared by the reaction of $SmCl_3$, K(Tp), and sodium *p*-X-benzoate (X = H, Cl, F, NO_2) in 1:1:2 ratio. These complexes have been characterized by elemental analysis, IR spectroscopy, thermogravimetry, optical properties, X-ray, and magnetic measurement studies. The X-ray structure shows that these complexes are isostructural, each samarium being seven-coordinate.[627]

Reaction of $[Sm(Tp^*)_2]$ with fluorenone, benzophenone, benzaldehyde azine, and phenantrenequinone gives monomeric $[Sm(Tp^*)_2X]$. Reaction of $[Sm(Tp^*)_2]$ with less bulky ligands such as benzaldehyde or pyrazine leads to dimetallic $[Sm(Tp^*)_2]_2[\mu\text{-}Y]$ (Y = dianion formed by C–C coupling of two reductively eliminated ligands).[628]

2.17.6. *Eu*

Two eight-coordinate Eu complexes $[Eu(Tp)_2(DPSO)_2]$ and $[Eu(Tp)_2(BPMU)_2]$ have been prepared and characterized also by X-ray diffraction analysis. Both derivatives are luminescent at 77 K.[629]

The synthesis and the X-ray single crystal studies of the dimeric heteroleptic eight-coordinate complexes $[Eu(O_2CPh)(Tp)_2]_2$ and $[Eu(O_2CPh)_2(Tp)]_2$ as also the structure of the known mononuclear eight-coordinated derivatives $[Eu(acac)(Tp)_2]$ and $[Eu(OC(Ph)CHC(Ph)O)(Tp)_2]$ have been reported. The molecular parameters of these compounds have been related to the Ln^{3+} ionic radius.[630]

Reduction of $[Eu(OTf)(Tp^*)_2]$ yielded the six-coordinate $[Eu(Tp^*)_2]$ with a trigonal antiprismatic geometry and planes of the κ^3–Tp* ligands parallel to one another.[615]

The Eu^{3+} ion of $[Eu(Tp)_3]$ is coordinated according to a single-crystal X-ray analysis in the first coordination sphere by nine N atoms in the shape of a tricapped trigonal prism, resulting in an effective crystal field (CF) of symmetry D_{3h}. The absorption spectrum of powdered $[Eu(Tp)_3]$ was recorded at approximately 5 K, and the luminescence spectrum at 77 K. On this basis an experimental CF splitting pattern has been derived.[600] The thermal behavior of [Nd(κ^3-Tp)$_3$] was investigated through TG/DTG and DSC analyses.[607]

The reaction of [Eu(Tp*)$_2$] with nitric oxide gave the seven-coordinate [Eu(Tp*)$_2$NO$_2$].[626]

2.17.7. *Gd*

[{Gd(Tp)$_2$}$_2${1,4-(O$_2$C)$_2$C$_6$H$_4$}] (Fig. 2.152), structurally characterized, contains two gadolinium centers linked by a ditopic 1,4-benzenedicarboxylate ligand.[630]

[Gd(DTBSQ)(Tp)$_2$]·2CHCl$_3$ has been obtained by metathetical reaction between the parent benzoato complex [Gd(OBz)(Tp)$_2$] and 3,5-di-tertbutylcatechol in alkaline methanol. The coordination sphere around gadolinium comprises six nitrogen atoms of Tp and two oxygen atoms of the semiquinone. Antiferromagnetic coupling has been observed. This compound has also been investigated by cyclic voltammetry.[631]

The dinuclear asymmetric [Gd$_2$(Tp)$_2$(DTBSQ)$_4$]·CHCl$_3$ consists of two Gd ions, one eight- and the other nine-coordinate. The magnetic properties of this compound have been investigated by magnetic susceptibility measurements and high-field electron paramagnetic resonance spectroscopy.[632]

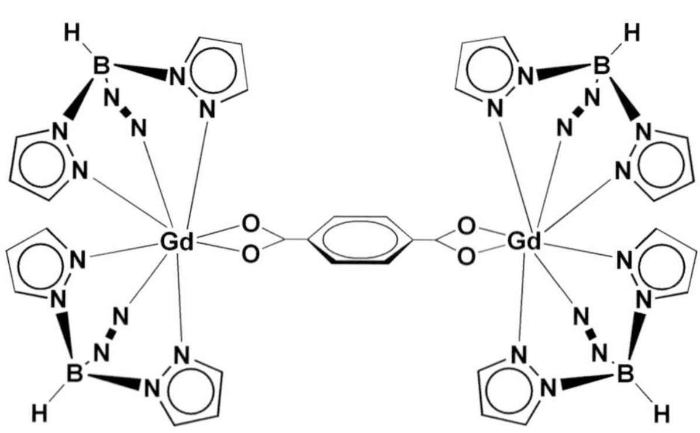

Fig. 2.152. [{Gd(Tp)$_2$}$_2${1,4-(O$_2$C)$_2$C$_6$H$_4$}].

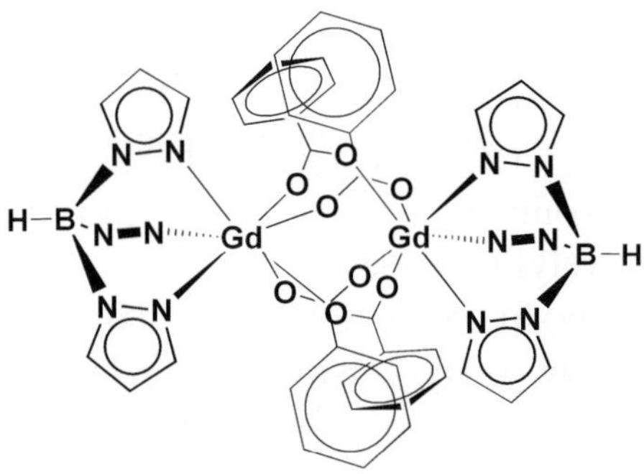

Fig. 2.153. $[Gd(\mu\text{-p-X-OBz})_2(Tp)]_2$.

When $GdCl_3$ reacts with 2 equiv. of K(Tp) and 2 equiv. of sodium p-X-benzoate (X = H, Cl), $[Gd(\mu\text{-p-X-OBz})_2(Tp)]_2$ (Fig. 2.153), which consists of a seven-coordinate tetrakis carboxylato-bridged dimetal unit with capping Tp ligands, is obtained. Luminescence properties of both complexes have been reported.[633]

2.17.8. *Tb*

When $TbCl_3$ reacted with 1 equiv. of Tp, and 2 equiv. of sodium *p*-X-benzoate (X = H, Cl, Br, NO_3) $[Tb(\mu\text{-}p\text{-X-OBz})_2(Tp)]_2$ is formed, whereas the reaction, in the presence of NaN_3, gave the tetranuclear complex $[Tb(\mu\text{-}N_3)(Tp)_2]_4$. The crystal structure of $[Tb(\mu\text{-}p\text{-X-OBz})_2(Tp)]_2$ (X = H, Cl) has been investigated by single-crystal X-ray study. These derivatives consist of a seven-coordinate tetrakis carboxylato-bridged dimetal unit with capping Tp ligands. Single-crystal X-ray analysis of $[(Tp)_2Tb(\mu\text{-}N_3)]_4$ indicates a cyclic 16-membered $[(Tb)_4(\mu\text{-}N_3)_4]$ unit with two Tp ligands bound to each of the Tb ion, resulting in a tetragonal antiprismatic N_8-coordination sphere. The luminescence properties of these complexes have also been investigated.[633]

The thermal behavior of the compounds $[Tb(\kappa^3\text{-}Tp)_3]$ was investigated through TG/DTG and DSC analyses.[607]

2.17.9. *Dy*

Stereospecific formation of the chiral heterodinuclear 3d–4f complexes (Λ–Δ)-[$(acac)_2$Cr(Ox)Dy$(Tp)_2$] (where (Λ–Δ) means the absolute configuration around the octahedral Cr and square antiprismatic Ln moiety, respectively) is reported.[634]

The dinuclear [Dy_2{μ-(*S*- or *RS*-pba)}$_4$$(Tp)_2$] has been reported. The skew bent B··· Dy ··· Dy··· B disposition leads to configurational chirality.[635]

2.17.10. *Ho*

[Ho(Trop)$(Tp)_2$] and [Ho(DTBSQ)$(Tp)_2$] have been synthesized and investigated by magnetization and susceptibility studies,[625] and [Ho(Trop)$(Tp)_2$] has also been structurally characterized. Mononuclear [Ho(*S*- or *RS*-pba)$(Tp)_2$] were also described.[635]

2.17.11. *Er*

Wong and coworkers[636] reported the synthesis of [Er(TPP)(Tp)] (TPP = 5,10,15,20-tetraphenylporphyrinate) with different substituents in the para position of the TPP phenyl group. These complexes were characterized by photoluminescence studies and single-crystal X-ray study.[636] Mononuclear [Er(*S*- or *RS*-pba)$(Tp)_2$] were also described.[635]

2.17.12. *Yb*

Reaction of YbI_2 with NaTp* gives the highly insoluble homoleptic [Yb$(Tp^*)_2$], structurally characterized as a trigonal antiprism.[615] Its reaction with nitric oxide gave the seven-coordinate [Yb$(NO_2)$$(Tp^*)_2$].[626] [Yb($\eta^2$-$O_2$)$(Tp)_2$] and [Yb($\eta^2$-$O_2$)$(Tp^*)_2$] have been prepared by the reactions of [Yb$(Tp)_2$] and [Yb$(Tp^*)_2$], respectively, with dioxygen in toluene at −75°C.[617]

Stereospecific formation of the chiral heterodinuclear 3d–4f complexes (Λ–Δ)-[$(acac)_2$Cr(Ox)Yb$(Tp)_2$] was demonstrated by comparing the NIR CD spectra in the 4f→4f transitions and/or the single-crystal X-ray analysis with those of a diastereomeric mixture of the mononuclear Λ-, and Δ-[Yb(Tp)$_2$(*S*-pba)] complexes.[634] The NIR CD spectrum of [(acac)$_2$Co(Ox)Yb(Tp)$_2$] compared with that of the related complexes suggests a stereospecific formation with configurational chirality around Yb^{III} ion.[623]

The synthesis and spectroscopic characterization of the monoporphyrinate trivalent lanthanide derivatives, bearing the monoanionic Tp, of general formulas [Yb(TPP)Tp] have been reported by Boncella *et al.*,[637] who also described their photophysical properties including absorption, emission, and transient absorption properties. Analogous Tm, Er, Ho, Nd, and Pr derivatives have been reported. All structures share the common feature that the metal ions are too large to fit in the cavity of the porphyrin ring and sit above the plane defined by the four N atoms of the porphyrin with the Tp ligand completing the coordination sphere of the distorted capped trigonal prism Yb.

Wong and coworkers reported[636] the synthesis of [Yb(TPP)(Tp)] (Fig. 2.154) and the structural characterization showing that seven nitrogen atoms are symmetrically coordinated to Yb ion. These complexes were also characterized by photoluminescence studies.[636]

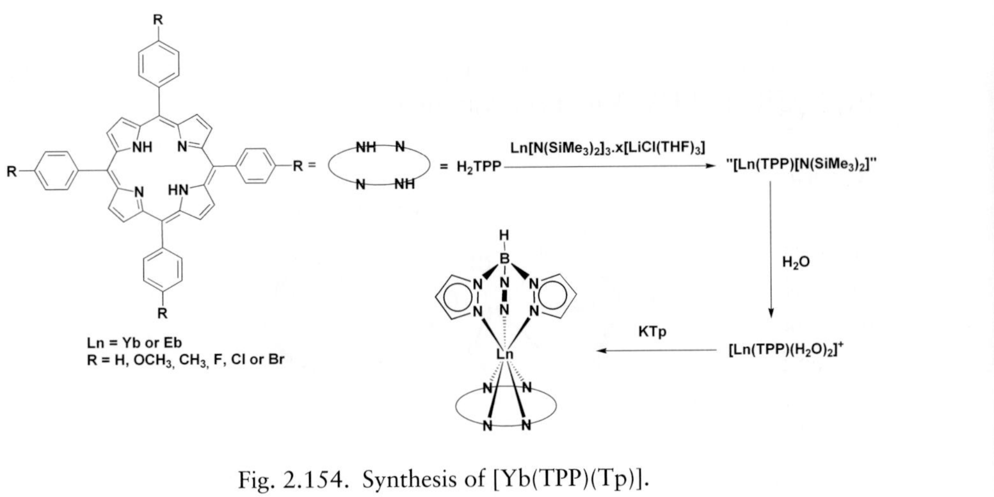

Fig. 2.154. Synthesis of [Yb(TPP)(Tp)].

$[Yb(Trop)(Tp)_2]$ and $[Yb(DTBSQ)(Tp)_2]$ have been synthesized by procedures analogous to that reported for Ho, Sm, Eu, Tb, Dy, Er, and Gd and studied by magnetization and susceptibility measurements. The Yb(III), Er(III), Tb(III), and Dy(III) complexes exhibit a dominating antiferromagnetic coupling.[625]

$[(Tp)_2Yb(\mu\text{-}C_6Cl_2O_4)Yb(Tp)_2]$ was synthesized and structurally characterized and reported to exhibit strong 4f–4f emission. Its structural and spectroscopic features were compared with a number of dinuclear complexes such as $[(acac)_2Cr^{III}(\mu\text{-}Ox)Ln^{III}(Tp)_2]$. Analogous Eu and Tb species were described in the same paper.[638]

The dinuclear $[Yb_2\{\mu\text{-}(S\text{- or }RS\text{-pba})\}_4(Tp)_2]$ was prepared and its NIR chiroptical properties were investigated. X-ray structural analysis confirmed a dinuclear structure with CH$\cdots\pi$ interactions and linear B$\cdots$Yb$\cdots$Yb$\cdots$B arrangements for the $[Yb(\mu\text{-}RS\text{-pba})_4Yb(Tp)_2]$. Analogous dinuclear Ho, Gd, Dy, and Nd species were reported.[635]

The molecular structure of $[Yb_2Cl_4(Tp)_2(THF)_2]\cdot C_7H_8$ (Fig. 2.155) is reported by Rabe *et al.*[639] This dimeric centrosymmetric complex exhibits a seven-coordinate Yb atom, bonded to three N atoms of the tridentate ligand, to the O atom of the neutral tetrahydrofuran ligand, and to one terminal as well as to two bridging Cl atoms.

2.17.13. *Lu*

$[(acac)_2Co(Ox)Lu(Tp)_2]$ has been prepared and characterized by FAB, IR, NMR, and UV–Vis spectroscopies.[623]

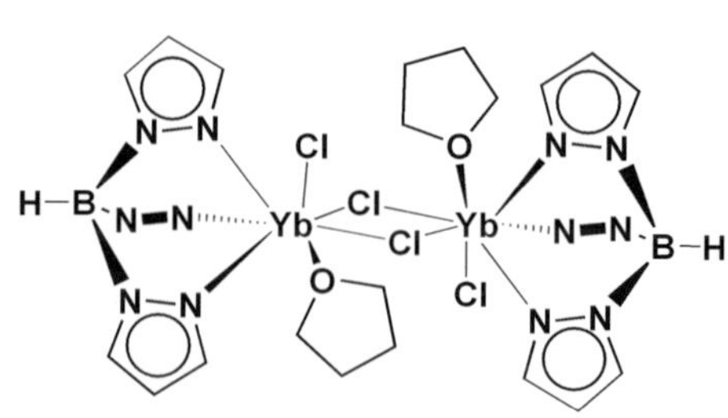

Fig. 2.155. The dimeric centrosymmetric $[Yb_2Cl_4(Tp)_2(THF)_2]$.

2.18. Actinides

2.18.1. *U*

Reaction between $[UI(Tp^*)_2]$ with KNR_2 ($R = C_6H_5$, $SiMe_3$) in THF yielded the monomeric trivalent actinide amide complexes $[U\{N(C_6H_5)_2\}(Tp^*)_2]$ and $[U\{N(SiMe_3)_2\}(Tp^*)_2]$ which have been fully characterized by spectroscopic methods and X-ray single crystal studies. In the solid state, both complexes show distorted pentagonal bipyramidal geometries. The U–NR_2 bond lengths in both complexes are the same, but in $[U\{N(SiMe_3)_2\}(Tp^*)_2]$ the greater steric demands of the $N(SiMe_3)_2$ ligand led to elongated U–$N_{(pz)}$ bonds, especially those opposite to the amido ligand.[640]

Takats *et al.*[641] examined the following reaction:

$$[UI_2(Tp^*)]\cdot(THF)_2(cr) + \tfrac{1}{2}I_2(soln) \rightarrow [UI_3(Tp^*)](soln) + 2THF(soln)$$

$$[UI_2(Tp^*)]\cdot(THF)_2(cr) + \tfrac{1}{2}{}^{t}BuO\text{-}O^{t}Bu(soln) \rightarrow [UI_2(Tp^*)(O\text{-}^{t}Bu)](soln) + 2THF(soln)$$

$$[UI_2(Tp^*)]\cdot(THF)_2(cr) + C_5H_{11}Cl(soln) \rightarrow [UClI_2(Tp^*)](soln) + \tfrac{1}{2}(C_5H_{11})_2(soln) + 2THF(soln)$$

$$[UI_2(Tp^*)]\cdot(THF)_2(cr) + \tfrac{3}{2}Cl_2(soln) \rightarrow [UCl_3(Tp^*)](soln) + I_2(soln) + 2THF(soln)$$

and presented a brief overview of the bond dissociation enthalpies of U(IV) complexes.[641]

Insertion of benzonitrile and acetonitrile into the U–C bond of $[UCl_2(CH_2SiMe_3)(Tp^*)]$ gave $[UCl_2\{NC(R)(CH_2SiMe_3)\}(Tp^*)]$ (R = Ph, Me). The octahedral $[UCl_2\{NC(Ph)(CH_2SiMe_3)\}(Tp^*)]$ has also been investigated by single-crystal X-ray diffraction study. The uranium amide complexes $[UCl_2(NR_2)(Tp^*)]$ (R = Et, Ph) have also been prepared and structurally characterized.[642]

K(pzTp) reacts with UCl_4 in THF giving $[UCl_2(pzTp)_2]$, which contains an eight-coordinated uranium center in a distorted square antiprismatic geometry. This complex reacts with sodium alkoxide,

NaS^{i}Pr, or Li(Me) yielding [UCl(OR)(pzTp)$_2$] (R = Et, ^{t}Bu, C_6H_4-o-Me, C_6H_2-2,4,6-Me$_3$), [U(O^{t}Bu)$_2$(pzTp)$_2$], [UCl(S^{i}Pr)(pzTp)$_2$], and [UCl(Me)(pzTp)$_2$]. [UCl(S^{i}Pr)(pzTp)$_2$] and [UCl(OC$_6$H$_2$-2,4,6-Me$_3$)(pzTp)$_2$], structurally characterized, are eight-coordinated, the first in a distorted square-antiprismatic and the latter in a bicapped triangular prismatic environment. Extended Huckel molecular orbital calculations have been used to understand the electronic properties of the ligand pzTp and relative stability of complexes containing U(pzTp)$_x$, fragments.[643]

Reaction of [U(NR$_2$)(Tp*)$_2$] (R = Ph, SiMe$_3$) with protic substrates such as 2,4,6-trimethylphenol (HOAr), pz*, 2-mercaptopyridine (HSC$_5$H$_4$N), and HC≡CPh afforded [U(OAr)(Tp*)$_2$], [U(Tp*)$_2$(pz*)] (Fig. 2.156), [U(η^2-SC$_5$H$_4$N)(Tp*)$_2$], and [U(C≡CPh)(Tp*)$_2$] respectively. Reaction of [U(Tp*)$_2$(NR$_2$)] with Me$_3$SnCl or Me$_3$SiBr gave [U(Tp*)$_2$Cl] and [U(Tp*)$_2$Br], respectively. Metathesis of [U(Tp*)$_2$I] with Na(Cp) yielded [U(κ^3-Tp*)(κ^2-Tp*)(η^5-Cp)] which, upon thermolysis, produced [U(Cp)$_3$(pz*)], pyrazabole, and [U(Tp*)(pz*)$_3$].[644]

Treatment of a CH$_2$Cl$_2$ solution of [UI$_4$(N≡CPh)$_4$] with K(Tp*) formed [UI$_3$(Tp*)], and is converted to [UI$_3$(Tp*)(NCMe)] by rinsing with MeCN. Addition of 2.2 equiv. of KTp* to a toluene solution of [UI$_4$(N≡CPh)$_4$] followed by heating at 95°C, led to the formation of [UI(pz*)(Tp*)]$_2$[μ-O] (Fig. 2.157), presumably by the hydrolytic cleavage of excess K(Tp*) by adventitious water. These three U(Tp*)

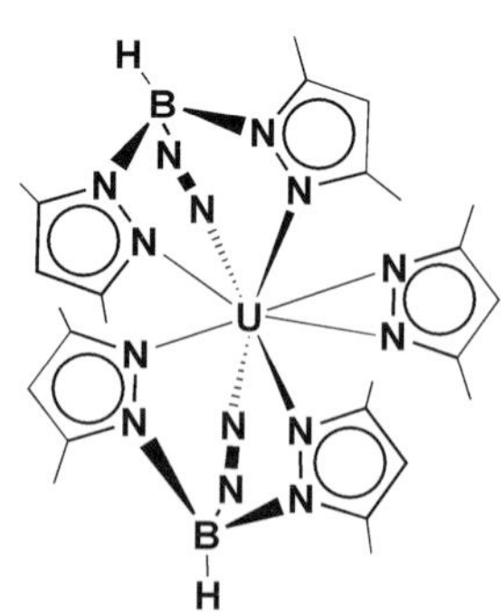

Fig. 2.156. [U(Tp*)$_2$(pz*)].

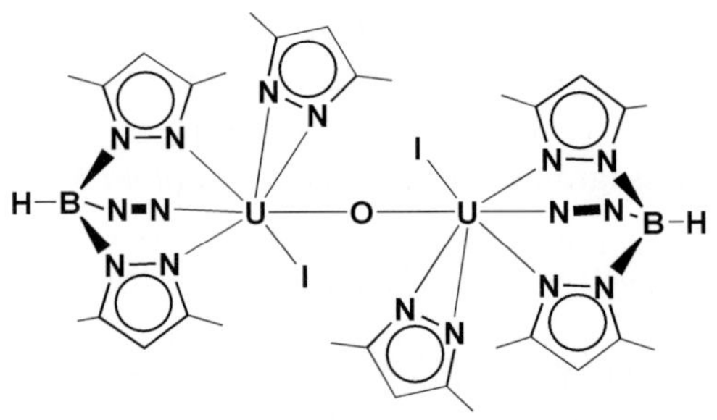

Fig. 2.157. [UI(pz*)(Tp*)]$_2$[μ-O].

complexes were structurally and spectroscopically (NMR, FT-IR, optical absorbance) characterized.[647]

References

1. Hernández L, Taboada S, D'Ornelas L, González T, Atencio R, *Acta Cryst* 2004, **E60**, m979.
2. Reglinski J, Garner M, Cassidy ID, Slavin PA, Spicer MD, Armstrong DR, *J Chem Soc Dalton Trans* 1999, 2119.
3. Joshi HK, Arvin ME, Durivage JC, Gruhn NE, Carducci MD, Westcott BL, Lichtenberger DL, Enemark JH, *Polyhedron* 2004, **23**, 429.
4. Sohrin Y, Matsui M, Hata Y, Hasegawa H, Kokusen H, *Inorg Chem* 1994, **33**, 4376.
5. Ballinas-López MG, Padilla-Martínez II, Martínez-Martínez FJ, Höpfl H, García-Báez EV, *Acta Cryst C Cryst Struct Commun* 2006, **62**, m132.
6. Blackwell J, Lehr C, Sun YM, Piers WE, Pearce-Batchilder SD, Zaworotko MJ, Young VG, *Can J Chem* 1997, **75**, 702.
7. Roitershtein D, Domingos Â, Pereira LCJ, Ascenso JR, Marques N, *Inorg Chem* 2003, **42**, 7666.
8. Long DP, Chandrasekaran A, Day RO, Bianconi PA, Rheingold AL, *Inorg Chem* 2000, **39**, 4476.
9. Kayal A, Kuncheria J, Lee SC, *Chem Commun* 2001, 2482.
10. Gazzi R, Perazzolo F, Sostero S, Ferrari A, Traverso O, *J Organomet Chem* 2005, **690**, 2071.

11. Antiñolo A, Carrillo-Hermosilla F, Corrochano AE, Fernández-Baeza J, Lanfranchi M, Otero A, Pellinghelli MA, *J Organomet Chem* 1999, **577**, 174.
12. Murtuza S, Casagrande OL, Jordan RF, *Polym Mat Sc Eng* 2001, **84**, 109.
13. Murtuza S, Casagrande Jr OL, Jordan RF, *Organometallics* 2002, **21**, 1882.
14. Cai H, Lam WH, Yu X, Liu X, Wu Z-Z, Chen T, Lin Z, Chen X-T, You X-Z, Xue Z, *Inorg Chem* 2003, **42**, 3008.
15. Lee H, Jordan RF, *J Am Chem Soc* 2005, **127**, 9384.
16. Sterzik A, Lönnecke P, Hey-Hawkins E, *Z Anorg Allg Chem* 2005, **631**, 2457.
17. Kuhl O, Blaurock S, Sieler J, Hey-Hawkins E, *Polyhedron* 2001, **20**, 2171.
18. Satish CD, Guraraj MV, Baygur SC, *Ind J Chem A Bio-inorg, Phys Theor Anal Chem* 2000, **39A**, 446.
19. Patiño P, D'Ornelas L, Hernández de Bastianoni L, Lara L, Urdaneta Y, Waters DN, *Vibrat Spectrosc* 1997, **15**, 163.
20. Hill AF, Smith MK, *Organometallics* 2007, **26**, 3900.
21. McLauchlan CC, McDonald KJ, *Acta Cryst* 2006, **E62**, m588.
22. McLauchlan CC, McDonald KJ, *Acta Cryst* 2005, **E61**, m2379.
23. Xing Y-H, Bai F-Y, Zhang S-J, Aoki K, *Jilin Huagong Xueyuan Xuebao* 2003, **20**, 6.
24. Zou W, Song J-Y, Xing Y-H, *Yingyong Huaxue* 2006, **23**, 94.
25. Xing Y-H, Bai F-Y, Aoki K, Sun Z, Ge MF, *Synth React Inorg, Met-Org Nano-Met Chem* 2007, **37**, 203.
26. Xing Y-H, Sun Z, Yuan H, Ge M, Niu S, Bai F-Y, *Chin Sc Bull* 2006, **51**, 661.
27. Kosugi M, Hikichi S, Akita M, Moro-oka Y, *Inorg Chem* 1999, **38**, 2567.
28. Li D, Parkin S, Wang G, Yee GT, Holmes SM, *Inorg Chem* 2006, **45**, 2773.
29. Brunker TJ, Cowley AR, O'Hare D, *Organometallics* 2002, **21**, 3123.
30. Brunker TJ, Green JC, O' Hare D, *Inorg Chem* 2003, **42**, 4366.
31. Berlinguette CP, Holm RH, *J Am Chem Soc* 2006, **128**, 11993.
32. Xing Y, Zhang Y, Sun Z, Ye L, Xu Y, Ge M, Zhang B, Niu S, *J Inorg Biochem* 2007, **101**, 36.
33. Etienne M, Carfagna C, Lorente P, Mathieu R, de Montauzon D, *Organometallics* 1999, 18, 3075.

34. Jaffart J, Etienne M, Maseras F, McGrady JE, Eisenstein O, *J Am Chem Soc* 2001, **123**, 6000.
35. Schorm A, Sundermeyer J, *Eur J Inorg Chem* 2001, 2947.
36. Pritchard HM, Etienne M, Vendier L, McGrady GS, *Organometallics* 2004, **23**, 1203.
37. Oulié P, Bréfuel N, Vendier L, Duhayon C, Etienne M, *Organometallics* 2005, **24**, 4306.
38. Teuma E, Etienne M, Donnadieu B, McGrady GS, *New J Chem* 2006, **30**, 409.
39. Jaffart J, Cole ML, Etienne M, Reinhold M, McGrady JE, Maseras F, *Dalton Trans* 2003, 4057.
40. Jaffart J, Etienne M, Reinhold M, McGrady JE, Maseras F, *Chem Commun* 2003, 876.
41. Besora M, Maseras F, McGrady JE, Oulié P, Dinh DH, Duhayon C, Etienne M, *Dalton Trans* 2006, 2362.
42. Seidel WW, Schaffrath M, Meel MJ, Hamilton T, Ariza SC, Pape T, *Z Naturforsch* 2007, **62b**, 791.
43. Hierso J-C, Etienne M, *Eur J Inorg Chem* 2000, 839.
44. Etienne M, Hierso J-C, Daff PJ, Donnadieu B, Dahan F, *Polyhedron* 2004, **23**, 379.
45. Brunker TJ, Green JC, O'Hare D, *Inorg Chem* 2002, **41**, 1701.
46. Rojas R, Valderrama M, We G, *Inorg Chem Commun* 2004, **7**, 1295.
47. De Alwis DCL, Schultz FA, *Inorg Chem* 2003, **42**, 3616.
48. Harris TD, Long JR, *Chem Commun* 2007, 1360.
49. Seino H, Arai Y, Ywata N, Nagao S, Mizobe Y, Hidai M, *Inorg Chem* 2001, **40**, 1677.
50. Hughes AK, Malget JM, Goeta AE, *J Chem Soc Dalton Trans* 2001, 1927.
51. Lee C-L, Wu Y-Y, Wu C-P, Chen J-D, Keng T-C, Wang J-C, *Inorg Chim Acta* 1999, **292**, 182.
52. Lee C-L, Yang P-Y, Su C-W, Chen J-D, Liou LS, Wang J-C, *Inorg Chim Acta* 2001, **314**, 147.
53. Mocella CJ, Delafuente DA, Keane JM, Warner GR, Friedman LA, Sabat M, Harman WD, *Organometallics* 2004, **23**, 3772.
54. Nagao S, Seino H, Hidai M, Mizobe Y, *Inorg Chim Acta* 2004, **357**, 4618.
55. Nemykin VN, Basu P, *Dalton Trans* 2004, 1928.
56. Nemykin VN, Basu P, *Inorg Chem* 2005, **44**, 7494.

57. Smith PD, Millar AJ, Young CG, Ghosh A, Basu P, *J Am Chem Soc* 2000, **122**, 9298.
58. Nemykin VN, Davie SR, Mondal S, Rubie N, Kirk ML, Somogyi A, Basu P, *J Am Chem Soc* 2002, **124**, 756.
59. Millar AJ, Doonan CJ, Laughlin LJ, Tiekink ERT, Young CG, *Inorg Chim Acta* 2002, **337**, 393.
60. Topaloğlu I, *Turk J Chem* 1998, **22**, 115.
61. Topaloğlu I, McCleverty JA, *Synth React Inorg, Met-Org Chem* 1997, **27**, 1205.
62. Topaloğlu I, McCleverty JA, *Synth React Inorg, Met-Org Chem* 1999, **29**, 1101.
63. Topaloğlu-Sozuer I, Dulger Irdem S, Jeffery JJ, Hamidov H, *J Coord Chem* 2005, **58**, 175.
64. Topaloğlu-Sozuer I, Gunyar A, Jeffery JJ, Hamidov H, *Trans Met Chem* 2004, **29**, 780.
65. Topaloğlu-Sozuer I, Gunyar A, Dulger-Irdem S, Baya M, Poli R, *Inorg Chim Acta* 2005, **358**, 3303.
66. Kowallick R, Jones AN, Reeves ZR, Jeffery JC, McCleverty JA, Ward MD, *New J Chem* 1999, **23**, 915.
67. Harden NC, Humphrey ER, Jeffery JC, Lee S-M, Marcaccio M, McCleverty JA, Rees LH, Ward MD, *J Chem Soc Dalton Trans* 1999, 2417.
68. Psillakis E, Shonfield PKA, Jouaiti A-A, Maher JP, McCleverty JA, Ward MD, *J Chem Soc Dalton Trans* 2000, 241.
69. Malaun M, Kowallick R, McDonagh AM, Marcaccio M, Paul RL, Asselberghs I, Clays K, Persoons A, Bildstein B, Fiorini C, Nunzi J-M, Ward MD, McCleverty JA, *J Chem Soc Dalton Trans* 2001, 3025.
70. McDonagh AM, Ward MD, McCleverty JA, *New J Chem* 2001, **25**, 1236.
71. Topaloğlu-Sozuer I, Dulger Irdem S, Jeffery JJ, Hamidov H, Senturk OS, *Z Naturforsch* 2005, **60b**, 15.
72. Topaloğlu-Sozuer I, Dulger-Irdem S, *Turk J Chem* 2005, **29**, 225.
73. Bayly SR, McCleverty JA, Ward MD, Gatteschi D, Totti F, *Inorg Chem* 2000, **39**, 1288.
74. Shonfield PKA, Behrendt A, Jeffery JC, Maher JP, McMcleverty JA, Psillakis E, Ward MD, Western C, *J Chem Soc Dalton Trans* 1999, 4341.
75. Behrendt A, Couchman SM, Jeffery JC, McCleverty JA, Ward MD, *J Chem Soc Dalton Trans* 1999, 4349.

76. Lee S-M, Marcaccio M, McCleverty JA, Ward MD, *Chem Mater* 1998, **10**, 3272.
77. Bayly SR, Humphrey ER, de Chair H, Paredes CG, Bell ZR, Jeffery JC, McCleverty JA, Ward MD, Totti F, Gatteschi D, Courric S, Steele BR, Screttas CG, *J Chem Soc Dalton Trans* 2001, 1401.
78. Stobie K, Bell ZR, Ward MD, McCleverty JA, *Collect Czech Chem Comm* 2001, **66**, 355.
79. Laye RH, Bell ZR, Ward MD, *New J Chem* 2003, **27**, 684.
80. Santos AM, Kühn FE, Bruus-Jensen K, Lucas I, Romão CC, Herdtweck E, *J Chem Soc Dalton Trans* 2001, 1332.
81. Kassim MB, Paul RL, Jeffery JC, McCleverty JA, Ward MD, *Inorg Chim Acta* 2002, **327**, 160.
82. Lantero DR, Glenn AG, Day CS, Welker ME, *Organometallics* 2003, **22**, 1998.
83. Schuster DM, White PS, Templeton JL, *Organometallics* 2000, **19**, 1540.
84. Li W, Jin G-X, *Wuji Huaxue Xuebao* 2005, **21**, 846.
85. Arrayás RG, Liebeskind LS, *J Am Chem Soc* 2003, **125**, 9026.
86. Zhang Y, Liebeskind LS, *J Am Chem Soc* 2006, **128**, 465.
87. Malinakova HC, Liebeskind LS, *Org Lett* 2000, **2**, 4083.
88. Gómez Arrayás R, Yin J, Liebeskind LS, *J Am Chem Soc* 2007, **129**, 1816.
89. Moretto AF, Liebeskind LS, *J Org Chem* 2000, **65**, 7455.
90. Zhang Y, Liebeskind LS, *J Am Chem Soc* 2005, **127**, 11258.
91. Joshi HK, Inscore FE, Schirlin JT, Dhawan IK, Carducci MD, Bill TG, Enemark JH, *Inorg Chim Acta* 2002, **337**, 275.
92. Inscore FE, Joshi HK, McElhaney AE, Enemark JH, *Inorg Chim Acta* 2002, **331**, 246.
93. Kirk ML, Peariso K, *Polyhedron* 2004, **23**, 499.
94. Helton ME, Gruhn NE, McNaughton RL, Kirk ML, *Inorg Chem* 2000, **39**, 2273.
95. Inscore FE, Knottenbelt SZ, Rubie ND, Joshi HK, Kirk ML, Enemark JH, *Inorg Chem* 2006, **45**, 967.
96. Joshi HK, Cooney JJA, Inscore FE, Gruhn NE, Lichtenberger DL, Enemark JH, *Proc Natl Ac Sc* 2003, **100**, 3719.
97. Uhrhammer D, Schultz FA, *Inorg Chem* 2004, **43**, 7389.
98. McElhaney AE, Inscore FE, Schirlin JT, Enemark JH, *Inorg Chim Acta* 2002, **341**, 85.

99. Young CG, Gable RW, Hill JP, George GN, *Eur J Inorg Chem* 2001, 2227.
100. Drew SC, Hill JP, Lane I, Hanson GR, Gable RW, Young CG, *Inorg Chem* 2007, **46**, 2373.
101. Drew SC, Young CG, Hanson GR, *Inorg Chem* 2007, **46**, 2388.
102. Graff JN, McElhaney AE, Basu P, Gruhn NE, Chang C-SJ, Enemark JH, *Inorg Chem* 2002, **41**, 2642.
103. Wadepohl H, Arnold U, Kohl U, Prizkow H, Wolf A, *J Chem Soc Dalton Trans* 2000, 3554.
104. Lalor FJ, O'Neill SA, *J Organomet Chem* 2003, **684**, 249.
105. Ferguson G, Lalor FJ, O'Neill SA, *Acta Cryst* 2003, **E59**, m644.
106. Wlodarczyk A, Coles SJ, Hursthouse MB, Malik KMA, Lieberman HF, *J Chem Soc, Dalton Trans* 1997, 2921.
107. Keane JM, Ding F, Sabat M, Harman WD, *J Am Chem Soc* 2004, **126**, 785.
108. Graham PM, Mocella CJ, Sabat M, Harman WD, *Organometallics* 2005, **24**, 911.
109. Meiere SH, Keane JM, Gunnoe TB, Sabat M, Harman WD, *J Am Chem Soc* 2003, **125**, 2024.
110. Liu W, You F, Mocella CJ, Harman WD, *J Am Chem Soc* 2006, **128**, 1426.
111. Mesaros EF, Pearson AJ, Schoffers E, Stone KC, Templeton JL, *Inorg Synth* 2004, **34**, 104.
112. Anderson S, Cook DJ, Hill AF, *Organometallics* 2001, **20**, 2468.
113. Cook DJ, Hill AF, *Organometallics* 2003, **22**, 3502.
114. Lim PJ, Slizys DA, Tiekink ERT, Young CG, *Inorg Chem* 2005, **44**, 114.
115. Kisała J, Białońska A, Ciunik Z, Kurek S, Wołowiec S, *Polyhedron* 2006, **25**, 3222.
116. Adams CJ, Bartlett IM, Boonyuen S, Connelly NG, Harding DJ, Hayward OD, McInnes EJL, Orpen AG, Quayle MJ, Rieger PH, *Dalton Trans* 2006, 3466.
117. Adams CJ, Bartlett IM, Carlton S, Connelly NG, Harding DJ, Hayward OD, Orpen AG, Patron E, Ray CD, Rieger PH, *Dalton Trans* 2007, 62.
118. Adams CJ, Connelly NG, Onganusorn S, *Dalton Trans* 2007, 1904.
119. Seidel WW, Ibarra Arias MD, Schaffrath M, Jahnke MC, Hepp A, Pape T, *Inorg Chem* 2006, **45**, 4791.

120. Farley RD, Hoefer P, Maher JP, McCleverty JA, Murphy DM, Rowlands CC, Ung VÂ, Ward MD, *Magn Reson Chem* 2002, **40**, 683.
121. Lutta ST, Kagwanja SM, *Trans Met Chem* 2001, **26**, 523.
122. Lutta ST, Kagwanja SM, *Trans Met Chem* 2000, **25**, 415.
123. Reis H, Raptis SG, Papadopoulos MG, *Phys Chem Chem Phys* 2001, **3**, 3901.
124. Gonçalves IS, Gamelas CA, Pereira CCL, Romão CC, *J Organomet Chem* 2005, **690**, 1718.
125. Ogura H, Walker FA, *Inorg Chem* 2001, **40**, 5729.
126. Dilworth JR, Gibson VC, Davies N, Redshaw C, White AP, Williams DJ, *J Chem Soc Dalton Trans* 1999, 2695.
127. Gibson VC, Redshaw C, Clegg W, Elsegood MRJ, Siemeling U, Türk T, *Polyhedron* 2004, **23**, 189.
128. Montilla F, Pastor A, Galindo A, *J Organomet Chem* 1999, **590**, 202.
129. Cho HY, Roh S-G, Jeong JH, *Polyhedron* 2002, **21**, 1211.
130. Fickert C, Nagel V, Kiefer W, Wahl G, Sundermeyer J, *J Mol Struct* 1999, **482–483**, 59.
131. Laughlin LJ, Eagle AA, George GN, Tiekink ERT, Young CG, *Inorg Chem* 2007, **46**, 939.
132. Bruce MI, Cole ML, Gaudio M, Skelton BW, White AH, *J Organomet Chem* 2006, **691**, 4601.
133. Fomitchev DV, McLauchlan CC, Holm RH, *Inorg Chem* 2002, 41, 958.
134. Zhang Y, Holm RH, *Inorg Chem* 2004, **43**, 674.
135. Berlinguette CP, Miyaji T, Zhang Y, Holm RH, *Inorg Chem* 2006, **45**, 1997.
136. Pesavento RP, Berlinguette CP, Holm RH, *Inorg Chem* 2007, **46**, 510.
137. Yamauchi T, Takagi H, Shibahara T, Akashi H, *Inorg Chem* 2006, **45**, 5429.
138. Sengar RS, Basu P, *Inorg Chim Acta* 2007, **360**, 2092.
139. Crane TW, White PS, Templeton JL, *Organometallics* 1999, **18**, 1897.
140. Jamison GM, White PS, Templeton JL, *Organometallics* 1991, **10**, 1954.
141. Stone KC, Jamison GM, White PS, Templeton JL, *Organometallics* 2003, **22**, 3083.
142. Feng SG, Templeton JL, *Organometallics* 1992, **11**, 2168.
143. Vogeley NJ, White PS, Templeton JL, *J Am Chem Soc* 2003, **125**, 12422.

144. Francisco LW, White PS, Templeton JL, *Organometallics* 1996, **15**, 5127.
145. Frohnapfel DS, Enriquez AE, Templeton JL, *Organometallics* 2000, **19**, 221.
146. Crane TW, White PS, Templeton JL, *Inorg Chem* 2000, **39**, 1081.
147. Wells MB, McConathy JE, White PS, Templeton JL, *Organometallics* 2002, **21**, 5007.
148. Jepsen AS, Vogeley NJ, White PS, Templeton JL, *J Organomet Chem* 2001, **617–618**, 520.
149. Enriquez AE, Templeton JL, *Organometallics* 2002, **21**, 852.
150. Cross JL, Crane TW, White PS, Templeton JL, *Organometallics* 2003, **22**, 548.
151. Cross JL, Garrett AD, Crane TW, White PS, Templeton JL, *Polyhedron* 2004, **23**, 2831.
152. Stone KC, Jamison GM, White PS, Templeton JL, *Inorg Chim Acta* 2002, **330**, 161.
153. Enriquez AE, White PS, Templeton JL, *J Am Chem Soc* 2001, **123**, 4992.
154. Stone KC, Onwuzurike A, White PS, Templeton JL, *Organometallics* 2004, **23**, 4255
155. Stone KC, White PS, Templeton JL, *J Organomet Chem* 2003, **684**, 13.
156. Vogeley NJ, Templeton JL, *Polyhedron* 2004, **23**, 311.
157. Vogeley NJ, White PS, Templeton JL, *Organometallics* 2004, **23**, 1378.
158. Garrett AD, Vogeley NJ, Warner JR, White PS, Templeton JL, *Organometallics* 2006, **25**, 1728.
159. Frohnapfel DS, White PS, Templeton JL, *Organometallics* 2000, **19**, 1497.
160. Lim PJ, Slizys DA, White JM, Young CG, Tiekink ERT, *Organometallics* 2003, **22**, 4853.
161. Eagle AA, Young CG, Tiekink ERT, *Aust J Chem* 2004, **57**, 269.
162. Eagle AA, George GN, Tiekink ERT, Young CG, *J Inorg Biochem* 1999, **76**, 39.
163. Eagle AA, Tiekink ERT, George GN, Young CG, *Inorg Chem* 2001, **40**, 4563.
164. Thomas S, Eagle AA, Sproules SA, Hill JP, White JM, Tiekink ERT, George GN, Young CG, *Inorg Chem* 2003, **42**, 5909.
165. Hill JP, Laughlin LJ, Gable RW, Young CG, *Inorg Chem* 1996, **35**, 3447.

166. Eagle AA, Gable RW, Thomas S, Sproules SA, Young CG, *Polyhedron* 2004, **23**, 385.
167. Eagle AA, Gable RW, Young CG, *Aust J Chem* 1999, **52**, 827.
168. Thomas S, Tiekink ERT, Young CG, *Inorg Chem* 2006, **45**, 352.
169. Sproules SA, Morgan HT, Doonan CJ, White JM, Young CG, *Dalton Trans* 2005, 3552.
170. Seino H, Iwata N, Kawarai N, Hidai M, Mizobe Y, *Inorg Chem* 2003, **42**, 7387.
171. Malarek MS, Logan BA, White JM, Young CG, *Organometallics* 2004, **23**, 4328.
172. Seidel WW, Schaffrath M, Pape T, *Chem Commun* 2006, 3999.
173. Seidel WW, Schaffrath M, Pape T, *Angew Chem Int Ed* 2005, **44**, 7798.
174. Seidel WW, Ibarra Arias MD, Schaffrath M, Bergander K, *Dalton Trans* 2004, 2053.
175. Bartlett IM, Carlton S, Connelly NG, Harding DJ, Hayward OD, Orpen AG, Ray CD, Rieger PH, *Chem Commun* 1999, 2403.
176. Adams CJ, Anderson KM, Connelly NG, Hardling DJ, Orpen AG, Patron E, Rieger PH, *Chem Commun* 2002, 130.
177. Weber L, Dembeck G, Boese R, Bläser D, *Organometallics* 1999, **18**, 4603.
178. Doufou P, Abboud KA, Boncella JM, *Inorg Chim Acta* 2003, **345**, 103.
179. Dewhurst RD, Hill AF, Rae AD, Willis AC, *Organometallics* 2005, **24**, 4703.
180. Dewhurst RD, Hill AF, Willis AC, *Chem Commun* 2004, 2826.
181. Dewhurst RD, Hill AF, Willis AC, *Organometallics* 2004, **23**, 1646.
182. Anderson S, Cook DJ, Hill AF, Malget JM, White AJP, Williams DJ, *Organometallics* 2004, **23**, 2552.
183. Mayr A, Ahn S, *Inorg Chim Acta* 2000, **300–302**, 406.
184. Weber L, Dembeck G, Lönneke P, Stammler H-G, Neumann B, *Organometallics* 2001, **20**, 2288.
185. Dewhurst RD, Hill AF, Smith MK, *Angew Chem Int Ed* 2004, **43**, 476.
186. Schwenzer B, Schleu J, Burzlaff N, Karl C, Fischer H, *J Organomet Chem* 2002, **641**, 134.
187. Stobie KM, Bell ZR, Munhoven TW, Maher JP, McCleverty JA, Ward MD, McInnes EJL, Totti F, Gatteschi D, *Dalton Trans* 2003, 36.

188. Graham PM, Meiere SH, Sabat M, Harman WD, *Organometallics* 2003, **22**, 4364.
189. Bassett KC, You F, Graham PM, Myers WH, Sabat M, Harman WD, *Organometallics* 2006, **25**, 435.
190. Todd MA, Grachan ML, Sabat M, Myers WH, Harman WD, *Organometallics* **2006**, **25**, 3948.
191. Delafuente DA, Myers WH, Sabat M, Harman WD, *Organometallics* 2005, **24**, 1876.
192. Lis EC, Delafuente DA, Lin Y, Mocella CJ, Todd MA, Liu W, Sabat M, Myears WH, Harman HW, *Organometallics* 2006, **25**, 5051.
193. Surendranath Y, Welch KD, Nash BW, Harman HW, Myers WH, Harman WD, *Organometallics* 2006, **25**, 5852.
194. Welch KD, Harrison DP, Lis Jr EC, Liu W, Salomon RJ, Harman WD, Myers WH, *Organometallics* 2007, **26**, 2791.
195. Harman WD, Trindle C, *J Comput Chem* 2005, **26**, 194.
196. Hong D, Zhang Y, Holm RH, *Inorg Chim Acta* 2005, **358**, 2303.
197. Xing YH, Aoki K, Bai FY, *Synth React Inorg Met-Org Chem* 2004, **34**, 1149.
198. Hossain F, Rigsby MA, Duncan CT, Milligan Jr PL, Lord RL, Baik M-H, Schultz FA, *Inorg Chem* 2007, **46**, 2596.
199. Brown SN, Mayer JM, *J Am Chem Soc* 1994, **116**, 2219.
200. Brown SN, Mayer JM, *J Am Chem Soc* 1996, **118**, 12119.
201. Domingos Â, Marçalo J, Paulo A, Pires de Matos A, Santos I, *Inorg Chem* 1993, **32**, 5114.
202. Paulo A, Domingos Â, Marçalo J, Pires de Matos A, Santos I, *Inorg Chem* 1995, **34**, 2113.
203. Trofimenko S, *Scorpionates, The Coordination Chemistry of Polypyrazolylborate Ligands*, Imperial College Press, 1999.
204. DuMez DD, Northcutt TO, Matano Y, Mayer JM, *Inorg Chem* 1999, **38**, 3309.
205. Matano Y, Northcutt TO, Brugman J, Bennett BK, Lovell S, Mayer JM, *Organometallics* 2000, **19**, 2781.
206. Bennett BK, Crevier TJ, DuMez DD, Matano Y, McNeil WS, Mayer JM, *J Organomet Chem* 1999, **591**, 96.
207. Gable KP, Brown EC, *J Am Chem Soc* 2003, **125**, 11018.
208. Gable KP, Chuawong P, Yokochi AFT, *Organometallics* 2002, **21**, 929.
209. Gable KP, Khownium K, Chuawong P, *Organometallics* 2004, **23**, 5268.

210. Seymore SB, Brown SN, *Inorg Chem* 2000, **39**, 325.
211. Lassen PR, Guy L, Karame I, Roisnel T, Vanthuyne N, Roussel C, Cao X, Lombardi R, Crassous J, Freedman TB, Nafie LA, *Inorg Chem* 2006, **45**, 10230.
212. Gunnoe TB, Sabat M, Harman WD, *J Am Chem Soc* 1999, **121**, 6499.
213. Gunnoe TB, Meiere SH, Sabat M, Harman WD, *Inorg Chem* 2000, **39**, 6127.
214. Gable KP, Zhuravlev FA, *J Am Chem Soc* 2002, **124**, 3970.
215. Paulo A, Domingos Â, Garcia R, Santos I, *Inorg Chem* 2000, **39**, 5669.
216. Faller JW, Lavoie AR, *Organometallics* 2000, **19**, 3957.
217. Middleditch M, Anderson JC, Blake AJ, Wilson C, *Inorg Chem* 2007, **46**, 2797.
218. Kückmann TI, Abram U, *Z Anorg Allg Chem* 2004, **630**, 783.
219. Barrado G, Doerrer L, Green MLH, Leech MA, *J Chem Soc Dalton Trans* 1999, 1061.
220. Gunnoe TB, Sabat M, Harman WD, *Organometallics* 2000, **19**, 728.
221. Valahovic MT, Gunnoe TB, Sabat M, Harman WD, *J Am Chem Soc* 2002, **124**, 3309.
222. Meiere SH, Ding F, Friedman LA, Sabat M, Harman WD, *J Am Chem Soc* 2002, **124**, 13506.
223. Friedman LA, Meiere SH, Brooks BC, Harman WD, *Organometallics* 2001, **20**, 1699.
224. Meiere SH, Brooks BC, Gunnoe TB, Carrig EH, Sabat M, Harman WD, *Organometallics* 2001, **20**, 3661.
225. Brooks BC, Meiere SH, Friedman LA, Carrig EH, Gunnoe TB, Harman WD, *J Am Chem Soc* 2001, **123**, 3541.
226. Meiere SH, Valahovic MT, Harman WD, *J Am Chem Soc* 2002, **124**, 15099.
227. Friedman LA, Sabat M, Harman WD, *J Am Chem Soc* 2002, **124**, 7395.
228. Myers WH, Welch KD, Graham PM, Keller A, Sabat M, Trindle CO, Harman WD, *Organometallics* 2005, **24**, 5267.
229. Keane JM, Chordia MD, Mocella CJ, Sabat M, Trindle CO, Harman WD, *J Am Chem Soc* 2004, **126**, 6806.
230. Schiffler MA, Friedman LA, Brooks BC, Sabat M, Harman WD, *Organometallics* 2003, **22**, 4966.
231. Bengali AA, Mezick BK, Hart MN, Fereshteh S, *Organometallics* 2003, **22**, 5436.

232. McNeil WS, DuMez DD, Matano Y, Lowell S, Mayer JM, *Organometallics* 1999, **18**, 3715.
233. Oliver JD, Mullica DF, Hutchinson BB, Milligan WO, *Inorg Chem* 1980, **19**, 165.
234. Sohrin Y, Kokusen H, Matsui M, *Inorg Chem* 1995, **34**, 3928.
235. Cho S-H, Wang D, Kim K, *Bull Korean Chem Soc* 1991, **12**, 107.
236. Armstrong WH, Lippard SJ, *J Am Chem Soc* 1983, **105**, 4837.
237. Armstrong WH, Spool A, Papaefthymiou GC, Frankel RB, Lippard SJ, *J Am Chem Soc* 1984, **106**, 3653.
238. Armstrong WH, Lippard SJ, *J Am Chem Soc* 1985, **107**, 3730.
239. King RB, Bond A, *J Am Chem Soc* 1974, **96**, 1334.
240. Anderson S, Hill AF, White AJP, Williams DJ, *Organometallics* 1998, **17**, 2665.
241. Anderson S, Hill AF, Slawin AMZ, White AJP, Williams DJ, *Inorg Chem* 1998, **37**, 594.
242. Cotton FA, Frenz BA, Shaver A, *Inorg Chim Acta* 1973, **7**, 161.
243. Bellachioma G, Cardaci G, Gramlich V, Macchioni A, Pieroni F, Venanzi LM, *J Chem Soc, Dalton Trans* 1998, 947.
244. Lescoüezec R, Vaissermann J, Lloret F, Julve M, Verdaguer M, *Inorg Chem* 2002, **41**, 5943.
245. Kim J, Han S, Cho I-K, Choi KY, Heu M, Yoon S, Suh BJ, *Polyhedron* 2004, **23**, 1333.
246. Li D, Parkin S, Wang G, Yee GT, Holmes SM, *Inorg Chem* 2006, **45**, 1951.
247. Wang S, Zuo J-L, Zhou H-C, Song Y, Song G, You X-Z, *Eur J Inorg Chem* 2004, 3681.
248. Wen H-R, Wang C-F, Song Y, Gao S, Zuo JL, You X-Z, *Inorg Chem* 2006, **45**, 8942.
249. Li D, Parkin S, Wang G, Yee CT, Prosvirin AV, Holmes SM, *Inorg Chem* 2005, **44**, 4903.
250. Liu W, Nfor EN, Li Y-Z, Zuo JL, You X-Z, *Inorg Chem Commun* 2006, **9**, 923.
251. Wang S, Zuo J-L, Zhou H-C, Choi HJ, Ke Y, Long JR, You XZ, *Angew Chem Int Ed* 2004, **43**, 5940.
252. Tao J-Q, Wang T-W, Gu Z-G, Zuo J-L, You X-Z, *Wuji Huaxue Xuebao* 2006, **22**, 2207.
253. Kim J, Han S, Pokhodnya KI, Migliori JM, Miller JS, *Inorg Chem* 2005, **44**, 6983.

254. Wang S, Zuo J-L, Zhou H-C, Song Y, You X-Z, *Inorg Chim Acta* 2005, **358**, 2101.
255. Wang S, Zuo JL, Gao S, Song Y, Zhou H-C, Zhang Y-Z, You X-Z, *J Am Chem Soc* 2004, **126**, 8900.
256. Gu Z-G, Yang Q-F, Liu W, Song Y, Li Y-Z, Zuo J-L, You X-Z, *Inorg Chem* 2006, **45**, 8895.
257. Wang S, Ferbinteanu M, Yamashita M, *Inorg Chem* 2007, **46**, 610.
258. Li D, Parkin S, Wang G, Yee GT, Clérac R, Wernsdorfer W, Holmes SM, *J Am Chem Soc* 2006, **128**, 4214.
259. Li D, Clérac R, Parkin S, Wang G, Yee GT, Holmes SM, *Inorg Chem* 2006, **45**, 5251.
260. Li D, Parkin S, Clérac R, Holmes SM, *Inorg Chem* 2006, **45**, 7569.
261. Wang C-F, Zuo J-L, Bartlett BM, Song Y, Long JR, You X-Z, *J Am Chem Soc* 2006, **128**, 7162.
262. Jiang L, Feng X-L, Lu T-B, Gao S, *Inorg Chem* 2006, **45**, 5018.
263. Liu W, Wang C-F, Li Y-Z, Zuo J-L, You X-Z, *Inorg Chem* 2006, **45**, 10058.
264. Park K, Holmes SM, *Phys Rev B* 2006, **74**, 224440.
265. Jiang L, Choi HJ, Feng X-L, Lu T-B, Long JR, *Inorg Chem* 2007, **46**, 2181.
266. Kim J, Han S, Lim JM, Choi K-Y, Nojiri H, Suh BJ, *Inorg Chim Acta* 2007, **360**, 2647.
267. Gu ZG, Liu W, Yang Q-F, Zhou X-H, Zuo J-L, You X-Z, *Inorg Chem* 2007, **46**, 3236.
268. Tyagi P, Li D, Holmes SM, Hinds BY, *J Am Chem Soc* 2007, **129**, 4929.
269. Sun Y-J, Yi L, Yang X, Liu Y, Cheng P, Liao D-Z, Yan S-P, Jiang Z-H, *Inorg Chim Acta* 2005, **358**, 396.
270. Wang S, Li YZ, Zuo JL, Cheng H, You XZ, *Acta Crystallogr* 2004, **C60**, m258.
271. Hong C-G, Zhou A-J, Tong M-L, *Acta Cryst* 2005, **E61**, m1774.
272. Turner JW, Schultz FA, *J Phys Chem B* 2002, **106**, 2009.
273. Graziani O, Toupet L, Hamon J-R, Tilset M, *J Organomet Chem* 2003, **669**, 200.
274. Anderson S, Hill AF, Ng YT, *Organometallics* 2000, **19**, 15.
275. Picraux LB, Smeigh AL, Guo D, McCusker JK, *Inorg Chem* 2005, **44**, 7846.
276. Cheng YH, Yeung LK, Lee HK, *J Inorg Biochem* 2003, **96**, 116.

277. Ferguson G, Restivo RJ, *J Chem Soc, Chem Commun* 1973, 847.
278. Bhambri S, Tocher DA, *Polyhedron* 1996, **15**, 2763.
279. McNair AM, Boyd DC, Mann KR, *Organometallics* 1986, **5**, 303.
280. Jalòn FA, Otero A, Rodriguez A, *J Chem Soc, Dalton Trans* 1995, 1629.
281. Corrochano AE, Jalòn FA, Otero A, Kubiki MM, Richard P, *Organometallics* 1997, **16**, 145.
282. Onishi M, Ikemoto K, Hiraki K, *Inorg Chim Acta* 1994, **219**, 3.
283. Hill AF, *J Organomet Chem* 1990, **395**, C35.
284. Alcock NW, Hill AF, Melling RP, *Organometallics* 1991, **10**, 3898.
285. Moreno B, Sabo-Etienne S, Chaudret B, Rodriguez-Fernandez A, Jalòn F, Trofimenko S, *J Am Chem Soc* 1994, **116**, 2635.
286. Halcrow MH, Chaudret B, Trofimenko S, *J Chem Soc Chem Commun* 1993, 465.
287. Rüba E, Gemel C, Slugovc C, Mereiter K, Schmid R, Kirchner K, *Organometallics* 1999, **18**, 2275.
288. Jiménez Tenorio MA, Jiménez Tenorio M, Puerta MC, Valerga P, *Organometallics* 2000, **19**, 1333.
289. Pavlik S, Mereiter K, Puchberger M, Kirchner K, *J Organomet Chem* 2005, **690**, 5497.
290. Pavlik S, Gemel C, Slugovc C, Mereiter K, Schmid R, Kirchner K, *J Organomet Chem* 2001, **617–618**, 301.
291. Pavlik S, Schmid R, Kirchner K, Mereiter K, *Monatshefte für Chemie* 2004, **135**, 1349.
292. Rüba E, Hummel A, Mereiter K, Schmid R, Kirchner K, *Organometallics* 2002, **21**, 4955.
293. Standfest-Hauser CM, Mereiter K, Schmid R, Kirchner K, *Eur J Inorg Chem* 2003, 1883.
294. Sanford MS, Valdez MR, Grubbs RH, *Organometallics* 2001, **20**, 5455.
295. Buriez B, Cook DJ, Harlow KJ, Hill AF, Welton T, White AJP, Williams DJ, Wilton-Ely JDET, *J Organomet Chem* 1999, **578**, 264.
296. Pavlik S, Mereiter K, Puchberger M, Kirchner K, *Organometallics* 2005, **24**, 3561.
297. Jayaprakash KN, Gunnoe TB, Boyle PD, *Inorg Chem* 2001, **40**, 6481.
298. Jayaprakash KN, Gillespie AM, Gunnoe TB, White DP, *Chem Commun* 2002, 372.
299. Conner D, Jayaprakash KN, Wells MB, Manzer S, Gunnoe TB, Boyle PD, *Inorg Chem* 2003, **42**, 4759.

300. Conner D, Jayaprakash KN, Gunnoe TB, Boyle PD, *Organometallics* 2002, **21**, 5265.
301. Jayaprakash KN, Conner D, Gunnoe TB, *Organometallics* 2001, **20**, 5254.
302. Conner D, Jayaprakash KN, Gunnoe TB, Boyle PD, *Inorg Chem* 2002, **41**, 3042.
303. Albertin G, Antoniutti S, Bortoluzzi M, Castro-Fojo J, Garcia-Fontán S, *Inorg Chem* 2004, **43**, 4511.
304. Beach NJ, Williamson AE, Spivak GJ, *J Organomet Chem* 2005, **690**, 4640.
305. Jimenéz-Tenorio MA, Jimenéz-Tenorio M, Puerta MC, Valerga P, *Inorg Chim Acta* 2000, **300–302**, 869.
306. Hartmann S, Winter RF, Brunner BM, Sarkar B, Knödler A, Hartenbach I, *Eur J Inorg Chem* 2003, 876.
307. Lo Y-H, Lin Y-C, Lee G-H, Wang Y, *Eur J Inorg Chem* 2004, 4616.
308. Wilson DC, Nelson JH, *J Organomet Chem* 2003, **682**, 272.
309. Burtscher D, Perner B, Mereiter K, Slugovc C, *J Organomet Chem* 2006, **691**, 5423.
310. Kremel S, Mereiter K, Slugovc C, Gemel C, Pfeiffer J, Schmid R, Kirchner K, *Monatshefte für Chemie* 2001, **132**, 551.
311. Gemel C, John R, Slugovc C, Mereiter K, Schmid R, Kirchner K, *J Chem Soc, Dalton Trans* 2000, 2607.
312. Standfest-Hauser CM, Mereiter K, Schmid R, Kirchner K, *Organometallics* 2004, **23**, 2194.
313. Onishi M, Yamaguchi M, Kumagae S, Kawano H, Arikawa Y, *Inorg Chim Acta* 2006, **359**, 990.
314. Slugovc C, Kirchner K, Mereiter K, *Acta Cryst* 2006, **E62**, m1356.
315. Huh S, Kim Y, Youm K-T, Jun M-J, *Polyhedron* 1999, **18**, 2625.
316. Feng Y, Lail M, Barakat KA, Cundari TR, Gunnoe TB, Petersen JL, *J Am Chem Soc*, 2005, **127**, 14174.
317. Feng Y, Lail M, Foley NA, Gunnoe TB, Barakat KA, Cundari TR, Petersen JL, *J Am Chem Soc* 2006, **128**, 7982.
318. Ng SM, Lam WH, Mak CC, Tsang CW, Jia G, Lin Z, Lau CP, *Organometallics* 2003, **22**, 641.
319. Feng Y, Gunnoe TB, Grimes TV, Cundari TR, *Organometallics* 2006, **25**, 5456.
320. Lee JP, Pittard KA, DeYonker NJ, Cundari TR, Gunnoe TB, Petersen JL, *Organometallics* 2006, **25**, 1500.

321. Albertin G, Antoniutti S, Bortoluzzi M, Zanardo G, *J Organomet Chem* 2005, **690**, 1726.
322. Yin C-Q, Li W, Fu X-b, Xu Z-t, Ng S-M, *Fenzi Cuihua* 2003, **17**, 22.
323. Yin C-Q, Xu Z-t, Yang S-Y, Ng SM, Wong KY, Lin Z, Lau CP, *Organometallics* 2001, **20**, 1216.
324. Yin C-Q, Feng Q-W, Chen Y, Bai Z-W, Li Z-Y, *Huaxue Xuebao* 2007, **65**, 722.
325. Ng SM, Lau CP, Fan M-F, Lin Z, *Organometallics* 1999, **18**, 2484.
326. Goj LA, Lail M, Pittard KA, Riley KC, Gunnoe TB, Petersen JL, *Chem Commun* 2006, 982.
327. Pittard KA, Cundari TR, Gunnoe TB, Day CS, Petersen JL, *Organometallics* 2005, **24**, 5015.
328. Arikawa Y, Nishimura Y, Kawano H, Onishi M, *Organometallics* 2003, **22**, 3354.
329. Arikawa Y, Nishimura Y, Ikeda K, Onishi M, *J Am Chem Soc* 2004, **126**, 3706.
330. Nishimura Y, Arikawa Y, Inoue T, Onishi M, *Dalton Trans* 2005, 930.
331. Arikawa Y, Asayama T, Onishi M, *J Organomet Chem* 2007, **692**, 194.
332. Lee JP, Jimenez-Halla JO, Cundari TR, Gunnoe TB, *J Organomet Chem* 2007, 692, 2175.
333. Pittard KA, Lee JP, Cundari TR, Gunnoe TB, Petersen JL, *Organometallics* 2004, **23**, 5514.
334. Lail M, Arrowood BN, Gunnoe TB, *J Am Chem Soc* 2003, **125**, 7506.
335. Lail M, Bell CM, Conner D, Cundari TR, Gunnoe TB, Petersen JL, *Organometallics* 2004, **23**, 5007.
336. Foley NA, Lail M, Lee JP, Gunnoe TB, Cundari TR, Petersen JL, *J Am Chem Soc* 2007, **129**, 6765.
337. Arrowood BN, Lail M, Gunnoe TB, Boyle PD, *Organometallics* 2003, **22**, 4692.
338. Leung CW, Zheng W, Wang D, Ng SM, Yeung CH, Zhou Z, Lin Z, Lau CP, *Organometallics* 2007, **26**, 1924.
339. Fung WK, Huang X, Man ML, Ng SM, Hung MY, Lin Z, Lau CP, *J Am Chem Soc* 2003, **125**, 11539.
340. Lail M, Gunnoe TB, Barakat KA, Cundari TR, *Organometallics* 2005, **24**, 1301.
341. Pavlik S, Puchberger M, Mereiter K, Kirchner K, *Eur J Inorg Chem* 2006, 4137.

342. Jimenez-Tenorio M, Palacios MD, Puerta MC, Valerga P, *J Mol Cat A: Chem* 2007, **261**, 64.
343. Liu SH, Xia H, Wan KL, Yeung RCY, Hu QY, Jia G, *J Organomet Chem* 2003, **683**, 331.
344. Man ML, Zhu J, Ng SM, Zhou Z, Yin C, Lin Z, Lau CP, *Organometallics* 2004, **23**, 6214.
345. Tanase T, Takeshita N, Inoue C, Kato M, Yano S, Sato K, *J Chem Soc Dalton Trans* 2001, 2293.
346. Huang L, Seward KY, Sullivan BP, Jones WE, Mecholsky JJ, Dressick WJ, *Inorg Chim Acta* 2000, **310**, 227.
347. Onishi M, Kumagae S, Asai K, Kawano H, Shigemitsu Y, *Chem Lett* 2001, 96.
348. Yuan P, Liu SH, Xiong W, Yin J, Yu G-a, Sung HY, Williams ID, Jia G, *Organometallics* 2005, **24**, 1452.
349. Slugovc C, Kirchner K, Mereiter K, *Acta Cryst E* 2005, **E61**, m1646.
350. Jiménez-Tenorio M, Palacios MD, Puerta MC, Valerga P, *Organometallics* 2005, **24**, 3088.
351. Yuan P, Yin J, Yu G-a, Hu Q, Liu SH, *Organometallics* 2007, **26**, 196.
352. Arikawa Y, Ikeda K, Asayama T, Nishimura Y, Onishi M, *Chem Eur J* 2007, **13**, 4024.
353. Shiu K-B, Chen J-Y, Yu S-J, Wang S-L, Liao F-L,Wang Y, Lee G-H, *J Organomet Chem* 2002, **648**, 193.
354. Freedman DA, Kruger S, Roosa C, Wymer C, *Inorg Chem* 2006, **45**, 9558.
355. Kumar AS, Tanase T, Iida M, *Langmuir* 2007, **23**, 391.
356. Rüba E, Simanko W, Mereiter K, Schmid R, Kirchner K, *Inorg Chem* 2000, **39**, 382.
357. Crevier TJ, Mayer JM, *J Am Chem Soc* 1998, **120**, 5595.
358. Crevier TJ, Mayer JM, *Angew Chem Int Ed* 1998, **37**, 1891.
359. Demadis KD, El-Samanody ES, Meyer TJ, White PS, *Polyhedron* 1999, **18**, 1587.
360. Demadis KD, El-Samanody ES, Coia GM, Meyer TJ, *J Am Chem Soc* 1999, **121**, 535.
361. Freedman DA, Gill TP, Blough AM, Koefod RS, Mann KR, *Inorg Chem* 1997, **36**, 95.
362. Bohanna C, Esteruelas MA, Gómez AV, López AM, Martínez MP, *Organometallics* 1997, **16**, 4464.
363. Steyn MM de V, Singleton E, Hietkamp S, Liles DC, *J Chem Soc, Dalton Trans* 1990, 2991.

364. Burns ID, Hill AF, White AJP, Williams DJ, Wilton-Ely JDET, *Organometallics* 1998, **17**, 1552.
365. Crevier TJ, Lovell S, Mayer JM, *Chem Commun* 1998, 2371.
366. Crevier TJ, Lowell S, Mayer JM, Rheingold AL, Guzei IA, *J Am Chem Soc* 1998, **120**, 6607.
367. Bennett BK, Pitteri SJ, Pilobello L, Lovell S, Kaminski W, Mayer JM, *J Chem Soc Dalton Trans* 2001, 3489.
368. Soper JD, Bennett BK, Lovell S, Mayer JM, *Inorg Chem* 2001, **40**, 1888.
369. Crevier TJ, Bennett BK, Soper JD, Bowman JA, Dehestani A, Horvath DA, Lovell S, Kaminsky W, Mayer JM, *J Am Chem Soc* 2001, **123**, 1059.
370. Huynh MHV, White PS, John KD, Meyer TJ, *Angew Chem Int Ed* 2001, **40**, 4049.
371. Soper JD, Rhile IJ, DiPasquale AG, Mayer JM, *Polyhedron* 2004, **23**, 323.
372. Soper JD, Saganic E, Weinberg D, Hrovat DA, Benedict JB, Kaminsky W, Mayer JM, *Inorg Chem* 2004, **43**, 5804.
373. Bennett BK, Saganic E, Lovell S, Kaminsky W, Samuel A, Mayer JM, *Inorg Chem* 2003, **42**, 4127.
374. Dehestani A, Kaminsky W, Mayer JM, *Inorg Chem* 2003, **42**, 605.
375. McCarthy MR, Crevier TJ, Bennett B, Dehestani A, Mayer JM, *J Am Chem Soc* 2000, **122**, 12391.
376. Wu A, Dehestani A, Saganic E, Crevier TJ, Kaminsky W, Cohen DE, Mayer JM, *Inorg Chim Acta* 2006, **359**, 2842.
377. El-Samanody E-S, Demadis KD, Meyer TJ, White PS, *Inorg Chem* 2001, **40**, 3677.
378. El-Samanody E-S, Demadis KD, Gallagher LA, Meyer TJ, White PS, *Inorg Chem* 1999, **38**, 3329.
379. Seymore SB, Brown SN, *Inorg Chem* 2002, **41**, 462.
380. Albertin G, Antoniutti S, Bacchi A, Bortoluzzi M, Pellizzi G, Zanardo G, *Organometallics* 2006, **25**, 4235.
381. Albertin G, Antoniutti S, Zanardo G, *J Organomet Chem* 2007, **692**, 3706.
382. Dickinson PW, Girolami GS, *Inorg Chem* 2006, **45**, 5215.
383. Castro-Rodrigo R, Esteruelas MA, López AM, Oliván M, Oñate E, *Organometallics* 2007, **26**, 4498.
384. Churchill MR, Gold K, Maw Jr CE, *Inorg Chem* 1970, **9**, 1597.
385. Fujihara T, Schönherr T, Kaizaki S, *Inorg Chim Acta* 1996, **249**, 135.

386. Hayashi A, Nakajima K, Nonoyama M, *Polyhedron* 1997, **16**, 4087.
387. Thompson JS, Sorrell T, Marks TJ, Ibers JA, *J Am Chem Soc* 1979, **101**, 4193.
388. Brunker TJ, Barlow S, O'Hare D, *Chem Commun* 2001, 2052.
389. Doren DJ, Konecny R, Theopold KH, *Cat Today* 1999, **50**, 669.
390. Hikichi S, Yoshizawa M, Sasakura Y, Komatsuzaki H, Moro-oka Y, Akita M, *Chem Eur J* 2001, **7**, 5011.
391. Sun Y-J, Shen W-Z, Cheng P, Yan S-P, Liao D-Z, Jiang Z-H, Shen P-W, *Inorg Chem Commun* 2002, **5**, 512.
392. Sun Y-J, Zhang LZ, Cheng P, Lin H-K, Yan S-P, Liao DZ, Jiang Z-H, Shen P-W, *J Mol Catal A: Chem* 2004, **208**, 83.
393. Trofimenko S, *J Am Chem Soc* 1969, **91**, 588.
394. Socol SM, Meek DW, *Inorg Chim Acta* 1985, **101**, L45.
395. Cocivera M, Ferguson G, Lalor FJ, Szczecinski P, *Organometallics* 1982, **1**, 1139.
396. Barrientos C, Ghosh CK, Graham WAG, Thomas MJ, *J Organomet Chem* 1990, **394**, C31.
397. Ball RG, Ghosh CK, Hoyano JK, McMaster AD, Graham WAG, *J Chem Soc Chem Commun*, 1989, 341.
398. Connelly NG, Emslie DJH, Metz B, Orpen AG, Quayle MJ, *Chem Commun* 1996, 2289.
399. Chauby V, Serra Le Berre C, Kalck P, Daran J-C, Commenges G, *Inorg Chem* 1996, **35**, 6354.
400. Wick DD, Jones WD, *Inorg Chem* 1997, **36**, 2723.
401. Wick DD, Northcutt TO, Lachicotte RJ, Jones WD, *Organometallics* 1998, **17**, 4484.
402. Circu V, Fernandes MA, Carlton L, *Inorg Chem* 2002, **41**, 3859.
403. Circu V, Fernandes MA, Carlton L, *Polyhedron* 2003, **22**, 3293.
404. Cao C, Wang T, Patrick BO, Love JA, *Organometallics* 2006, **25**, 1321.
405. Van Rooy S, Cao C, Patrick BO, Lam A, Love JA, *Inorg Chim Acta* 2006, **359**, 2918.
406. Connelly NG, Emslie DJH, Geiger WE, Hayward OD, Linehan EB, Quayle MJ, Rieger PH, *J Chem Soc Dalton Trans* 2001, 670.
407. Chauby V, Daran J-C, Serra-Le Berre C, Malbosc F, Kalck P, Delgado Gonzalez O, Haslam CE, Haynes A, *Inorg Chem* 2002, **41**, 3280.
408. Herberhold M, Eibl S, Milius W, Wrackmeyer B, *Z Anorg Allg Chem* 2000, **626**, 552.

409. Geiger WE, Ohrenberg NC, Yeomans B, Connelly NG, Emslie DJH, *J Am Chem Soc* 2003, **125**, 8680.
410. Connelly NG, Emslie DJH, Hayward OD, Orpen AG, Quayle MJ, *J Chem Soc, Dalton Trans* 2001, 875.
411. Malbosc F, Chauby V, Serra-Le Berre C, Etienne M, Daran J-C, Kalck P, *Eur J Inorg Chem* 2001, 2689.
412. Webster CE, Hall MB, *Inorg Chim Acta* 2002, **330**, 268.
413. Eckert J, Webster CE, Hall MB, Albinati A, Venanzi LM, *Inorg Chim Acta* 2002, **330**, 240.
414. Carmona D, Viguri F, Lahoz FJ, Oro LA, *Inorg Chem* 2002, **41**, 2385.
415. Carmona E, Cingolani A, Marchetti F, Pettinari C, Pettinari R, Skelton BW, White AH, *Organometallics* 2003, **22**, 2820.
416. Albinati A, Bucher UE, Gramlich V, Renn O, Rüegger H, Venanzi LM, *Inorg Chim Acta* 1999, **284**, 191.
417. Misumi Y, Seino H, Mizobe Y, *J Organomet Chem* 2006, **691**, 3157.
418. Nagao S, Seino H, Hidai M, Mizobe Y, *Dalton Trans* 2005, 3166.
419. Seino H, Yoshikawa T, Hidai M, Mizobe Y, *Dalton Trans* 2004, 3593.
420. Boaretto R, Ferrari A, Merlin M, Sostero S, Traverso O, *J Photochem Photobiol A: Chem* 2000, **135**, 179.
421. Boaretto R, Paolucci G, Sostero S, Traverso O, *J Mol Catal A Chem* 2003, **204–205**, 253.
422. Yeston JS, McNamara BK, Bergman RG, Moore CB, *Organometallics* 2000, **19**, 3442.
423. Nicasio MC, Paneque M, Pérez PJ, Pizzano A, Poveda ML, Rey L, Sirol S, Taboada S, Trujillo M, Monge A, Ruiz C, Carmona E, *Inorg Chem* 2000, **39**, 180.
424. Paneque M, Pérez PJ, Pizzano A, Poveda ML, Taboada S, Trujillo M, Carmona E, *Organometallics* 1999, **18**, 4304.
425. Paneque M, Sirol S, Trujillo M, Gutiérrez-Puebla E, Monge MA, Carmona E, *Angew Chem Int Ed* 2000, **39**, 218.
426. Paneque M, Sirol S, Trujillo M, Carmona E, Gutiérrez-Puebla E, Monge MA, Ruiz C, Malbosc F, Serra-Le Berre C, Kalck P, Etienne M, Daran JC, *Chem Eur J* 2001, **7**, 3868.
427. Bowden AA, Hughes RP, Lindner DC, Incarvito CD, Liable-Sands LM, Rheingold AL, *J Chem Soc Dalton Trans* 2002, 3245.
428. Nagao S, Saito N, Kojima A, Seino H, Mizobe Y, *Bull Chem Soc Jap* 2005, **78**, 1641.
429. Northcutt TO, Wick DD, Vetter AJ, Jones WD, *J Am Chem Soc* 2001, **123**, 7257.

430. Wick DD, Reynolds KA, Jones WD, *J Am Chem Soc* 1999, **121**, 3974.
431. Vetter AJ, Rieth RD, Jones WD, *Proc Natl Ac Sc* 2007, **104**, 6957.
432. Vetter AJ, Jones WD, *Polyhedron* 2004, **23**, 413.
433. Heinekey DM, Oldham Jr WJ, *J Am Chem Soc* 1994, **116**, 3137.
434. Paneque M, Poveda ML, Taboada S, *J Am Chem Soc* 1994, **116**, 4519.
435. Oldham Jr WJ, Heinekey DM, *Organometallics* 1997, **16**, 467.
436. Gutiérrez-Puebla E, Monge A, Nicasio MC, Pérez PJ, Poveda ML, Rey L, Ruíz C, Carmona E, *Inorg Chem* 1998, **37**, 4538.
437. Ball RG, Ghosh CK, Hoyano JK, McMaster AD, Graham WAG, *J Chem Soc Chem Commun* 1989, 341.
438. Fernandez MJ, Rodriguez MJ, Oro LA, *Polyhedron* 1991, **14**, 1595.
439. Tanke RS, Crabtree RH, *Inorg Chem* 1989, **28**, 3444.
440. Gutiérrez E, Monge A, Nicasio MC, Poveda ML, Carmona E, *J Am Chem Soc* 1994, **116**, 791.
441. Alvarado Y, Daff PJ, Pérez PJ, Poveda ML, Sánchez-Delgado R, Carmona E, *Organometallics* 1996, **15**, 2192.
442. Boutry O, Poveda ML, Carmona E, *J Organomet Chem* 1997, **528**, 143.
443. Gutiérrez-Puebla E, Monge A, Nicasio MC, Pérez PJ, Poveda ML, Carmona E, *Chem Eur J* 1998, **4**, 2225.
444. Wiley JS, Heinekey DM, *Inorg Chem* 2002, **41**, 4961.
445. O'Connor JM, Closson A, Gantzel P, *J Am Chem Soc* 2002, **124**, 2434.
446. Paneque M, Poveda ML, Salazar V, Carmona E, Ruiz-Valero C, *Inorg Chim Acta* 2003, **345**, 367.
447. Paneque M, Poveda ML, Carmona E, Salazar V, *Dalton Trans* 2005, 1422.
448. Paneque M, Poveda ML, Santos LL, Salazar V, Carmona E, *Chem Commun* 2004, 1838.
449. Álvarez E, Paneque M, Petronilho AG, Poveda ML, Santos LL, Carmona E, Mereiter K, *Organometallics* 2007, **26**, 1231.
450. Wiley JS, Oldham Jr WJ, Heinekey DM, *Organometallics* 2000, **19**, 1670.
451. Tellers DM, Bergman RG, *J Am Chem Soc* 2000, **122**, 954.
452. Tellers DM, Bergman RG, *Organometallics* 2001, **20**, 4819.
453. Tellers DM, Bergman RG, *Can J Chem* 2001, **79**, 525.
454. Ilg K, Paneque M, Poveda ML, Rendón N, Santos LL, Carmona E, Mereiter K, *Organometallics* 2006, **25**, 2230.
455. Paneque M, Poveda ML, Salazar V, Gutiérrez-Puebla E, Monge A, *Organometallics* 2000, **19**, 3120.

456. Paneque M, Posadas CM, Poveda ML, Rendón N, Mereiter K, *Organometallics* 2007, **26**, 1900.
457. Álvarez E, Gómez M, Paneque M, Posadas CM, Poveda ML, Rendón N, Santos LL, Rojas-Lima S, Salazar V, Mereiter K, Ruiz C, *J Am Chem Soc* 2003, **125**, 1478.
458. Paneque M, Posadas CM, Poveda ML, Rendón N, Mereiter K, *Organometallics* 2007, **26**, 3120.
459. Álvarez A, Paneque M, Poveda ML, Rendón N, *Angew Chem Int Ed* 2006, **45**, 474.
460. Lara P, Paneque M, Poveda ML, Salazar V, Santos LL, Carmona E, *J Am Chem Soc* 2006, **128**, 3512.
461. Salazar V, Suárez-Castillo OR, Padilla R, Macías PJC, Méndez-Rojas MA, Tamariz J, Benavides A, *Organometallics* 2006, **25**, 172.
462. Paneque M, Poveda ML, Santos LL, Carmona E, Lledos A, Ujiaque G, Mereiter K, *Angew Chem Int Ed* 2004, **43**, 3708–3711.
463. Webster CE, Singleton DA, Szymanski MJ, Hall MB, Zhao C, Jia G, Lin Z, *J Am Chem Soc* 2001, **123**, 9822.
464. Elliott PIP, Haslam CE, Spey SE, Haynes A, *Inorg Chem* 2006, **45**, 6269.
465. Li J, Djurovich PI, Alleyne BD, Tsyba I, Ho NN, Bau R, Thompson ME, *Polyhedron* 2004, **23**, 419.
466. Padilla-Martínez II, Poveda ML, Carmona E, Monge MA, Ruiz-Valero C, *Organometallics* 2002, **21**, 93.
467. López-Linares F, Agrifoglio G, Labrador Á, Karam A, *J Mol Catal A Chem* 2004, **207**, 115.
468. Inoue H, Shitagaki S, Seo S, Ohsawa N, *PCT Int Appl* 2006, WO2006-JP30547420006314.
469. Seo S, Inoue H, Ohsawa N, *PCT Int Appl* 2006, WO2006-JP306377 20060322.
470. Trofimenko S, *J Am Chem Soc* 1967, **89**, 3170.
471. Trofimenko S, *J Am Chem Soc* 1967, **89**, 6288.
472. Lehmkuhl H, Näser J, Mehler G, Keil T, Danowski F, Benn R, Mynott R, Schroth G, Gabor B, Krüger C, Betz P, *Chem Ber* 1991, **124**, 441.
473. Gutiérrez E, Hudson SA, Monge A, Nicasio MC, Paneque M, Ruiz C, *J Organomet Chem* 1998, **551**, 215.
474. Desrochers PJ, Cutts RW, Rice PK, Golden ML, Graham JB, Barclay TM, Cordes AW, *Inorg Chem* 1999, **38**, 5690.
475. Hikichi S, Sasakura Y, Yoshizawa M, Ohzu Y, Moro-oka Y, Akita M, *Bull Chem Soc Jpn* 2002, **75**, 1255.

476. Desrochers PJ, Telser J, Zvyagin SA, Ozarowski A, Krzystek J, Vicic DA, *Inorg Chem* 2006, **45**, 8930.
477. Santi R, Romano AM, Sommazzi A, Grande M, Bianchini C, Mantovani G, *J Mol Catal A Chem* 2005, **229**, 191.
478. Uehara K, Hikichi S, Akita M, *Chem Lett* 2002, **31**, 1198.
479. Sun Y-J, Shen W-Z, Cheng P, Yan S-P, Liao D-Z, Jiang Z-H, Shen P-W, *Polyhedron* 2004, **23**, 211.
480. Sun Y-J, Cheng P, Yan S-P, Jiang Z-H, Liao D-Z, Shen P-W, *Inorg Chem Commun* 2000, **3**, 289.
481. Desrochers PJ, LeLievre S, Johnson RJ, Lamb BT, Phelps AL, Cordes AW, Gu W, Cramer SP, *Inorg Chem* 2003, **42**, 7945.
482. Grande M, Romano AM, Bianchini C, Mantovani G, Santi R, Sommazzi A, *PCT Int Appl* 2003, PIXXD2 WO 2003070737.
483. Li D, Ruschman C, Parkin S, Clérac R, Holmes SM, *Chem Commun* 2006, 4036.
484. Bielawski J, Hodgkins TG, Layton WJ, Niedenzu K, Niedenzu PM, Trofimenko S, *Inorg Chem* 1986, **25**, 87.
485. Ohkita K, Kurosawa H, Hasegawa T, Shirafuji T, Ikeda I, *Inorg Chim Acta* 1992, **198–200**, 275.
486. Canty AJ, Jin H, Roberts AS, Traill PR, Skelton BW, White AH, *J Organomet Chem* 1995, **489**, 153.
487. Canty AJ, Jin H, *J Organomet Chem* 1998, **565**, 135.
488. Canty AJ, Traill PR, *J Organomet Chem* 1992, **435**, C8.
489. Onishi M, Hiraki K, Konda H, Ishida Y, Ohama Y, Uchibori Y, *Bull Chem Soc Jpn* 1986, **59**, 201.
490. Onishi M, Hiraki K, Itoh T, Ohama Y, *J Organomet Chem* 1983, **254**, 381.
491. Canty AJ, Minchin NJ, Engelhardt LM, Skelton BW, White AH, *J Chem Soc, Dalton Trans* 1986, 645.
492. Das MK, Niedenzu K, Roy S, *Inorg Chim Acta* 1988, **150**, 47.
493. Do HO, Lee JH, Kim H, Park S, *J Coord. Chem* 2001, **53**, 143.
494. Kläui W, Turkowski B, Wunderlich H, *Z Anorg Allg Chem* 2001, **627**, 2397.
495. Kläui W, Turkowski B, Borisovna Chenskaya T, *Z Anorg Allg Chem* 2001, **627**, 2609.
496. Cámpora J, Palma P, del Rio D, Carmona E, Graiff C, Tiripicchio A, *Organometallics* 2003, **22**, 3345.
497. Cámpora J, Palma P, del Rio D, López JA, Valerga P, *Chem Commun* 2004, 1490.

498. Cámpora J, Palma P, del Rio D, López JA, Alvarez E, Connelly NG, *Organometallics* 2005, **24**, 3624.
499. Kujime M, Hikichi S, Akita M, *Organometallics* 2001, **20**, 4049.
500. Binotti B, Bellachioma G, Cardaci G, Macchioni A, Zuccaccia C, Foresti E, Sabatino P, *Organometallics* 2002, **21**, 346.
501. Clark HC, Manzer LE, *J Chem Soc, Chem Commun* 1973, 870.
502. Clark HC, Manzer LE, *Inorg Chem* 1974, **13**, 1291.
503. Clark HC, Manzer LE, *Inorg Chem* 1974, **13**, 1996.
504. Manzer LE, Meakin PZ, *Inorg Chem* 1976, **15**, 3117.
505. Davies BW, Payne NC, *Inorg Chem* 1974, **13**, 1843.
506. King RB, Bond A, *J Am Chem Soc* 1974, **96**, 1338.
507. Roth S, Ramamoorthy V, Sharp PR, *Inorg Chem* 1990, **29**, 3345.
508. Canty AJ, Jin H, Skelton BW, White AH, *Aust. J. Chem* 1999, **52**, 417.
509. Wick DD, Goldberg KI, *J Am Chem Soc* 1999, **121**, 11900.
510. Reinartz S, White PS, Brookhart M, Templeton JL, *Organometallics* 2000, **19**, 3748.
511. Haskel A, Keinan E, *Organometallics* 1999, **18**, 4677.
512. Reinartz S, Baik M-H, White PS, Brookhart M, Templeton JL, *Inorg Chem* 2001, **40**, 4726.
513. Iron MA, Lo HC, Martin JML, Keinan E, *J Am Chem Soc* 2002, **124**, 7041.
514. Reinartz S, White PS, Brookhart M, Templeton JL, *Organometallics* 2001, **20**, 1709
515. Reinartz S, White PS, Brookhart M, Templeton JL, *J Am Chem Soc* 2001, **123**, 6425.
516. Reinartz S, White PS, Brookhart M, Templeton JL, *J Am Chem Soc* 2001, **123**, 12724.
517. Norris CM, Reinartz S, White PS, Templeton JL, *Organometallics* 2002, **21**, 5649.
518. West NM, Reinartz S, White PS, Templeton JL, *J Am Chem Soc* 2006, **128**, 2059.
519. Norris Kostelansky C, MacDonald MG, White PS, Templeton JL, *Organometallics* 2006, **25**, 2993.
520. Reinartz S, White PS, Brookhart M, Templeton JL, *Organometallics* 2000, **19**, 3854.
521. Jensen MP, Wick DD, Reinartz S, White PS, Templeton JL, Goldberg KI, *J Am Chem Soc* 2003, **125**, 8614.
522. Norris CM, Templeton JL, *Organometallics* 2004, **23**, 3101.
523. Haskel A, Keinan E, *Tetrahedron Letters* 1999, **40**, 7861.

524. Fekl U, van Eldik R, Lowell S, Goldberg KI, *Organometallics* 2000, **19**, 3535.
525. Reinartz S, Brookhart M, Templeton JL, *Organometallics* 2002, **21**, 247.
526. Lo HC, Iron MA, Martin JML, Keinan E, *Chem Eur J* 2007, **13**, 2812.
527. MacDonald MG, Norris Kostelansky C, White PS, Templeton JL, *Organometallics* 2006, **25**, 4560.
528. Ruiz J, Florenciano F, Rodríguez V, De Haro C, López G, Pérez J, *Eur J Inorg Chem* 2002, 2736.
529. Murphy A, Hathaway BJ, King TJ, *J Chem Soc, Dalton Trans* 1979, 1646.
530. Kitajima N, Moro-oka Y, Uchida A, Sasada Y, Ohashi Y, *Acta Crystallogr.* 1988, **C44**, 1876.
531. Bruce MI, Ostazewski APP, *J Chem Soc, Dalton Trans* 1973, 2433.
532. Kitajima N, Koda T, Moro-oka Y, *Chem Letters* 1988, **17**, 347.
533. Gioia Lobbia G, Pettinari C, Santini C, Colapietro M, Cecchi P, *Polyhedron* 1997, **16**, 207.
534. Gioia Lobbia G, Pettinari C, Marchetti F, Bovio B, Cecchi P, *Poly hedron* 1996, **15**, 881.
535. Kilner CA, McInnes EJL, Leech MA, Beddard GS, Howard JAK, Mabbs FE, Bridgeman AJ, Halcrow MA, *Dalton Trans* 2004, 236.
536. Di Nicola C, Marchetti F, Monari M, Pandolfo L, Pettinari C, *Inorg Chem Commun* 2008, in press.
537. Díaz-Requejo MM, Belderrain TR, Nicasio MC, Pérez PJ, *Organometallics* 2000, **19**, 285.
538. Pellei M, Pettinari C, Santini C, Skelton BW, Somers N, White AH, *J Chem Soc Dalton Trans* 2000, 3416.
539. Pettinari C, Marchetti F, Santini C, Pettinari R, Drozdov A, Troyanov S, Battiston GA, Gerbasi R, *Inorg Chim Acta* 2001, **315**, 88.
540. Acosta A, Zink JI, Cheon J, *Inorg Chem* 2000, **39**, 427.
541. Santini C, Pellei M, Gioia Lobbia G, Fedeli D, Falcioni G, *J Inorg Biochem* 2003, **94**, 348.
542. Santini C, Pellei M, Gioia Lobbia G, Alidori S, Berrettini M, Fedeli D, *Inorg Chim Acta* 2004, **357**, 3549.
543. Mandal GC, Sarkhel P, Poddar RK, Bermejo E, Sevillano P, Castineiras A, *Ind J Chem Inorg Bio-Inorg Phys Theor Anal Chem* 2001, **40A**, 630.
544. Thomas AM, Mukherjee A, Saha MK, Nethaji M, Chakravarty AR, *Ind J Chem Inorg Bio-Inorg Phys Theor Anal Chem* 2004, **43A**, 1626.

545. Wang S, Li YZ, Zuo JL, You XZ, *Acta Crystallogr E* 2004, **E60**, m376.
546. Aullón G, Gorun SM, Alvarez S, *Inorg Chem* 2006, **45**, 3594.
547. de la Lande A, Gérard H, Moliner V, Izzet G, Reinaud O, Parisel O, *J Biol Inorg Chem* 2006, **11**, 593.
548. Fujisawa K, Miyashita Y, Yamada Y, Okamoto K-i, *Bull Chem Soc Jpn* 2001, **74**, 1065.
549. Braga AAC, Maseras F, Urbano J, Caballero A, Díaz-Requejo MM, Pérez PJ, *Organometallics* 2006, **25**, 5292.
550. Abu Salah OM, Bruce MI *J Organomet Chem* 1975, **87**, C15.
551. Bruce MI, Walsh JD, *Aust J Chem* 1979, **32**, 2753.
552. Santini C, Gioia Lobbia G, Pettinari C, Pellei M, Valle G, Calogero S, *Inorg Chem* 1998, **37**, 890.
553. Effendy, Gioia Lobbia G, Pettinari C, Santini C, Skelton BW, White AH, *Inorg Chim Acta* 2000, **308**, 65.
554. Humphrey ER, Reeves Z, Jeffery JC, McCleverty JA, Ward MD, *Polyhedron* 1999, **18**, 1335.
555. Effendy, Gioia Lobbia G, Pettinari C, Santini C, Skelton BW, White AH, *Inorg Chim Acta* 2000, **298**, 146.
556. Effendy, Gioia Lobbia G, Marchetti F, Pellei M, Pettinari C, Pettinari R, Santini C, Skelton BW, White AH, *Inorg Chim Acta* 2004, **357**, 4247.
557. Effendy, Hanna JV, Pettinari C, Pettinari R, Santini C, Skelton BW, White AH, *Inorg Chem Commun* 2007, **10**, 571.
558. Gioia Lobbia G, Hanna JV, Pellei M, Pettinari C, Santini C, Skelton BW, White AH, *Dalton Trans* 2004, 951.
559. Effendy, Gioia Lobbia G, Pellei M, Pettinari C, Santini C, Skelton BW, White AH, *Inorg Chim Acta* 2001, **315**, 153.
560. Borkett NF, Bruce MI, *Inorg Chim Acta* 1975, **12**, L33.
561. Borkett NF, Bruce MI, Walsh JD, *Aust J Chem* 1980, **33**, 949.
562. Vicente J, Chicote MT, Guerriero R, Herber U, Bautista D, *Inorg Chem* 2002, **41**, 1870.
563. Yang K-W, Wang Y-Z, Huang Z-X, Sun J, *Polyhedron* 1997, **16**, 1297.
564. Looney A, Parkin G, *Inorg Chem* 1994, **33**, 1234.
565. Kumar PNV, Marynick DS, *Inorg Chem* 1993, **32**, 1857.
566. Vahrenkamp H, *Acc Chem Res* 1999, **32**, 589.
567. Lipton AS, Wright TA, Bowman MK, Reger DL, Ellis PD, *J Am Chem Soc* 2002, **124**, 5850.

568. McWhinnie WR, Monsef-Mirzai Z, Perry MC, Shaikh N, Hamor TA, *Polyhedron* 1993, **12**, 1193.
569. Reger DL, Myers SM, Mason SS, Rheingold AL, Haggerty BS, Ellis PD, *Inorg Chem* 1995, **34**, 4996.
570. Reger DL, Collins JE, Rheingold AL, Liable-Sands LM, *Inorg Chem* 1999, **38**, 3235.
571. Parkin G, *Science* 2004, **305**, 1117.
572. Resa I, Carmona E, Gutierrez-Puebla E, Monge A, *Science* 2004, **305**, 1136.
573. Reger DL, Mason SS, Rheingold AL, *J Am Chem Soc* 1993, **115**, 10406.
574. Gioia Lobbia G, Bonati F, Cecchi P, Pettinari C, *Gazz Chim Ital* 1991, **121**, 355.
575. Han R, Looney A, Parkin G, *J Am Chem Soc* 1989, **111**, 7276.
576. Reger DL, Ding Y, *Organometallics* 1993, **12**, 4485.
577. Green JC, Suter JL, *J Chem Soc Dalton Trans* 1999, 4087.
578. Reger DL, Mason SS, Rheingold AL, Ostrander RL, *Inorg Chem* 1994, **33**, 1803.
579. Kläui W, Peters W, Liedtke N, Trofimenko S, Rheingold AL, Sommer RD, *Eur J Inorg Chem* 2001, 693.
580. Kläui W, Liedtke N, Peters W, *Magn Reson Chem* 1999, **37**, 867.
581. Fraser A, Piggott B, *J Chem Soc Dalton Trans* 1999, 3483.
582. Craven E, Mutlu E, Lundberg D, Temizdemir S, Dechert S, Brombacher H, Janiak C, *Polyhedron* 2002, **21**, 553.
583. Filippou AC, Portius P, Kociok-Köhn G, Albrecht V, *J Chem Soc Dalton Trans* 2000, 1759.
584. Reger DL, Coan PS, *Inorg Chem* 1996, **35**, 258.
585. Dungan CH, Maringgele W, Meller A, Niedenzu K, Noth H, Serwatowska J, Serwatowski J, *Inorg Chem* 1991, **30**, 4799.
586. Lee SK, Nicholson BK, *J Organomet Chem* 1986, **309**, 257.
587. Gioia Lobbia G, Bonati F, Cecchi P, Lorenzotti A, Pettinari C, *J Organomet Chem* 1991, **403**, 317.
588. Nicholson BK, *J Organomet Chem* 1984, **265**, 153.
589. Oshiki T, Mashima K, Kawamura S, Tani K, Kitaura K, *Bull Soc Chem Jap* 2000, **73**, 1735.
590. Wang J-t, He H-y, Xu Y-m, *Heteroatom Chem* 1998, **5**, 479.
591. Reger DL, Huff MF, Rheingold AL, Haggerty BS, *J Am Chem Soc* 1992, **114**, 579.

592. Reger DL, Wright TD, Smith MD, Rheingold AL, Kassel S, Concolino T, Rhagitan B, *Polyhedron* 2002, **21**, 1795.
593. Dodds CA, Kennedy AR, Reglinski J, Spicer MD, *Inorg Chem* 2004, **43**, 394.
594. Stainer MVR, Takats J, *Inorg Chem* 1982, **21**, 4050.
595. Stainer MVR, Takats J, *J Am Chem Soc* 1983, **105**, 410.
596. Moffat WD, Stainer MVR, Takats J, *Inorg Chim Acta* 1987, **139**, 75.
597. Moss MAJ, Jones CJ, *Polyhedron* 1989, **8**, 555.
598. Apostolidis C, Carvalho A, Domingos Â, Kanellakopulos B, Maier R, Marques N, Pires de Matos A, Rebizant J, *Polyhedron* 1999, **18**, 263.
599. Onishi M, Nagaoka N, Hiraki K, Itoh K, *J Alloys Comp* 1996, **236**, 6.
600. Apolistidis C, Rebizant J, Walter O, Kanellakopulos B, Reddmann H, Amberger H-D, *Z Anorg Allg Chem* 2002, **628**, 2013.
601. Kunrath FA, Casagrande Jr, OL, Toupet L, Carpentier J-F, *Polyhedron* 2004, **23**, 2437.
602. Roitershtein D, Domingos Â, Pereira LCJ, Ascenso JR, Marques N, *Inorg Chem* 2003, **42**, 7666.
603. Domingos Â, Elsegood MRJ, Hillier AC, Lin G, Liu SY, Lopes I, Marques N, Maunder GH, McDonald R, Sella A, Steed JW, Takats J, *Inorg Chem* 2002, **41**, 6761.
604. Galler JL, Goodchild S, Gould J, McDonald R, Sella A, *Polyhedron* 2004, **23**, 253.
605. Amberger H-D, Reddmann H, Apostolidis C, Kanellakopulos B, *Z Anorg Allg Chem* 2003, **629**, 147.
606. Reddmann H, Apostolidis C, Walter O, Amberger H-D, *Z Anorg Allg Chem* 2006, **632**, 1405.
607. Miranda Jr P, Aricó EM, Máduar MF, Matos JR, de Carvalho CAA, *J Alloys Comp* 2002, **344**, 105.
608. Singh UP, Kumar R, *J Mol Struct* 2007, **837**, 214.
609. Lopes I, Lin GY, Domingos A, McDonald R, Marques N, Takats J, *J Am Chem Soc* 1999, **121**, 8110.
610. Amberger HD, Reddman H, Jank S, Lopes MI, Marques N, *Eur J Inorg Chem* 2004, 98.
611. Lopes I, Hillier AC, Liu SY, Domingos Â, Ascenso J, Galvão A, Sella A, Marques N, *Inorg Chem* 2001, **40**, 1116.
612. Reddmann H, Apostolidis C, Walter O, Rebizant J, Amberger H-D, *Z Anorg Allg Chem* 2005, **631**, 1487.

613. Christou V, Salata OV, Shipley C, *PCT Int Appl* 2000, WO 99-GB3201 19990924, WO 00/18851.
614. Lopes I, Monteiro B, Lin G, Domingos Â, Marques N, Takats J, *J Organomet Chem* 2001, **632**, 119.
615. Hillier AC, Zhang XW, Maunder GH, Liu SY, Eberspacher TA, Metz MV, McDonald R, Domingos Â, Marques N, Day VW, Sella A, Takats J, *Inorg Chem* 2001, **40**, 5106.
616. Onishi M, Yamaguchi H, Shimotsuma H, Hiraki K, Nagaoka J, Kawano H, *Chem Lett* 1999, **28**, 573.
617. Deng DL, Zhang Y-H, Dai C-Y, Zeng H, Ye C-Q, Hage R, *Inorg Chim Acta* 2000, **310**, 51.
618. Lin G, McDonald R, Takats J, *Organometallics* 2000, **19**, 1814.
619. Onishi M, Kayano K-h, Inada K-i, Yamaguchi H, Nagaoka J, Arikawa Y, Takatani T, *Inorg Chim Acta* 2004, **357**, 4091.
620. Hillier AC, Liu S-Y, Sella A, Elsegood MRJ, *Inorg Chem* 2000, **39**, 2635.
621. Hillier AC, Sella A, Elsegood MRJ, *J Organomet Chem* 2002, **664**, 298.
622. Hillier AC, Liu SY, Sella A, Elsegood MRJ, *J Alloys Comp* 2000, **303–304**, 83.
623. Md Subhan A, Sanada T, Suzuki T, Kaizaki S, *Inorg Chim Acta* 2003, **353**, 263.
624. Md Subhan A, Suzuki T, Kaizaki S, *J Chem Soc Dalton Trans* 2002, 1416.
625. Caneschi A, Dei A, Gatteschi D, Poussereau S, Sorace L, *Dalton Trans* 2004, 1048.
626. Maunder GH, Russo MR, Sella A, *Polyhedron* 2004, **23**, 2709.
627. Singh UP, Kumar R, Upreti S, *J Mol Struct* 2007, **831**, 97.
628. Lopes I, Dias R, Domingos Â, Marques N, *J Alloys Comp* 2002, **344**, 60.
629. Carvalho A, Domingos Â, Isolani PC, Marques N, Pires de Matos A, Vicentini G, *Polyhedron* 2000, **19**, 1707.
630. Lawrence RG, Hamor TA, Jones CJ, Paxton K, Rowley NM, *J Chem Soc Dalton Trans* 2001, 2121.
631. Caneschi A, Dei A, Gatteschi D, Sorace L, Vostrikova K, *Angew Chem Int Ed* 2000, **39**, 246.
632. Dei A, Gatteschi D, Massa CA, Pardi LA, Poussereau S, Sorace L, *Chem Eur J* 2000, **6**, 4580.

633. Singh UP, Tyagi S, Sharma CL, Görner H, Weyhermüller T, *J Chem Soc Dalton Trans* 2002, 4464.
634. Md Subhan A, Suzuki T, Kaizaki S, *J Chem Soc Dalton Trans* 2001, 492.
635. Md Subhan A, Suzuki T, Fuyuhiro A, Kaizaki S, *Dalton Trans* 2003, 3785.
636. He H, Guo J, Zhao Z, Wong W-K, Wong W-Y, Lo W-K, Li K-F, Luo L, Cheah K-W, *Eur J Inorg Chem* 2004, 837.
637. Foley TJ, Harrison BS, Knefely AS, Abboud KA, Reynolds JR, Schanze KS, Boncella JM, *Inorg Chem* 2003, **42**, 5023.
638. Md Subhan A, Kawahata R, Nakata H, Fuyuhiro A, Tsukuda T, Kaizaki S, *Inorg Chim Acta* 2004, **357**, 3139.
639. Rabe GW, Zhang-Presse M, Riederer FA, Rheingold AL, *Acta Crystallogr* 2006, **E62**, m3149.
640. Antunes MA, Ferrence GM, Domingos Â, McDonald R, Burns CJ, Takats J, Marques N, *Inorg Chem* 2004, **43**, 6640.
641. Leal JP, Marques N, Takats J, *J Organomet Chem* 2001, **632**, 209.
642. Silva M, Antunes MA, Dias M, Domingos Â, Cordeiro dos Santos I, Marçalo J, Marques N, *Dalton Trans* 2005, 3353.
643. Campello MPC, Domingos Â, Galvão A, Pires de Matos A, Santos I, *J Organomet Chem* 1999, **579**, 5.
644. Antunes MA, Domingos Â, Cordeiro dos Santos I, Marques N, Takats J, *Polyhedron* 2005, **24**, 3038.
645. Enriquez AE, Scott BL, Neu MP, *Inorg Chem* 2005, **44**, 7403.

Chapter 3

Homoscorpionates — Second Generation

3.1. Tp^{Me}

$[Cu(Tp^{Me})_2]$ prepared from the reaction of $CuCl_2$ with $K(Tp^{Me})$ has been characterized by magnetic susceptibilities, fast atom bombardment and IR spectrometry, and X-ray single-crystal study.[1] Copper(I) complexes containing water soluble phosphines, $[Cu(Tp^{Me})(L)]$ (L = 4- or 2-(diphenylphosphane)benzoic acid), have been characterized by IR in the solid state, and by NMR (^1H and $^{31}P\{^1H\}$) and electrospray mass spectroscopy in solution. Chemiluminescence technique was used to evaluate their superoxide scavenging activity.[2]

In $[Ag(Tp^{Me})(P^iBu_3)]$ the silver atom is in a distorted tetrahedral coordination geometry. Variable temperature $^{31}P\{^1H\}$ NMR studies and ESI MS data on this compound suggested the existence in solution of species such as $[Ag(P^iBu_3)_2]^+$ upon dissociation of the Tp^{Me} ligand, and also the formation of dimeric species of the form $[Ag_2(Tp^{Me})(P^iBu_3)_2]^+$.[3]

3.2. Tp^{4Me}

$[Cu(Tp^{4Me})_2]$ has been prepared by following the same procedure reported for $[Cu(Tp^{Me})_2]$.[1]

3.3. pz^0Tp^{Me}

$[Cu(pz^0Tp^{Me})_2]$ has been characterized by magnetic susceptibilities, fast atom bombardment and IR spectrometry, and X-ray single-crystal study.[1] The solid state and solution properties of $[\{Ag(pz^0Tp^{Me})\}(dppf)]$ have been investigated through analytical and spectroscopic measurements.[4]

3.4. Tp^{Me_3} ($\equiv Tp^{*Me}$)

$[Zn(Tp^{Me_3})_2]$ and the pseudotetrahedral $[(Bp^{Me_3})Zn(\mu\text{-}pz^{Me_3})_2\text{-}Zn(Bp^{Me_3})]$ are present in equal amounts in a crystal reported by Reger *et al.*[5] The structure of $[Zn(Tp^{Me_3})_2]$ is a trigonally distorted octahedron very similar to the structure of $[Zn(Tp^*)_2]$[6,7] suggesting that the introduction of the third methyl group on the pyrazolyl ring does not impact on the structure.[8] $[Fe(Tp^x)_2]$ complexes have been investigated by Reger *et al.* to study the effect of substituents, intramolecular ligand distortions, and intermolecular supramolecular structures on the spin-state crossover (SCO) behavior. The molecular structure of $[Fe(Tp^{Me_3})_2]$, which remains high-spin when the temperature is lowered, reveals that this complex has an intramolecular ring twist distortion that is not observed in analogous complexes exhibiting a SCO at low temperatures, thus indicating that this distortion greatly influences the properties of these complexes. It has been reported that the use of Tp^{Me_3} can lead to the successful isolation of thermally unstable high-valent dinuclear bis(μ-oxo) complexes $[(Tp^{Me_3})M(\mu\text{-}O_2)M(Tp^{Me_3})]$ (M = Ni or Co).[9] Synthesis of half-sandwich type complexes $[M(Tp^{Me_3})(\kappa^2\text{-}O,O'\text{-}L)]$ (M = Ni or Co; L = NO_3, OAc) has been achieved by dropwise addition of a THF solution of $Na(Tp^{Me_3})$ to an excess amount of ML_2 dissolved in MeOH.[10]

3.5. Tp^{*Br} ($\equiv Tp^{Me_2Br}$)

Several Tp^{*Br} complexes have been also described. In $[M(Tp^{*Br})\text{-}(\kappa^2\text{-}(O,O')\text{-}NO_3)]$ (M = Ni or Co), characterized by X-ray crystallography, the metal centers have a six-coordinated octahedral geometry.[10]

The heterobimetallic complexes $[(Tp^{*Br})MCo(CO)_4]$ (M = Ni, Co) and $[(Tp^{*Br})NiRu(Cp)(CO)_2]$ have been also reported. DFT calculations have been performed for the model complexes $[(Tp)NiCo(CO)_4]$, $[(Tp)NiCo(PH_3)(CO)_3]$, and $[(Tp)NiRu(Cp)(CO)_2]$.[11] The high-spin triplet, xenophilic complex, $[(Tp^{*Br})Ni–RuCp(CO)_2]$, upon interaction with $2e^-$-donors gave the diamagnetic adducts $[(Tp^{*Br})Ni–RuCp(\mu\text{-}CO)_2(L)]$.[12]

3.6. $Tp^{*Cl}(\equiv Tp^{Me_2Cl})$

The electronic structure of $K(Tp^{*Cl})$ has been investigated using gas-phase photoelectron spectroscopy and DFT. The X-ray crystal structures of $K(Tp^{*Cl})$ and $(\mu\text{-}H_2O)_3[K(Tp^{*Cl})]_2$ have been also reported.[13] $[Cu(Tp^{*Cl})_2]$ has been characterized by magnetic susceptibilities, fast atom bombardment and IR spectrometry, and X-ray single-crystal study.[1] Desrochers *et al.* reported the X-ray single-crystal study of $[Ni(Tp^{*Cl})_2]$ in which the ligand imposes pseudooctahedral coordination environment.[14]

When $[Rh(Tp^{*Cl})(CO)_2]$ was irradiated in NH^iPr_2 a single diastereomer of the C–H activated metallacycle $[Rh(Tp^{*Cl})H\{CH_2\text{-}CH(CH_3)NH^iPr\}]$ was formed. This process is reminiscent of the heteroatom (oxygen) assistance of C–H and C–F bond activation of pentafluoroanisole by a Rh(Cp) derivative.[15] $[Rh(Tp^{*Cl})(CO)_2]$ reacts with $CHCl_3$ to give $[Rh(Tp^{*Cl})Cl(CHCl_2)(CO)]$ upon oxidative addition of a C–Cl bond. Further reaction with NH^iPr_2 gives the aminocarbene complex $[Rh(Tp^{*Cl})Cl_2(CHN^iPr_2)]$.[16]

3.7. Tp^{*Et}

The reaction of nitric oxide with the $[Ln(Tp^{*Et})_2]$ (Ln = Sm, Eu, Yb) yielded the corresponding eight-coordinate $[Ln(Tp^{*Et})_2NO_2]$. The single-crystal X-ray diffraction study of $[Sm(Tp^{*Et})_2NO_2]$, fluxional in solution even at the lowest accessible temperatures (NMR), reveals a dodecahedral metal center.[17]

3.8. PhTp^{Me}

The synthesis, structural, magnetic, and Mössbauer spectral properties of [Fe(PhTp^{Me})$_2$] are described by the Reger group. A distorted octahedral structure is found at both 294 and 90 K. The absence of a SCO in [Fe(PhTp^{Me})$_2$] is the result of its molecular structure in which the three pyrazolyl rings of each tridentate ligand are highly twisted away from an ideal C_{3v} coordination symmetry.[18] On the other hand the closely related complexes [Fe{(*p*-R–C$_6$H$_4$)Tp^{Me}}$_2$] (R = I, –*C*≡C–H, –*C*≡C–SiMe$_3$, or –*C*≡C–C$_6$H$_5$) exhibit electronic spin-state changes on cooling.

3.9. Tp$^{^iPr}$

Monomeric amide or aryloxide derivatives [Ca(X)(Tp$^{^iPr}$)] (X = CH[CMeNC$_6$H$_3$-2,6-^{i}Pr$_2$]$_2$, CH[CMeNC$_6$H$_4$-2-OMe]$_2$) have been prepared and characterized and employed for the ring-opening polymerizations of lactide.[19,20] [Ca(Tp$^{^iPr}$)N(SiMe$_3$)$_2$]·THF has been synthesized and structurally characterized.[21]

[Co(Tp$^{^iPr}$)SCN] has been reported and investigated by ^{1}H NMR and IR spectroscopy.[22]

[MoO$_2$X(Tp$^{^iPr}$)] (X = OC$_6$H$_4$^{s}Bu-2, OC$_6$H$_4$R-3 (R = Me, ^{t}Bu), OC$_6$H$_4$R-4 (R = Br, Et, OMe), OC$_6$H$_3$^{t}Bu$_2$-3,5, OC$_6$H$_3$^{t}BuMe-2,4, S^{n}Bu, S^{s}Bu) have been obtained by metathesis of [MoO$_2$Cl(Tp$^{^iPr}$)] with phenol/NEt$_3$, or thiol/NEt$_3$. These diamagnetic complexes were characterized by analytical, spectroscopic and mass spectroscopic investigations. Reaction between [MoO$_2$Cl(Tp$^{^iPr}$)] and P(O^{i}Pr$_3$) produced [MoOCl(Tp$^{^iPr}$)](μ-O)[MoO(OH)(Tp$^{^iPr}$)], which shows corner shared bioctahedral structures with a single bridging oxo ligand linking [MoOX(Tp$^{^iPr}$)]. Oxygen atom transfer from [MoO$_2$Cl(Tp$^{^iPr}$)] to PPh$_3$ in py or Lut yields [MoOCl(Tp$^{^iPr}$)(py)] and [MoOCl(Tp$^{^iPr}$)(Lut)], respectively.[23]

Reaction of [MoO$_2$(OPh)(Tp$^{^iPr}$)] with PEt$_3$ in MeCN yields [MoO(OPh)(Tp$^{^iPr}$)(OPEt$_3$)], which upon further reaction with propylene sulfide generated [MoOS(OPh)(Tp$^{^iPr}$)]. Aerobic chromatographic workup of the reaction products afforded the dinuclear

$[Mo(Tp^{iPr})]_2(\mu\text{-}O)(\mu\text{-}S_2)$.[24] *cis*-$[MoO_2(OAr)(Tp^{iPr})]$ (OAr = phenolate or naphtholate) are obtained by metathesis of $[MoO_2Cl(Tp^{iPr})]$ with $HOAr/NEt_3$ in dichloromethane. These complexes display distorted octahedral geometries, with a *cis* arrangement of terminal oxo ligands. Maximal frontal steric congestion is observed in the 2-phenolate derivatives, identified as precursors for strictly monomeric (solid and solution state) oxosulfido-Mo(VI) counterparts.[25] Atom transfer reactions have been employed to convert $[MoO_2(OAr)(Tp^{iPr})]$ into monomeric *cis*-oxosulfido-Mo(VI) and dimeric μ-disulfido-Mo(V) species $[MoOS(OAr)\text{-}(Tp^{iPr})]_n$ (OAr = phenolate or naphtholate derivative; $n = 1$ and 2). The monomeric species $[MoOS(OAr)(Tp^{iPr})]$ (**I**, Fig. 3.1) contain distorted octahedral *cis*-oxosulfido-Mo(VI) centers whereas the dimeric complexes $[MoOS(OAr)(Tp^{iPr})]_2$ undergo S–S bond cleavage forming monomeric oxosulfido-Mo(VI) species in solution. Hydrolysis of the oxosulfido-Mo(VI) complexes results in the formation of $[MoO(Tp^{iPr})]_2(\mu\text{-}S_2)(\mu\text{-}O)$ (**II**, Fig. 3.1).[26]

The unstable $[MoO(SR)(Tp^{iPr})(NCMe)]$ complexes, obtained by reaction of $[MoO_2(SR)(Tp^{iPr})]$ with PPh_3, react with CO in toluene

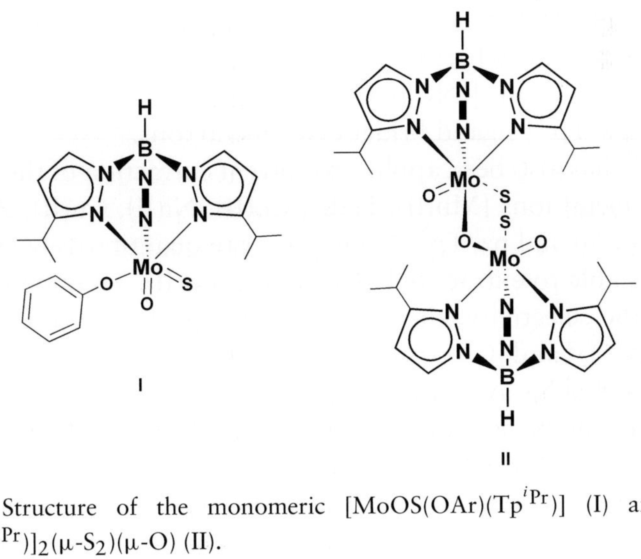

I

II

Fig. 3.1. Structure of the monomeric $[MoOS(OAr)(Tp^{iPr})]$ (I) and dimeric $[MoO(Tp^{iPr})]_2(\mu\text{-}S_2)(\mu\text{-}O)$ (II).

yielding [MoO(SR)($Tp^{^iPr}$)(CO)] (R = Et, ^{i}Pr, Ph, *p*-Tol, Bn). The crystal structure of [MoO(S^{i}Pr)($Tp^{^iPr}$)(CO)] revealed a distorted octahedral geometry. The distorted octahedral *cis*-[MoO_2(SR)($Tp^{^iPr}$)] (R = Et,^{i}Pr, Ph, *p*-Tol, Bn, ^{t}Bu) are formed by metathesis of [MoO_2Cl($Tp^{^iPr}$)] with HSR/NEt_3 in CH_2Cl_2.[27]

[MoOX(OPR_3)($Tp^{^iPr}$)] (X = Cl, phenolate or thiolate) complexes have been isolated from the reaction of [MoO_2X($Tp^{^iPr}$)] with phosphines (PEt_3, $PMePh_2$, PPh_3). The X-ray crystal structures of [MoOX($Tp^{^iPr}$)($OPEt_3$)] (X = OC_6H_4-2-^{s}Bu, S^{n}Bu), [MoO(OPh)($Tp^{^iPr}$)($OPMePh_2$)], and [MoOCl($Tp^{^iPr}$)($OPPh_3$)] have been determined. The electronic structures and stabilities of the complexes have been probed by computational methods.[28] Also, the distorted octahedral [MoO(OPh)($OPEt_3$)($Tp^{^iPr}$)] has been isolated and characterized by X-ray crystallography.[29]

3.10. $pz^0Tp^{^iPr}$

[CuCl($pz^0Tp^{^iPr}$)($Hpz^{5\text{-}^iPr}$)], [$NiNO_3$($pz^0Tp^{^iPr}$)($Hpz^{5\text{-}^iPr}$)], [MX(pz^0-$Tp^{^iPr}$)] (M = Ni, Cu; X = Cl, NO_3), and [M($pz^0Tp^{^iPr}$)$_2$] complexes have been obtained depending on the ligand to metal ratio employed in the synthesis.[30] In [NiCl($pz^0Tp^{^iPr}$)], the ligand $pz^0Tp^{^iPr}$ is three-coordinate whereas a κ^2-binding mode was found in [Ni($pz^0Tp^{^iPr}$)$_2$] and in [$(NCS)_2$Co($pz^0Tp^{^iPr}$)Co($pz^0Tp^{^iPr}$)Co$(NCS)_2$], where the $pz^0Tp^{^iPr}$ ligand bridges two metal ions.[30]

$pz^0Tp^{^iPr}$ has also been applied to solvent extraction of the first row transition metal ions [Mn(II), Fe(II), Co(II), Ni(II), Cu(II), Zn(II)].[31] Although pzTp and pz^0Tp^{Me} complexes are quantitatively extracted, $pz^0Tp^{^iPr}$ is able to extract only Cu(II) and Zn(II). To understand the origin of the selectivity, the structures of the extracted compounds [Zn($pz^0Tp^{^iPr}$)$_2$] and [Cu($pz^0Tp^{^iPr}$)(O_2CCH_3)] have been investigated by single-crystal X-ray study. [Zn($pz^0Tp^{^iPr}$)$_2$] has a tetrahedral geometry, each ligand being bidentate. [Cu($pz^0Tp^{^iPr}$)(O_2CCH_3)] has a distorted square-pyramidal geometry, $pz^0Tp^{^iPr}$ acting in the tridentate fashion and the acetate in the bidentate fashion. The bulky isopropyl groups in $pz^0Tp^{^iPr}$ likely prevent octahedral geometry and extraction of the other metal ions.[31]

3.11. $Tp^{^iPr,4Br}$

The syntheses of the four-coordinate tetrahedral $[MCl(Tp^{^iPr,4Br})]$ complexes (M = Fe, Mn) were described by O'Hare and coworkers.[32] Reaction of $CrCl_2(MeCN)_2$ with $Tl(Tp^{^iPr,4Br})$ gave $[Cr(\kappa^3\text{-}Tp^{^iPr,4Br*})(\kappa^2\text{-}Tp^{^iPr,4Br})]$, five-coordinate, in which the κ^3-ligand had isomerized to $Tp^{^iPr,4Br*}$. Variable temperature magnetic susceptibility measurements on this complex were also performed. In benzene solution $[Cr(\kappa^3\text{-}Tp^{^iPr,4Br*})(\kappa^2\text{-}Tp^{^iPr,4Br})]$ decomposes to $[Cr(\kappa^3\text{-}Tp^{^iPr,4Br*})_2][Tp^{^iPr,4Br}]$ which contains a pseudooctahedral Cr(III) cation with both ligands in the isomerized form and an uncoordinated $Tp^{^iPr,4Br}$.[32] Mechanochemical synthesis of $[CoCl(Tp^{^iPr,4Br})]$ has been reported.[33]

3.12. $Tp^{^iPr_2}$

$Tp^{^iPr_2}$ continues to be the second generation homoscorpionate ligand most used. $[Ni(\mu\text{-}O)(Tp^{^iPr_2})]_2$, prepared from $[Ni(\mu\text{-}OH)(Tp^{^iPr_2})]_2$ and H_2O_2, is able to mediate, in excess of H_2O_2, intramolecular oxygenation of the ^{i}Pr substituents of $Tp^{^iPr_2}$ to give enolate groups (Fig. 3.2).[34] A new O_2-activation method via a higher valent bis(μ-oxo) dimetal species, resulting from extensive electron transfer, has been proposed by Akita and Moro-oka on the basis of bioinorganic studies on the dioxygen complexes supported by $Tp^{^iPr_2}$.[35]

Heterobimetallic xenophilic complexes $[(Tp^{^iPr_2})MCo(CO)_4]$ (M = Mn, Ni, Co, Fe), containing an unsupported polar metal–metal

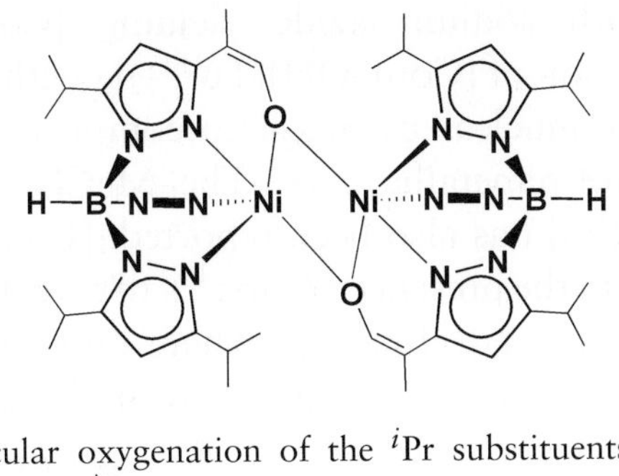

Fig. 3.2. Intramolecular oxygenation of the ^{i}Pr substituents of $Tp^{^iPr_2}$ has been observed in $[Ni(\mu\text{-}OH)(Tp^{^iPr_2})]_2$.

bond, have been prepared by the reaction of the labile cationic species $[M(Tp^{^iPr_2})(NCMe)_3]^+$ with PPN$[Co(CO)_4]$. The reaction of $[(Tp^{^iPr_2})MCo(CO)_4]$ with MeCN produced the ion pair $[M(Tp^{^iPr_2})(NCMe)_3][Co(CO)_4]$. $[(Tp^{^iPr_2})NiCo(CO)_4]$ and $[(Tp^{^iPr_2})NiCo(CO)_3(PPh_3)]$ have been structurally characterized.[36]

Coordinatively unsaturated 15e$^-$ alkynyl-$[Co(C{\equiv}C{-}R)(Tp^{^iPr_2})]$ and aryl-$[Co(Ar)(Tp^{^iPr_2})]$ cobalt complexes have been synthesized by dehydrative condensation of the hydroxo complex $[Co(\mu\text{-}OH)_2(Tp^{^iPr_2})]_2$ with 1-alkyne and arylation of $[CoCl(Tp^{^iPr_2})]$ with Grignard reagents. The $\{Co(Tp^{^iPr_2})\}$ and hydrocarbyl fragments are bonded through σ-bonding interaction and π-interaction, but back-donation is not significant. The Co–C bonds are polarized so as to be readily protonated even by moisture to give the corresponding hydrocarbons. $[Co(C{\equiv}C{-}COOMe)(Tp^{^iPr_2})]$ is found to catalyze specific linear trimerization of methyl propiolate.[37]

$[M(Tp^{^iPr_2})(NCMe)_3]$OTf (M = Ni, Co) were prepared by chloride abstraction from the corresponding chloro complexes $[MCl(Tp^{^iPr_2})]$ with AgOTf in MeCN. They can be readily transformed in $[M(Tp^{^iPr_2})(L)_n]$OTf by reaction with N- and P-donors $[NC(CH_2)_2CN$, *p*-NCC_6H_4CN, py, bipy, 4,4′-bipy, $Ph_2P(CH_2)_m$-PPh_2 $(m = 2, 3)]$.[38]

Mononuclear $[Co(X)(Tp^{^iPr_2})]$ (X = NO_3 and OBz) and binuclear $[Co(Tp^{^iPr_2})]_2(\mu\text{-}X)(\mu\text{-}OBz)$ (X = OH, N_3) complexes have been prepared and characterized by Singh *et al.*[39] $[Co(NO_3)(Tp^{^iPr_2})]$, structurally characterized, was converted to $[Co(\mu\text{-}OH)(Tp^{^iPr_2})]_2$ by the reaction with aqueous NaOH, and to $[Co(Tp^{^iPr_2})]_2(\mu\text{-}OH)(\mu\text{-}OBz)$ upon interaction with benzoic acid. The latter compound reacts with sodium azide yielding $[Co(Tp^{^iPr_2})]_2(\mu\text{-}N_3)(\mu\text{-}OBz)$. The reaction of $[Co(\mu\text{-}OH)(Tp^{^iPr_2})]_2$ with excess of benzoic acid gave the mononuclear carboxylato complex $[Co(OBz)(Tp^{^iPr_2})]$ upon breaking the bimetallic core. The MeCN derivative $[Co(\mu\text{-}OBz)(Tp^{^iPr_2})(MecN)]$ has also been reported. $[Co(\mu\text{-}F\text{-}OBz)(Tp^{^iPr_2})]$ has been oxidized in the presence of $Hpz^{^iPr_2}$ to give $[Co(F\text{-}OBz)\{HB(3\text{-}OCMe_2\text{-}5\text{-}^iPrpz)(pz^{^iPr_2})_2\}(Hpz^{^iPr_2})]$, where only one methyl carbon of isopropyl group on the pyrazole ring is oxidized and coordinated to the cobalt center.[40]

Reaction of $[M(\mu\text{-OH})(Tp^{iPr_2})]_2$ (M = Co, Ni) with active methylene compounds CH_2XY (X/Y=CN/COPh; CN/CN; $COCH_2$-CMe_2CH_2CO; COMe/COOMe; COOMe/COOMe) gave N/O-bound enolato complexes with monomeric chelated $[M(\text{–XCHY})(Tp^{iPr_2})]$ or dimeric cyclic structure $[M(Tp^{iPr_2})]_2(\mu\text{-XCHY})_2$. $[M(\mu\text{-OH})_2(Tp^{iPr_2})]_2$ reacts with methylcyanoacetate giving the κ^2-carboxylato complex $[M(\kappa^2\text{-OOCCH}_2\text{CN})(Tp^{iPr_2})]$. $[M(Tp^{iPr_2})]_2(\mu\text{-XCHY})_2$ (M = Co, Ni) reacts with benzaldehyde affording the metallacyclic products $[(Tp^{iPr_2})Co\{\text{–O–C(Ph)–CH(CN)–Co}(Tp^{iPr_2})\text{–CH–(CN)–C(Ph)–O–}\}]$ (Fig. 3.3) and $[(Tp^{iPr_2})Ni\{\text{–O–C(Ph)–C(CN–H)–CHPh–C(CN)–C(Ph)–O–}\}]$.[41]

The dinuclear $[(Tp^{iPr_2})MCo(CO)_4]$ (M = Ni, Co, Fe, Mn) have been prepared by the reaction of cationic complexes $[M(Tp^{iPr_2})(NCMe)_3]PF_6$ with the corresponding metalates. The tetrahedral, high-spin $M(Tp^{iPr_2})$ moiety and the coordinatively saturated carbonyl–metal fragments are connected by a M–M interaction.[11]

$[MR(Tp^{iPr_2})]$ (R = η^3-allyl, M = Fe, Co, Ni; R = η^3-prenyl, M = Ni; R = η^1-*p*-methylbenzyl, M = Fe, Co, Fe; R = η^1-R-naphthylmethyl, M = Co, R = η^1-ethyl M = Co, Fe) have been prepared by the reaction of corresponding precursors, $[ML(Tp^{iPr_2})]$, with the corresponding Grignard reagents. The thermally unstable η^1-ethylnickel derivative was characterized via conversion to the acyl derivative $[Ni(CO)C(=O)Et(Tp^{iPr_2})]$. The η^1-hydrocarbyl complexes are tetrahedral, highly coordinatively unsaturated species with 14 (M = Fe) and 15 (M = Co) valence electrons. Carbonylation of

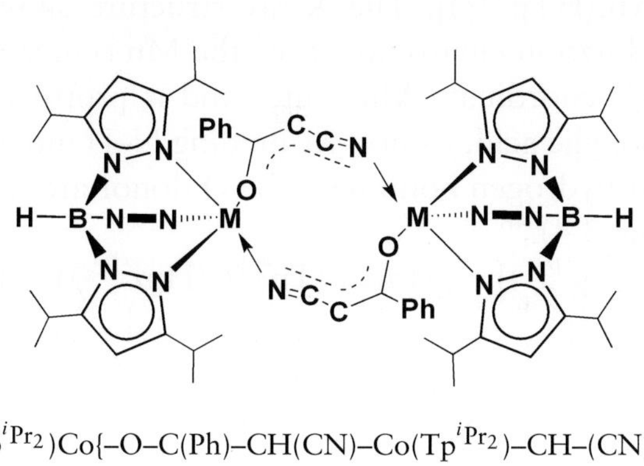

Fig. 3.3. $[(Tp^{iPr_2})Co\{\text{–O–C(Ph)–CH(CN)–Co}(Tp^{iPr_2})\text{–CH–(CN)–C(Ph)–O–}\}]$.

[MR(Tp^{iPr_2})] led to acyl-carbonyl species [M(CO)$_n$C(=O)R(Tp^{iPr_2})] (M/n = Fe/2, Co/1). Reaction of phenylacetylene with the ethyl complexes resulted in the insertion of C≡C bond into the M–C bond or protonolysis of the M–C bond to afford the acetylide complex [Fe(C≡CPh)(Tp^{iPr_2})], which undergoes carbonylation to [Fe(C≡CPh)(Tp^{iPr_2})$(CO)_2$]. Magnetic susceptibility measurements on these species were also performed.[42]

Reaction of a vanadium(IV) hydroxo complex containing the Tp^{iPr_2} ligand with O_2 gave, upon reductive O_2 activation, [V(O)-(Tp^{iPr_2})(η^2-O_2)(L)], the first example of a peroxo vanadium complex from O_2. The six-coordinated distorted octahedral [V$(O)_2$(Tp^{iPr_2})-(pz^{iPr_2})]·THF and the seven-coordinated pentagonal-bipyramidal [V(O)(Tp^{iPr_2})(η^2-O)(pz^{iPr_2})]·THF have been structurally characterized.[43] $VOCl_2(MeCN)_2(H_2O)$ reacts with Tp^{iPr_2} yielding the octahedral [V(O)(Cl)(Tp^{iPr_2})L] (L = Hpz^{iPr_2} or py). Its hydrolysis produces [V(O)(OH)(OH_2)(Tp^{iPr_2})], which dimerizes through intermolecular hydrogen-bonding interaction between the hydroxo and the aqua ligand forming an $H_3O_2^-$ bridging ligand; [(Tp^{iPr_2})V(O)(μ-OH)(μ-pz^{iPr_2})V(O)(μ-OH)(μ-pz^{iPr_2})V(O)(Tp^{iPr_2})] is a trinuclear complex that consists of two octahedral {V(O)Tp^{iPr_2}} fragments and of a distorted trigonal-bipyramidal non-Tp^{iPr_2} supported VO^{2+} center.[44]

A side-on peroxo complex, [Mn(η^2-O_2)(Tp^{iPr_2})(2-Meim)], unable to oxygenate triphenylphosphine and ethyl vinyl ether, has been prepared by the reaction of [Mn(O)(Tp^{iPr_2})]$_2$ with H_2O_2 and 2-Meim.[45]

[Mn(μ-OH)(Tp^{iPr_2})]$_2$ reacts with benzoic acid giving [(Tp^{iPr_2})-Mn(μ-OBz)$_3$Mn(HTp^{iPr_2})]. The X-ray structure shows an unsymmetrical coordination environment for the Mn centers: one Tp^{iPr_2} is bound to a five-coordinate Mn center and is protonated by one of the three carboxylic acid, its non-M-binding N–H moiety forming an intramolecular hydrogen bond with the O donor atom of a carboxylate ligand (Fig. 3.4).[46]

[Mn(FOBz)(Tp^{iPr_2})], [Mn(Cl)(Tp^{iPr_2})(Hpz^{iPr_2})], and [Mn_2(μ-FOBz)$_3$(Tp^{iPr_2})(Hpz^{iPr_2})$_2$] were reported and characterized,[47] the latter two being also structurally characterized and tested for their superoxide dismutase activity.

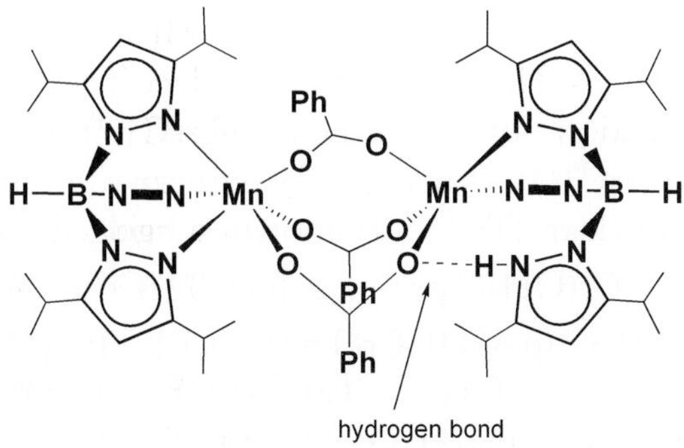

Fig. 3.4. $[(Tp^{iPr_2})Mn(\mu\text{-}OBz)_3Mn(Tp^{iPr_2})]$.

$[CrCl(Tp^{iPr_2})(Hpz^{iPr_2})]$ reacts with benzyl chloride yielding the corresponding trivalent dichlorochromium species $[CrCl_2(Tp^{iPr_2})\text{-}(Hpz^{iPr_2})]$. The reaction of $[CrCl(Tp^{iPr_2})(L)]$ ($L = Hpz^{iPr_2}$, py or bipy) with CH_2Cl_2 gives the methylene inserted derivative $[CrCl_2(Tp^{iPr_2})(CH_2L)]$, via a methylene carbene intermediate (Fig. 3.5). This is supported by cyclopropanation of styrene with CH_2Cl_2 in the presence of $[Cr(\mu\text{-}Cl)(Tp^{iPr_2})]_2$. The complex $[CrCl(Tp^{iPr_2})(py)]$ induces reductive coupling of benzaldehyde to give a 3,4-diphenyl-2,5-dioxachromacyclopent-3-ene, $[Cr(Tp^{iPr_2})\{\kappa^2\text{-}OC(Ph){=}C(Ph)\text{–}O\}]$. The complexes exhibit a variety of coordination geometries: $[Cr(\mu\text{-}Cl)(Tp^{iPr_2})]_2$ is a square-pyramidal five-coordinate complex, as is $[CrCl(Tp^{iPr_2})(py)]$; $[CrCl(Tp^{iPr_2})(Hpz^{iPr_2})]$ is trigonal bipyramidal, whereas $[CrCl(Tp^{iPr_2})(bipy)]$ is octahedral.[48]

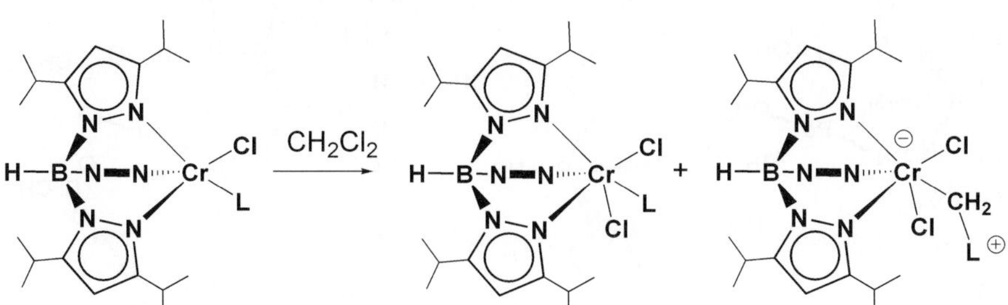

Fig. 3.5. The reaction of $[CrCl(Tp^{iPr_2})(L)]$ ($L = Hpz^{iPr_2}$, py or bipy) with CH_2Cl_2 gave methylene inserted derivatives $[CrCl_2(Tp^{iPr_2})(CH_2L)]$.

The methylene fragment $[CrCl_2(Tp^{iPr_2})(CH_2L)]$ is trapped by nucleophiles such as Hpz, py, and olefin.[49] The oxidative dehydrative condensation of $[Cr(OH)_2(OH_2)(Tp^{iPr_2})]_2$ with H_2O_2 yields $[Cr(O_2)_2(Tp^{iPr_2})]$, whereas O_2-oxidative addition to Cr(II) species gave dinuclear $\{Tp^{iPr_2}Cr\}$ moieties bridged by (CrO_4^{2-}) or $(Cr_2O_7^{2-})$ ions. $[Cr(OH)(Tp^{iPr_2})]_2$, $[(Tp^{iPr_2})Cr(\mu\text{-}OH)(\mu\text{-}pz^{iPr})Cr(Tp^{iPr_2})]_2$, $[(Tp^{iPr_2})Cr(\mu\text{-}OH)_3Cr(Tp^{iPr_2})(Cl)]_2$, $[Cr(OH)_2(Tp^{iPr_2})]_2$, $[Cr(O_2)_2(Tp^{iPr_2})]$, $[(Tp^{iPr_2})TpCr(\mu\text{-}OH\cdots H_2O)_2(\mu\text{-}CrO_4)Cr(Tp^{iPr_2})]_2$, and $[(Tp^{iPr_2})Cr(H_2O)(\mu Cr_2O_7)Cr(H_2O)(Tp^{iPr_2})]_2$ have been structurally characterized.[50]

Mono and dinuclear hydroxopalladium complexes $[Pd(OH)(\kappa^2\text{-}Tp^{iPr_2})(py)]$ and $[(\mu\text{-}OH)_2\{Pd(\kappa^2\text{-}Tp^{iPr_2})(H_2O)\}_2]$ have been prepared by alkaline hydrolysis of $[PdCl(\kappa^2\text{-}Tp^{iPr_2})(py)]$ and $[Pd(\mu\text{-}Cl)(\kappa^2\text{-}Tp^{iPr_2})(H_2O)]_2$. Dehydrative condensation of $[Pd(OH)(\kappa^2\text{-}Tp^{iPr_2})(py)]$ with protic substrates HA gives a series of substituted products $[Pd(A)(\kappa^2\text{-}Tp^{iPr_2})(py)]$, whereas the reaction of $[(\mu\text{-}OH)_2\{Pd(\kappa^2\text{-}Tp^{iPr_2})(H_2O)\}_2]$with acetic acid affords the mononuclear $[Pd(OAc)_2(\kappa^2\text{-}Tp^{iPr_2}\text{-}H)(HOAc)]$ (Fig. 3.6).[51]

$[Pd(OOH)(\kappa^2\text{-}Tp^{iPr_2})(PPh_3)]$, containing a hydrogen-bonding interaction between the OOH moiety and the uncoordinated pyrazolyl ring (Fig. 3.7), has been described by Akita *et al.*[52]

When $[Pd(OOH)(\kappa^2\text{-}Tp^{iPr_2})(py)]$ was treated with a catalytic amount of a Cu(II) species, peroxidation of the methine

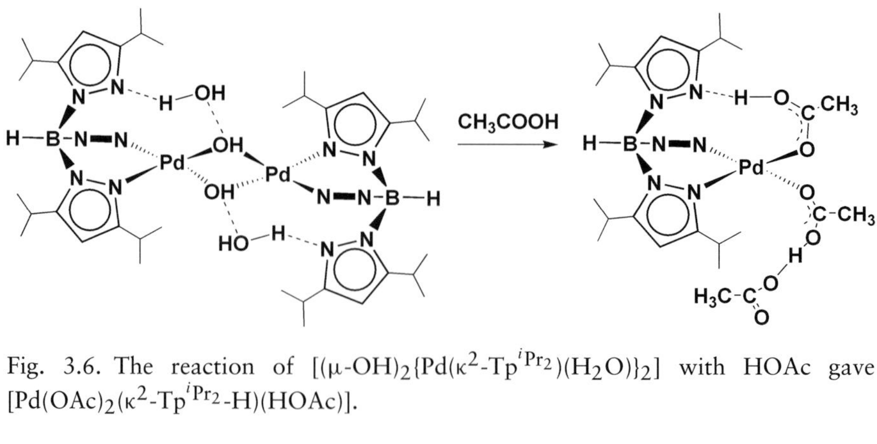

Fig. 3.6. The reaction of $[(\mu\text{-}OH)_2\{Pd(\kappa^2\text{-}Tp^{iPr_2})(H_2O)\}_2]$ with HOAc gave $[Pd(OAc)_2(\kappa^2\text{-}Tp^{iPr_2}\text{-}H)(HOAc)]$.

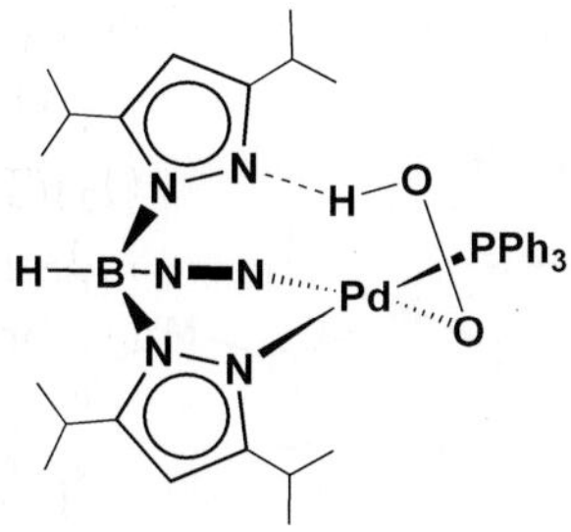

Fig. 3.7. $[Pd(OOH)(\kappa^2\text{-}Tp^{iPr_2})(PPh_3)]$.

moiety of an isopropyl group occurred, yielding a cyclic alkylperoxopalladium derivative.[53] $[Pd(OH)(Tp^{iPr_2})(py)]$ reacts with hydroperoxides (X(OOH); X = H, ^{t}Bu) giving $[Pd(OOH)(\kappa^2\text{-}Tp^{iPr_2})(py)]$ and $[Pd(OO^tBu)(\kappa^2\text{-}Tp^{iPr_2})(py)]$. PPh_3 also displaces py in $[Pd(OOH)(\kappa^2\text{-}Tp^{iPr_2})(py)]$ forming $[Pd(OOH)(\kappa^2\text{-}Tp^{iPr_2})(PPh_3)]$. Condensation of $[Pd(OH)(Tp^{iPr_2})(py)]$ with $[Pd(OOH)(\kappa^2\text{-}Tp^{iPr_2})(py)]$ gave $[(Tp^{iPr_2})(py)Pd(\mu\text{-}OO)Pd(Tp^{iPr_2})(py)]$, while condensation of $[(PdTp^{iPr_2})_2(\mu\text{-}OH)_2]$ with $[Pd(OOH)(\kappa^2\text{-}Tp^{iPr_2})(py)]$ afforded $[(Tp^{iPr_2})(py)Pd(\mu\text{-}OO)Pd(Tp^{iPr_2})]$. The OOH moiety in $[Pd(OOH)(\kappa^2\text{-}Tp^{iPr_2})(PPh_3)]$ interacts with the noncoordinated nitrogen atom of the pendant pyrazolyl group through hydrogen bond. The O_2 moieties in these complexes add across carbon–heteroatom multiple bonds of acetonitrile and acetaldehyde to give $[(Tp^{iPr_2})Pd\{OOC(Me){=}N\}Pd(Tp^{iPr_2})(py)]$ and $[Pd\{OC({=}O)\text{-}CH_3\}(Tp^{iPr_2})(py)]$. The hydroxo and hydroperoxo species are able to convert methyl vinyl ether and styrene to corresponding methyl ketones.[54]

Reaction of $[Pd(OOH)(\kappa^2\text{-}Tp^{iPr_2})(py)]$ with $[Ni(\mu\text{-}OH)(Tp^{iPr_2})]_2$ results in the dehydrogenation of an isopropyl group of the Tp^{iPr_2} ligand to give heterobimetallic di-μ-hydroxo complexes bearing the 3-isopropenyl-substituted Tp ligand $[HB(pz^{iPr_2})_2(pz^{3\text{-isopropenyl-}5\text{-}^iPr})]$ (Fig. 3.8). Similar dehydrogenation is observed in the reaction of hydroperoxo palladium complex with $[Co(\mu\text{-}OH)(Tp^{iPr_2})]_2$. The dehydrogenated products are characterized by spectroscopic and single-crystal X-ray studies.[55]

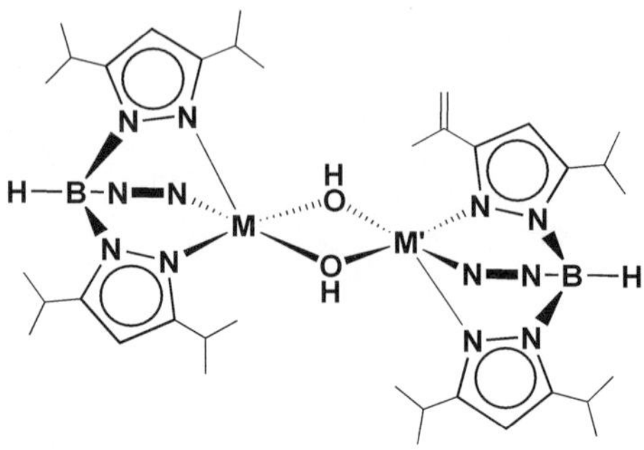

Fig. 3.8. Heterobimetallic complexes containing the ligand [HB(pz^{iPr_2})$_2$-(pz$^{3\text{-isopropenyl-5-iPr}}$)] obtained upon dehydrogenation of an isopropyl group in Tp^{iPr_2}.

Akita *et al.* compared the structural and spectroscopic features of a series of [Rh(Tp^{iPr_2})(Ph$_2$PXPPh$_2$)] complexes: the factor determining the hapticity of the Tp^{iPr_2} ligand (κ^2 vs κ^3) was traced to the conformation of the rhodadiphosphacycle. A folded RhP$_2$X conformation hinders coordination of the pendant pz^{iPr_2} group due to steric repulsion with the bridging X group leading to a κ^2-Tp^{iPr_2} species. A κ^3-coordination is found in complexes with a flat RhP$_2$X conformation (Fig. 3.9).[56]

[Rh(Tp^{iPr_2})(COE)(MeCN)], obtained by the reaction of [RhCl-(COE)$_2$]$_2$ with K(Tp^{iPr_2}), reacts with diphosphines (L$_2$ = dppm, dppe, or dppp) or CO (L) giving square planar [Rh(Tp^{iPr_2})(L$_2$)]

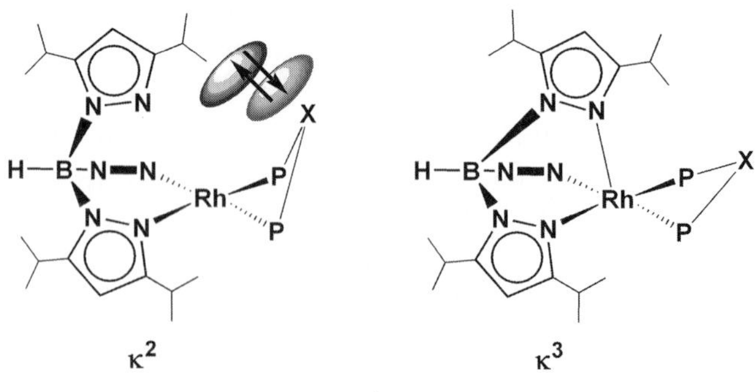

Fig. 3.9. Different hapticity of Tp^{iPr_2} depends on rhodadiphosphacycle conformation.

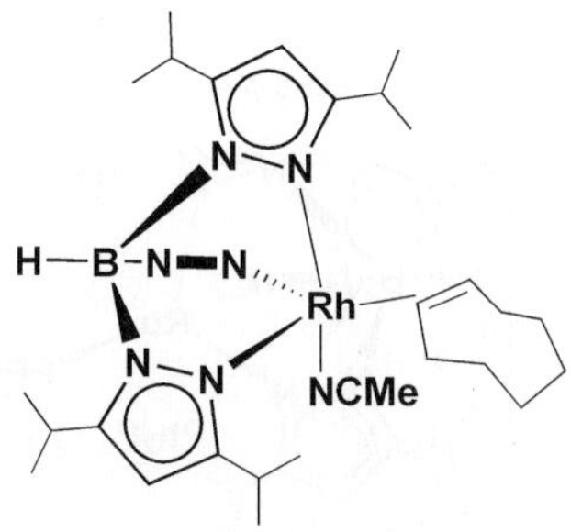

Fig. 3.10. Oxidative addition of XY to [Rh(Tp^{iPr_2})(COE)(MeCN)] generates [Rh(Tp^{iPr_2})(X)(Y)(MeCN)] species.

upon displacement of both COE and MeCN. [Rh(Tp^{iPr_2})-(COE)(MeCN)] undergoes oxidative addition of XY (XY = I_2, SiH_2Et_2, SiH_3Ph) yielding [Rh(Tp^{iPr_2})(X)(Y)(MeCN)] (Fig. 3.10). κ^2-Tp^{iPr_2} vs κ^3-Tp^{iPr_2} interconversion in [Rh(Tp^{iPr_2})(L_2)] species becomes slower as the chelate ring size of L_2 increases.[57]

The square-planar [Rh(Tp^{iPr_2})(dppe)], stored under an O_2 atmosphere, afforded also [Rh(OOH)(Tp^{iPr_2})(Hpz^{iPr_2})] in which the OOH functionality results from the ring-opening protonation of the O_2 ligand in [Rh(μ-O_2)(Tp^{iPr_2})(Hpz^{iPr_2})]. [Rh(μ-OOH)(Tp^{iPr_2})(pz)(Hpz)] and the cyclometalated [Rh(Tp^{iPr_2})H(*O*, C^{Ph}-O-dppe-O)] (Fig. 3.11) have been also reported.[58]

O_2-treatment of [Rh(Tp^{iPr_2})(dppe)], in the presence of Hpz^{iPr_2}, gives the κ^2-peroxometal complex, [Rh(κ^2-O_2)(Tp^{iPr_2})(Hpz^{iPr_2})], which has been converted to [Rh(μ-OOH)(Tp^{iPr_2})(pz)(Hpz)] when

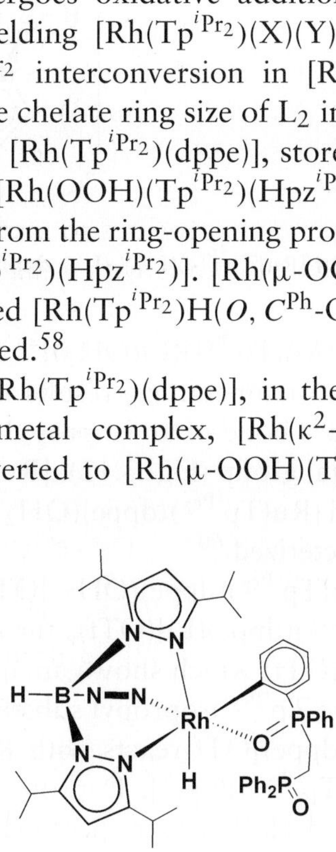

Fig. 3.11. Structure of the cyclometalated [Rh(Tp^{iPr_2})H(O,C^{Ph}-O-dppe-O)].

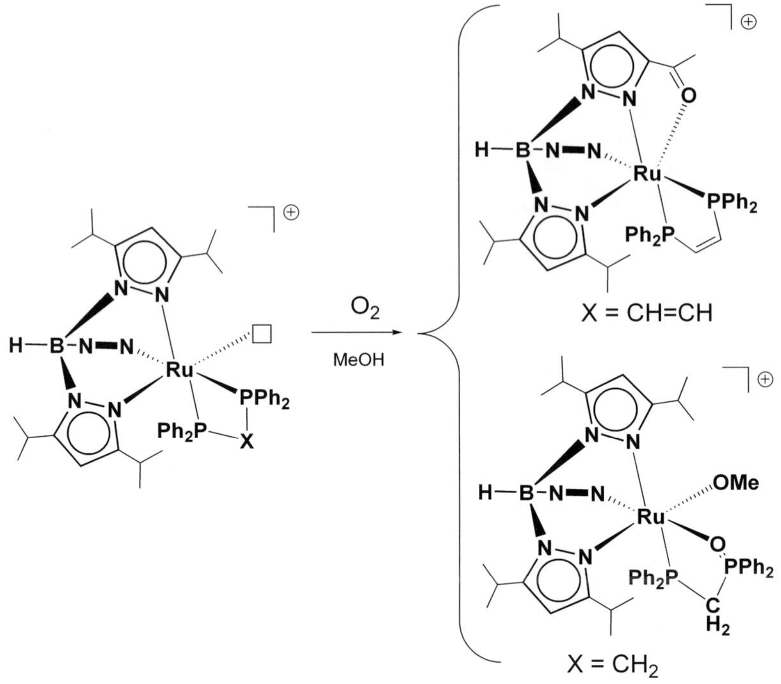

Fig. 3.12. Reaction of $[Ru(Tp^{^iPr_2})(L)]^+$ (L = dppm or dppe) with O_2.

reacted with Hpz.[59] $[Ru(Tp^{^iPr_2})(L)(OH_2)]^+$ (L = dppene), when exposed to O_2 at room temperature, results in the oxidative cleavage of the $C(sp^3)$–$C(sp^3)$ bond of an isopropyl group in the $Tp^{^iPr_2}$ to give an acetyl functional group (Fig. 3.12). $[Ru(OMe)(Tp^{^iPr_2})$(dppm-O)]OTf (Fig. 3.12) and $[Ru(Tp^{^iPr_2})(dppe)(OH_2)](OTf)$ have been also synthesized and characterized.[60]

The reaction of $[Ru(Tp^{^iPr_2})(dppe)(OH_2)]$OTf with L (L = acetone or N_2) gives $[Ru(Tp^{^iPr_2})(dppe)(L)](OTf)$, through the formation of $[Ru(\kappa^4\text{-}Tp^{^iPr_2})(dppe)](OTf)$ which shows an agostic interaction with a methyl C-bond of the $Tp^{^iPr_2}$ isopropyl substituent (Fig. 3.13).[61]

$[Ru(Tp^{^iPr_2})(OH_2)(dppe)]$OTf reacts with KOH yielding the six-coordinate $[Ru(OH)(Tp^{^iPr_2})(dppe)]$. Deprotonation of $[Ru(Tp^{^iPr_2})$-(OH_2)(bipy)]OTf affords the hydroxo species $[Ru(OH)(Tp^{^iPr_2})(OH_2)$-(bipy)]. $[Ru(OH)(Tp^{^iPr_2})(dppe)]$ is basic enough to be condensed with acidic substrates such as carboxylic acid. Thus, addition of acetic acid produced the acetato complex $[Ru(\kappa^1\text{-}OAc)(Tp^{^iPr_2})(dppe)]$

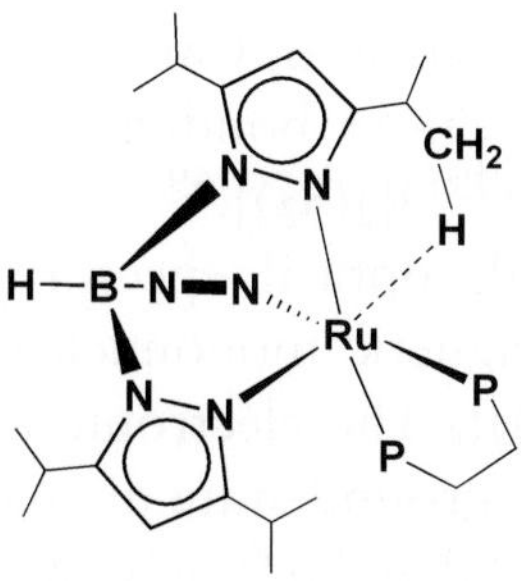

Fig. 3.13. Structure of the cation of $[Ru(\kappa^4\text{-}Tp^{^iPr_2})(dppe)](OTf)$.

through the formation of the intermediate $[Ru(Tp^{^iPr_2})(OH_2)(dppe)]$-OAc. Excess of acetic acid yields $[Ru(Tp^{^iPr_2})\{H(OAc)_2\}(dppe)]$. $[Ru(Tp^{^iPr_2})(\kappa^1\text{-}OCOC_6H_4Cl)(dppe)]$ and the intermediate $[Ru(Tp^{^iPr_2})(OH_2)(dppe)(\kappa^1\text{-}OCOC_6H_4Cl)]$ have also been reported and structurally characterized.[62]

O_2-treatment of $[Ru(OH_2)_{3-n}(L)_n(Tp^{^iPr_2})(OTf)(THF)_x]$ (L = bipy, py, THF) produced $[Ru(OH)(Tp^{^iPr_2})(bipy)](OTf)]$, $[Ru(OH)(OTf)(Tp^{^iPr_2})(OH_2)(py)(THF)_2]$ and the dinuclear mixed aquo-hydroxo complexes bridged by the Ru–O···H···O–Ru hydrogen-bonding interactions, $[(Tp^{^iPr_2})Ru^{II}(OH)_2(OH_2)_4(OTf)Ru^{III}(Tp^{^iPr_2})]$ and $[(RuTp^{^iPr_2})_2(OH)_2(OTf)_2(OH_2)_4]$.[63]

The five- and four-coordinate complexes, $[Cu(SMeIm)(Tp^{^iPr_2})]$ and $[Cu(SCPh_3)(Tp^{^iPr_2})]$, respectively, have been investigated by UV–Vis, MCD, Raman, EPR, and X-ray absorption spectroscopy. These compounds have been compared to red and blue copper centers in metalloproteins.[64] $[Cu(SCPh_3)(Tp^{^iPr_2})]$ has been also investigated by MCD, XAS, and resonance Raman studies combined with DFT calculations, using $[Cu(SCH_3)(Tp)]$ as a model.[65] The square-pyramidal $[Cu(NO_3)(Tp^{^iPr_2})]$ has been prepared and characterized by X-ray, IR, UV–Vis, and EPR spectroscopies. The structural properties of $[Cu(NO_3)(Tp^{^iPr_2})]$ and $[\{Cu(Tp^{^iPr_2})\}_2(\mu\text{-}OH)_2]$ have been related to the substituents in the 3-position.[66] $[Cu(OAr)(Tp^{^iPr_2})]$ (Ar = C_6H_4-F, 2,6-$Me_2C_6H_3$, 2,6-$^tBuC_6H_3$), models for the reaction intermediates of tyrosinase catalyzed oxidation of phenols, have been prepared and structurally characterized.[67] $[Cu(\eta^2\text{-}ONO)(Tp^{^iPr_2})]$ has been also synthesized, spectroscopically characterized, and investigated by DFT calculations.[68] Raman spectroscopy and DFT calculations have been applied to the study of Cu–S_2 bonding interactions in the side-on μ-η^2:η^2-bridged [{Cu(Tp^{iPr_2})}$_2$(S_2)].[69]

Magnetic, vibrational, optical spectra and DFT studies have been reported for diamagnetic mononuclear side-on Cu^{II}-superoxo complex [Cu(O_2)(Tp^{iPr_2})]. The electronic nature of the lowest singlet/triplet states and the ground-state diamagnetism of this complex has been also described.[70] [Cu(Tp^{iPr_2})(NCMe)] and [Cu(Tp^{iPr_2})(CO)] were prepared and fully characterized by IR, far-IR, ^{1}H NMR, and ^{13}C NMR spectroscopy. The structure of [Cu(Tp^{iPr_2})(NCMe)] has been determined by X-ray crystallography.[71] ^{63}Cu NMR spectroscopic studies of [Cu(Tp^{iPr_2})(PPh_3)], [Cu(Tp^{iPr_2})(MeCN)], and [Cu(Tp^{iPr_2})] have been reported. Temperature dependence of ^{63}Cu NMR signals for these copper-(I) complexes show that a quadrupole relaxation process is much more significant to their ^{63}Cu NMR line widths than a ligand exchange process.[72] Spectroscopic studies combined with calculations are used to describe the electronic structure and vibrational properties of mononuclear four-coordinate end-on alkylperoxo and hydroperoxo Cu(II) complexes.[73]

The crystal structure of [Cu($OOCMe_2Ph$)(Tp^{iPr_2})] has been also reported.[73] Two thiolato Cu redox pairs containing Tp^{iPr_2} (models for blue Cu proteins) have been reported by the Fujisawa group, which discusses also the electron-transfer processes and the structure of K[Cu(SC_6F_5)(Tp^{iPr_2})] and [Zn(SMeIm)(Tp^{iPr_2})],[74] as well as synthesis and characterization by UV–Vis absorption, EPR, NMR, far-IR, and FT-Raman spectroscopies and X-ray of the distorted tetrahedral [M(SC_6F_5)(Tp^{iPr_2})] (M = Cu, Mn, Fe, Co, Ni, and Zn).[75] [M(SC_6F_5)(Tp^{iPr_2})] has been also investigated by a combination of absorption, MCD, resonance Raman, and S K-edge X-ray absorption spectroscopies and DFT calculations. The geometry variations and bond energies have been related to the nature of the metal–ligand bonding interactions.[76]

Hikichi reported the characterization of a series of reactive bis(μ-oxo)dimetal(III) complexes (metal = Cu, Fe) containing the Tp^{iPr_2} ligands.[9]

The diazene complexes $[(Tp^{iPr_2})CuNH{:}NHCu(Tp^{iPr_2})]$ and $[(Tp^{iPr_2})CuMeN{:}NMeCu(Tp^{iPr_2})]$ have been synthesized by the addition of hydrazines to $[Cu(Tp^{iPr_2})(\mu\text{-}OH)]_2$. The diazene complexes were characterized by X-ray crystallography as well as ^{1}H NMR, UV/visible absorption, IR, far-IR, and resonance Raman spectroscopy.[77]

Fujisawa *et al.* reported $[M(Cl)(Tp^{iPr_2})]$ complexes (M = Zn, Cd and Hg) and analyzed bond-length contraction through single-crystal X-ray diffraction and metal–halogen stretching vibrations.[78]

$K(Tp^{iPr_2})$ reacts with $[UI_3(THF)_4]$ in THF giving $[UI_2(\kappa^3\text{-}Tp^{iPr_2})(THF)_x]$ ($x = 2$–3) which decomposes in solution, the main decomposition product being $[UI_2(\kappa^3\text{-}Tp^{iPr_2})(\kappa^1\text{-}pz^{iPr_2}H)]$.[79] $[UI_3(THF)_4]$ reacts with 1 equiv. of $K(Tp^{iPr_2})$ in the presence of L giving $[UI_2(Tp^{iPr_2})(L)_x]$ (L = $OPPh_3$, $x = 1$; L = py, $x = 2$; L = $Hpz^{tBu,Me}$, $x = 2$; L = bipy, $x = 1$).

$[UI_2(Tp^{iPr_2})(THF)_{2-3}]$ has been isolated by reacting $[UI_3(THF)_4]$ with 1 equiv. of $K(Tp^{iPr_2})$. $[UI_2(Tp^{iPr_2})(OPPh_3)]$ displays distorted octahedral coordination geometry, while $[UI_2(Tp^{iPr_2})(py)_2]$ and $[UI_2(Tp^{iPr_2})(bipy)_2]$ display distorted pentagonal bipyramid and capped octahedral geometries, respectively.[80]

3.13. Tp^{iPr_2Br}

$[Pd(OH)(\kappa^2\text{-}Tp^{iPr_2,Br})(py)]$ has been prepared by alkaline hydrolysis of $[PdCl(\kappa^2\text{-}Tp^{iPr_2,Br})(py)]$.[51]

3.14. Tp^{4Br}

$[Cu(Tp^{4Br})(PR_3)]$[81,82] have been reported (PR_3 = 4 or 2-(diphenylphosphane)benzoic acid or tris(*m*-sulfonatophenyl) posphine trisodium salt) and characterized by FT-IR in the solid state, and by NMR (^{1}H and $^{31}P\{^1H\}$) and electrospray mass spectrometry (ESI-MS) in solution. $[Cu\{PPh_2(4\text{-}C_6H_4COOH)\}(Tp^{4Br})]$ shows an interesting dimeric supramolecular assembly. These complexes present a considerable superoxide scavenging activity,[81] which has been investigated by NMR, IR, and ESI MS spectroscopy.[82]

3.15. Tp^{tBu}

[Tl(Tp^{tBu})] exhibits photoluminescence properties. Comparison is made with [Tl(Tp^{Ph})], [Tl(Tp^{Tol})], and [Tl(Tp^{Th})].[83]

Amide and alkoxide coordination complexes of calcium supported by β-diiminato and bulky Tp^x complexes have been reported by Chisholm and Gallucci[21] together with their activity in lactide ring-opening polymerization. [Ca(Tp^{tBu})N($SiMe_3$)$_2$], [Ca(Tp^{tBu})(O-2,6-^{i}$Pr_2C_6H_3$)]·THF, and [Ca(Tp^{tBu})(O-2,6-^{i}$Pr_2C_6H_3$)] were synthesized and structurally characterized.[21] [Ca(OC_6H_3-2,6-^{i}Pr_2)(Tp^{tBu})]·(PO) has been isolated and structurally characterized and shown to be inert to the polymerization of PO.[20]

[FeCl(Tp^{tBu})], prepared from the reaction of Tp^{tBu} with $FeCl_2$, is able to catalyze the epoxidation of stilbenes to epoxides, whereas [Fe(Tp^{tBu})](OTf) exhibits efficient catalytic activity with high selectivity and reaction rate for epoxidation by using O_2 with isobutyraldehyde as co-reductant.[84]

Complexes [M(OR)(Tp^{tBu})] (M = Zn or Mg, R = Et, $SiMe_3$) have been prepared and the zinc species employed in the ring-opening polymerization of *L*−, *rac*-, and *meso*-lactide.[85]

The reaction of [Cd_2(THF)$_5$](BF_4)$_4$ with 2 equiv. of [Tl(Tp^{tBu})] gives the intermediate [Cd(Tp^{tBu})][BF_4] that reacts with potassium thiocyanate to yield the tetrahedral monomeric [Cd(NCS)(Tp^{tBu})] as a consequence of sterically demanding 3-*tert*-butyl groups.[86] [Cd(Et_2NCS_2)(Tp^{tBu})] is a distorted trigonal bipyramid, an unusual geometry for a molecule containing a tridentate Tp^{tBu} ligand.[87]

The synthesis and characterization of [Ca(Tp^{tBu})$_2$] has been also reported.[88]

3.16. pz^0Tp^{tBu}

Variable temperature ^{1}H NMR spectra of [Pd(η^3-2-methallyl)-(pz^0Tp^{tBu})] were recorded and its stereochemically nonrigid motions were investigated. The four *tert*-butyl groups appear as 1:1:1:1 and 2:1:1 set of resonances at low and high temperature, respectively, whereas *syn* and *anti* protons of the η^3-2-methallyl moiety resonated separately without their mutual interconversion. The single-crystal X-ray study confirmed κ^2-coordination of (pz^0Tp^{tBu}). The temperature-dependent spectral change has been assigned to the rotation of the uncoordinated pyrazolyl group near palladium, i.e. axial one, about the boron–nitrogen bond axis.[89]

3.17. $PhTp^{tBu}$

[MgR($PhTp^{tBu}$)] (R = Me, Et) and [ZnEt($PhTp^{tBu}$)] have been prepared by the reaction of [Tl($PhTp^{tBu}$)] with MR_2 (M = Mg, Zn), structurally characterized and compared with [MMe(Tp^{tBu})]. The phenyl substituent on boron has an almost negligible effect on the metal–carbon bond lengths.[90]

3.18. $^tBuTp^{tBu}$ and $^tBuTp^{iPr}$

When $FeCl_2(THF)_{1.5}$ reacts with the sterically hindered thallium salt of $^tBuTp^{tBu}$, *trans*-$FeCl_2(Hpz^{tBu})_4$ is formed,[91] whereas the same reaction carried out with [Tl($^tBuTp^{iPr}$)] gave [FeCl($^tBuTp^{iPr}$)], structurally characterized. The reaction of the latter compound with MeLi, followed by carbonylation gave [Fe($^tBuTp^{iPr}$)$(CO)_2${C(=O)Me}] whereas the reaction with RMgX resulted in the formation of [MgX($^tBuTp^{iPr}$)].[92]

3.19. $Tp^{tBu,Me}$

Open-shell cobalt(III) imido complex [Co=NAd($Tp^{tBu,Me}$)] is a stable representative of a broader class of multiply bonded terminal [Co=X(Tp^x)] species (for example, X = NR); the existence of [Co=NAd($Tp^{tBu,Me}$)] suggests that other isoelectronic embodiments (X = O, S, PR, CR_2) may also be accessible and possess some — albeit fleeting — stability. Nevertheless, the facile transformation of [Co=NAd($Tp^{tBu,Me}$)] by a C–H activation into the complex **III** in Fig. 3.14 demonstrates the inherent reactivity of these compounds.[93]

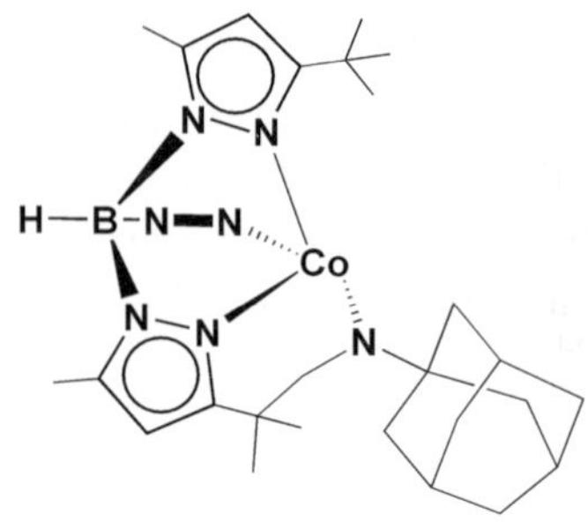

III

Fig. 3.14. Product derived from the C–H activation in [Co=NAd(Tp$^{^tBu,Me}$)].

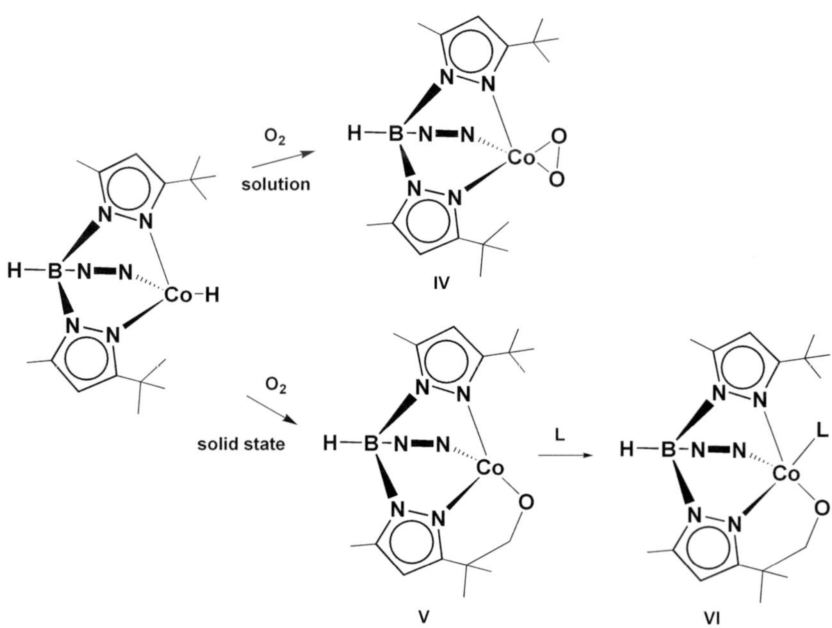

Fig. 3.15. Reaction of [CoH(Tp$^{^tBu,Me}$)] toward O_2.

Reaction of [CoH(Tp$^{^tBu,Me}$)] with excess O_2 in solution (Fig. 3.15) yielded [Co(O_2)(Tp$^{^tBu,Me}$)] (**IV**) and [Co(OH)(Tp$^{^tBu,Me}$)]. When [CoH(Tp$^{^tBu,Me}$)] was exposed to O_2 the complex **V**, containing a functionalized pyrazole in which a H of a *tert*-butyl group has been replaced by the oxygen atom bonded to cobalt, also formed. This complex interacts with 1 equiv. of $Hpz^{^tBu,Me}$ yielding the five-coordinate trigonal bipyramid complex **VI**, structurally characterized, in which the neutral pyrazole occupies an axial position. These reactions proceed via a spectroscopically observable hydroperoxide $[CoOOH(Tp^{^tBu,Me})]$.[94]

When $[Co(O_2)(Tp^{^tBu,Me})]$ reacted with NO a mixture of $[Co(NO_2)(Tp^{^tBu,Me})]$ and $[Co(NO_3)(Tp^{^tBu,Me})]$ are formed, whereas reaction of $[Co(Tp^{^tBu,Me})(N_2)]$ with NO yielded the paramagnetic cobalt nitrosyl species $[Co(NO)(Tp^{^tBu,Me})]$, which, exposed to O_2 in solution, yielded $[Co(NO_2)(Tp^{^tBu,Me})]$. The same reaction carried out in the solid state produced $[Co(NO_3)(Tp^{^tBu,Me})]$. A labile peroxynitrite has been indicated as an intermediate in this reaction.[95] On the other hand, reaction of $[Co(Tp^{^tBu,Me})(N_2)]$ in THF with 1 equiv. of Me_3SiN_3 yielded $[Co(Tp^{^tBu,Me@})N(H)SiMe_3]$, in which the cobalt atom is bound to three nitrogen atoms of $Tp^{^tBu,Me}$, a trimethylsilylamido group, and a methylene carbon from one of the *tert*-butyl substituents of $Tp^{^tBu,Me}$ (Fig. 3.16). The latter complex reacts with excess 2-propanol forming $[Co(Tp^{^tBu,Me})(O^iPr)]$ and acetone, upon alcoholysis of the amide, followed by β-hydrogen atom transfer from coordinated *iso*-propoxide to the alkyl group, and finally reaction of the resulting $[Co(Tp^{^tBu,Me})(O{=}CMe_2)]$ with an additional equivalent of alcohol. $[Co{=}NSiMe_3(Tp^{R,R'})]$ is a highly reactive intermediate capable of abstracting hydrogen atoms from alkyl groups. The isoelectronic $[Co{=}O(Tp^{R,R'})]$ are very reactive species as well, and their

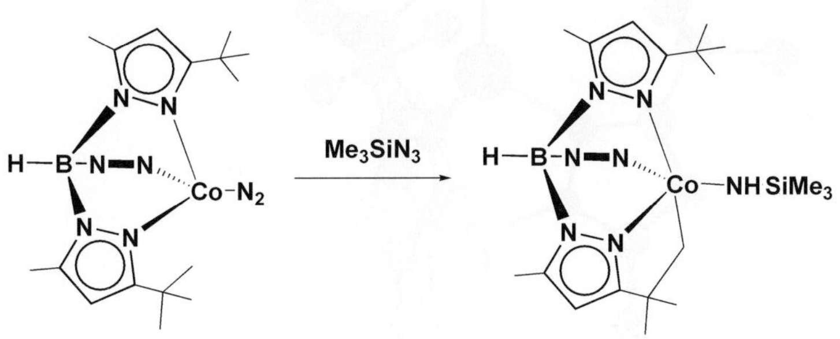

Fig. 3.16. Synthesis of $[Co(Tp^{^tBu,Me})N(H)SiMe_3]$, a compound in which a bond between Co and a methylene carbon from the ^{t}Bu group is present.

generation from O_2 and use in the transformations of external substrates is a significant challenge.[96]

$[NiCl(Tp^{^tBu,Me})]$ has been investigated as a catalyst toward ethylene polymerization.[97] The oxidation of $[Cr(OH)(Tp^{^tBu,Me})(pz^{^tBu,Me})]$ in CH_2Cl_2 was investigated electrochemically.[98]

The zinc hydroxide complex $[Zn(OH)(Tp^{^tBu,Me})]$ is protonated by $(C_6F_5)_3B(OH_2)$ to yield the aqua derivative $[Zn(OH_2)(Tp^{^tBu,Me})][HOB(C_6F_5)_3]$, which has been structurally characterized by X-ray diffraction (Fig. 3.17). The protonation is reversible, and treatment of $[Zn(OH_2)(Tp^{^tBu,Me})]^+$ with Et_3N regenerates $[Zn(OH)(Tp^{^tBu,Me})]$. $[Zn(OH_2)(Tp^{^tBu,Me})]^+$ is inert toward CO_2, whereas $[Zn(OH)(Tp^{^tBu,Me})]$ is in rapid equilibrium with the bicarbonate complex $[ZnOC(O)OH(Tp^{^tBu,Me})]$.[99]

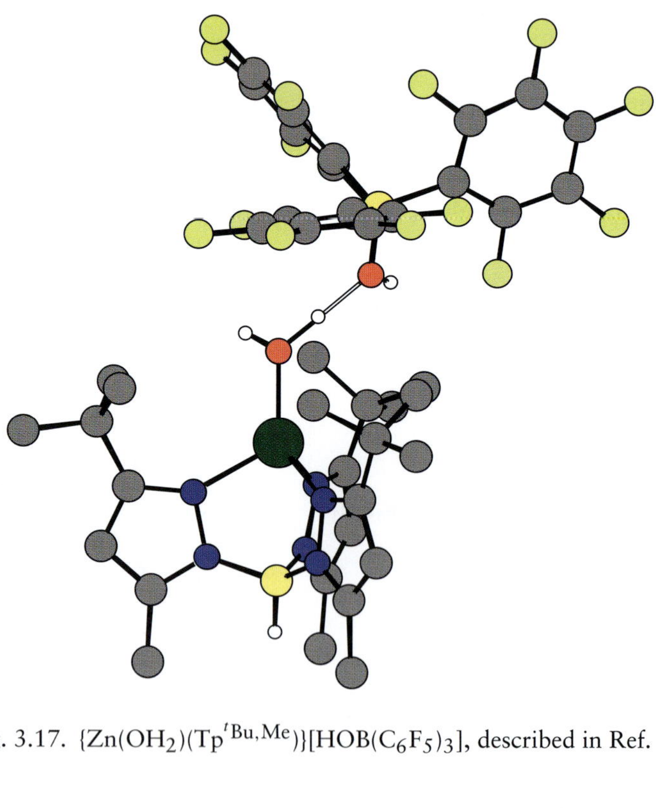

Fig. 3.17. $\{Zn(OH_2)(Tp^{^tBu,Me})\}[HOB(C_6F_5)_3]$, described in Ref. 99.

The cobalt hydroxide complex $[Co(OH)(Tp^{^tBu,Me})]$ is likewise protonated by $(C_6F_5)_3B(OH_2)$ to yield the aqua derivative $[Co(OH_2)(Tp^{^tBu,Me})][HOB(C_6F_5)_3]$, isostructural with the zinc complex. Complexes $[M(OH_2)(Tp^{^tBu,Me})][HOB(C_6F_5)_3]$ (M = Zn, Co) exhibit a hydrogen bonding interaction between the metal-aqua and boron hydroxide moieties. This hydrogen-bonding interaction may be viewed as analogous to that between the aqua ligand and Thr-199 at the active site of carbonic anhydrase. In addition to the structural similarities between the zinc and cobalt complexes $[Zn(OH)(Tp^{^tBu,Me})]$ and $[Co(OH)(Tp^{^tBu,Me})]$, and between $[Zn(OH_2)(Tp^{^tBu,Me})]^+$ and $[Co(OH_2)(Tp^{^tBu,Me})]^+$, DFT (B3LYP) calculations demonstrated that the p*Ka* value of $[Zn(OH_2)(Tp^x)]^+$ is similar to that of $[Co(OH_2)(Tp^x)]^+$. $[Co(OH)(Tp^{^tBu,Me})]$ reacts with CO_2 to give the bridging carbonate complex $[Co(Tp^{^tBu,Me})]_2(\mu\text{-}\eta^1,\eta^2\text{-}CO_3)$. In $[Zn(Tp^{^tBu,Me})]_2(\mu\text{-}\eta^1,\eta^1\text{-}CO_3)$, the carbonate bridges in an unidentate manner to both zinc centers.[100]

A solution of $[Zn(OH)(Tp^{^tBu,Me})]$ in alcohols (ROH = Me, Et, ^{i}Pr) is able to transfer hydride to the NAD^+ model 10-methylacridinium perchlorate. By employing deuterium labeling studies it has been demonstrated that the source of the hydride is the B–H group of $Tp^{^tBu,Me}$. The $Tp^{^tBu,Me}$ ligand acts as a hydride donor in the reaction of $[Zn(OH_2)(Tp^{^tBu,Me})]^+[HOB(C_6F_5)_3]^-$ with MeOH, giving $[ZnH(Tp^{^tBu,Me})]$ (Fig. 3.18). This fact demonstrated a caveat for the

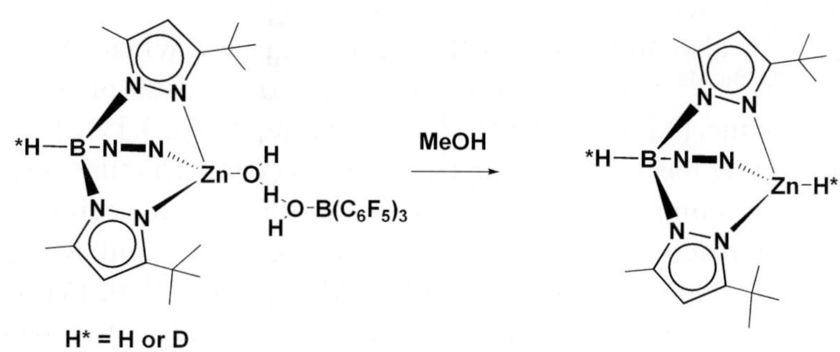

Fig. 3.18. $Tp^{^tBu,Me}$ acts as hydride donor in the reaction of $[Zn(OH)_2(Tp^{^tBu,Me})]^+$ with $[HOB(C_6F_5)_3]$ which yields $[ZnH(Tp^{^tBu,Me})]$.

often assumed inertness of the B–H group in Tp^x ligands, especially in the presence of reactive cationic species.[101]

Complexes $[ZnX(Tp^{'Bu,Me})]$ (X = Br, Cl, and OH) have been investigated by low-temperature solid-state ^{67}Zn NMR spectroscopy. The value of the quadrupole coupling constant C_q for the zinc increased monotonically with the electronegativity of X, e.g. $Br < Cl \ll OH$.[102]

$[Zn(SH)(Tp^{'Bu,Me})]$ does not react with esters, phosphates, or CO_2, e.g., for thiolytic cleavage reactions, whereas acidic organic X–OH compounds (carboxylic acids, trinitrophenol, hexafluoroacetylacetone) replace the SH groups with the formation of $[Zn(OX)(Tp^{'Bu,Me})]$.[103] The reaction of $[Zn(OH)(Tp^{'Bu,Me})]$ with alcohols has been followed by combined experimental and computational investigations in order to determine the factors influencing the formation of tetrahedral alkoxide complexes. The $[Zn(OR)(Tp^{'Bu,Me})]$ derivatives have been found unstable toward hydrolysis.[104]

$Tp^{'Bu,Me}$ has been also used to prepare ternary zinc complexes with triorganodiphosphates $(R_3POP)^-$ and diorganodiphosphates (R_2POP^{2-}). Monodentate and bis-monodentate attachment of the diphosphates, with tetrahedral zinc in a ZnN_3O environment, have been identified in $[Zn(R_3POP)(Tp^{'Bu,Me})]$ and $[(Tp^{'Bu,Me})Zn(R_2POP)Zn(Tp^{'Bu,Me})]$.[105]

IMOMM(Becke3LYP)MM3 calculations on $[CuCl(Tp^{'Bu})]$ and $[CuCl(Tp^{'Bu,Me})]$ show that the nonsymmetrical arrangement observed around the Cu center is controlled by a Jahn–Teller effect, and not from the sterical demands of the $Tp^{'Bu,X}$.[106]

$Tp^{'Bu,Me}$ also stabilizes Ln(II) hydrido species. When $[Yb(CH_2SiMe_3)(Tp^{'Bu,Me})(THF)]$ is exposed to a hydrogen atmosphere (1000 psi) in pentane, $[Yb(\mu\text{-}H)(Tp^{'Bu,Me})]_2$ is obtained (Fig. 3.19). The deuterium analog has been also prepared. The crystal structure revealed a dinuclear complex containing five-coordinate Yb centers. This compound reacts with $HN(SiMe_3)_2$, $HC{\equiv}CSiMe_3$, and $Me_3SiC{\equiv}CC{\equiv}CSiMe_3$ and CO yielding $[YbN(SiMe_3)_2(Tp^{'Bu,Me})]$, $[Yb(\mu\text{-}\eta^1\text{-}\eta^2\text{-}HC{\equiv}CSiMe_3)(Tp^{'Bu,Me})]$ and $[Yb(Me_3SiC{\equiv}CCH{=}CSiMe_3)(Tp^{'Bu,Me})]$, respectively.[107]

Reaction of $[Yb(I)(Tp^{'Bu,Me})(THF)]$ with K(OMs) yielded $[Yb(OMs)(Tp^{'Bu,Me})(THF)]$. Reaction of $[Yb(\mu\text{-}H)_2(Tp^{'Bu,Me})]$ with

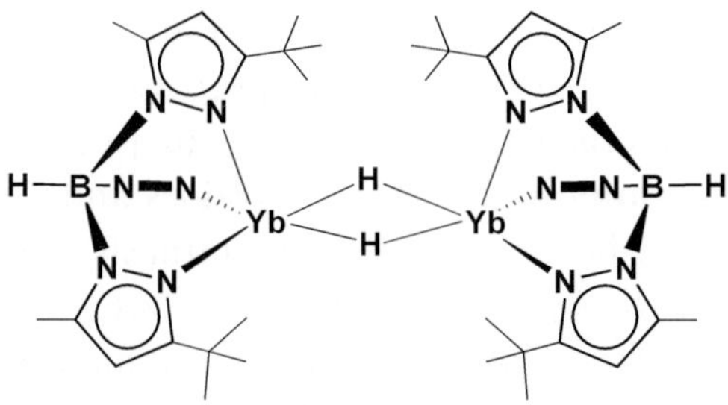

Fig. 3.19. $[Yb(\mu\text{-}H)(Tp^{^tBu,Me})]_2$.

HOMs in benzene gave the monomeric solvent-free $[(Tp^{^tBu,Me})$-$Yb(OMs)]$ that, however, formed $[Yb(OMs)(Tp^{^tBu,Me})(THF)]$ upon dissolution in THF. Finally, $[Yb(\mu\text{-}H)(Tp^{^tBu,Me})]_2$ reacts with dpm affording the related $[Yb(Tp^{^tBu,Me})(dpm)]$. These complexes have been fully characterized by spectroscopy and X-ray single-crystal study. In $[Yb(OMs)(Tp^{^tBu,Me})(THF)]$ and $[Yb(Tp^{^tBu,Me})(dpm)]$ the metal is five-coordinate (trigonal bipyramid). $[Yb(OMs)(Tp^{^tBu,Me})]$, four-coordinate, exhibits a distorted tetrahedral arrangement.[108]

Reaction of $[Yb(\mu\text{-}H)(Tp^{^tBu,Me})]_2$ with cyclopentadiene (C_5H_6) and trimethylsilyl cyclopentadiene ($C_5H_5SiMe_3$) resulted in the formation of $[Yb(C_5H_4R)(Tp^{^tBu,Me})]$ (R = H, $SiMe_3$) characterized by multinuclear NMR spectroscopy and the latter also by X-ray, which evidenced well-separated monomeric units with the $Tp^{^tBu,Me}$ in an unusual distortion: one of the pyrazolyl rings is rotated in such a way as to bring both pyrazolyl nitrogens in bonding contact with ytterbium (Fig. 3.20).[109]

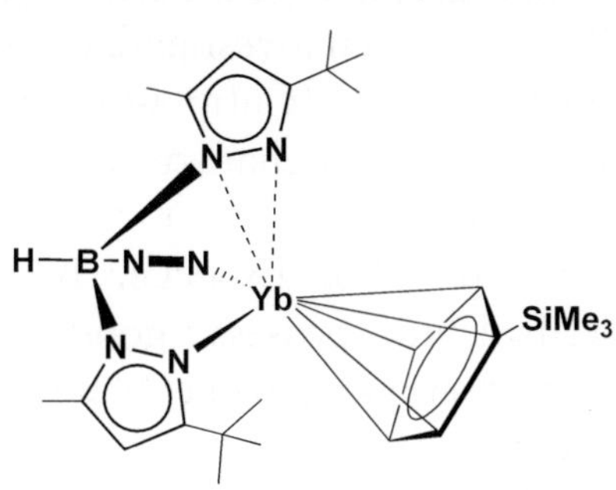

Fig. 3.20. $[Yb(C_5H_4SiMe_3)(Tp^{^tBu,Me})]$.

An overview of recent work from the Takats laboratory deals with divalent {Ln($Tp^{^tBu,Me}$)} complexes and preparation, molecular structure, NMR characteristics and reactivity of [Yb(μ-H)($Tp^{^tBu,Me}$)]$_2$. Reactions of this compound with Lewis bases and Lewis acids, such as the perfluoroaryl boranes, σ-bond metathesis reactions generally involving acidic H–X bonds, but also an example of C–Si bond cleavage, and insertion chemistry, primarily involving alkynes have been described. The molecular structures of representative complexes have been determined by X-ray crystallography.[110]

The reaction of [Yb(E)(THF)($Tp^{^tBu,Me}$)] (E = I, CH_2SiMe_3) with $ImMe_4$ has been also investigated. $ImMe_4$ acts as a simple Lewis base toward [YbI($Tp^{^tBu,Me}$)(THF)] resulting in substitution of the THF and formation of [YbI($ImMe_4$)($Tp^{^tBu,Me}$)]. Reaction with [$YbCH_2SiMe_3$($Tp^{^tBu,Me}$)(THF)], in addition to displacement of THF, gave metalation of one of the N–CH_3 substituents of $ImMe_4$ and formation of [Yb($ImMe_4$)(CH_2N)(C(CH_3)C(CH_3)-N(CH_3)C)($Tp^{^tBu,Me}$)].[111]

3.20. $Tp^{^tBu,^iPr}$

The trigonal pyramidal [Cu(NO_3)($Tp^{^tBu,^iPr}$)] and the tetrahedral [Cu(OH)($Tp^{^tBu,^iPr}$)] have been prepared and characterized by X-ray, IR, UV–Vis, and EPR spectroscopies.[66] [Cu($Tp^{^tBu,^iPr}$)(NCMe)] and [Cu($Tp^{^tBu,^iPr}$)(CO)] were prepared and fully characterized by IR, far-IR, ^1H NMR, and ^{13}C NMR spectroscopy. The structure of [Cu($Tp^{^tBu,^iPr}$)(CO)][71] and [CuCl($Tp^{^tBu,^iPr}$)][112] were determined by X-ray crystallography. [Cu($OOCMe_2Ph$)($Tp^{^tBu,^iPr}$)] and [Cu(OO^tBu)($Tp^{^tBu,^iPr}$)] have been investigated by spectroscopic studies combined with calculations.[73] [Cu(η^2-ONO){$Tp^{^tBu,^iPr}$}] has been also synthesized, spectroscopically characterized, and investigated by DFT calculations.[68]

Complexes [Mn(X)($Tp^{^tBu,^iPr}$)] (X = Cl, Br, NO_3) have been reported and described to possess good activity for ethylene polymerization and ethylene-R-olefin copolymerization. The [Mn(Cl)($Tp^{^tBu,^iPr}$)]/Al(^{i}Bu)$_3$/[Ph_3C][B(C_6F_5)$_4$] system is also active toward propylene and gives isotactic polypropylene.[113] Spectral properties and electronic structure of the four-coordinate high-spin

$[Fe(OO^tBu)(Tp^{tBu,iPr})]^+$ have been investigated[114] and compared to the six-coordinate high-spin $[Fe(6\text{-}Me_3TPA)(OH_x)(OO^tBu)]^{x+}$ (x = 1 or 2).[115]

The synthesis of $[Zn(OH)(Tp^{tBu,iPr})]$, which has a structure very similar to the active site of the enzyme carbonic anhydrase, was described. $[Zn(L)(Tp^{tBu,iPr})]$ (HL = uracil or halouracil) were also reported. The halogenated uracil complexes were injected in Dalton's lymphoma tumor system in mice and [Zn(5-fluorouracilate)-$(Tp^{tBu,iPr})]$ exhibits significant antitumor activity. This complex also promotes the hydrolysis of various esters and the maximum rate of hydrolysis is found with *p*-nitrophenylacetate.[116]

3.21. Tp^{Ph}

$[Tl(Tp^{Ph})]$ shows photoluminescence due to the metal-centered sp triplet of the Tl^+ ion. This compound has been compared with $[Tl(Tp^{tBu})]$, $[Tl(Tp^{Tol})]$, and $[Tl(Tp^{Th})]$ and it has been reported that, in the solid state, the emission wavelength depends on the substituent on the Tp ligand, since the Tl···Tl distance varies with R.[83] The structure of $[Tl(Tp^{Ph})]$ reported by Ruman and coworkers[117] shows the ligand coordinated in a κ^3-fashion. The compound crystallizes as centrosymmetric dimer in which weak CH/π intra- and inter-dimer interactions are responsible for arrangement in the crystal structure. $[MgEt(Tp^{Ph})(THF)]$ has been also structurally characterized.[85]

The high-spin $[CoR(Tp^{Ph})]$ has been investigated. Tp^{Ph} provides intermediate steric hindrance for the central metal ion, which is able to bond two other groups as the thiocyanate and Hpz or MeOH yielding $[Co(NCS)(L)(Tp^{Ph})]$ (L = Hpz or MeOH), or two oxygen atoms from lactate giving $[Co(O_2\text{-lactate})(Tp^{Ph})]$. $[Co(NCS)(Hpz)(Tp^{Ph})]$ has been structurally characterized.[118]

Complexes $[Co(X)(Tp^{Ph})(CH_3OH)_m]\cdot nCH_3OH$ ($X = N_3$, $m = 1$, $n = 2$; X = NCS, $m = 0$, $n = 0$) have been synthesized and used as catalysts in the bicarbonate dehydration reaction. The coordination geometries of these Co(II) complexes in solution are five-coordinated trigonal bipyramidal as revealed by the spectroscopic measurements.[119]

$[Cu(Tp^{Ph})(L)](ClO_4)$ (L = bipy, phen, dpq, dppz) have been prepared by the reaction of copper(II) acetate hydrate with $K(Tp^{Ph})$ and L in CH_2Cl_2. The complexes have a square-pyramidal coordination geometry in which two nitrogens of L and two nitrogens of the Tp^{Ph} ligand occupy the basal plane and one nitrogen of Tp^{Ph} binds at the axial site. $[Cu(Tp^{Ph})(dpq)](ClO_4)$ and $[Cu(Tp^{Ph})(dppz)](ClO_4)$ display distortion from the square-pyramidal geometry. The phenyl groups of Tp^{Ph} form a bowl-shaped structure that encloses the {CuL} moiety. The steric encumbrance is greater for the bipy and phen ligands compared to that for dpq and dppz.[120] On the other hand, the steric hindrance due to Tp^{Ph} in $[Cu(L)(Tp^{Ph})](ClO_4)$ (L = dpq, dppz) leads to efficient cleavage of supercoiled DNA to its relaxed form upon exposure to red light as a result of protection of the photosensitizer in the molecular bowl of the $\{Cu(Tp^{Ph})\}$ moiety, which generates singlet oxygen as the reactive species in a type-II process.[121]

$[CuL(Tp^{Ph})]$ (HL = 5-*tert*butylsalicylaldehyde, 5-*tert*butyl-3-methylsulfanylsalicylaldehyde or 5-*tert*butyl-3-phenylsulfanyl-salicylaldehyde) complexes exhibit a square-pyramidal coordination geometry. They also undergo a ligand-based oxidation in CH_2Cl_2, that is, chemically reversible by voltammetry. The resultant phenoxyl radical products, however, decomposed rapidly at low temperatures in bulk solution likcly duc to intramolecular steric repulsions between the phenoxide *tert*butyl substituents, and a phenyl group of Tp^{Ph}.[122]

$[Cu(L)(Tp^{Ph})]BF_4$ (L = 1-(pyrid-2-yl)-3-(2′,5′-dimethoxyphenyl)-pyrazole) has been electrochemically investigated. It has been concluded that the intramolecular π–π interaction existing in the complex (Fig. 3.21) has no effect on the lifetime of an aryl radical cation.[123]

Half-sandwich copper complexes $[CuX(Tp^{Ph})]$ ($X = OH^-$, N_3^-, NCS^-) have been synthesized as model for carbonic anhydrase and the structure of $[CuNCS(Tp^{Ph})]$ has been determined by X-ray diffraction analysis. The apparent dehydration rate constant k^d_{obs} varies linearly with total Cu(II) concentration, and the catalytic activity of the model complexes $[CuX(Tp^{Ph})]$ decreases on going from OH^- to N_3^- and finally to NCS^-.[124] The EPR-silent species $[Cu(L)(Tp^{Ph})]^+$ (HL = 2-hydroxy-3-methylsulfanyl-5-methylbenzaldeyde) exhibits a UV–V IS–NIR spectrum that shows analogies with that of active

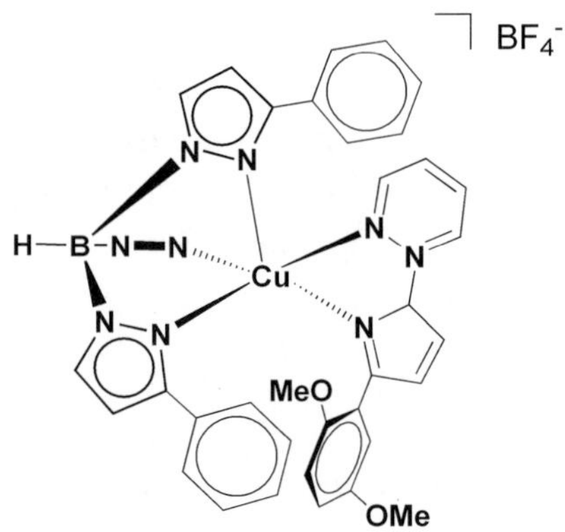

Fig. 3.21. [Cu(1-(pyrid-2-yl)-3-(2′,5′-dimethoxyphenyl)pyrazole)(Tp^{Ph})]BF_4.

galactose oxidase. The structure of [Cu(L)(Tp^{Ph})]$^+$ (HL = 2-hydroxy-3-methylsulfanyl-5-methylbenzaldeyde) shows a square-pyramidal Cu(II) center with a N_3O_2 donor set.[125] Other [Cu(L)(Tp^{Ph})] derivatives, where L is a deprotonated hydroxobenzaldeyde-type ligand, have been reported and characterized, also square-pyramidal with a CuN_3O_2 or CuN_4O coordination sphere. The UV–Vis spectra of these species in CH_2Cl_2 show the presence of tetragonal copper(II) center. Cyclic voltammetry studies have been performed together with spectrochemical characterization.[126]

The reaction of K(Tp^{Ph}) with $Cu(OAc)_2{\cdot}H_2O$, $CuCl_2$, and $Cu(BF_4)_2{\cdot}6H_2O$ has been investigated. [Cu(OAc)(Tp^{Ph})] converts in solution to the square pyramidal [Cu(OAc)(Tp^{Ph})(Hpz^{Ph})] likely due to a hydrolytic B–N cleavage.[127] [CuCl(Tp^{Ph})(Hpz^{Ph})], [Fe(NO_3)$_2$(Tp^{Ph})], [Fe(SCN)(Tp^{Ph})(THF)], [Cu(SCN)(Tp^{Ph})(Hpz^{Ph})], and [Co(Tp^{Ph})(Bp^{Ph})] have been also reported and characterized by X-ray.[128]

Synthesis, characterization, and ethylene polymerization behavior of [MCl_3(Tp^{Ph})] (M = Zr, or Ti) are reported. The results have been compared with those found for analogous $Tp^{^tBu}$ and Tp^{Np} derivatives.[129]

[Pd(Tp^{Ph})(p-Tol)(PPh_3)] has been reported and used for the synthesis of alternating co-oligomers of carbon monoxide and norbornadiene.[130]

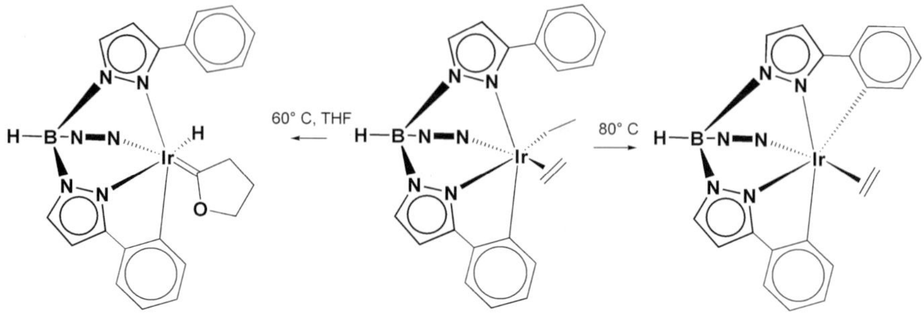

Fig. 3.22. Reactivity of (OC-6-25)-[Ir{κ^4-(N,N',N'',C^{Ph})-Tp^{Ph}}(Et)(η^2-ethene)].

The ligand Tp^{Ph} acts as pentadentate in the compound [Ir{κ^5-($N,N',N'',C^{Ph},C^{Ph'}$)-Tp^{Ph}}(η^2-C_2H_4)] that has been obtained by thermal activation of a mixture of [IrCl(COE)$_2$]$_2$ with [Tl(Tp^{Ph})] and ethene through the formation of the intermediate [Ir{κ^4-(N,N',N'',C^{Ph})-Tp^{Ph}}Et(η^2-C_2H_4)].[131] [Ir(Tp^{Ph})(olefin)$_2$] has been also prepared and its capability to activate C–H bonds discussed by Carmona and coworkers.[132] (OC-6-25)-[Ir{κ^4-(N,N',N'',C^{Ph})-Tp^{Ph}}(Et)(η^2-ethene)] was formed diastereoselectively which, upon heating, can be converted exclusively to (OC-6-24)-[Ir{κ^5-($N,N',N'',C^{Ph},C^{Ph'}$)-Tp^{Ph}}(η^2-ethene)]. (OC-6-25)-[Ir{κ^4-(N,N',N'',C^{Ph})-Tp^{Ph}}(Et)(η^2-ethene)] with THF gave (OC-6-35)-[Ir{κ^3-(N,N',N'')-Tp^{Ph}}H(dihydrofuran-2(3H)-ylidene)] (Fig. 3.22). The reactivity of [Ir(Tp^{Ph})(butadiene)] was also investigated by the same authors.[132]

The compound [Ir(κ^3-Tp^{Ph})(η^4-isoprene)] has been synthesized and showed to effect the regioselective, double C–H activation of a variety of ether and amine substrate by allowing their precoordination to the metal center. The Fischer-type carbenes in Fig. 3.23 have been produced by cyclometalation-β-elimination (or α-elimination) reaction.[133]

[Zn(L)(Tp^{Ph})]BF_4 (L = 1-{pyrid-2-yl}-3-{2′,5′-dimethoxyphenyl}-pyrazole) has been obtained in low-to-moderate yields. [Zn(L)(Tp^{Ph})]BF_4 is a strongly distorted square-pyramidal coordination compound which undergoes a fluxional process in solution (VT NMR) involving libration of the ligand L within one "wedge" of

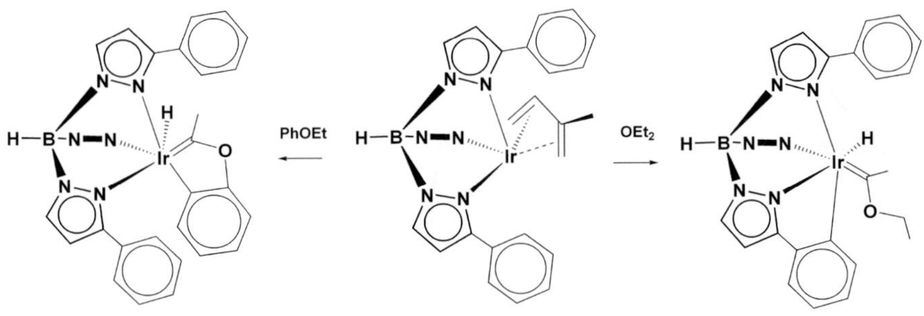

Fig. 3.23. Fischer-type carbenes prepared from [Ir(κ^3-Tp^{Ph})(η^4-isoprene)] and described in Ref. 133.

the Tp^{Ph} tripod. The electrochemistry of this complex has been also reported.[134]

$CdMe_2$ reacts with Tl(Tp^{Ph}) forming [CdMe(Tp^{Ph})], which upon reaction with thiotoluene and mercaptopyridine yields [CdS(*p*-$C_6H_4CH_3$)(Tp^{Ph})] and [Cd(SNC_6H_4)(Tp^{Ph})], respectively.[87] The thermodynamic parameters for the binding of a series of epoxides to [Cd(OAc)(Tp^{Ph})] have been determined via ^{113}Cd NMR spectroscopy.[135]

[YCl_2(Tp^{Ph})(THF)] has been prepared and characterized by NMR by Long *et al.*[136]

3.22. Tp^{PhF}

[Zn(L)(Tp^{PhF})]BF_4 (L = 1-{pyrid-2-yl}-3-{2′,5′-dimethoxyphenyl}-pyrazole) has been obtained in low-to-moderate yields. The electrochemistry of this complexes has been also reported.[134]

3.23. Tp^{PhCl}

[Rh(κ^2-Tp^{PhCl})$(CO)_2$] and [$(CO)_2$Rh(μ:κ^2,κ^1-Tp^{PhCl})Rh(Cl)$(CO)_2$] have been prepared and characterized by spectroscopy and X-ray crystallography. [Rh(κ^2-Tp^{PhCl})$(CO)_2$] exhibits three different isomeric forms with different Rh environments due to the uncoordinated third pyrazolyl arm. In one isomer a pz^x group was rotated to equatorial position; in the other two forms it occupies axial positions with

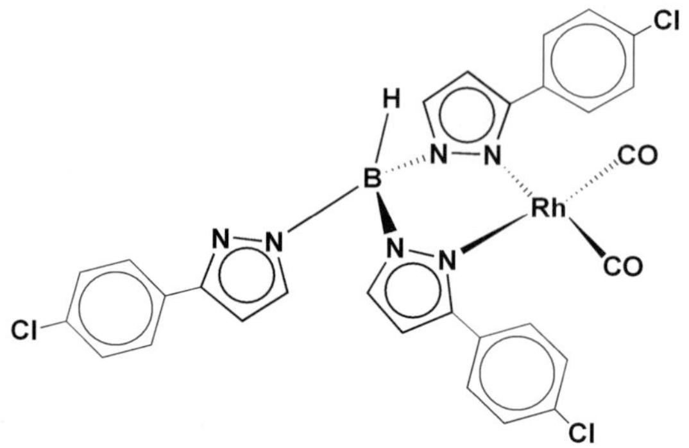

Fig. 3.24. The square-planar (SP2) structure of $[Rh(\kappa^2\text{-}Tp^{PhCl})(CO)_2]$.

different orientation. Two square-planar coordination geometries (one with the third pz^x ring over the square-planar (SP1) and the other inverted with the free pyrazolyl arm away from the square-planar (SP2)) and one square pyramidal geometry (SPy) have been identified (Fig. 3.24). In $[(CO)_2Rh(\mu{:}\kappa^2,\kappa^1\text{-}Tp^{PhCl})Rh(Cl)(CO)_2]$, Tp^{PhCl} acts as a bridging non-symmetric ligand.[137]

3.24. $Tp^{Ph,Me}$

Complexes $[M(Tp^{Ph,Me})(NCMe)_3](OTf)$ (M = Ni, Co) were prepared by chloride abstraction from the corresponding chloro complexes $[MCl(Tp^{Ph,Me})]$ with Ag(OTf) in MeCN. They can be readily transformed in $[M(Tp^{Ph,Me})(L)_n](OTf)$, by reaction with N- and P-donors $[NC(CH_2)_2CN$, *p*-NCC_6H_4CN, py, bipy, 4,4′-bipy, $Ph_2P(CH_2)_mPPh_2$ ($m = 2, 3$)].[38]

The tetrahedral $[CoCl(Tp^{Ph,Me})]$ reacts with L (L = hydroxypyridinone, hydroxypyridinethione, pyrone, and thiopyrone) giving $[Co(L)(Tp^{Ph,Me})]$. Single-crystal X-ray studies show that the $[Co(L)(Tp^{Ph,Me})]$ is pentacoordinate with a general tendency toward square pyramidal geometry.[138] The molecular structure of $[Co(NO_3)(Tp^{Ph,Me})(THF)]$ has been also reported.[139] Mechanochemical synthesis of $[NiCl(Tp^{Ph,Me})]$ has been developed, which enabled the preparation of metal complexes of the types [NiCl(Tp^x)] and [MCl(Tp^x)]·Hpz^x (Hpz^x is generated by hydrolysis of the corresponding Tp^x ligand). Analogous Co and Mn species have been reported.[33] [ML($Tp^{Ph,Me}$)]·$Hpz^{Ph,Me}$ (M = Ni, Co and Mn; HL = 2,6-di-*tert*-butyl-4-carboxyphenol) were prepared and structurally (Ni and Co) characterized. Oxidation of [ML($Tp^{Ph,Me}$)]·$Hpz^{Ph,Me}$ (M = Ni, Co) gave coordinated phenoxyl radical and diphenoquinone, whereas only quinone formed in the case of Mn complex and free 2,6-di-*tert*-butyl-4-carboxyphenol.[140] Ruman *et al.* also reported the synthesis and structure of [Co($Tp^{Ph,Me}$)$_2$].[141] [Mn(OAc)($Tp^{Ph,Me}$)($Hpz^{Ph,Me}$)] has a five-coordinate manganese center and it has been shown to possess SOD activity.[47] Reaction of K($Tp^{Ph,Me}$) with copper(II) acetate in methanol at room temperature gives the distorted square pyramidal [Cu(O_2CCH_3)($Tp^{Ph,Me}$)].[142]

Reaction of [Zn(OH)($Tp^{Ph,Me}$)] with CS_2 resulted in the quantitative formation of [Zn(SH)($Tp^{Ph,Me}$)] and COS; in the presence of 1 equiv. of MeOH the reaction yielded [Zn(SC(S)OMe)($Tp^{Ph,Me}$)], supporting the existence of the four-center intermediate [Zn(SC(S)OH)($Tp^{Ph,Me}$)] found when the reaction was computed at the B3LYP/6 − 311 + G* level.[143]

[Zn(SR)($Tp^{Ph,Me}$)] (R = C_2H_5, $^iC_4H_9$, $CH_2C_6H_5$, C_6H_5), [Zn(HCys)(OEt)(NAc)($Tp^{Ph,Me}$)] and [($Tp^{Ph,Me}$)Zn-(HCys)-Zn-($Tp^{Ph,Me}$)] have been prepared and structurally characterized. Methylation of the thiolates could also be achieved with dimethylsulfate and trimethylsulfonium iodide.[144]

[Zn(H)($Tp^{Ph,Me}$)], obtained from the reaction of [Zn(F)($Tp^{Ph,Me}$)] with Et_3SiH, and [Zn(OH)($Tp^{Ph,Me}$)] undergo insertion reactions with heterocumulenes. CO, CS_2, and RCNS insert into the Zn–H bond giving [Zn(OCH(O))($Tp^{Ph,Me}$)], [Zn(SCH(S))($Tp^{Ph,Me}$)], and [Zn(SCH(NR))($Tp^{Ph,Me}$)], respectively. [Zn(OH)($Tp^{Ph,Me}$)] reacts with RCNS yielding [Zn(SC(O)NHR)($Tp^{Ph,Me}$)]. In the presence of EtOH the reaction of [Zn(X)($Tp^{Ph,Me}$)] with RNCS yields [Zn-(SC(NR)OEt)($Tp^{Ph,Me}$)]. These compounds have been characterized by X-ray crystallography, supporting a four-center reaction mechanism proposed for hydrolytic zinc enzymes.[145] [Zn(OH)-($Tp^{Ph,Me}$)] reacts with simple aminoacids to form [Zn(AB)($Tp^{Ph,Me}$)]

complexes, in which AB is a O,N-chelate ligand (AB = Gly, Val, Leu, Met, Phe, Trop). Analogously, aminoacid–nitroanilides were deprotonated at the N atom yielding [Zn(NN)($Tp^{Ph,Me}$)] (NN = chelating ligands: Gly-Nit, Ala-Nit, Leu-Nit). When terminal amino functions were missing, pure carboxylate coordination resulted, as in [Zn(GlyPheBoc)($Tp^{Ph,Me}$)] and [Zn(AspPheOMe)($Tp^{Ph,Me}$)]. The binuclear complexes [($Tp^{Ph,Me}$)Zn(AA)Zn($Tp^{Ph,Me}$)] (AA = Asp, Glu, Tyr, and Cys) have been also reported.[146]

[Zn(OR)($Tp^{Ph,Me}$)] (R = Me, Et, ^{i}Pr) undergoes acid–base reactions with protic compounds whose acidity is higher than that of the corresponding alcohol ROH. The reaction of [Zn(OR)($Tp^{Ph,Me}$)] with MeCN gave [Zn(CH_2CN)($Tp^{Ph,Me}$)] (**VII**, Fig. 3.25).[147] The reaction with NH acids cyanamide, trifluoroacetamide, pyrazoles, and saccharin yields corresponding complexes containing deprotonated acid (as the saccharinate **VIII** in Fig. 3.25).

[Zn(SH)($Tp^{Ph,Me}$)] has been prepared by the reaction of [Zn(OH)($Tp^{Ph,Me}$)] with H_2S. [Zn(SH)($Tp^{Ph,Me}$)] does not react with esters, phosphates, or CO_2; whereas X–OH (carboxylic acids, trinitrophenol, hexafluoroacetylacetone) replace the SH groups with the formation of [Zn(OX)($Tp^{Ph,Me}$)]. [Zn(SH)($Tp^{Ph,Me}$)] is quite reactive toward alkylation with methyl iodide, yielding [ZnI($Tp^{Ph,Me}$)]

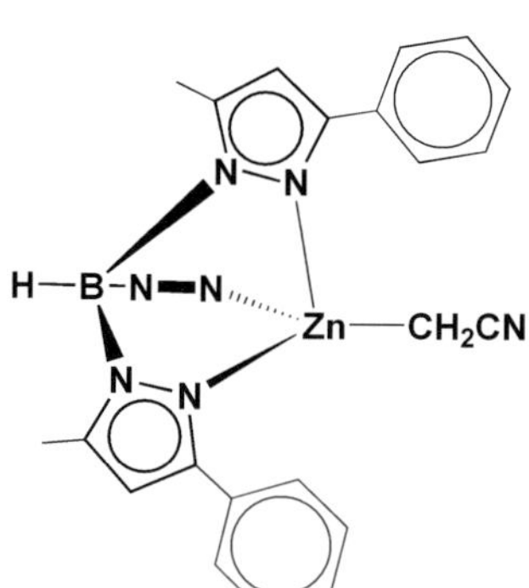

VII

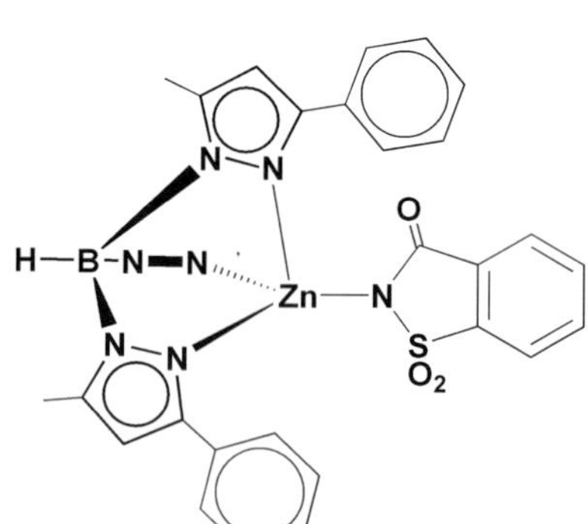

VIII

Fig. 3.25. [Zn(CH_2CN)($Tp^{Ph,Me}$)] (VII) and [Zn(saccharinate)($Tp^{Ph,Me}$)] (VIII).

and CH_3SH. Methylation of [Zn(SH)($Tp^{Ph,Me}$)] is a clean second-order reaction.[103] [Zn(OMe)($Tp^{Ph,Me}$)] does not induce the cleavage of nonactivated esters, phosphoesters, lactones or lactams but induces cleavage the P–O–P linkage of tetraalkylpyrophosphates.[148] β-lactams and β-amino acids react with [Zn(OH)($Tp^{Ph,Me}$)] yielding complexes containing N-bound β-lactamide ligands and semibidentate carboxylate ligands with a NH···O hydrogen bond respectively. Likewise the 6-aminopenicillanic and 7-aminocephalosporanic acid are incorporated as carboxylate ligands. β-Lactams bearing nitrophenyl or acyl substituents at the nitrogen atoms are opened hydrolytically by [Zn(OH)($Tp^{Ph,Me}$)]. Only when trifluoroacetyl is the N-substituent does hydrolytic cleavage occur at the external amide bond, yielding the free β-lactam and [Zn(O_2CCF_3)($Tp^{Ph,Me}$)].[149] [Zn(OH)($Tp^{Ph,Me}$)] reacts with functional thiols (HL) containing an additional donor function (for example COOH, COOR, NH_2, NHR, OH) in a position favorable for chelation, yielding [Zn(L)($Tp^{Ph,Me}$)] species (Fig. 3.26) which persist in solution.[150]

While a solution of [Zn(OH)($Tp^{Ph,Me}$)] in MeOH contains only traces of [Zn(OMe)($Tp^{Ph,Me}$)], the reaction of [Zn(OH)($Tp^{Ph,Me}$)] with trifluoroethanol and hexafluoro-2-propanol easily yields [Zn(OCH_2CF_3)($Tp^{Ph,Me}$)] and [Zn($OCH(CF_3)_2$)($Tp^{Ph,Me}$)].[151] [Zn-(OR)($Tp^{Ph,Me}$)] species (R = Me, Et, ^{i}Pr, and CH_2CH_2F) are accessible from [Zn(H)($Tp^{Ph,Me}$)] and the corresponding alcohol. A relation

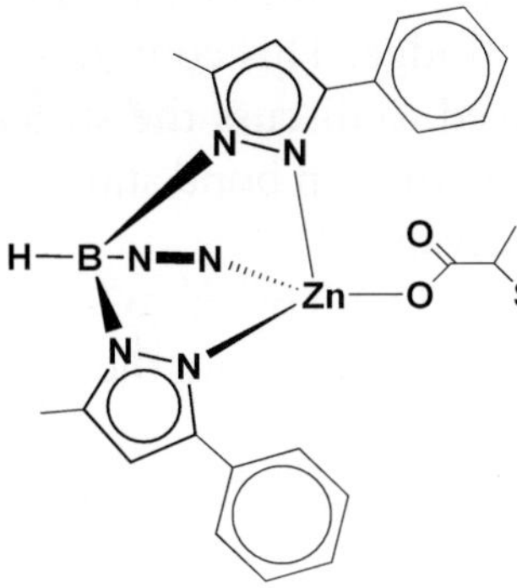

Fig. 3.26. A stable [Zn(carboxylate-thiol)($Tp^{Ph,Me}$)] species, a model for zinc enzyme inhibitors.

of [Zn(alkoxides)(Tp^x)] to the function of the zinc enzyme alcoholdehydrogenase exists in the reaction of [Zn(OCH(CH_3)$_2$)($Tp^{Ph,Me}$)] with aromatic aldehydes, which yields acetone and the corresponding benzyl oxides [Zn(OCH_2Ar)($Tp^{Ph,Me}$)]. CO_2, CS_2, isothiocyanates, and isocyanate have been inserted into the Zn–OR bonds, yielding [Zn(OC(O)OMe)($Tp^{Ph,Me}$)], [Zn(SC(S)OR)($Tp^{Ph,Me}$)], [Zn(SC(NR′)OR)($Tp^{Ph,Me}$)], and [Zn(NRCOOMe)($Tp^{Ph,Me}$)].[151] [Zn(OH)($Tp^{Ph,Me}$)] interacts with acetohydroxamic acid, 3-mercapto-2-butanone, *N*-(methyl)mercaptoacetamide, β-mercaptoethanol, 3-mercapto-2-propanol, and 3-mercapto-2-butanol yielding [Zn(ZBG)($Tp^{Ph,Me}$)] (ZBG = zinc-binding group), all structurally characterized. These complexes can be employed as models for thiol-derived matrix metalloproteinase (MMP) inhibitors.[152] Analogous reaction was performed with 1-hydroxy-2(1*H*)-pyridinone, 3-hydroxy-2(1*H*)-pyridinone, 3-hydroxy-1-methyl-2(1*H*)-pyridinone, 3-hydroxy-1,2-dimethyl-4(1*H*)-pyridinone, 1-hydroxy-2(1*H*)-pyridinethione, and 3-hydroxy-2-methyl-4-pyrone; all [Zn(ZBG)($Tp^{Ph,Me}$)] complexes have been investigated by X-ray, IR, and ^{1}H NMR.[153] [Zn(OH)($Tp^{Ph,Me}$)] and [CoCl($Tp^{Ph,Me}$)] were combined with HL (HL = 3-hydroxy-2H-pyran-2-one (3,2-pyrone), 3-hydroxy-4H-pyran-4-one (3,4-pyrone), and tropolone to form the corresponding [M(L)($Tp^{Ph,Me}$)] complexes (M = Zn, Co). [M(3,2-pyronate)($Tp^{Ph,Me}$)] (M = Co and Zn) have been structurally characterized.[154]

Reaction of Zn(ClO_4)$_2$ with $Tp^{Ph,Me}$ yields [Zn_2(H_3O_2)($Tp^{Ph,Me}$)$_2$]ClO_4 (Fig. 3.27) in which two {Zn($Tp^{Ph,Me}$)} moieties are linked through a $H_3O_2^-$ bridge. This compound, structurally characterized, has been employed to discuss the structure of the hydrolytic zinc enzymes that utilize hydrogen bond stabilized water nucleophiles to yield peptide bond cleavage.[155]

The dinuclear [Zn_2(H_3O_2)($Tp^{Ph,Me}$)$_2$]BPh_4 and [Zn($Tp^{Ph,Me}$)(H_2O)$_3$]BPh_4 were also reported and the former was structurally characterized.[156]

The reaction of [Zn(OH)($Tp^{Ph,Me}$)] and [Zn(OH_2)($Tp^{Ph,Me}$)]$^+$ with chelating Schiff's base ligands shows that protic ligands HL (HL = *N*-propyl-1-(5-methyl-2-imidazolyl)methanimine, *N*-propyl-1-(4-imidazolyl)-methanimine, and *N*-propyl-1-(2-imidazolyl)-methanimine) react with [Zn(OH)($Tp^{Ph,Me}$)] giving [ZnL($Tp^{Ph,Me}$)];

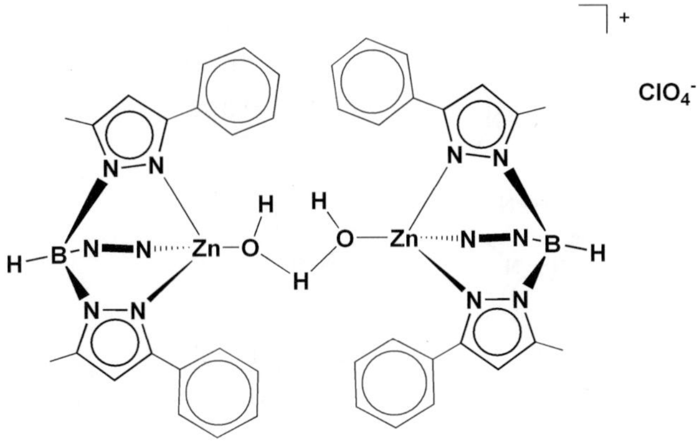

Fig. 3.27. $[Zn_2(H_3O_2)(Tp^{Ph,Me})_2]ClO_4$.

whereas reactions with neutral aprotic ligands L′ (L′ = *N*-propyl-1-(1-methyl-2-imidazolyl)methanimine and *N*-propyl-1-(2-thiazolyl)-methanimine) give the corresponding $[ZnL'(Tp^{Ph,Me})]^+$ complexes. Although the phenolic group of N-propyl-1-(2-hydroxyphenyl)methanimine is protic, this ligand forms a cationic four-coordinate derivative containing an intraligand hydrogen bond. Protic ligands form five-membered chelates. The ^1H NMR resonance frequencies of the chelating ligands typically shift upfield upon coordination to the zinc center, due to ring current effects from the phenyl groups of the $Tp^{Ph,Me}$ ligand.[157]

The interaction of nine nucleobases (thymine, uracil, dihydrouracil, cytosine, adenine, guanine, diaminopurine, xanthine, hypoxanthine), in their deprotonated form, with zinc and with themselves in $[Zn(nucleobase)(Tp^{Ph,Me})]$ has been investigated. Selected compounds have been characterized by single-crystal X-ray study.[158] Gelinski *et al.* reported the synthesis and characterization of two zinc complexes containing the ligand $Tp^{Ph,Me}$ and glutathione derivatives (Fig. 3.28).[159]

The reversible uptake of CO_2 by $[Zn(OH)(Tp^{Ph,Me})]$ in alcoholic solutions gave $[Zn(alkylcarbonate)(Tp^{Ph,Me})]$ complexes.[160] The condensation of $[Zn(OH)(Tp^{Ph,Me})]$ to nucleobases was also investigated by Badura and Vahrenkamp.[161]

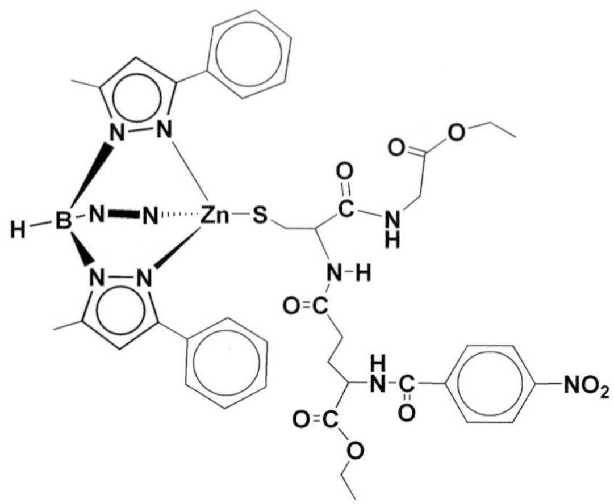

Fig. 3.28. [Zn($Tp^{Ph,Me}$)(glutathionate)].

[Rh(COD)($Tp^{Ph,Me}$)] has been investigated by low-temperature NMR spectroscopy; two rotational isomers originate from slow rotation of the dangling $pz^{Ph,Me}$.[162] [Rh$(CO)_2$($Tp^{Ph,Me}$)] has been prepared by the reaction of K($Tp^{Ph,Me}$) with [Rh(acac)$(CO)_2$] in benzene. When the reaction was carried out in the presence of PPh_3, [Rh(CO)(PPh_3)($Tp^{Ph,Me}$)] is formed. Both complexes, square-planar, have been structurally characterized and their fluxional behavior has been investigated by VT ^{1}H NMR spectroscopy. The $\kappa^2 \leftrightarrow \kappa^3$ conversion for [Rh$(CO)_2$($Tp^{Ph,Me}$)] is fast on the NMR time-scale even at $-90°$C, whereas [Rh(CO)(PPh_3)($Tp^{Ph,Me}$)] is a unique example of two well-separated $\kappa^2 \leftrightarrow \kappa^3$ equilibria.[163]

3.25. Tp^{Ph_2}

The reaction of Tp^{Ph_2} with $CuCl_2$ affords a mixture of [Cu(Cl)(Tp^{Ph_2})] and [CuCl(Tp^{Ph_2})(Hpz^{Ph_2})]. Tp^{Ph_2} also forms [Cu$(Tp^{Ph_2})_2$] in high yield.[127]

The structure of the aliphatic α-keto carboxylate complex [Fe^{II}($O_2CC(O)CH_3$)(Tp^{Ph_2})], and of the carboxylate complexes [Fe^{II}(OBz)(Tp^{Ph_2})] and [Fe^{II}(OAc)(Tp^{Ph_2})(Hpz^{Ph_2})], was determined by single-crystal X-ray diffraction, all being five-coordinate centers.

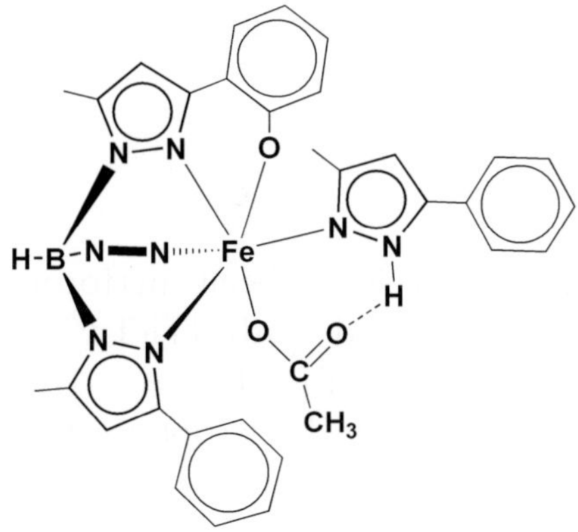

Fig. 3.29. Structure of the octahedral iron(III) phenolate product $[Fe^{III}(OAc)(Tp^{Ph_2^*})(Hpz^{Ph_2})]$.

Both the α-keto carboxylate and the carboxylate complexes react with dioxygen resulting in the hydroxylation of a single *ortho* phenyl position of the Tp^{Ph_2} ligand. The oxygenation products were characterized spectroscopically, and the structure of the octahedral iron(III) phenolate product $[Fe^{III}(OAc)(Tp^{Ph_2^*})(Hpz^{Ph_2})]$ was established by X-ray diffraction (Fig. 3.29). The reaction of the α-keto carboxylate model compounds with O_2 produces the phenolate product with concomitant oxidative decarboxylation of the α-keto acid. Isotope labeling studies show that $^{18}O_2$ ends up in the $Tp^{Ph_2^*}$ phenolate oxygen and the carboxylate derived from the α-keto acid.[164]

When $K(Tp^{Ph_2})$ reacts with hydrated $CuCl_2$ and equimolar amount of $Na(L^1)\cdot H_2O$ (HL^1 = 2-hydroxy-5-methyl-1,4-benzoquinone), $[Cu(L^1)(Tp^{Ph_2})]$ is formed, which contains a regular square-pyramidal copper center.[165]

The polymerization of 4-phenoxyphenol was performed by the use of the tyrosinase model complex $[Cu(Cl)(Tp^{Ph_2})]$. Very little of C–C coupling dimers were afforded with the $Cu(Tp^{Ph_2})$ catalyst in toluene or THF.[166]

$[Co(NO_3)(Tp^{Ph_2})]$ has been reported to contain a pentacoordinate cobalt center from a tridentate scorpionate and a bidentate nitrate.[167] $[Ni(Cl)(Tp^{Ph_2})(Hpz^{Ph_2})]$ and $[Ni(X)(Tp^{Ph_2})]$ (X = Br, NO, NO_3,

OAc, acac) have been synthesized and the acetate also structurally characterized.[168]

3.26. Tp^{Tol}

Photoluminescence studies have been performed on $[Tl(Tp^{Tol})]$ that has been also compared with $[Tl(Tp^{Ph})]$, $[Tl(Tp^{Th})]$, and $[Tl(Tp^{tBu})]$.[83] $[Zn(L)(Tp^{Tol})]BF_4$ (L = 1-{pyrid-2-yl}-3-{2′,5′-dimethoxyphenyl}pyrazole) has been obtained in low-to-moderate yield. The electrochemistry of this complex has been also reported.[134]

3.27. Tp^{Ms} and Tp^{Ms*}

When $[Tl(Tp^x)]$ (Tp^x = Tp^{Ms} or Tp^{Ms*}) reacts with $NiCl_2{\cdot}6H_2O$, $[NiCl(Tp^{Ms})]$ and $[NiCl(Tp^{Ms*})]$ are formed in good yields. $[NiCl(Tp^{Ms*})]$ undergoes an isomerization process, giving $[NiCl(Tp^{Ms**})]_2$ (Tp^{Ms**} = $HB(pz^{Ms})(pz^{5Ms})_2$. Treatment of $[NiCl(Tp^{Ms})]$ and $[NiCl(Tp^{Ms*})]$ with alkylaluminum cocatalysts such as MAO and trimethylaluminum in toluene generates active catalysts for ethylene oligomerization.[169]

The synthesis, molecular structures, and olefin polymerization behavior of the sterically crowded tris(pyrazolyl)borate complexes $[ZrCl_3(Tp^{Ms})]$, $[ZrCl_3(Tp^{Ms*})]$, $[ZrCl_2(Tp^{Ms**})]_2(\mu\text{-}O)$, and $[HfCl_3(Tp^{Ms*})]$ (Fig. 3.30) are reported by Michiue and Jordan[170]

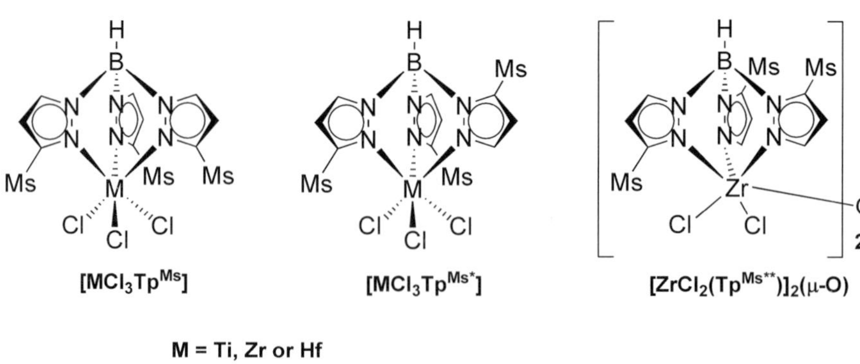

Fig. 3.30. $[MCl_3(Tp^{Ms})]$, $[MCl_3(Tp^{Ms*})]$, and $[ZrCl_2(Tp^{Ms**})]_2(\mu\text{-}O)$ complexes.

who also described the molecular structures of $Tl(Tp^{Ms})$ and $Tl(Tp^{Ms*})$.

The synthesis and characterization of $[ZrCl_3(Tp^{Ms*})]$ and its performance in polymerizing ethylene have been described by Casagrande and coworkers.[171] In $[TiCl_3(Tp^{Ms})]\cdot 2CH_2Cl_2$, the Ti atom is coordinated by a κ^3-Tp^{Ms} ligand.[172] The reaction of $[TiCl_3(Tp^{Ms*})]$ with $K(O^tBu)$ in toluene yields $[Ti^{III}Cl_3(Tp^{Ms*})K]_2$ (Fig. 3.31). An X-ray diffraction study of this compound reveals a unique dimeric structure in which two $[Ti^{III}Cl_3(Tp^{Ms*})]^-$ anions are linked by two K^+ cations.[173] In the same paper the authors reported the successful synthesis of the alkoxides in Fig. 3.32. The solid-state structure of $[TiCl_3(Tp^{Ms*})]$ has been also described.[174]

When $[RuH(COD)(NH_2N(CH_3)_2)_3]BPh_4$ reacts with $[Tl(Tp^{Ms})]$, $[Ru(Tp^{Ms})(COD)H]$ is formed. $[Ru(Tp^{Ms})(COD)H]$ interacts with

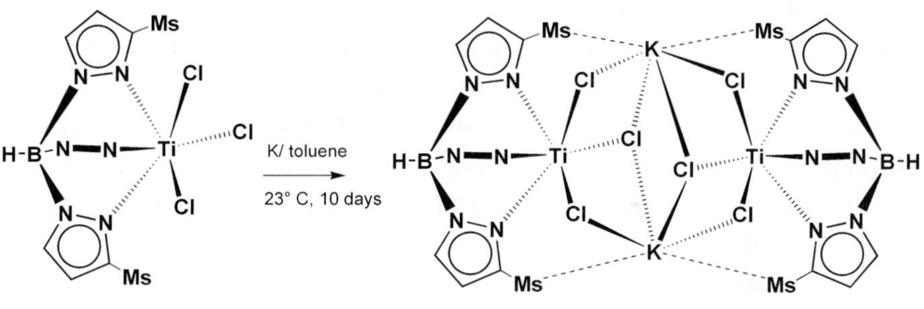

Fig. 3.31. Synthesis of $[Ti^{III}Cl_3(Tp^{Ms*})K]_2$.

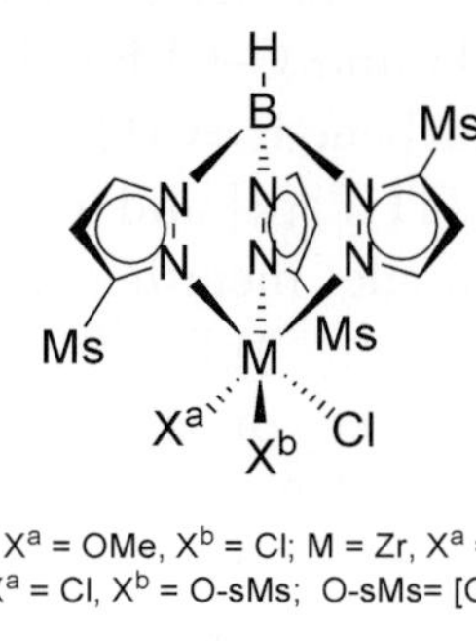

M = Ti, X^a = Cl, X^b = O-sMs; M = Zr, X^a = OMe, X^b = Cl; M = Zr, X^a = Cl, X^b = OMe; M = Zr, X^a = O^tBu, X^b = Cl; M = Zr, X^a = Cl, X^b = O^tBu; M = Zr, X^a = Cl, X^b = O-sMs; O-sMs= [O-2,4,6-$^tBu_3$-Ph]).

Fig. 3.32. The $[TiX_2Cl(Tp^x)]$ complexes reported in Ref. 173.

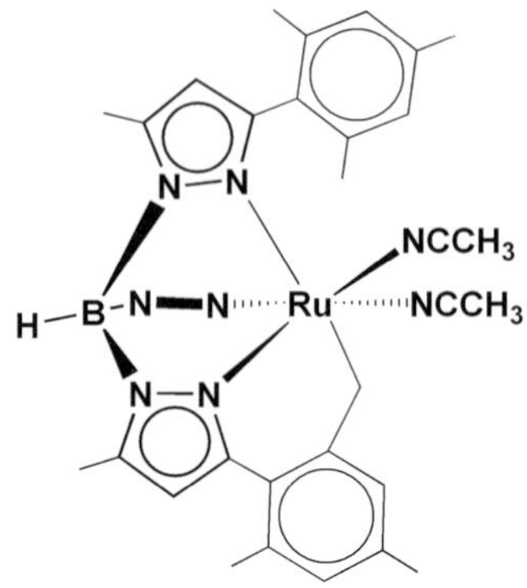

Fig. 3.33. [Ru($Tp^{Ms@}$)($NCCH_3$)$_2$] containing a Tp^{Ms} ligand bonded in a κ^4-N_3, C fashion.

CO affording the acyl complex [Ru(Tp^{Ms})(CO)$_2$C(O)C_8H_{13}] via intramolecular olefin insertion followed by CO addition and insertion. On the other hand, reaction of [Ru(Tp^{Ms})(COD)H] with MeCN was accompanied by C–H activation of one of the orthomethyl groups of the ligand's mesityl substituents and H_2 elimination, producing [Ru($Tp^{Ms@}$)($NCCH_3$)$_2$] as the final product (Fig. 3.33: $Tp^{Ms@}$ has been sometimes employed by the authors to indicate a metalated Tp^{Ms} ligand). The molecular structures of the three complexes have been determined by single-crystal X-ray diffraction methods.[175]

[Ir($Tp^{Ms*@}$)(N_2)], which contains a pentadentate, doubly metalated 3-mesityl substituted tris(pyrazolyl)borate ligand, was obtained according to the reaction scheme depicted in Fig. 3.34. The compound induces the cleavage of C–H and C–Cl bonds of CH_2Cl_2 to yield a highly electrophilic chlorocarbene Ir=C(H)Cl complex.[176]

Reaction between [YCl_3(THF)$_{3.5}$] and 1 or 2 equiv. of Tl(Tp^{Ms*}) leads in both cases to single metathesis, giving a mixture of [YCl_3(Tp^{Ms*})Tl] and [YCl_2(Tp^{Ms**})] [Tp^{Ms**} = HB(pz^{Ms})(pz^{5Ms})$_2^-$], that results from the transfer of a second mesityl group to the 5-position of the pyrazolyl ring. The solid-state structure of [YCl_3(Tp^{Ms*})Tl]$_2$ shows a unique "ate" dimeric structure with the Tl^+ cations coordinated by two μ^2- and two μ^3-bridging Cl atoms as well as two η^3-Tp^{Ms*} ligands.[177]

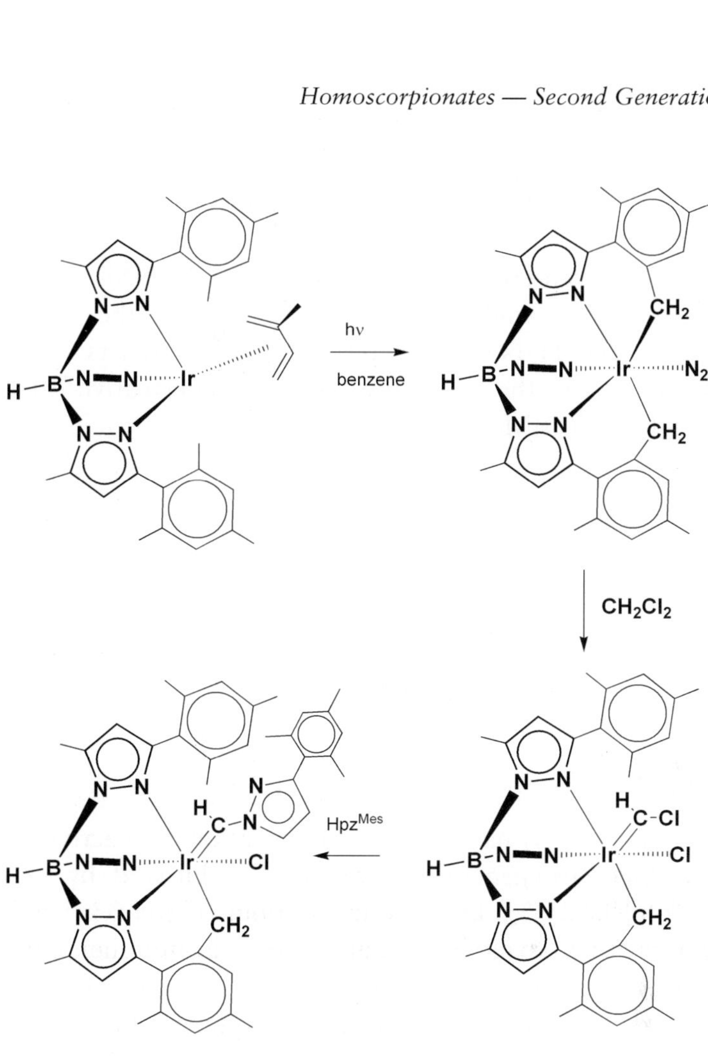

Fig. 3.34. Synthesis of [Ir(Tp$^{Ms*@}$)N_2)] and formation of a chlorocarbene complex upon cleavage of C–H and C–Cl in CH_2Cl_2.

Synthesis and reactivity of ytterbium(II) compounds with the sterically demanding Tp^{Ms,R} (R = H or Me) and Tp^{Ms*,R} ligands are described by Lopes *et al.*[178]

UCl_4 reacts with 1 equiv. of [Tl(Tp^{Ms})] yielding [UCl_3(Tp^{Ms*})] due to isomerization of Tp^{Ms}. Also a fortuitous synthesis of [UCl_3(Tp^{Ms})] is reported and its crystal structure determined. Upon addition of THF to [UCl_3(Tp^{Ms*})], [UCl_3(Tp^{Ms*})]·THF is formed. Derivatization of [UCl_3(Tp^{Ms*})] with $KN(SiMe_3)_2$ and li($C_6H_4CH_2NMe_2$-*o*) gave [$UCl_2(NSiMe_3)_2$(Tp^{Ms*})] and

$[UCl_2(C_6H_4CH_2NMe_2\text{-}o)(Tp^{Ms^*})]$, respectively. The U center has been found in octahedral configuration with the exception of the seven-coordinate $[UCl_2(C_6H_4CH_2NMe_2\text{-}o)(Tp^{Ms^*})]$. Nitriles and isocyanides do not insert into the U–C bond of $[UCl_2(C_6H_4CH_2NMe_2\text{-}o)(Tp^{Ms^*})]$, which reacts with acetone leading to the uranium aldolate derivative $[UCl_2(\eta^2\text{-}OC(Me_2)_2CH_2C(O)Me)(Tp^{Ms^*})]$.[179]

3.28. $MeTp^{Ms}$

Dias *et al.* reported that high-quality polyaniline can be prepared catalytically in MeCN using aniline, HCl, H_2O_2, and the copper catalyst $[CuCl(MeTp^{Ms})]$.[180]

3.29. $Tp^{Cum,Me}$

$[Zn(OH)(Tp^{Cum,Me})]$ has been investigated for its reactivity toward substrates that are generally hydrolytically cleaved by $[Zn(OH)(Tp^x)]$ complexes.[151] β-lactams and β-amino acids react with $[Zn(OH)(Tp^{Cum,Me})]$ yielding complexes containing N-bound β-lactamide ligands and semibidentate carboxylate ligands with a NH···O hydrogen bond, respectively. Likewise the 6-aminopenicillanic and 7-aminocephalosporanic acid are incorporated as carboxylate ligands. β-Lactams bearing nitrophenyl or acyl substituents at the nitrogen atoms are opened hydrolytically by $[Zn(OH)(Tp^{Cum,Me})]$. Only with trifluoroacetyl as the N-substituent does the hydrolytic cleavage occur at the external amide bond, yielding the free β-lactam and $[Zn(Tfa)(Tp^*)]$.[149]

$[Zn(OR)(Tp^{Cum,Me})]$ (R = Me and ^{i}Pr) has been prepared by the reaction of $[ZnH(Tp^{Cum,Me})]$ with the corresponding alcohol ROH.[151]

Upon reaction of $[Zn(OH)(Tp^{Cum,Me})]$ with trichloroacetaldehyde and pentafluorobenzaldehyde complexes $[Zn(O\text{-}(CHOH)R)(Tp^{Cum,Me})]$, containing the aldehyde ligands as the deprotonated aldehyde hydrates, have been obtained. [Zn(OH)-

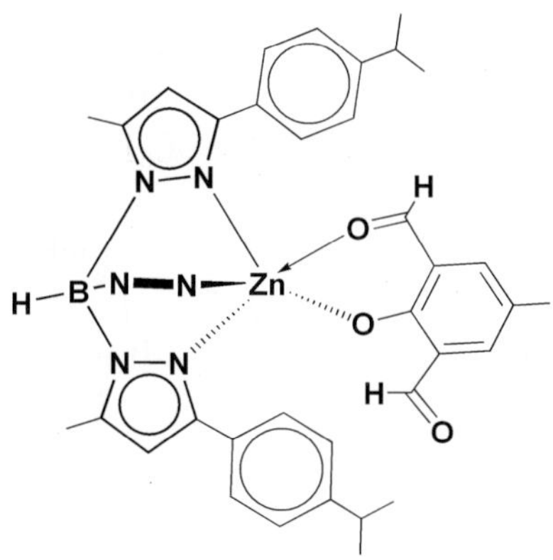

Fig. 3.35. The phenolate complex obtained from the reaction of $[Zn(OH)(Tp^{Cum,Me})]$ with 2,6-diformylphenol.

$(Tp^{Cum,Me})]$ reacts with 2-formylphenol, 2,6-diformylphenol, and 2-formylthiophenol undergoing condensation and formation of phenolate (Fig. 3.35) and thiophenolate species.[181]

$[Zn(OCHMe_2)(Tp^{Cum,Me})]$ is oxidized by *N*-benzylnicotinamide chloride to acetone and $[ZnCl(Tp^{Cum,Me})]$. 2,3-Dichloro-5,6-dicyanobenzoquinone oxidizes $[Zn(phenolates)(Tp^{Cum,Me})]$ with *o*-hydroxymethyl substituents to the corresponding zinc phenolates with *o*-formyl substituents. Benzaldehydes with electronegative substituents (F, NO_2, CF_3) are reduced by *N*-benzyldihydronicotinamide in the presence of $[Zn(OH)(Tp^{Cum,Me})]$ to the corresponding benzyl alcohols which are isolated as $[Zn(OCH_2Ar)(Tp^{Cum,Me})]$.[182]

The interaction of several nucleobases with themselves in $[Zn(nucleobase)(Tp^{Cum,Me})]$ has been investigated. With exception of the guanine, the complexes $[Zn(base)(Tp^{Cum,Me})]$ have been isolated in all cases. Selected compounds have been characterized by single-crystal X-ray study.[158] $[Zn(nucleobase)Tp^x]$ shows in solution (VT NMR) self-association and base pairing with external nucleobase (Fig. 3.36). The self-association was studied quantitatively for both $[Zn(hypoxanthinate)(Tp^{Cum,Me})]$ and $[Zn(thyminate)(Tp^{Cum,Me})]$. $[Zn(thyminate)(Tp^{Cum,Me})]$:9-ethyladenine, $[Zn(cytosinate)(Tp^{Cum,Me})]$:9-isobutylguanine, and

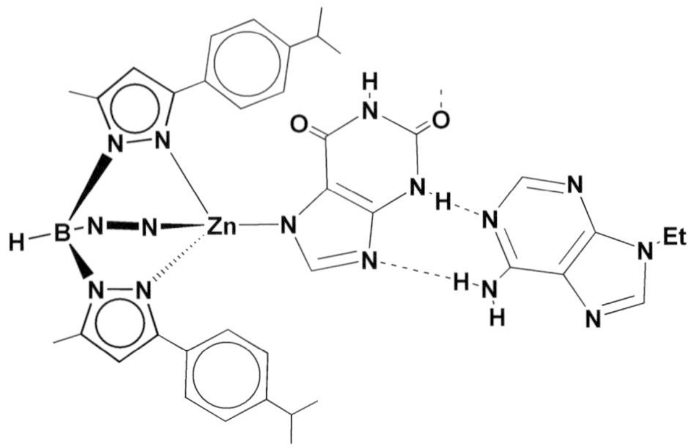

Fig. 3.36. Self-association observed in solution of [Zn(nucleobase)Tp^x] species.

[Zn(xanthinate)($Tp^{Cum,Me}$)]:9-ethyladenine were investigated by single-crystal X-ray study. In the solid state, extended base pairing leads to quartet and polymer arrangements.[183]

Spectroscopic and magnetic studies have been carried out by Shultz on a series of [Cu(NN-SQ)($Tp^{Cum,Me}$)] derivatives (Fig. 3.37) that have been considered as potential building blocks for the formation of spin-diverse molecule-based magnetic materials.[184]

The synthesis of the derivative [Zn($Tp^{Cum,Me}$)]$_2$L, shown in Fig. 3.38, has been reported. This compound has $S = 1/2$ that has

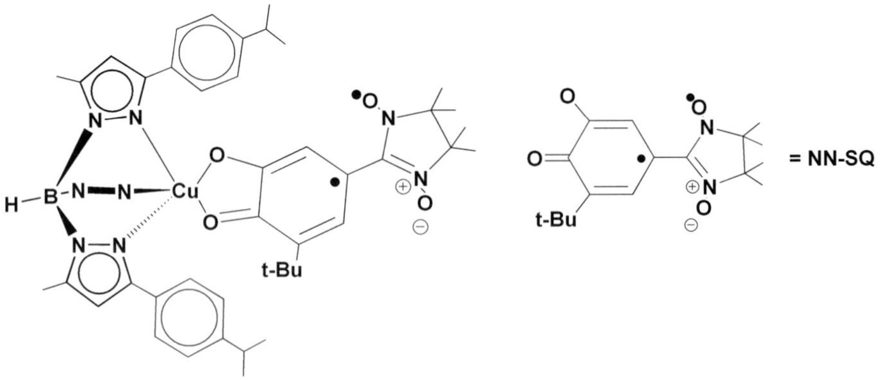

Fig. 3.37. [Cu(NN-SQ)($Tp^{Cum,Me}$)], a building block for the formation of spin-diverse molecule-based magnetic materials.

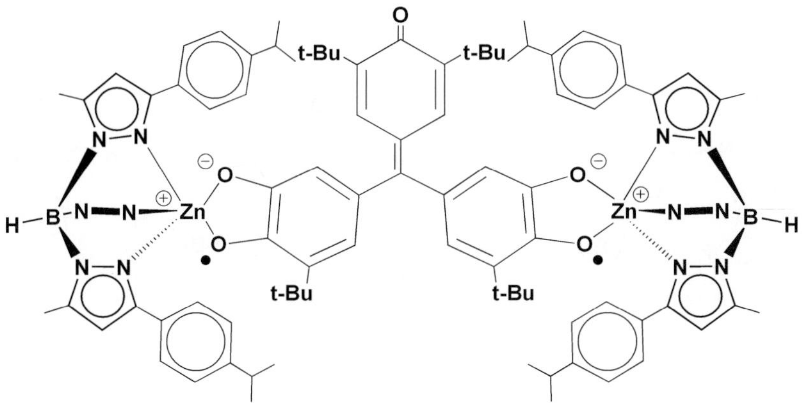

Fig. 3.38. The biradical $[Zn(Tp^{Cum,Me})]_2L$ reported in Ref. 185.

been ascribed to a tautomerization of a reaction intermediate. The biradical species in Fig. 3.38 has been prepared by oxidation with PbO_2.[185]

Synthesis and molecular structure of two bis(zinc-dioxolene) complexes have been also described.[186] The $[\{Zn(Tp^{Cum,Me})\}_2(L{-}H_2)]$ derivative in Fig. 3.39 has two protonated catecholate ligands, whereas $[\{Zn(Tp^{Cum,Me})\}_2(L{-}H)]$ has one protonated catecholate and one semiquinone ligand.

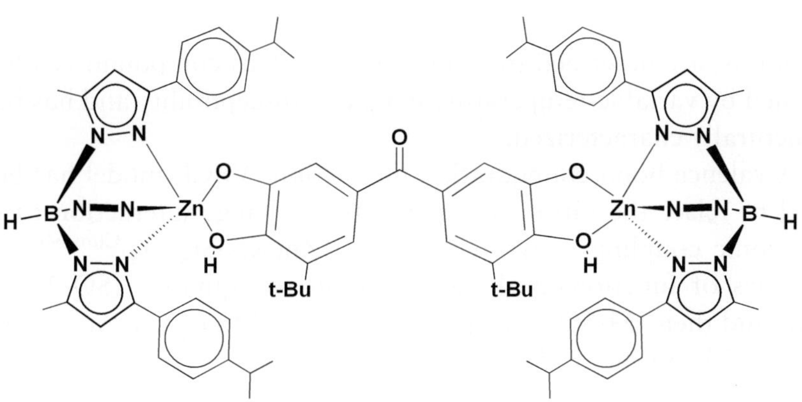

Fig. 3.39. $[\{Zn(Tp^{Cum,Me})\}_2(L{-}H_2)]$ in which both catecholate ligands are protonated.

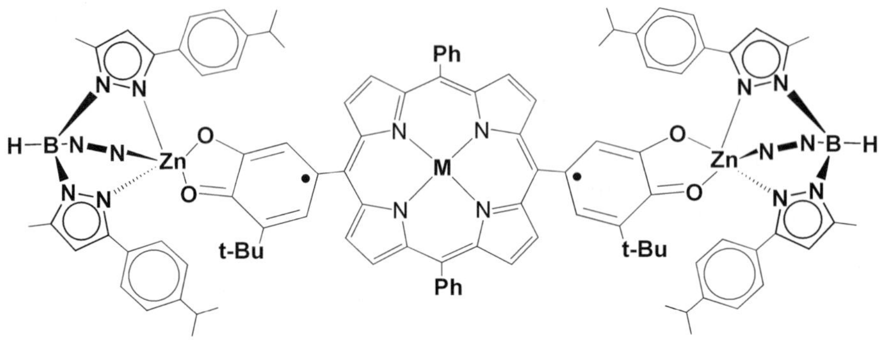

Fig. 3.40. Two [Zn($Tp^{Cum,Me}$)] moieties linked through a bis(semiquinone) porphyrin system.

The [Zn($Tp^{Cum,Me}$)] moiety can be employed to investigate the dependence of product spin state on tautomerization of bis(catechol) ligands.[187] The isostructural series of biradical complexes based on bis(semiquinone) and containing [Zn($Tp^{Cum,Me}$)] moieties has been investigated by X-ray crystal structures, variable-temperature electron paramagnetic resonance spectroscopy, zero-field splitting parameters, and variable-temperature magnetic susceptibility measurements in order to evaluate molecular conformation and electron-spin exchange coupling.[188]

The syntheses and EPR spectra of the porphyrin complex in Fig. 3.40, substituted with two radical groups bound to *trans*-meso positions, are described by Shultz *et al.*[189] This compound has been studied by variable-temperature magnetic susceptibility and has been structurally characterized.

A valence bond configuration interaction (VBCI) model has been used to relate the intraligand magnetic exchange interaction to the electronic coupling matrix element in [Zn(NN-SQ)($Tp^{Cum,Me}$)].[190] A series of functionalized radical anion semiquinone (SQ-Ar) ligands and their [M(SQ-Ar)($Tp^{Cum,Me}$)] (M = Mn or Cu, Ar = C_6H_4-NO_2, C_6H_4OMe, $C_6H_4{}^tBu$, etc.) derivatives were prepared and characterized.[191]

[NiCl($Tp^{Cum,Me}$)] has been prepared and tested toward ethylene polymerization.[97]

3.30. Tp^{Cy} and $Tp^{Cy,4Br}$

The ^{1}H, ^{13}C, and ^{15}N NMR spectroscopic properties of [Tl(Tp^{x})] (Tp^{x} = Tp^{Cy} and Tp^{Cy,4Br}) have been reported.[192] Tp^{Cy} has been employed to synthesize [Cu(L)(Tp^{Cy})]BF_4 (L = 1-(pyrid-2-yl)-3-(2′, 5′-dimethoxyphenyl)pyrazole).[123] Tp^{Cy} reacts with an equimolar quantity of $CuCl_2$ yielding a mixture of [Cu(Cl)(Tp^{Cy})] and [Cu(Cl)(Tp^{Cy})(Hpz^{Cy})]. With Cu(BF_4)$_2$·6H_2O, the reaction gives [Cu(Tp^{Cy})(Hpz^{Cy})$_2$]BF_4 which exhibits a square-pyramidal environment with two N–H···FBF_3 and one Cl_3C–H···FBF_3 hydrogen bond; whereas the reaction of $CuCl_2$ with 2 equiv. of Tp^{Cy} affords the square-pyramidal [Cu(pz^{Cy})(Hpz^{Cy})(Tp^{Cy})].[127] Tp^{Cy} forms with Cu, also [Cu(κ^2-Tp^{Cy})$_2$]·2CH_2Cl_2 that contains a four-coordinate Cu(II) ion lying on a crystallographic inversion center, giving rise to a near-regular square planar stereochemistry. A nonagostic contact has been found between the Cu and the ligand B–H group.[193] [NiCl(Tp^{Cy})] has been synthesized mechanochemically.[33]

3.31. Tp^{Tn}

Photoluminescence studies have been performed on [Tl(Tp^{Tn})] and compared with those for [Tl(Tp^{Ph})], [Tl(Tp^{Tol})], and [Tl(Tp^{tBu})].[83]

[RuH(Tp^{Tn})(COD] has been prepared from the reaction of [RuH(NH_2NMe_2)$_3$(COD)]BPh_4 with Tp^{Tn} which acts as a κ^3-*N*,*N*′, *BH* tridentate donor due to a BH···Ru agostic interaction. Reaction between [RuH(Tp^{Tn})(COD)] and L (L = PMe_3, PMe_2Ph, CN^{t}Bu) gave [RuH(Tp^{Tn})(L)(COD)] in which L replaces one of the coordinated N atoms, leaving intact the stated BH···Ru interaction. [RuH(Tp^{Tn})(PMe_3)(COD)] exhibits a distorted octahedral environment with the Tp^{Tn} coordinated through only one pyrazole ring and an agostic BH group, the two hydrides being mutually *trans*. The molecular structure of the model compound [RuH(Tp)(PH_3)(H_2C=CH_2)$_2$] has been investigated by *ab initio* calculations.[194]

[Ir(Tp^{Tn})(olefin)$_2$] has been prepared and its capability to activate C–H bonds is discussed by Carmona and coworkers[132] who discussed also the reactivity of [Ir(Tp^{Tn})(butadiene)].

3.32. $Tp^{Tn,Me}$

Hydrotris(5-methyl-3-(2-thienyl)pyrazol-1-yl)borate ($Tp^{Tn,Me}$) is coordinated in the solid state in κ^3-mode to Mo in both $[Et_4N][Mo(Tp^{Tn,Me})(CO)_3]$ and $[Mo(Tp^{Tn,Me})(CO)_3(\eta^3\text{-allyl})]$.[195]

3.33. Tp^{An}

$[Mo(NO)(Cl)(NHC_6H_4\text{-}p\text{-}N(CH_3)_2)(Tp^{An})]$ and $[Mo(NO)(Cl)(NHC_6H_4\text{-}p\text{-}CH_3)(Tp^{An})]$ have been investigated by X-ray. They have a characteristic bowl-like molecular shape with a columnar packing arrangement with molecular interpenetration between neighboring columns. Due to the potential application as liquid crystal materials, $[Mo(Tp^{An})(NO)(Cl)(NHC_6H_4\text{-}p\text{-}C_nH_{2n+1})]$ (n = 6, 10, 12, 14, 16) complexes have been also prepared (Fig. 3.41).[196]

Systems based on donor/acceptor organometallic groups connected by π-conjugated diene $[Mo(NO)(Cl)(Tp^{An})]$–R (R = C_6H_4Fc, C_6H_3R'–X=X–$C_6H_3R''Fc$, O_2N–C_6H_4–CH=CH–C_6H_4Fc) have been prepared and their nonlinear optical properties were investigated. At the macroscopic level the SHG and THG responses were measured on corona-poled spin-coated films, and at the molecular level cubic hyperpolarizability data were obtained from solution.

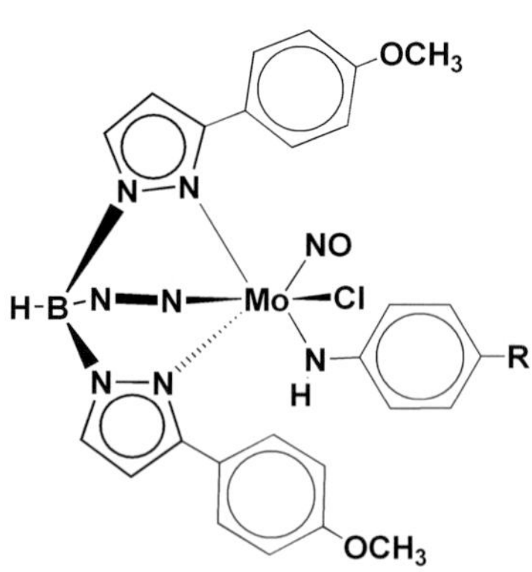

Fig. 3.41. $[Mo(Tp^{An})(NO)(Cl)(NHC_6H_4\text{-}p\text{-}R)]$ (R = C_nH_{2n+1}, n = 6, 10, 12, 14, 16).

The most relevant results on the THG response were from the use of $[Mo(NO)(Cl)(Tp^{An})]$ as acceptor group.[197] Agulló-López and coworkers[198] described the synthesis and characterization of $[Mo(NO)(Cl)\{(Y\text{-}\pi\text{-bridge})(4\text{-Fc})\}(Tp^{An})]$ (Y = NH or O; bridge = C_6H_4, $C_6H_3R{-}N{=}N{-}C_6H_3R'$ (R = R′ = H; R = CH_3, R′ = H; R = R′ = CH_3) or $C_6H_4{-}CH{=}CH{-}C_6H_4$). The role of the diverse bridging structures on the susceptibilities, including different lengths, as well as methyl substituents and linking (C=C vs N=N) groups have been evaluated. The acceptor $[Mo(NO)(Cl)(Y)(Tp^{An})]$ and nitro groups have been compared in compounds having the same ferrocenyl donor group and polarizable bridging system. $[Mo(NO)(Cl)(Tp^{An})\{HN{-}C_6H_3(3\text{-}CH_3)[N{=}NC_6H_3(5\text{-}CH_3){-}]_2(4\text{-}Fc)\}]$ has been synthesized, characterized and its second-order nonlinear optical properties were studied,[199] whereas the third-order nonlinear optical susceptibilities of push–pull organobimetallic Mo/Fe compounds of the type $[Mo(Tp^{An})(NO)(Cl)\{(HN\text{-}\pi\text{-bridge})(4\text{-}Fc)\}]$ (bridge = C_6H_4, $C_6H_3(3\text{-}R)\text{-}4\text{-}[N{=}N{-}C_6H_3(3\text{-}R')]$ [R = R′ = H; R = CH_3, R′ = H; R = R′R′ = CH_3], $C_6H_4\text{-}4\text{-}[CH{=}CH{-}C_6H_4]$) ("straight" compounds) and $[Mo(Tp^{An})(NO)(Cl)\{HNC_6H_3(3\text{-}CH_3)\}][\text{-}N{=}N\text{-}C_6H_3(5\text{-}CH_3)\text{-}2\text{-}\{4\text{-}Fc\}]$ ("bent" compound) have been measured in both chloroform solutions and dispersed into spincoated poly(methylmethacrylate) films. Highest values of $|\gamma|$ (5×10^{-33} esu) have been obtained for "straight" aza-bridged derivatives having the $[Mo(Tp^{An})(NO)(Cl)]$ fragment as acceptor group. They are comparable to those reported for highly conjugated planar phthalocyanines and other related compounds.[200]

$[Cu(OAc)(Tp^{An})]$ and $[CuCl(Tp^{An})(Hpz^{An})]$ have been prepared and characterized by IR, ESR and mass spectra and CV measurements.[201] Mechanochemical synthesis of $[NiCl(Tp^{An})]\cdot(Hpz^{An})$ has been also reported.[33] $K(Tp^{An})$ was reacted with MX_2 (M = Zn, Cd; X=Cl, OAc) affording $[MX(Tp^{An})\cdot nL]$, which were characterized by elemental analysis, IR, and ^1H NMR. The single-crystal structure of $[ZnCl(Tp^{An})(Hpz^{An})]$ exhibits a distorted tetrahedral zinc coordination environment.[202]

$[Rh(NBD)(Tp^{An})]$, $[Rh(COD)(Tp^{An})]$, and $[Rh(CO)_2(Tp^{An})]$ have been prepared and characterized by IR and NMR (solution). In all

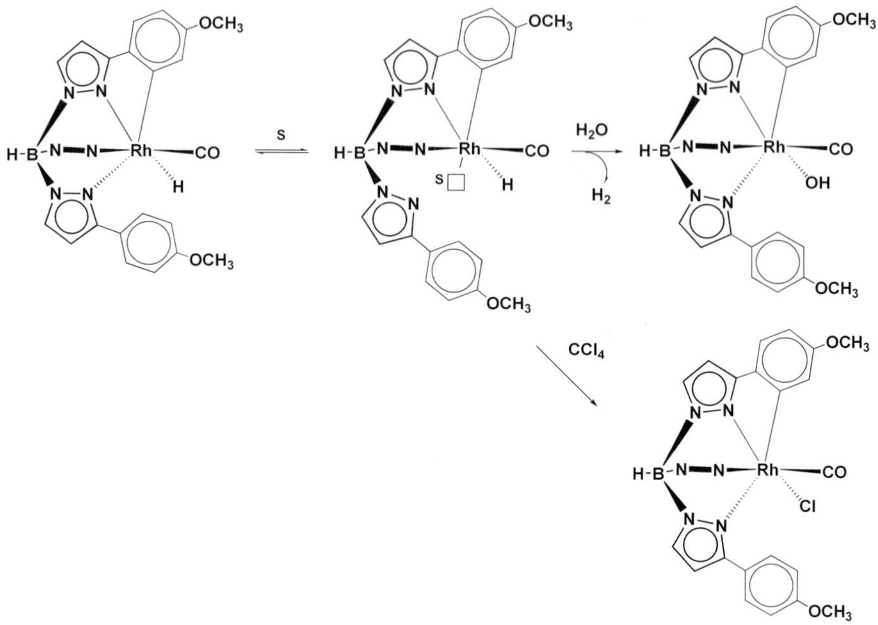

Fig. 3.42. Evolution of the [Rh(CO)H($Tp^{An@}$)] in different solvents.

cases the ligand Tp^{An} acts in κ^2-bonding mode, two different isomers containing an uncoordinated pyrazolyl ring being detected. [Rh(NBD)(Tp^{An})] and [Rh$(CO)_2$(Tp^{An})] are four-coordinate in the solid state. Photochemical irradiation of [Rh$(CO)_2$(Tp^{An})] affords an aryl hydride complex by intramolecular cyclometalation involving an ortho C–H bond of one *p*-anisyl substituents (Fig. 3.42). Evolution of the hydride species to a chloro complex occurs through a dehydrochlorination process.[203]

3.34. $Tp^{An,Me}$

[CuCl($Tp^{An,Me}$)($Hpz^{Me,An}$)] has been prepared and characterized by IR spectra, ESR and mass spectroscopy and CV measurements,[201] as have [CuCl($Tp^{An,Me}$)($Hpz^{Me,An}$)]·CH_2Cl_2 and [Cu(OAc)($Tp^{An,Me}$)-($Hpz^{Me,An}$)]·CH_2Cl_2 for which a pseudoaxial symmetrical $\{d_{x^2-y^2}\}^1$ configuration at copper is indicated. The four-coordinate [ZnCl-

$(Tp^{An,Me})$] and [Zn(OAc)($Tp^{An,Me}$)] were prepared by reacting ZnX_2 (X = Cl or OAc) with $Tp^{An,Me}$. [CdCl($Tp^{An,Me}$)($Hpz^{Me,An}$)] and [Cd(OAc)($Tp^{An,Me}$)($Hpz^{Me,An}$)]·CH_2Cl_2 are five-coordinated, with the cadmium center in a distorted square pyramidal geometry and a N atom of the tridentate $Tp^{An,Me}$ occupying the apical position.[204]

3.35. Tp^{Np}

Synthesis, characterization, and ethylene polymerization behavior of [$TiCl_3$(Tp^{Np})] are reported and the catalytic properties are compared with those found for analogous Tp^{tBu} and Tp^{Ph} derivatives. It has been observed that the highest activity is achieved by using the more sterically open catalyst precursor [$TiCl_3$(Tp^{Np})].[129]

3.36. $Tp^{Ad,iPr}$

The geometric and electronic structures of the mononuclear [Cu(O_2)($Tp^{Ad,iPr}$)] complex have been evaluated using Cu K- and L-edge X-ray absorption spectroscopy (XAS) studies in combination with valence bond configuration interaction (VBCI) simulations and spin-unrestricted broken symmetry density functional theory (DFT) calculations.[205]

3.37. Tp^{CF_3}

[Cu(Tp^{CF_3})(C_2H_4)] has been prepared by the reaction of Na(Tp^{CF_3}) with Cu(OTf) in the presence of ethylene.[206] [Cu(Tp^{CF_3})(PR_3)] has been synthesized from the reaction of CuCl with PR_3 (PR_3 = 4-(diphenylphosphane)benzoic acid) and K(Tp^{CF_3}). The superoxide scavenging activity of this complex, investigated by NMR, IR, and ESI MS spectroscopy, has been reported.[82]

[Fe(Tp^{CF_3})$_2$], distorted octahedral, has been prepared and characterized by ^{57}Fe Mossbauer spectroscopy and also investigated by X-ray absorption spectroscopy and its features are compared with those of the analogous Tp^{Me} derivatives. At variance with Tp^{Me}

derivative the high spin state for $[Fe(Tp^{CF_3})_2]$ is the most stable one. $[Fe(Tp^{CF_3})_2]$ has been compared with $[Fe(Tp^{R})_2]$ (R = 4-Me, 4-Br) and with $[Fe(pz^0Tp^{Me})_2]$.[207]

In $[Ag(Tp^{CF_3})(P^iBu_3)]$, the silver atom is in a distorted tetrahedral coordination geometry. This compound has been investigated by $^{31}P\{^1H\}$ NMR variable temperature NMR studies and ESI MS spectroscopy.[3]

3.38. $Tp^{CF_3,Me}$

Gorun and coworkers[208] reported that replacing all C–H bonds by C–F bonds in the proximity of a copper center, as in $[Cu_2(Tp^{CF_3,Me})_2]$, yields a hemocyanin spectroscopic model which also binds O_2. This compound partially dissociates in CD_2Cl_2, whereas reaction with MeCN yields $[Cu(Tp^{CF_3,Me})(MeCN)]\cdot[Cu_2(Tp^*)_2]$, having the same topology as $[Cu(Tp^{CF_3,Me})]_2$. This complex forms the peroxo complex $[Cu(Tp^{CF_3,Me})_2O_2]$ stable only at low temperature. The potassium salt $K(Tp^{CF_3,Me})$ has been also described and used to synthesize the distorted octahedral geometry $[Co(\kappa^2\text{-}NO_3)(\kappa^3\text{-}Tp^{CF_3,Me})(NCMe)]$. Catalytic oxidation of cyclohexane using this complex and cumene hydroperoxide as oxygen donor yields a mixture of cyclohexanol and cyclohexanone.[209] Also, $[Mn(\eta^2\text{-}NO_3)(\kappa^3\text{-}Tp^{CF_3,Me})(NCMe)]$ exhibits a N_4O_2 donor set which defines a distorted octahedral geometry. $[CuTp^{CF_3,Me}]_2$ binds O_2 at 25°C in CH_2Cl_2 to give $[(CuTp^{CF_3,Me})_2(O_2)]$. NMR studies in the same solvent show that, in the absence of O_2, $[CuTp^{CF_3,Me}]_2$ is in equilibrium with the mononuclear intermediate $[Cu(Tp^{CF_3,Me})]$. Addition of acetone produces $[Cu(Tp^{CF_3,Me})(acetone)]$ that transforms in $[(CuTp^{CF_3,Me})_2(O_2)]$ when the [acetone]/[Cu] ratio is below 200. In contrast, $[CuTp^{CF_3,Me}]_2$ at [acetone]/[Cu] > 300 yields crystals of $[Cu(Tp^{CF_3,Me})(lactate)]$ which have a square pyramidal geometry resembling that of $[Cu(Tp^{CF_3,Me})(acetate)]$. The one-pot formation of lactate form acetone appears unprecedented: this is the first example of an external substrate oxygenated aerobically using a robust Cu complex whose metal environmnet exhibits no C–H bonds.[210] The stable copper complex $[(CuTp^{CF_3,Me})_2(O_2)]$, amorphous upon slow evaporation, results in the formation of crystals that reveal a crystallographically unique $[Cu(Tp^{CF_3,Me})O]$ fragment which generates a disordered dinuclear complex. The system exhibits a strong antiferromagnetic coupling. The Cu···Cu and O–O distances are, respectively, shorter and longer compared to the corresponding distances in oxyHc and μ-η^2-η^2 peroxo models. The elongation and contraction of the average O–O and Cu···Cu contacts could be due to the presence of a small amount of cocrystallized $[(CuTp^{CF_3,Me})_2(OH)_2]$.[211]

Tp^x ligands are often compared with Cp^x because of their identical charge, number of donated electrons, and analogous facial coordinating geometry, the six-electron donation of Tp^x being, however, of sigma type. The X-ray structure of $[(Tp^{CF_3,Me})CuK(\mu_4\text{-}CO_3)K(Tp^{CF_3,Me})]_2$ shows for the first time an unprecedented η^5-$Tp^{CF_3,Me}$ potassium interaction. Both K–F and K-μ_4-CO_3^{2-} interactions stabilize the aggregate, as suggested by the lack of hexanuclear aggregation and K incorporation in the absence of fluorine groups or when the carbonate is replaced by the acetate. $[Cu(OAc)(Tp^{CF_3,Me})]$ retains a Cu coordination sphere similar to that present in the $[Cu(Tp^{CF_3,Me})]$ subset of $[(Tp^{CF_3,Me})CuK(\mu_4\text{-}CO_3)K(Tp^{CF_3,Me})]_2$. This novel mode of coordination suggests that Tp ligands might function not only as σ-donors but also as Cp-like π-donors.[212]

3.39. $Tp^{(CF_3)_2}$

The electronic structure of $K(Tp^{(CF_3)_2})$ has been investigated using gas-phase photoelectron spectroscopy and DFT calculations. Only slight differences in the ionizations from the frontier orbitals in the photoelectron spectra of Tp* and $Tp^{(CF_3)_2}$ have been observed. The experimentally observed first ionization energy of $Tp^{(CF_3)_2}$ is stabilized by ~2.0 eV relative to Tp* due to the strong electron withdrawing effect of the trifluoromethyl groups. The photoelectron spectroscopic studies of $Na(Tp^{(CF_3)_2})$ further confirm the assignments of ionizations from σ_N orbitals for $Tp^{(CF_3)_2}$ associated with the *a* and *e* sets in C_3 symmetry.[13]

[In(Tp$^{(CF_3)_2}$)] and [Ga(Tp$^{(CF_3)_2}$)] have been prepared, characterized, and compared with analogous compounds containing nonfluorinated ligands. The mononuclear [In(Tp$^{(CF_3)_2}$)] has been also structurally characterized, its In–N bond distances being longer than the corresponding distances in nonfluorinated analogs. In the monomeric [Sn(OTf)(Tp$^{(CF_3)_2}$)]·Hpz$^{(CF_3)_2}$, one Hpz$^{(CF_3)_2}$ is H-bonded to a coordinated OTf group (Fig. 3.43).[213]

[Ag(Tp$^{(CF_3)_2}$)(η^2-toluene)] reacts with [(^{n}Pr)$_2$ATI]GeN$_3$ or [(^{n}Pr)$_2$ATI]SnN$_3$ in dichloromethane giving stable [Ag(Tp$^{(CF_3)_2}$)]-M(N$_3$)[(^{n}Pr)$_2$ATI] (M = Ge, Sn) complexes that exhibit unsupported silver–germanium and silver–tin bonds[214]; [Ag(Tp$^{(CF_3)_2}$)]{(OAc)$_2$-CN$_2$} (Fig. 3.44) contains an essentially unperturbed diazoalkane moiety with Ag–O rather than Ag–N interactions. The stability of such complex suggests that metal–oxygen bonded species play an important role in metal-mediated activation of diazoketones.[215]

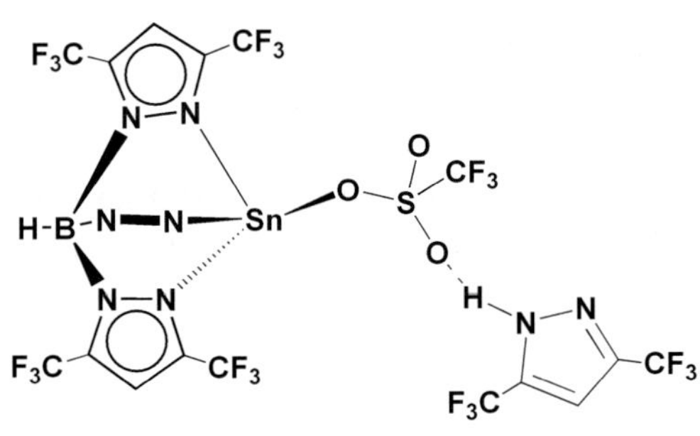

Fig. 3.43. [Sn(OTf)(Tp$^{(CF_3)_2}$)]·Hpz$^{(CF_3)_2}$.

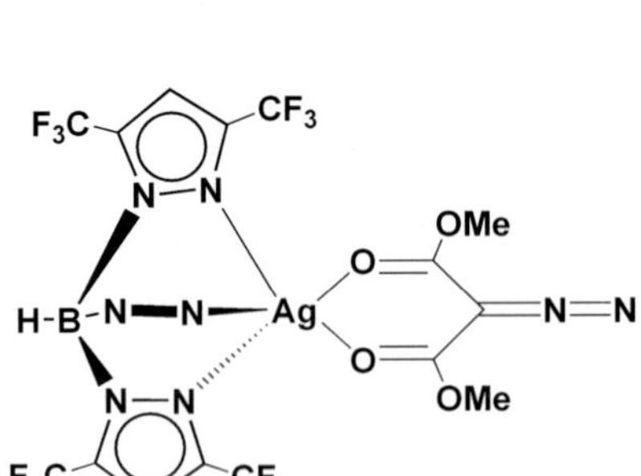

Fig. 3.44. [Ag(Tp$^{(CF_3)_2}$)]{(OAc)$_2$CN$_2$}.

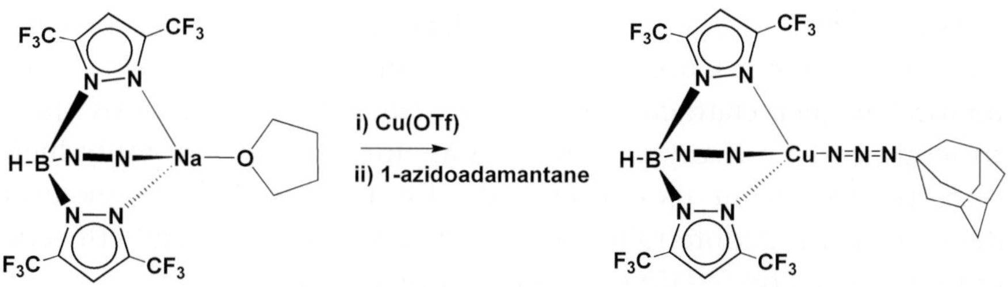

Fig. 3.45. Synthesis of [Cu(Tp$^{(CF_3)_2}$)(NNN(1-Ad))].

The reaction of [Na(Tp$^{(CF_3)_2}$)(THF)] with Cu(OTf) and 1-azidoadamantane (NNN(1-Ad)) yields [Cu(Tp$^{(CF_3)_2}$)(NNN(1-Ad))] in which Cu is coordinated to the terminal nitrogen atom of (NNN(1-Ad)) (Fig. 3.45). [Ag(Tp$^{(CF_3)_2}$)(NNN(1-Ad))] has been obtained from the reaction of [Ag(Tp$^{(CF_3)_2}$)(THF)] with (NNN(1-Ad)). Different from the copper analogs, the silver is coordinated to the alkylated nitrogen atom of the azidoadamantane ligand.[216]

[Ag(Tp$^{(CF_3)_2}$)(η^2-toluene)] reacts with [(Me)$_2$ATI]GeCl giving the 1:1 adduct [Tp$^{(CF_3)_2}$Ag]←GeCl[(Me)$_2$ATI], containing a silver–germanium bond. This compound underwent metathesis with AgOTf, yielding the [Tp$^{(CF_3)_2}$Ag]←Ge(OTf)[(Me)$_2$ATI] that has been also structurally characterized.[217] Tp$^{(CF_3)_2}$ is also useful for the isolation of the tetrahedral ethylene oxide adduct [Ag(Tp$^{(CF_3)_2}$)(OC_2H_4)] and of the related sulfide [Ag(Tp$^{(CF_3)_2}$)(SC_3H_6)]. Syntheses, spectroscopic, and structural characterization for both species have also been presented.[218]

The activation of alkyl halides via a silver-catalyzed carbene insertion process has been performed by using [Ag(Tp$^{(CF_3)_2}$)] complexes. This transformation results in the formation of a new sp^3–sp^3 carbon–carbon bond and the net migration of a halide atom. Interestingly, the silver salt of the nonfluorinated ligand [Ag(Tp*)] did not provide any product resulting from C–Cl insertion or from the elimination pathway.[219] [Ag(Tp$^{(CF_3)_2}$)(OSMe$_2$)] is a monomeric tetrahedral silver complex with an O-bonded dimethylsulfoxide ligand. [Ag(Tp$^{(CF_3)_2}$)(OSMe$_2$)] and the related [Ag(Tp$^{(CF_3)_2}$)(THF)] show good antibacterial activity.[220]

$[Cu(Tp^{(CF_3)_2})(C_2H_4)]$ has been prepared by the reaction of $Na(Tp^{(CF_3)_2})$ with Cu(OTf) in the presence of ethylene. This compound has been characterized by ^1H, ^{13}C, ^{19}F NMR, and IR spectroscopy and by X-ray crystallography and displayed notably high stability toward air oxidation and ethylene loss. X-ray structural data show a pseudotetrahedral copper ion and η^2-bonded ethylene units.[204] $[Cu(Tp^{(CF_3)_2})(NCMe)]$ has been also reported.[221]

$[Al(Tp^{(CF_3)_2})Me_2]$ and $[Zn(Tp^{(CF_3)_2})Et]$ complexes can be prepared by the reaction of $[Ag(Tp^{(CF_3)_2})(THF)]$ with Me_3Al and Et_2Zn. The scorpionate ligand coordinates to the aluminum center in κ^2-fashion, while it acts as a κ^3 tripodal donor to zinc.[222]

3.40. $Tp^{CF_3,Ph}$

The sodium salt of the fluorinated $Tp^{CF_3,Ph}$ ligand has been synthesized from the corresponding pyrazole and $NaBH_4$. $Tp^{CF_3,Ph}$ yields stable adducts with oxygen-based donors like THF and water. $[Na(Tp^{CF_3,Ph})(THF)]$ is monomeric in the solid state, whereas $[Na(Tp^{CF_3,Ph})(H_2O)]_n$ exhibits a polymeric chain structure. These sodium compounds show intra and/or intermolecular sodium–fluorine interactions. $[Na(Tp^{CF_3,Ph})(H_2O)]_n$ and $[Na(Tp^{(CF_3)_2})(OEt_2)]$ show similar CF···Na contacts. $[Cu(Tp^{CF_3,Ph})(CO)]$ was synthesized from $[Na(Tp^{CF_3,Ph})(THF)]$ with Cu(OTf) in the presence of CO. The IR stretching frequency data show relatively high υ(CO) value ($2103\,cm^{-1}$) in accordance with the presence of highly electrophilic copper site.[223] $[Cu(Tp^{CF_3,Ph})(C_2H_4)]$ has been prepared by the reaction of the corresponding sodium salts with Cu(OTf) in the presence of ethylene.

3.41. $MeTp^{CF_3}$

The thallium derivative $Tl(MeTp^{CF_3})$ has been synthesized via a two-step process using the corresponding pyrazole, $Li[MeBH_3]$, and thallium(I) acetate. Reaction of $[Tl(MeTp^{CF_3})]$ with CuBr in the presence of ethylene gave $[Cu(MeTp^{CF_3})(C_2H_4)]$ which reacts with $[(Bn)_2ATI]SnCl$ to yield $[(MeTp^{CF_3})Cu]\leftarrow Sn(Cl)[(Bn)_2ATI]$, containing an unsupported Cu(I)–Sn(II) bond.[224]

3.42. Tp^{4Bo}

The hydrotris(indazolyl)borate ligand Tp^{4Bo} has been indicated to be prone to supramolecular assembly which, in the Co and Ni derivatives of formula $[M(Tp^{4Bo})_2]$, is determined by aromatic π–π and C–H···π interactions generating noncovalent framework structures, about 30% porous.[225] $[Cu(Tp^{4Bo})_2]$ exhibits a pseudooctahedral geometry with a Jahn–Teller distortion. It has been synthesized and characterized by IR, UV/Vis, and MS. The ligand field strength of the Tp^{4Bo} ligand is slightly higher than that of the normal Tp ligand. The molecular and crystal structure of this compound, with different solvents of crystallization, has been elucidated by X-ray crystallography. It exhibits the usual pseudooctahedral D_{3d}-symmetrical geometry for two tripodal ligands with a Jahn–Teller distortion. The binary complexes $[M(Tp^{4Bo})_2]$ (M = Fe, Co, Ni, Zn), having the usual pseudooctahedral D_{3d}-symmetrical geometry, have been also synthesized and characterized by NMR, IR, UV/Vis, MS, and Mössbauer spectroscopy. They present large noncovalent framework structures with voids and channels ranging in volume from 19% to 39% of the unit cell volume, the crystal packing being controlled mainly by C–H···π interactions from the indazolyl moiety.[226]

The molecular metal–ligand arrangement and crystal packing of $[Tl(Tp^{4Bo})]$ have been compared with that of [Tl(Tp)]. In $[Tl(Tp^{4Bo})]$ the molecular units are arranged in pairs through indazolyl π–π stacking.[227]

A short route to prepare a ruthenium complex, described as an organometallic molecular turnstile, with a pentaphenyl substituted cyclopentadienyl and Tp^{4Bo} was reported by Rapenne and coworkers[228]: heating [Ru(1,2,3,4,5-penta(*p*-bromophenyl)-cyclopentadiene)$(CO)_2$Br] with 2 equiv. of $K(Tp^{4Bo})$ for 24 h in freshly distilled THF resulted in displacement of both bromide and CO ligands and attachment of the scorpionates. The X-ray structure of this species (Fig. 3.46) confirmed the coordination of both ligands with Tp^{4Bo} binding in a facial tripodal mode (i.e. κ^3-N,N',N'').

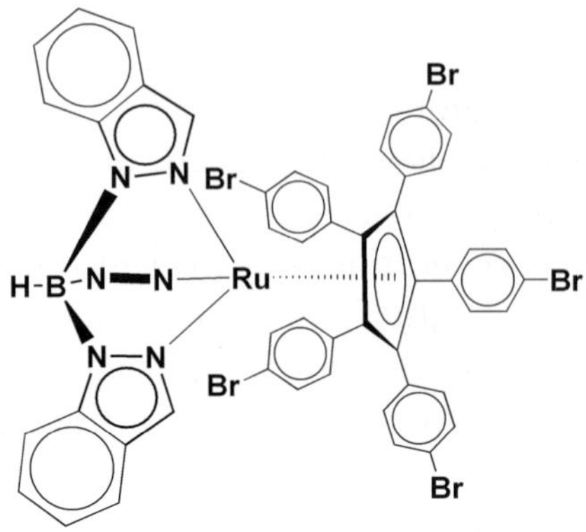

Fig. 3.46. [Ru(Tp^{4Bo})(1,2,3,4,5-penta(*p*-bromophenyl)cyclopentadiene)].

3.43. Tp^{4Bo,3Me}

Reaction of [$NbCl_3(MeOCH_2CH_2OMe)(PhC{\equiv}CMe)$] with K(Tp^{4Bo,3Me}) yields [$NbCl_2$(Tp^{4Bo,3Me})(PhC≡CMe)]. The four-electron donor alkyne sits in the molecular mirror plane and the restricted rotation of the alkyne ligand allows the observation of an equilibrium between two rotamers.[229]

3.44. Tp^{Bo,7Me}

[RuH(Tp^{Bo,7Me})(COD)] has been prepared from the reaction of [$RuH(NH_2NMe_2)_3(COD)]BPh_4$ with Tp^{Bo,7Me}. Reaction between [RuH(Tp^{Bo,7Me})(COD)] and L (L = PMe_3, PMe_2Ph, CN-*t*-Bu) gave [RuH(Tp^{Bo,7Me})(L)(COD)].[194]

3.45. Tp$^{Bo4Me,7^iPr}$

Single-site achiral and chiral C_3-symmetric complexes [Zn(OR)(κ^3-Tp$^{Bo4Me,7^iPr}$)] (Tp$^{Bo4Me,7^iPr}$ = HB((7*R*)-^{i}Pr-(4*R*)-Me-4,5,6,7-tetrahydro-2H-indazolyl)$_3$, R = Et, ^{t}Bu, Ph or $SiMe_3$) have been prepared and employed in the ring-opening polymerization of *L*−, *rac*-, and *meso*-lactide in CH_2Cl_2 at 25°C. The molecular structure of [ZnMe(κ^3-Tp$^{Bo4Me,7^iPr}$)] has been reported.[85]

3.46. pz^0Tp^{Cpr} and Tp^{Cpr}

In the crystal of $[Fe(pz^0Tp^{Cpr})_2]\cdot(CH_3OH)$ two independent molecules are present in the unit cell, both of which are high spin; only one of these high-spin iron(II) sites, the site with the lesser ring-twist distortion, is observed to be low-spin iron(II) in the 90 K structure. Magnetic and Mössbauer measurements on $[Fe(pz^0Tp^{Cpr})_2]$ and $[Fe(Tp^{Cpr})_2]$ indicate that both complexes exhibit a partial SCO from fully high-spin iron(II) at higher temperatures to a 50:50 high-spin/low-spin mixture of iron(II) below 100 K. A comparison of the Mössbauer spectral properties of these two complexes with those of $[Fe(Tp^{Me})_2]$ has been also performed.[8]

The ^1H, ^{13}C, and ^{15}N NMR spectroscopic properties of $[Tl(Tp^x)]$ ($Tp^x = Tp^{Cpr}$ and pz^0Tp^{Cpr}) ($[Tl(Tp^{Cpr})]$ is as shown in Fig. 3.47) have been also reported and compared with those of the analogous $[Tl(Tp^{Cpe})]$ and $[Tl(Tp^{Cbu})]$. The data acquired in solution allow understanding the data recorded in the solid state. The ^{205}Tl–^{15}N and ^{205}Tl–^{13}C coupling constants measured in the solid state were reported. Among those, a very large coupling constant (in the range 194–282 Hz) has been measured on the carbon to position 4 of the pyrazole ring in several compounds and particularly for the cyclobutyl derivative $Tl(Tp^{Cbu})$. The large value could be due to direct 4-C–H···Tl type interaction.[192]

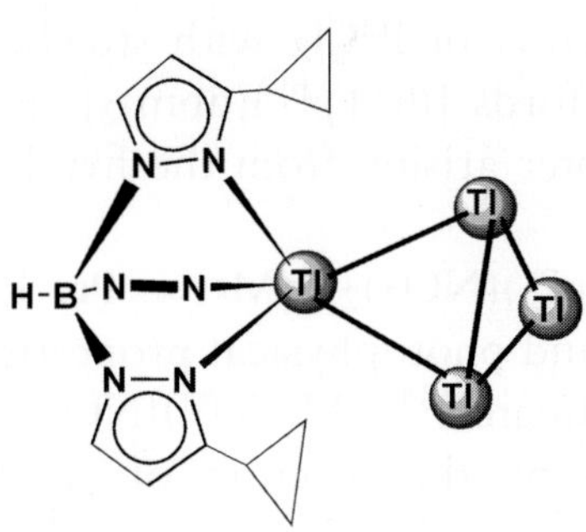

Fig. 3.47. $[Tl(Tp^{Cpr})]_4$ crystallizes in a very unusual structure. The four thallium atoms form a regular tetrahedron, each Tl being linked to the three N atoms of the scorpionate; three Tp^{Cpr} are not shown for clarity.

3.47. Tp^{py}

Tp^{py} complexes of Mn(II), Co(II), and Zn(II) have been prepared and structurally characterized. In the mononuclear $[Co(Tp^{py})][PF_6]\cdot CH_2Cl_2$, all three bidentate arms of the ligand are coordinated to the Co center in a relatively strain-free manner to give a trigonal prismatic coordination geometry. In contrast, in $[Mn_4(Tp^{py})_4][PF_6]_4\cdot 4MeCN\cdot Et_2O$ and $[Zn_4(Tp^{py})_4][PF_6]_3[OH]\cdot 12EtOH$, the $[M_4(Tp^{py})_4]^{4+}$ cationic units are tetrahedral clusters. Each Tp^{py} coordinates one bidentate arm to each of three metal ions in a $\kappa^2:\kappa^2:\kappa^2$ coordination mode, such that each ligand caps one triangular face of the metal tetrahedron. This trinucleating coordination mode, and the 1:1 correspondence of octahedral metal ions and hexadentate ligands, necessarily results in the formation of a tetrahedral cluster in which all four tris(chelate) metal centers have the same chirality. Thus, the mononucleating κ^6 coordination mode results when the metal ion can tolerate a trigonal prismatic geometry, whereas the trinucleating $\kappa^2:\kappa^2:\kappa^2$ mode occurs when the metal ions are octahedral. The monomeric and tetrameric forms are retained in solution and do not interconvert, as indicated by spectral data.[230] The crystal structure of ternary 10-coordinated lanthanide complexes $[Tb(Tp^{py})(dbm)_2]$ and $[Eu(Tp^{py})(Trop)(NO_3)]$ has been reported.[231] $[Nd(Tp^{Py})(Bp^{Py})][Nd(Hpz^{py})(NO_3)_4]$, obtained from the reaction of Tp^{Py} with $Nd(NO_3)_3$, exhibits a 10-coordinate cation $[Nd(Tp^{Py})(Bp^{Py})]$ from interleaved hexadentate and tetradentate ligands.[232] Reaction of $PrCl_3$ with stoichiometric quantities of Hdbm and Tp^{2py} affords $[Pr(Tp^{py})(dbm)_2]$ (Fig. 3.48) with a 10-coordinate metal center arising from the five bidentate arms of the anionic ligands.[233]

Complexes $[M(Tp^{py})(NO_3)_2]$ (M = Pr, Er) were synthesized and their structural and photophysical properties reported. All complexes are 10-coordinate.[234] $[M(Tp^{py})(NO_3)_2]$ (M = Nd, Yb) have been prepared by the reaction of equimolar quantities of the hydrated $M(NO_3)_3$ and $K(Tp^{py})$ in MeOH at room temperature. $[M(Tp^{py})_2][BPh_4]$ was prepared by the reaction of hydrated MCl_3 with 2 equiv. of $K(Tp^{py})$ in MeOH, followed by precipitation

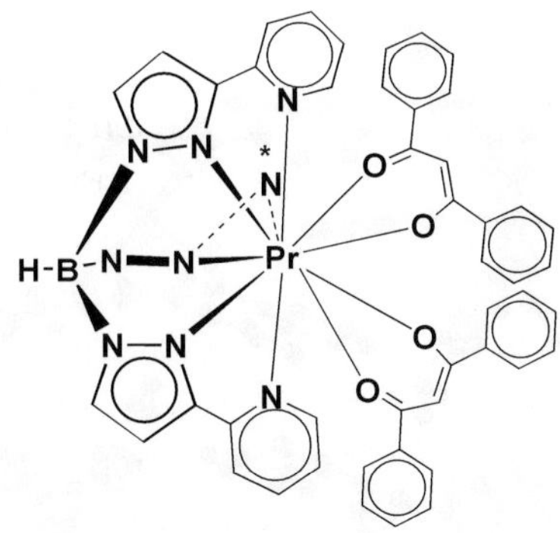

Fig. 3.48. The 10-coordinate [Pr(Tp^{py})(dbm)$_2$] (the labeled N* indicates the pyridyl group bonded to third pyrazolyl ring).

with aqueous $NaBPh_4$; finally [M(Tp^{py})(dbm)$_2$] was prepared by the reaction of the hydrated metal chloride, K(Tp^{py}), and Hdbm, in a 1:1:2 ratio, in MeOH. These complexes have been investigated also by near-IR spectroscopy and [Nd(Tp^{py})(NO$_3$)$_2$]·Et$_2$O, [Nd(Tp^{py})$_2$][BPh$_4$]0.6(Et$_2$O)·0.4(Me$_2$CO) also by X-ray diffraction studies.[235]

In [Eu(Tp^{py})(dbm)$_2$], eight-coordinate, the Tp^{py} ligand acts only as a tetradentate chelate. [Nd(Tp^{py})(dbm)$_2$] on recrystallization rearranged to [Nd(Tp^{py})$_2$][Nd(dbm)$_4$], which contains a 12-coordinate [Nd(Tp^{py})$_2$]$^+$ cation (Fig. 3.49).[236]

3.48. Tp^{py,Me}

[Zn(SH)(Tp^{py,Me})] has been synthesized by the reaction of the corresponding zinc-hydroxide complex with H_2S. Acidic organic X–OH compounds (carboxylic acids, trinitrophenol, hexafluoroacetylacetone) replace the SH groups with formation of [Zn(OX)(Tp^{py,Me})].[103]

3.49. Tp^{py*}

Ward and coworkers reported the synthesis of a chiral C$_3$-symmetric hexadentate tripodal ligand tris[3-{2-(pinene[4,5]pyridyl)}pyrazolyl]-

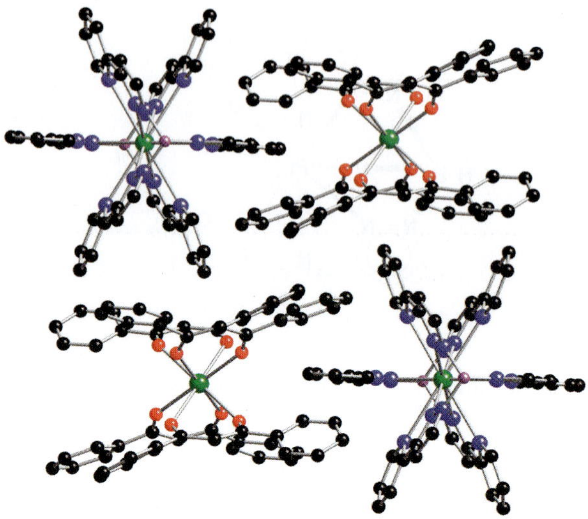

Fig. 3.49. Structure of $[Nd(Tp^{py})_2][Nd(dbm)_4]$, which contains a 12-coordinate $[Nd(Tp^{py})_2]^+$ cation.

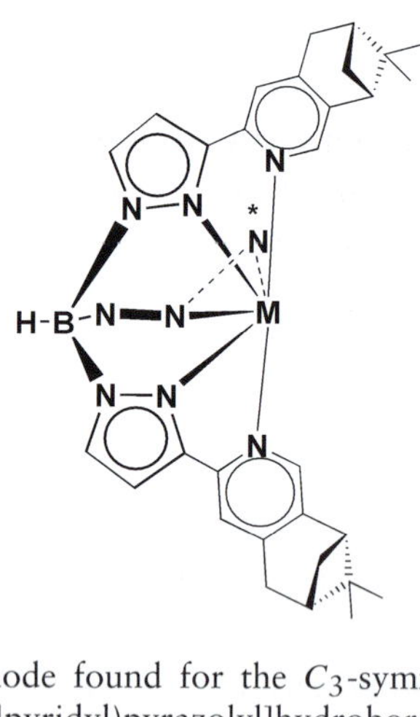

Fig. 3.50. Coordination mode found for the C_3-symmetric hexadentate tripodal ligand tris[3-{2-(pinene[4,5]pyridyl)pyrazolyl]hydroborate (Tp^{py^*}).

hydroborate (Tp^{py^*}) (Fig. 3.50) isolated as its Tl salt, the Tl center being coordinated mainly by the three pyrazolyl N donors in a pyramidal arrangement. $[Tb(Tp^{py^*})(NO_3)_2]\cdot 2CH_2Cl_2$ is also reported and structurally characterized.[237]

3.50. BisTp^{Py}

The 12-dentate compartmental ligand hexakis{[3-(2-pyridyl)-pyrazol]-1yl}diboran(IV)ate (BisTp^{Py}) has two hexadentate tris (pyrazolyl)borate-based cavities, linked back-to-back by a B–B bond, which have been reported.[238] [Ln(BisTp^{Py})(NO_3)$_2$]$_2$ (Ln = La or Gd) is dinuclear with each metal center being 10-coordinate from a tripodal hexadentate ligand cavity and two bidentate nitrates.

3.51. Tp$^{4\text{-pic,Me}}$

[Zn(SH)(Tp$^{4\text{-pic,Me}}$)] has been synthesized by the reaction of the corresponding zinc-hydroxide with H_2S. Acidic organic X–OH compounds (carboxylic acids, trinitrophenol, hexafluoroacetylacetone) replace the SH groups with formation of [Zn(OX)(Tp$^{4\text{-pic,Me}}$)].[103]

3.52. Tp^{camph}

The synthesis and characterization of the optically active hydrido-(4,5,6,7-tetrahydro-7,8,8-trimethyl-4,7-methanoindazolyl)borate in the form of its sodium salt Na(Tp^{camph}) are reported by Babbar *et al.*[239] The Cu^I/Tp^{camph} and Cu^{II}/Tp^{camph} systems catalyze the cyclopropanation of styrene and ethyl diazoacetate with ee's of up to 40% and yield of up to 67%.

3.53. Tp^{Menth} and Tp^{Mementh}

The solution and solid state structural features of the optically pure [Rh(Tp^{Menth})(CO)$_2$] have been reported, Tp^{Menth} being found in both κ^2- and κ^3-coordination modes. The optically active Tp^{Menth} and Tp^{Mementh} are able to control the stereoselectivity of the C–H bond activation by their coordinated rhodium center. Irradiation of [Rh(Tp^{Menth})(CO)$_2$] under N_2 resulted in the generation of an 85:15 mixture of diastereomeric alkyl hydrides due to intramolecular cyclometalation reaction involving the methyl substituents on the ligand isopropyl group (Fig. 3.51).[240]

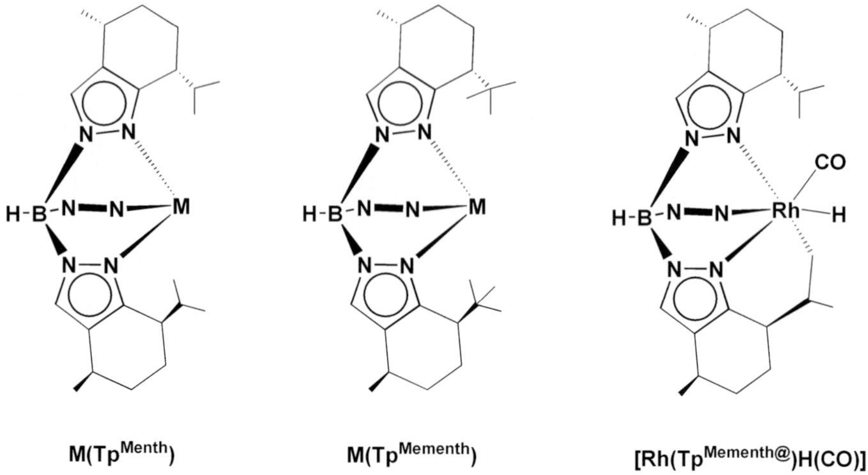

Fig. 3.51. Metal derivatives of Tp^{Menth} and $Tp^{Mementh}$ and (right) the Rh complex resulting from intramolecular cyclometalation.

3.54. $Tp^{2,4(OMe)_2Ph}$

Monohydride complexes $[RuH(Tp^{2,4(OMe)_2Ph})(COD)]$ have been prepared from the reaction of $[RuH(NH_2NMe_2)_3(COD)]BPh_4$ with $Tp^{2,4(OMe)_2Ph}$.[194]

3.55. $HB(tz)_3$

A 2D layer of water molecules has been stabilized between organic layers of the nickel(II) chelate complex $[Ni\{HB(tz)_3\}_2]$, the structure of which has been determined by neutron diffraction studies at 278, 263, and 20 K.[241] $HB(tz)_3$ reacts with main group metals leading to eight-coordinate complexes $[Ca(HB(tz)_3)_2(H_2O)_2]$, $[Sr\{HB(tz)_3\}_2(H_2O)_2]$, and $[Pb\{HB(tz)_3\}_2(H_2O)_2]$. In these complexes the two $HB(tz)_3$ ligands coordinate in a "bent" arrangement to allow for the coordination of the two aqua ligands. In the synthesis of $[Pb\{HB(tz)_3\}_2(H_2O)_2]$ an intermediate of the form $[Pb\{\mu_3\text{-}HB(tz)_3\}(NO_3)H_2O]$ has been found and structurally characterized. Long M–N bonds indicate a more ionic bonding to the modified $HB(tz)_3$ when compared to analogous Tp complexes.[242]

$[Cu\{HB(tz)_3\}(PR_3)_x]$ (x = 1 or 2) has been prepared from the reaction of CuCl with PR_3 (including 4-(diphenylphosphane)benzoic acid),[82] and $K[HB(tz)_3]$ and investigated by NMR, IR, and ESI MS spectroscopy. Chemiluminescence techniques were used to evaluate the superoxide scavenging activity of these new copper complexes. The structurally authenticated arrays fall into two different types: (a) discrete mononuclear $[Cu\{HB(tz)_3\}(PR_3)]$ with a $CuP(N_3)$ coordination sphere, (b) one-dimensional polynuclear $[Cu\{(tz)BH(tz)(tz)\}(PR_3)_2]_{(\infty|\infty)}$ with a four-coordinate $CuP_2(N_2)$ coordination sphere in which one of the ligand tz rings is uncoordinated and the other two bridge through N(4) to successive copper atoms (Fig. 3.52).[243]

$[Fe\{HB(tz)_3\}_2]ClO_4{\cdot}2MeCN$, octahedral, has been employed as a molecular building block. The noninterpenetrating 3D coordination polymer framework of $[Fe\{HB(tz)_3\}_2][Rh_2(OAc)_4]_3{\cdot}xS$ (S = solvent) has been prepared and characterized by single-crystal X-ray structure analysis.[244]

Silver(I) derivatives containing monodentate tertiary phosphines PR_3 (R = Ph, Bn, *o*-Tol, *m*-Tol, *p*-Tol) and $HB(tz)_3$ have been also

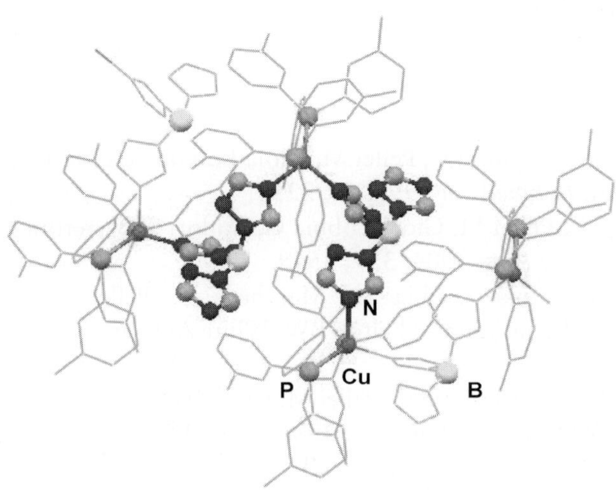

Fig. 3.52. $[Cu\{(tz)BH(tz)(tz)\}(PR_3)_2]_{(\infty|\infty)}$ with a four-coordinate $CuP_2(N_2)$ coordination sphere in which one of the ligand tz rings is uncoordinated and the other two bridge through N(4) to successive copper atoms.

prepared. All the compounds are soluble in chlorinated solvents and are nonelectrolytes in CH_2Cl_2 and acetone solutions. Owing to the presence of the methyl substituents on the phosphine ligand, the complex [Ag{HB(tz)$_3$}{P(*o*-Tol)$_3$}], as expected, is mononuclear.[245]

Hydrotris(1,2,4-triazolyl)borate organic salts (imidazolium, triazolium, and tetrazolium) were synthesized in high yield by metathesis reaction with [Ag{HB(tz)$_3$}]. These borate salts melt at relatively low temperature to give thermally stable ionic liquids.[246]

3.56. HB(btz)$_3$

Silver(I) derivatives [Ag{HB(btz)$_3$}(PR$_3$)$_x$] (x ranging from 1 to 3), have been prepared from the reaction of $AgNO_3$ with PR_3 (R = Ph, *o*-Tol, *m*-Tol, *p*-Tol, Bn) and K{HB(btz)$_3$} (Hbtz = 1,2,3-benzotriazole). Solid state and solution properties of all complexes have been investigated by analytical and spectroscopic measurements (IR, ^{1}H, ^{31}P NMR), the ^{1}H and ^{31}P NMR spectra being interpreted in terms of equilibria that involve mono and dinuclear complexes. [Ag{HB(btz)$_3$}(PPh$_3$)$_3$]·(1/2H$_2$O) has been characterized by single-crystal X-ray studies, the HB(btz)$_3$ ligand being unidentate in an $NAgP_3$ coordination environment.[247]

References

1. Santini C, Pettinari C, Pellei M, Gioia Lobbia G, Pifferi A, Camalli M, Mele A, *Polyhedron* 1999, **18**, 2255.
2. Santini C, Pellei M, Gioia Lobbia G, Alidori S, Berrettini M, Fedeli D, *Inorg Chim Acta* 2004, **357**, 3549.
3. Effendy, Gioia Lobbia G, Marchetti F, Pellei M, Pettinari C, Pettinari R, Santini C, Skelton BW, White AH, *Inorg Chim Acta* 2004, **357**, 4247.
4. Effendy, Gioia Lobbia G, Pellei M, Pettinari C, Santini C, Skelton BW, White AH, *Inorg Chim Acta* 2001, **315**, 153.
5. Reger DL, Wright TD, Smith MD, Rheingold AL, Rhagitan B, *J Chem Crystallogr* 2000, **30**, 665.
6. Looney A, Han R, Gorrell IB, Cornebise M, Yoon K, Parkin G, Rheingold AL, *Organometallics* 1995, **14**, 274.

7. Yang K-W, Wang Y-Z, Huang Z-X, Sun J, *Polyhedron* 1997, **16**, 1297.
8. Reger DL, Gardinier JR, Elgin JD, Smith MD, Hautot D, Long GJ, Grandjean F, *Inorg Chem* 2006, **45**, 8862.
9. Hikichi S, Yoshizawa M, Sasakura Y, Akita M, Moro-oka Y, *J Inorg Biochem* 1999, **74**, 31.
10. Hikichi S, Sasakura Y, Yoshizawa M, Ohzu Y, Moro-oka Y, Akita M, *Bull Chem Soc Jpn* 2002, **75**, 1255.
11. Uehara K, Hikichi S, Inagaki A, Akita M, *Chem Eur J* 2005, **11**, 2788.
12. Uehara K, Hikichi S, Akita M, *Chem Lett* 2002, **12**, 1198.
13. Joshi HK, Arvin ME, Durivage JC, Gruhn NE, Carducci MD, Westcott BL, Lichtenberger DL, Enemark JH, *Polyhedron* 2004, **23**, 429.
14. Desrochers PJ, Brown JR, Arvin ME, Jones GD, Vivic DA, *Acta Cryst* 2005, **E61**, m1455.
15. Teuma E, Malbosc F, Pons V, Serra-Le Berre C, Jaud J, Etienne M, Kalck P, *J Chem Soc Dalton Trans* 2001, 2225.
16. Teuma E, Malbosc F, Etienne M, Daran J-C, Kalck P, *J Organomet Chem* 2004, **689**, 1763.
17. Maunder GH, Russo MR, Sella A, *Polyhedron* 2004, **23**, 2709.
18. Reger DL, Elgin JD, Smith MD, Grandjean F, Rebbouh L, Long GJ, *Polyhedron* 2006, **25**, 2616.
19. Chisholm MH, Gallucci JC, Phomphrai K, *Inorg Chem* 2002, **41**, 2785.
20. Chisholm MH, Gallucci JC, Phomphrai K, *Inorg Chem* 2004, **43**, 6717.
21. Chisholm MH, Gallucci JC, Phomphrai K, *Chem Commun* 2003, 48.
22. Ruman T, Ciunik Z, Wołowiec S, *Eur J Inorg Chem* 2003, 89.
23. Millar AJ, Doonan CJ, Laughlin LJ, Tiekink ERT, Young CG, *Inorg Chim Acta* 2002, **337**, 393.
24. Smith PD, Slizys DA, George GN, Young CG, *J Am Chem Soc* 2000, **122**, 2946.
25. Doonan CJ, Millar AJ, Nielsen DJ, Young CG, *Inorg Chem* 2005, **44**, 4506.
26. Doonan CJ, Nielsen DJ, Smith PD, White JM, George GN, Young CG, *J Am Chem Soc* 2006, **128**, 305.
27. Malarek MS, Evans DJ, Smith PD, Bleeker AR, White JM, Young CG, *Inorg Chem* 2006, **45**, 2209.
28. Millar AJ, Doonan CJ, Smith PD, Nemykin VN, Basu P, Young CG, *Chem Eur J* 2005, **11**, 3255.

29. Smith PD, Millar AJ, Young CG, Ghosh A, Basu P, *J Am Chem Soc* 2000, **122**, 9298.
30. J. Kisała, Ciunik Z, Drabent K, Ruman T, Wołowiec S, *Polyhedron* 2003, **22**, 1645.
31. Kitano T, Sohrin Y, Hata Y, Mukai H, Wada H, Ueda K, *Polyhedron* 2004, **23**, 283.
32. Brunker TJ, Hascall T, Cowley AR, Rees LH, O'Hare D, *Inorg Chem* 2001, **40**, 3170.
33. Kolotilov SV, Addison AW, Trofimenko S, Dougherty W, Pavlishchuk VV, *Inorg Chem Commun* 2004, **7**, 485.
34. Hikichi S, Yoshizawa M, Sasakura Y, Komatsuzaki H, Akita M, Moro-oka Y, *Chem Lett* 1999, **9**, 979.
35. Akita M, Moro-oka Y, *Catal Today* 1998, **44**, 183.
36. Uehara K, Hikichi S, Akita M, *Organometallics* 2001, **20**, 5002.
37. Yoshimitsu S-i, Hikichi S, Akita M, *Organometallics* 2002, **21**, 3762.
38. Uehara K, Hikichi S, Akita M, *J Chem Soc Dalton Trans* 2002, 3529.
39. Singh UP, Babbar P, Sharma AK, *Inorg Chim Acta* 2005, **358**, 271.
40. Singh UP, Aggarwal V, Sharma AK, *Inorg Chim Acta* 2007, **360**, 3226
41. Kujime M, Hikichi S, Akita M, *Inorg Chim Acta* 2003, **350**, 163.
42. Shirasawa N, Nguyet TT, Hikichi S, Moro-oka Y, Akita M, *Organometallics* 2001, **20**, 3582.
43. Kosugi M, Hikichi S, Akita M, Moro-oka Y, *J Chem Soc Dalton Trans* 1999, 1369.
44. Kosugi M, Hikichi S, Akita M, Moro-oka Y, *Inorg Chem* 1999, **38**, 2567.
45. Singh UP, Sharma AK, Hikichi S, Komatsuzaki H, Moro-oka Y, Akita M, *Inorg Chim Acta* 2006, **359**, 4407.
46. Singh UP, Singh R, Hikichi S, Akita M, Moro-oka Y, *Inorg Chim Acta* 2000, **310**, 273.
47. Singh UP, Sharma AK, Tyagi P, Upreti S, Singh RK, *Polyhedron* 2006, **25**, 3628.
48. Sugawara K-i, Hikichi S, Akita M, *J Chem Soc Dalton Trans* 2002, 4514.
49. Sugawara K-i, Hikichi S, Akita M, *Chem Lett* 2001, **30**, 1094.
50. Sugawara K-i, Hikichi S, Akita M, *Dalton Trans* 2003, 4346.
51. Akita M, Miyaji T, Muroga N, Mock-Knoblauch C, Adam W, Hikichi S, Moro-oka Y, *Inorg Chem* 2000, **39**, 2096.
52. Akita M, Miyaji T, Hikichi S, Moro-oka Y, *Chem Lett* 1999, **8**, 813.
53. Kujime M, Hikichi S, Akita M, *Chem Lett* 2003, **32**, 486.

54. Miyaji T, Kujime M, Hikichi S, Moro-oka Y, Akita M, *Inorg Chem* 2002, **41**, 5286.
55. Kujime M, Hikichi S, Akita M, *Dalton Trans* 2003, 3506.
56. Akita M, Hashimoto M, Hikichi S, Moro-oka Y, *Organometallics* 2000, **19**, 3744.
57. Ohta K, Hashimoto M, Takahashi Y, Hikichi S, Akita M, Moro-oka Y, *Organometallics* 1999, **18**, 3234.
58. Takahashi Y, Hashimoto M, Hikichi S, Akita M, Moro-oka Y, *Angew Chem Int Ed* 1999, **38**, 3074.
59. Takahashi Y, Hashimoto M, Hikichi S, Moro-oka Y, Akita M, *Inorg Chim Acta* 2004, **357**, 1711.
60. Takahashi Y, Hikichi S, Akita M, Moro-oka Y, *Chem Commun* 1999, 1491.
61. Takahashi Y, Hikichi S, Akita M, Moro-oka Y, *Organometallics* 1999, **18**, 2571.
62. Akita M, Takahashi Y, Hikichi S, Moro-oka Y, *Inorg Chem* 2001, **40**, 169.
63. Takahashi Y, Hikichi S, Moro-oka Y, Akita M, *Polyhedron* 2004, **23**, 225.
64. Basumallick L, DeBeer George S, Randall DW, Hedman B, Hodgson KO, Fujisawa K, Solomon EI, *Inorg Chim Acta* 2002, **337**, 357.
65. Randall DW, DeBeer George S, Hedman B, Hodgson KO, Fujisawa K, Solomon EI, *J Am Chem Soc* 2000, **122**, 11620.
66. Fujisawa K, Kobayashi T, Fujita K, Kitajima N, Moro-oka Y, Miyashita Y, Yamada Y, Okamoto K, *Bull Chem Soc Jpn* 2000, **73**, 1797.
67. Fujisawa K, Iwata Y, Kitajima N, Higashimura H, Kubota M, Miyashita Y, Yamada Y, Okamoto K, Moro-oka Y, *Chem Lett* 1999, **28**, 739.
68. Lehnert N, Cornelissen U, Neese F, Ono T, Noguchi Y, Okamoto K-i, Fujisawa K, *Inorg Chem* 2007, **46**, 3916.
69. Chen P, Fujisawa K, Helton ME, Karlin KD, Solomon EI, *J Am Chem Soc* 2003, **125**, 6394.
70. Chen P, Root DE, Campochiaro C, Fujisawa K, Solomon EI, *J Am Chem Soc* 2003, **125**, 466.
71. Fujisawa K, Ono T, Ishikawa Y, Amir N, Miyashita Y, Okamoto K-i, Lehnert N, *Inorg Chem* 2006, **45**, 1698.
72. Kujime M, Kurahashi T, Tomura M, Fujii H, *Inorg Chem* 2007, **46**, 541.

73. Chen P, Fujisawa K, Solomon EI, *J Am Chem Soc* 2000, **122**, 10177.
74. Fujisawa K, Fujita K, Takahashi T, Kitajima N, Moro-oka Y, Matsunaga Y, Miyashita Y, Okamoto K-i, *Inorg Chem Commun* 2004, 7, 1188.
75. Matsunaga Y, Fujisawa K, Ibi N, Miyashita Y, Okamoto K-i, *Inorg Chem* 2005, **44**, 325.
76. Gorelsky SJ, Basumallick L, Vura-Weis J, Sarangi R, Hodgson KO, Hedman B, Fujisawa K, Solomon EI, *Inorg Chem* 2005, **44**, 4947.
77. Fujisawa K, Lehnert N, Ishikawa Y, Okamoto K-i, *Angew Chem Int Ed* 2004, **43**, 4944.
78. Fujisawa K, Matsunaga Y, Ibi N, Amir N, Miyashita Y, Okamoto K-i, *Bull Soc Chem Jpn* 2006, **79**, 1894.
79. Maria L, Domingos Â, Santos I, *Inorg Chem Commun* 2003, **6**, 58.
80. Maria L, Domingos Â, Galvão A, Ascenso J, Santos I, *Inorg Chem* 2004, **43**, 6426.
81. Pellei M, Gioia Lobbia G, Santini C, Spagna R, Camalli M, Fedeli D, Falcioni G, *Dalton Trans* 2004, 2822.
82. Santini C, Pellei M, Gioia Lobbia G, Fedeli D, Falcioni G, *J Inorg Biochem* 2003, **94**, 348.
83. Kunkely H, Vogler A, *Chem Phys Lett* 2000, **327**, 162.
84. Chen J, Woo LK, *J Organomet Chem* 2000, **601**, 57.
85. Chisholm MH, Eilerts NW, Huffman JC, Iyer SS, Pacold M, Phomphrai K, *J Am Chem Soc* 2000, **122**, 11845.
86. Reger DL, Wright TD, Smith MD, Rheingold AL, Kassel S, Concolino T, Rhagitan B, *Polyhedron* 2002, **21**, 1795.
87. Reger DL, Wright TD, Smith MD, *Inorg Chim Acta* 2002, **334**, 1.
88. Harder S, Brettar J, *Angew Chem Int Ed* 2006, **45**, 3474.
89. Onishi M, Yamaguchi M, Nishimoto E, Itoh Y, Nagaoka J, Umakoshi K, Kawano H, *Inorg Chim Acta* 2003, **343**, 111.
90. Kisko JL, Fillebeen T, Hascall T, Parkin G, *J Organomet Chem* 2000, **596**, 22.
91. Graziani O, Toupet L, Hamon J-R, Tilset M, *Inorg Chim Acta* 2002, **341**, 127.
92. Graziani O, Toupet L, Tilset M, Hamon J-R, *Inorg Chim Acta* 2007, **360**, 3083.
93. Shay DT, Yap GPA, Zakharov LN, Rheingold AL, Theopold KH, *Angew Chem Int Ed* 2005, **44**, 1508.
94. Thyagarajan S, Incarvito CD, Rheingold AL, Theopold KH, *Chem Commun* 2001, 2198.

95. Thyagarajan S, Incarvito CD, Rheingold AL, Theopold KH, *Inorg Chim Acta* 2003, **345**, 333.
96. Thyagarajan S, Shay DT, Incarvito CD, Rheingold AL, Theopold KH, *J Am Chem Soc* 2003, **125**, 4440.
97. Santi R, Romano AM, Sommazzi A, Grande M, Bianchini C, Mantovani G, *J Mol Cat A: Chem* 2005, **229**, 191.
98. Van Kirk CC, Qin K, Theopold KH, Evans DH, *J Electroanal Chem* 2004, **565**, 185.
99. Bergquist C, Parkin G, *J Am Chem Soc* 1999, **121**, 6322.
100. Bergquist C, Fillebeen T, Morlok MM, Parkin G, *J Am Chem Soc* 2003, **125**, 6189.
101. Bergquist C, Koutcher L, Vaught AL, Parkin G, *Inorg Chem* 2002, **41**, 625.
102. Lipton AS, Bergquist C, Parkin G, Ellis PD, *J Am Chem Soc* 2003, **125**, 3768.
103. Rombach M, Vahrenkamp H, *Inorg Chem* 2001, **40**, 6144.
104. Bergquist C, Storrie H, Koutcher L, Bridgewater BM, Friesner RA, Parkin G, *J Am Chem Soc* 2000, **122**, 12651.
105. Müller-Hartmann A, Vahrenkamp H, *Eur J Inorg Chem* 2000, 2371.
106. Cucurull-Sánchez L, Maseras F, Lledós A, *Inorg Chem Commun* 2000, **3**, 590.
107. Ferrence GM, McDonald R, Takats J, *Angew Chem Int Ed* 1999, **38**, 2233.
108. Morisette M, Haufe S, McDonald R, Ferrence GM, Takats J, *Polyhedron* 2004, **23**, 263.
109. Ferrence GM, McDonald R, Morissette M, Takats J, *J Organomet Chem* 2000, **596**, 95.
110. Ferrence GM, Takats J, *J Organomet Chem* 2002, **647**, 84.
111. Ferrence GM, Arduengo AJ, Jockisch A, Kim H-J, McDonald R, Takats J, *J Alloys Comp* 2006, **418**, 184.
112. Fujisawa K, Tada N, Ishikawa Y, Higashimura H, Miyashita Y, Okamoto KI, *Inorg Chem Commun* 2004, **7**, 209.
113. Nabika M, Seki Y, Miyatake T, Ishikawa Y, Okamoto K-i, Fujisawa K, *Organometallics* 2004, **23**, 4335.
114. Lehnert N, Fujisawa K, Solomon EI, *Inorg Chem* 2003, **42**, 469.
115. Lehnert N, Ho RYN, Que Jr L, Solomon EI, *J Am Chem Soc* 2001, **123**, 12802.
116. Tyagi S, Sharma CL, Mahendra SM, Singh UP, *Ind J Chem Inorg Bio-Inorg Phys Theor An Chem* 2004, **43A**, 83.

117. Ciunik Z, Ruman T, Łukasiewicz M, Wołowiec S, *J Mol Struct* 2004, **690**, 175.
118. Łukasiewicz M, Ciunik Z, Wołowiec S, *Polyhedron* 2001, **20**, 2119.
119. Sun Y-J, Zhang LZ, Cheng P, Lin H-K, Yan S-P, Liao D-Z, Jiang Z-H, Shen P-W, *Biophys Chem* 2004, **109**, 281.
120. Dhar S, Reddy PAN, Nethaji M, Mahadevan S, Saha MK, Chakravarty AR, *Inorg Chem* 2002, **41**, 3469.
121. Dhar S, Chakravarty AR, *Inorg Chem* 2005, **44**, 2582.
122. Silvestre I, Wolowska J, McInnes EJL, Kilner CA, Halcrow MA, *Inorg Chim Acta* 2005, **358**, 1337.
123. Liu X, Chia LML, Kilner CA, Yellowlees LJ, Thornton-Pett M, Trofimenko S, Halcrow MA, *Chem Commun* 2000, 1947.
124. Sun Y-J, Zhang LZ, Cheng P, Lin H-K, Yan S-P, Sun W, Liao DZ, Jiang Z-H, Shen P-W, *Inorg Chem* 2003, **42**, 508.
125. Halcrow MA, Chia LML, Liu X, McInnes EJL, Yellowlees LJ, Mabbs FE, Davies JE, *Chem Commun* 1998, 2465.
126. Halcrow MA, Chia LML, Liu X, McInnes EJL, Yellowlees LJ, Mabbs FE, Scowen IJ, McPartlin M, Davies JE, *J Chem Soc Dalton Trans* 1999, 1753.
127. Chia LML, Radojevic S, Scowen IJ, McPartlin M, Halcrow MA, *J Chem Soc Dalton Trans* 2000, 133.
128. Siemer CJ, Meece FA, Armstrong WH, Eichhorn DM, *Polyhedron* 2001, **20**, 2637.
129. Gil MP, Casagrande Jr OL, *J Organomet Chem* 2004, **689**, 286.
130. Novikova EV, Belov GP, Klaui W, Solotnov AA, *Vysokomolekulyarnye Soedineniya Seriya A i Seriya B* 2003, **45**, 1605.
131. Slugovc C, Mereiter K, Trofimenko S, Carmona E, *Chem Commun* 2000, 121.
132. Slugovc C, Mereiter K, Trofimenko S, Carmona E, *Helv Chim Acta* 2001, **84**, 2868.
133. Slugovc C, Mereiter K, Trofimenko S, Carmona E, *Angew Chem Int Ed* 2000, **39**, 2158.
134. Chia LML, Wheatley AEH, Feeder N, Davies JE, *Polyhedron* 2000, **19**, 109.
135. Darensbourg DJ, Billodeaux DR, Perez LM, Helerow MA, *Organometallics* 2004, **23**, 5286.
136. Long DP, Chandrasekaran A, Day RO, Bianconi PA, Rheingold AL, *Inorg Chem* 2000, **39**, 4476.

137. Criado R, Cano M, Campo JA, Heras JV, Pinilla E, Torres MR, *Polyhedron* 2004, **23**, 301.
138. Jacobsen FE, Breece RM, Myers WK, Tierney DL, Cohen SM, *Inorg Chem* 2006, **45**, 7306.
139. Ruman T, Ciunik Z, Goclan A, Łukasiewicz M, Wołowiec S, *Polyhedron* 2001, **20**, 2965.
140. Yakovenko AV, Kolotilov SV, Addison AW, Trofimenko S, Yap GPA, Lopushanskaya V, Pavlishchuk VV, *Inorg Chem Commun* 2005, **8**, 932.
141. Ruman T, Ciunik Z, Szklanny E, Wołowiec S, *Polyhedron* 2002, **21**, 2743.
142. He H-S, *Acta Cryst* 2006, **E62**, m2016.
143. Bräuer M, Anders E, Sinnecker S, Rombach M, Brombacher H, Vahrenkamp H, *Chem Commun* 2000, 647.
144. Brand U, Rombach M, Seebacher J, Vahrenkamp H, *Inorg Chem* 2001, **40**, 6151.
145. Rombach M, Brombacher H, Vahrenkamp H, *Eur J Inorg Chem* 2002, 153.
146. Rombach M, Gelinsky M, Vahrenkamp H, *Inorg Chim Acta* 2002, **334**, 25.
147. Brombacher H, Vahrenkamp H, *Inorg Chem* 2004, **43**, 6054.
148. Brombacher H, Vahrenkamp H, *Inorg Chem* 2004, **43**, 6050.
149. Gross F, Vahrenkamp H, *Inorg Chem* 2005, **44**, 4433.
150. Tekeste T, Vahrenkamp H, *Inorg Chem* 2006, **45**, 10799.
151. Brombacher H, Vahrenkamp H, *Inorg Chem* 2004, **43**, 6042.
152. Puerta DT, Cohen SM, *Inorg Chem* 2002, **41**, 5075.
153. Puerta DT, Cohen SM, *Inorg Chem* 2003, **42**, 3423.
154. Jacobsen FE, Lewis JA, Heroux KJ, Cohen SM, *Inorg Chim Acta* 2007, **360**, 264.
155. Puerta DT, Cohen SM, *Inorg Chim Acta* 2002, **337**, 459.
156. Ibrahim MM, Pérez Olmo C, Tekeste T, Seebacher J, He G, Maldonado Calvo JA, Böhmerle K, Steinfeld G, Brombacher H, Vahrenkamp H, *Inorg Chem* 2006, **45**, 7493.
157. He H, Puerta DT, Cohen SM, Rodgers KR, *Inorg Chem* 2005, **44**, 7431.
158. Badura D, Vahrenkamp H, *Inorg Chem* 2002, **41**, 6013.
159. Gelinsky M, Vogler R, Vahrenkamp H, *Inorg Chim Acta* 2003, **334**, 230.
160. Peter A, Vahrenkamp H, *Z Anorg Allg Chem* 2005, **631**, 2347.

161. Badura D, Vahrenkamp H, *Eur J Inorg Chem* 2003, 89.
162. Ruman T, Ciunik Z, Trzeciak AM, Wołowiec S, Ziólkowski JL, *Organometallics* 2003, **22**, 1072.
163. Moszner M, Wołowiec S, Trösch A, Vahrenkamp H, *J Organomet Chem* 2000, **595**, 178.
164. Mehn MP, Fujisawa K, Hegg EL, Que Jr L, *J Am Chem Soc* 2003, **125**, 7828.
165. Foster CL, Liu X, Kilner CA, Thornton-Pett M, Halcrow MA, *J Chem Soc Dalton Trans* 2000, 4563.
166. Higashimura H, Kubota M, Shiga A, Fujisawa K, Moro-oka Y, Uyama H, Kobayashi S, *Macromolecules* 2000, **33**, 1986.
167. Ruman T, Ciunik Z, Mazurek J, Wołowiec S, *Eur J Inorg Chem* 2002, 754.
168. Harding DJ, Harding P, Adams H, Tuntulani T, *Inorg Chim Acta* 2007, **360**, 3335.
169. Kunrath FA, de Souza RF, Casagrande Jr OL, Brooks NR, Young Jr VG, *Organometallics* 2003, **22**, 4739.
170. Michiue K, Jordan RF, *Organometallics* 2004, **23**, 460.
171. Furlan LG, Gil MP, Casagrande OL, *Macromol Rapid Commun* 2000, **21**, 1054.
172. Michiue K, Steele I, Casagrande OL, Jordan RF, *Acta Cryst* 2006, **E62**, m2297.
173. Michiue K, Murtuza S, Jordan RF, *Polym Mat Sci Eng* 2002, **86**, 95.
174. Michiue K, Jordan RF, *Macromolecules* 2003, **36**, 9707.
175. Pariya C, Incarvito CD, Rheingold AL, Theopold KH, *Polyhedron* 2004, **23**, 439.
176. López JA, Mereiter K, Paneque M, Poveda ML, Serrano O, Trofimenko S, Carmona E, *Chem Commun* 2006, 3921.
177. Kunrath FA, Casagrande Jr OL, Toupet L, Carpentier J-F, *Eur J Inorg Chem* 2004, 4803.
178. Lopes I, Dias R, Domingos Â, Marques N, *J All Compds* 2002, **344**, 60.
179. Silva M, Domingos Â, Píres de Matos A, Marques N, Trofimenko S, *J Chem Soc Dalton Trans* 2000, 4628.
180. Dias HVR, Rajapakse RMG, Krishantha DMM, Fianchini M, Wang X, Elsenbaumer RL, *J Mater Chem* 2007, **17**, 1762.
181. Walz R, Ruf M, Vahrenkamp H, *Eur J Inorg Chem* 2001, 139.
182. Walz R, Vahrenkamp H, *Inorg Chim Acta* 2001, **314**, 58.
183. Badura D, Vahrenkamp H, *Inorg Chem* 2002, **41**, 6020.
184. Depperman EC, Bodnar SH, Vostrikova KE, Shultz DA, Kirk ML, *J Am Chem Soc* 2001, **123**, 3133.

185. Shultz DA, Bodnar SH, *Inorg Chem* 1999, **38**, 591.
186. Shultz DA, Bodnar SH, Kumar RK, Lee H, Kampf JW, *Inorg Chem* 2001, **40**, 546.
187. Shultz DA, Bodnar SH, Kumar RK, Kampf JW, *J Am Chem Soc* 1999, **121**, 10664.
188. Shultz DA, Fico Jr RM, Bodnar SH, Kumar RK, Vostrikova KE, Kampf JW, Boyle PD, *J Am Chem Soc* 2003, **125**, 11761.
189. Shultz DA, Mussari CP, Ramanathan KK, Kampf JW, *Inorg Chem* 2006, **45**, 5752.
190. Kirk ML, Shultz DA, Depperman EC, Brannen CL, *J Am Chem Soc* 2007, **129**, 1937.
191. Shultz DA, Sloop JC, Coote T-A, Beikmohammadi M, Kampf J, Boyle PD, *Inorg Chem* 2007, **46**, 273.
192. Claramunt RM, Sanz D, Santa María MD, Elguero J, Trofimenko S, *J Organomet Chem* 2004, **689**, 463.
193. Halcrow MA, Kilner CA, Thornton-Pett M, *Acta Crystallogr* 2001, **C57**, 711.
194. Caballero A, Gómez-de la Torre F, Jalón FA, Manzano BR, Rodriguez AM, Trofimenko S, Sigalas MP, *J Chem Soc Dalton Trans* 2001, 427.
195. Kisała J, Białońska A, Ciunik Z, Kurek S, Wołowiec S, *Polyhedron* 2006, **25**, 3222.
196. Campo JA, Cano M, Heras JV, Lagunas MC, Pinilla E, Torres MR, *Polyhedron* 2001, **20**, 2997.
197. Rojo G, Agulló-López F, Campo JA, Cano M, Lagunas MC, Heras JV, *Synth Met* 2001, **124**, 201.
198. López-Garabito C, Campo JA, Heras JV, Cano M, Rojo G, Agulló-López F, *J Phys Chem B* 1998, **102**, 10698.
199. Campo JA, Cano M, Heras JV, Lopez-Garabito C, Pinilla E, Torres R, Rojo G, Agulló-Lopez F, *J Mater Chem* 1999, **9**, 899.
200. Rojo G, Agulló-López F, Campo JA, Heras JV, Cano M, *J Phys Chem B* 1999, **103**, 11016.
201. Deng Y, Ding T-Z, Li Y, *Synth React Inorg Met-Org Nano-Met Chem* 2002, **32**, 219.
202. Deng Y, Wang RJ, Sun SQ, Feng YP, Ding TZ, *Chem J Chin Univers Chin Ed* 2001, **22**, 764.
203. Santa María MD, Claramunt RM, Campo JA, Cano M, Criado R, Heras JV, Ovejero P, Pinilla E, Torres MR, *J Organomet Chem* 2000, **605**, 117.
204. Deng Y, Wang R-J, Ding T-Z, Li Y, Sun S-Q, Feng Y-P, Zhao Y-F, *Polyhedron* 2001, **20**, 291.

205. Sarangi R, Aboelella N, Fujisawa K, Tolman WB, Hedman B, Hodgson KO, Solomon EI, *J Am Chem Soc* 2006, **128**, 8286.
206. Dias HVR, Lu H-L, Kim H-J, Polach SA, Goh TKHH, Browning RG, Lovely CJ, *Organometallics* 2002, **21**, 1466.
207. Cecchi P, Berrettoni M, Giorgetti M, Gioia Lobbia G, Calogero S, Stievano L, *Inorg Chim Acta* 2001, **319**, 67.
208. Hu Z, Williams RD, Tran D, Spiro TG, Gorun SM, *J Am Chem Soc* 2000, **122**, 3556.
209. Gorun SM, Hu Z, Stibrany RT, Carpenter G, *Inorg Chim Acta* 2000, **297**, 383.
210. Diaconu D, Hu Z, Gorun SM, *J Am Chem Soc* 2002, **124**, 1564.
211. Hu Z, George GN, Gorun SM, *Inorg Chem* 2001, **40**, 4812.
212. Hu Z, Gorun SM, *Inorg Chem* 2001, **40**, 667.
213. Dias HVR, Jin W, *Inorg Chem* 2000, **39**, 815.
214. Dias HVR, Ayers AE, *Polyhedron* 2002, **21**, 611.
215. Dias HVR, Polach SA, *Inorg Chem* 2000, **39**, 4676.
216. Dias HVR, Polach SA, Goh S-K, Archibong EF, Marynick DS, *Inorg Chem* 2000, **39**, 3894.
217. Dias HVR, Wang Z, *Inorg Chem* 2000, **39**, 3890.
218. Dias HVR, Wang Z, *Inorg Chem* 2000, **39**, 3724.
219. Dias HVR, Browning RG, Polach SA, Diyabalanage HVK, Lovely CJ, *J Am Chem Soc* 2003, **125**, 9270.
220. Dias HVR, Batdorf KH, Fianchini M, Diyabalanage HVK, Carnahan S, Mulcahy R, Rabiee A, Nelson K, Van Waasbergen LG, *J Inorg Biochem* 2006, **100**, 158.
221. Balili MNC, Pintauer T, *Acta Cryst* 2007, **E63**, m988.
222. Dias HVR, Jin W, *Inorg Chem* 2003, **42**, 5034.
223. Dias HVR, Goh TKHH, *Polyhedron* 2004, **23**, 273.
224. Dias HVR, Wang X, Diyabalanage HVK, *Inorg Chem* 2005, **44**, 7322.
225. Janiak C, Temizdemir S, Dechert S, *Inorg Chem Commun* 2000, **33**, 271.
226. Janiak C, Temizedmir S, Dechert S, Deck W, Girgsdies F, Heinze J, Kolm MJ, Scharmann TG, Zipffel OM, *Eur J Inorg Chem* 2000, 1229.
227. Craven E, Mutlu E, Lundberg D, Temizdemir S, Dechert S, Brombacher H, Janiak C, *Polyhedron* 2002, **21**, 553.
228. Carella A, Jaud J, Rapenne G, Launay J-P, *Chem Commun* 2003, 2434.
229. Oulié P, Teichert J, Vendier L, Dablemont C, Etienne M, *New J Chem* 2006, **30**, 679.

230. Paul RL, Amoroso AJ, Jones PL, Couchman SM, Reeves ZR, Rees LH, Jeffery JC, McCleverty JA, Ward MD, *J Chem Soc Dalton Trans* 1999, 1563.
231. Ward MD, McCleverty JA, Mann KLV, Jeffery JC, Motson GR, Hurst J, *Acta Cryst* 1999, **C55**, 2055.
232. Bell ZR, Motson GR, Jeffery JC, McCleverty JA, Ward MD, *Polyhedron* 2001, **20**, 2045.
233. Davies GM, Adams H, Ward MD, *Acta Cryst* 2005, **C61**, m221.
234. Davies GM, Adams H, Pope SJA, Faulkner S, Ward MD, *Photochem Photobiol Sci* 2005, **4**, 829.
235. Beeby A, Burton-Pye BP, Faulkner S, Motson GR, Jeffery JC, McCleverty JA, Ward MD, *J Chem Soc Dalton Trans* 2002, 1923.
236. Davies GM, Aarons RJ, Motson GR, Jeffery JC, Adams H, Faulkner S, Ward MD, *Dalton Trans* 2004, 1136.
237. Motson GR, Mamula O, Jeffery JC, McCleverty JA, Ward MD, von Zelewsky A, *J Chem Soc Dalton Trans* 2001, 1389.
238. Armaroli N, Accorsi G, Barigelletti F, Couchman SM, Fleming JS, Harden NC, Jeffery JC, Mann KLV, McCleverty JA, Rees LH, Starling SR, Ward MD, *Inorg Chem* 1999, **38**, 5769.
239. Babbar P, Brunner H, Singh UP, *Ind J Chem Sect A Inorg Bio-Inorg Phys Theor Anal Chem* 2001, **40**, 225.
240. Keyes MC, Young Jr VC, Tolman WB, *Organometallics* 1996, **15**, 4133.
241. Janiak C, Scharmann TG, Mason SA, *J Am Chem Soc* 2002, **124**, 14010.
242. Janiak C, Temizdemir S, Scharmann TG, Schmalstieg A, Demtschuk J, *Z Anorg Allg Chem* 2000, **626**, 2053.
243. Gioia Lobbia G, Pellei M, Pettinari C, Santini C, Skelton BW, Somers N, White AH, *Dalton Trans* 2002, 2333.
244. Youm KT, Kim MG, Ko J, Jun M-J, *Angew Chem Int Ed* 2006, **45**, 4003.
245. Gioia Lobbia G, Pellei M, Pettinari C, Santini C, Skelton BW, White AH, *Inorg Chim Acta* 2005, **358**, 1162.
246. Zeng Z, Twamley B, Shreeve JM, *Organometallics* 2007, **26**, 1782.
247. Gioia Lobbia G, Pellei M, Pettinari C, Santini C, Skelton BW, White AH, *Inorg Chim Acta* 2005, **358**, 3633.

Chapter 4

Heteroscorpionates

4.1. General Considerations

This chapter will describe heteroscorpionates (Fig. 4.1), i.e. ligands of general structure RR′(Bp^x) in which R = or ≠ R′ = H, alkyl, or aryl, or a pyrazolyl group different from those forming the B(μ-pz^x)M ring.[1–3] The following class of heteroscorpionates can be individuated:

a. H_2Bp^x (≡ $H_2B(pz^x)_2$).
b. R_2Bp^x (≡ $R_2B(pz^x)_2$) (R = alkyl, aryl or halogen).
c. RR′Bp^x (≡ $R_2B(pz^x)_2$) (R = H, alkyl, aryl or halogen, R′ = group containing heteroatoms (O, S, NR″)).
d. R(Az)Bp^x (≡ R(Az)B(pz^x)$_2$) (R = H, alkyl, aryl; HAz = azole-type ligand).
e. $H_2B(pz^x)(pz^y)$ where pz^x and pz^y are different pyrazolyl groups.

4.2. Specific Ligands

4.2.1. *Bp^x*

Bp^x ligands are characterized by the presence of the BH_2 group which although renders the Bp^x ligands more hydrolytically labile than their Tp^x counterparts. The Bp^x ligands are able to establish an agostic

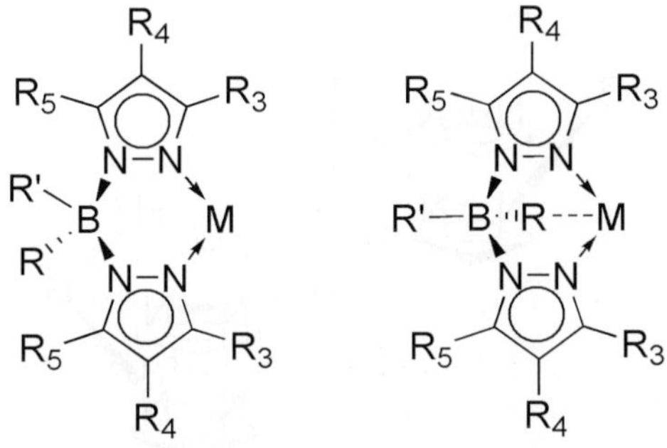

I II

Fig. 4.1. Heteroscorpionates can coordinate the metal in chelating bidentate fashion (I) or through the pseudoaxial R group (II).

B–H–M bond with many metals. The residual reducing power of the BH_2 group makes the Bp^x complexes of easily reducible cations, such as silver(I) and palladium(II), unstable. They can be used in some instances for organic reductions.

4.2.1.1. *Bp and Bp**

The Bp ligand is the simplest of all the heteroscorpionate ligands. It is basically bidentate, forming the typical boat-shaped $H_2B(\mu\text{-pz})_2M$ ring, in which the pseudoaxial B–H may form an agostic bond with the metal.[4,5] $[M(Bp)_2]$ complexes of first row transition metals are usually square planar or tetrahedral.[6,7] Some octahedral anionic $[M(Bp)_3]^-$ species of low stability can be also isolated.[8,9] Five-coordinate anionic species $[MX(Bp)_2]^-$ (M = Cr, Mn; X = halides or pseudohalides) have been also known.[10,11]

The molecular structure of $[Tl(Bp)]_2$, compared with those of analogous $[Tl(Bp^x)]_2$ derivatives, demonstrates how the existence of Tl $\cdots$ Tl interaction (Fig. 4.2) is highly dependent on the nature of the pyrazolyl substituents. The $Tl^I \cdots Tl^I$ separation in $[Tl(Bp)]_2$ is the shortest with respect to those found in the other $[Tl(Bp^x)]_2$.[12]

Pressure effect studies on the spin crossover behavior of the mononuclear compounds $[Fe(Bp)_2(bipy)]$ and $[Fe(Bp)_2(phen)]$ have been performed.[13]

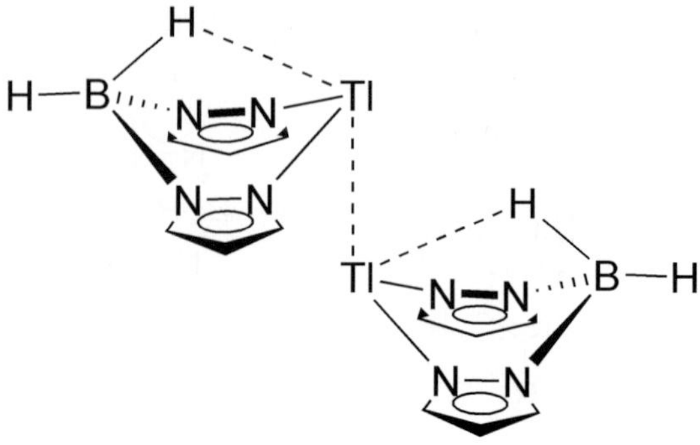

Fig. 4.2. Tl···Tl interaction in $[Tl(Bp)]_2$.

Reaction of $CuCl_2$ and $Ni(OAc)_2$ with Na(Bp) yielded $[Cu(Bp)_2]$ and $[Ni(Bp)_2]$, respectively, which have been characterized by IR, thermal analyses, and X-ray diffraction study.[14]

The square pyramidal (μ-O)di-[VO(Bp*)] has been prepared and characterized by IR, electronic and ESR spectroscopic, and magnetic susceptibility measurements. A kinetic study of its capability to oxidize BuOH has been also reported.[15]

Two new pentanuclear complexes $(NEt_4)_2[MS_4(CuBp)_4]$ (M = Mo, W) have been synthesized and characterized by IR, UV–Vis, ^1H, and ^{13}C NMR spectra, and the structure of the tungsten complex $(NEt_4)_2[WS_4(CuBp)_4]$ has been also determined by single-crystal X-ray diffraction. In the pentanuclear dianion $[WS_4(CuBp)_4]$, having a crystallographic S_4 symmetry, the coordination environment around each copper atom is distorted tetrahedral with two N donors of the bidentate Bp ligands and two S donors of the WS_4 core, representing a structural model for some copper proteins.[16] The pentanuclear cluster $(NEt_4)_2[MoS_4(CuBp)_4]$ has also been also prepared and its NLO properties are investigated. The structure of this complex has been determined by single-crystal X-ray diffraction. The coordination environment around each copper atom is distorted tetrahedral with two N donors of the bidentate bis(pyrazolyl)borate ligands and two S donors of the MoS_4 group.[17] Two copper complexes $(NEt_4)_2[MS_4(CuBp^*)_2]\cdot X$ (M = Mo, X = acetone; M = W, X = MeCN) have also been prepared and characterized by IR,

UV-Vis, ^1H and ^{13}C NMR spectroscopy. They are isostructural and exhibit a distorted tetrahedral coordination environment around each copper atom from two N atoms of Bp* and two S donor of the central MS_4 group.[18]

The complex $[Et_4N]_2[(Bp^*)CuMoS_4Cu_2(\mu\text{-}Bp^*)Cu_2MoS_4Cu(Bp^*)]$ — obtained in acetone by the reaction of $[MoS_4(CuCl)_3]^{2-}$ with Bp* — crystallizes as centrosymmetric dimeric anion, Bp* acting both as bridging and terminal ligand (Fig. 4.3), and Cu being in two different coordination environments, one three-coordinate and the other four-coordinate. $[Et_4N]_2[(Bp^*)CuMoS_4Cu_2(\mu\text{-}Bp^*)Cu_2MoS_4Cu(Bp^*)]$ is, however, unstable in MeCN where it generates the square-planar $[Cu(Bp^*)_2]$ complex (Fig. 4.4).[19]

$Zn(Bp)_2$ [20] and $Cd(Bp^*)_2$ [21] have been structurally characterized, both zinc and cadmium centers being in a pseudotetrahedral environment.

$[Pd(Me)(Bp^x)(L)]$ (Bp^x = Bp or Bp*; L = PPh_3, PCy_3) have been synthesized and characterized. The reaction of $[Pd(Me)(Bp^x)(L)]$ with

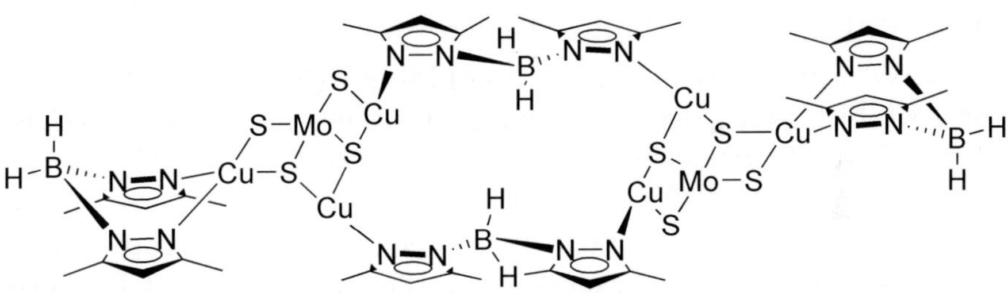

Fig. 4.3. The centrosymmetric dimeric anion $[(Bp^*)CuMoS_4Cu_2(\mu\text{-}Bp^*)Cu_2MoS_4Cu(Bp^*)]$.

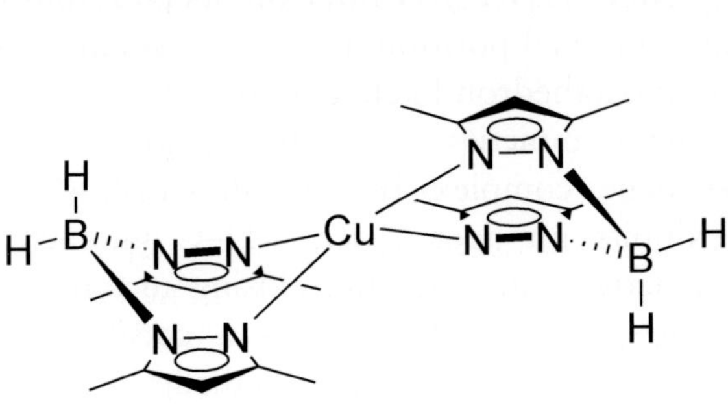

Fig. 4.4. The square-planar $[Cu(Bp^*)_2]$.

CO produces [Pd(C(=O)Me)(Bp^x)(L)]. The distorted square-planar [Pd(C(=O)Me)(Bp)(PPh_3)] has been also structurally characterized.[22] [$PdMe_3$(Bp)(L)] (L = py-d_5, PMe_2Ph) have been synthesized in solution by oxidative addition of iodomethane to [$PdMe_2$(Bp)]$^-$ at −70°C followed by the addition of L. The Pd(IV) complexes reductively eliminate ethane above 0°C. The Pt(IV) analogs have also been isolated.[23]

[Mo(η^2-C(O)$CHMe_2$)(Bp)(CO)(PMe_3)$_2$] can be obtained by the reaction of KH with [Mo(η^2-C(O)CH_3)(Bp)(CO)(PMe_3)$_2$] followed by the addition of MeI. [Mo(η^2-C(O)$CHMe_2$)(Bp)(CO)(PMe_3)$_2$] evolves gradually at room temperature to [Mo(H)(Bp)$(CO)_2$(PMe_3)$_2$] and propene. [Mo(η^2-C(O)$CHMe_2$)(Bp^x)(CO)(PMe_3)$_2$] (Bp^x = Bp or Bp*) reacts with chelating diphosphines $R_2PCH_2CH_2PR_2$ (R = Me or Et) yielding [Mo(η^2-C(O)$CHMe_2$)(Bp^x)(CO)($R_2PCH_2CH_2PR_2$)].[24]

[W(≡CR)Br$(CO)_2$(NC_5H_4Me4)$_2$] (R = $C_6H_3Me_2$-2,6,C_6H_2-Me_3-2,4,6) reacts with K(Bp) yielding two interconverting isomers of [W(≡CR)$(CO)_2$($NC_5H_4Me_4$)(Bp)], one isomer structurally characterized for R = $C_6H_3Me_2$-2,6. In the absence of light, the reaction of [W(≡CR)Br$(CO)_2$(CNR′)$_2$] with K(Bp) gives [W(≡CR)$(CO)_2$(CNR′)(Bp)]. Under mild photolytic conditions, the thermally unstable ketenyl complex [W(η^2-OCCR)-(CO)(CNR′)$_2$(Bp)] has been reported. The picoline ligand in [W(≡CR)Br$(CO)_2$(pic)(Bp)] is readily replaced by PMe_2Ph to provide [W(≡CR)$(CO)_2$(PMe_2Ph)(Bp)]. The reaction of either [W(≡CR)$(CO)_2$(PMe_2Ph)(Bp)] or [W(≡CR)$(CO)_2$(pic)(Bp)] (R = $C_6H_2Me_3$-2,4,6) with excess PMe_2Ph gives [W(≡CR)(CO)-(PMe_2Ph)$_2$(Bp)].[25]

[Ru(CO){(*p*-MeC_6H_4)$_3$P}$_2$(H)Bp] exhibits two *trans*-disposed tri-*p*-tolylphoshines in axial position, the overall geometry of the metal being a distorted octahedron likely due to steric congestion between the bulky phosphino moieties and the Bp ligand.[26]

The benzylidene complex [Ru=CHPh(Cl)(Bp)(C_8H_{14})(PCy_3)], octahedral, exhibits an agostic Ru···H–C interaction and displays catalytic activity for the ring closing metathesis of diethylmalonate in the presence of CuCl as a cocatalyst in refluxing toluene.[27] The crystal structure of the octahedral hydrido compound [Ru(Bp)(H)(CO)($AsPh_3$)$_2$]·H_2O has been reported.[28]

The synthesis and the coordination chemistry of [Rh(Bp^x)(C_2H_4)$_2$] (Bp^x = Bp or Bp*) has been described by Carmona and coworkers.[29] When [Rh(Bp^x)(C_2H_4)$_2$] reacts with PR_3 or a N-donor ligand (L), the formation of [Rh(Bp^x)(C_2H_4)(L)] is observed. Excess of PR_3 produces the disubstituted 16 electron complex [Rh(Bp^x)(PR_3)$_2$]. When [Rh(Bp^x)(C_2H_4)$_2$] reacts with L′ (L′ = CO or CN^{t}Bu), [Rh(Bp^x)(L′)$_2$] is formed. Single-crystal studies indicated square-planar structures for [Rh(Bp*)(C_2H_4)$_2$] and [Rh(Bp)(PPh_3)$_2$].

[Rh(Bp*)(CO)(py)] has been prepared from the reaction of [Rh(Bp*)(CO)$_2$] with py. The reaction of [Rh(Bp*)(CO)(py)] with MeI gives a Rh carbene complex [HB(pz*)$_2$(O(Me)C=Rh(H)(I)(py))] (Fig. 4.5), upon addition of MeI to the Rh–C bond concomitant with hydride transfer from B to Rh. Thermolysis of this compound induces migration of the Rh hydride to the α-carbon to give [HB(pz*)$_2$(O(Me)HCRh(I)(py))].[30]

[Rh(Bp)(CO)L] (L = P(NC_4H_4)$_3$, PPh_3, PCy_3, P(C_6H_4OMe-4)$_3$) were prepared by exchange of the acac in [Rh(acac)(CO)L]. The spectroscopic and electrochemical properties as well as X-ray data of [Rh(Bp)(CO)L] complexes were investigated. The cyclic voltammetric results indicate that the Bp behaves as a much stronger electron donor than acac and a value of the Lever EL ligand parameter identical to that of the pyrazolate ligand is proposed for Bp^x and Tp^x. All complexes are very attractive precursors for hydroformylation catalysts, and aldehydes were obtained with all complexes without extra phosphine as cocatalyst.[31]

trans-[ReO_2(py)$_4$]Cl reacts with Bp^x (Bp^x = Bp and Bp*) yielding *trans,trans*-[ReO_2(py)$_4$][ReO_2(Bp^x)$_2$]. These compounds react with MClMe$_3$ (X = Si, Sn) giving [ReO(OSiMe_3)(Bp^x)$_2$]

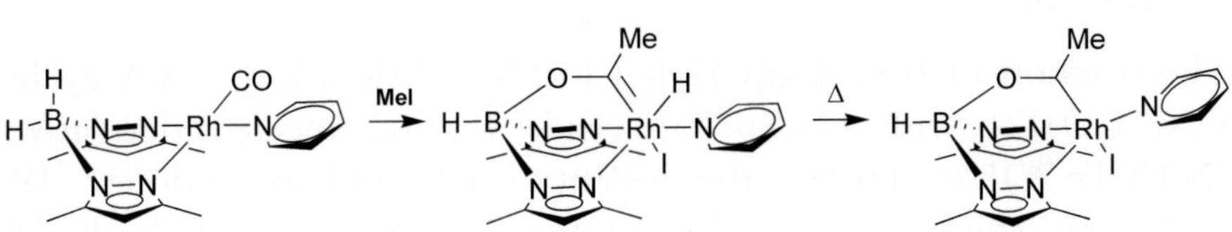

Fig. 4.5. Synthesis of the Rh–carbene complex [HB(pz*)$_2$(O(Me)C=Rh(H)(I)(py))]. Its thermolysis induces the formation of [HB(pz*)$_2$(O(Me)HCRh(I)(py))].

and [ReO(OSnMe$_3$)(Bp)$_2$]. *trans*-[ReO$_2$(py)$_4$][ReO$_2$(Bp)$_2$] and [ReO(OSiMe$_3$)(Bp)$_2$] have been also structurally characterized.[32]

[Ag(Bp^x)(PR$_3$)$_2$] (Bp^x = Bp and Bp*) (R$_3$ = Ph$_3$, Cy$_3$, Bn$_3$, *m*-Tol$_3$, *p*-Tol$_3$, EtPh$_2$, and MePh$_2$, have been prepared from the reaction of K(Bp^x) with AgNO$_3$ and PR$_3$. These compounds are stable, soluble in chlorinated solvents, and nonelectrolytes in CH$_2$Cl$_2$ and acetone. When P(*o*-Tol)$_3$ was employed as ancillary ligand, [Ag(Bp^x)(P(*o*-Tol)$_3$)] was formed.[33] Also, silver(I) derivatives containing bidentate phosphines and Bp^x (Bp^x = Bp or Bp*) have been prepared from the one-pot reaction of the heteroscorpionates with AgNO$_3$ and dppm, dppe, dppp, dppb or dppf. Their solid state and solution properties have been investigated through analytical and spectroscopic measurements. The structure of [{Ag(Bp)}(dppf)] has been determined by single-crystal X-ray studies, the bis(pyrazolyl)borate being always bidentate with the silver(I) center in distorted tetrahedral geometry.[34]

The synthesis of [ZrCl(η-C$_8$H$_8$)(κ^2-Bp)] has been reported by Hill and Smith, who described that the reaction of [ZrCl$_2$(η-C$_8$H$_8$)(THF)] with Bp* gave only the organometallic oxide cluster [Zr$_4$(μ-O)$_4$-(μ-Cl)$_2$(Cl)$_2$(η-C$_8$H$_8$)$_4$].[35] Some deprotonated diindolylmethanes have been employed to coordinate Ti and Zr and the compounds obtained have been compared with analogous bis(pyrazolyl)borates derivatives.[36]

4.2.1.2. *Bp^{iPr_2}*

[Cr(Bp^{iPr_2})$_2$] and [CrCl(Bp^{iPr_2})] were identified as by-products in the metathesis reaction of M(Tp^{iPr_2}) with Cr(II) salts.[37]

4.2.1.3. *Bp^{tBu}*

The reaction of *trans*-[Ni(CH$_2$R)Cl(PMe$_3$)$_2$] (R = SiMe$_3$, CMe$_2$Ph, CN, CO$_2$Et) with [Tl(Bp^{tBu})] yields the unexpected derivative [NiCl(Tp^{tBu})] containing the tridentate κ^3-scorpionate donor. IR and X-ray studies confirm the unexpected conversion of the Bp^{tBu} into Tp^{tBu} due to a degradation of the Bp ligands by Lewis or Brönsted-acid-induced cleavage of the B–N bond. On the other hand

$[Ni(Bp^{^tBu})_2]$ has been obtained when the reaction of $[Tl(Tp^{^tBu})]$ was carried out with $[NiCl_2(py)_4]$. $[Ni(Bp^{^tBu})_2]$ and $[Ni(Bp^{^tBu,Me})_2]$ contain a tridentate κ^3-*N*,*N*-*H*-coordination mode of the scorpionate ligand. Finally, using $MCl_2{\cdot}nH_2O$ (M = Fe, Co) as starting reagents, $[M(Bp^{^tBu})_2]$ complexes, isostructural with $[Ni(Bp^{^tBu})_2]$, have been obtained and structurally characterized.[38]

4.2.1.4. $Bp^{^tBu,Me}$

$[ZnX(Bp^{^tBu,Me})]$ (X = Me, I) have be easily obtained from the reaction of ZnX_2 with the corresponding Tl salt. The tetrahedral $[Cd(\mu\text{-}I)(Bp^{^tBu,Me})]_2$ has been also prepared and structurally characterized.[39]

The tetrahedral complexes $[AlMe_2(Bp^{^tBu,Me})]$ and $[GaMe_2(Bp^{^tBu,Me})]$ have been synthesized from the reaction of $AlMe_3$ and $GaMe_3$ with $K(Bp^{^tBu,Me})$. Single-crystal X-ray study indicated that the most noticeable difference in these very similar structures is the C–M–C bond angle.[40]

$[UI_2(Bp^{^tBu,Me})(THF)_2]$ and $[UI_2(Bp^{^tBu,Me})(OPPh_3)_2]$ have been synthesized and their crystal structure was determined by single-crystal X-ray diffraction. The monomeric $[UI_2(Bp^{^tBu,Me})(THF)_2]$ and $[UI_2(Bp^{^tBu,Me})(OPPh_3)_2]$ exhibit a seven coordinate U^{III} center from two pyrazolyl nitrogen, two iodide, two oxygens (THF and $OPPh_3$), and a B–H $\cdots$ U agostic interaction.[41]

4.2.1.5. $Bp^{^tBu,^iPr}$

$[ZnX(Bp^{^tBu,^iPr})]$ (X = Me, I) has been prepared and structurally characterized, the zinc center being in a trigonal planar coordination environment.[39]

4.2.1.6. Bp^{Ph}

The tetrahedral $[Co(Bp^{Ph})_2]$ has been synthesized and characterized by X-ray diffraction.[42] Reaction between $K(Bp^{Ph})$ and $Zn(OAc)_2$ resulted in the formation of the distorted tetrahedral $[Zn(Bp^{Ph})_2]$, isolated and crystallized in THF at -10°C.[43]

4.2.1.7. Bp^{2py}

Reaction of Bp^{2py} with cobalt(II) or nickel(II) salts gave the cyclic derivatives $[M_8(Bp^{2py})_{12}X]X_3$ (M = Co, Ni; X = ClO_4, PF_6) (Fig. 4.6), in which one anion is encapsulated in the central cavity.[44]

Ward and coworkers[45] also reported the synthesis of $[M(Bp^{2py})_2(NO_3)]$ (M = Nd, Yb) and the crystal and molecular structure of the Nd derivative. $[Ln(Bp^{2py})(dbm)_2]$ (Ln = Pr, Nd, Eu, Gd, Tb, Er, Yb) complexes were also prepared and structurally characterized, all being eight-coordinate, from one tetradentate ligand Bp^{2py} and two bidentate dbm ligands. Photophysical studies have been also performed on the whole series.[46] The complexes $[Ln(Bp^{2py})_2(NO_3)]$ (M = Pr,[47] Er[47] and Tb,[48] Eu[48] and Gd[48]) were synthesized and their structural and photophysical properties were studied. All complexes are ten-coordinate. Crystallographic analysis has shown that for the smaller Er(III) ion, steric congestion at the metal center results in two of the Er–N_{py} distances being particularly long, which does not occur with the larger Pr(III) ion, that is better able to accommodate 10-fold coordination. The emission properties of the metal derivatives have been rationalized in terms of subtle variations in the steric and electronic properties of the ligands.

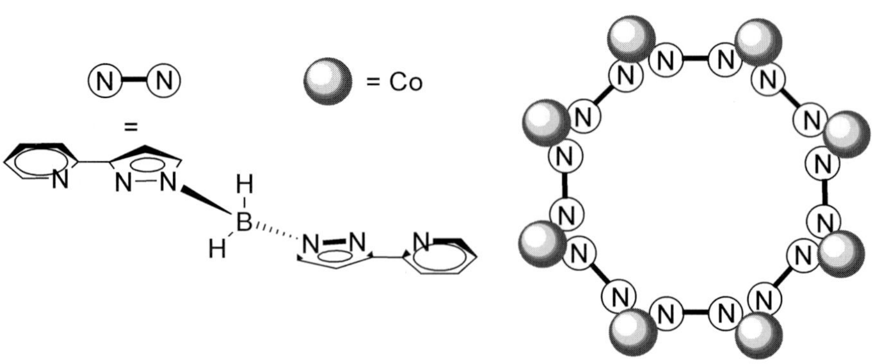

Fig. 4.6. The cyclic structure of $[M_8(Bp^{2py})_{12}X]X_3$.

4.2.1.8. Bp^{4Bo}

[Tl(Bp^{4Bo})] exhibits Tl-Bp^x units with pronounced Tl-π_{azolyl} interaction between neighboring molecules, and extended 3D and 1D structures through the bridging action of the poly(pyrazolyl)borate ligand between three symmetry related thallium centers.[49]

4.2.1.9. $Bp^{(CF_3)_2}$

Reaction of [RuH(COD)($Bp^{(CF_3)_2}$)] with 2 equiv. of PR_3 under 3 bar of H_2 produces [RuH(H_2)($Bp^{(CF_3)_2}$)(PR_3)$_2$] (R = Cy; R = ^{i}Pr). X-ray structural analysis of [RuH(H_2)($Bp^{(CF_3)_2}$)(P^iPr_3)$_2$] confirms a κ^2-*N*,*H*-bonding mode of $Bp^{(CF_3)_2}$. In the absence of dihydrogen, the monohydride complex [RuH($Bp^{(CF_3)_2}$)(COD)(PCy_3)] is formed.[50] Upon pressurization to 3 bar H_2, [RuH($Bp^{(CF_3)_2}$)(COD)(PCy_3)] converts into [RuH($Bp^{(CF_3)_2}$)(H_2)(PCy_3)]. Addition of 2 equiv. of PPh_3 or P(pyl)$_3$ to [RuH($Bp^{(CF_3)_2}$)(COD)] under 3 bar of H_2 leads to the formation of the corresponding hydrido complexes [RuH($Bp^{(CF_3)_2}$)(PR_3)$_2$] (R = Ph, pyl). Under similar conditions, [RuH($Bp^{(CF_3)_2}$)(COD)] reacts with 2 equiv. of ^{t}BuNH$_2$ to produce the amine adduct [RuH($Bp^{(CF_3)_2}$)(^{t}BuNH$_2$)$_2$]. By the addition of 1 equiv. of MeI to the latter compound, [RuI($Bp^{(CF_3)_2}$)(^{t}BuNH$_2$)$_2$] is isolated and characterized by an X-ray structural determination. [RuCl($Bp^{(CF_3)_2}$)(^{t}BuNH$_2$)$_2$] can be prepared by stirring a CH_2Cl_2 solution of [RuH($Bp^{(CF_3)_2}$)(^{t}BuNH$_2$)$_2$]. In the absence of dihydrogen, [RuH($Bp^{(CF_3)_2}$)(COD)] reacts with a large excess of ^{t}BuNH$_2$ to give [RuH($Bp^{(CF_3)_2}$)(COD)(^{t}BuNH$_2$)] in which a κ^2-*N*,*H* coordination of the $Bp^{(CF_3)_2}$ ligand is found. When the reaction was carried out under N_2 atmosphere, [RuH($Bp^{(CF_3)_2}$)(^{i}Pr$_2$NH)]$_2$(μ^2-N_2) has been isolated and characterized by single-crystal X-ray determination. Addition of mesitylene to [RuH($Bp^{(CF_3)_2}$)(COD)] produces the arene complex [RuH($Bp^{(CF_3)_2}$)(η^6-$C_6H_3Me_3$)]. The X-ray structure determination of this compound gave conclusive evidence for the κ^2-*N*,*N* bonding mode of the $Bp^{(CF_3)_2}$ ligand. When exposing [RuH($Bp^{(CF_3)_2}$)(COD)] to a CO atmosphere, [Ru($Bp^{(CF_3)_2}$)(COC_8H_{13})(CO)$_2$] is isolated. Analogous complexes have been obtained using the nonfluorinated complex [RuH(Bp*)(COD)] as starting material. The variation of the hapticity of the Bp ligand plays a crucial role in this chemistry, whereas the CF_3 substituents seem to have a limited influence.[50]

Dias and Lu demonstrated that it is possible to synthesize Bp^x metal complexes (metal ≠ alkali metal) directly from metal tetrahydroborates. The reaction of $[CuBH_4(PPh_3)_2]$ with $Hpz^{(CF_3)_2}$ at *ca.* 180°C yields $[Cu(Bp^{(CF_3)_2})(PPh_3)_2]$ in which $Bp^{(CF_3)_2}$ is coordinated to Cu through only one nitrogen atom, one of the hydrogens on boron showing a weak interaction with the metal (Fig. 4.7).[51]

Synthesis, structure, and photoluminescence properties are reported for the three-coordinate mononuclear and dinuclear $[M(Bp^{(CF_3)_2})(2,4,6\text{-collidine})]$ (M = Cu, Ag). The copper compound in the solid state exhibits bright blue phosphorescence at room temperature. Both compounds show blue phosphorescence at 77 K.[52] The ground-state and low-lying excited electronic states in $[M(Bp^{(CF_3)_2})(2,4,6\text{-collidine})]$ have been studied using DFT approach.[53]

Thermally stable copper(I) complexes containing phosphorus and nitrogen based donors, alkyne and alkene ligands, in detail $[Cu(Bp^{(CF_3)_2})(PPh_3)]$, $[Cu(Bp^{(CF_3)_2})(NCMe)]$, and $[Cu_2(Bp^{(CF_3)_2})_2(1,5\text{-COD})]$ were synthesized by using $K(Bp^{(CF_3)_2})$, Cu(OTf),

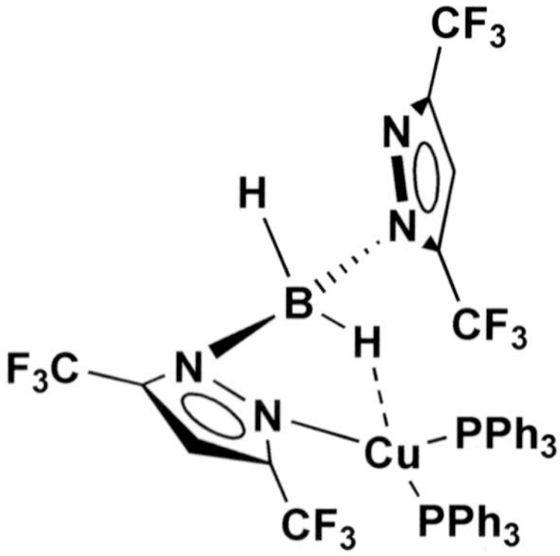

Fig. 4.7. In $[Cu(Bp^{(CF_3)_2})(PPh_3)_2]$ the heteroscorpionate acts as a bidentate *N,H*-donor.

and PPh_3, NCMe or 1,5-COD. $[Cu(Bp^{(CF_3)_2})(HC{\equiv}CPh)]$ and $[Cu(Bp^{(CF_3)_2})(H_2C{=}CHPh)]$ were also reported. X-ray data show that $[Cu(Bp^{(CF_3)_2})L]$ ($L{=}PPh_3$, $HC{\equiv}CPh$, $H_2C{=}CHPh$) and $[Cu_2(Bp^{(CF_3)_2})_2(1,5\text{-COD})]$ have three-coordinate, trigonal planar copper sites, with $Bp^{(CF_3)_2}$ in the common boat conformation.[54] $[Cu(Bp^{(CF_3)_2})L]$ (L = olefin: COE, Van, Clsty, Tevs, Fn) have been prepared one-pot by Pampaloni *et al.*[55] The structures of $[Cu(Bp^{(CF_3)_2})L]$ (L = COE, Van, Tevs, Fn) have been determined by X-ray diffraction methods. The carbonylation reactions of $[Cu(Bp^{(CF_3)_2})L]$ (L = COE, Van, Clsty, Tevs) have been also gas-volumetrically investigated.

4.2.1.10. $Bp^{CF_3,Me}$

Gorun *et al.*[56] reported the synthesis and characterization of the potassium salt $K(Bp^{CF_3,Me})$. In the blue-purple $[Cu(Bp^{CF_3,Me})_2]$ the N_4 donor set defines a symmetry imposed tetragonal geometry.

4.2.2. R_2Bp^x

The R_2Bp^x ligands generally provide more stable derivatives, due to the absence of reducing BH_2 groups. Agostic interactions are also rare, whereas the R groups are able to shield effectively the coordinated metals from other ligands.

4.2.2.1. Et_2Bp

$[AlEt_2(Et_2B(\mu\text{-pz})_2)]$, synthesized by the reaction of $AlEt_2Cl$ with $K(Et_2Bp)$, exhibits the metal center eclipsed with respect to the plane defined by the four nitrogen atom. This complex reacts smoothly with O_2 to form alkyl(alkoxy)aluminum species. The dimeric five-coordinate compound $[Et_2B(\mu\text{-pz})_2Al(\mu\text{-OEt})Et]_2$ has been also reported (Fig. 4.8).[57]

Thompson and coworkers[58] reported the synthesis and structural characterization of $[Pt(Et_2Bp)(2\text{-}(2,4\text{-difluorophenyl})pyridyl)]$ (Fig. 4.9).

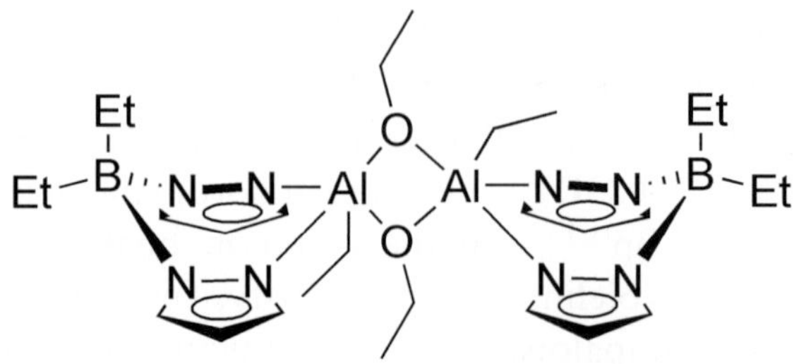

Fig. 4.8. $[Et_2B(\mu\text{-pz})_2Al(\mu\text{-OEt})Et]_2$.

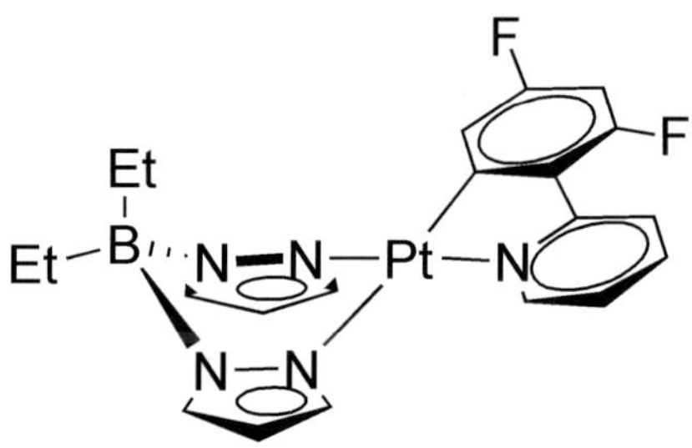

Fig. 4.9. $[Pt(Et_2Bp)(2\text{-}(2,4\text{-difluorophenyl})pyridyl)]$.

4.2.2.2. *Ph_2Bp*

Thomas and Peters[59] compare the mononuclear platinum complexes of Ph_2Bp and $[Ph_2B(CH_2PPh_2)_2]^-$(Ph_2BP_2). The relative *trans* influence of these ligands has been assessed by comparison of the structural and spectroscopic (NMR) data of $[PtMe_2(Ph_2Bp)][NBu_4]$ and $[Pt(Me)_2(Ph_2BP_2)][ASN]$. $[Pt(Me)(Ph_2Bp)(NCCH_3)]$, $[Pt(Me)(Ph_2Bp)(CO)]$, and $[Pt(Me)(Ph_2Bp)(P(C_6F_5)_3)]$ have been also reported. The CO stretching frequency in $[Pt(Me)(Ph_2Bp)(CO)]$ suggests that the charged borate moiety renders the R_2Bp^x ligands more electron-donating than their neutral analogs. Upon protonation or methide abstraction in benzene at room temperature, $[PtMe_2(Ph_2Bp)][NBu_4]$ rapidly activates two molecules of benzene to generate $[PtPh_2(Ph_2Bp)][NBu_4]$ (Fig. 4.10). The double C–H activation reaction can be halted by the addition of MeCN to trap the intermediate $[Pt(Ph)(Ph_2Bp)(NCMe)]$. $[PtMe_2(Ph_2Bp)][NBu_4]$ also

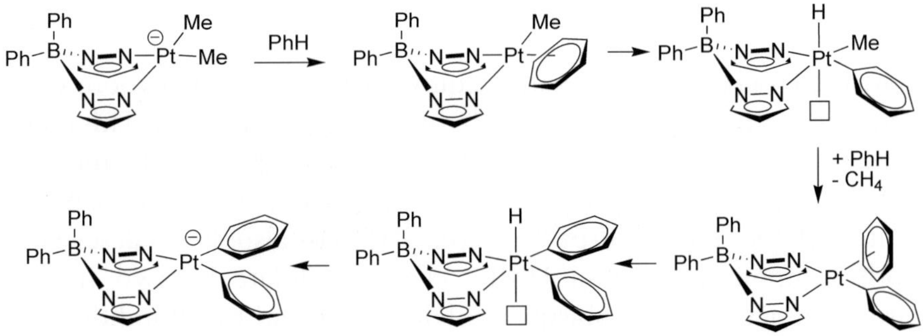

Fig. 4.10. $[PtMe_2(Ph_2Bp)][NBu_4]$ is able to rapidly activate two molecules of benzene giving $[PtPh_2(Ph_2Bp)][NBu_4]$.

displays reactivity toward the benzylic C–H bonds of mesitylene at room temperature to form $[Pt(CH_2C_6H_3(CH_3)_2)(Ph_2Bp)(Hpz)]$.[59]

In the six-coordinate $[U(Ph_2Bp)_3]$ the metal is in a trigonal prismatic geometry around the uranium center.[41]

4.2.2.3. *BBN(pz)*$_2$

[Ni{BBN(pz)$_2$}(*o*-Tol)(PPh$_3$)] has been synthesized by the reaction of [Ni(Br)(*o*-Tol)(PPh$_3$)$_2$] with K{BBN(pz)$_2$}. When the same reaction was carried out in the presence of moisture, the nickel complex [Ni{BBN(pz)(OH)}(*o*-Tol)(PPh$_3$)] was formed. The square planar [Ni{BBN(pz)$_2$}(*o*-Tol)(PPh$_3$)] and [Ni(Tp^{Ph})(*o*-Tol)(PPh$_3$)] show very similar activity in the copolymerization of CO and ethene. This seems to indicate that the borane hydrogen and the noncoordinated pyrazolyl ring are not essential for the catalysis. [Ni{BBN(pz)(OH)}-(*o*-Tol)(PPh$_3$)] is inactive.[60]

4.2.3. *$H_2B(pz^x)(pz^y)$*

4.2.3.1. *$H_2B(pz^{4Me})(pz^*)$*

Agrifoglio and Capparelli[61] reported $[M\{H_2B(pz^{4Me})(pz^*)\}_2]$ (M = Co, Zn), in which M is tetrahedrally coordinated to two bidentate ligands. The molecules display C_1 symmetry.

4.2.3.2. *$H_2B(pz^{^tBu,Me})(pz^{Me,^tBu})$ and $H_2B(pz^{^tBu,Me})(pz^*)$*

When $[UI_3(THF)_4]$ reacts with 2 equiv. of $K(Bp^{^tBu,Me})$ at room temperature, $[UI\{\kappa^3\text{-}H(\mu\text{-}H)B(pz^{^tBu,Me})(pz^{Me,^tBu})\}_2]$ is formed. This complex, stabilized by two asymmetric heteroscorpionate ligands, is obtained upon an isomerization process promoted *in situ* by the metal center. $[UI\{\kappa^3\text{-}H(\mu\text{-}H)B(pz^{^tBu,Me})(pz^*)\}_2]$ has also been described. Both compounds contain U(III) seven-coordinate by two tridentate asymmetric dihydrobis(pyrazolyl)borates and by an iodide. The geometry around the atom is strongly distorted and can be described as a pentagonal bipyramid. In the attempted crystallization, the complex $[UI(\mu\text{-}I)\{\kappa^3\text{-}Bp^{^tBu,Me}\}(Hpz^{^tBu,Me})]$ has been obtained in which the U(III) center, seven-coordinate, exhibits a 4 : 3 tetragonal base-trigonal geometry.[62]

4.2.3.3. *$H_2B(pz^*)(pz^{Ph_2})$*

Ruman *et al.*[63] reported the synthesis of $H_2B(pz^*)(pz^{Ph_2})$ and also of its high-spin cobalt(II) complexes. The molecular structure of $[Co(Bp^{Ph_2})(H_2B(pz^*)(pz^{Ph_2}))]$ and $[Co(H_2B(pz^*)(pz^{Ph_2}))_2]$ were established on the basis of a single-crystal X-ray study. $[Co(Bp^{Ph_2})(H_2B(pz^*)(pz^{Ph_2}))]$ contains a six-coordinate metal center with five Co–N(pyrazolyl) bonds and a short Co–H (attached to the boron atom of the scorpionate), whereas $[Co(H_2B(pz^*)(pz^{Ph_2}))_2]$ is distorted tetrahedral from two bidentate $H_2B(pz^*)(pz^{Ph_2})$ ligands.

4.2.4. *R(Az)Bp^x*

This section will cover Bp^x ligands, definitely tridentate due to the presence of a third pyrazolyl group different from pz^x. The $R(Az)Bp^x$ includes also those where Az is a methimazolyl group.

4.2.4.1. *$(pz^{Ph_2})Bp^*$ and $(pz^*)Bp^{Ph_2}$*

The two heteroscorpionates $(pz^{Ph_2})Bp^*$ and $(pz^*)Bp^{Ph_2}$ have been employed to prepare high-spin cobalt(II) complexes. The molecular structure of the six-coordinate $[Co(NO_3)\{(pz^{Ph_2})Bp^*\}(THF)]$ and of

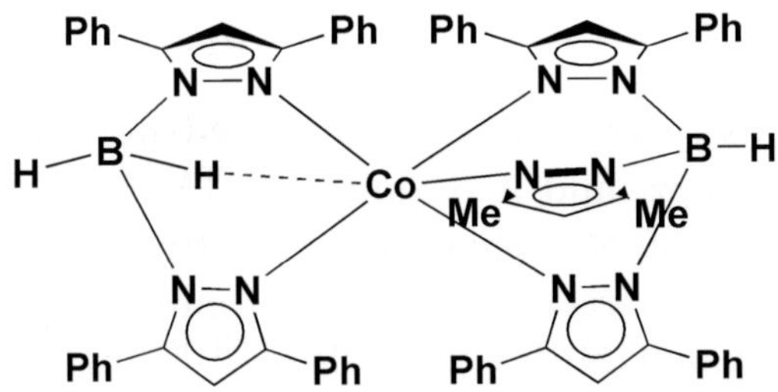

Fig. 4.11. $[Co(Bp^{Ph_2})\{(pz^*)Bp^{Ph_2}\}]$, in which Bp^{Ph_2} is bonded in the κ^3-N_2,H coordination mode.

$[Co(Bp^{Ph_2})\{(pz^*)Bp^{Ph_2}\}]\cdot 2C_6H_{14}$ (Fig. 4.11) have been reported. The scorpionate ligands are coordinated in the classical κ^3-coordination mode in all complexes. In the nitrate derivative, the coordination sphere is completed by three oxygen donor atoms from NO_3 and THF; whereas in the heteroleptic complex, Bp^{Ph_2} acts as a κ^3-N_2, H-donor.[64]

4.2.4.2. $(pz^{^iPr,Ph})Bp^{Ph,^iPr}$ *and* $(pz^{Ph,^iPr})Bp^{^iPr,Ph}$

Condensation of 5(3)-^{i}Pr-3(5)-Phpz with $NaBH_4$ led to a mixture of tris(pyrazolyl)borate ligands: the homoscorpionate $Tp^{Ph,^iPr}$ and the heteroscorpionates $(pz^{^iPr,Ph})Bp^{Ph,^iPr}$ and $(pz^{Ph,^iPr})Bp^{^iPr,Ph}$ have been prepared and employed to coordinate {Co(NCS)} groups. The molecular structures of the five-coordinate $[Co(SCN)(Tp^{Ph,^iPr})(THF)]$ and $[Co(SCN)\{(pz^{Ph,^iPr})Bp^{^iPr,Ph}\}(THF)]$ were also reported.[65]

4.2.4.3. $(pz^{5-^iPr})Bp^{Ph,Me}$ *and* $(pz^{5-^iPr})Bp^{Ph_2}$

Condensation of $NaBH_4$ with $Hpz^{3(5)-^iPr}$ and either $Hpz^{5(3)-Me-3(5)-Ph}$ or Hpz^{Ph_2} gave heteroscorpionate ligands $(pz^{5-^iPr})Bp^{Ph,Me}$ and $(pz^{5-^iPr})Bp^{Ph_2}$, respectively. These tripodal ligands react with CoX_2 salts yielding $[Co\{(pz^{5-^iPr})Bp^{Ph,Me}\}_2]$ and $[CoSCN\{(pz^{5-^iPr})\ Bp^{Ph,Me}\}(THF)]$, both structurally characterized, in which the heteroscorpionate ligand is bound in an η^3 fashion. These compounds as well as $[Co(SCN)\{(pz^{5-^iPr})Bp^{Ph_2}\}]$ were investigated by ^1H NMR and IR spectroscopy.[66]

4.2.4.4. $(pz^*)Bp^{^iPr,Ph}$

$(pz^*)Bp^{^iPr,Ph}$ has been prepared by condensation of 2.1 equiv. of 5(3)^{i}Pr-3(5)-Phpz and 0.9 equiv. of Hpz* with $NaBH_4$. Its reaction with $Co(SCN)_2$ gave [Co(SCN){(pz*)$Bp^{^iPr,Ph}$}THF] and it is structurally characterized. This compound was converted thermally into [Co(SCN){HB($pz^{^iPr,Ph}$)($pz^{Ph,^iPr}$)(pz*)}THF]$_2$ a dimer (Fig. 4.12) obtained in a crystalline form from toluene/hexane. In [Co(SCN){HB($pz^{^iPr,Ph}$)($pz^{Ph,^iPr}$)(pz*)}THF]$_2$, two chiral scorpionate ligands of opposite boron-centered configurations are coordinated to cobalt(II) ions in a κ^3-fashion. This chiral scorpionate ligand is the first example of a borate with fixed boron-centered chirality.[66]

The structure of [Rh{(pz*)$Bp^{^iPr,Ph}$}(NBD)] has been also reported, the scorpionate acting in the κ^3-tridentate form. The same ligand was found to be coordinate in κ^2-fashion in [Rh{(pz*)$Bp^{^iPr,Ph}$}(COD)] by variable temperature 1D and 2D ^{1}H NMR measurements. The intramolecular exchange between coordinated and uncoordinated 3-Ph-5-^{i}Prpz rings occurs with $\Delta G^{\neq} = 42.2$ kJ/mol as it was calculated from the ^{1}H NMR studies.[67]

4.2.4.5. $(pz^*)Bp^{Ph,Me}$ and $(pz^{Me,Ph})Bp^{Ph,Me}$

When [Na{(pz*)$Bp^{Ph,Me}$}] reacts with $Co(NCS)_2$, [Co(NCS){(pz*)-$Bp^{Ph,Me}$}(THF)]·THF is formed. Heating these cobalt complexes at 428 K in propylbenzene resulted in a borotropic shift of $pz^{Ph,Me}$ moiety and subsequent formation of the chiral

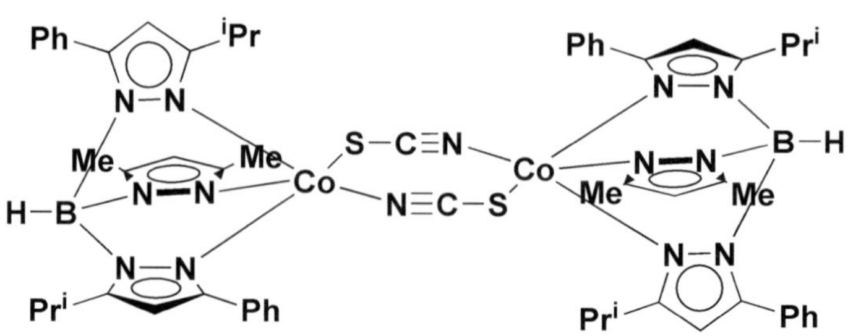

Fig. 4.12. [Co(SCN){HB($pz^{^iPr,Ph}$)($pz^{Ph,^iPr}$)(pz*)}THF]$_2$, containing two chiral scorpionate ligands of opposite boron-centered configurations.

$[Co(NCS)\{HB(pz^{Ph,Me})(pz^{Me,Ph})(pz^*)\}]$ and $[Co(NCS)\{(pz^{Me,Ph})Bp^{Ph,Me}\}]$.[68]

The structure of $[Tl\{(pz^*)Bp^{Ph,Me}\}]$ in which the ligand is coordinated in a κ^3-fashion is reported by Ruman and coworkers.[69] This compound crystallizes as centrosymmetric dimer in which weak CH/π intra- and inter-dimer interactions are responsible for the arrangement in the crystal structure.

The homoleptic $[M\{(pz^{Me,Ph})Bp^{Ph,Me}\}_2]$ (M = Co, Cu) have been prepared and their structure is elucidated crystallographically.[70] $(pz^{Me,Ph})Bp^{Ph,Me}$ has been also employed to synthesize the four-coordinate $[Co(NCS)\{(pz^{Me,Ph})Bp^{Ph,Me}\}(THF)] \cdot THF$, which contains a high-spin metal center. This complex readily underwent conversion to the five-coordinate $[Co(OOCCH(OH)CH_3)\{(pz^{Me,Ph})Bp^{Ph,Me}\}]$ and to the six-coordinate $[Co(Tp)\{(pz^{Me,Ph})\ Bp^{Ph,Me}\}]$.[71]

4.2.4.6. $(pz^{Et_2})Bp^{Ph,Me}$

$[Na\{(pz^{Et_2})Bp^{Ph,Me}\}]$, $[Co(NCS)\{(pz^{Et_2})Bp^{Ph,Me}\}(THF)]$, and $[Co(Tp)\{(pz^{Et_2})Bp^{Ph,Me}\}]$ were synthesized and structurally characterized. The steric hindrance imposed by $(pz^{Et_2})Bp^{Ph,Me}$ was probed with ^1H NMR by *in situ* conversions of $[Co(NCS)\{(pz^{Et_2})Bp^{Ph,Me}\}(CD_3OD)]$ into $[Co\{(pz^{Et_2})Bp^{Ph,Me}\}(lactate)]$ and $[Co(Tp)\{(pz^{Et_2})Bp^{Ph,Me}\}]$.[72]

$(pz^{Et_2})Bp^{Ph,Me}$ has been demonstrated to be able to coordinate to rhodium(I) in a κ^2-fashion giving $[Rh\{(pz^{Et_2})Bp^{Ph,Me}\}(LL)$ (LL = COD or $(CO)_2$). While one of the $pz^{Ph,Me}$ groups and the nonsterically hindered pz^{Et_2} are coordinated through the N2 atoms, the axially appended $pz^{Ph,Me}$ is in side-on conformation. Variable-temperature ^1H NMR measurements have been employed to determine the energy of activation of the exchange between the coordinated and the uncoordinated pyrazolyl groups. Analogous $(pz^*)Bp^{Ph,Me}$–and $(pz^{Me,Ph})Bp^{Ph,Me}$–Rh derivatives have been reported.[73]

4.2.4.7. $(pz)Bp^{^tBu_2}$

$[Tl\{(pz)Bp^{^tBu_2}\}]$, prepared by the reaction of $LiBH_4$ with a mixture of Hpz and $Hpz^{^tBu_2}$, followed by a treatment with Tl(OAc), structurally characterized, proves that steric environment about a metal center may be modulated in a symmetric manner.[74]

4.2.4.8. $(pz^{^tBu})Bp^{(CF_3)_2}$

$K\{(pz^{^tBu})Bp^{(CF_3)_2}\}$, $[Tl\{(pz^{^tBu})Bp^{(CF_3)_2}\}]$, $[ZnCl\{(pz^{^tBu})Bp^{(CF_3)_2}\}]$, and $[Zn(OSiMe_3)\{(pz^{^tBu})Bp^{(CF_3)_2}\}]$ have been reported and characterized by Chisholm *et al.*[75]

4.2.4.9. $(pz^{^tBu})Bp^{Ph}$

The high-spin four-coordinate complexes $[Co(R)(pz^{^tBu}Bp^{Ph})]$ have been reported (R = unidentate NCS or lactate) and $[Co(SCN)\{(pz^{^tBu})Bp^{Ph}\}]$ has been structurally characterized.[76]

4.2.4.10. $(pz^{^tBu_2})Bp^{Ph}$

$(pz^{^tBu_2})Bp^{Ph}$ has been employed in the synthesis of the four-coordinate $[Co(NCS)\{(pz^{^tBu_2})Bp^{Ph}\}]$.[71]

4.2.4.11. $(mt^{^tBu})Bp$

Zn–ethyl-, –benzyl-, –phenyl-, and –*p*-nitrophenyl-thiolate complexes containing the ligand $(mt^{^tBu})Bp$ (Fig. 4.13) have been reported and their alkylation by MeI investigated.[77]

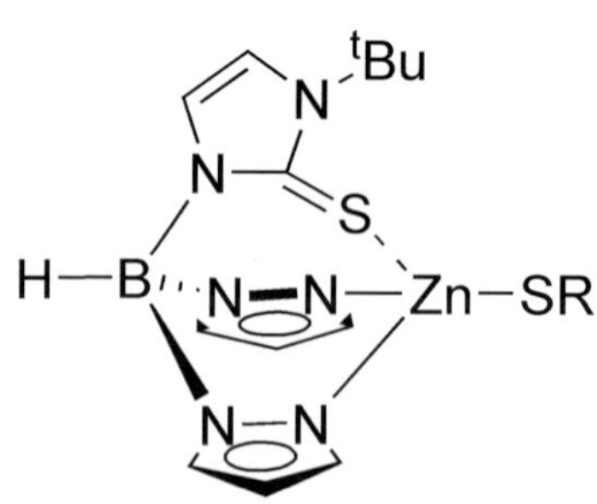

Fig. 4.13. $[Zn(SR)\{(mt^{^tBu})Bp\}]$ (R = Et, Bn, Ph, *p*-NO_2Ph).

4.3. Pyrazaboles

Organoboron polymers containing transition metals (Ni, Pd, Pt) and pyrazaboles were obtained by Cu-catalyzed coupling polymerization. The Ni- and Pt-containing polymers showed MLCT (metal-to-ligand charge transfer) absorption.[78]

The complex $[(\eta^1:\eta^1\text{-Indz})(\mu\text{-B})Et_2]_2$ obtained by the treatment of indazole with BEt_3 in toluene has an *anti* dimeric molecular structure (Fig. 4.14).[79]

4.4. $H_2B(Az)_2$ (Az = tz or btz)

$[Cu\{H_2B(tz)_2\}(PR_3)_x]$ (x = 1 or 2) have been prepared from the reaction of CuCl with PR_3 and $K\{H_2B(tz)_2\}_3$ and investigated by NMR, IR, and ESI MS spectroscopy. The structurally authenticated arrays fall into two different types: (i) discrete mononuclear $[Cu\{H_2B(tz)_2\}(PR_3)_2]$ with a $CuP_2(N_2)$ coordination sphere; (ii) one-dimensional polynuclear $[Cu\{(tz)BH_2(tz)\}(PR_3)]_{(\infty|\infty)}$ with a four-coordinate $CuP_2(N_2)$ coordination sphere due to a coordination by way of both N(2) and N(4) in one of the heterocyclic rings (Fig. 4.15).[80]

Silver(I) derivatives containing monodentate tertiary phosphines PR_3 (R = Ph, Bn, *o*-Tol, *m*-Tol, *p*-Tol) and $H_2B(tz)_2$ have been also reported. All the compounds are soluble in chlorinated solvents and nonelectrolytes in CH_2Cl_2 and acetone solutions. $[Ag\{H_2B(tz)_2\}(PPh_3)_2]$ and $[Ag\{H_2B(tz)_2\}\{P(m\text{-Tol})_3\}_2]$ are simple

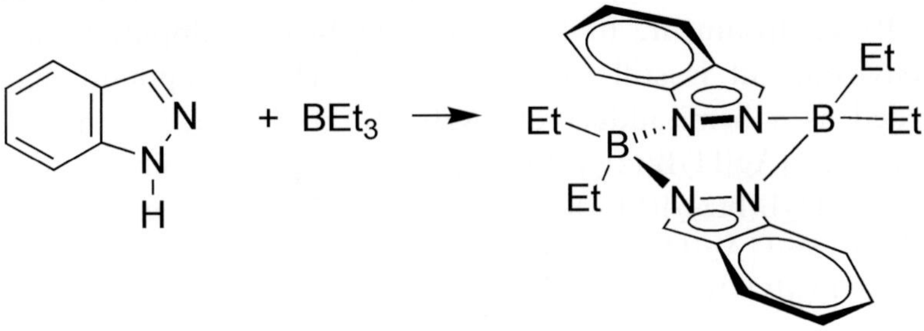

Fig. 4.14. Reaction of BEt_3 with indazole gives $[(\eta^1:\eta^1\text{-Indz})(\mu\text{-B})Et_2]_2$.

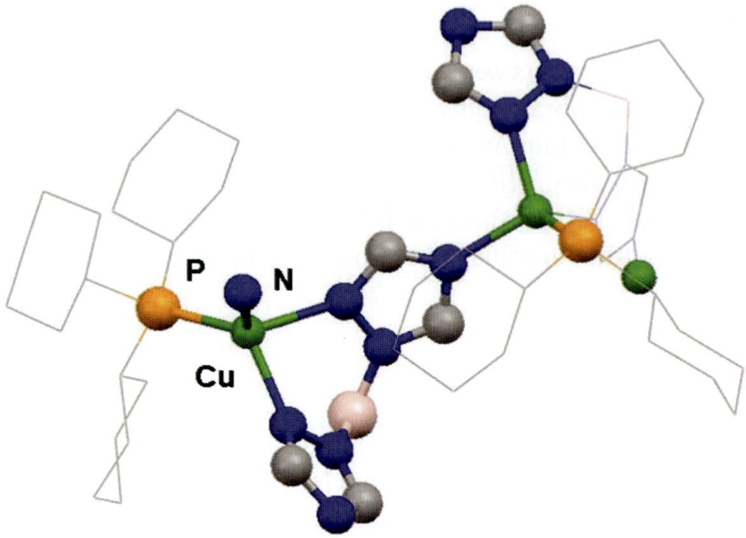

Fig. 4.15. The polynuclear $[Cu\{(tz)BH_2(tz)\}(PCy_3)]_{(\infty|\infty)}$.

mononuclear arrays, in which the silver atoms are lying in four-coordinate N_2AgP_2 environments. In $[Ag\{H_2B(tz)_2\}\{P(p\text{-Tol})_3\}]$, the silver environment is still four-coordinate but $PAgN_3$, utilizing the coordinating capability of one of the additional (exo-)ring nitrogens not only to complete the four-coordinate array about the silver but, necessarily, to link successive asymmetric units into a single-stranded polymer.[81] Also analogous species containing bidentate diphosphines have been reported. $[Ag\{H_2B(tz)_2\}(dppf)]$ is a simple mononuclear array, the silver atom lying in four-coordinate N_2AgP_2 environment. In $[Ag_3(NO_3)\{H_2B(tz)_2\}_2(dppm)_2]$, the silver environment is still four-coordinate. This compound may be regarded as a hemi-adduct of $[Ag\{H_2B(tz)_2\}(dppm)]$ with silver(I) nitrate, thus $[Ag\{H_2B(tz)_2\}(dppm)]$: $AgNO_3$ (2:1).[82] Silver(I) derivatives $[Ag\{H_2B(btz)_2\}(PR_3)_x]$ ($x = 1$ or 2, R = Ph, *o*-Tol, *m*-Tol, *p*-Tol, Bn) and $[Ag\{H_2B(btz)_2\}]_2(L)$ (L = dppe or dppf) have been also reported. $[Ag\{H_2B(btz)_2\}]_2(dppf)$ is a dimer, the dppf bridging the two silver atoms and $H_2B(btz)_2$, which chelates one silver, bridging to the second also.[83]

References

1. Trofimenko S, *J Am Chem Soc* 1966, **88**, 1842.
2. Trofimenko S, *J Am Chem Soc* 1967, **89**, 3170.
3. Trofimenko S, *J Am Chem Soc* 1967, **89**, 6288.
4. Trofimenko S, *Inorg Chem* 1970, **9**, 2493.
5. Cotton FA, Stanislowski AG, *J Am Chem Soc* 1974, **96**, 5074.
6. Trofimenko S, *Inorg Synth* 1970, **12**, 99.
7. Bernarducci E, Schwindinger WF, Hughey IV JL, Krogh-Jespersen K, Schugar HJ, *J Am Chem Soc* 1981, **103**, 1686.
8. Liang Z, Marshall AG, Marçalo J, Marques N, Pires de Matos A, Santos I, Well DA, *Organometallics* 1991, **10**, 2794.
9. Mani F, *Inorg Chim Acta* 1986, **117**, L1.
10. DiVaira M, Mani F, *J Chem Soc Dalton Trans* 1990, **191**.
11. Bencini A, DiVaira M, Mani F, *J Chem Soc Dalton Trans* 1991, **41**.
12. Ghosh P, Rheingold AL, Parkin G, *Inorg Chem* 1999, **38**, 5464.
13. Galet A, Gaspar AB, Agusti G, Muñoz MC, Levchenko G, Real JA, *Eur J Inorg Chem* 2006, 3571.
14. Xing YH, Yuan HQ, Zhang BL, Zhang YH, Niu SY, Bai FY, *J Coord Chem* 2007, **60**, 163.
15. Gogoi PK, Borborah A, *Nat Acad Sci Lett* 2003, **26**, 150.
16. Sadr MH, Clegg W, Bijhanzade HR, *Polyhedron* 2004, **23**, 637.
17. Sadr MH, Sardroodi JJ, Zare D, Brooks NR, Clegg W, Song Y, *Polyhedron* 2006, **25**, 3285.
18. Beheshti A, Clegg W, Sadr MH, *Inorg Chim Acta* 2002, **335**, 21.
19. Beheshti A, Clegg W, Sadr MH, *Polyhedron* 2001, **20**, 179.
20. Reger DL, Wright TD, Smith MD, Rheingold AL, Rhagitan B, *J Chem Crystallogr* 2000, **30**, 665.
21. Reger DL, Wright TD, Rheingold AL, Rhagitan B, *J Chem Crystallogr* 2001, **31**, 93.
22. Díaz-Requejo MM, Nicasio MC, Belderrain TR, Pérez PJ, Puerta MC, Valerga P, *Eur J Inorg Chem* 2000, 1359.
23. Canty AJ, Patel J, Pfeffer M, Skelton BW, White AH, *Inorg Chim Acta* 2002, **327**, 20.
24. Contreras L, Pizzano A, Sánchez L, Carmona E, *J Organomet Chem* 1999, **582**, 3.
25. Hill AF, Malget JM, White AJP, Williams DJ, *Eur J Inorg Chem* 2004, **818**.
26. Huh S, Park YJ, Lough AJ, Jun M-J, *Acta Cryst* 2000, **C56**, 416.

27. Patel J, Jackson RW, Serelis AK, *Inorg Chim Acta* 2004, **357**, 2374.
28. Huh S, Kim Y, Park S, Park T-J, Jun M-J, *Acta Cryst* 1999, **C55**, 850.
29. Baena MJ, Reyes ML, Rey L, Carmona E, Nicasio MC, Pérez PJ, Gutierrez E, Monge A, *Inorg Chim Acta* 1998, **273**, 244.
30. Yeston JS, Bergman RG, *Organometallics* 2000, **19**, 2947.
31. Trzeciak AL, Borak B, Ciunik Z, Ziołkowski JJ, Guedes Da Silva FMC, Pombeiro AJL, *Eur J Inorg Chem* 2004, 1411.
32. Reddy KR, Domingos Â, Paulo A, Santos I, *Inorg Chem* 1999, **38**, 4278.
33. Effendy, Gioia Lobbia G, Pellei M, Pettinari C, Santini C, Skelton BW, White AH, *J Chem Soc Dalton Trans* 2000, 2123.
34. Effendy, Gioia Lobbia G, Pellei M, Pettinari C, Santini C, Skelton BW, White AH, *Inorg Chim Acta* 2001, **315**, 153.
35. Hill AF, Smith MK, *Organometallics* 2007, **26**, 3900.
36. Mason MR, Ogrin D, Fneich B, Barnard TS, Kirschbaum K, *J Organomet Chem* 2005, **690**, 157.
37. Sugawara K-i, Hikichi S, Akita M, *J Chem Soc Dalton Trans.* 2002, 4514.
38. Belderraín TR, Paneque M, Carmona E, Gutiérrez-Puebla E, Monge MA, Ruiz-Valero C, *Inorg Chem* 2002, **41**, 425.
39. Dowling CM, Parkin G, *Polyhedron* 2001, **20**, 285.
40. Dowling CM, Parkin G, *Polyhedron* 1999, **18**, 3567.
41. Maria L, Campello MP, Domingos Â, Santos I, Andersen R, *J Chem Soc Dalton Trans* 1999, 2015.
42. Siemer CJ, Meece FA, Armstrong WH, Eichhorn DM, *Polyhedron* 2001, **20**, 2637.
43. Darensbourg DJ, Yarbrough JC, *J Organomet Chem* 2000, **614–615**, 305.
44. Jones PL, Byrom KJ, Jeffery JC, McCleverty JA, Ward MD, *Chem Commun* 1997, 1361.
45. Beeby A, Burton-Pye BP, Faulkner S, Motson GR, Jeffery JC, McCleverty JA, Ward MD, *J Chem Soc Dalton Trans* 2002, 1923.
46. Davies GM, Aarons RJ, Motson GR, Jeffery JC, Adams H, Faulkner S, Ward MD, *Dalton Trans* 2004, 1136.
47. Davies GM, Adams H, Pope SJA, Faulkner S, Ward MD, *Photochem Photobiol Sci* 2005, **4**, 829.
48. Armaroli N, Accorsi G, Barigelletti F, Couchman SM, Fleming JS, Harden NC, Jeffery JC, Mann KLV, McCleverty JA, Rees LH, Starling SR, Ward MD, *Inorg Chem* 1999, **38**, 5769.

49. Craven E, Mutlu E, Lundberg D, Temizdemir S, Dechert S, Brombacher H, Janiak C, *Polyhedron* 2002, **21**, 553.
50. Rodriguez V, Atheaux I, Donnadieu B, Sabo-Etienne S, Chaudret B, *Organometallics* 2000, **19**, 2916.
51. Dias HVR, Lu, H-L, *Inorg Chem* 2000, **39**, 2246.
52. Omary MA, Rawashdeh-Omary MA, Diyabalanage HVK, Dias HVR, *Inorg Chem* 2003, **42**, 8612.
53. Gahungu G, Zhang J, *Chem Phys Lett* 2005, **410**, 302.
54. Dias HVR, Richey SA, Diyabalanage HVK, Thankamani J, *J Organomet Chem* 2005, **690**, 1913.
55. Pampaloni G, Peloso R, Graiff C, Tiripicchio A, *Dalton Trans* 2006, 3576.
56. Gorun SM, Hu Z, Stibrany RT, Carpenter G, *Inorg Chim Acta* 2000, **297**, 383.
57. Lewiński J, Zachara J, Gós P, Grabska E, Kopeó T, Madura I, Marciniak W, Prowotorow I, *Chem Eur J* 2000, **6**, 3215.
58. Ma B, Li J, Djurovich PI, Yousufuddin M, Bau R, Thompson ME, *J Am Chem Soc* 2005, **127**, 28.
59. Thomas CM, Peters JC, *Organometallics* 2005, **24**, 5858.
60. Kläui W, Turkowski B, Rheinwald G, Lang H, *Eur J Inorg Chem* 2002, 205.
61. Agrifoglio G, Capparelli MV, *J Chem Crystallogr* 2005, **35**, 95.
62. Maria L, Domingos Â, Santos I, *Inorg Chem* 2003, **42**, 3323.
63. Ruman T, Ciunik Z, Mazurek J, Woł owiec S, *Eur J Inorg Chem* 2002, 754.
64. Ruman T, Ciunik Z, Goclan A, Łukasiewicz M, Wołowiec S, *Polyhedron* 2001, **20**, 2965.
65. Ruman T, Ciunik Z, Wołowiec S, *Eur J Inorg Chem* 2003, 2475.
66. Ruman T, Ciunik Z, Wołowiec S, *Eur J Inorg Chem* 2003, 89.
67. Ruman T, Ciunik Z, Wołowiec S, *Polyhedron* 2004, **23**, 219.
68. Ruman T, Łukasiewicz M, Ciunik Z, Wołowiec S, *Polyhedron* 2001, **20**, 2551.
69. Ciunik Z, Ruman T, Łukasiewicz M, Wołowiec S, *J Mol Struct* 2004, **690**, 175.
70. Ruman T, Ciunik Z, Szklanny E, Wołowiec S, *Polyhedron* 2002, **21**, 2743.
71. Łukasiewicz M, Ciunik Z, Ruman T, Skóra M, Wołowiec S, *Polyhedron* 2001, **20**, 237.

72. Ruman T, Ciunik Z, Wołowiec S, *Polyhedron* 2003, **22**, 581.
73. Ruman T, Ciunik Z, Trzeciak AM, Wołowiec S, Ziołkowski JJ, *Organometallics* 2003, **22**, 1072.
74. Ghosh P, Churchill DG, Rubinshtein M, Parkin G, *New J Chem* 1999, **23**, 961.
75. Chisholm MH, Eilerts NW, Huffman JC, Iyer SS, Pacold M, Phomphrai K, *J Am Chem Soc* 2000, **122**, 11845.
76. Łukasiewicz M, Ciunik Z, Wołowiec S, *Polyhedron* 2001, **20**, 2119.
77. Rombach M, Seebacher J, Ji M, Zhang G, He G, Ibrahim MM, Benkmil B, Vahrenkamp H, *Inorg Chem* 2006, **45**, 4571.
78. Matsumoto F, Chujo Y, *Pure and App Chem* 2006, **78**, 1407.
79. Cortes-Llamas SA, Hernández-Pérez JM, Hô M, Muñoz-Hernández MÁ, *Organometallics* 2006, **25**, 588.
80. Gioia Lobbia G, Pellei M, Pettinari C, Santini C, Skelton BW, Somers N, White AH, *Dalton Trans* 2002, 2333.
81. Gioia G Lobbia, Pellei M, Pettinari C, Santini C, Skelton BW, White AH, *Inorg Chim Acta* 2005, **358**, 1162.
82. Gioia G Lobbia, Pellei M, Pettinari C, Santini C, Skelton BW, White AH, *Polyhedron* 2005, **24**, 181.
83. Gioia Lobbia G, Pellei M, Pettinari C, Santini C, Skelton BW, White AH, *Inorg Chim Acta* 2005, **358**, 3633.

Chapter 5

Scorpionates Based on Thioimidazolones as Sulfur Donors and Related Ligands

5.1. $\{R'B[C_3H_2N_2S(R)]_3\} \equiv Tm^R$ (R = substituent on the heterocyclic ring N)

5.1.1. *Tm^{Me}(≡ Tm) derivatives*

Reglinski and coworkers reported in 1996 the synthesis of the S_3-donor Tm (Fig. 5.1) by the reaction of $[BH_4]^-$ with 2-mercaptoimidazole in the melt, a procedure similar to that used in the preparation of tris(pyrazolyl)borates.[1] Tm can be considered as a soft analog of Tp designed to maintain the tripodal geometry around the boron, allowing the replacement of the three nitrogen donor atoms by three sulfur (thione) donor atoms, thus providing a complementary soft, tridentate, face capping ligand system. It was compared with Tp by X-ray analysis and *ab initio* calculations: Na(Tm) is essentially salt-like with discrete anions and hydrated sodium cations. *Ab initio* calculations have been performed on sodium and copper derivatives of this ligands.[2] The synthesis and structure of $[Na(\kappa^3\text{-}(S,S,H)\text{-Tm})(\mu\text{-DMF})]_2$ and $[Na_4(\kappa^3\text{-Tm})_2(OH_2)_{13}(DMF)_2][(Tm)_2]$ demonstrated that sulfur can preferentially bind to sodium irrespective of the presence of more apposite donor species such as DMF. The structure of the noncoordinated Tm has been also reported.[3] The synthesis and crystal structure of $[Tl(Tm)_2]^+$ cation, in which the thallium ion is coordinated by six sulfur thione donors, in a regular

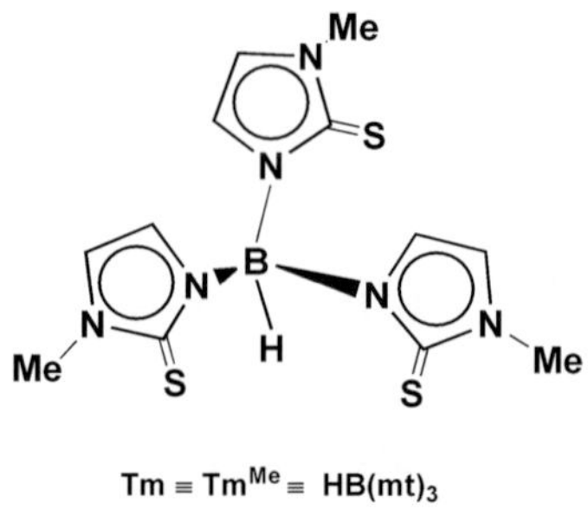

Fig. 5.1. Tm, a soft analog of Tp.

octahedral environment was also described.[4] The reaction of excess Tm with $ZnBr_2$ leads to the preferential formation of the 1:1 complex [ZnBr(Tm)].[1]

Reaction of MX_2 with Tm gave $[M(Tm)_2]$ (M = Fe, Ni), found in a trigonally distorted octahedral geometry by a S_6 donor set. Spectroscopic properties indicate that the Tm ligand generates a weak ligand field, with *Dq* between that of H_2O and Cl^-. $[M(Tm)_2]$ displays a classical paramagnetic behavior and $[Fe(Tm)_2]$ is a high-spin complex. Reaction of $NiX_2{\cdot}6H_2O$ with [Tl(Tm)] results also in a small quantity of $[Ni(Tm)_2]X$.[5] The paramagnetic nickel(II) complexes [Ni(Tm)(dppe)]X (X = Cl, Br) have also been prepared.[6]

Reaction of Tm with cobalt halides leads to the formation of paramagnetic pseudotetrahedral [CoX(Tm)] (X = Cl, Br, I), [CoBr(Tm)] being structurally characterized. Mass spectrometry reveals the presence of fragments $[Co(Tm)_2]^+$ and $[Co_2X(Tm)_2]^+$ in solution. Aerial oxidation leads to the formation of the $[Co(Tm)_2]^+$ cation, crystallographically characterized as the BF_4^-, ClO_4^-, Br^-, and I^- salts. The mixed-sandwich complex [Co(Cp*)(Tm)]I has been also isolated and characterized. Electrochemical measurements indicated that $[Co^{III}(Tm)]$ species are more difficult to reduce than their Tp and Cp analogs.[7] $[Mn(Tm)(CO)_3]$ has been also reported.[8]

The $[Ag(Tm)(PR_3)]$ derivatives (PR_3 = PCy_3, PBn_3, $(P(p\text{-Tol})_3)$, PMs_3 or $PEtPh_2$) were prepared through the one-pot reaction of AgX with PR_3 and M(Tm) (M = Na or K). The same reaction carried out with $PMePh_2$ gave $[Ag(Tm)(PMePh_2)_2]$, whereas $P(m\text{-Tol})_3$ formed the likely dinuclear $[Ag_2(Tm)_2\{P(m\text{-Tol})_3\}]$ complex.[9]

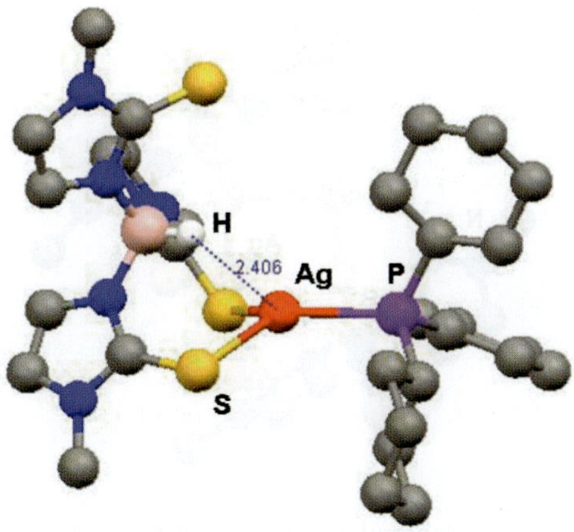

Fig. 5.2. In the [Ag(Tm){PCy_3}] complex Tm acts as S_2, *H*-donor.

dppe and dppf reacted with the [Ag(Tm)] moiety giving, respectively, [Ag(Tm)(dppe)] and [Ag(Tm)(dppf)]. The solid state structure of [Ag(Tm){PCy_3}] shows the silver atom tri-coordinate with the donor Tm acting in the S_2-bidentate form, but with a short Ag–B contact distance suggesting the existence of a two-electron B–H···Ag agostic interaction (Fig. 5.2). $[Ag(Tm)]_2$ was obtained in absence of PR_3 donors and structurally characterized as $[Ag(Tm)]_2{\cdot}2CHCl_3$, a binuclear species accompanied by a pair of chloroform solvent molecules, possessing a four-membered $Ag(\mu\text{-}S)_2Ag$ ring (Fig. 5.3).[10]

[Cu(Tm)(PR_3)] ($PR_3 = PPh_3, PCy_3$, P(*p*-Tol)$_3$, P(*m*-Tol)$_3$, $PEtPh_2$, and $PMePh_2$) were synthesized from Tm, CuX, and the corresponding P-donor. No adduct was obtained when $AsPh_3$, $SbPh_3$, PBn_3, and P(*o*-Tol)$_3$ were employed as ancillary ligands in the reaction of Tm with CuCl, $[Cu(Tm)]_n$ being the unique product always obtained. $[Cu(Tm)]_n$ reacts with py to yield [Cu(Tm)(py)]. The copper complexes were characterized through analytical and spectral (IR, ^1H and ^{31}P NMR) measurements. In [Cu(Tm)(P(*m*-Tol)$_3$)] and [Cu(Tm)(P(*p*-Tol)$_3$)] the copper atom is four-coordinate with the donor Tm behaving as an S_3-, face-capping ligand.[11]

The divalent closed shell metal complexes $[Ca(Tm)_2]{\cdot}6H_2O$, (**I**, Fig. 5.4), $[Ba(Tm)_2](H_2O)_2$ (**II**, Fig. 5.4) and $[Hg_4Cl_4(Tm)_4]$ were prepared and characterized by single-crystal X-ray diffraction. These

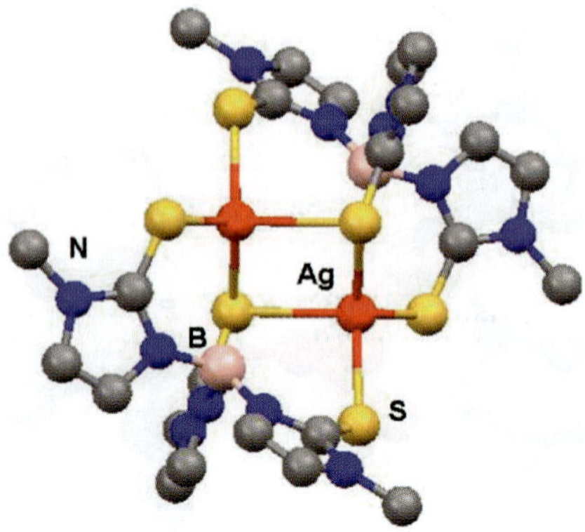

Fig. 5.3. The four-membered $Ag(\mu\text{-}S)_2Ag$ ring in $[Ag(Tm)]_2{\cdot}2CHCl_3$.

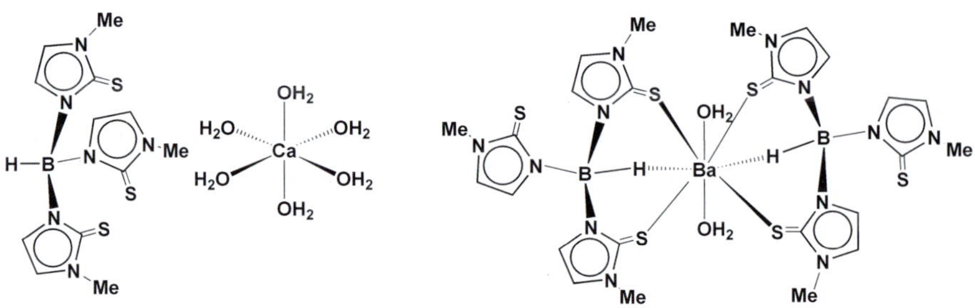

I II

Fig. 5.4. $[Ca(Tm)_2]{\cdot}6H_2O$ (I) (only one Tm is displayed) and $[Ba(Tm)_2](H_2O)_2$ (II).

three complexes exhibit different ligand binding modes, Tm acting as a noncoordinating anion for Ca(II) only engaging in H bonding with the metal bound water molecules, binding via two sulfur atoms to Ba(II) as well as a Ba$\cdots$H–B agostic interaction and finally forming a mercury cluster, with an interstitial chloride in the center of a tetrahedral arrangement of metal ions.[12]

[MX(Tm)] (M = Zn, Cd and Hg; X = Cl, Br or I) were obtained and characterized by NMR and mass spectrometry. X-ray single-crystal studies of [ZnX(Tm)] (X = Cl, Br, I) revealed a symmetrical coordination mode of the ligand. The structures of [MBr(Tm)] (M = Cd, Hg) are isomorphous, Tm being symmetrically coordinated to the metal, and the halide significantly displaced from the

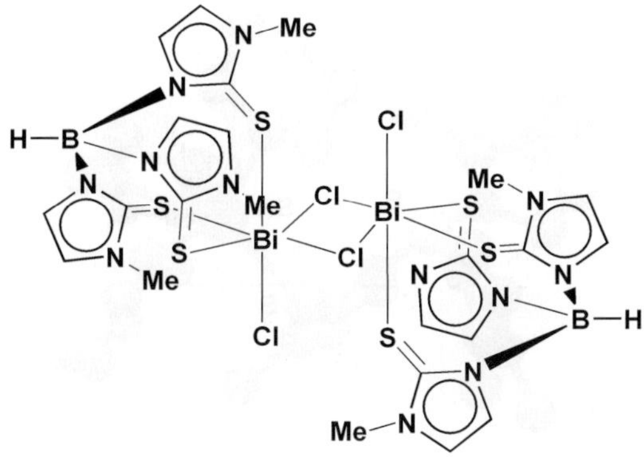

Fig. 5.5. The dinuclear [BiCl(μ-Cl)Tm]$_2$.

approximate C_3 axis of the {M(Tm)} unit, yielding a quasi-trigonal pyramidal geometry.[13]

In the reaction of $BiCl_3$ with 3 equiv. of Tm, crystals of [BiCl(μ-Cl)Tm]$_2$ (Fig. 5.5) and a bulk solid consistent with [Bi(Cl)Tm$_2$] are found. The latter compound reacted with 2 equiv. of Tp yielding [Bi(Tm)$_2$][NaTp$_2$]·4CH_2Cl_2·Cl(CH_2)$_2$Cl, also structurally characterized.[14]

A variety of structural motifs were obtained by reaction of the main-group element halide with Na(Tm), in detail: [In(κ^3-Tm)$_2$]$^+$, [Sb(κ^3-Tm)(κ^1-Tm)X] (X = Br, I) and the iodo-bridged [M(κ^3-Tm)I(μ$_2$-I)]$_2$ (M = Bi, Sb). Synthesis and structure of [Bi(Tm)$_2$]NO_3 and [Sb(κ^3-Tm)(κ^2-Tm)][Tm] were also reported. While in a number of Tm complexes of As, Sn, and Bi, the stereochemical influence of the lone-pairs is negligible, in the Sb complexes a significant influence of the nonbonded electron pair is found.[15]

Arsenic(III) and tin(IV) adducts [E(κ^3-Tm)$_2$]$^{n+}$ (E = As, $n = 1$; E = Sn, $n = 2$) have been prepared by Dodds *et al.*[16] Attempts to prepare the tin(II) adduct produced the tin(IV) species [Sn(κ^3-Tm)$_2$][Tm]$_2$ even when the reaction was carried out under inert atmospheres.[16] The reaction of Tm with $SnRCl_3$ (R = Me, ^{n}Bu or Ph), SnR_2Cl_2 (R = Me, Et, ^{n}Bu or Ph), and SnR_3Cl (R = Me, ^{n}Bu, Ph or Cy) acceptors afforded [SnRCl$_2$(Tm)], [SnR$_2$Cl(Tm)], and [SnR$_3$(Tm)] species, respectively, investigated by ^{1}H and ^{119}Sn NMR spectroscopy. The

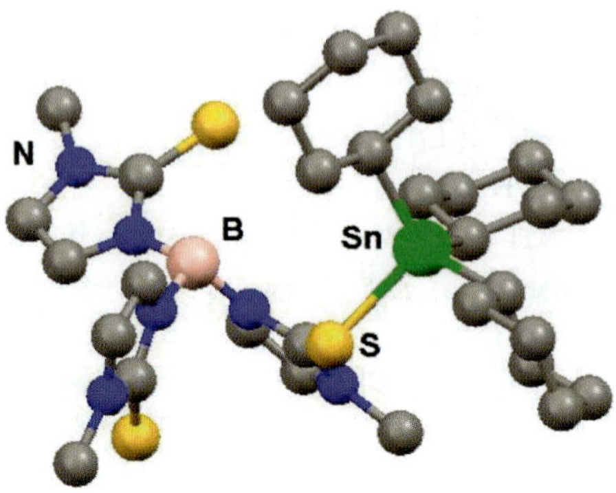

Fig. 5.6. In $[Sn(Cy)_3(Tm)]$, the ligand Tm acts as a monodentate S-donor.

triorganotin(IV) derivatives are tetrahedral both in solution and in the solid state, as confirmedz by X-ray. $[Sn(Cy)_3(Tm)]$ contains the Tm coordinated in a monodentate S-donor fashion (Fig. 5.6).[17]

Synthesis, structure, and spectroscopic properties of $[Te(\kappa^2\text{-}Tm)_2]$, displaying a square planar geometry at the Te atom and two distinct ligand conformations, have been reported by Reglinski's group.[18]

The reactions of $[MCl_4(\eta\text{-}C_5H_5)]$ (M = Nb, Ta) with $[SnPh_3(Tm)]$ yielded $[MCl_3(\eta\text{-}C_5H_5)\{\kappa^2\text{-}S,S'\text{-}ClBm\}]$ containing the new ligand chlorobis(methimazolyl)borate ClBm. Both Nb and Ta complexes as also the triorganotin species $[SnPh_3(Tm)]$ have been structurally characterized.[19] Tm-molybdenum stannyl complexes have been synthesized by oxidative addition of $[SnPh_3(Tm)]$ to Mo(0) precursors or trough reactions of halostannanes with $Na[Mo(CO)_3(Tm)]$.[20]

The octahedral $[M({=}NR)Cl_2(Tm)]$ (M = Nb, Ta; R = $C_6H_3{}^iPr_2$-2,6) have been prepared from the reactions of $[Nb({=}NR)Cl_3(dme)]$ and $[Ta({=}NR)Cl_3(THF)_2]$ with Tm. Formation of these complexes demonstrates that Tm is also compatible with "hard" metals in high oxidation states.[21]

[MCl(Tm)] (M = Mn, Fe, Co, Ni), upon reaction with NaI, afford [MI(Tm)]. The structures of $[M(\mu\text{-}Cl)(Tm)]_2$ (M = Mn, Ni) and $[Ni(\mu\text{-}Br)(Tm)]_2$ have been reported, Tm occupying the tripodal coordination site of the metal ions, yielding a square pyramidal or trigonal bipyramidal coordination geometry.[22]

The reactions of $[RuCl_2(HMB)]_2$ with Tm gave [Ru(Tm)(HMB)]Cl, characterized by diffraction analysis and cyclic voltammetric measurements.[23] $[RuCl(Tm)(DMSO)_2]$, [Ru(Tm)(*p*-cymene)]Cl], [RuCp(Tm)], and $[Mn(Tm)(CO)_3]$ have been synthesized, and [Ru(Tm)(*p*-cymene)]Cl and $[Mn(Tm)(CO)_3]$ structurally characterized. The Tm ligand acts as a tripod through its sulfur donors in all complexes except in [RuCp(Tm)], where bidentate coordination and formation of a 16-electron species are found.[24]

$[Ru^{III}Cp^*(Tm)]X$ ($X = Cl, PF_6$) and $[Ru^{II}Cp^*(Tm)]$ were prepared by the reaction of Tm with $[Ru^{III}Cp^*Cl_2]_2$ and $[Ru^{II}Cp^*(OMe)]_2$, respectively. Both complexes contain a tridentate κ^3-S,S',S'-Tm ligand. However, they undergo an isomerization reaction in solution (Fig. 5.7), where the sulfur on one mt group is displaced by the hydrogen bonded to the boron yielding a κ^3-H,S',S'-coordination.[25]

$[Mo(\eta^3\text{-allyl})(CO)_2(Tm)]$ and $[WI(CO)_3(Tm)]$ (Fig. 5.8) have been reported and indicated as general entry points for a wide range of group VI organometallic complexes. Their crystal structures have been determined and comparisons are made with the homologous Tp and Cp complexes.[26]

$[Mo(O)_2Cl(Tm)]$ has been prepared and investigated as a model for the metalloenzyme sulfite oxidase. This complex undergoes oxygen atom transfer chemistry and performs the primary function of the enzyme, e.g., sulfite oxidation.[27] $[Mo(O)_2Cl(Tm)]$, $[Mo(O)Cl_2(Tm)]$, $[Mo(O)Cl(Tm)]_2O$, and $[MoO(Tm)(CNC(Me)_2)]PF_6$ have been also

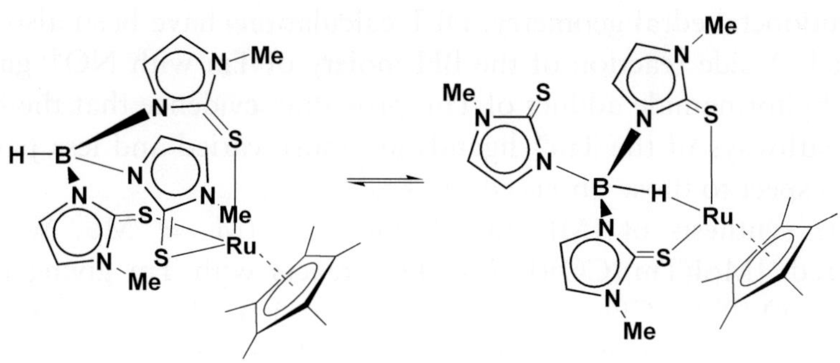

Fig. 5.7. The isomerization reaction found in solution for $[Ru^{III}Cp^*(Tm)]^{n+}$ ($n = 0, 1$).

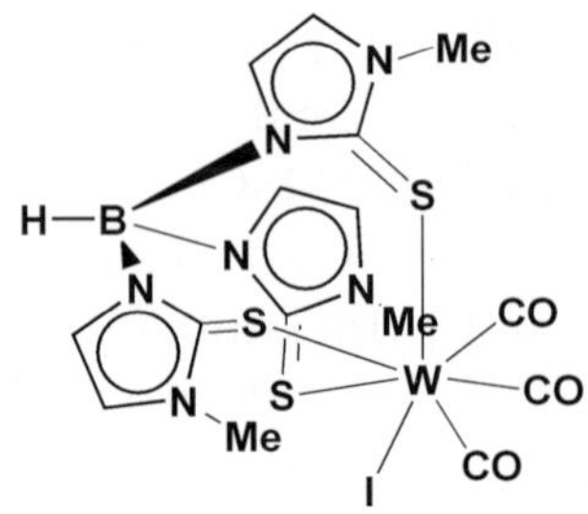

Fig. 5.8. $[WI(CO)_3Tm]$.

synthesized and characterized by X-ray crystallography, cyclic voltammetry, and EPR.[28]

$[Mo(\eta^2\text{-}SCNMe_2)(CO)_2(Tm)]$ has been prepared from the reaction of $[Mo(\eta^2\text{-}SCNMe_2)Cl(CO)_2(TMEDA)]$ with Tm or through interaction of $Na[Mo(CO)_x(Tm)]$ with Me_2NCSCl.[29]

$[W(\equiv CR)(CO)_2(Tm)]$ derivatives (R = N^iPr_2, C_4H_3S-2, $C\equiv CCMe_3$, C_6H_4Me-4, $C_6H_3Me_2$-2,6, and $C_5H_4Mn(CO)_3$) have been also described. The crystal structure of $[W(\equiv CR)(CO)_2(Tm)]$ (R = N^iPr_2, C_4H_3S-2, C_6H_4Me-4) reveals a progressive distortion of the alkylidyne ligand from octahedral coordination at tungsten, a chiral C_3-symmetric conformation for the $B(S)_3W$ cage being found.[30]

$[M(Tm)(CO)_2(NO)]$ (M = Mo, W) have been synthesized by the reaction of $[M(Tm)(CO)_3]^-$ with $NOBF_4$. These complexes adopt a pseudooctahedral geometry. DFT calculations have been also performed. A side reaction of the BH moiety of Tm with NO^+ gave a dimethylformamide adduct of Tm, providing evidence that the reaction pathways of the Tm^R ligands are more varied and less passive with respect to those observed for Tp.[31]

The synthesis of [MI(Tm)(CO)(alkyne)] (M = Mo, W) was reported. $[MoI(Tm)(CO)(PhC\equiv CPh)]$ reacts with Tm giving $[Mo(Tm)_2(CO)(PhC\equiv CPh)]$ in which the two Tm adopt two different noninterconverting coordination modes, κ^3-S_3 and κ^1-S.[32]

$[PtMe_3I]_4$ reacted with Tm affording the octahedral organometallic complex $[PtMe_3(Tm)]$.[33]

Santos and coworkers[34] described the synthesis of the building block *fac*-[^{99}Tc(Tm)(CO)$_3$], as well as its biological evaluation and application on the design of radiopharmaceuticals for the targeting of 5-HT1A receptors.

5.1.2. *Tm^{Ph}*

Tm^{Ph} has been used to synthesize [Zn(Tm^{Ph})SPh], which provides a good structural model for zinc enzymes possessing [Zn(Cys)$_4$] motifs.[35] [Zn(Tm^{Ph})SCH$_2$C(O)N(H)Ph], also featuring a tetrahedral [ZnS$_4$] motif, may be obtained by the reaction of Zn(NO$_3$)$_2$ with [Li(Tm^{Ph})] and Li[SCH$_2$C(O)N(H)Ph]. The thiolate ligand in [Zn(Tm^{Ph})SCH$_2$C(O)N(H)Ph] is subject to alkylation, the reaction with MeI yielding PhN(H)C(O)CH$_2$SMe and [ZnI(Tm^{Ph})]. A mechanistic investigation of thiolate alkylation has been reported.[36] The tetrahedral [Zn(Tm^{Ph})(Hmt^{Ph})][ClO$_4$] has been synthesized by the reaction of [Li(Tm^{Ph})] with Zn(ClO$_4$)$_2$ in the presence of Hmt^{Ph}. X-ray diffraction studies demonstrated that the mt^{Ph} ligands are almost orthogonal to the Zn–S–C plane, the orthogonal arrangement observed for [Zn(Tm^{Ph})(Hmt^{Ph})]$^+$ being a consequence of the existence of a N–H$\cdots$O hydrogen bond between the N–H group of the Hmt^{Ph} ligand and MeOH. The Hmt^{Ph} ligand of [Zn(Tm^{Ph})(Hmt^{Ph})]$^+$ may be immediately displaced by I$^-$ to give [ZnI(Tm^{Ph})].[37]

[Zn(OH)(Tm^{Ph})], synthesized by the reaction of [Li(Tm^{Ph})] with Zn(ClO$_4$)$_2$ in the presence of OH$^-$, has been indicated as a good structural model for the zinc enzyme 5-aminolevulinate dehydratase.[38]

Lead readily displaces the zinc in [ZnX(Tm^{Ph})] derivatives. [ZnI(Tm^{Ph})] reacts rapidly with Pb(ClO$_4$)$_2\cdot x$H$_2$O giving [Pb(Tm^{Ph})](ClO$_4$) that can be prepared independently also by the reaction of [Li(Tm^{Ph})] with Pb(ClO$_4$)$_2$. [Pb(Tm^{Ph})](ClO$_4$) possesses a monomeric trigonal pyramidal [Pb(Tm^{Ph})]$^+$ cation with lead at the apex, this structure being very similar to the active site of Pb^{II}ALAD.[39] Parkin and coworkers[40] provided, by using [Pb(Tm^{Ph})](ClO$_4$) as model complex, the first detailed, molecular insights into how lead binding to proteins is involved in gene expression which could directly account for some of the severe developmental problems known to result from lead poisoning.

Reaction of [Li(Tm^{Ph})] with $Tl(ClO_4)_3$ yields [Tl(Tm^{Ph})$_2$][ClO_4]. [Tl(Tm^{Ph})]$_2$ and [Tl(Tm^{Ph})$_2$][ClO_4] have been structurally characterized by X-ray diffraction. All the mercaptoimidazolyl groups are coordinated to thallium in the six-coordinate octahedral cation of [Tl(Tm^{Ph})$_2$][ClO_4].[41]

[Pb(Tm^{Ph})$_2$], prepared by the reaction of $Pb(ClO_4)_2$ with 2 equiv. of [Li(Tm^{Ph})], possesses an unusual structure with an inverted η^4-coordination mode for the (Tm^{Ph}) ligand which includes a Pb···H–B interaction (Fig. 5.9).[42]

In [M(Tm^{Ph})$_2$] (M = Fe, Co) Tm^{Ph} binds the metal through only two sulfur atoms, the "tetrahedral" metal centers interacting, however, also with the two B–H groups. Comparison of [M(Tm^{Ph})$_2$] (M = Fe, Co) with the related [M(Tp^{Ph})$_2$] counterparts indicates that Tm^R favors lower primary coordination numbers in divalent metal complexes.[43]

[MCl(Tm^{Ph})] (M = Mn, Fe, Co, Ni), upon reaction with NaI, afforded [MI(Tm^{Ph})]. The structures of [MCl(Tm^{Ph})] (M = Mn, Fe, Co, Ni) and [MI(Tm^{Ph})] (M = Mn, Co) have been reported. Reaction of [Fe(Tm^{Ph})$_2$] with excess $FeCl_2$ yielded [FeCl(Tm^{Ph})].[22]

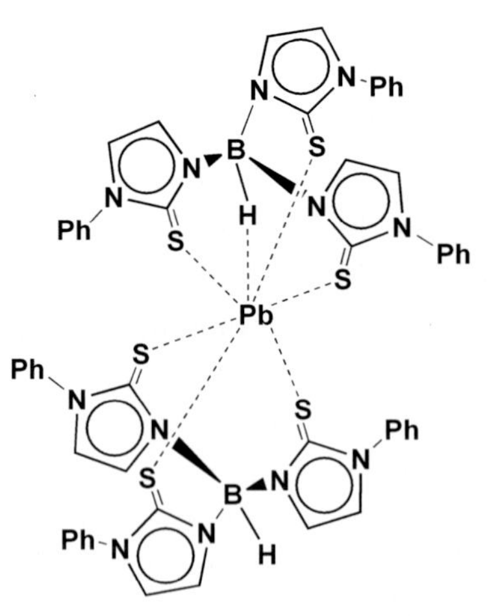

Fig. 5.9. In [Pb(Tm^{Ph})$_2$] a Tm^{Ph} ligand exhibits an inverted η^4-coordination mode.

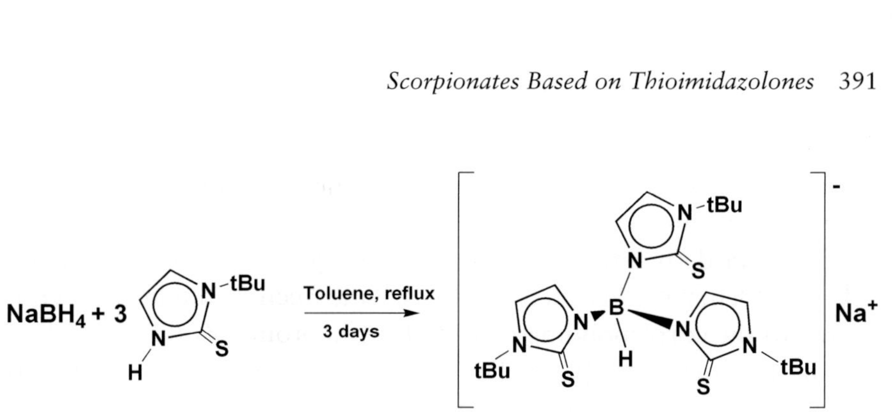

Fig. 5.10. Synthesis of Tm^{tBu} in toluene.

5.1.3. Tm^{tBu}

Rabinovich and coworkers[44] reported an alternative preparation (Fig. 5.10) of the bulky Tm^{tBu} together with convenient syntheses of [MBr(Tm^{tBu})] (M = Zn, Cd, Hg), structurally characterized. The isostructural complexes display a distorted tetrahedral structure, with the heavier members of the group exhibiting larger Br–M–S and smaller S–M–S angles. The nickel complexes [Ni(dppe)(Tm^{tBu})]X (X = Cl, Br) have been also described.[6]

Parkin and coworkers[45] reported a series of hydrochalcogenido and phenylchalcogenolate complexes prepared from [ZnMe(Tm^{tBu})], all supported by Tm^{tBu} ligation, in detail [Zn(EH)(Tm^{tBu})] (E = S, Se) and [Zn(EPh)(Tm^{tBu})] (E = O, S, Se, Te). [M(Tm^{tBu})(SCH_2-C(O)N(H)Ph)] (M = Zn, Cd) have been also obtained via reaction of [M(Me)(Tm^{tBu})] with PhN(H)C(O)CH_2SH.[46] [Zn(Tm^{tBu})-(SCH_2C(O)N(H)Ph)] mimics the function of the Ada DNA repair protein by undergoing alkylation with MeI to give [ZnI(Tm^{tBu})] and $MeSCH_2$C(O)N(H)Ph.

Tm^{tBu} has reacted with zinc salts yielding the tetrahedral complexes [ZnX(Tm^{tBu})] (X = F, Cl, ONO_2) and [Zn(Tm^{tBu})$_2$]. The labile [Zn($OClO_3$)Tm^{tBu}] was also described as a good starting material for the incorporation of substrates relevant for the chemistry of horse liver alcohol dehydrogenase, e.g. the addition of ethanol produced [Zn(Tm^{tBu})EtOH]·ClO_4·EtOH. Pyridine-2-carbaldehyde and salicylic aldehyde were also incorporated. (R-pyCH_2OH) yielded the trinuclear [Zn_3(Tm^{tBu})$_2$(R-pyCH_2O)$_2$](ClO_4)$_2$ with bridging Tm and pyridylmethoxide ligands. Zn(Tm^{tBu}) complexes have been shown to promote both the dehydrogenation of 2-propanol and the hydrogenation of pentafluorobenzaldehyde.[47] [ZnSR(Tm^{tBu})] complexes have been reported to reproduce the S_3Zn-SR coordination of thiolate-alkylating enzyme. Selected thiolates with methyl iodide were converted to the corresponding methyl thioethers and [ZnI(Tm^{tBu})].[48]

The paramagnetic [CoX(Tm^{tBu})] (X = Cl, Br, I), the mononuclear [Co(Tm^{tBu})(PPh_3)]X, and the dinuclear compounds [Co_2-(Tm^{tBu})$_2$X]Y have been also described.[49] [MCl(Tm^{tBu})] (M = Mn, Fe, Co, Ni) upon reaction with NaI afford [MI(Tm^{tBu})]. The structures of [MI(Tm^{tBu})] (M = Fe, Co, Ni) have been reported.[22]

Melnick and Parkin[50] reported a functional model for mercury detoxification by organomercurial lyase MerB. The alkyl compound [HgR(Tm^{tBu})] (R = Me or Et) reacts with phenylthiol (PhSH) yielding [HgSPh(Tm^{tBu})] and RH. [HgR(Tm^{tBu})] exist as linear two-coordinate complexes in the solid state, whereas in solution a rapid equilibrium between [HgR(κ^2-Tm^{tBu})] and [HgR(κ^3-Tm^{tBu})] occurs.

[Au(Tm^{tBu})] and [Au(Tm^{tBu})(PPh_3)] have been synthesized and structurally characterized and compared with their copper and silver analogous [M(Tm^{tBu})] and [M(Tm^{tBu})(PPh_3)] (M = Cu, Ag). Comparative structural analyses with the known methyl-substituted analog [Ag(Tm)] and various [M(Tm)(PR_3)] derivatives (M = Cu, Ag) have been carried out with the aim to evaluate the steric and electronic effects of the bulky *tert*-butyl substituents.[51]

The trinuclear species [Au_3(Tm^{tBu})$_2$]X (X = Cl, BF_4, PF_6, SbF_6) have been prepared and fully characterized. The X-ray single crystal study of [Au_3(Tm^{tBu})$_2$]SbF_6 indicated a cationic complex [Au_3(Tm^{tBu})$_2$]$^+$ in which a triangular array (Fig. 5.11) of three gold atoms is coordinated simultaneously by two Tm^{tBu} ligands.[52]

5.1.4. *Tm^Cum^*

Tm^{Cum} (Cum = C_6H_4-*p*-$CH(CH_3)_2$) gave tetrahedral complexes [ZnX(Tm^{Cum})] (X = F, Cl, ONO_2). [Zn($OClO_3$)(Tm^{Cum})], [Zn-(ONit)(Tm^{Cum})][47], and [ZnSR(Tm^{Cum})] complexes have been also reported.[48]

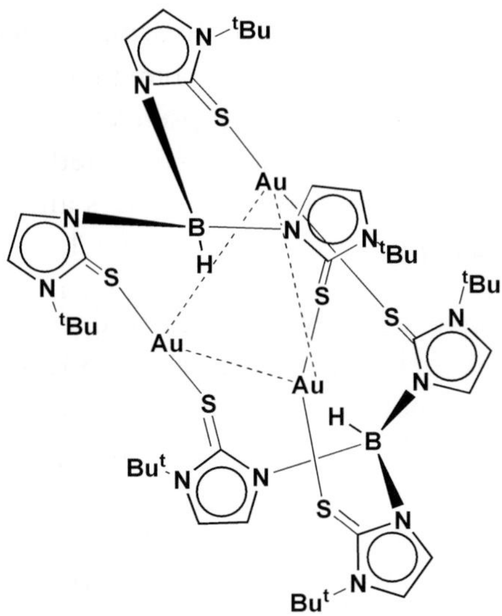

Fig. 5.11. The $[Au_3(Tm^{'tBu})_2]^+$cation.

$[MCl(Tm^{Cum})]$ (M = Mn, Fe, Co, Ni) upon reaction with NaI afford $[MI(Tm^{Cum})]$. $[MX(Tm^{Cum})]$ (M = Mn, Fe, Co, Ni; X = N Cl, I) have been also structurally characterized.[22]

5.1.5. *Tm^p-Tol^*

The cadmium thiolate complexes $[Cd(Tm^{p\text{-Tol}})(SR)]$ (R = Bn, Ph, *p*-Tol, C_6F_5) were prepared by the reaction of $[CdBr(Tm^{p\text{-Tol}})]$ with equimolar amounts of the corresponding thallium thiolates. $[CdSPh(Tm^{p\text{-Tol}})]$ is a four-coordinate species exhibiting a distorted tetrahedral geometry in the solid state.[53]

$Tm^{p\text{-Tol}}$ has been also employed to prepare $[MBr(Tm^{p\text{-Tol}})]$ (M = Zn, Cd), characterized by analytical and spectroscopic techniques, including ESI MS[54] and in the synthesis of mono and dinuclear Zn-OH species. The Zn-OH_2, Zn-OH, Zn-OH-Zn, and Zn-O_2-H_3-Zn coordination motifs were found.[55] $[Zn(SR)(Tm^{p\text{-Tol}})]$ complexes have been also reported.[48] The tetrahedral $[Zn(Tm^{p\text{-Tol}})$-$(Hmt^{p\text{-Tol}})][ClO_4]$ has been prepared by the reaction of $[Li(Tm^{p\text{-Tol}})]$ with $Zn(ClO_4)_2$ in the presence of $Hmt^{p\text{-Tol}}$.[37]

[Ni(Tm$^{p\text{-Tol}}$)(dppe)]X (X = Cl, Br) complexes have been prepared by the reaction of NiX_2 with Tm in the presence of dppe. [Ni(Tm$^{p\text{-Tol}}$)(dppe)]Cl shows a weak pseudoaxial Ni···H–B interaction and an overall square pyramidal geometry.[6]

5.1.6. *Tm^{R}(R = Et, Bn)*

Tm^{Et} and Tm^{Bn} have been synthesized and employed to examine the racemization of their C_3-symmetric complexes with both four- and six-coordinate metals. The four-coordinate [ZnCl(Tm^{Et})], [CdBr(Tm^{Et})], [HgCl(Tm^{Et})], [Cu(Tm^{Et})PPh_3], [Ag(Tm^{Et})PPh_3], [ZnCl(Tm^{Bn})] and the six-coordinate [Ru(Tm^{Et})(*p*-cymene)]Cl, [Ru(Tm^{Et})(*p*-cymene)]PF_6, and [Mn(Tm^{Et})$(CO)_3$] were synthesized and the crystal structures of [ZnCl(Tm^{Et})], [CdBr(Tm^{Et})], [Cu(Tm^{Et})PPh_3], and [Ru(Tm^{Et})(*p*-cymene)]Cl reported. The diastereotopic nature of the ethyl and benzyl hydrogen atoms in the ligands allows the enantiomeric forms of these complexes to be distinguished by ^{1}H NMR spectroscopy. The racemization mechanism has been investigated also with *ab initio* DFT calculations.[56]

Tm^{Bn} has been used to prepare [MBr(Tm^{Bn})] (M = Zn, Cd), that have been characterized by analytical and spectroscopic techniques, including ESI MS and also by single-crystal X-ray diffraction.[54]

5.1.7. *Tm^{Ms}*

The tris(2-mercapto-1-mesitylimidazolyl)borate ligand, Tm^{Ms}, reacts with $Zn(ClO_4)_2{\cdot}6H_2O$ yielding [Zn(Tm^{Ms})(HOMe)]$^+$, a stable monomeric tetrahedral zinc–methanol complex which shows analogies with the alcohol intermediate in the catalytic cycle of the mechanism of action of liver alcohol dehydrogenase.[57]

5.1.8. *Tm^{Xyl}*

[ZnSR(Tm^{Xyl})] (R = alkyl) complexes have been prepared to reproduce the S_3Zn–SR coordination environment of thiolate-alkylating enzyme. The most reactive thiolate, [ZnSEt(Tm^{Xyl})], reacts slowly with trimethyl phosphate yielding methyl–ethylthioether and [Zn(Tm^{Xyl})OPO$(OMe)_2$].[48]

5.1.9. *RTm (R = Ph or Me)*

A good method for the synthesis of [Li(PhTm)] has been described by the Reglinski group, and the structure of the methanol-solvated [Li(OHMe)$_4$][PhTm] determined. The syntheses and characterization of [M(PhTm)(PR$_3$)] (M = Cu, Ag, Au; R = Et, Ph) have been reported, the pseudotetrahedral [Cu(PhTm)(PPh$_3$)], the trigonal planar [Ag(PhTm)(PEt$_3$)], and the linear [Au(PhTm)(PEt$_3$)] being structurally characterized. Reaction of $CuCl_2$ with [Li(PhTm)] results in the formation of insoluble [CuCl(PhTm)]. A copper(I) dimeric complex, [Cu$_2$(PhTm)$_2$], is formed by ultra-slow diffusion of the reactants $CuCl_2$ and [Li(PhTm)].[58]

[Li(RTm)] (R = Me or Ph) have been prepared by Santos *et al.*[59] in refluxing toluene with Li[RBH$_3$]. Their reaction with (NEt$_4$)$_2$[Re(CO)$_3$Br$_3$] gave [Re(RTm)(CO)$_3$] in which RTm acts as S$_3$-donor. These compounds were characterized by NMR spectroscopy and [Re(MeTm)(CO)$_3$] also by X-ray single-crystal study.[59]

5.1.10. *MeimTm$^{\alpha MeBn}$*

A chiral tris(methimazolyl)borate ligand has been prepared by Bailey *et al.*[60] by the activation of tris(dimethylamino)borane toward reaction with a chiral methimazole (1-(*S*)-α-methylbenzyl-2-mercaptoimidazolyl) by *N*-methylimidazole (Fig. 5.12). Its coordination to [RuCl$_2$(*p*-cymene)]$_2$ provides a single diastereomer complex in which the chirality of the methimazolyl substituents

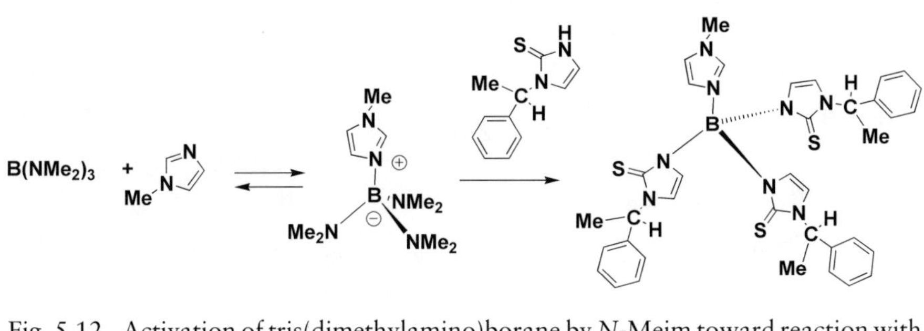

Fig. 5.12. Activation of tris(dimethylamino)borane by *N*-Meim toward reaction with the chiral methimazole (1-(*S*)-α-methylbenzyl-2-mercaptoimidazolyl) gave a chiral tris(methimazolyl)borate.

dictates the chirality of the bicyclo[3.3.3]cage formed by the ligand, upon coordination to the metal.

5.2. $\{HR'B[C_3H_2N_2S(R)]_2\} \equiv RBm^{R'}$

5.2.1. *Bm^{Me}(≡Bm) derivatives*

The sodium salt of Bm has been prepared and employed to obtain homoleptic complexes $[M(Bm)_2]$ (M = Zn, Cd, Hg). $[Cd(Bm)_2]$ shows a distorted tetrahedral $[MS_4]$ central core with two weak vicinal M···H–B bonds.[61] [Cd(SR)(Bm)] (R = Ph, *p*-Tol, C_6F_5) have been synthesized by an analogous procedure. [Cd(S-*p*-C_6H_4Me)(Bm)] exhibits a dimeric structure in the solid state, in which each five-coordinate cadmium atom displays a trigonal-bipyramidal geometry.[62]

[Li(Bm)], upon reaction with pzH yielded [Li(pzBm)], which reacts with ZnI_2 forming [ZnI(pzBm)], a compound that can be obtained also through the reaction of the three-coordinate zinc complex [ZnI(Bm)] with pzH. The molecular structures of both [ZnI(pzBm)] and [ZnI(Bm)] have been also reported; in the former, the ligand binds the zinc via its nitrogen and sulfur donors in a manner which resembles the active site of LADH.[63]

Bm reacts with ZnX_2 (X = Me, I) yielding [ZnMe(Bm)] and [ZnI(Bm)], each of which possesses 3-center–2-electron [Zn···H–B] interactions (Fig. 5.13). A redistribution reaction of [ZnMe(Bm)] in $CHCl_3$ yielded the homoleptic tetrahedral complex $[Zn(Bm)]_2$.[64]

[Tl(Bm)], polymeric, has been prepared through a one-pot metathesis of Tl(OAc) with Bm.[65]

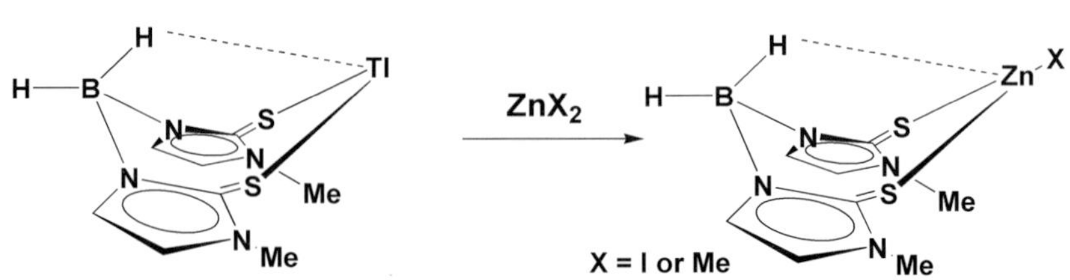

Fig. 5.13. The 3-center–2-electron [Zn···H–B] interactions in [ZnX(Bm)].

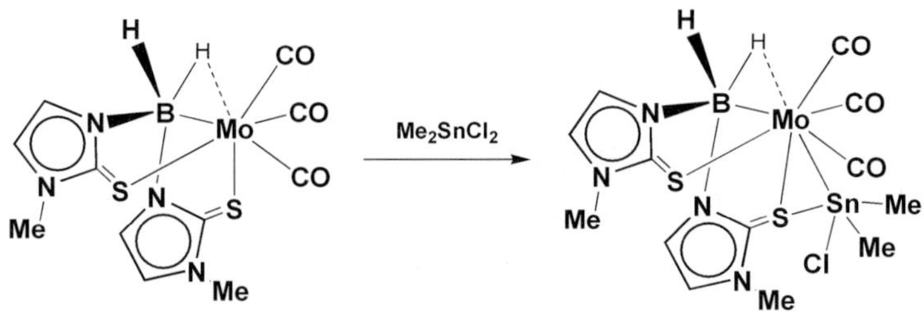

Fig. 5.14. Synthesis of $[Mo(CO)_3(SnMe_2Cl)(Bm)]$.

A Bm-containing molybdenum stannyl complex has been synthesized by oxidative addition of $[Me_2Sn(Bm)]Cl$ to Mo(0) precursors (Fig. 5.14) or through reactions of halostannanes with Na[Mo-$(CO)_3(Bm)$]. The crystal structure of $[Mo(CO)_3(SnMe_2Cl)(Bm)]$ in which Bm bridges both molybdenum and tin centers is also reported.[26]

A poly(methimazolyl)borato complex of group VI, [Ti(=NCMe$_3$)-$(Bm)_2$] has been obtained from the reaction of Bm with [Ti-(=NCMe$_3$)$Cl_2(py)_3$], the two Bm ligands bonded to Ti, acting as κ^2-*S*,*S*- and κ^3-*H*,*S*,*S*-donor.[66]

The paramagnetic nickel(II) complex [Ni(Bm)(dppe)]Cl exhibits a weak pseudoaxial Ni···H–B interaction and an overall square pyramidal geometry.[6] $[Mn(Bm)(CO)_3]$ has been also prepared and characterized.[8]

The reaction of $[M(\equiv CC_6H_2Me_3\text{-}2,4,6)X(CO)_2(L)_2]$ (M = Mo, W; X = Cl, Br; L = py, Hpz*) with Bm provides [Mo($\equiv CC_6H_2Me_3$-2,4,6)$(CO)_2$(κ^3-*H*,*S*,*S*′-Bm)] and [W(CO)(κ^2-*S*,*S*′-Bm)(κ^3-*H*,*S*,*S*′-Bm)].[67]

[Mo(η^2-SCNMe$_2$)$(CO)_2$(Bm)] (Fig. 5.15) has been prepared from the reactions of [Mo(η^2-SCNMe$_2$)Cl$(CO)_2$(TMEDA)] and Bm. An agostic interaction B–H···Mo is found both in solution and in the solid state.[29]

The synthesis and characterization of [M(κ^3-*H*,*S*,*S*′-Bm)$(CO)_3$] (M = Re, ^{99}Tc, ^{99m}Tc), exhibiting an agostic B–H···M bond and obtained directly by reacting $[M(OH_2)_3(CO)_3]^+$ (M = Re, Tc) with Bm, have been reported by Santos *et al.*[68] These complexes represent unique examples of structurally characterized Re

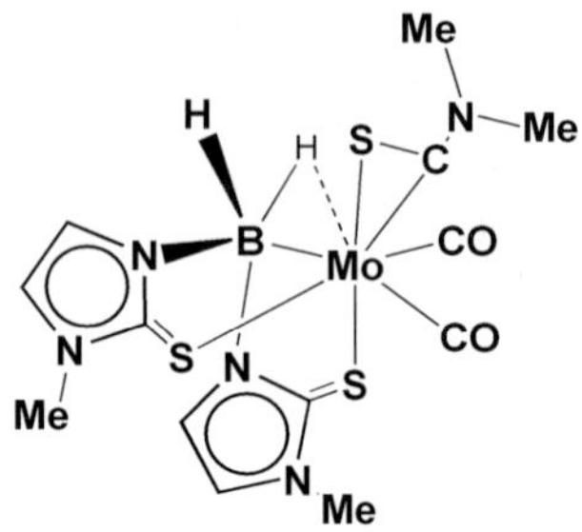

Fig. 5.15. $[Mo(\eta^2\text{-}SCNMe_2)(CO)_2(Bm)]$.

or Tc carbonyl complexes exhibiting B–H···M agostic interactions. Complexes $[Re(Bm)(CO)_3(L)]$ (L = imH, ^{t}BuNC, PPh_3, 4-NMe_2py) have been also prepared through reactions kinetically unfavorable, the cleavage of the B–H···M agostic interaction needing a large excess of substrates. The imH derivative has been structurally characterized.

The building block *fac*-$[^{99}Tc(Bm)(CO)_3]$ has been synthesized and applied to the design of radiopharmaceuticals for the targeting of 5-HT_{1A} receptors have been reported.[34]

The square planar [Rh(Bm)(COD)] was prepared and reacted with CO or $CNC_6H_3Me_2$-2,6 to yield the derivatives $[Rh(Bm)L_2]$ (L = CO, $CNC_6H_3Mc_2$-2,6). The crystal structure of [Rh(Bm)(COD)] reveals long B–H···Rh interactions not present in solution.[69] The reaction of Na(Bm) with $[RhCl(CO)(PPh_3)_2]$ gave $[Rh(Bm)(CO)(PPh_3)]$. When Bm reacted with $[RhCl(CS)(PPh_3)_2]$, C,C-phosphaboration of the thiocarbonyl ligand occurs, $[RhH(PPh_3)\{\eta^2(C,S),\kappa^2\text{-}S',S''\text{-}SC(PPh_3)BH(mt)_2\}]$ being formed.[70]

The reaction of $[ZrCl_2(THF)(\eta\text{-}C_8H_8)]$ with Bm gave $[ZrCl(Bm)(\eta\text{-}C_8H_8)]$, in which the metal is electronically unsaturated despite the capability of Bm to increase the metal coordination number.[71]

The reaction of $[RuCl_2(HMB)]_2$ with Bm gave [Ru(Bm)(HMB)]Cl, whereas the reaction of $[Ru(OMe)(Cp^*)]_2$ with Bm yielded $[RuCp^*(Bm)]$. Single-crystal X-ray diffraction analyses and cyclic voltammetric measurements were carried out on both complexes.[23]

$[PtMe_3I]_4$ reacted with Bm affording $[PtMe_3(Bm)]$ which upon further reaction with $[PtMe_3I]_4$ gave the sulfur-bridged dinuclear complex $[Me_3Pt(\mu\text{-}Bm)PtMe_3I]$ (Fig. 5.16).[33]

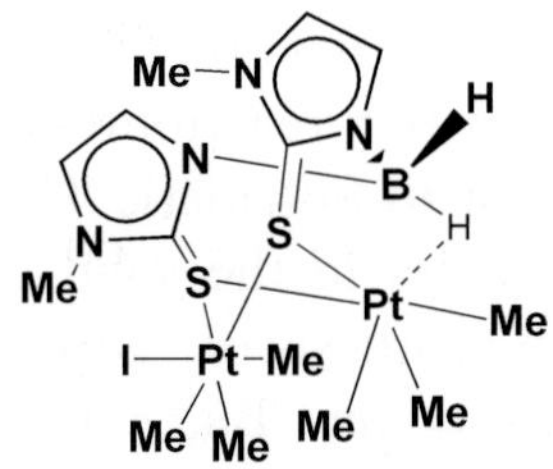

Fig. 5.16. $[Me_3Pt(\mu\text{-Bm})PtMe_3I]$.

The Au atom in $[Au(Bm)(Ph_3P)]$ is coordinated by two S (thione) atoms and one Ph_3P ligand in a distorted trigonal environment. The two Au–S distances are significantly different and the thioimidazolyl rings are approximately perpendicular to each other.[72]

The synthesis and structural characterization of $[U(Bm)_2\text{-}(THF)_3][BPh_4]$, in which the ligand acts as tridentate through the two thione sulfurs and one agostic hydrogen, have been reported by Santos and coworkers.[73] $[U(Bm)_2(L')_3][BPh_4]$ complexes (L' = py or MeCN) have been easily obtained by stirring $[U(Bm)_2(THF)_3][BPh_4]$ in the corresponding solvents.[73]

5.2.2. *Bm^Ms*

Bm^{Ms} has been prepared and its interaction with ZnX_2 (X = Me, I, NO_3), yielding $[Zn(X)(Bm^{Ms})]$, is investigated.[64]

5.2.3. *Bm^R* (R = Bn, *^tBu*, *p-Tol*)

The sodium salts of $Na(Bm^R)$ (R = Bn, ^{t}Bu, *p*-Tol) have been prepared and employed to obtain homoleptic complexes $[M(Bm^R)_2]$ (M = Zn, Cd, Hg). X-ray diffraction studies of $[M(Bm^{^tBu})_2]$ (M = Zn, Cd, Hg) indicate the presence of distorted tetrahedral $[MS_4]$ central cores with two weak vicinal M···H–B bonds. In the case of zinc, in the presence of bulky ligands, as in $[Zn(Bm^{^tBu})_2]$, an unexpected expansion in the coordination number from four to six is found.[61]

[Tl(Bm^{R})] (R = Bn, ^{t}Bu, *p*-Tol)] derivatives have been conveniently prepared in a one-pot reaction by metathesis of Tl(OAc) with Na(Bm^{R}). In contrast to polymeric Tl(Bm), Tl($Bm^{^tBu}$) consists of discrete dimeric molecules in the solid state.[65] Rabinovich and coworkers[6] reported the synthesis of [Ni($Bm^{^tBu}$)(dppe)]X (X = Cl, Br). [Mn(Bm^{R})$(CO)_3$] (R = Bn, ^{t}Bu, *p*-Tol) have been synthesized and fully characterized. The presence of 3-center–2-electron Mn···H–B interactions in these species, both in solution and in the solid state, has been investigated using a combination of IR, NMR, and X-ray single-crystal study.[8]

Santos and coworkers[34] described the synthesis of the building block *fac*-[M($Bm^{Bu\text{-}pip}$)$(CO)_3$] and of *fac*-[M{H_2B(mt)($mt^{Bu\text{-}pip}$)}-$(CO)_3$] (M = Re or ^{99}Tc) and their application on the design of radiopharmaceuticals for the targeting of 5-HT1A receptors has been reported.

5.2.4. *pzBm*

[Zn$(pzBm)_2$], [Co$(pzBm)_2$], and [Cd$(pzBm)_2$] have been synthesized and structurally characterized. [Zn$(pzBm)_2$] exhibits a tetrahedral {ZnS_4} structure in which only the sulfur donors coordinate to zinc, [Co$(pzBm)_2$] exhibits a trigonal-bipyramidal {CoS_3NH} structure in which one of the pz groups and the B–H are coordinated to cobalt, and [Cd$(pzBm)_2$] possess a six-coordinate {CdS_4H_2} coordination environment in which both B–H groups interact with the cadmium center.[74]

The trivalent complex [Co$(pzBm)_2$]I exhibits octahedral coordination, the two pzBm being coordinated in the κ^3-*N*,S_2 fashion mode.[43]

5.2.5. *[($pz^{Ph,Me}$)Bm^{R}]* (R = o-*An, Cum, Xyl)*

The *N*,*S*,*S*-donor [($pz^{Ph,Me}$)$Bm^{o\text{-}An}$] yielded [Zn{($pz^{Ph,Me}$)$Bm^{o\text{-}An}$}-(EtOH)]ClO_4, first zinc complex containing a *N*,*S*,*S*-ligation and a coordinated ethanol molecule as in the active center of horse liver alcohol dehydrogenase.[75]

The S_2,*N*-donor ligands $(pz^{Ph,Me})Bm^R$ (R = Cum, Xyl) were employed for the synthesis of mono- and di-nuclear Zn-hydroxo species. The new Zn-OH–Zn motif was realized by using the tripodal NS_2-donor $(pz^{Ph,Me})Bm^R$.[55] $[Zn(OClO_3)\{(pz^{Ph,Me})Bm^{Xyl}\}]$ decomposed in attempts to replace the perchlorate ion by *p*-nitrophenolate in CH_2Cl_2 solution.[76] $[Zn(NO_3)\{(pz^{Ph,Me})Bm^R\}]$ (R = *o*-An, Xyl) and $[Zn\{(pz^{Ph,Me})Bm^R\}(ClO_4)]$ are converted into [Zn(tripod)(thiolate)], including the homocysteinate. Methylation with MeI converts $[Zn\{(pz^R)Bm^R\}thiolate]$ to $[ZnI\{(pz^R)Bm^{R'}\}]$ and the corresponding thioethers CH_3SR, including methionine. The alkylation is faster in $[Zn(N_2S\text{-donor})(thiolate)]$ than for $[Zn(NS_2\text{-donor})(thiolate)]$ and $[Zn(N_3\text{-donor})(thiolate)]$.[77]

5.2.6. $pz^{Ph,Me}Bm^{o\text{-}An}$

Zn–ethyl-, –benzyl-, –phenyl-, and –*p*-nitrophenyl-thiolate complexes containing the ligand $pz^{Ph,Me}Bm^{o\text{-}An}$ (**I**, Fig. 5.17) have been widely reported and their alkylation by MeI investigated.[78]

5.2.7. *PhBm*

$[Mn(PhBm)(CO)_3]$ has been synthesized and fully characterized,[8] whereas the building block *fac*-$[^{99}Tc(PhBm)(CO)_3]$ has been applied on the design of radiopharmaceuticals for the targeting of 5-HT1A

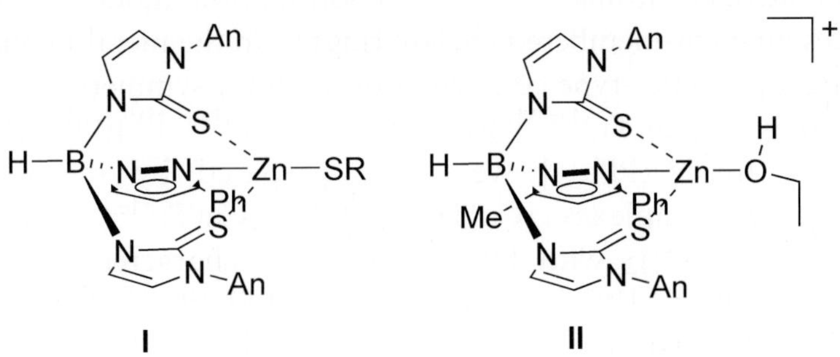

Fig. 5.17. $[Zn(SR)(pz^{Ph,Me}Bm^{o\text{-}An})]$ (I) and $[Zn(pz^{Ph,Me}Bm^{o\text{-}An})(EtOH)]$-$ClO_4$ (II).

receptors.[34] [M(PhBm)(PPh_3)] (M = Cu, Ag), containing PhBm, adventitiously formed, have been crystallographically characterized, both species being in a trigonal planar primary coordination sphere, with a secondary M···H–B interaction.[58] [U(PhBm)$_2$(THF)$_3$][BPh_4] and [U(PhBm)$_2$(L)$_3$][BPh_4] (L = py or MeCN) have been also reported by Santos and coworkers.[73]

5.3. mt(Tm)

K{mt(Tm)} was prepared and characterized in solution and in the solid state as two independent very similar enantiomorphs in the asymmetric unit. The packing gives rise to big holes between mt(Tm) filled by clusters of K^+ and H_2O.[79]

5.4. Thioxotriazolylborates (≡ Tr^R)

Reaction of 5-thioxo-4,5-dihydro-3,4-dimethyl-1,2,4-triazole with $NaBH_4$ provides the N_3/S_3-donor ambidentate tripod ligand hydridotris(thioxotriazolyl)borate $Tr^{Me,Me}$. Its coordinating ability toward different metals has been investigated. [Na(H_2O)$_6$]-[Na($Tr^{Me,Me}$)$_2$], [Na($Tr^{Me,Me}$)(DMF)$_3$]·Et_2O, [Bi($Tr^{Me,Me}$)$_2$]Cl·H_2O, [Sn($Tr^{Me,Me}$)Cl_3], and [Mn($Tr^{Me,Me}$)(CO)$_3$] have been also characterized by X-ray crystallography which shows the ambidentate character of the ligand it which with the softer metals bismuth and tin exhibit sulfur coordination, while it binds sodium and manganese(I) via the nitrogen atoms. In the S_3-coordination mode the ligand produces an eight-membered chelate rings with the metal-ligand unit adopting a propeller-type conformation with C_3-symmetry.[80]

Complexes [Cu($Tr^{Mes,Me}$)(PPh_3)], [Cu($Tr^{Me,o\text{-}py}$)(PPh_3)], and [Cu{($pz^{o\text{-}py}$)Br^{Mes}}(PPh_3)] were also prepared by the reaction of dinuclear complexes [Cu($Tr^{Mes,Me}$)]$_2$, [Cu($Tr^{Me,o\text{-}py}$)]$_2$, and [Cu{($pz^{o\text{-}py}$)Br^{Mes}}]$_2$ with PPh_3. They were characterized by ^1H, ^{13}C, and ^{31}P NMR spectroscopy and ESI-mass spectrometry. The molecular structure of [Cu($Tr^{Mes,Me}$)(PPh_3)] (Fig. 5.18) and of [Cu($Tr^{Me,o\text{-}py}$)(PPh_3)] is reported. These compounds are not fluxional in solution, whereas [Cu{($pz^{o\text{-}py}$)Br^{Mes}}(PPh_3)] is subject to an equilibrium between two species, the major species consisting of the

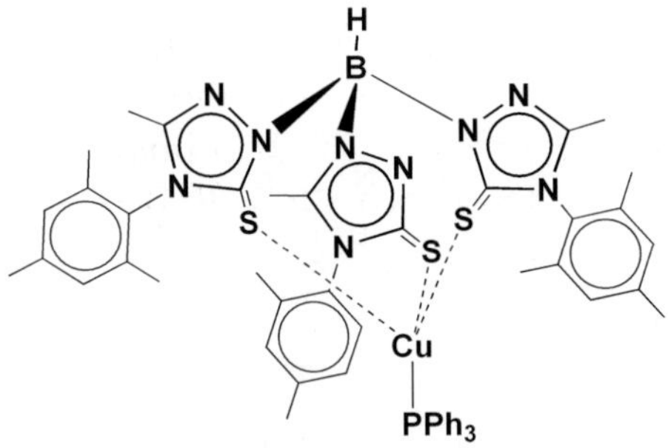

Fig. 5.18. $[Cu(Tr^{Mes,Me})(PPh_3)]$.

$\{(pz^{o\text{-}py})Br^{Mes}\}$ ligand bounded to Cu(I) in the κ^3-S_2H fashion with two Cd-S bonds and a Cd$\cdots$H–B interaction. $[Cu(Tr^{Mes,Me})(Tu)]$ was also prepared and structurally characterized.[81]

The complex $[Ni(Tr^{Me,o\text{-}py})_2]$ was synthesized and characterized by UV–Vis spectroscopy, ESI-mass spectrometry, and X-ray crystallography. It possesses a slightly distorted octahedral geometry (N_4S_2 coordination), the metal center being coordinated by two ligands that act as *S,N/N′*-donor ligands.[82] Connelly and coworkers reported the synthesis and the molecular structure of $[Rh(Tr^{Me,Et})(CO)(PPh_3)]$ through oxidative addition of the B–H bond, provides an unprecedented route to rhodaboratrane.[83]

5.5. Tz and Tbz

The synthesis of the Tl salts of hydrotris(mercaptothiazolyl)borate (Tz) and hydrotris(mercaptobenzothiazolyl)borate (Tbz) has been reported by the Reglinski group. These species can be used as convenient ligand transfer reagents. $[Tl(Tbz)]_\infty$ exhibits a bridging thione linking Tl(Tbz) units (Fig. 5.19) into zig-zag one-dimensional polymeric chains.[84]

5.6. Tdz

The tris(mercaptothiadiazolyl)borate ligand (Tdz) has been reported by Gardinier.[85] It has been defined as a Janus scorpionate, a hybrid

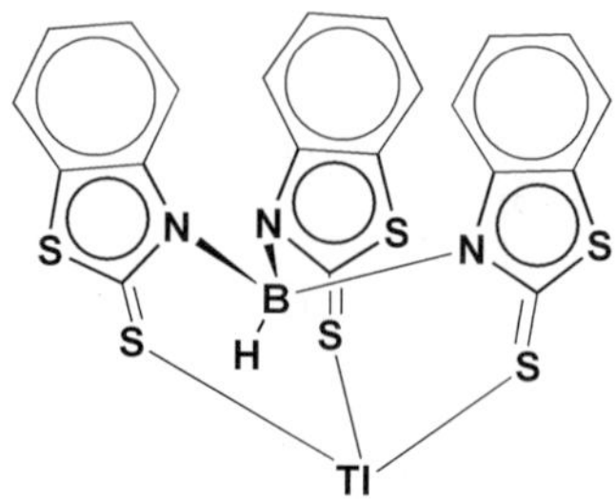

Fig. 5.19. The Tl(Tbz) unit which assembles to zig-zag one-dimensional polymeric chains.

of Tp and Tm. Tdz has been investigated for the controlled solid state organization of homo- and hetero-metallic alkali metal coordination polymers as Na(Tdz) with both acentric and centric chains. Tdz exhibits an extraordinary coordination capacity and versatility. It is able to bind from two up to five metal cations as in its thallium(I) salt [Tl(Tdz)]·DMF, a model for metal–surface binding (Fig. 5.20).[86]

5.7. [Bse^{Me}], [Tse^{Me}], and [Tse^{Ms}]

The ligands bis- and tris(2-seleno-1-methylimidazolyl)hydroborate ({Bse^{Me}} and {Tse^{Me}}) have been described by Parkin's group.[87] The structural comparison of [ZnI{Bse^{Me}}]$_2$ and [ZnI(Bm)] demonstrates

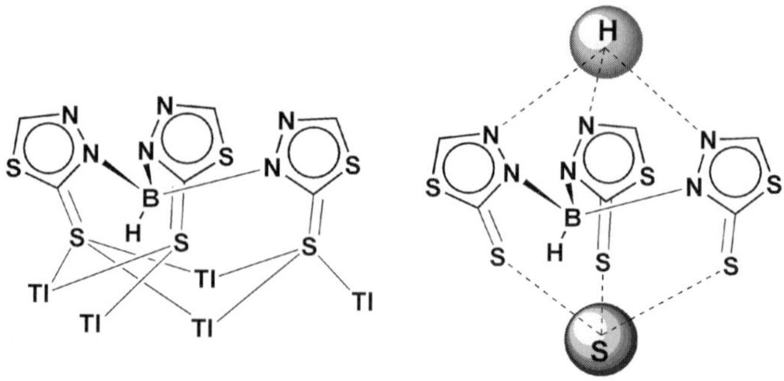

Fig. 5.20. [Tl(Tdz)]·DMF, a model for metal-surface binding. Tdz is able to bond hard (H) and soft (S) species.

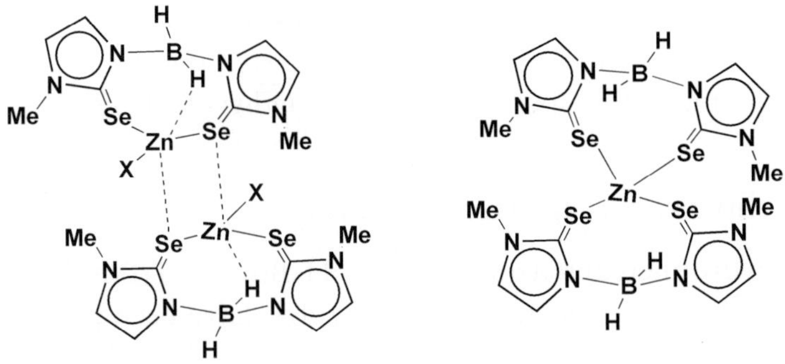

Fig. 5.21. $[ZnI\{Bse^{Me}\}]_2$ (III) and $[Zn\{Bse^{Me}\}_2]$ (IV).

that the seleno ligand exhibits a greater tendency than the mercapto one to bridge two metal centers (Fig. 5.21).

Comparison of $[Re(Tse^{Me})(CO)_3]$ and $[Re(Tse^{Mes})(CO)_3]$ suggests that Tse^{Me} is more electron donating and less sterically demanding than the Tse^{Ms} ligand. On the other hand, comparison of the IR spectroscopic data of $[Re(Tse^{Mes})(CO)_3]$ with those of a variety of related complexes $[Re(L)(CO)_3]$ demonstrates that the Tse^{Mes} ligand is more strongly electron donating than Cp, Cp*, Tp, Tp*, and Tse^{Me} ligands.[88]

The linear nickel nitrosyl compound supported by tridentate selenium ligands $[Ni\{Tse^{Mes}\}NO]$ has been prepared and structurally characterized by a X-ray diffraction study. Comparison was made with [Ni(Tp*)NO].[89]

5.8. mt^RBp

The anionic bis(pyrazolyl)(thioimidazolyl)borate ligands mt^RBp (R = ^{t}Bu or ^{i}Pr) were prepared by reaction of Tp with the corresponding thioimidazoles in the melt at 150°C. They were used in the synthesis of zinc complexes such as $[ZnCl\{(mt^{^tBu})Bp\}]$ and $[ZnSC_6H_4\text{-}p\text{-}Cl\{(mt^{^iPr})Bp\}]$.[90] Several zinc–thiolate (aromatic and cysteine) complexes containing the same donors (mt^R)Bp (R = ^{t}Bu or ^{i}Pr) have been reported. Single-crystal X-ray studies confirmed the ZnN_2S_2 coordination. Reaction with methyl iodide converted the thiolates into the corresponding iodide species and free thioethers CH_3SR.[91]

5.9. Trihydromercaptoazolylborates

The trihydro(mercaptoazolyl)borates $H_3B(mt)$, $H_3B(mt^{Bupip})$, and $H_3B(bzt)$ (bzt = 2-mercaptobenzothiazolyl) have been prepared and employed to stabilize the $[M(CO)_3]^+$ core (M $={}^{99}$Tc, Re). The complexes *fac*-$[M\{\kappa^3\text{-}H(\mu\text{-}H)_2B(mt)\}(CO)_3]$ (M $={}^{99}$Tc, Re), *fac*-$[Re\{\kappa^3\text{-}H(\mu\text{-}H)_2B(mt^{Bupip})\}(CO)_3]$, and *fac*-$[Re\{\kappa^3\text{-}H(\mu\text{-}H)_2B(bzt)\}(CO)_3]$ have been synthesized and characterized. The soft scorpionates are coordinated to the metal in unique κ^3-*H*,*H′*,*S* fashion, as confirmed by X-ray crystallography. The trihydro(mercaptoazolyl)borate technetium tricarbonyl complexes are suitable for the design of CNS receptor ligand radiopharmaceuticals as exemplified due to their lipophilicity, small-size, and easy functionalization with adequate biomolecules.[92]

5.10. Metallaboratranes: M → B(mt)$_3$

The treatment of the complex $[Ru(CH{=}CHCPh_2OH)Cl(CO)(PPh_3)_2]$ with Tm gave the ruthenaboratrane complex $[Ru\{B(mt)_3\}(CO)(PPh_3)]$ in which the ruthenium atom adopts a distorted octahedral coordination despite the constraints of bis-chelation (Fig. 5.22).[93]

Hill proposed an unambiguous electron-counting notation for metallaboratranes: (M → B). This notation follows the tradition often encountered in denoting compounds with metal–metal multiple bonds and allows a correct assignment of oxidation number in these complexes.[94]

The reaction of $[Os(Ph)Cl(CO)(PPh_3)_2]$ with Tm results in benzene elimination and formation of the first osmaboratrane $[Os\{B(mt)_3\}(CO)(PPh_3)](Os \rightarrow B)$, showing a somewhat long *trans*-annular dative bond between osmium(0) and boron(III) (Fig. 5.23).[95]

The reaction of $[Rh(Ph)Cl_2(PPh_3)_2]$ with Tm gave $[RhCl\{B(mt)_3\}(PPh_3)](Rh \rightarrow B)$, the first example of rhodaboratrane (Fig. 5.24).[96]

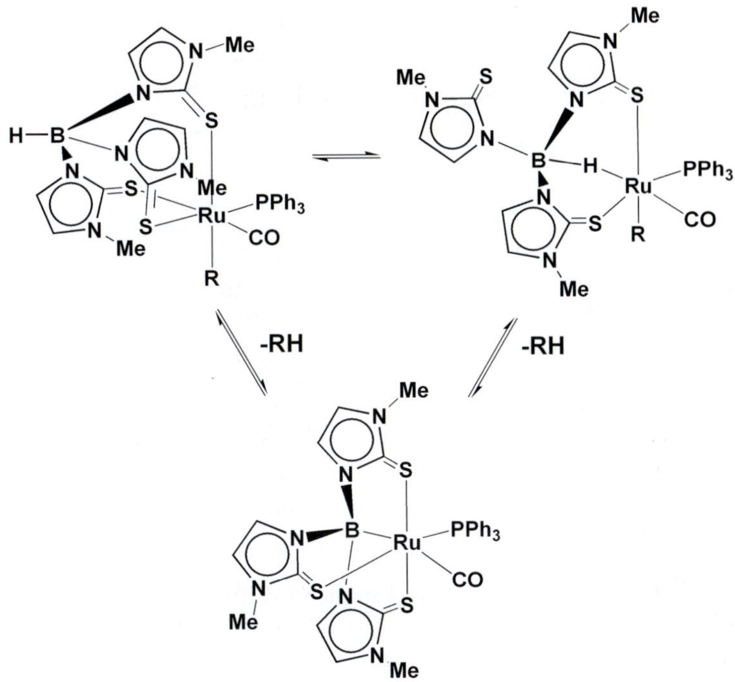

Fig. 5.22. Formation of the metallaboratrane $[Ru\{B(mt)_3\}(CO)(PPh_3)]$.

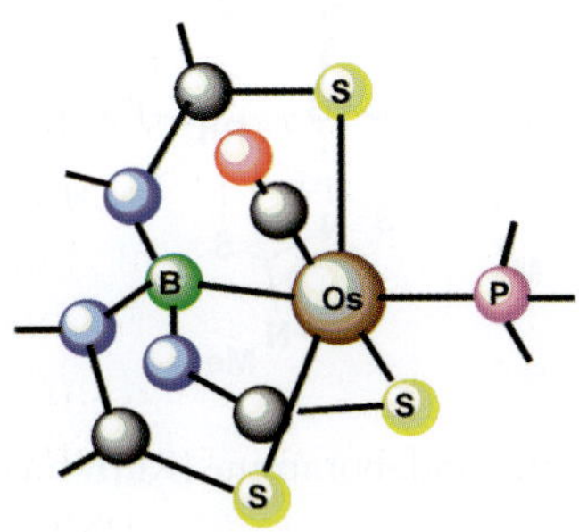

Fig. 5.23. Coordination environment of Os in the osmaboratrane complex $[Os\{B(mt)_3\}(CO)(PPh_3)](Os \rightarrow B)$.

Other rhodaboratrane complexes, e.g. $[RhH\{B(mt)_3\}(PPh_3)]$ and $[Rh\{B(mt)_3\}(\eta^4\text{-}C_8H_{12})]Cl$, have been prepared, but also isonitrile (Fig. 5.25), phosphine, and dialkyldithiocarbamate species, in detail: $[Rh\{B(mt)_3\}(CNR)(PPh_3)]Cl$ (R $=$ ^{t}Bu, $C_6H_3Me_2$-2,6,

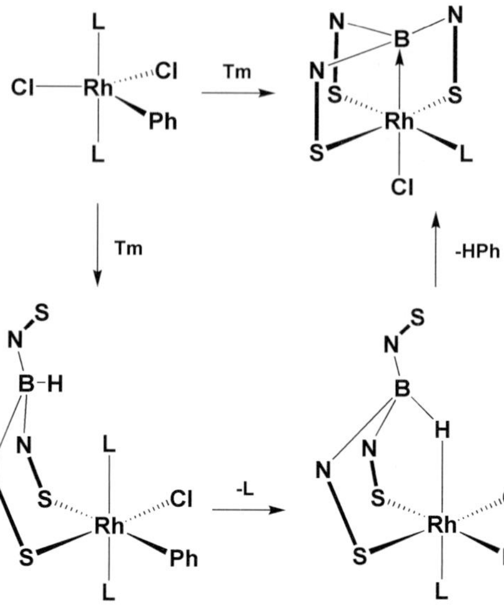

Fig. 5.24. Formation of the rhodaboratrane $[RhCl\{B(mt)_3\}(PPh_3)](Rh \rightarrow B)$.

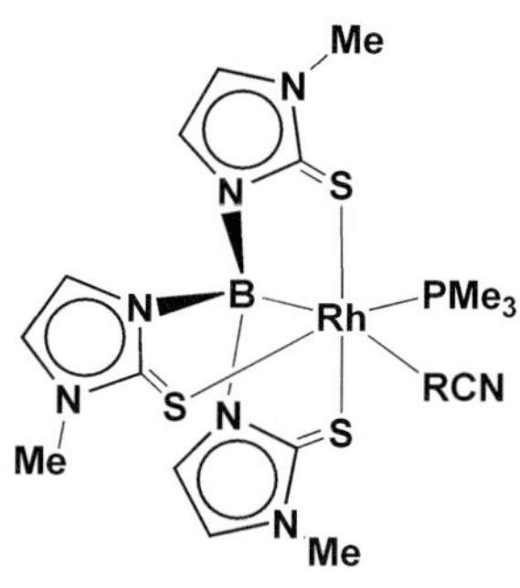

Fig. 5.25. The cationic rhodaboratrane $[Rh\{B(mt)_3\}(CNR)(PPh_3)]^+$.

$C_6H_2Me_3$-2,4,6), $[Rh\{B(mt)_3\}(PMe_3)_n(PPh_3)_{2-n}]Cl$ ($n = 0, 1, 2$), and $[Rh\{B(mt)_3\}(S_2CNEt_2)]Cl$. The structures of $[Rh\{B(mt)_3\}(L)(L')]Cl$ ($L = CN^tBu$, $CN(C_6H_3Me_2$-2,6), $L = PPh_3$; $L = L' = PMe_3$) were also reported.[97]

From the reaction of $[Rh\{B(mt)_3\}(\eta^4\text{-}C_8H_{12})]Cl$ with an excess of Tm the binuclear metallaboratrane $[Rh_2\{B(mt)_3\}_2\{\kappa^2\text{-}S,S'\text{-Tm}\}]Cl$ is formed. This compound is also a decomposition product of rhodium tris(methimazolyl)borate complexes.[98]

The reaction of $[RuHCl(CO)(PPh_3)_3]$ or $[RuH(NCMe)_2(CO)(PPh_3)_2]ClO_4$ with Tm provide the octahedral $[RuH(CO)(PPh_3)\{\kappa^3\text{-}H,S,S\text{-Tm}\}]$, able to hydrogenate ethynylbenzene to styrene with the formation of the ruthenaboratrane $[Ru(CO)(PPh_3)\{B(mt)_3\}]$-$(Ru \rightarrow B)$.[99]

The structure of the first cobaltaboratrane $[\{B(mt^{tBu})_3\}Co(PPh_3)]BPh_4$ has been reported by Rabinovich and coworkers.[100]

Iridium and rhodium complexes with $M \rightarrow B$ dative bonds, namely $[Ir\{\kappa^3\text{-}B,S,S\text{-}B(mt^R)_3\}(CO)(PPh_3)H]$ ($R = {}^tBu$, Ph) and $[M\{\kappa^4\text{-}(B(mt^{tBu})_3)\}(PPh_3)Cl]$ (M = Rh, Ir), have been prepared through the reactions of $Ir(PPh_3)_2(CO)Cl$ with Tm^{tBu} or Tm^{Ph} and also from the reaction of $[M(PPh_3)Cl(COD)]$ with Tm^{tBu}, respectively, and structurally characterized.[101] The Vaska's complex $IrCl(CO)(PPh_3)_2$ reacts with 1 equiv. of Na(Tm) or Na(Bm) yielding also the iridaboratranes $[IrH(CO)(PPh_3)\{\kappa^3\text{-}B,S,S\text{-}B(mt)_2R\}](Ir \rightarrow B)$ (R = mt, H), in which a transannular metal $\rightarrow$ boron bond is supported by two methimazolyl groups.[102]

It has been reported that dichloromethane, frequently used in the reaction of poly(methimazolyl)borates, is not an innocent solvent, but is able to form slowly heterocyclic salts (Fig. 5.26) $[H_2C(mt)_2BR_2]Cl$, ($BR_2 = BH_2$, BH(mt), 9-borabicyclononyl) which have been characterized by X-ray single-crystal studies.[103]

The ferraboratrane $[Fe\{\kappa^4\text{-}(B(mt^{tBu})_3)\}(CO)_2]$ is the first example of complex having a retrodative $Fe \rightarrow B$ bond. It has been prepared by

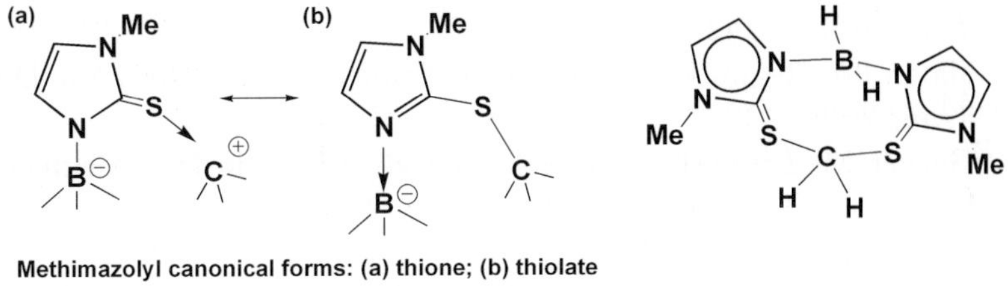

Fig. 5.26. Interaction of CH_2Cl_2 with poly(methimazolyl)borates.

the reaction of [FeCl(Tm^{tBu})] with $LiCH_2SiMe_3$ and treatment with CO.[104]

The dinuclear complex [Pd{μ-κ^1,κ^3-B(mt^{tBu})$_3$}]$_2$, exhibiting a Pd → B dative bond, was prepared by the reaction of Tm^{tBu} with Pd(OAc)$_2$; [Pd{μ-κ^1,κ^3-B(mt^{tBu})$_3$}]$_2$ reacts with PMe_3 giving the mononuclear boratrane derivative [Pd{μ-κ^1,κ^3-B(mt^{tBu})$_3$}-(PMe_3)].[105]

5.11. Applications

Reglinski *et al.*[106] reported the use of Tm and Tz anions as surface modifiers for colloidal silver particles for the production of novel soft coating. The use of these species in surface-enhanced resonance Raman spectra has been described.

Tripod ligands of the pyrazolyl/thioimidazolyl-borate offering N_2S, NS_2, and S_3-donor sets were employed in the study of biologically relevant alkylations of the R-thiolate zinc complexes (R = ethyl-, benzyl-, phenyl-, and *p*-nitrophenyl).[78] A review recently described the use of poly(mercaptoimidazolyl)-borate in the chelation of cadmium and mercury.[107]

References

1. Garner M, Reglinski J, Cassidy I, Spicer MD, Kennedy AR, *J Chem Soc Chem Commun* 1996, 1975.
2. Reglinski J, Garner M, Cassidy ID, Slavin PA, Spicer MD, Armstrong DR, *J Chem Soc Dalton Trans* 1999, 2119.
3. Biernat A, Schwalbe M, Wallace D, Reglinski J, Spicer MD, *Dalton Trans* 2007, 2242.
4. Slavin PA, Reglinski J, Spicer MD, Kennedy AR, *J Chem Soc Dalton Trans* 2000, 239.
5. Garner M, Lewinski K, Pattek-Janczyk A, Reglinski J, Sieklucka B, Spicer MD, Szaleniec M, *Dalton Trans* 2003, 1181.
6. Alvarez HM, Tanski JM, Rabinovich D, *Polyhedron* 2004, **23**, 395.
7. Dodds CA, Lehmann M-A, Ojo JF, Reglinski J, Spicer MD, *Inorg Chem* 2004, **43**, 4927.

8. Graham LA, Fout AR, Kuehne KR, White JL, Mookherji B, Marks FM, Yap GPA, Zakharov LN, Rheingold AL, Rabinovich D, *Dalton Trans* 2005, 171.
9. Santini C, Pettinari C, Gioia Lobbia G, Spagna R, Pellei M, Vallorani F, *Inorg Chim Acta* 1999, **285**, 81.
10. Effendy, Gioia Lobbia G, Pettinari C, Santini C, Skelton BW, White AH, *Inorg Chim Acta* 2000, **308**, 65.
11. Gioia Lobbia G, Pettinari C, Santini C, Somers N, Skelton BW, White AH, *Inorg Chim Acta* 2001, **319**, 15.
12. Çetin A, Ziegler CJ, *Dalton Trans* 2006, 1006.
13. Cassidy I, Garner M, Kennedy AR, Potts GBS, Reglinski J, Slavin PA, Spicer MD, *Eur J Inorg Chem* 2002, 1235.
14. Reglinski J, Spicer MD, Garner M, Kennedy AR, *J Am Chem Soc* 1999, **121**, 2317.
15. Dodds CA, Reglinski J, Spicer MD *Chem Eur J* 2006, **12**, 931.
16. Dodds CA, Jagoda M, Reglinski J, Spicer MD, *Polyhedron* 2004, **23**, 445.
17. Santini C, Pellei M, Gioia Lobbia G, Pettinari C, Drozdov A, Troyanov S, *Inorg Chim Acta* 2001, **325**, 20.
18. Dodds CA, Kennedy AR, Reglisnki J, Spicer MD, *Inorg Chem* 2004, **43**, 394.
19. Hill AF, Smith MK, *Chem Commun* 2005, 1920.
20. Foreman MR St-J, Hill AF, Smith MK, Tshabang N, *Organometallics* 2005, **24**, 5224.
21. Hill AF, Rae AD, Smith MK, *Inorg Chem* 2005, **44**, 7316.
22. Senda S, Ohki Y, Hirayama T, Toda D, Chen J-L, Matsumoto T, Kawaguchi H, Tatsumi K, *Inorg Chem* 2006, **45**, 9914.
23. Kuan SL, Leong WK, Goh LY, Webster RD, *J Organomet Chem* 2006, **691**, 907.
24. Bailey PJ, Lorono-Gonzales DJ, McCormack C, Parsons S, Price M, *Inorg Chim Acta* 2003, **354**, 61.
25. Kuan SL, Leong WK, Goh LY, Webster RD, *Organometallics* 2005, **24**, 4639.
26. Garner M, Lehmann M-A, Reglinski J, Spicer MD, *Organometallics* 2001, **20**, 5233.
27. Wallace D, Gibson LT, Reglinski J, Spicer MD, *Inorg Chem* 2007, **46**, 3804.
28. Tran BL, Carrano CJ, *Inorg Chem* 2007, **46**, 5429.

29. Foreman MRSt-J, Hill AF, Tshabang N, White AJP, Williams DJ, *Organometallics* 2003, **22**, 5593.
30. Foreman MRSt-J, Hill AF, White AJP, Williams DJ, *Organometallics* 2003, **22**, 3831.
31. Schwalbe M, Andrikopoulos PC, Armstrong DR, Reglinski J, Spicer MD, *Eur J Inorg Chem* 2007, 1351.
32. Hill AF, Tshabang N, Willis AC, *Eur J Inorg Chem* 2007, 3781.
33. Crossley IR, Hill AF, Willis AC, *Organometallics* 2005, **24**, 4889.
34. Garcia R, Gano L, Maria L, Paulo A, Santos I, Spies H, *J Biol Inorg Chem* 2006, **11**, 769.
35. Bridgewater BM, Fillebeen T, Friesner RA, Parkin G, *J Chem Soc Dalton Trans* 2000, 4494.
36. Morlok MM, Janak KE, Zhu G, Quarless DA, Parkin G, *J Am Chem Soc* 2005, **127**, 14039.
37. Docrat A, Morlok MM, Bridgewater BM, Churchill DG, Parkin G, *Polyhedron* 2004, **23**, 481.
38. Bridgewater BM, Parkin G, *Inorg Chem Commun* 2001, **4**, 126.
39. Bridgewater BM, Parkin G, *J Am Chem Soc* 2000, **122**, 7140.
40. Magyar JS, Weng T-C, Stern CM, Dye DF, Rous BW, Payne JC, Bridgewater BM, Mijovilovich A, Parkin G, Zaleski JM, Penner-Hahn JE, Godwin HA, *J Am Chem Soc* 2005, **127**, 9495.
41. Kimblin C, Bridgewater BM, Hascall T, Parkin G, *J Chem Soc Dalton Trans* 2000, 1267.
42. Bridgewater BM, Parkin G, *Inorg Chem Commun* 2000, **3**, 534.
43. Kimblin C, Churchill DG, Bridgewater BM, Girard JN, Quarless DA, Parkin G, *Polyhedron* 2001, **20**, 1891.
44. White JL, Tanski JM, Rabinovich D, *J Chem Soc Dalton Trans* 2002, 2987.
45. Melnick JG, Docrat A, Parkin G, *Chem Commun* 2004, 2870.
46. Melnick JG, Zhu G, Buccella D, Parkin G, *J Inorg Biochem* 2006, **100**, 1147.
47. Tesmer M, Shu M, Vahrenkamp H, *Inorg Chem* 2001, **40**, 4022.
48. Ibrahim MM, Seebacher J, Steinfeld G, Vanhrenkamp H, *Inorg Chem* 2005, **44**, 8531.
49. Mihalcik DJ, White JL, Tanski JM, Zakharov LN, Yap GPA, Incarvito CD, Rheingold AL, Rabinovich D, *Dalton Trans* 2004, 1626.
50. Melnick JG, Parkin G, *Science* 2007, **317**, 225.

51. Patel DV, Mihalcik DJ, Kreisel KA, Yap GPA, Zakharov LN, Kassel WS, Rheingold AL, Rabinovich D, *Dalton Trans* 2005, 2410.
52. Patel DV, Kreisel KA, Yap GPA, Rabinovich D, *Inorg Chem Commun* 2006, **9**, 748.
53. Bakbak S, Incarvito CD, Rheingold AL, Rabinovich D, *Inorg Chem* 2002, **41**, 998.
54. Bakbak S, Bhatia VK, Incarvito CD, Rheingold AL, Rabinovich D, *Polyhedron* 2001, **20**, 3343.
55. Ibrahim MM, Pérez Olmo C, Tekeste T, Seebacher J, He G, Maldonado Calvo JA, Böhmerle K, Steinfeld G, Brombacher H, Vahrenkamp H, *Inorg Chem* 2006, **45**, 7493.
56. Bailey PJ, Dawson A, McCormack C, Moggach SA, Oswald IDH, Parsons S, Rankin DWH, Turner A, *Inorg Chem* 2005, **44**, 8884.
57. Kimblin C, Bridgewater BM, Churchill DG, Parkin G, *Chem Commun* 1999, 2301.
58. Dodds CA, Garner M, Reglinski J, Spicer MD, *Inorg Chem* 2006, **45**, 2733.
59. Garcia R, Paulo A, Domingos Â, Santos I, *Dalton Trans* 2003, 2757.
60. Bailey PJ, McCormack C, Parsons S, Rudolphi F, Sanchez Perucha A, Wood P, *Dalton Trans* 2007, 476.
61. Alvarez HM, Tran TB, Richter MA, Alyounes DM, Rabinovich D, Tanski JM, Krawiec M, *Inorg Chem* 2003, **42**, 2149.
62. Philson LA, Alyounes DM, Zakharov LN, Rheingold AL, Rabinovich D, *Polyhedron* 2003, **22**, 3461.
63. Kimblin C, Hascall T, Parkin G, *Inorg Chem* 1997, **36**, 5680.
64. Kimblin C, Bridgewater BM, Hascall T, Parkin G, *J Chem Soc Dalton Trans* 2000, 891.
65. Alvarez HM, Gillespie PA, Gause CD, Rheingold AL, Golen JA, Rabinovich D, *Polyhedron* 2004, **23**, 617.
66. Hill AF, Smith MK, *Dalton Trans* 2006, 28.
67. Abernethy RJ, Hill AF, Neumann H, Willis AC, *Inorg Chim Acta* 2005, **358**, 1605.
68. Garcia R, Paulo A, Domingos Â, Santos I, Ortner K, Alberto R, *J Am Chem Soc* 2000, **122**, 11240.
69. Crossley IR, Hill AF, Humphrey ER, Smith MK, *Organometallics* 2006, **25**, 2242.
70. Crossley IR, Hill AF, Willis AC, *Organometallics* 2007, **26**, 3891.
71. Hill AF, Smith MK, *Organometallics* 2007, **26**, 3900.

72. Mohamed AA, Rabinovich D, Fackler JP, *Acta Crystallogr* 2002, **E58**, m726.
73. Maria L, Domingos Â, Santos I, *Inorg Chem* 2001, **40**, 6863.
74. Kimblin C, Bridgewater BM, Churchill DG, Hascall T, Parkin G, *Inorg Chem* 2000, **39**, 4240.
75. Seebacher J, Shu M, Vahrenkamp H, *Chem Commun* 2001, 1026.
76. He G, Vahrenkamp H, *Z Anorg Allg Chem* 2007, **633**, 51.
77. Ibrahim MM, He G, Seebacher J, Benkmil B, Vahrenkamp H, *Eur J Inorg Chem* 2005, 4070.
78. Rombach M, Seebacher J, Ji M, Zhang G, He G, Ibrahim MM, Benkmil B, Vahrenkamp H, *Inorg Chem* 2006, **45**, 4571.
79. Soares LF, Menezes DC, Silva RM, Doriguetto AC, Ellena J, Mascarenhas YP, Castellano EC, *Polyhedron* 2004, **23**, 205.
80. Bailey PJ, Lanfranchi M, Marchiò L, Parsons S, *Inorg Chem* 2001, **40**, 5030.
81. Gennari M, Lanfranchi M, Marchiò L, Pellinghelli MA, Tegoni M, Cammi R, *Inorg Chem* 2006, **45**, 3456.
82. Gennari M, Giannetto M, Lanfranchi M, Marchiò L, Pellinghelli MA, Tegoni M, *Polyhedron* 2004, **23**, 1829.
83. Blagg RJ, Claramunt JPH, Connelly NG, Haddow MF, Orpen AG, *Chem Commun* 2006, 2350.
84. Ojo JF, Slavin PA, Reglinski J, Garner M, Spicer MD, Kennedy AR, Teat SJ, *Inorg Chim Acta* 2001, **313**, 15.
85. Silva RM, Gwengo C, Lindeman SV, Smith MD, Gardinier JR, *Inorg Chem* 2006, **45**, 10998.
86. Gardinier JR, Silva RM, Gwengo C, Lindeman SV, *Chem Commun* 2007, 1524.
87. Landry VK, Buccella D, Pang K, Parkin G, *Dalton Trans* 2007, 866.
88. Minoura M, Landry VK, Melnick JG, Pang K, Marchiò L, Parkin G, *Chem Commun* 2006, 3990.
89. Landry VK, Pang K, Quan SM, Parkin G, *Dalton Trans* 2007, 820.
90. Benkmil B, Ji M, Vanhrenkamp H, *Inorg Chem* 2004, **43**, 8212.
91. Ji M, Benkmil B, Vahrenkamp H, *Inorg Chem* 2005, **44**, 3518.
92. Maria L, Paulo A, Santos IC, Santos I, Kurz P, Spingler B, Alberto R, *J Am Chem Soc* 2006, **128**, 14590.
93. Hill AF, Owen GR, White AJP, Williams DJ, *Angew Chem Int Ed* 1999, **38**, 2759.
94. Hill AF, *Organometallics* 2006, **25**, 4741.

95. Foreman MRSt-J, Hill AF, White AJP, Williams DJ, *Organometallics* 2004, **23**, 913.
96. Crossley IR, Foreman MRSt-J, Hill AF, White AJP, Williams DJ, *Chem Commun* 2005, 221.
97. Crossley IR, Hill AF, Willis AC, *Organometallics* 2006, **25**, 289.
98. Crossley IR, Hill AF, Humphrey ER, Willis EC, *Organometallics* 2005, **24**, 4083.
99. Foreman MRSt-J, Hill AF, Owen GR, White AJP, Williams DJ, *Organometallics* 2003, **22**, 4446.
100. Mihalcik DJ, White JL, Tanski JM, Zakharov LN, Yap GPA, Incarvito CD, Rheingold AL, Rabinovich D, *Dalton Trans* 2004, 1626.
101. Crossley IR, Hill AF, Willis AC, *Organometallics* 2005, **24**, 1062.
102. Landry VK, Melnick JC, Buccella D, Pang K, Ulichny JC, Parkin G, *Inorg Chem* 2006, **45**, 2588.
103. Crossley IR, Hill AF, Humphrey ER, Smith MK, Tshabang N, Willis AC, *Chem Commun* 2004, 1878.
104. Figueroa JS, Melnick JG, Parkin G, *Inorg Chem* 2006, **45**, 7056.
105. Pang K, Quan SM, Parkin G, *Chem Commun* 2006, 5015.
106. Reglisnki J, Spicer MD, Ojo JF, McAnally GD, Skórska A, Smith SJ, Smith WE, *Langmuir* 2003, **19**, 6336.
107. Rabinovich D, *Struct Bond* 2006, **120**, 143.

Chapter 6

Scorpionates Based on Thioethers as Sulfur Donors

In 1994, Riordan and coworkers[1] have prepared monoanionic, acyclic borates of the form $B(CH_2SR)_4^-$, potentially C_{3v}-symmetric face-capping ligands. These molecules can be considered sulfur analogs of the versatile poly(pyrazolyl)borates. The thioether ligands described in this chapter provide a softer coordination sphere for metal ions and, sometimes, a different metal atom environment.

6.1. $\{B(CH_2SCH_3)_4\}$ ($\equiv$ RTt)

$[B(CH_2SCH_3)_4]^-$, ($CH_3SCH_2Tt \equiv$ RTt), (Fig. 6.1) can be readily prepared in good yield from the *in situ* reaction of an excess of $LiCH_2SCH_3$ with $BF_3 \cdot Et_2O$ in THF at $-78°C$. The anion is precipitated from aqueous solutions as $[Bu_4N]^+$ salt. Addition of 1 equiv. of $[Bu_4N][RTt]$ to a THF solution of $Mo(C_7H_8)(CO)_3$ results in $[Bu_4N][Mo(RTt)(CO)_3]$ for which the diffraction study confirmed the tridentate, face-capping nature of the ligand. Protonation of $[Bu_4N][Mo(RTt)(CO)_3]$ with $HBF_4 \cdot Et_2O$ in THF resulted in immediate formation of the air-sensitive Mo hydride $[Mo(H)(RTt)(CO)_3]$ (Fig. 6.2). Reaction of 2 equiv. of $[Bu_4N][RTt]$ with $[Fe(H_2O)_6][BF_4]_2$ over several hours in THF gave $[Fe(RTt)_2]$.[1]

The synthesis and characterization of $[M(RTt)_2]$ (M = Fe, Co, Ni) complexes have been reported. RTt provides tridentate, face-capping

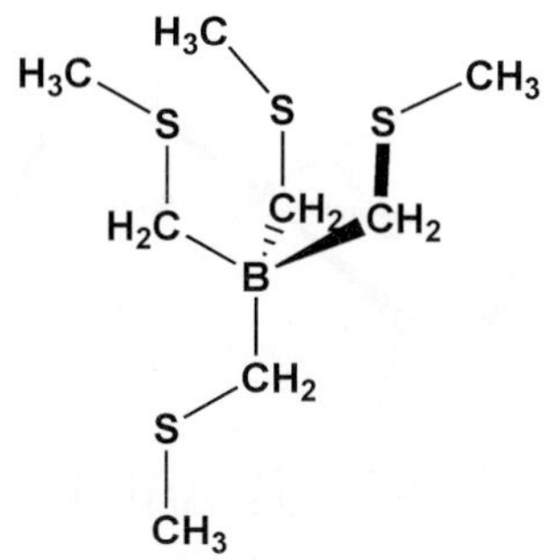

Fig. 6.1. Structure of $[B(CH_2SCH_3)_4]$ (RTt), a sulfur analog of Tp.

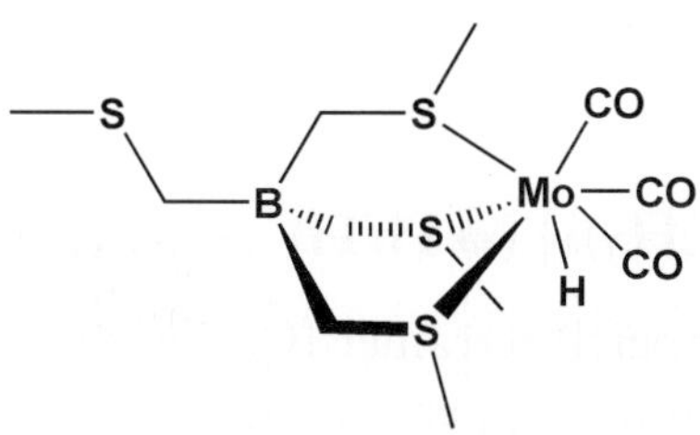

Fig. 6.2. The air-sensitive $[Mo(H)(RTt)(CO)_3]$.

coordination to the divalent metal ion. The Fe(II) complex shows spin-crossover behavior both in solution and in the solid state.[2]

6.2. $\{B(C_4H_3S)_4\}$ (≡ RTt^{Tn})

The anion tetrakis(2-thienyl)borate, $[B(C_4H_3S)_4]^-$ (≡ RTt^{Tn}), which is the thiophene analog of the polythioether ligand RTt, has been reported. RTt^{Tn} does not coordinate $Mo(CO)_3$ when reacted with $[Mo(C_7H_8)(CO)_3]$ also if the sulfur atoms are oriented to coordinate the metal in a pyramidal η^1 sulfur-bound mode. Approximate molecular orbital calculations were used to compare the metal–ligand interactions in RTt and RTt^{Tn} species. The results indicated that the magnitude and polarizability of the electronic charge density of the lone pairs on the sulfur atoms dictate the coordination strength of

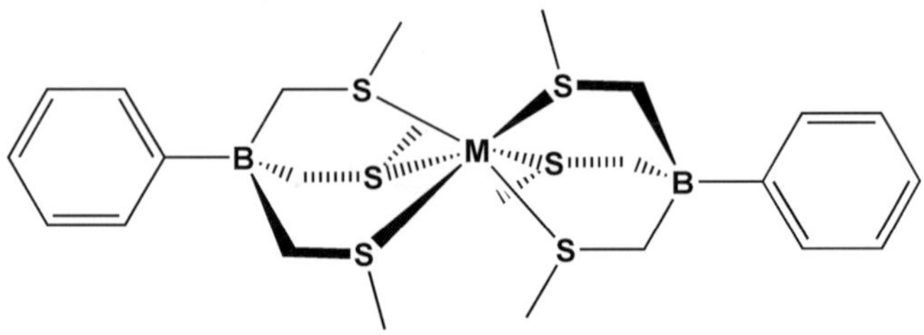

Fig. 6.3. The octahedral [M(PhTt)$_2$].

the ligands. The conjugation of the sulfur lone pair electrons with adjacent π bonds in the ligands decreases the Lewis basicity of the molecule.[3]

6.3. {PhB(CH$_2$SCH$_3$)$_3$} (≡ PhTt)

The homoleptic copper(I) tetramer [Cu(PhTt)]$_4$, structurally characterized, shows the four copper ions nearly coplanar in an approximately tetrahedral environment; [Cu(PhTt)]$_4$ is stable toward O_2, H_2O, and CO and reacts with PPh$_3$, pyridine, or [NEt$_4$]SPh to yield the corresponding mononuclear [Cu(PhTt)L] species (L = PPh$_3$, py, [NEt$_4$]SPh).[4] The synthesis and structural characterization of [M(PhTt)$_2$] (M = Fe, Co, Ni) have been also described, these compounds being iso-structural of [M(RTt)$_2$] (Fig. 6.3).[2]

6.4. {PhB(CH$_2$S(^{t}Bu))$_3$} (≡ PhTt^{tBu})

The synthesis and the structure of [Tl(PhTt^{tBu})], having a one-dimensional extended structure, and two halide complexes, [MCl(PhTt^{tBu})] (M = Co, Ni), have been described. Reactivity studies of [NiCl(PhTt^{tBu})], also structurally characterized, toward CO, NO, and THF have been discussed.[5] [ZnBr(PhTt^{tBu})] provides entry into [ZnX(*S*$_3$-donor)] complexes and its metathesis with KSPh yielded the phenylthiolato complex, [Zn(SPh)(PhTt^{tBu})], which represents a structural mimic of the homocysteine ligated form of the enzyme.[6]

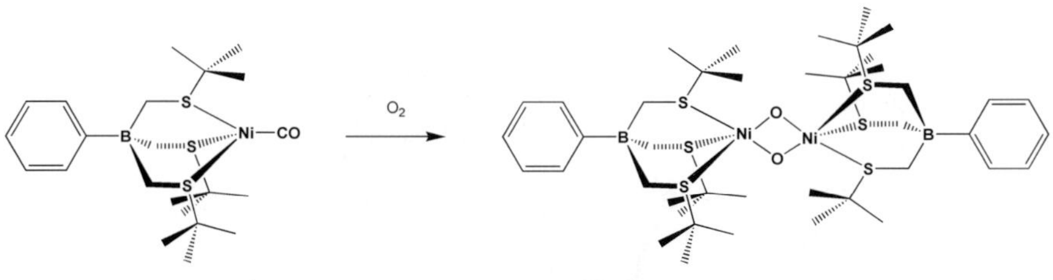

Fig. 6.4. Oxygenation of [Ni(PhTt$^{t\mathrm{Bu}}$)(CO)] gives [Ni(μ-O)(PhTt$^{t\mathrm{Bu}}$)]$_2$.

Reaction of [CoCl(PhTt$^{t\mathrm{Bu}}$)] with $MgMe_2$ or LiMe in THF resulted in the formation of [CoMe(PhTt$^{t\mathrm{Bu}}$)], whereas [NiCl(PhTt$^{t\mathrm{Bu}}$)] reacts with the same organometallic compounds giving the diamagnetic complex [Ni(η^2-CH_2S^tBu)(PhTt$^{t\mathrm{Bu}}$)] which has been characterized by X-ray diffraction. Addition of PMe_3 or CO gave the stable [Ni(PhTt$^{t\mathrm{Bu}}$)(L)] complexes (L = CO or PMe_3) characterized by X-ray diffraction. In [Ni(PhTt$^{t\mathrm{Bu}}$)(CO)] the CO is located on the molecular pseudo-three-fold axis.[7] Oxygenation of [Ni(PhTt$^{t\mathrm{Bu}}$)(CO)] in toluene yielded the thermally sensitive species [Ni(μ-O)(PhTt$^{t\mathrm{Bu}}$)]$_2$ (Fig. 6.4).[8]

[Ni(CO)(PhTt$^{t\mathrm{Bu}}$)] has been described as a possible model for key Ni–CO reaction intermediates in the acetyl-CoA synthase catalytic cycle.[9] Spectroscopic and DFT electronic structure computational studies on [Ni_2(μ-O)$_2$(PhTt$^{t\mathrm{Bu}}$)$_2$] provide insight into the possible role of the (PhTt$^{t\mathrm{Bu}}$) thioether ligand in the formation of the oxo complex.[10] The copper(I) complex [Cu(PhTt$^{t\mathrm{Bu}}$)]$_\infty$ was obtained from the reaction of [Cu(MeCN)$_4$][BF_4] with [Cs(PhTt$^{t\mathrm{Bu}}$)]. [Cu(PhTt$^{t\mathrm{Bu}}$)]$_\infty$, in MeCN and in the presence of PPh_3, converts to [Cu(PhTt$^{t\mathrm{Bu}}$)(PPh_3)].[11]

Addition of [Ni(CO)(PhTt$^{t\mathrm{Bu}}$)] to a Cu(N–N) complex resulted in the gradual formation of the mixed-metal complex reported in Fig. 6.5 containing a [Cu(μ-O)$_2$Ni]$^{2+}$ core.[12]

A series of electronically and coordinatively unsaturated four-coordinate organocobalt(II) complexes, [Co(R)(PhTt$^{t\mathrm{Bu}}$)] (R = Me, Et, Ph, Bn), have been prepared by reaction of [CoCl(PhTt$^{t\mathrm{Bu}}$)] with the corresponding MgR_2 reagent and characterized spectroscopically and structurally. [Co(η^3-allyl)(PhTt$^{t\mathrm{Bu}}$)] is a low-spin species.

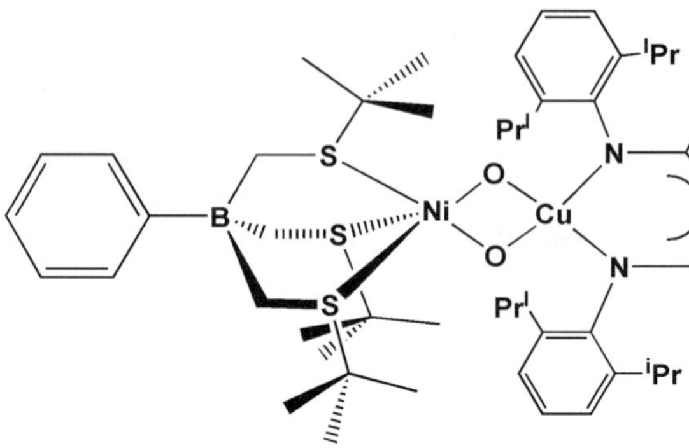

Fig. 6.5. The mixed-metal complex [(PhTt^{tBu})Ni(μ-O)$_2$Cu(N–N)] reported in Ref. 12.

Reaction of [Co(R)(PhTt^{tBu})] with CO yielded the acyl adducts [Co(CO)C(O)(R)(PhTt^{tBu})] (R=Me, Et, Ph). These five-coordinate, low-spin complexes possess square pyramidal stereochemistry with a thioether occupying the apical position. [Co(R)(PhTt^{tBu})] (R = Bn or allyl), upon reaction with CO, yields [Co(CO)$_2$(PhTt^{tBu})]. [Ni(η^3-allyl)(κ^2-PhTt^{tBu})] shows no reactivity with CO under similar conditions. [Co(R)(PhTt^{tBu})] also reacts with NO, giving [Co(NO)$_2$(κ_2-PhTt^{tBu})].[13]

The two thiolate-bridged nickel–copper complexes in Fig. 6.6 have been prepared by reaction of [N$_2$S$_2$]Ni-type complexes with [Cu(CO)(PhTt^{tBu})], the latter available via the carbonylation of [Cu(MeCN)(PhTt^{tBu})]. These complexes can be employed as model for the catalytic site of the acetyl coenzyme A synthase.[14]

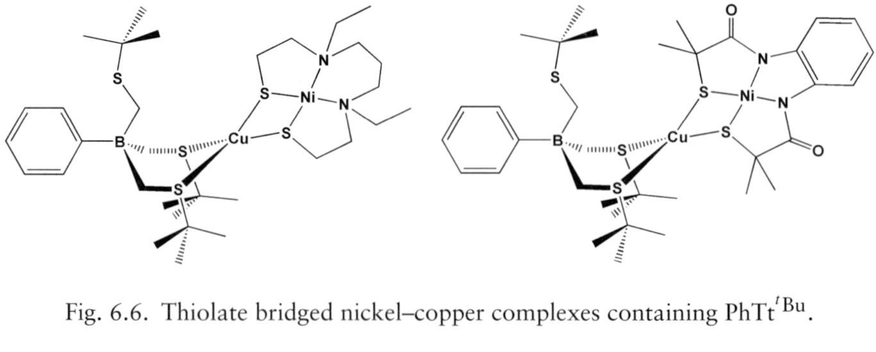

Fig. 6.6. Thiolate bridged nickel–copper complexes containing PhTt^{tBu}.

6.5. {PhB(CH_2S(*p*-Tol))$_3$} (≡ PhTt^{pTol})

Copper(I) complexes of the tridentate thioether ligand (PhTt^{pTol}) have been prepared by reaction of its Bu_4N salt with [Cu(MeCN)$_4$][BF_4]. When the reaction was carried out in MeCN in absence of ancillary ligands, the one-dimensional extended structure [Cu(PhTt^{pTol})] was obtained; whereas in the presence of PPh_3, the mononuclear tetrahedral complex [Cu(PhTt^{pTol})(PPh_3)] afforded.[11]

6.6. {PhB(CH_2S(1-adamantyl))$_3$} (≡ PhTt^{Ad})

The ligand phenyltris{1-adamantyl(thiomethyl)}borate PhTt^{Ad}, has been employed for the preparation of four-coordinate nickel(I) complexes: [Ni(PhTt^{Ad})(CO)], [Ni(PhTt^{Ad})(PMe_3)], and [Ni(η^2-CH_2SAd)(κ^2-PhTt^{Ad})]. The latter is a degradation product resulting from attempted reduction of the nickel(II) precursor [NiCl(PhTt^{Ad})].[15] [NiCl(PhTt^{Ad})] reacted with O_2 generating a 1:1 species identified as a side-on dioxygen adduct [Ni(O_2)(PhTt^{Ad})], able to oxidize PPh_3 to $OPPh_3$, NO to NO_3^-, and [Ni(CO)(PhTt$^{^tBu}$)] to the nonsymmetric [(PhTt$^{^tBu}$)Ni(μ-O)$_2$Ni(PhTt$^{^tBu}$)] dimer.[16]

6.7. {Ph(pz)B(CH_2S(^{t}Bu))$_2$} (≡ Ph(pz)Bt$^{^tBu}$) and {Ph(pz$^{^tBu}$)B(CH_2S(^{t}Bu))$_2$} (≡ Ph(pz$^{^tBu}$)Bt$^{^tBu}$)

[ZnX(S_2N–donor)] (X = Br, CH_3, SPh) species were prepared using the new mixed-donor ligands, {Ph(pz)Bt$^{^tBu}$} and {Ph(pz$^{^tBu}$)Bt$^{^tBu}$}. Protonolysis of [Zn(Me){Ph(pz$^{^tBu}$)Bt$^{^tBu}$}] by PhSH in toluene yielded [Zn(SPh){Ph(pz$^{^tBu}$)Bt$^{^tBu}$}], a synthetic analog of the homocysteine ligated form of cobalamin-independent methionine synthase.[6]

6.8. {Ph(pz$^{^tBu}$)Bt$^{^tBu}$}

[Zn(Me){Ph(pz$^{^tBu}$)Bt$^{^tBu}$}] has been employed in the synthesis of [Zn-(SC_6H_4*o*-NHC(O)^{t}Bu){Ph(pz$^{^tBu}$)Bt$^{^tBu}$}], [Zn(SC_6H_4*o*-NDC(O)^{t}Bu)-{Ph(pz$^{^tBu}$)Bt$^{^tBu}$}], and [Zn(SC_6H_4*p*-NHC(O)^{t}Bu){Ph(pz$^{^tBu}$)Bt$^{^tBu}$}].

$[Ph(pz^{^tBu})Bt^{^tBu}]$, transformed in $[Ph(pz^{^tBu})Bt^{^tBu}]H$, reacts with $CdCl_2$ giving $[CdCl\{Ph(pz^{^tBu})Bt^{^tBu}\}]$ which upon treatment with KSPh yielded $[CdSPh\{Ph(pz^{^tBu})Bt^{^tBu}\}]$.[17]

6.9. $\{Ph_2B(CH_2SCH_3)_2\}$ ($\equiv Ph_2Bt$)

Reaction between 2 equiv. of $[Bu_4N]Ph_2Bt$ and $Ni(BF_4)_2{\cdot}6H_2O$ in acetone resulted in precipitation of a deep red solid, $[Ni(Ph_2Bt)_2]$, structurally characterized, containing the Ni ion on a crystallographic inversion center which renders *trans* thioethers metrically equivalent and ensures a planar ligation sphere.[18] Ph_2Bt is also able to form the tetrameric species $[Cu(Ph_2Bt)]_4$, structurally characterized, containing both terminal and bridging thioether ligation, and the mononuclear tetrahedral species $[Cu(Ph_2Bt)(PPh_3)_2]$.[11] The reactivity of $[Ni(Ph_2Bt)_2]$ has been investigated, $[Ni(Ph_2Bt)_2(L)_2]$ (L = MeCN or py) and $[Bu_4N][NiBr_2(Ph_2Bt)_2]$ being formed.[19]

6.10. $\{Ph_2B(CH_2SPh)_2\}$ ($\equiv Ph_2Bt^{Ph}$)

The adducts $[Cu(Ph_2Bt^{Ph})(L)_2]$ (L = MeCN or PPh_3) have been prepared by the reaction of $[Bu_4N][Ph_2Bt^{Ph}]$ with $[Cu(MeCN)_4][BF_4]$ in the absence or in the presence of PPh_3.[11]

6.11. $\{Ph_2B(CH_2S(^tBu))_2\}$ ($\equiv Ph_2Bt^{^tBu}$)

Reaction of $[Cs(Ph_2Bt^{^tBu})]$ with $NiCl_2$ gave an unexpected organometallic derivative $[Ni(\eta^2\text{-}CH_2S^tBu)(Ph_2Bt^{^tBu})]$, resulting from B–C bond rupture.[19]

6.12. $\{Et_2B(CH_2SCH_3)_2\}$ ($\equiv Et_2Bt$)

The ligand Et_2Bt, prepared as $[Bu_4N]$ salt, was employed to synthesize $[Cu(Et_2Bt)(PPh_3)_2]$, structurally characterized.[11]

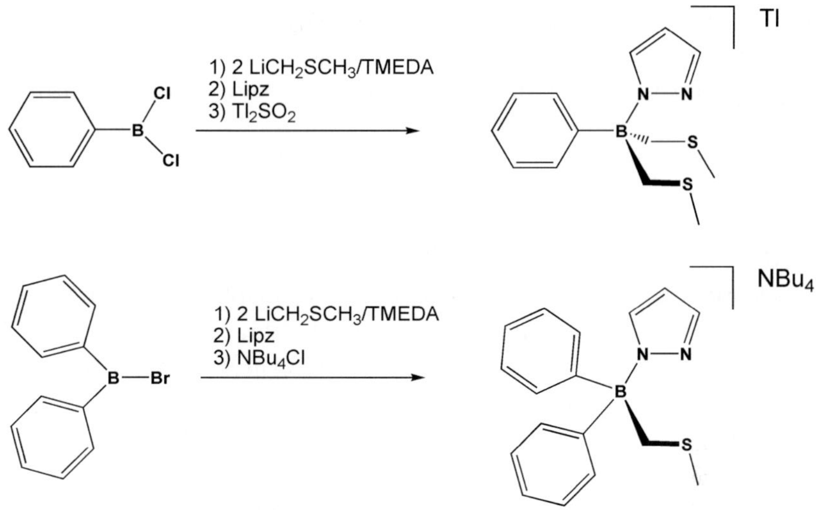

Fig. 6.7. Synthesis of {Ph(pz)B(CH_2SCH_3)$_2$} (≡ Ph(pz)Bt) and {Ph_2(pz)B(CH_2-SCH_3)}.

6.13. {Ph(pz)B(CH_2SCH_3)$_2$} (≡ Ph(pz)Bt) and {Ph_2(pz)$_2$B(CH_2SR)} (R=Me or ^{t}Bu)

Two new borate ligands, {Ph(pz)B(CH_2SCH_3)$_2$} (≡ Ph(pz)Bt) and {Ph_2(pz)B(CH_2SCH_3)} have been synthesized according to the reactions in Fig. 6.7. They provide the mixed donor sets [*N*, S_2] and [*N*, *S*], respectively, to transition metal ions. The homoleptic complexes [M{Ph(pz)Bt}$_2$] (M = Fe, Co, Ni) and [M{Ph_2(pz)B(CH_2SCH_3)}$_2$] (M = Co, Ni, Zn) have been described, and [Ni{Ph(pz)Bt}$_2$] and [Co{Ph_2(pz)B(CH_2SCH_3)}$_2$] also structurally characterized.[20] The ligands {Ph_2B(pz)(CH_2SR)} (R = Me or ^{t}Bu) react with Ni salts, yielding exclusively the *cis* stereoisomers of [Ni{Ph_2B(pz)(CH_2SR)}$_2$].[19]

References

1. Ge P, Haggerty BS, Rheingold AL, Riordan CG, *J Am Chem Soc* 1994, **116**, 8406.
2. Ohrenberg C, Ge P, Schebler P, Riordan CG, Yap GPA, Rheingold AL, *Inorg Chem* 1996, **35**, 749.

3. Sargent AL, Titus EP, Riordan CG, Rheingold AL, Ge P, *Inorg Chem* 1996, **35**, 7095.
4. Ohrenberg C, Saleem MM, Riordan CG, Yap GP, Verma AK, Rheingold AL, *Chem Commun* 1996, 1081.
5. Schebler PJ, Riordan CG, Guzei IA, Rheingold AL, *Inorg Chem* 1998, **37**, 4754.
6. Chiou S-J, Innocent J, Riordan CG, Lam K-C, Liable-Sands L, Rheingold AL, *Inorg Chem* 2000, **39**, 4347.
7. Schebler PJ, Mandimutsira BS, Riordan CG, Liable-Sands LM, Incarvito CD, Rheingold AL, *J Am Chem Soc* 2001, **123**, 331.
8. Mandimutsira BS, Yamarik JL, Brunold TC, Gu W, Cramer SP, Riordan CG, *J Am Chem Soc* 2001, **123**, 9194.
9. Craft JL, Mandimutsira BS, Fujita K, Riordan CG, Brunold TC, *Inorg Chem* 2003, **42**, 859.
10. Schenker R, Mandimutsira BS, Riordan CG, Brunold TC, *J Am Chem Soc* 2002, **124**, 13842.
11. Ohrenberg C, Liable-Sands LM, Rheingold AL, Riordan CG, *Inorg Chem* 2001, **40**, 4276.
12. Aboelella NW, York JT, Reynolds AM, Fujita K, Kinsinger CR, Cramer CJ, Riordan CG, Tolman WB, *Chem Commun* 2004, 1716.
13. DuPont JA, Coxey MB, Schebler PJ, Incarvito CD, Dougherty WG, Yap GPA, Rheingold AL, Riordan CG, *Organometallics* 2007, **26**, 971.
14. Krishnan R, Voo JK, Riordan CG, Zahkarov L, Rheingold AL, *J Am Chem Soc* 2003, **125**, 4422.
15. Fujita K, Rheingold AL, Riordan CG, *Dalton Trans* 2003, 2004.
16. Fujita K, Schenker R, Gu W, Brunold TC, Cramer SP, Riordan CG, *Inorg Chem* 2004, **43**, 3324.
17. Chiou S-J, Ge P, Riordan CG, Liable-Sands LLM, Rheingold AL, *Chem Commun* 1999, 159.
18. Ge P, Riordan CG, Yap GPA, Rheingold AL, *Inorg Chem* 1996, **35**, 5408.
19. Ge P, Rheingold AL, Riordan CG, *Inorg Chem* 2002, **41**, 1383.
20. Chiou S-J, Riordan CG, Rheingold AL, *Proc Natl Acad Sci* 2003, **100**, 3695.

Chapter 7

Scorpionates Based on Diaryl and Dialkylphosphine as Donors

Scorpionates based on phosphines (phosphinoborates) are a new emerging class of ligands almost exclusively developed by Peters' group.[1] They are successfully used in coordination chemistry and catalysis and represent a progress with respect to phosphine ligands due to easy modification of their electronic and steric properties. Convenient synthetic procedures for the synthesis of diaryl- and dialkyl-(phosphino)borates have been reviewed and the synthesis of [Tl{$PhB(CH_2PR_2)_3$}] (the abbreviation system used by Peters will be used in the whole chapter), a useful transmetalation reagent, has been described in detail.[2] Metal complexes supported by (phosphino)borates can be described as zwitterionic complexes containing a cationic metal center chelated by anionic donor ligands. In contrast to poly(pyrazolyl)borates, the phosphinoborates do not exhibit an accepted means for delocalizing the charge of the borate by resonance.

7.1. {$PhB(CH_2PR_2)_3$}($\equiv PhBP_3^R$)

7.1.1. *$PhBP_3$*($\equiv PhBP_3^{Ph}$)

[Li(TMEDA)][$PhBP_3$] has been prepared by addition of 3 equiv. of [Li(TMEDA)][CH_2PPh_2] to $BPhCl_2$. The tripodal anionic $PhBP_3$ has been employed in the synthesis of a neutral iridium silylene complex

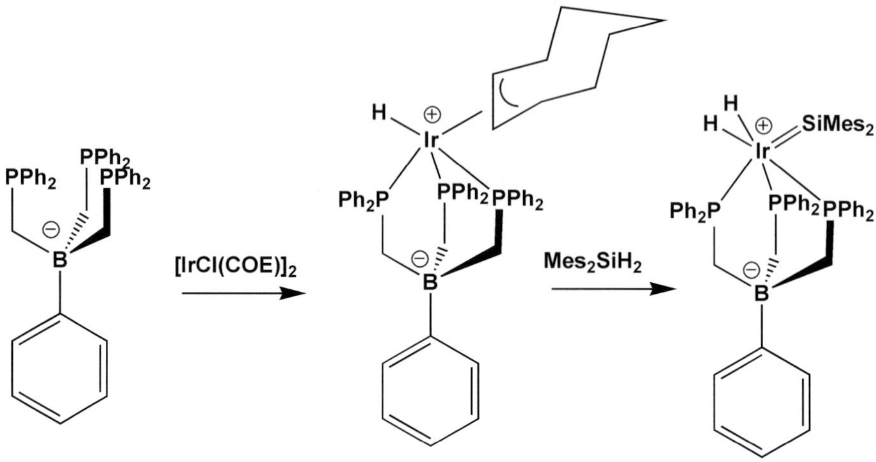

Fig. 7.1. Synthesis of the neutral iridium silylene complex [Ir=SiMes$_2$(H)$_2$(PhBP$_3$)].

[Ir=SiMes$_2$(H)$_2$(PhBP$_3$)] by interaction of the cyclooctenyl complex [Ir(H)(η^3-C$_8$H$_{13}$)(PhBP$_3$)] with Mes$_2$SiH$_2$ (Fig. 7.1).[1]

Other [Ir=SiR$_2$(H)$_2$(PhBP$_3$)] complexes (R = Ph, Et, Me) have been reported. Analogously, the germylene complex [Ir=GeMes$_2$-(H)$_2$(PhBP$_3$)] has been prepared. Primary silanes react with [Ir(H)(η^3-C$_8$H$_{13}$)(PhBP$_3$)] giving [Ir=Si(R)(*c*-C$_8$H$_{15}$)(H)$_2$(PhBP$_3$)] (R = Me or 2,4,6-triisopropylphenyl). R$_3$SiH reacts with [Ir(H)(η^3-C$_8$H$_{13}$)-(PhBP$_3$)] affording [IrH$_3$(SiR$_3$)(PhBP$_3$)] (R = Et or Me).[3] The allyl complexes [Ir(H)(η^3-C$_8$H$_{13}$)(PhBP$_3$)] and [Ir(H)(η^3-C$_3$H$_5$)(PhBP$_3$)] can be employed as precursors to the dihalides derivatives [IrX$_2$-(PhBP$_3$)] (X = Cl or I). [IrMe$_2$(PhBP$_3$)] and [IrCl$_2$(ClBP$_3$)] were also synthesized and fully characterized. Addition of CO to [Ir(H)(η^3-C$_3$H$_5$)(PhBP$_3$)] gave [Ir(CO)$_2$(PhBP$_3$)], whereas reaction of [Ir(H)(η^3-C$_8$H$_{13}$)(PhBP$_3$)] with H$_2$ generated [Ir(H$_2$)(COE)(PhBP$_3$)] in THF and [IrH$_2$(PhBP$_3$)]$_2$ in benzene. [Ir(H)(η^3-C$_8$H$_{13}$)(PhBP$_3$)], upon reaction with PMePh$_2$, gave [Ir(H$_2$)(PMePh$_2$)(PhBP$_3$)], which upon protonation with acids gave [Ir(H$_3$)(PMePh$_2$)(PhBP$_3$)]. Several C–H activation reactions have been observed (Fig. 7.2).[4]

Peters and coworkers also reported the homoleptic [Tl(PhBP$_3$)], structurally characterized,[5] and an anomalous low-spin cobalt complex, [CoI(PhBP$_3$)], which exhibits a distorted tetrahedral

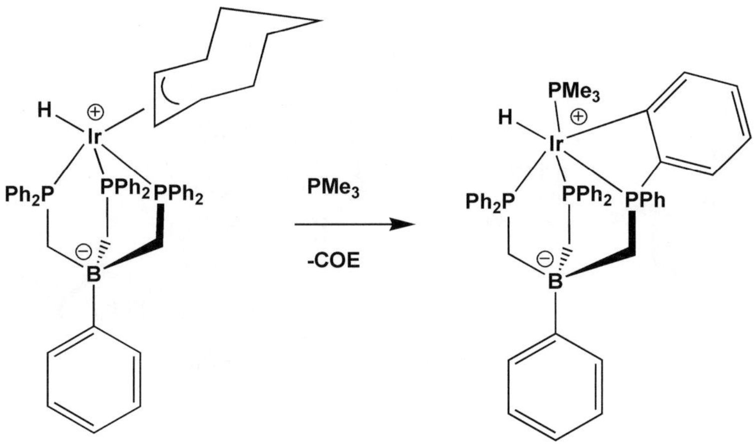

Fig. 7.2. C–H activation by using $[Ir(H)(\eta^3\text{-}C_8H_{13})(PhBP_3)]$.

geometry. It has been proposed that the ground-state electronic configuration of $[CoI(PhBP_3)]$ arises likely from a strong axial distortion, geometrically enforced by $PhBP_3$. Addition of PMe_3 to $[CoI(PhBP_3)]$ gave $[CoI(PhBP_3)(PMe_3)]$ which underwent smooth reduction by Na amalgam affording $[Co(PhBP_3)(PMe_3)]$. The latter complex reacts with (*p*-Tol)azide forming the nitrene complex $[Co(\equiv N\text{-}p\text{-}Tol)(PhBP_3)]$. The imido functionality can be transferred to carbon monoxide giving OCN-(*p*-Tol) and $[Co(PhBP_3)(CO)_2]$. Finally the reaction of $[Co(PhBP_3)(PMe_3)]$ with diphenylcarbene gave $[Co{=}N{=}N{=}CPh_2(PhBP_3)]$.[6] $[CoI(PhBP_3)]$ is an useful precursor to the corresponding dimers bromide and chlorides $[Co(\mu\text{-}X)(PhBP_3)]_2$ (Fig. 7.3). Stoichiometric oxidation of $[CoI(PhBP_3)]$ gave the high-spin complex $[CoI\{PhB(CH_2P(O)Ph_2)_2(CH_2PPh_2)\}]$. Finally reaction of $[CoI(PhBP_3)]$ with Tl(OAr) (Ar = 2,6-Me_2Ph) afforded the high-spin complex $[Co(O\text{-}2,6\text{-}Me_2Ph)(PhBP_3)]$.[7] The pseudotetrahedral d^7 complex $[Co(OSiPh_3)(PhBP_3)]$ exhibits thermally induced spin-crossover.[8] An investigation of the structural and magnetic properties of four-coordinate $[Co(X)(PhBP_3)]$ complexes (X = OC_6F_5, $OSiPh_3$,...) has been also reported.[9]

The four-coordinate complex $[Fe(PPh_3)(PhBP_3)]$, obtained by reaction of $[FeCl(PhBP_3)]$ with PPh_3 in sodium amalgam, undergoes

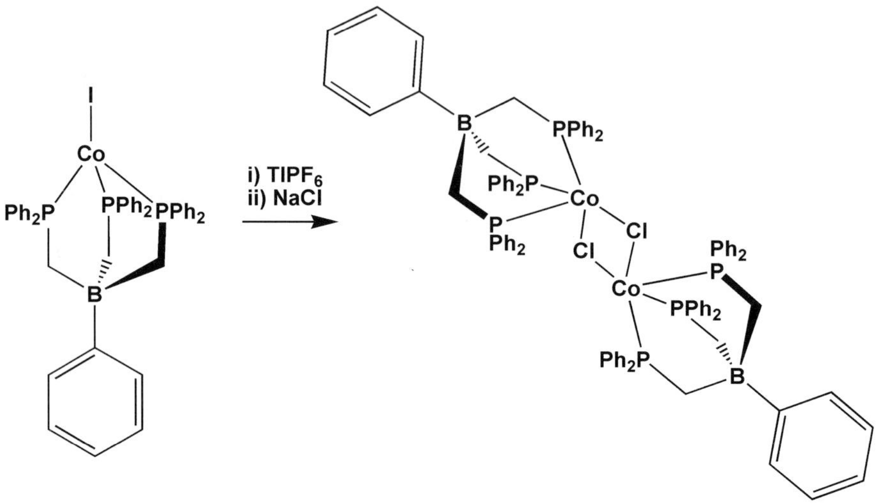

Fig. 7.3. Synthesis of the dimer $[Co(\mu\text{-}Cl)(PhBP_3)]_2$.

rapid oxidation by aryl azide giving a mononuclear terminally bonded Fe(III) imide $[Fe(\equiv N\text{-}p\text{-}Tol)(PhBP_3)]$. This complex at room temperature upon CO addition quantitatively releases isocyanate (O=C=N-*p*-Tol) and forms the distorted square-pyramidal dicarbonyl by-product $[Fe(CO)_2(PhBP_3)]$.[10] $[Fe(\equiv N\text{-}p\text{-}Tol)(PhBP_3)]$, upon exposure to 1 atm of hydrogen pressure at room temperature, features both partial and complete hydrogenolysis of the Fe≡NR linkage (Fig. 7.4).[11]

Sodium amalgam reduction of the ferromagnetically coupled dimer, $[Fe(\mu\text{-}1,3\text{-}N_3)(PhBP_3)]_2$ yields the diamagnetic bridging nitride species $[\{Fe(PhBP_3)\}_2(\mu\text{-}N)][Na(THF)_5]$. The diamagnetic imide complex $[Fe\equiv N(1\text{-}Ad)(PhBP_3)][^nBu_4N]$ has been also prepared by sodium amalgam reduction of its low-spin precursor $[Fe(\equiv N(1\text{-}Ad))\text{-}(PhBP_3)]$. The diamagnetic $[Fe(\equiv N(1\text{-}Ad))\{PhBP_3\}][^nBu_4N]$ provides a bona fide example of a pseudotetrahedral iron(II) center in a low-spin ground-state configuration.[12] $[\{Fe(PhBP_3)\}_2(\mu\text{-}N)]\text{-}[Na(THF)_5]$, which consists of two antiferromagnetically coupled high-spin Fe(II) centers, and its neutral one-electron oxidation

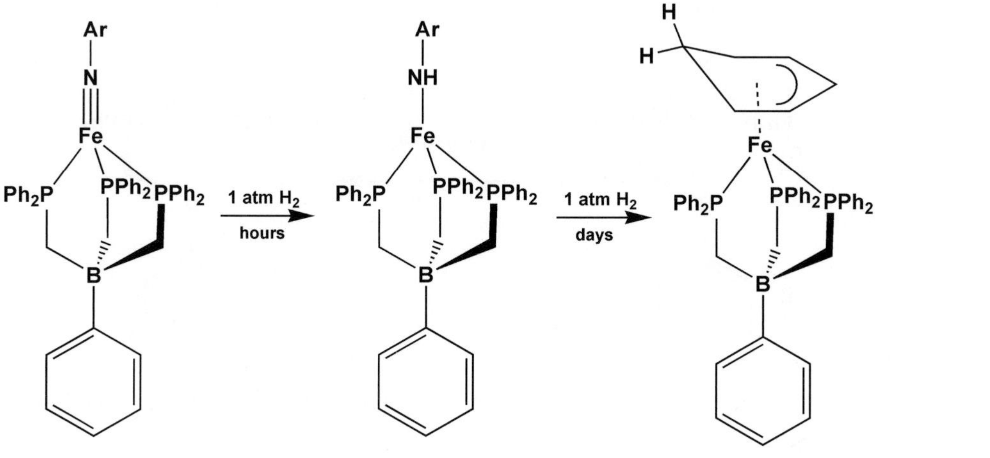

Fig. 7.4. [Fe(≡N–Ar)($PhBP_3$)], upon exposure to H_2, exhibits both partial and complete hydrogenolysis of the Fe≡NR linkage.

product [{Fe($PhBP_3$)}$_2$(μ-N)] are able to mediate heterolytic H_2 activation yielding the di-iron imide-hydride [{Fe($PhBP_3$)}(μ-NH)-(μ-H){Fe($PhBP_3$)}]. [{Fe($PhBP_3$)}$_2$(μ-N)] reacts with 5 equiv. of CO giving both [Fe$(CO)_2$($PhBP_3$)] and [Fe$(CO)_2$(NCO)($PhBP_3$)].[13]

The pseudotetrahedral complexes [NiCl($PhBP_3$)] and [NiI($PhBP_3$)] were prepared by the metalation of [Tl($PhBP_3$)] with [$NiCl_2$$(Ph_3P)_2$] and NiI_2, respectively. [NiCl($PhBP_3$)] is a precursor to [Ni(N_3)-($PhBP_3$)], [Ni($OSiPh_3$)($PhBP_3$)], [Ni(O-*p*-^{t}Bu-Ph)($PhBP_3$)], and [Ni(S-*p*-^{t}Bu-Ph)($PhBP_3$)], characterized by X-ray diffraction (XRD) as pseudotetrahedral monomers in the solid state. [NiCl($PhBP_3$)] reacts readily with oxygen to form the four-electron-oxidation product, [NiCl{PhB(CH_2P(O)Ph_2)$_2$(CH_2PPh_2)}]. Chemical reduction of [NiCl($PhBP_3$)], using Na/Hg or other potential one-electron reductants, generates a product that arises from partial ligand degradation, [Ni(η^2-CH_2PPh_2)($PhBP_3$)] (Fig. 7.5); however, in the presence of L (L = PPh_3, CN^{t}Bu) the reaction produces [Ni(PPh_3)(L)]. [Ni($PhBP_3$)-(NO)] is prepared by the reaction of [NiCl($PhBP_3$)] with excess NO or by treating [Ni($PhBP_3$)(PPh_3)] with stoichiometric NO. [Ni(κ^2-$PhBP_3$)$(CO)_2$][^{n}Bu_4N] was obtained by reacting [Tl($PhBP_3$)] with [Ni$(PPh_3)_2$$(CO)_2$] in the presence of [^{n}Bu_4N]Br.[14]

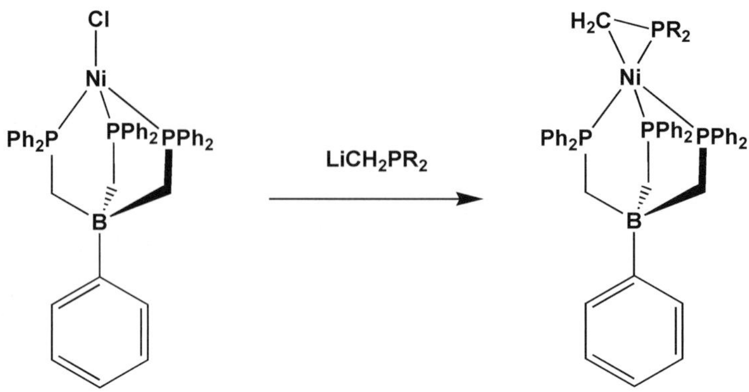

Fig. 7.5. Reduction of [NiCl(PhBP$_3$)] with LiCH$_2$PR$_2$ gives [Ni(η^2-CH$_2$PPh$_2$)-(PhBP$_3$)].

Reaction of [PhBP$_3$][^{n}Bu$_4$N] with platinum methyl precursors gave [PtMe$_3$(PhBP$_3$)] and [PtMe$_2$(κ^2-PhBP$_3$)][^{n}Bu$_4$N], also structurally characterized. [PtMe$_3$(PhBP$_3$)] is very thermally and chemically robust. In contrast, the square planar [PtMe$_2$(κ^2-PhBP$_3$)][^{n}Bu$_4$N] demonstrated reactivity at both the Pt and the unbound phosphine. The reaction of [^{n}Bu$_4$N][PtMe$_2$(κ^2-PhBP$_3$)] with elemental sulfur gave [PtMe$_2$(κ^2-PhB(CH$_2$P(S)Ph$_2$)(CH$_2$PPh$_2$)$_2$)][^{n}Bu$_4$N] whereas the reaction with BH$_3$·SMe$_2$ formed [PtMe$_2$(κ^2-PhB(CH$_2$P(BH$_3$)-Ph$_2$)(CH$_2$PPh$_2$)$_2$)][^{n}Bu$_4$N] which, with additional BH$_3$·SMe$_2$, gave [Pt(μ-H){(κ^2-PhB(CH$_2$P(BH$_3$)Ph$_2$)(CH$_2$PPh$_2$)$_2$)}]$_2$[^{n}Bu$_4$N]$_2$.[15]

[Li(TMEDA)][PhBP$_3$] reacts with SnCl$_2$ yielding [SnCl(PhBP$_3$)] which may be de-halogenated to give the η^3-complex [Sn(PhBP$_3$)]-[PF$_6$].[16]

7.1.2. *PhBP*$_3^{i\text{Pr}}$

Fourteen-electron, coordinatively unsaturated, Fe(II) silyl complexes containing PhBP$_3^{i\text{Pr}}$, have been prepared and their reactivity investigated. These silyl complexes can undergo redox processes to generate isolable Fe(I) species with concomitant loss of the silyl ligand. The paramagnetic [FeBr(PhBP$_3^{i\text{Pr}}$)] reacted with KSi(SiMe$_3$)$_3$ in benzene yielding a diamagnetic product in which a η^5 coordination of a silylated benzene moiety to the iron center is found (Fig. 7.6). When

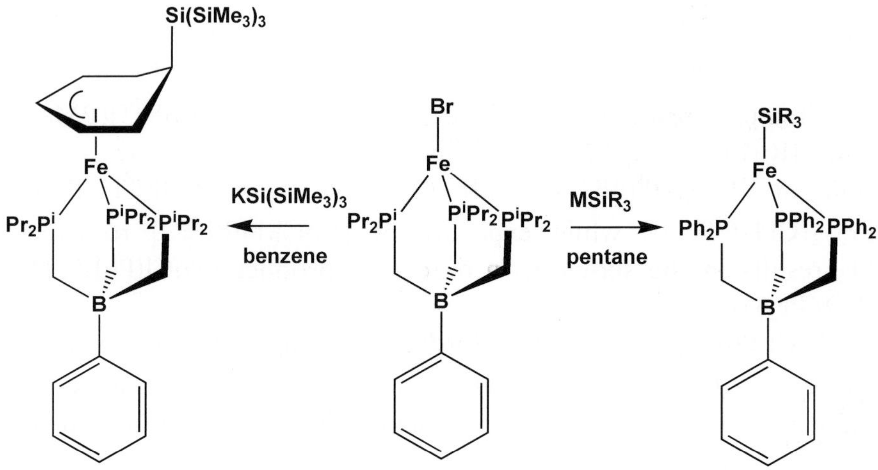

Fig. 7.6. [FeBr(PhBP$_3^{iPr}$)] reacts with KSi(SiMe$_3$)$_3$ in benzene forming a 6-exocyclohexadienyl complex, whereas the same reaction carried out in pentane yields [Fe(SiR$_3$)(PhBP$_3^{iPr}$)].

[FeBr(PhBP$_3^{iPr}$)] reacts with KSi(SiMe$_3$)$_3$ in pentane, the paramagnetic [FeSi(SiMe$_3$)$_3$(PhBP$_3^{iPr}$)] affords, whereas [FeBr(PhBP$_3^{iPr}$)] reacts with either Ph$_3$SiLi(THF)$_3$ or Ms$_2$SiHLi(THF)$_2$ in pentane, leading to the formation of the corresponding silyl and hydrosilyl complexes [Fe(SiPh$_3$)(PhBP$_3^{iPr}$)] and [Fe(SiMes$_2$H)(PhBP$_3^{iPr}$)], respectively (Fig. 7.6).[17]

A short review summarizes recent advances in the synthesis and reactivity of iron imides and nitrides complexes containing phosphino-borate ligands PhBP$_3^R$.[18]

The Tl(I) complex [Tl(PhBP$_3^{iPr}$)] has been synthesized and characterized in the solid-state by X-ray diffraction analysis. This precursor is an effective transmetalating agent used in the preparation of the monomeric, four-coordinate, high-spin derivatives [FeCl(PhBP$_3^{iPr}$)] and [CoX(PhBP$_3^{iPr}$)] (X = Cl, I). The 16-electron complexes [Ru(μ-Cl)(PhBP$_3^{iPr}$)]$_2$ and [Ru(μ-Cl)(PhBP$_3$)]$_2$ have been prepared for comparison. [MCl(PhBP$_3^{iPr}$)] reacted with excess CO to afford [MCl(PhBP$_3^{iPr}$(CO))] (M = Fe, Co). Reaction of [CoI(PhBP$_3^{iPr}$)] with excess CO resulted in [CoI(PhBP$_3^{iPr}$)(CO)$_2$].

$[Ru(\mu\text{-}Cl)(PhBP_3^{iPr})]_2$ and $[Ru(\mu\text{-}Cl)(PhBP_3)]_2$ bound CO readily, giving $[RuCl(PhBP_3^{iPr})(CO)_2]$ and $[RuCl(PhBP_3)(CO)_2]$, which, upon reaction with PMe_3, formed $[RuCl(PhBP_3^{iPr})(PMe_3)]$ and $[RuCl(PhBP_3)(PMe_3)]$, respectively. Stoichiometric oxidation of $[CoCl(PhBP_3^{iPr})]$ with O_2 gave $[CoCl(PhB(CH_2P(O)^i\text{-}Pr_2)_2(CH_2P^iPr_2))]$ while exposure of $[CoCl(PhBP_3^{iPr})]$ to excess O_2 results in the six-electron oxidation product $[CoCl\{PhB(CH_2\text{-}P(O)^iPr_2)_3\}]$.[19]

$[FeCl(PhBP_3^{iPr})]$ reacts with Li(dbabh) giving the high-spin amide species $[Fe(dbabh)(PhBP_3^{iPr})]$ which is thermally unstable and shows clean first-order decay when warmed at room temperature forming $[Fe{\equiv}N(PhBP_3^{iPr})]$. This compound exhibits a propensity to undergo bimolecular condensation to $[Fe(PhBP_3^{iPr})]_2(\mu\text{-}N_2)$ via nitride coupling.[20]

Complexes $[MnX(PhBP_3^{iPr})]$ (X = Cl, I) have been prepared and characterized by X-ray diffraction, SQUID magnetometry, and EPR spectroscopy. $[MnX(PhBP_3^{iPr})]$ complexes serve as a precursor to the pseudotetrahedral $[Mn(Y)(PhBP_3^{iPr})]$ (Y = N_3, CH_2Ph, Me, $NH(2,6\text{-}^iPr_2C_6H_3)$, dbabh, 1-Ph(isoindolate)). The two Mn(I) species $[(PhBP_3^{iPr})Tl\text{–}MnBr(CO)_4]$ and $[Mn(PhBP_3^{iPr})(CN^tBu)_3]$ have been also reported.[21]

The allyl complexes $[IrH(\eta^3\text{-}C_8H_{13})(PhBP_3^{iPr})]$ and $[IrH(\eta^3\text{-}C_3H_5)(PhBP_3^{iPr})]$ were reported. $[IrH(\eta^3\text{-}C_8H_{13})(PhBP_3^{iPr})]$ reacted with secondary silanes (H_2SiR_2, with R = Et, Ph) to give silyl-capped trihydride complexes $[IrH_3(SiHR_2)(PhBP_3^{iPr})]$. This complex underwent H/D exchange with D_2 to incorporate deuterium into both the Ir–H and Si–H bonds, whereas its reaction with PMe_3 resulted in formation of $[Ir(H)_2(PhBP_3^{iPr})(PMe_3)]$. Complexes $[Ir(H)_2(PhBP_3^{iPr})(L)]$ (L = PMe_3, PH_2Cy, CO) have been synthesized from the direct reaction of $[IrH(\eta^3\text{-}C_8H_{13})(PhBP_3^{iPr})]$ with L. $[Ir(\kappa^2\text{-}PhBP_3^{iPr})(PMe_3)_2]$ and $[Rh(PhBP_3^{iPr})(CO)_2]$ were also reported. Variable-temperature ^1H and ^{31}P NMR experiments have been carried out on these complexes. $[Ir(\kappa^2\text{-}PhBP_3^{iPr})(PMe_3)_2]$ reacted with H_2 to give $[IrH_2\text{-}(\kappa^2\text{-}PhBP_3^{iPr})(PMe_3)_2]$. The reaction of $[Ir(\kappa^2\text{-}PhBP_3^{iPr})(PMe_3)_2]$ with

1 equiv. of Ph_2SiH_2 resulted in loss of a ligand arm to give the bis(phosphino)borane complex [Rh(H)$_2$(SiHPh$_2$)(PhBP$_2^{iPr}$)(PMe$_3$)]. [Rh(PhBP$_3^{iPr}$)(CO)$_2$] reacted with H_2 in the presence of 1 equiv. of Me$_3$NO to give [Rh(H)$_2$(PhBP$_3^{iPr}$)(CO)].[22]

[Fe(PhBP$_3^{iPr}$)] and related cobalt systems coordinate, activate, and promote the methylation of nitrogen at the fourth binding site. A THF solution of [MX(PhBP$_3^{iPr}$)] (M = Fe or Co, X = Cl or I) stirred under a blanket of nitrogen in the presence of Mg^0, produced the anionic, paramagnetic [Fe(PhBP$_3^{iPr}$)(N$_2$)]-[MgCl(THF)$_2$] and [Co(PhBP$_3^{iPr}$)(N$_2$)]$_2$[Mg(THF)$_4$]. Gentle oxidation by Cp_2Fe^+ in THF yielded the neutral, dinuclear N_2-bridged products [Fe(PhBP$_3^{iPr}$)]$_2$(μ-N$_2$) and [Co(PhBP$_3^{iPr}$)]$_2$(μ-N$_2$). Their exposure to sodium amalgam produced, respectively, the mixed-valence (M^0/M^I)complexes [Fe(PhBP$_3^{iPr}$)]$_2$(μ-N$_2$)[Na(THF)$_6$] and [Co(PhBP$_3^{iPr}$)]$_2$(μ-N$_2$)[Na(THF)$_6$]. [M(PhBP$_3^{iPr}$)]$_2$(μ-N$_2$) complexes undergo rapid oxidation upon addition of either tolyl or adamantyl azide to afford trivalent imide products such as [Fe(≡NAd)(PhBP$_3^{iPr}$)] and [Co(≡N-*p*-Tol)(PhBP$_3^{iPr}$)]. The diazenido complex [(PhBP$_3^{iPr}$)Fe–N=NMe] has been also reported.[23]

[NiCl(PhBP$_3^{iPr}$)] has been obtained by the metalation of [Tl(PhBP$_3^{iPr}$)] with [NiCl$_2$(dme)]. [NiCl(PhBP$_3^{iPr}$)] reacts with Li(dbabH) to provide the three-coordinate [Ni(κ^2-PhBP$_3^{iPr}$)(dbabh)]. Reduction of [NiCl(PhBP$_3^{iPr}$)] in the presence of PMe$_3$ and CN^{t}Bu afforded [Ni(PhBP$_3^{iPr}$)(PMe$_3$)] and [Ni(PhBP$_3^{iPr}$)(CN^{t}Bu)]. [Ni(κ^2-PhBP$_3^{iPr}$)(CO)$_2$][ASN] was obtained by reacting [Tl(PhBP$_3^{iPr}$)] with [Ni(Ph$_3$P)$_2$(CO)$_2$] in the presence of [ASN]Br.[14]

7.2. [Ph$_2$B(CH$_2$PR$_2$)$_2$]$^-$(≡ Ph$_2$BP$_2^R$)

7.2.1. *Ph$_2$BP$_2$(≡ Ph$_2$BP$_2^{Ph}$)*

Thomas and Peters[24] reported the synthesis of [Ph$_2$B(CH$_2$-PPh$_2$)$_2$][ASN] (Ph$_2$B(CH$_2$PPh$_2$)$_2$ = Ph$_2$BP$_2$). Its reaction with [PtMe$_2$(COD)] and [PtMe(Ph)(COD)] yields the anionic Pt(II) compounds [PtMe$_2$(Ph$_2$BP$_2$)][ASN] and [PtMe(Ph)(Ph$_2$BP$_2$)][ASN], respectively. Protonation of [$PtMe_2(Ph_2BP_2)$][ASN] gave [PtMe-(Ph_2BP_2)(THF)].

A series of structurally similar, but charge-differentiated, platinum complexes has been prepared using the bidentate phosphine ligand Ph_2BP_2, $Ph_2Si(CH_2PPh_2)_2$, and dppp. Complexes supported by anionic Ph_2BP_2 are shown to be more electron-rich than those supported by $Ph_2Si(CH_2PPh_2)_2$ and dppp. A study of the temperature dependence of the rate of THF self-exchange between neutral, formally zwitterionic [Pt(Me)(Ph_2BP_2)(THF)] and its cationic relative [Pt(Me)(Ph_2SiP_2)(THF)][$B(C_6F_5)_4$] has been carried out. Extended thermolysis of neutral [Pt(Ph)(Ph_2BP_2)(THF)] results primarily in a disproportionation into the complex molecular salt [$PtPh_2(Ph_2BP_2)$][Pt(Ph_2BP_2)$(THF)_2$]. The bulky phosphine adduct [Pt(Me)(Ph_2BP_2){$P(C_6F_5)_3$}] also undergoes thermolysis in benzene.[25]

Analogous reaction between [Ph_2BP_2][ASN] and [$PdMe_2$-(TMEDA)] produces [$PdMe_2(Ph_2BP_2)$][ASN] which can be protonated by the ammonium salt [HN^iPr_2Et][BPh_4] to generate the zwitterion [PdMe(Ph_2BP_2)(THF)]. This solvento adduct converts under CO and ethylene gas to the unstable product [Pd(C(O)Me)(Ph_2BP_2)(CO)]. Exposure of high concentrations of [PdMe(Ph_2BP_2)(THF)] to an atmosphere of CO produces the dimeric [Pd(Ph_2BP_2)]$_2$ (Fig. 7.7). An alternative method for cleanly generating dimeric species results from exposure of [PdMe(Ph_2BP_2)(THF)] to ethylene gas in the absence of CO. Monitoring this transformation under ethylene at low temperature establishes one observable intermediate, assigned as an insertion product, [Pd(C_2H_5)(Ph_2BP_2)($H_2C{=}CH_2$)].[26]

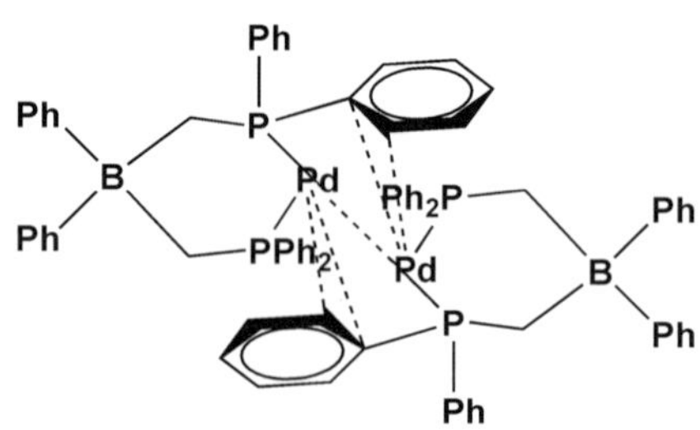

Fig. 7.7. The dimeric compound [Pd(Ph_2BP_2)]$_2$.

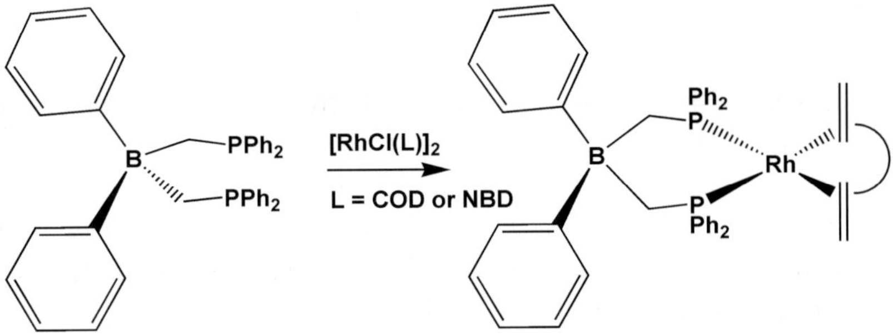

Fig. 7.8. Synthesis of $[Rh(Ph_2BP_2)(diolefin)]$ complexes.

$[Ph_2BP_2][ASN]$ reacts with $[RhCl(L)]_2$ (L = COD or NBD) giving $[Rh(L)(Ph_2BP_2)]$ (Fig. 7.8) which upon interaction with PMe_3 yields $[Rh(Ph_2BP_2)(PMe_3)_2]$. The related MeCN adduct $[Rh(Ph_2BP_2)(MeCN)_2]$ has been also reported. $[Rh(COD)(Ph_2BP_2)]$ reacts with CO giving $[Rh(Ph_2BP_2)(CO)_2]$ which underwent single substitution by PMe_3, PPh_3, and *N*-heterocyclic carbenes to produce $[Rh(Ph_2BP_2)(CO)(PR_3)]$ (R = Me or Ph) and $[Rh(Ph_2BP_2)$-$(CO)(carbene)]$. These zwitterionic phosphane rhodium systems are

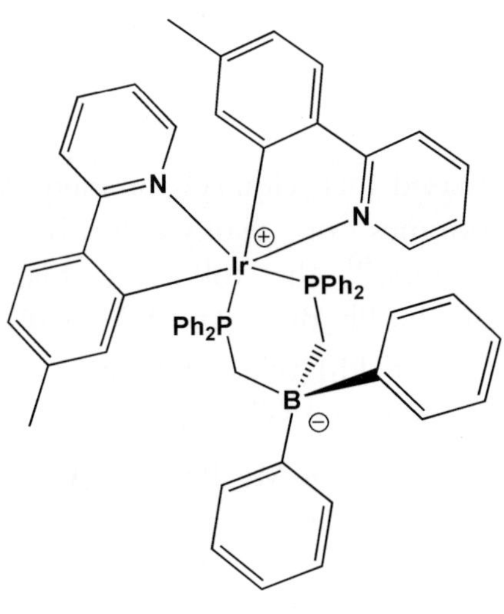

Fig. 7.9. The $[Ir(Ph_2BP_2)(2\text{-}p\text{-tolylpyridyl})_2]$ complex described in Ref. 28.

promising catalysts for H–E (E = C, Si, B) addition reactions to olefins.[27]

The [Ir(Ph_2BP_2)(tpy)$_2$] has been also reported and structurally characterized (Fig. 7.9).[28]

7.2.2. *(p-^{t}BuPh)$_2$BP$_2$(≡ (p-^{t}BuPh)$_2$BP$_2^{Ph}$), (3,5-Me$_2$Ph)$_2$BP$_2^{tBu}$, Ph$_2$BP$_2^{iPr}$, and PhBP$_2^{tBu}$*

The bis(phosphino)borate [(*p*-^{t}BuPh)$_2$BP$_2$][ASN] reacts with CuI giving the 1:1 adduct [CuI{(p-^{t}BuPh)$_2$BP$_2$}][ASN]. CuCl reacts with [Li{(3,5-Me$_2$Ph)$_2$BP$_2^{tBu}$}] in THF to afford [CuCl{(3,5-Me$_2$Ph)$_2$BP$_2^{tBu}$}][Li(THF)$_4$] along with a minor by-product, the three-coordinate [CuCl{(3,5-Me$_2$Ph)$_2$BP$_2^{tBu}$}]. An analogous neutral complex, [CuCl(PhBP$_2^{tBu}$)], has been prepared directly by reacting CuCl with the neutral PhBP$_2^{tBu}$. Reaction of [CuBr(Me$_2$S)] with [Tl{(3,5-Me$_2$Ph)$_2$BP$_2^{tBu}$}] provides the dimeric [Cu{(3,5-Me$_2$Ph)$_2$BP$_2^{tBu}$}]$_2$(μ-Br)Tl. [Cu(Ph$_2$BP$_2^{iPr}$)(MeCN)] and [Cu{(3,5-Me$_2$Ph)$_2$BP$_2^{tBu}$}(MeCN)] have been also described. The latter compound is a useful precursor to [Cu{(3,5-Me$_2$Ph)$_2$BP$_2^{tBu}$}(L)] species (L = PMe$_2$Ph, S=PMe$_3$, Lut, CN^{t}Bu).[29]

7.2.3. *Other dialkyl- and diarylphosphino-borates*

The reaction of lithiated methyldiaryl- or methyldialkylphosphines with diarylchloroboranes or dialkylchloroboranes is reported by Thomas and Peters,[30] the following bis(phosphino)borates being easily obtained: Ph$_2$BP$_2$, (*p*-MePh)$_2$BP$_2$, (*p*-^{t}BuPh)$_2$BP$_2$, (*p*-MeOPh)$_2$BP$_2$, (*p*-CF$_3$Ph)$_2$BP$_2$, Cy$_2$BP$_2$, Ph$_2$BP$_2^{p\text{-}tBuPh}$, (*p*-MeOPh)$_2$BP$_2^{\{p\text{-}tBuPh\}}$, Ph$_2$BP$_2^{\{p\text{-}CF_3Ph\}}$, Ph$_2$BP$_2^{(BH_3)(Me)_2}$, Ph$_2$BP$_2^{(S)(Me)_2}$, Ph$_2$BP$_2^{iPr}$, Ph$_2$BP$_2^{tBu}$, and (*m*,*m*-Me$_2$Ph)$_2$BP$_2^{tBu}$. Selected diarylphosphines were employed for the chelation of PtMe$_2$ moieties. The electronic effects of substituted bis(phosphino)borates on the carbonyl stretching frequency of neutral platinum alkyl carbonyl complexes were investigated by infrared spectroscopy.

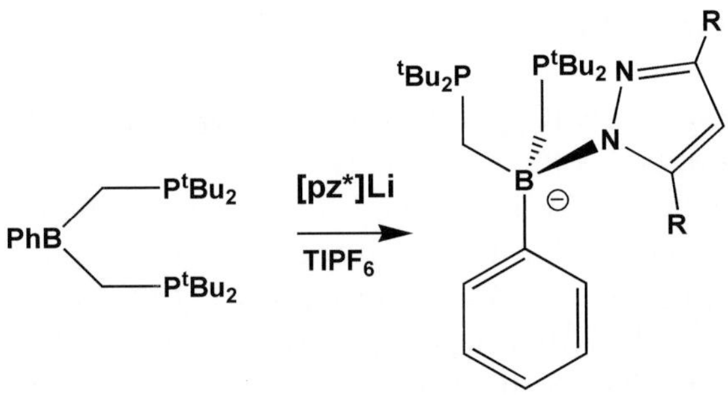

Fig. 7.10. Synthesis of $PhBP_2^{tBu}(pz^R)$ (R = H or Me).

7.3. $PhBP_2^{tBu}(pz)$

Reaction of [Li(pz)] with $PhBP_2^{tBu}$ followed immediately by salt metathesis with $TlPF_6$ (Fig. 7.10) leads to the clean formation of $[Tl\{PhBP_2^{tBu}(pz)\}]$. Metathesis of $[Tl\{PhBP_2^{tBu}(pz)\}]$ with $FeCl_2$ leads to the clean formation of the high-spin ($S = 2$) complex $[FeCl\{PhBP_2^{tBu}(pz)\}]$. One-electron reduction of $[FeCl\{PhBP_2^{tBu}(pz)\}]$ with sodium/mercury amalgam in the presence of excess PMe_3 generates the high-spin d^7 precursor $[Fe\{PhBP_2^{tBu}(pz)\}(PMe_3)]$. The reaction between this complex and 2 equiv. of 1-AdN_3 generates the expected Fe(III) imide $[Fe(\equiv NAd)\{PhBP_2^{tBu}(pz)\}]$ which could be chemically oxidized with $[Fc][BAr_F]$ at low temperature (−50°C) in THF solution to generate a $[Fe^{IV}(\equiv NAd)\{PhBP_2^{tBu}(pz)\}][BAr_F]$. Analogous complexes containing the pz* moiety have been reported.[31]

References

1. Peters JC, Feldman JD, Tilley TD, *J Am Chem Soc* 1999, **121**, 9871.
2. Peters JC, Thomas JC, *Inorg Synth* 2004, **34**, 8.
3. Feldman JD, Peters JC, Tilley TD, *Organometallics* 2002, **21**, 4065.
4. Feldman JD, Peters JC, Tilley TD, *Organometallics* 2002, **21**, 4050.
5. Shapiro IR, Jenkins DM, Thomas JC, Day MW, Peters JC, *Chem Commun* 2001, 2152.
6. Jenkins DM, Betley TA, Peters JC, *J Am Chem Soc* 2002, **124**, 11238.

7. Jenkins DM, Di Bilio AJ, Allen MJ, Betley TA, Peters JC, *J Am Chem Soc* 2002, **124**, 15336.
8. Jenkins DM, Peters JC, *J Am Chem Soc* 2003, **125**, 11162.
9. Jenkins DM, Peters JC, *J Am Chem Soc* 2005, **127**, 7148.
10. Brown SD, Betley TA, Peters JC, *J Am Chem Soc* 2003, **125**, 322.
11. Brown SD, Peters JC, *J Am Chem Soc* 2004, **126**, 4538.
12. Brown SD, Peters JC, *J Am Chem Soc* 2005, **127**, 1913.
13. Brown SD, Mehn MP, Peters JC, *J Am Chem Soc* 2005, **127**, 13146.
14. MacBeth CE, Thomas JC, Betley TA, Peters JC, *Inorg Chem* 2004, **43**, 4645.
15. Thomas JC, Peters JC, *Polyhedron* 2004, **23**, 489.
16. Barney AA, Heyduk AF, Nocera DG, *Chem Commun* 1999, 2379.
17. Turculet L, Feldman JD, Tilley TD, *Organometallics* 2003, **22**, 4627.
18. Mehn MP, Peters JC, *J Inorg Biochem* 2006, **100**, 634.
19. Betley TA, Peters JC, *Inorg Chem* 2003, **42**, 5074.
20. Betley TA, Peters JC, *J Am Chem Soc* 2004, **126**, 6252.
21. Lu CC, Peters JC, *Inorg Chem* 2006, **45**, 8597.
22. Turculet L, Feldman JD, Tilley TD, *Organometallics* 2004, **23**, 2488.
23. Betley TA, Peters JC, *J Am Chem Soc* 2003, **125**, 10782.
24. Thomas JC, Peters JC, *J Am Chem Soc* 2001, **123**, 5100.
25. Thomas JC, Peters JC, *J Am Chem Soc* 2003, **125**, 8870.
26. Lu CC, Peters JC, *J Am Chem Soc* 2002, **124**, 5272.
27. Betley TA, Peters JC, *Angew Chem Int Ed* 2003, **42**, 2385.
28. Li J, Djurovich PI, Alleyne BD, Yousufuddin M, Ho NN, Thomas JC, Peters JC, Bau R, Thompson ME, *Inorg Chem* 2005, **44**, 1713.
29. Thomas JC, Peters JC, *Polyhedron* 2004, **23**, 2901.
30. Thomas JC, Peters JC, *Inorg Chem* 2003, **42**, 5055.
31. Thomas CM, Mankad NP, Peters JC, *J Am Chem Soc* 2006, **128**, 4956.

Chapter 8

Application

8.1. General Considerations

Some recent examples of applications of scorpionate ligands will be presented in this chapter. Most of the new synthesized compounds have been described in previous chapters. This will be mostly a summary of different applicative aspects.

From their genesis, polypyrazolylborates have found applications in many areas of chemistry such as the modeling of active sites of metalloenzymes, perspective materials, organic synthesis, homogenous catalysis, or as "spectator ligands" in the development of reactive metal centers. A number of miscellaneous items will also be discussed in this chapter.

Recent research on transition metal complexes containing Tp^x ligands which ranges from bioinorganic chemistry of dioxygen complexes to organometallic systems, has been reviewed by Akita and Hikichi.[1] The versatile Tp^x ligands can be employed in a wide variety of inorganic studies, because the coordination properties can be finely tuned in a systematic manner by choosing appropriate substituents on the pyrazolyl rings to serve the purpose of the study (e.g. stabilization, activation, functionalization). For example, Tp^x acts as a tetrahedral enforcer giving coordinatively unsaturated tetrahedral species with first row transition metals which can be interconverted to five- and six-coordinate species via coordination/dissociation of donor(s).

The Tp^x ligands can mimic the coordination environment created by three imidazolyl groups from histidine residues, which is frequently found in the active sites of metalloenzymes. Higher valent bis(μ-oxo) species, $[(Tp^x)M(\mu\text{-}O)_2M(Tp^x)]$ via O–O cleavage of $[(Tp^x)M(\mu\text{-}\eta^2{:}\eta^2\text{-}O_2)M(Tp^x)]$ intermediates, but also peroxo, hydroperoxo, and alkylperoxo species, active species undergoing oxidative C–C cleavage reaction, stable hydrocarbyl complexes, and dinuclear xenophilic complexes, $[(Tp^x)M\text{–}M'L_n]$, are all relevant to chemical and biological processes, most of which are associated with transition metal catalytic species.

8.2. Catalysis

8.2.1. *Polymerization and oligomerization*

Ti complexes containing Tp or Tp* have been investigated as polymerization precatalysts for ethylene, propylene, and styrene. A patent on olefin polymerization catalysts and polymerization of olefins has been presented by Michiue[2] which describes the use of $[TiCl_3(Tp^{Ms*})]$ as cocatalyst and its reaction with K in toluene.

Activation of $[TiCl_3(Tp^x)]$ by MAO exhibits a greater catalytic activity in ethylene polymerization for $[TiCl_3(Tp)]$ than for the corresponding $[TiCl_3(Cp)]$ analog. Mono substitution of a OPh for one Cl ligand in $[TiCl_3(Tp^*)]$ does not alter the activity. Titanium species containing Tp and Tp* have been also demonstrated to yield atactic polymers from styrene.[3] More recently, a series of $[TiCl_n(OCH_3)_{3-n}(Tp^x)]$ (Tp^x = Tp or Tp*, n = 1, 2, 3) has been employed for ethylene polymerization. $[Ti(OCH_3)_{3-n}Cl_n(Tp)]$ (n = 1, 2) activated with MAO or MMAO showed higher catalytic activity than $[Ti(OCH_3)_{3-n}Cl_n(Tp^*)]$ (n = 1, 2). The ethylene polymerization seems to be determined by steric effects: the polymers produced with use of $[Ti(OCH_3)_{3-n}Cl_n(Tp)]$ (n = 1, 2) have generally a higher content of chain branching groups than the polymers produced using $[TiCl_3(Tp)]$.[4] In general, $[TiCl_2(OR)(Tp)]$ (R = Me, Et, ^{i}Pr, ^{n}Bu) and $[TiCl_3(Tp)]$ complexes, activated with very low concentration of MAO, showed very high activity in the ethylene polymerization. The precatalyst activity decreases in the following order: $[TiCl_2(OEt)(Tp)] > [TiCl_2(O^nBu)(Tp)] \approx [TiCl_2(O^iPr)(Tp)] > [TiCl_2(OMe)(Tp)] > [TpTiCl_3(Tp)]$.[5] $[TiCl_2(OR)(Tp)]$ (R = Me, Et, ^{i}Pr, ^{n}Bu), combined with MAO, produced active catalysts for styrene polymerization. These pre-catalysts show higher activities than $[TiCl_3(Tp)]$. The activity and syndiospecificity of $[TiCl_2(OR)(Tp)]$ are affected by R nature.[6] The same research has been extended to complexes containing other Tp^x ligands, a number of $[TiCl_3(Tp^R)]$ and $[TiCl_2(OR)(Tp^R)]$ complexes being employed for ethylene polymerization under MAO activation conditions ($Tp^x = Tp^*$, Tp or BuTp, Tp^{Ms^*}, Tp^{Ms}). It has been reported that the activity of the $[TiCl_3(Tp^x)]$/MAO system varies in the following order: $[TiCl_3(Tp^{Ms^*})] > [TiCl_3(Tp^{Ms})] > [TiCl_3(Tp^*)] > [TiCl_3(Tp)] > [TiCl_3(BuTp)] > [TiCl_2(O^tBu)(Tp^*)] > [TiCl_2(O\text{-}2\text{-}^tBu\text{–}C_6H_4)(Tp^*)]$. It has been reported that the activity of $[TiCl_3(Tp^x)]$/MAO is very sensitive to the steric properties of the Tp^x ligands. The highest activity is presented by moderately crowded catalysts containing bulky substituents at the 3 position of two of the three pyrazolyl ring.[7] $[TiCl_3(Tp^*)]$ and $[TiCl_3(Tp)]$ always in the presence of MAO, have been employed in the polymerization of ethylene with a blend of INPETUS®, a Ziegler–Natta catalyst. The blends with Tp complexes showed lowest activities with respect to analogous Cp^x complexes.[8]

$[TiCl_3(Tp^{Ms})]$ and $[TiCl_3(Tp^{Ms^*})]$ in the presence of MAO at 65°C have been employed in copolymerization of ethylene/1-hexene. For both catalysts it has been found that the productivities decrease with increasing 1-hexene concentration. The highest productivities and 1-hexene incorporations were achieved when $[TiCl_3(Tp^{Ms^*})]$ was used as a catalyst, which in accordance with X-ray photoelectron spectroscopy seems to present a stronger cationic character after MAO treatment.[9] The ethylene polymerization using $[TiCl_3(Tp^{Ms})]$ and $[TiCl_3(Tp^{Ms^*})]$ was also performed in toluene or hexane at 60°C in the presence of MAO or TiBA/MAO (1:1) as cocatalysts.[10] The activity of $\{TiTp^{Ms}\}$ and $\{TiTp^{Ms}\}$ systems toward olefin polymerization has been compared with that of analogous Tp, Tp^*, and ^{n}BuTp complexes.[11] Branched polyethylene/linear polyethylene blends were prepared using the combined $[Ni(\alpha\text{-diimine})Cl_2]$ and $[TiCl_3(Tp^{Ms^*})]$ catalysts supported *in situ* on MAO. The polymerization reactions were performed in toluene at two different polymerization temperatures.[12] Branched polyethylene/high-density polyethylene blends were also prepared using the same homogeneous binary catalytic system activated with MAO and/or TiBA in hexane at two different polymerization temperatures (30°C and 55°C) and by varying the nickel loading molar fraction.[13]

The living radical polymerization of styrene initiated by epoxide radical ring opening was also investigated by using the piano-stool complex [$TiCl_3$(Tp)]. Comparison was made with half-sandwich metallocenes. On the basis of a combination of steric and electronic properties, the ligands rank as Cp ≥ Ind > Cp* > Tp. The polymerization is mediated by the reversible termination of the growing chains with Ti(III) species derived from Zn reduction of parent Ti(IV) derivatives. The poor performance observed for [$TiCl_3$(Tp)] is likely due to over-reduction.[14] Ethylene was also polymerized using a combination of [$NiCl_2$(α-diimine)] (diimine = 1,4-bis(2,6-diisopropylphenyl)-acenaphthenediimine), and [$TiCl_3(Tp^{Ms*})$] compounds in the presence of MAO at 308°C.[15] The immobilization of soluble catalyst [$TiCl_3(Tp^{Ms*})$] on silica and MAO-modified silicas yields active supported catalysts for ethylene polymerization.[16] The ethylene polymerization behavior of a series of titanium(IV) and vanadium(V) complexes [$MXCl_2(Tp^x)$] (M = Ti, $Tp^x = Tp^{Ms*}$, X = Cl; M = Ti, $Tp^x = Tp^{Ms}$, X = Cl; M = V, $Tp^x = Tp^{Ms*}$, X = N^tBu; M = V, $Tp^x = Tp^{Ms*}$, X = N-2,6-$^iPr_2$-C_6H_3) was investigated in toluene and in hexane using MAO or TiBA/MAO (1:1) as activators, and varying the polymerization temperature.[17] Pronounced effects of Tp^x ligand structure on ethylene polymerization activity observed for [$TiCl_3(Tp^x)$]/MAO catalysts (Tp^{Ms*} > Tp^{Ms} ≫ uncrowded Tp^x ligands) are also operative in [$ZrCl_3(Tp^x)$] and [$Ti^{III}Cl_3(Tp^x)$]$^-$ catalysts.[18] A series of hybrid-supported catalysts was prepared by combining [$ZrCl_2(^iBuCp)_2$] and [$TiCl_3(Tp^{Ms*})$] complex sequentially grafted onto MAO modified silica. The resulting catalysts were evaluated in terms of catalyst activity and polymer properties.[19]

Zr complexes containing Tp or Tp* have been investigated as polymerization precatalysts for ethylene, propylene, and styrene. Activation of these complexes by MAO exhibits a greater catalytic activity in ethylene polymerization for $[ZrCl_3(Tp)]$ than for the corresponding $[TiCl_3(Cp)]$ analog. $[ZrCl_3(Tp^*)]$ is more active than $[ZrCl_3(Tp)]$. Tp and Tp* Ti species have been demonstrated in the same work to yield atactic polymers from styrene.[3] $[ZrCl_2(Cp)(Bp)]$ and $[ZrCl_2(Cp)(Tp)]$ were also tested as ethene and propene polymerization catalysts.[20]

$[V(L)Cl_2(Tp^{Ms^*})]$ (L = N^tBu; L = O) were *in situ* supported onto SiO_2 and onto MAO and trimethylaluminum. All catalyst systems were shown to be active in ethylene polymerization. The systems were stable at different [Al]/[V] molar ratios and polymerization temperatures.[21] Branched polyethylene/high-density polyethylene blends were prepared using the combined $[NiCl_2(\alpha\text{-diimine})]$ and $[VCl_2(Tp^{Ms^*})(N^tBu)]$ catalysts. The polymerization reactions were performed in hexane or toluene at three different polymerization temperatures (0°C, 30°C and 50°C) and several nickel molar fractions, using MAO as cocatalyst.[22] Tp^{Ms}- and Tp^{Ms^*}-imido vanadium (V) were immobilized onto a series of inorganic supports: All the systems were shown to be active in ethylene polymerization in the presence of MAO or TiBA/MAO mixture.[23]

Linear low-density polyethylene with different branching contents was prepared from ethylene, using a combination of catalyst precursors $[NiCl(Tp^{Ms})]$ and $[ZrCl_2Cp_2]$ activated with MAO/TMA (1:1) in toluene at 0°C,[24] and also, without the addition of an α-olefin co-monomer, using a combination of the catalyst precursors $[NiCl(Tp^{Ms})]$ and $ZrCl_2Cp_2$/SMAO-4, by varying the nickel loading mole fraction.[25]

The $[Mn(Tp^{^tBu,^iPr})]Cl/Al(^iBu)_3/[Ph_3C][B(C_6F_5)_4]$ system is active toward propylene polymerization giving isotactic polypropylene.[26]

The synthesis and structure of alternating co-oligomers of carbon monoxide and norbornadiene produced in the presence of $[Pd(Tp^{Ph})(p\text{-Tol})(PPh_3)]$ have been investigated by Novikova *et al.*[27]

A vinyl compound–olefin copolymer containing polar groups can be prepared in the presence of a catalyst containing alkylaluminum and a $\{Cr(Tp^x)\}$ complex.[28]

The catalytic activity toward ethylene polymerization of $[NiCl(Tp^{^tBu,Me})]$ and of $[Pd(\eta^3\text{-allyl})(Tp^*)]$,[29] prepared following a literature method,[30] has been tested. The latter showed little activity with either Lewis acid or MAO activator.

At high temperatures, $[Ru(CH_3)(Tp)(CO)(NCMe)]$ acts as catalyst for the production of polystyrene. The dependence of polystyrene molecular weight on benzene/cumene molar ratios suggests a radical polymerization mechanism. The polymerization of methyl methacrylate, in the presence of $[Ru(CH_3)(Tp)(CO)(NCMe)]$ with carbon tetrachloride or methyl dichloroacetate, has been observed at 90°C.[31]

[Rh(Tp)(COD)] and [Rh(Bp)(COD)] are good catalysts for phenylacetylene polymerization in ionic liquids and in CH_2Cl_2, in the presence of tetraammonium halides ($[R_4N]X$, R = Bu, Et; X = Cl, Br).[32]

$[Ca(X)(Tp^{^iPr})]$ (X = $CH[CMeNC_6H_3\text{-}2,6\text{-}^iPr_2]_2$, $CH[CMeNC_6H_4\text{-}2\text{-}OMe]_2$) species have been employed for the ring-opening polymerizations of lactide.[33]

A new concept, the "radical-controlled" oxidative polymerization of phenols catalyzed by a tyrosinase model complex, has been proposed. The polymerization of 4-phenoxyphenol was performed by using the tyrosinase model complex $[Cu(Cl)Tp^{Ph_2}]$.[34]

Dias *et al.*[35] reported that high-quality polyaniline can be prepared catalytically in MeCN using aniline, HCl, H_2O_2, and the copper catalyst $[CuCl(MeTp^{Ms})]$. Controlled, copper-catalyzed functionalization of polyolefins by using $[Cu(Tp^x)L]$ (L = MeCN) complexes has been reported by Pérez and coworkers.[36]

8.2.2. *Carbene and nitrene transfer*

Complexes of general formula $[Cu(Tp^x)]$ ($Tp^x = Tp^{Br_3}$, Tp^{Cy}, Tp^{Ms}) have been employed as catalysts for the insertion of diethyl diazoacetate into CH bonds of a variety of hydrocarbons and ethers, in moderate to high yields: tertiary, secondary, and primary C–H bonds have been functionalized. The presence of electron-withdrawing groups in the Tp^x ligands seems to be crucial for the activity and selectivity of the catalyst.[37] In particular, the results of the olefin cyclopropanation reaction with a series of *in situ* generated copper(I) complexes of general formula $[Cu(Tp^x)]$ ($Tp^x = Tp^{Ph}$, $Tp^{Ph,4-Me}$, $Tp^{Ph,4-Pr}$, $Tp^{\alpha-Nt}$, and Tp^{Ms}) indicated a remarkable diastereoselectivity for the $[Cu^I(Tp^{Ms})]$ catalyst in the styrene cyclopropanation (*cis:trans* selectivity >98:2). This behavior can be related to the restriction on phenyl rotation that the methyl groups enforce in the mesityl case. The mesityl ring can be essentially orthogonal with respect to the pyrazolyl plane, providing the smallest catalytic pocket, which may be responsible for high *cis* selectivity.[38] The *in situ* prepared $[Cu(Tp^x)]$ ($Tp^x = Tp^{Ph}\ Tp^{\alpha-Np}$, $Tp^{Ph,4-Me}$, and Tp^{Ms}) also catalyze the olefin cyclopropanation reaction using ethyldiazoacetetate as carbene source, and high values of activity and selectivity toward the *cis* isomers have been obtained for styrene, α-methylstyrene, 1-hexene, 1-octene, vinyl acetate, *n*-butyl vinyl ether, 2,5-dimethyl-2,4-hexadiene, and 3,3-dimethyl-1-butene. Kinetic studies suggested that the homoscorpionate ligand acts in a dihapto form during the catalytic process.[39]

Cu(I) catalysts containing Tp^x (Tp^x = Tp, Tp*) have been also used to efficiently convert equimolar mixtures of ethyl diazoacetate and alkynes into cyclopropenes at room temperature (Fig. 8.1 and Table 8.1).[40]

$[Cu(Tp^x)]$ catalyzes also the addition of the carbene fragment $:CHCO_2Et$ to furans under mild conditions to give different cyclopropanes and dienes with ratios depending on the Tp^x ligand,

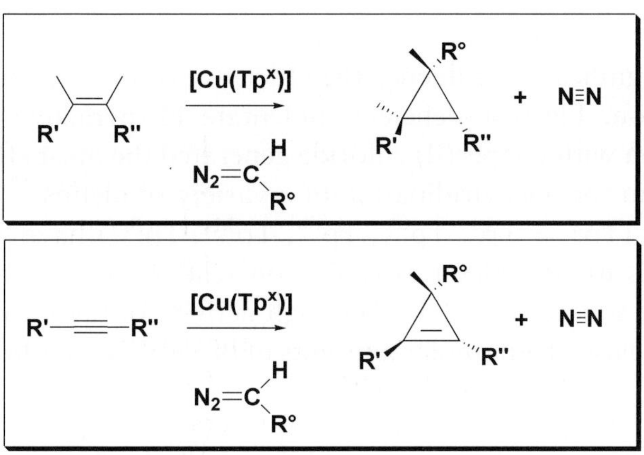

Fig. 8.1. $[Cu(Tp^x)]$-catalyzed cyclopropanation and cyclopropenation reactions.

Table 8.1. Cyclopropenation of hex-3-yne with [Cu(Tp^x)] catalysts.

Tp^x	Yield	Tp^x	Yield
Tp	32	Tp$^{\alpha-Np}$	27
Tp*	82	Tp^{Ms}	31
Tp^{tBu}	85	Tp^{ClPh}	52
Tp^{Np}	95	Tp^{An}	43
Tp^{Cy}	97	Tp^{Tol}	49
Tp^{Ph}	21		

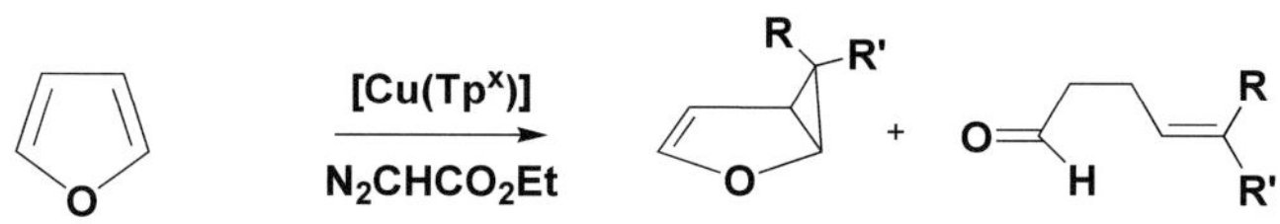

Fig. 8.2. [Cu(Tp^x)] catalyzes the addition of the carbene fragment $CHCO_2Et$ to furans.

therefore inducing the control of the selectivity in this transformation (Fig. 8.2).[41]

In situ generated [Cu(Tp^x)] complexes are also effective for the aziridination of olefins. A significant influence of the combination of the starting copper oxidation state and the hapticity of the poly(pyrazolyl)borate ligand on the efficiency of the reaction has been observed.[42] Handy *et al.*[42,43] reported that electronic features of Tp^x and Bp^x significantly influence the efficiency of the copper-catalyzed aziridination. Electron-deficient, bidentate bis(pyrazolyl)borates in conjunction with copper(II) chloride generated the most effective catalyst system for the aziridination of a variety of olefins.[43] [Cu(Tp^x)] complexes (Tp^x = Tp*, Tp^{Cy}, Tp^{Ph}, Tp^{Ms}, Tp^{Br_3}) have been tested as catalysts for the olefin aziridination reaction using PhI=NTs as the nitrene source (Fig. 8.3). The high activity of these catalysts has allowed the use of equimolar amounts of PhI=NTs and olefins, avoiding the request of excess olefin.[44]

[Cu(Tp*)(C_2H_4)] has been employed in combination with $N_2CHCO_2CH_2CH_3$ and $C_6H_5INSO_2C_6H_4CH_3$ (PhI=NTs) in cyclopropanation and aziridination of a series of *para*-substituted styrenes.

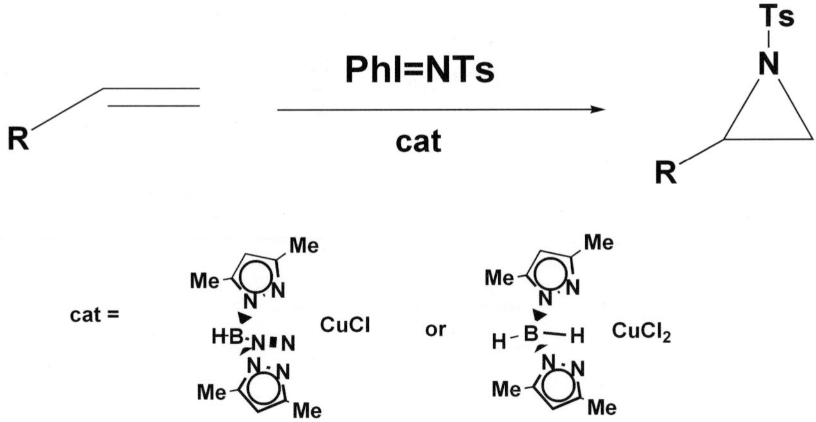

Fig. 8.3. [Cu(Tp^x)] catalyzes the olefin aziridination reaction.

An electrophilic metal–carbene complex intermediate has been proposed in the cyclopropanation reaction, whereas a paramagnetic copper nitrene species, which behaves as an electrophilic, nitrogen-centered radical, is proposed as the intermediate for the aziridination reaction.[45] [Cu(Tp$^{(CF_3)_2}$)(C_2H_4)] is a good aziridination catalyst, readily converting a variety of olefins into the corresponding *N*-tosyl aziridines.[46]

A series of catalysts of general formula [Cu(Tp^x)] (Tp^x = Tp, Tp*, Tp^{Np}, Tp^{Cy}, Tp^{Ph}, and Tp^{Ms}), by using ethyl diazoacetate as carbene source, catalyzes the insertion of :$CHCO_2Et$ into the C–H bonds of cycloalkanes and cyclic ethers in moderate to high yields. The influence of the groups attached to the Tp^x pyrazolyl rings on the activity of the [Cu(Tp^x)] catalysts has been also investigated and the results discussed.[47] The mesityl derivative provided the highest degree of conversion.

[Cu(Tp^{Ms})] catalyzes the addition of the :$CHCO_2Et$ unit (derived from N_2HCHCO_2Et) to benzene which yields a cycloheptatriene ring, in analogy with the Büchner reaction. When alkyl groups are linked to the aromatic rings, the selectivity of the reaction can be oriented toward insertion into an alkyl C–H bond by using [Cu(Tp^{Ms})].[48] On the other hand, [Cu(Tp*)] efficiently catalyzes the insertion of

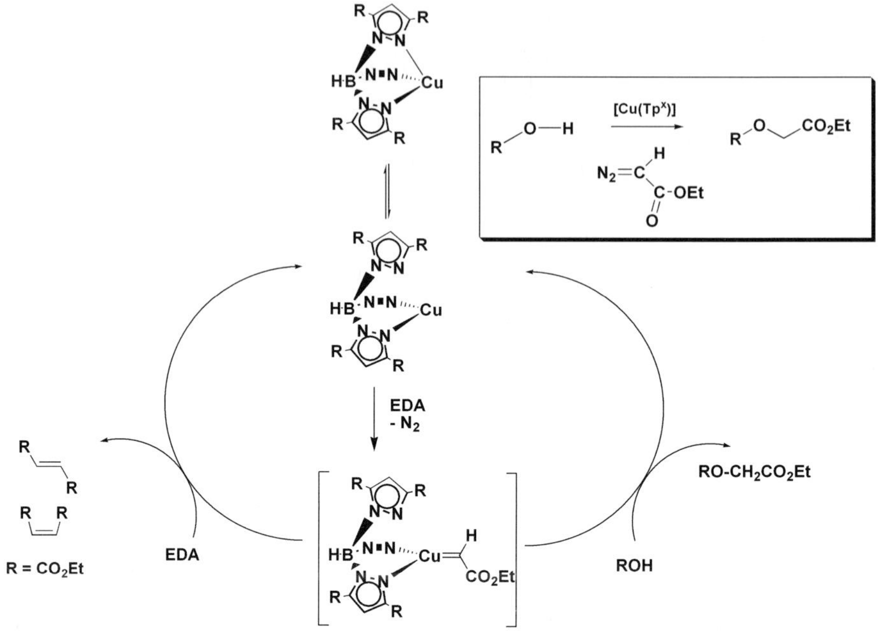

Fig. 8.4. [Cu(Tp^{x})] catalyzes the insertion of the carbene fragment :$CHCO_2Et$, derived from ethyldiazoacetate into the OH bond of saturated and unsaturated alcohols under mild conditions.

the carbene fragment :$CHCO_2Et$ into the OH bond of saturated and unsaturated alcohols under mild conditions (Fig. 8.4). When unsaturated alcohols were employed, the reaction proceeded selectively toward the insertion products. [Cu(Tp^{x})] (Tp^{x} = Tp^{Ms}, Tp^{Br_3}, Tp^{Cy}) have been also examined as catalysts for the same reaction and the results related to the Tp^{x} substituents.[49]

The [Cu(Bp)] system has been employed to investigate kinetics of the ethyl diazoacetate decomposition reaction in the presence or absence of olefin. The available data have allowed the proposition for a copper-mediated olefin cyclopropanation reaction. It has been proposed that the real catalyst is a 14-electron species, independent of the nature of the ligand bonded to the copper center.[50]

Silver derivatives containing Tp$^{(CF_3)_2}$ catalyze the carbene insertion into carbon–hydrogen bonds of cyclic and acyclic hydrocarbons at room temperature. [Ag(Tp$^{(CF_3)_2}$)(THF)] is more effective with respect to the related copper complex [Cu(Tp$^{(CF_3)_2}$)(THF)] for hydrocarbon activation via this route.[51] [Ag(Tp$^{(CF_3)_2}$)(THF)] catalyzes the addition of ethyl diazoacetate to benzene rings, providing norcaradienes, which undergo electrocyclization to provide the corresponding cycloheptatriene[52] and also effectively catalyzes carbene transfer to allylic and propargylic halides, leading to the formation of α-haloacetate derivatives.[53]

8.2.3. *Oxidation*

Copper(I) complexes containing Tp^x ligands exhibit catalytic activity toward the epoxidation reaction, as they did with the olefin cyclopropanation and aziridination reactions. These complexes can operate under both homogeneous and heterogeneous conditions. [Cu(Tp*)], one of the first homoscorpionates complexes, has been employed as catalyst for the transformation of styrene into styrene oxide using ozone as the oxidising agent. This copper complex can be supported on silica-gel, acting as heterogeneous catalysts in water (Fig. 8.5).[54,55]

[Cu(Bp)] has been employed as catalyst for the transformation of styrene into styrene oxide using ozone as the oxidizing agent.[55]

The one-pot formation of [Cu(Tp^{CF_3CH_3})(lactate)] from acetone has been observed: this is the first example of an external substrate

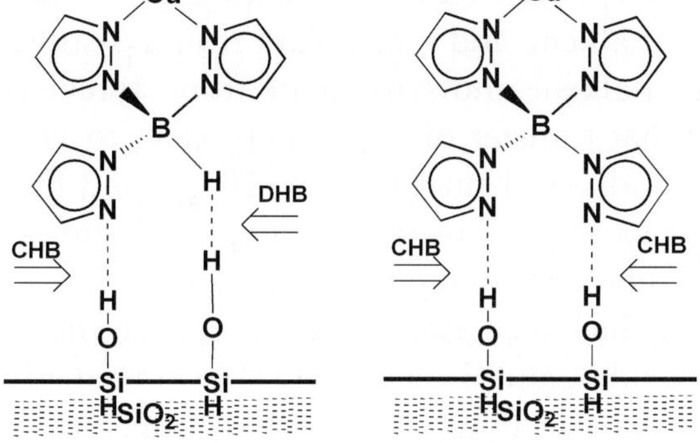

Fig. 8.5. Complex–support interaction: formation of classical hydrogen bonds (CHB) and nonclassical dihydrogen bonds (DHB).

oxygenated aerobically using a robust Cu complex whose metal environment exhibits no C–H bonds.[56]

Catalytic oxidation of cyclohexane using $[Co(\eta^2\text{-}NO_3)(\kappa^3\text{-}Tp^{CF_3,CH_3})(NCMe)]$ and cumene hydroperoxide yields a mixture of cyclohexanol and cyclohexanone.[57]

Reaction of $[MoO_2Cl_2(THF)_2]$ (X = Cl or Br) with Tp yielded $[MoO_2Cl(Tp)]$. The turnover frequency of this complex in olefin epoxidation, with *t*-butyl hydroperoxide as oxidizing agent, is reported in the middle of the range observed for $[MoO_2X_2L_2]$ complexes (X = halide, L = N-donor ligands).[58]

$[ReO_3(Tp^*)]$ and $[ReO_3(pzTp)]$, reduced *in situ* with PPh_3 or $P(OEt)_3$, have been demonstrated to catalyze efficiently O atom transfer from epoxides to stoichiometric phosphorus reductant in the temperature range 75–105°C. It has been reported that the reaction is stereospecific and the rate is greater with *cis*- than with *trans*-alkenes. The use of Tp* is advantageous due to the fact that the tris(pyrazolyl)borate ligand provides a critical tool for tuning the reaction through the use of the steric interactions.[59] [M(Tp)CO(L)] (M = Re, Mo, W) promises greater economy and safety as dearomatizing agents.[60] Both $[Re(O)(OH)_2(Tp^*)]$ and $[Re(O)(OEt)(OH)(Tp^*)]$ are efficient catalysts for O atom transfer. The kinetic behavior of the catalytic system (epoxide: Re complexes = 20) is complex, the majority of rhenium existing as *syn*-$[Re(O)(diolate)(Tp^*)]$, formed by ring expansion of the epoxide. $[Re(O)(OEt)(OH)(Tp^*)]$ with excess *cis*-stilbene oxide and styrene oxide, in the absence of reductant, yielded a 4:1 mixture of alkene and *syn*-diolate from *cis*-stilbene oxide or a 5.5:1 mixture of alkene and *syn*-diolate from styrene oxide.[61] It has been reported that transfer of O from epoxides to PPh_3 is catalyzed by $[ReO_3(Tp^*)]$ also in benzene at 75–105°C and that the reaction tolerates a wide variety of functional groups including ketones, esters, nitriles, ethers, silyl ethers, and phthalimides.[62]

$[FeCl(Tp^{'Bu})]$ has been employed to catalyze the epoxidation of stilbenes to epoxides, and $[Fe(Tp^{'Bu})]OTf$ exhibited efficient catalytic activity with high selectivity and reaction rate for epoxidation, by using O_2 with isobutyraldehyde as co-reductant.[63]

The oxygenation of $[Rh(Tp^{'Pr})(dppe)]$ produces $[Rh(Tp^{'Pr})(O_2)(Hpz^{'Pr})]$ and $[Rh(OOH)(Tp^{'Pr})(pz^{'Pr})(Hpz^{'Pr})]$ can be considered as the first structural evidence for the key steps of the O_2 activation by a transition metal species and conversion of molecular oxygen into a hydroperoxo species.[64]

8.2.4. *Miscellaneous catalysis*

In situ regioselective quinoline hydrogenation with transition metal complexes, such as [Rh(COD)Cl_2], [Ir(COD)Cl_2], [Ir$(COE)_2Cl_2$], and [Ru$(MeCN)_4Cl_2$] stabilized by Tp and Tp* ligands via one-pot reaction has been described. The results have been compared with those obtained with [Ir(Tp)$(C_2H_4)_2$]. The system [Rh(COD)Cl_2]/Tp is the most efficient and regioselective catalyst precursor for this reaction.[65]

[Ru(Tp)(PPh_3)$(MeCN)_2$]PF_6 has been employed as catalyst to produce 1-iodo-2-naphthol in DMF and 2-iodobenzo[d]oxepin in benzene from 1-(2′-iodoethynylphenyl)-2-propyloxirane. The solvent-dependent chemoselectivity has been ascribed to a solution equilibrium between ruthenium-π-iodoalkyne and ruthenium-2-iodovinylidene intermediates.[66] The same ruthenium phosphine complex has been efficiently employed as catalyst in the nucleophilic addition of water, alcohols, aniline, acetylacetone, pyrroles, and dimethyl malonate to unfunctionalized enediynes that yielded functionalized benzene products in good yields (Fig. 8.6).[67] [Ru(Tp)-(PPh_3)$(MeCN)_2$]PF_6 has been also found very active in catalytic benzannulation of 1-phenyl-2-ethynylbenzenes in dichloroethane to afford phenanthrene.[68]

[Ru(Tp)(PPh_3)$(MeCN)_2PF_6$], [Ru(Tp)(PPh_3)(MeCN)Cl], and [Ru-(Tp)$(PPh_3)_2$Cl] catalyze an efficient cleavage of organic alkynes in which ethynyl alcohol is split into alkene and CO.[69] [Ru(Tp)(PPh_3)-$(MeCN)_2PF_6$] reacts also with (*o*-ethynyl)phenyl epoxides yielding 2-naphthols or 1-alkylidene-2-indanones, very selectively. The reaction intermediate has hypothesized a ruthenium-π-ketene species that can be trapped efficiently by alcohol to give an ester compound.[70] [Ru(Tp)(PPh_3)$(MeCN)_2PF_6$] catalyzes cyclo-isomerization of *cis*-3-en-1-ynes to cyclopentadiene and analogous derivatives through a 1,5-sigmatropic shift of Ru-vinylidene intermediate.[71]

[RuCl(Tp)$(PPh_3)_2$] acts as a catalyst in the reaction between terminal alkynes and hydrazine to yield nitriles. The reaction occurs

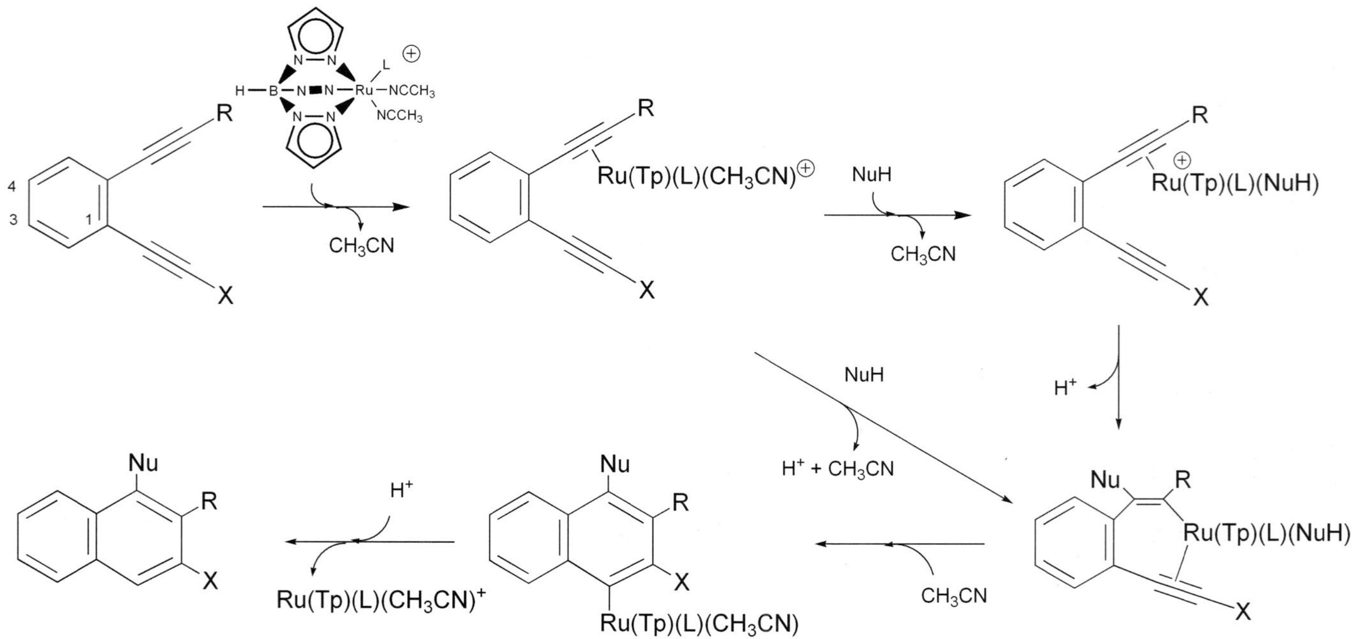

Fig. 8.6. The nucleophilic addition of water, alcohols, pirroles, and Hacac to unfunctionalized enediynes yields functionalized benzene derivatives.

through an intermolecular attack of nitrogen at the α-carbon atom of (vinylidene)metal intermediates generated from terminal alkynes,[72] whereas [RuCl(Tp)$(py)_2$] was found to be most active between a number of catalysts for the isomerization of *cis*-stilbene oxide to its *trans*-isomer without formation of by-product.[73]

[{Ru(Tp)(L)}$_2$(μ-N_2)][PF_6]$_2$ and [Ru(Tp)(methacrolein)(L)][PF_6] are catalysts for the solvent-free regio and enantioselective Diels–Alder reactions between methacrolein and Cp or Cp*. [{Ru(Tp)(L)}$_2$(μ-N_2)][PF_6]$_2$ also catalyzes the 1,3-dipolar cycloaddition reaction between methacrolein and benzylidenephenylamine *N*-oxide to yield 5-methyl-2-*N*-3-diphenylisoxazolidine-5-carbaldehyde.[74]

The σ-enynyl complex [Ru{C(Ph)=C(Ph)C≡CPh}(Tp)(PMe^{i}Pr$_2$)] is able to catalyze the regioselective cyclization of α,ω-alkynoic acids to yield endocyclic enol lactones.[75]

[Rh(Tp*)$(PPh_3)_2$] and [Rh(Tp*)(COD)] are active catalysts for alkyne hydrophosphinylation,[76] and [Rh(Tp*)(COD)] also for dehydrogenative coupling of 4,4,5,5-tetramethyl-1,3,2-dioxaborolane with arenes.[77]

[Ir(Tp*)(η^4-2,3-dimethylbutadiene)] was investigated for selective deuteration of norbornene derivatives.[78]

[PtMe(Tp)], a new alkyne protecting group, can be used under a variety of reaction conditions, including basic and acidic media and the environment required for catalytic hydrogenation and for chromate oxidation reaction.[79]

[UI_2(Bp$^{^t\text{Bu},\text{Me}}$)]·2THF has been described to be an efficient Lewis acid catalyst for Diels–Alder reactions performed in mild conditions and with low amounts of catalysts. The replacement of an iodide in UI_3 by a dihydrobispyrazolylborate ligand, however, does not increase activity or selectivity.[80]

8.3. Enzyme Modeling

Much of the work in this area concerns modeling zinc-based enzymes. Scorpionates can be considered as the most popular ligands able to mimic the enzyme behavior.

8.3.1. *Vanadium*

Vanadium haloperoxidases are enzymes able to catalyze the oxidation of halides to corresponding hypohalous acids, which then readily undergo halogenation of organic substrates or conversion of hydrogen peroxide to singlet oxygen and generation of halides. $[VO(O_2)(Tp)(pzH)]$ and $[VO(O_2)(pzTp)(pzH)]$ were synthesized and characterized and $[VO(O_2)(Tp)(pzH)]$ indicated as model for haloperoxidase enzymes.[81]

8.3.2. *Molybdenum*

$[Mo(Tp^*)(E)(bdt)]$, $[Mo(Tp^*)(E)(tdt)]$, $[Mo(Tp^*)(E)(bdtCl_2)]$[82,83] (E = O, NO), and $[MoO(qdt)(Tp^*)]$[84] have been investigated as models for various pyranopterin Mo enzyme active sites, including sulfite oxidase. Solution redox potentials and heterogeneous electron transfer rate constants for these species have been also reported.[85] The interactions between the sulfur π-orbitals of arene dithiolates and high-valent Mo in $[MoO(Tp^*)(bdt)]$ have been investigated by gas-phase photoelectron spectroscopy and DFT methods in order to understand the properties of the active site of pyranopterin Mo–W enzymes.[86] Temperature-dependent measurements of potential and electron-transfer rate constants are also reported for electrochemical reduction of a series of $[MoO(Tp^*)(X,Y)]$ complexes.[87] The molecular and electronic structures of the SO active site $[MoO(Tp^*)(bdt)]$ have been also reported.[88]

The double cubane $[(Tp)_2Mo_2Fe_6S_8(PEt_3)_4]$ system has been investigated as model for the reactivity of the nitrogenase P^N cluster. Four distinct processes for the conversion of double cubanes to P^N-type clusters have been documented.[89]

8.3.3. *Tungsten*

Eagle *et al.* have investigated the biomimetic chemistry of tungsten–dithiolene complexes containing Tp*.[90]

$[(Tp^*)WFe_3S_4Cl_3]^-$ reacts with a system composed of excess Et_3P, BH_4^- and HS^- yielding $[(Tp^*)_2W_2Fe_5S_9Na(SH)(MeCN)]^{3-}$ which exhibits a core topology similar to the P^N cluster of nitrogenase.[91]

8.3.4. *Manganese*

The superoxide dismutase activity of [Mn(FOBz)(Tp^{iPr_2})], [Mn(Cl)-(Tp^{iPr_2})(Hpz^{iPr_2})], and [Mn_2(μ-FOBz)$_3$(Tp^{iPr_2})(Hpz^{iPr_2})$_2$] has been investigated.[92]

Dehydrative condensation of the dinuclear [Mn(μ-O)(Tp^{iPr_2})]$_2$ with H_2O_2 in the presence of 2-Meim yielded the peroxomanganese(III) complex [Mn(η^2-O_2)(Tp^{iPr_2})(2-MeIm)]$_2$. Formation of this stable peroxomanganese complex may mimic the essential role of the "distal" histidine residue in the hemoglobin/myoglobin, i.e. the hydrogen-bonding interaction between imidazolyl N–H group and the dioxygen ligand increases the stability of the M–O_2 framework.[93]

8.3.5. *Iron*

Catechol dioxygenases are nonheme iron enzymes which catalyze the oxidative cleavage of catechols. [Fe(Cat)(Tp*)(Hpz*)] [Cat = catecholate = $C_6H_4O_2$, 3,5-^{t}Bu_2, $C_6H_2O_2$, $C_6Cl_4O_2$) have been prepared and their reactivity toward O_2 investigated.[94]

α-Ketoglutarate-dependent dioxygenases constitute a large class of nonheme iron-containing enzymes that are essential for the biosynthesis of a diverse array of compounds. Their action mechanism is not well known and in order to achieve this goal (Fig. 8.7), the aliphatic α-keto carboxylate, [Fe^{II}(O_2CC(O)CH_3)(Tp^{Ph_2})], and the carboxylate

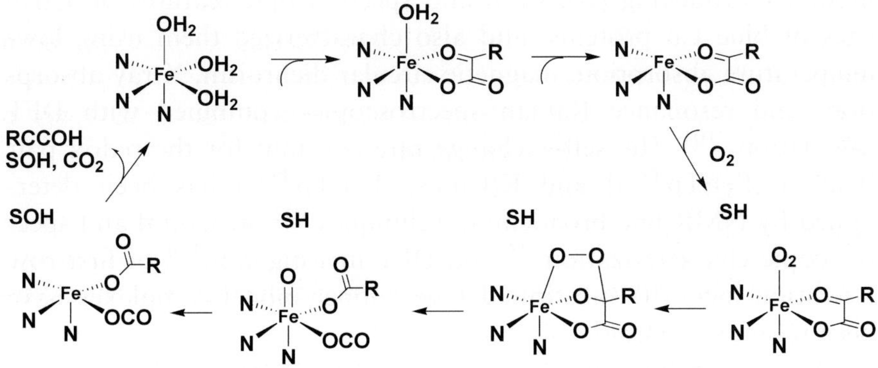

Fig. 8.7. Proposed mechanism for hydroxylation of phenyl ring with α-keto acid precursor.

complexes, $[Fe^{II}(OBz)(Tp^{Ph_2})]$ and $[Fe^{II}(OAc)(Tp^{Ph_2})(Hpz^{Ph_2})]$, were synthesized.[95]

$[Fe(Tp^{iPr_2})(OOPtn)]$ which is relevant as a model for potential intermediates in pterin-dependent hydroxylases, has been reported and its structure and reactivity investigated.[96]

8.3.6. *Cobalt*

The dehydration kinetic measurements of HCO_3^-, catalyzed by $[Co(Tp^*)(N_3)]$ and $[Co(Tp^*)(NCS)(MeOH)_2]$ have been recently reported.[97] $[Co(X)(Tp^{Ph})(CH_3OH)_m]\cdot nCH_3OH$ ($X = N_3$, $m = 1$, $n = 2$; $X = NCS$, $m = 0$, $n = 0$) have been used as catalysts in the bicarbonate dehydration reaction.[98]

8.3.7. *Nickel*

Desrochers *et al.* reported monomeric five-coordinate [Ni(*l*-cysteine)(Tp*)], analogous to the nickel component of the active site in hydrogenase enzymes.[99]

8.3.8. *Copper*

Fujisawa and coworkers[100] reported in 1992 the copper(II)thiolato complexes $[Cu(SC_6F_5)(Tp^{iPr_2})]$ and $[Cu(SCPh_3)(Tp^{iPr_2})]$, synthetic models reproducing structural and spectroscopic features of active sites of blue Cu proteins, and also characterized them using low-temperature absorption, magnetic circular dichroism, X-ray absorption, and resonance Raman spectroscopies combined with DFT calculations.[101] The self-exchange rate constant for the redox pair $[Cu^{II}(SC_6F_5)(Tp^{iPr_2})]$ and $K[Cu^{I}(SC_6F_5)(Tp^{iPr_2})]$ has been determined by NMR line-broadening techniques.[102] Structural and spectroscopic characterization[103] and DFT investigation[104] of first-row transition metal(II) substituted blue copper model complexes have been recently reported.

Oxygen binding, activation, and reduction constitute processes carried out by copper enzymes. Solomon reported spectroscopic and theoretical studies of mononuclear copper alkyl and hydroperoxo complexes.[105] Mononuclear Cu(II)-superoxo species are also likely involved in the chemistry at several Cu protein active sites, such as Cu/Zn superoxide dismutase, dopamine monooxygenase, peptidylglycine α-hydroxylating monooxygenase. $[Cu(Tp^{tBu,iPr})(DMF)]$, $[Cu(O_2)(Tp^{tBu,iPr})]$, $[Cu(Tp^{Ad,iPr})(DMF)]$, and $[Cu(O_2)(Tp^{Ad,iPr})]$ have been investigated in order to elucidate the electronic structure of the diamagnetic mononuclear side-on Cu^{II}–superoxo complex.[106] Superoxide–Cu^{II} and peroxide–Cu^{II} bonding in $[Cu(O_2)(Tp^{Ad,iPr})]$ have been also compared through X-ray absorption edge spectroscopy and computational studies.[107] $[\{Cu(Tp^{iPr_2})\}_2(S_2)]$, both in the side-on and end-on $Cu_2(S_2)$ cores, have been prepared and compared with the peroxide analogous.[108] Quantum chemistry investigations on a series of tripodal CuN_3 cuprous complexes interacting with O_2, have been performed by de la Lande *et al.*[109] The structures of these complexes, the coordination mode (η^2 side-on or η^1 end-on) of O_2 to Cu^{II}, and the charge transfer from the metal to dioxygen have been investigated. It is noteworthy that $[Cu(Tp^{CF_3,CH_3})_2(O_2)]$ exhibits Cu···Cu and O–O distances, respectively, shorter and longer with respect to the corresponding distances observed in oxyHc and μ–η^2–η^2 peroxo models.[110] Aullón and coworkers[111] have investigated by DFT methods, the effect of the Tp^x substituents on the structure of dinuclear $[Cu_2(\mu\text{-}O)_2(Tp^x)_2]$ derivatives, analogs of the active site of oxyhemocyanin. They reported that the type of bridging oxygen, peroxo, or bisoxo is strongly influenced by the nature and position of the R substituents because of variable substituent–bridging oxygen interactions, as well as electronic effects. The electronic effects of the substituents at the 5 position are not significant. The energy profiles of compounds with several substitution patterns clearly indicate that upon fluorination of position 3 of the pyrazolyl rings (either directly or by attaching a fluoroalkyl group) it becomes harder to split the O–O bond, i.e. fluorination makes the bonded peroxo groups closer to free dioxygen.

The structures and the O–O and M–O bonding characters of a series of reported side-on (η^2) 1:1 metal complexes of O_2, including those reported in Fig. 8.8 are analyzed by DFT. Comparison of the

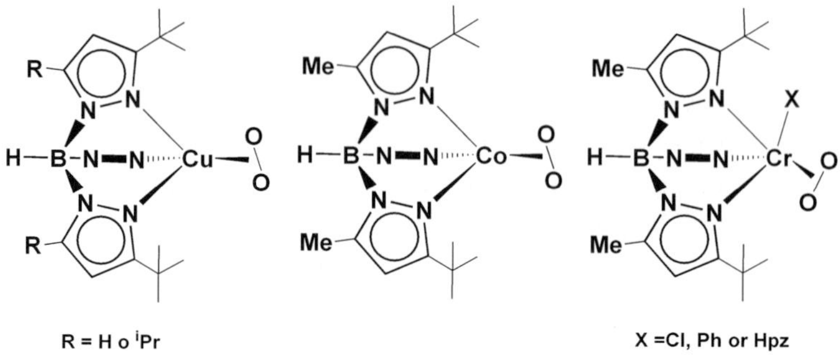

Fig. 8.8. Side-on $(Tp^x)M{:}\eta^2\text{-}O_2$ 1:1 complexes.

calculated and experimental systems with respect to O–O bond distance, O–O stretching frequency, and O–O and M–O bond orders gave insights into subtle influences relevant to O_2 activation processes in biology and catalysis.[112]

Other model complexes relevant to enzymes have been described as the EPR-silent species $[Cu(L)(Tp^{Ph})]^+$ (HL = 2-hydroxy-3-methylsulfanyl-5-methylbenzaldeyde), showing spectroscopic similarity with the galactose oxidase,[113] or $[Cu(OAr)(Tp^{^iPr_2})]$ (Ar = C_6H_4-4-F; 2,6-$Me_2C_6H_3$), 2,6-($Bu_2C_6H_3$)-C-t) described as a model of intermediates of tyrosinase catalyzed oxidation of phenols,[114] and $[Cu(Tp^{CO_2Et,Me})(H_2O)_2]ClO_4$ as a good structural model for many of the enzymes of the vicinal oxygen chelate superfamily.[115]

Pérez and coworkers[116] have reported that in the reaction of $[Cu(Tp^x)]$ complexes in solution with dioxygen at room temperature two different behaviors can occur: (i) reaction with O_2 or (ii) lack of any transformation. This fact has been associated with the electron density at the metal center. An experimental study has been carried out to evaluate the effects of the substituents on the $[Cu(Tp^x)]$ activation of O_2.

8.3.9. *Zinc*

The construction of suitable synthetic analogs of zinc enzymes is an important process: the ligand design and its choice constitute an important step in order to obtain coordination environments analogous to that enforced by the unique topology of a protein. Tripodal ligands, in which X, Z, and Y donor groups are attached to a tetrahedral boron or carbon center, enforce the "facial binding" and directly influence the steric environment about the metal center upon appropriate choice of their substituents (Fig. 8.9). In the past 10 years both $Tp^{RR'}$ and Tm^{R} have received considerable attention with respect to modeling zinc enzymes and studies have been directed mainly to mimic carbonic anhydrase, liver alcohol dehydrogenase, 5-aminolevulinate dehydratase, and Ada DNA repair protein behavior.

Most of the work in this field has been reviewed by Parkin in 2004.[117]

Tp^{x} have been employed as ligands in carbonic anhydrase models (Fig. 8.10).

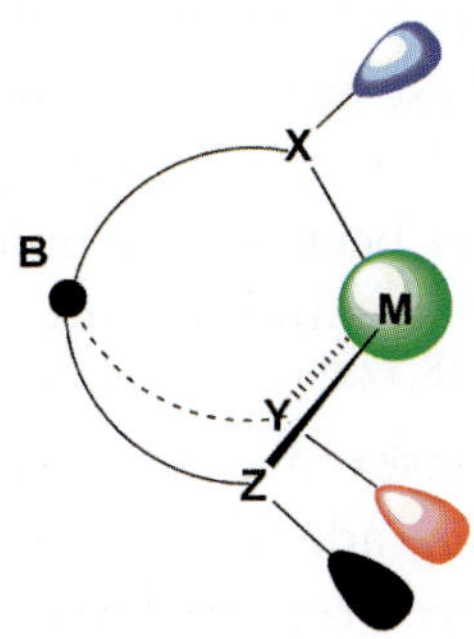

Fig. 8.9. Tripodal ligands in which X, Z, and Y donor groups are attached to a tetrahedral boron or carbon center enforce the "facial binding."

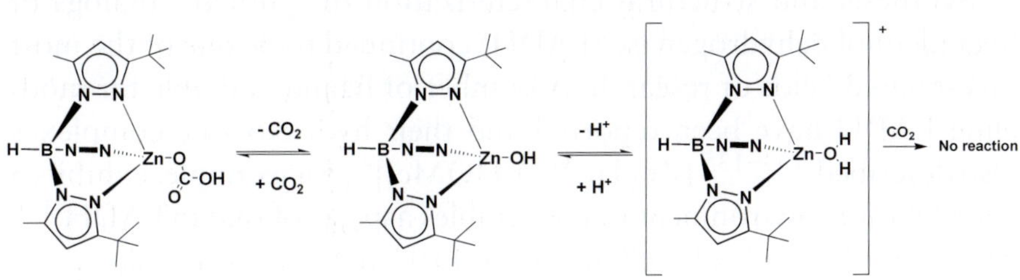

Fig. 8.10. $[Zn(OH)(Tp^{^tBu,Me})]$ reacts rapidly with CO_2, whereas $[Zn(OH_2)(Tp^{^tBu,Me})]^+$ does not.

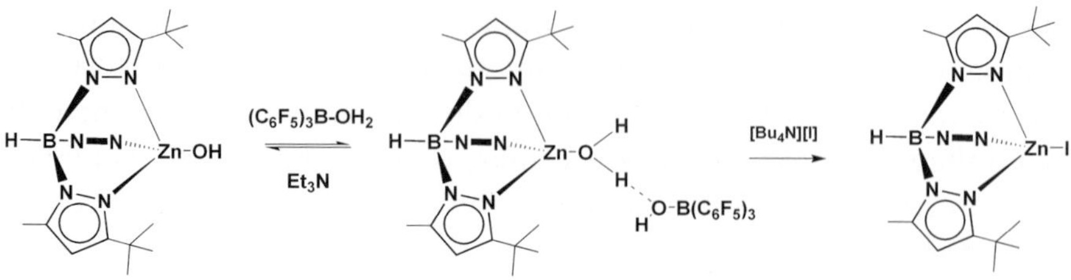

Fig. 8.11. $(C_6F_5)_3B(OH)_2$ is able to protonate the hydroxide ligand in $[Zn(OH)(Tp^{^tBu,Me})]$. Reaction of $[\{Zn(Tp^{R,Me})\}_2(H_3O_2)]^+$ with I^- gives [Zn(I)-$(Tp^{^tBu,Me})$].

$[ZnOH(Tp^x)]$ species (including Tp^x incorporating hydrogen-bonding accepting ester substituents) have been synthesized and widely characterized.[118–121] It is worth noting that $(C_6F_5)_3$-$B(OH)_2$[122] is capable of protonating the hydroxide ligand in [Zn-$(OH)(Tp^{^tBu,Me})$] to yield the aqua complex $[Zn(OH_2)(Tp^{^tBu,Me})]$-$[HOB(C_6F_5)_3]$ (Fig. 8.11).[123] Theoretical calculations have been performed for both of these species,[124] and their molecular structure compared with dinuclear complexes $[\{Zn(Tp^{R,Me})\}_2(H_3O_2)]^+$. $[Zn(OH)(Tp^{Ar,Me})]$[125] has been also employed to synthesize zinc–nucleobase complexes[126,127] and to investigate the reactivity of aminoacids toward zinc.[128,129]

Matrix metalloproteinases are a recently discovered class of zinc enzymes able to degrade extracellular matrix components. $[Zn(OH)Tp^{Cum,Me})]$ has been reported to cleave activated amides and ester.[130] The reaction of $[Zn(OH)(Tp^{Ph,Me})]$ with acetohydroxamic acid has been also described,[131] as well as protonolis of zinc thiolate bond in $[Zn(SPh)(Tp^{Ph})]$.[132]

Synthesis and structural characterization of synthetic analogs of liver alcohol dehydrogenase (LADH) continued to be one of the most investigated fields of research. A number of ligands suitable for modeling LADH have been reported and their hydroxo zinc complexes also described.[133–135] $[Zn(Tm^{Mes})(HOMe)]^+$, for example, exhibits a coordination environment that resembles aspects of that in LADH.[136] The ethanol complex $[Zn(Tm^{^tBu})(HOEt)]^+$ has been also isolated.[137] $[NS_2]$ donor ligands, that feature thioether donors, also provide coordination environments that mimic the active site of LADH.[138,139]

Several papers appeared in literature which report on the mechanism of action of LADH, showing examples for the formation of alkokides[140–142] and hydride transfer.[143] Zinc enzyme modeling with zinc complexes containing polar pyrazolylborate ligands have been reported by Vahrenkamp and coworkers.[144]

The tris(mercaptoimidazolyl) ligand Tm^{Ar} has been employed to emulate the coordination environment of 5-aminolevulinate dehydratase (ALAD). Several $[ZnX(Tm^{Ar})]$ have been described[145–147] that also have helped in the knowledge of the mechanism of action of ALAD. To study the lead poisoning that is associated with lead interaction with ALAD, $[PbX(Tm^{Ar})]$ complexes have been also reported.[148]

A structural analog for the Cys_4Zn motif of the ADA protein is constituted by the thiolate complex $[ZnSPh(Tm^{Ph})]$.[132] Related cadmium complexes have been also described.[149] These studies are also pertinent to other enzymes that feature cysteine thiolate alkylation such as methionine synthase, methanol:coenzyme methyltransferase, farnesyl transferase.

Influence of the metal on the structure, composition, and reactivities of synthetic analogs, scorpionate-containing, of zinc enzymes has been also investigated.[150–152]

8.3.10. *Mercury*

Melnick and Parkin[153] reported a functional model for mercury detoxification by organomercurial lyase *MerB*. The interaction of the alkyl compound $[HgR(Tm^{^tBu})]$ with phenylthiol has been investigated.

8.4. C–H Activation

Jones reported an overview of the use of tris(pyrazolyl)borate rhodium complexes for the activation of arene and alkane C–H bonds, in particular detailing the fundamental studies with $[Rh(CNR)(R)H(Tp^*)]$. Selected example is reported in Fig. 8.12.[154] Jones and coworkers[155] also carried out some experiments in order to verify the involvement of

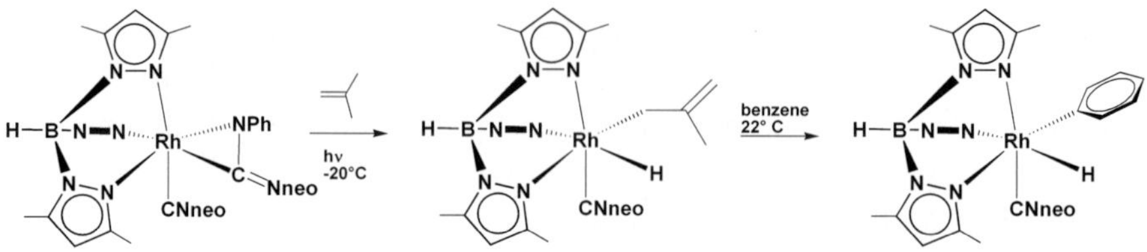

Fig. 8.12. Synthesis of [Rh(CNR)(R)H(Tp*)] and its activation of arene C–H bonds.

alkane σ-complexes in oxidative addition/reductive elimination reaction of [Rh(L)(R)H(Tp*)] (L = $CNCH_2CMe_3$). The isopropyl hydride complex [Rh(L)($CHMe_2$)H(Tp*)] has been found to rearrange to the *n*-propyl hydride derivative [Tp*Rh(L)($CH_2CH_2CH_3$)H] through an intramolecular reaction (Fig. 8.13). The same behavior has been found for the *sec*-butyl complex. These reactions have been investigated also by using the corresponding deuteride complexes.

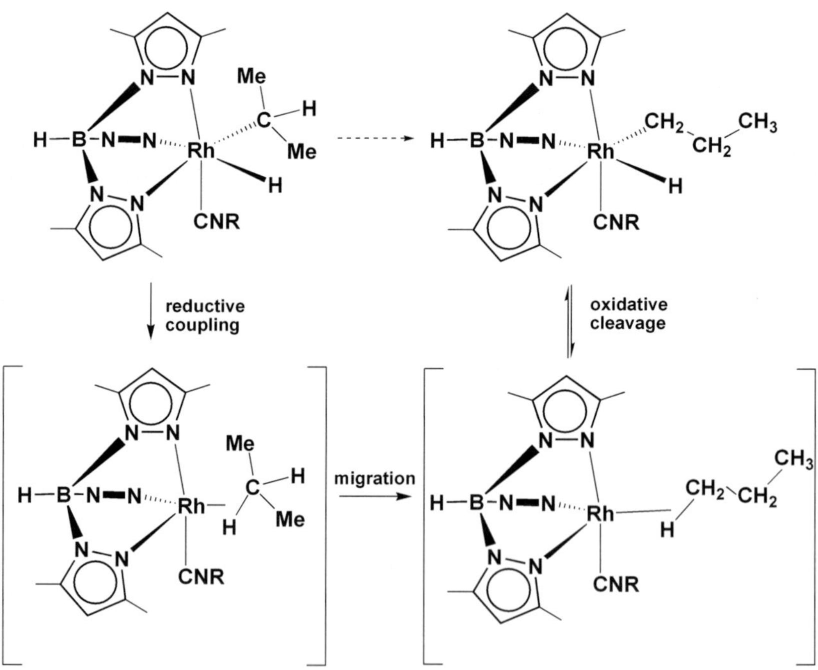

Fig. 8.13. The isopropyl hydride complex [Rh(L)($CHMe_2$)H(Tp*)] rearranges to the *n*-propyl hydride [Rh(L)($CH_2CH_2CH_3$)H(Tp*)] through an intramolecular reaction.

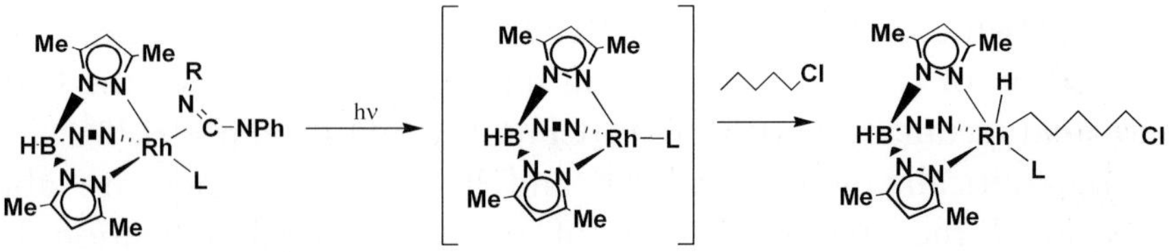

Fig. 8.14. [Rh(CNR)(Tp*)(L)] reacts with chloro-substituted alkanes giving primary C–H oxidative addition products.

[Rh(CNneopentyl)(Tp*)(L)] reacts with chloro-substituted alkanes giving primary C–H oxidative addition products (Fig. 8.14). If the alkane has chlorine substitution β to the methyl group, then activation is followed by rapid β-elimination to give [RhHCl(Tp*)(L)] and the corresponding olefin.[156–158]

The strength of the {M*}–R bond for a number of hydrocarbyl groups R (M* = [Rh(H)(Tp)(MeCN)]) has been also analyzed with DFT-based energy decomposition scheme.[159]

[Rh(Tp*)(C_2H_4)$_2$] reacts with L giving [Rh(CH=CH_2)(C_2H_5)-(Tp*)(L)] (L = MeCN or py), which upon heating at 60°C converts to [Rh(Tp*)(C_2H_4)(L)] (L = MeCN) and is able to induce the activation of one of the C–H bonds in C_6H_6. The adducts [Rh(Tp*)(C_2H_4)(PR_3)] (R = Me or Et), are efficient reagents for the C–H activation of C_6H_6, py or thiophene.[160] In the latter case, rupture of the C–S bond is also detected, although the C–S bond activation complexes are thermodynamically disfavored with respect to those derived from the cleavage of one of the R–C–H bonds.

[Rh(Tp*)(PPh_3)$_2$] catalyzes the hydrothiolation of a range of alkynes with both aryl and alkyl thiols. The reactions with alkyl thiols proceeded with excellent regioselectivity, providing convenient access to branched alkyl vinyl sulfides.[161] [Rh(Tp*)(PPh_3)$_2$] reacts with dichloromethane giving [RhCl(H)$_2$(PPh_3)$_2$(Hpz*)] and [$RhCl_2$(H)(PPh_3)$_2$(Hpz*)], and with C_6F_5SH giving [(PPh_3)$_2$Rh(μ-SC_6F_5)$_2$Rh(SC_6F_5)(H)(PPh_3)(Hpz*)].[162]

Kinetic data for the reaction of [Rh(Tp*)(CO)$_2$] with cyclohexane in xenon indicated the presence of the same transient IR bands (1978 and 1996 cm^{-1}) that decay with formation of the Rh alkyl hydride product [Rh(cyclohexyl)(H)(Tp*)(CO)], absorbing at 2032 cm^{-1}.[163] Irradiation of [Rh(Tp^{*Cl})$(CO)_2$] in diisopropylamine gave the C–H activated metallacycle [RhH{$CH_2CH(CH_3)NH^iPr$}(Tp^{*Cl})].[164]

The optically active Tp^{Menth} and $Tp^{Mementh}$ ligands are able to control the stereoselectivity of the C–H bond activation by their coordinated rhodium center. Irradiation of [Rh(Tp^{Menth})$(CO)_2$] under N_2 resulted in generation of an 85:15 mixture of diastereomeric alkyl hydrides resulting from intramolecular cyclometalation reaction involving the methyl substituents on the ligand isopropyl group.[165] Photochemical irradiation of [Rh$(CO)_2$(Tp^{An})] affords an aryl hydride complex by intramolecular cyclometalation involving an ortho C–H bond of one *p*-anisyl substituents.[166]

[Rh(Tp*)(COD)] has been employed as catalyst in the dehydrogenative coupling of 4,4,5,5-tetramethyl-1,3,2-dioxaborolane with arene. Several pyrazolylborates have been employed and it has been found that the catalytic activity is affected by substituents on the pz rings.[77]

[Rh(Tp*)(η^4-1,5-COD)], photoisomerizes to [Rh(Tp*)(η^4-1,3-COD)] through an intramolecular (3,4) hydrogen shift of the coordinated 1,5-COD likely involving an allylrhodium(III) intermediate.[167] Photolysis of [Rh(Tp*)(η^4-1,5-COD)] in benzene at 400 nm gives [Rh(Tp*)(η^4-1,3-COD)], which upon selected photolysis at 336 nm in the presence of $P(OMe)_3$, CH_3OH, and ^{t}Bu-acrylate can be used to produce [Rh(H)(C_6H_5)(Tp*){$P(OMe)_3$}], [Rh$(H)_2$(Tp*)(CO)], and [Rh(C_6H_5)(Tp*)($CH_2CH_2COO^tBu$)], respectively.[168]

The unsaturated [Ir$(C_6H_5)_2$(Tp*)] fragment, generated *in situ* either from the N_2 complex [Ir$(C_6H_5)_2$(Tp*)(N_2)] or from the mild heating of [Ir$(C_2H_4)_2$(Tp*)] in C_6H_6 produces, under argon, the double α-C–H bond activation of THF (and other cyclic ethers) with formation of the corresponding cyclic carbene. This reaction is general for other {Ir^{III}Tp} precursors and a variety of related cyclic ethers.[169] For example, [Ir$(C_2H_4)_2$(Tp*)] activates successfully C–H bonds in two molecules of benzene to generate a mixture of [Ir$(C_6H_5)_2$(Tp*)(N_2)] and [Ir(Tp*)$(C_6H_5)_2$]$_2$(μ-N_2). [Ir(Tp*)$(C_2H_4)_2$] activates the OCH_2–functionality of THF, yielding a Fischer carbene compound [Tp*(H)(Bu)Ir=C(CH_2)O].[170] In fact,

$\{Ir^{III}Tp^{x}\}$ fragments based on different auxiliary tris(pyrazolyl)borate ligands have been employed to activate even ethers like Et_2O, $MeOCH_2CH_2OMe$, $MeOBu^n$, and others of low Lewis basicity, that very seldom participate in rearrangements of this type.[171–176] An overview of this work, concentrating on the activation of aliphatic ethers and of anisole by $\{Ir^{III}Tp^{x}\}$ complexes, has been presented by Carmona *et al.*[177]

More recently also the formation and cleavage of C–H, C–C, and C–O bonds of ortho-methyl-substituted anisoles by $\{Ir^{III}Tp^{x}\}$ fragments have been reported (Fig. 8.15).[178]

A new route to iridabenzenes, involving the coupling of the primary alkene $MeCH{=}CH_2$ with iridacycles (Fig. 8.16) has been also described. Transient iridium alkylidenes are likely key intermediates of this rearrangement that involves the formation of new carbon–carbon bonds.[179]

The iridium fragment $[Ir(C_6H_5)_2(Tp^*)]$ also mediates the coupling of benzene and the methyl ethers $MeOBu^t$, $MeOBu^n$, and $MeOCH_2CH_2OMe$, to give the benzyl ethers $C_6H_5CH_2OR$ (Fig. 8.17). Mechanistic studies have allowed the trapping of some key intermediates and elucidated not only the nature of the elementary

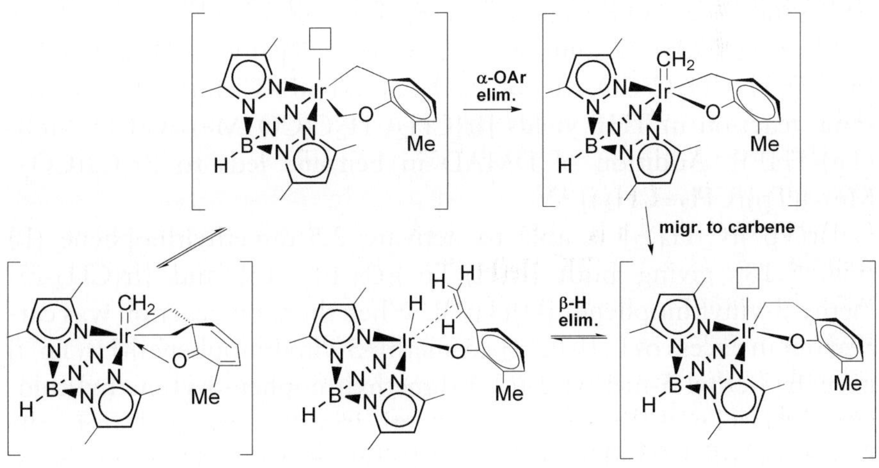

Fig. 8.15. Formation and cleavage of C–H, C–C, and C–O bonds of ortho-methyl-substituted anisoles by $\{Ir^{III}Tp^{x}\}$ fragments.

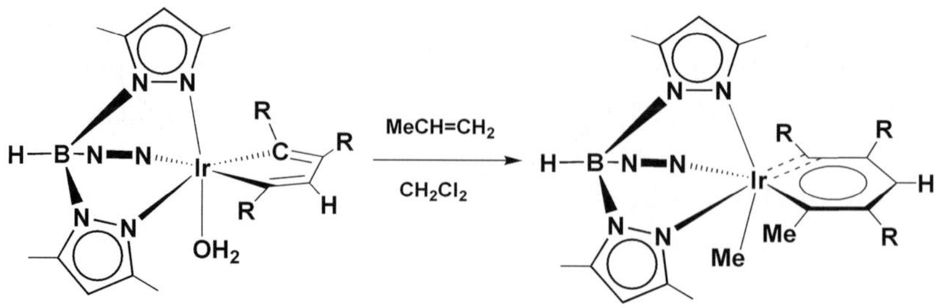

Fig. 8.16. A route to iridabenzenes involving the coupling of the primary alkene $MeCH{=}CH_2$ with iridacycles.

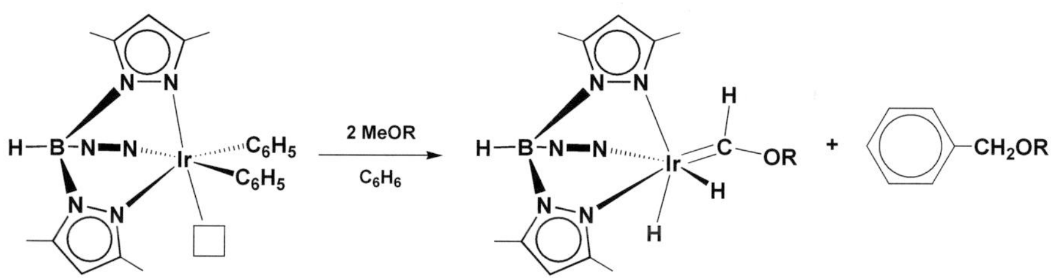

Fig. 8.17. The iridium fragment $[Ir(C_6H_5)_2(Tp^*)]$ mediates the coupling of benzene and MeOR.

steps involved in the global reaction but also the precise sequence with which they occur in the complex reaction manifold.[180]

Heating a MeCN solution of $[Ir(C_2H_4)_2(Tp)]$ with DMAD leads to $[Ir\{CH_2CH_2C(CO_2Me){=}C(CO_2Me)\}(Tp)(NCMe)]$ whereas the same reaction in THF yields $[Ir\{CH_2CH_2C(CO_2Me){=}C(CO_2Me)\}$-(Tp)(THF)]. Addition of DMAD in benzene leds to $[Ir\{C_4(CO_2$-$Me)_4\}(Tp)(CH_2{=}CH_2)]$.[181]

$[Ir(Tp^*)(C_2H_4)_2]$ is able to activate 2,5-dimethylthiophene (L) (Fig. 8.18) giving both $[Ir(H)(Tp^*)(C_3\text{-}L)(S\text{-}L)]$ and $[Ir\{CH_2$-(5-methy-3-vinylthiophene)}(H)(Tp*)]. When the same reaction was carried out in excess of L, $[Ir\{CH_2$-(5-methy-3-vinylthiophene)}(H)(Tp*)] gave [Ir{3-ethyl-5-methyl-2-((2,5-dimethylthiophen-3-yl)methyl)thiophenyl}(H)(Tp*)].[182] sp^2 C–H activation is needed for the formation of $[Ir(H)(Tp^*)(C_3\text{-}L)(S\text{-}L)]$, whereas $[Ir\{CH_2$-(5-methy-3-vinylthiophene)}(H)(Tp*)] afforded upon activation of an sp^3 C–H bond and formation of a new C–C bond.[183]

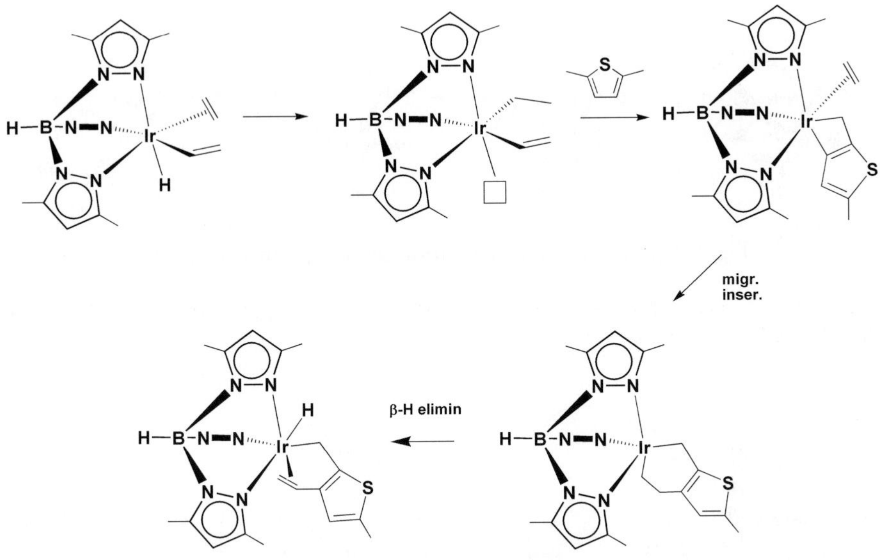

Fig. 8.18. $[Ir(Tp^*)(C_2H_4)_2]$ activation of 2,5-dimethylthiophene.

[Ir(Tp)COD] has been employed as catalyst in the dehydrogenative coupling of 4,4,5,5-tetramethyl-1,3,2-dioxaborolane with arene.[177]

Reaction of $[Ru(Tp^{Ms})(COD)H]$ with MeCN was accompanied by C–H activation of a ligand orthomethyl group of the mesityl substituent and H_2 elimination.[184]

Theoretical calculations on the metathesis process, $[MR(\eta^2\text{-H–}CH_3)(Tp)(PH_3)] \rightarrow [M(CH_3)(\eta^2\text{-H–R})(Tp)(PH_3)]$ (M = Fe, Ru, and Os; R = H and CH_3), have been systematically carried out to clarify the reaction mechanisms.[185]

Hydrocarbon C–H bond activation with $[Pt(Tp^x)]$ complexes has been widely investigated by Templeton's group.[186] It has been hypothesized that the first step in alkane C–H activation by platinum(II) scorpionates involves coordination of the alkane to a coordinatively unsaturated metal center to form a σ-bound alkane intermediate Pt(R–H) which is thought to interconvert rapidly with a $Pt^{IV}(R)(H)$ oxidative addition product (Fig. 8.19). $[PtR(H)_2(Tp^x)]$ species have been isolated and structurally characterized. When

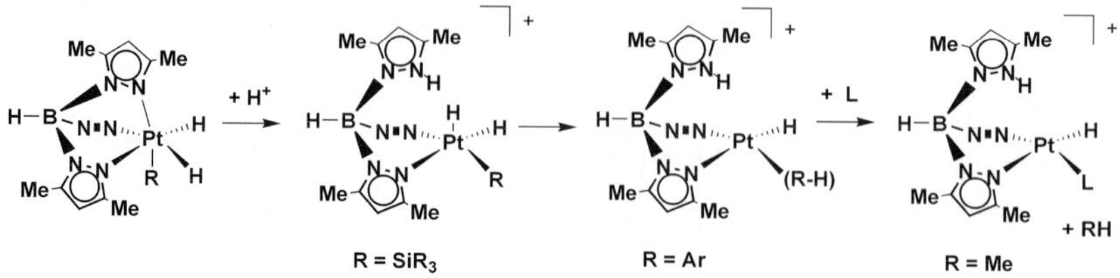

Fig. 8.19. Alkane C–H activation by platinum(II) reagents.

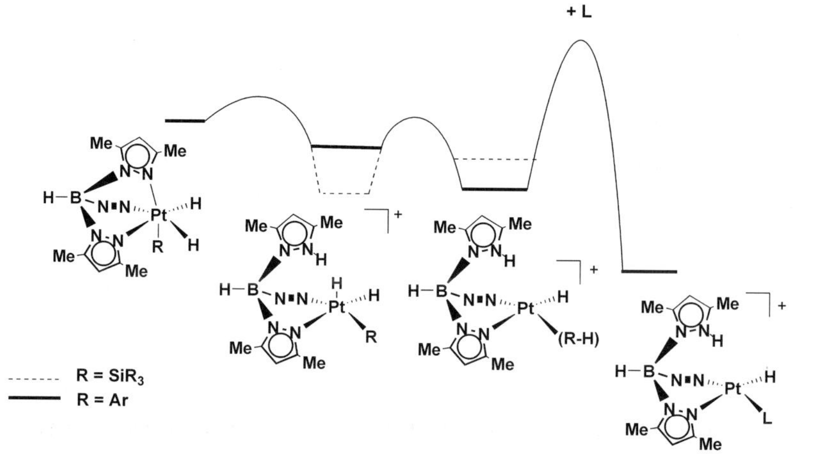

Fig. 8.20. When R = SiR_3 and C_6H_5 rather than R = alkyl in the low-temperature protonation of $[PtR(H)_2(Tp^*)]$, isolation of analogs of proposed intermediates in C–H bond activation by Pt(II) complexes is possible.

R = SiR_3, five-coordinate Pt(IV) complexes of the type $[Pt(H)_2(SiR_3)(\kappa^2\text{-HTp}^x)][BAr_F]$ formed.[187] A Pt(II) arene adduct prior to replacement of the Ar–H ligand was achieved in the form of Pt(II) η^2–arene adduct (Fig. 8.20).[188,189]

When nitrile ligands were added at low-temperature to Pt(II) η^2-benzene complexes $[Pt(\eta^2\text{-}C_6H_6)(R)(\kappa^2\text{-HTp}^*)][BAr_F]$ (R = H, Ph), reversible trapping of the five-coordinate C–H oxidative addition product was observed, $[Pt(C_6H_5)(H)(R)(NCR')(\kappa^2\text{-HTp}^*)][BAr_F]$ (R = Ph, H; R′ = Me, CMe_3, *p*-XC_6H_4) being formed. At higher temperature, irreversible loss of benzene occurred from the Pt(II) η^2-benzene forming Pt(II) nitrile derivatives $[Pt(R)(NCR')(\kappa^2\text{-HTp}^*)][BAr_F]$.[190]

Table 8.2. Effects of catalysts strucure on the silylation of benzene.

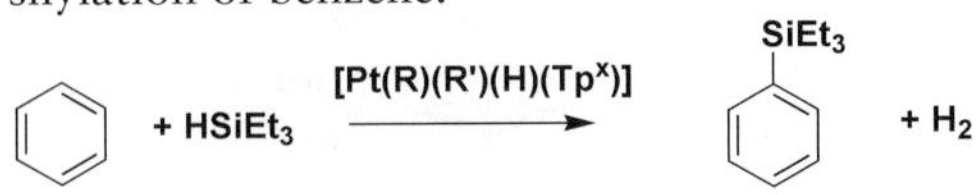

Entry	Tp^x	R	R′	Yield (%)
1	Tp	Me	Me	72
2	Tp^*	Me	Me	86
3	Tp^*	Me	H	6
4	Tp^*	Ph	Ph	78

The reactivity of $[PtR_3(Tp^*)]$ derivatives for the dehydrogenative couplings of silanes with hydrocarbons is reported. The effect of catalyst structure on the silylation of benzene has been reported (Table 8.2).[191]

8.5. New Materials

Quaternary ammonium surfactants (dodecyl-, tetradecyl-, and hexadecyltrimethylammonium or didodecyldimethylammonium) with Tp anion were prepared and characterized also by measurement of their Kraft temperatures and critical aggregation concentrations. ^1H NMR spectroscopy suggested that the single-chain surfactants containing the dodecyl ammonium form small aggregates in water. The X-ray structure analysis of the dodecyl derivative was performed.[192]Azolium poly(1,2,4-triazolyl)borate salts melt at relatively low temperatures to give thermally stable ionic liquids. Some of them exhibit conductivities close to that of KCl in $MeCN/H_2O$ mixtures.[193]

Metal complexes containing Tp^x ligands have been employed for forming films by MOCVD in semiconductor devices (based on $(Ba,Sr)TiO_3$ fabrication). They were the object of a patent by Uhlenbrock and Vaartstra.[194]

Rapenne and Vives[195] reported some technomimetic molecules designed to simulate macroscopic objects at the molecular level. The molecule in Fig. 8.21 has been synthesized starting from pentaphenyl

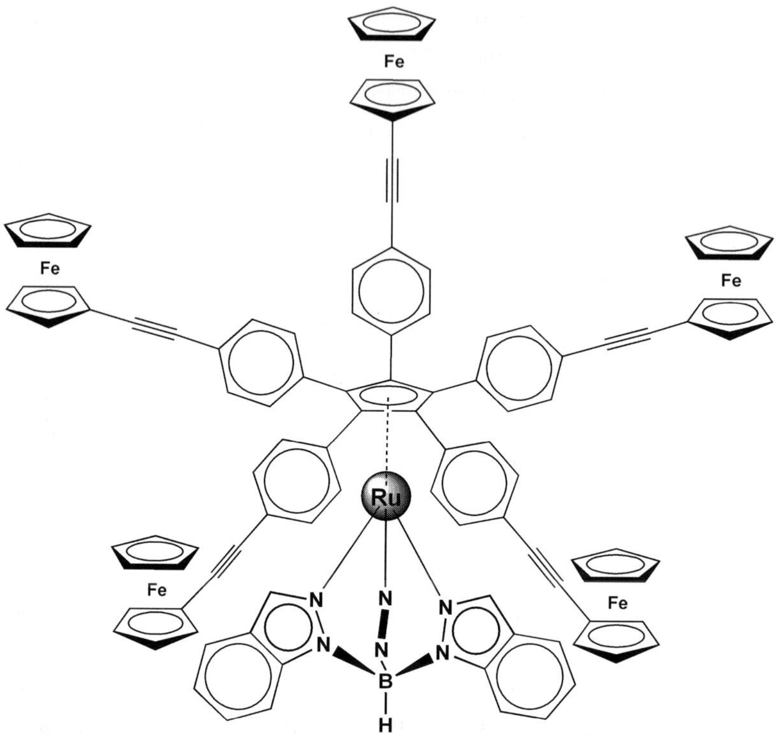

Fig. 8.21. A technomimetic molecule designed to imitate macroscopic objects at the molecular level.

cyclopentadiene, brominated selectively in the *para* positions and at the saturated carbon of the cyclopentadiene ring,[195] coordinated to ruthenium by reaction with $Ru_3(CO)_{12}$, and therefore, introducing the tris(indazolyl)borate ligand. The last step involves the coupling of five ethynyl ferrocene electroactive groups.[196]

Cascade hole-injection and effective electron-blocking/exciton confinement for light-emitting diodes (LEDs) have been reported using the blue phosphorescent emitter iridium bis(4′,6′-difluorophenylpyridinato)(pzTp).[197]

Interest toward spin-state crossover in iron(II) complexes is increasing because of potential technological applications,[198,199] such as "intelligent" contrast agents for biomedical imaging, as temperature/pressure threshold indicators, and as optical elements in display devices. The molecular and supramolecular organization of Tp^x complexes of iron(II) can have a profound impact on their electronic spin-state crossover properties.[200–202] For example, magnetic and Mössbauer spectral measurements on $[Fe(pz^0Tp^{Cpr})_2]$ and $[Fe(Tp^{Cpr})_2]$ indicate that both complexes exhibit a partial SCO from fully high-spin iron(II) at higher temperatures. In all complexes in which a phase transition is observed, this change dominates the SCO behavior. Comparison of the Mössbauer spectral properties of these two complexes and of $[Fe(Tp^{Me})_2]$ with that of other complexes reveals correlations between the Mössbauer-effect isomer shift and the average Fe–N bond distance and between the quadrupole splitting and the average Fe–N–N–B intraligand dihedral torsion angles and the distortion of the average N–Fe–N intraligand bond angles.[203]

Molecular materials with second-order nonlinear optical (NLO) properties can be successfully applied in optoelectronics technology.[204] Cano and his group have reported the second-order NLO response of novel bimetallic push–pull systems designed by connecting metallic fragments as donor (for example ferrocenyl groups) and acceptor groups (for example $[MoCl(Tp^{An})(NO)]$) through a π-conjugated system. They also investigated the second and third harmonic generations of the bimetallic system.[205–208]

Bp and Tp were investigated as corrosion inhibitors for iron.[209]

Thin film electroluminescent device structures containing $[Eu(Tp^*)_2]$, which has a high quantum efficiency in the solid state, exhibits bright orange luminescence.[210] Luminescent platinum bis(pyrazolyl)borate complexes bearing orthometalated phenyl pyridines have been also synthesized and characterized.[211]

Boncella and coworkers[212,213] reported photoluminescent and electroluminescent properties of near-infrared (near-IR) emitting lanthanide monoporphyrinate complexes, [Yb(TPP)(Tp)] (Fig. 8.22) blended into conjugated and nonconjugated polymer hosts. They found complete quenching of the alkoxysubstituted poly(*p*-phenylene) host fluorescence (i.e. $>95\%$) at 5 mol% of [Yb(TPP)-(Tp)]. Host quenching is accompanied by sensitization of near-IR emission from the lanthanide complex. The spectral data analysis suggests that the dominant mechanism operating in the EL devices

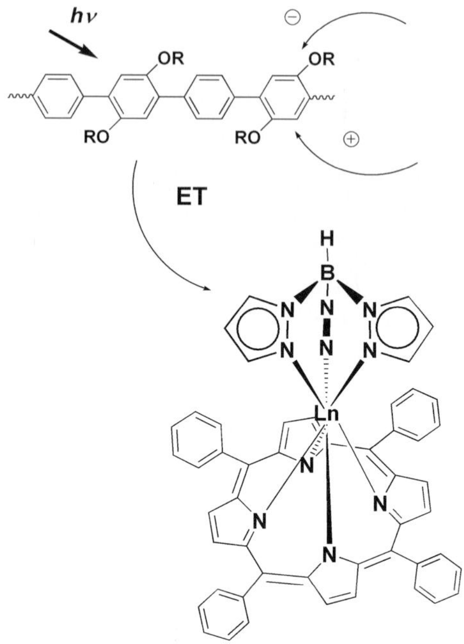

Fig. 8.22. The dominant mechanism operating in the EL devices involves the [Yb(TPP)(Tp)] complex as the charge-transport material.

involves the [Yb(TPP)(Tp)] complex as the charge-transport material. The near-IR emission has been investigated also for other [Ln(TPP)Tp] complexes (Ln = Er, Ho, Nd).

[Yb(TPP)Tp] was blended with the polymers poly(methylmethacrylate), poly(*N*-butylmethacrylate), and with poly(bisphenol-A-carbonate). Near-IR emitting polymeric light-emitting diodes (PLEDs) were fabricated by spin-coating of films of the [Yb(TPP)Tp]/polymer blends on ITO glass substrates. The best device characteristics concerning the near-IR irradiance and EL efficiency was obtained for the [Yb(TPP)Tp]/PMMA blend. Flexible PLEDs can be fabricated from the [Yb(TPP)Tp]/PBMA blend.[214]

$[Mg(Tp)]_2$ has been tested as a precursor for the MOCVD of borate phase thin films. $[Mg(Tp)]_2$ provides constant evaporation rates even for very long deposition times and, hence, highly homogeneous films of carefully controlled thickness.[215]

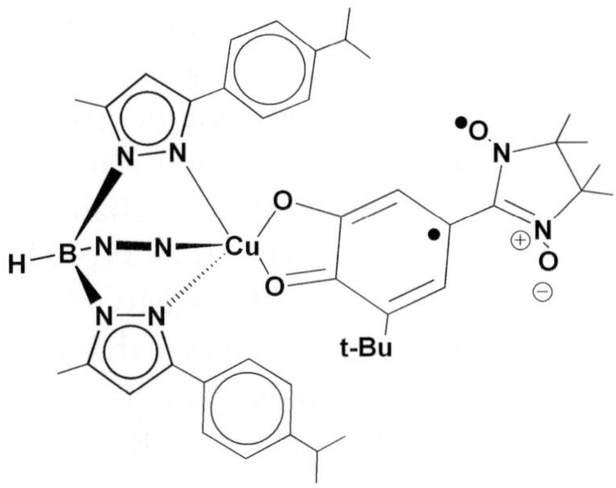

Fig. 8.23. A $[Cu(L)(Tp^{Cum,Me})]$ derivative, potential building block for the formation of spin-diverse molecule-based magnetic materials.

Volatility studies, electrospray mass spectra, and IR in vapor phase have been carried out for a series of $[Cu(Tp^x)(PR_3)]$. XRD data of copper films obtained from MOCVD experiments on the same derivatives have been also reported.[216]

Some studies have been carried out by Shultz on a series of $[M(L)(Tp^{Cum,Me})]$ derivatives (M = Cu, Mn, Zn) as that depicted in Fig. 8.23, that have been considered as potential building blocks for the formation of spin-diverse molecule-based magnetic materials.[217–220]

The three-coordinate mononuclear and dinuclear $[M(Bp^{(CF_3)_2})(2,4,6\text{-collidine})]$ (M = Cu, Ag) exhibit blue-pyrazole-base structured emissions with short phosphorescence lifetimes, due to an internal heavy-metal effect.[221]

One of the most recent synthetic challenges in the field of molecule-based magnetism is to create nano-sized superparamagnetic molecular clusters (single-molecule magnets = SMMs) having a large ground spin and a large negative anisotropy and exhibiting very slow relaxation of the magnetization below critical temperature.[222,223] SMMs have potential applications in areas including information data storage,[224] quantum computing,[225] and single-electron electronic devices.[226] The building block *fac*-$[Fe(Tp)(CN)_3]$,[227–229] able to direct the formation of new cyanide-bridged compounds with interesting structures and magnetic properties, has been employed in the synthesis of a number of heteronuclear cyano-bridged complexes showing SMM behavior. For example, $(Bu_4N)[(Tp)Fe(CN)_3]$ reacts with Cu, Co, and Ni cations, giving four one-dimensional (1D) heterobimetallic cyano-bridged chain complexes of squares, $[Fe_2^{III}(Tp)_2(CN)_6Cu(CH_3OH){\cdot}2CH_3OH]_n$, $[(Tp)_2Fe_2^{III}(CN)_6Cu(DMF){\cdot}DMF]_n$, $[Fe_2^{III}(Tp)_2(CN)_6M(CH_3OH)_2{\cdot}2CH_3OH]_n$ (M = Co, Ni). The copper complexes exhibit intrachain ferromagnetic coupling and single-chain magnets behavior, whereas the nickel and cobalt derivatives show metamagnetic behavior.[230] Also the incorporation of low-spin iron(III) centers into the face-centered cubic cluster $[(Tp)_8(H_2O)_6Cu_6^{II}Fe_8^{III}(CN)_{24}]^{4+}$ led to single-molecule magnet behavior, despite the apparent *Oh* symmetry of its core structure.[231] *fac*-$[Fe(Tp)(CN)_3]$ has been employed as building block also toward a $[Cu(bipy)_2]^{2+}$ core giving $[Fe(Tp)(CN)_3Cu(bipy)_2]ClO_4{\cdot}CH_3OH$ which exhibits ferromagnetic coupling between Cu and Fe.[232] If a tridentate blocking ligand is bonded to the Cu^{II} centers, the systems generates a trigonal bipyramidal cluster, $[Tp_2(Me_3tacn)_3Cu_3Fe_2(CN)_6]^{4+}$, in which the reduced symmetry affords a significantly increased anisotropy barrier.[233]

8.6. Metal Extraction

The separation of first-series transition metal and Cd^{++} ions has been investigated by solvent extraction using Tp^{Me} and pz^0Tp^{Me}. The results were compared with the X-ray single-crystal studies of the extracted species: $[M(Tp^{Me})_2]$ (M = Co, Zn, Cd, and Mn) and $[M(pz^0Tp^{Me})_2]$ (M = Mn, Fe, Co, Ni, and Zn).[234] Separation of first-series transition metal ions (Mn^{2+}, Fe^{2+}, Co^{2+}, Ni^{2+}, Cu^{2+}, and Zn^{2+}) and Cd^{2+} has been studied by solvent extraction using Tp^{Me}, pz^0Tp^{Me}, Tp^{iPr} and pz^0Tp^{iPr}. The selectivity is principally governed by the stability of the $[M(Tp^x)_2]$ complexes. Tp^{iPr} and pz^0Tp^{iPr} destabilize the $[M(Tp^x)_2]$ complex when small metal ions, such as Ni^{2+}, were employed. The stability of $[M(pz^0Tp^{iPr})_2]$ markedly decreases for large metal ions.[235] Separation of first-series transition metal ions {Mn(II), Fe(II), Co(II), Ni(II), Cu(II) Zn(II) and Cd(II)} has been studied by solvent extraction also using Tp and pzTp. These ligands quantitatively extract all the studied metal ions. Tp is one of the most powerful extractants of these metal ions, since it extracts them from acidic aqueous solution (pH 2.1). pzTp extracts large metal ions, such as Mn(II) and Cd(II), around higher pH compared to Tp. Also, the extraction constant of pzTp for Cu(II) is low and does not conform to the order of the Irving–Williams series. The different selectivity of these ligands cannot be explained on the basis of the basicity of the donor atoms. To explore the origin of the selectivity, X-ray crystal structures of the six extracted species have been determined. The complexes $[Mn(Tp)_2]$, $[Mn(pzTp)_2]$, $[Co(pzTp)_2]$, $[Ni(pzTp)_2]$, and $[Zn(pzTp)_2]$ have distorted octahedral geometries with each ligand having tridentate coordination. The complex $[Cu(pzTp)_2]$ has square bipyramidal geometries. Examination of the dimensions of the above-mentioned complexes has revealed that although the Tp complexes are highly C_3 symmetric, the pzTp complexes for Mn(II), Cu(II), and Cd(II) are largely distorted. These results suggest that Tp is sterically efficient, and the increase in steric energy with complex formation is sufficiently low. On the other hand, because of the intraligand contact between pyrazolyl rings, pzTp accompanies substantial steric energy increases in coordination to Mn(II), Cu(II), and Cd(II), decreasing the stability of the complexes and resulting in unique selectivity.[236]

The extraction of Cu(II), Ni(II), Co(II), Cd(II), Zn(II), Pb(II), and Mn(II) with K(Bp) in dichloromethane has been studied by Shukla and Rao[237] and the extraction constants compared with those obtained with dibenzoylmethane, thenoyltrifluoroacetone, and 1-phenyl-3-methyl-4-benzoyl-5-pyrazolone.

8.7. Miscellaneous Studies

Gatteschi and coworkers used the homologous series of compounds with the diamagnetic tropolonato ligand, $[Ln(Trop)(Tp)_2]$, and the paramagnetic semiquinone ligand, $Ln(DTBSQ)(Tp)_2$ (Ln = Sm(III), Eu(III), Gd(III), Tb(III), Dy(III), Ho(III), Er(III), or Yb(III) to obtain information on the crystal-field effects in paramagnetic rare earth ions. The X-ray crystal structure of a new form of tropolonate derivative is presented, which shows, as expected, a marked similarity with the structure of the semiquinonate derivative. Details of the coordination sphere of [Ho(Trop)$(Tp)_2$], evidenced the geometry intermediate between square antiprismatic and bicapped trigonal prismatic. The [Ln(Trop)$(Tp)_2$] derivatives were also used as a reference for the qualitative determination of crystal-field effects in the exchange coupled semiquinone derivatives. Through magnetization and susceptibility measurements, this empirical diamagnetic substitution method evidenced for Tb(III), Dy(III), Er(III), and Yb(III) derivatives a dominating antiferromagnetic coupling. The increased antiferromagnetic contribution compared to other radical–rare earth metal complexes formed by nitronyl nitroxide ligands may be related to the increased donor strength of the semiquinone ligand.[238]

[VO(acac)(Tp)], [VO(acac)(Tp*)], and [VO(acac)(Tp*)]·CH_3CN were prepared and characterized by IR, UV–vis, and NMR techniques. The toxicity of these complexes was tested and it has been reported that [VO(acac)(Tp*)]·CH_3CN has feasibility to cure diabetes.[239,240] Cytotoxic effects of [VO(O_2)(Tp)(pzH)] and [VO(O_2)(pzTp)(pzH)] on 3T3 cell proliferation have been also reported.[81]

A chemiluminescence study has demonstrated the superoxide scavenging activity of some [Cu(Tp^x)(PR_3)] complexes.[241,242] Selected complexes were investigated as antioxidants in mitigating oxidative DNA damage.

Highly fluorinated Tp^x were tested for their antimicrobial activity against several bacterial species. [Ag(THF)($Tp^{(CF_3)_2}$)] and [Na(THF)-($Tp^{(CF_3)_2}$)] seem highly effective at inhibiting the growth of two different species of Gram-positive bacteria.[243]

Rhenium and technetium complexes anchored by scorpionates (Fig. 8.24) have been also widely investigated for their potential use in radiopharmaceutical applications.[244]

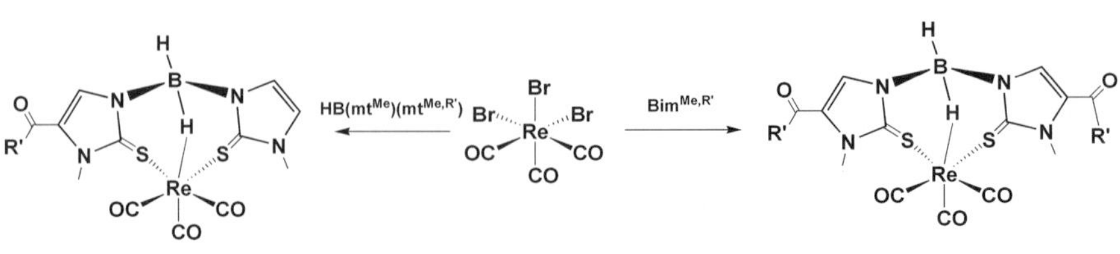

Fig. 8.24. Rhenium complexes anchored by scorpionates investigated for their potential use in radiopharmaceutical applications.

Krzystek *et al.*[245] described a methodology based on backward wave oscillator sources in high frequency and field EPR which has been applied to the study of the complex $[Co(N_3)(Tp^{tBu})]$.

References

1. Akita M, Hikichi S, *Bull Soc Chem Jap* 2002, **75**, 1657.
2. Michiue K, Jordan RF, CODEN: JKXXAF JP 2005239910 A2 20050908 Patent written in Japanese. Application: JP 2004-52712 20040227, Jpn Kokai Tokkyo Koho 2005.
3. Nakazawa H, Ikai S, Imaoka K, Kai Y, Yano T, *J Mol Catal A: Chem* 1998, **132**, 33.
4. Karam A, Jimeno M, Lezama J, Catarí E, Figueroa A, Rojas de Gascue B, *J Mol Catal A: Chem* 2001, **176**, 65.
5. Karam A, Casas E, Catari E, Pekerar S, Albornoz A, Méndez B, *J Mol Catal A: Chem* 2005, **238**, 233.
6. Karam A, Pastran J, Casas E, Méndez B, *Polym Bull* 2005, **55**, 11.
7. Murtuza S, Casagrande Jr, OL, Jordan RF, *Organometallics* 2002, **21**, 1882.
8. López-Linares F, Díaz Barrios A, Ortega H, Karam A, Agrifoglio G, Gonzalez E, *J Mol Catal A: Chem* 2002, **179**, 87.
9. Gil MP, dos Santos JHZ, Casagrande Jr, OL, *Macromol Chem Phys* 2001, **202**, 319.
10. Gil MP, dos Santos JHZ, Casagrande Jr, OL, Simplicio LMT, da Rocha ZN, *J Mol Catal Chem* 2005, **238**, 96.
11. Murtuza S, Casagrande Jr, OL, Jordan RF, *Polym Mat Sc Eng* 2001, **84**, 109.
12. Junges F, de Souza RF, dos Santos JHZ, Casagrande Jr, OL, *Macromol Mater Eng* 2005, **290**, 72.
13. Kunrath FA, Mauler RS, de Souza RF, Casagrande Jr, OL, *Macromol Chem Phys* 2002, **203**, 2058.
14. Asandei AD, Moran IW *J Polym Sci Part A: Polym Chem* 2005, **43**, 6039.
15. Kunrath FA, de Souza RF, Casagrande Jr, OL, *Macromol Rapid Commun* 2000, **21**, 277.
16. Gil MP, dos Santos JHZ, Casagrande Jr, OL, *J Mol Catal A: Chem* 2004, **209**, 163.
17. Casagrande ACA, Gil MP, Casagrande Jr, OL, *J Braz Chem Soc* 2005, **16**, 1283.

18. Michiue K, Murtuza S, Jordan RF, *Polym Mat Sci Eng* 2002, **86**, 295.
19. Pires GP, Gil MP, Rohrman JA, Stedile FC, Casagrande Jr, OL, dos Santos JHZ, Sano T, *J App Polym Sc* 2006, **99**, 2002.
20. Janiak C, Lange KCH, Scharmann TG, *Appl Organomet Chem* 2000, **14**, 316.
21. Casagrande ACA, dos Anjos PS, Gamba D, Casagrande Jr, OL, Dos Santos JHZ, *J Mol Catal A: Chem* 2006, **255**, 19.
22. Furlan LG, Casagrande Jr, OL, *J Braz Chem Soc* 2005, **16**, 1248.
23. Casagrande ACA, Tavares TT Da R, Kuhn MCA, Casagrande Jr, OL, dos Santos JHZ, Teranishi T, *J Mol Catal A: Chem* 2004, **212**, 267.
24. Furlan LG, Kunrath FA, Mauler RS, de Souza RF, Casagrande Jr, OL, *J Mol Catal A: Chem* 2004, **214**, 207.
25. Kuhn MCA, da Silva JL, Casagrande ACA, Mauler RS, Casagrande Jr OL, *Macromol Chem Phys* 2006, **207**, 827.
26. Nabika M, Seki Y, Miyatake T, Ishikawa Y, Okamoto K-i, Fujisawa K, *Organometallics* 2004, **23**, 4335.
27. Novikova EV, Belov GP, Klaui W, Solotnov AA, *Vysokomolekulyarnye Soedineniya Seriya Ai Seriya B* 2003, **45**, 1605.
28. Nozaki K, Kondo F, Patent. Application: JP 2004-290985 20041004, 2006.
29. Santi R, Romano AM, Sommazzi A, Grande M, Bianchini C, Mantovani G, *J Mol Catal A: Chem* 2005, **229**, 191.
30. Trofimenko S, Calabrese JC, Thompson JS, *Inorg Chem* 1987, **26**, 1507.
31. Arrowood BN, Lail M, Gunnoe TB, Boyle PD, *Organometallics* 2003, **22**, 4692.
32. Trzeciak AM, Ziólkowski JJ, *Appl Organomet Chem* 2004, **18**, 124.
33. Chisholm MH, Gallucci JC, Phomphrai K, *Inorg Chem* 2004, **43**, 6717.
34. Higashimura H, Kubota M, Shiga A, Fujisawa K, Moro-oka Y, Uyama H, Kobayashi S, *Macromolecules* 2000, **33**, 1986.
35. Dias HVR, Rajapakse RMG, Krishantha DMM, Fianchini M, Wang X, Elsenbaumer RL, *J Mater Chem* 2007, **17**, 1762.
36. Díaz-Requejo MM, Wehrmann P, Leatherman MD, Trofimenko S, Mecking S, Brookhart M, Pérez PJ, *Macromolecules* 2005, **38**, 4966.
37. Caballero A, Díaz-Requejo MM, Belderraín TR, Nicasio MC, Trofimenko S, Pérez PJ, *Organometallics* 2003, **22**, 4145.
38. Díaz-Requejo MM, Belderraín TR, Trofimenko S, Pérez PJ, *J Am Chem Soc* 2001, **123**, 3167.

39. Díaz-Requejo MM, Caballero A, Belderraín TR, Nicasio MC, Trofimenko S, Pérez PJ, *J Am Chem Soc* 2002, **124**, 978.
40. Díaz-Requejo MM, Mairena MA, Belderraín TR, Nicasio MC, Trofimenko S, Pérez PJ, *Chem Commun* 2001, 1804.
41. Caballero A, Díaz-Requejo MM, Trofimenko S, Belderrain TR, Pérez PJ, *J Org Chem* 2005, **70**, 6101.
42. Handy ST, Czopp M, *Org Lett* 2001, **3**, 1423.
43. Handy ST, Ivanow A, Czopp M, *Tetrahedron Lett* 2006, **47**, 1821.
44. Mairena MA, Díaz-Requejo MM, Belderrain TR, Nicasio MC, Trofimenko S, Pérez PJ, *Organometallics* 2004, **23**, 253.
45. Díaz-Requejo MM, Pérez PJ, Brookhart M, Templeton JL, *Organometallics* 1997, **16**, 4399.
46. Dias HVR, Lu H-L, Kim H-J, Polach SA, Goh TKHH, Browning RG, Lovely CJ, *Organometallics* 2002, **21**, 1466.
47. Díaz-Requejo MM, Belderraín TR, Nicasio MC, Trofimenko S, Pérez PJ, *J Am Chem Soc* 2002, **124**, 896.
48. Morilla ME, Díaz-Requejo MM, Belderrain TR, Nicasio MC, Trofimenko S, Pérez PJ, *Organometallics* 2004, **23**, 293.
49. Morilla ME, Molina MJ, Diaz-Requejo MM, Belderrain TR, Nicasio MC, Trofimenko S, Peréz PJ, *Organometallics* 2003, **22**, 2914.
50. Díaz-Requejo MM, Belderrain TR, Nicasio MC, Prieto F, Pérez PJ, *Organometallics* 1999, **18**, 2601.
51. Dias HVR, Browning RG, Richey SA, Lovely CJ, *Organometallics* 2004, **23**, 1200.
52. Lovely CJ, Browning RG, Badarinarayana V, Dias HVR, *Tetrahedron Lett* 2005, **46**, 2453.
53. Krishnamoorthy P, Browning RG, Singh S, Sivappa R, Lovely CJ, Dias HVR, *Chem Commun* 2007, 731.
54. Díaz-Requejo MM, Belderrain TR, Nicasio MC, Pérez PJ, *Organometallics* 2000, **19**, 285.
55. Díaz-Requejo MM, Belderraín TR, Pérez PJ, *Chem Commun* 2000, 1853.
56. Diaconu D, Hu Z, Gorun SM *J Am Chem Soc* 2002, **124**, 1564.
57. Gorun SM, Hu Z, Stibrany RT, Carpenter G, *Inorg Chim Acta* 2000, **297**, 383.
58. Santos AM, Kühn FE, Bruus-Jensen K, Lucas I, Romão CC, Herdtweck E, *J Chem Soc, Dalton Trans* 2001, 1332.
59. Gable KP, Brown EC, *Organometallics* 2000, **19**, 944.
60. Harman WD, Trindle C, *J Comput Chem* 2005, **26**, 194.

61. Gable KP, Brown EC, *J Am Chem Soc* 2003, **125**, 11018.
62. Gable KP, Brown EC, *Synlett* 2003, **14**, 2243.
63. Chen J, Woo LK, *J Organomet Chem* 2000, **601**, 57.
64. Takahashi Y, Hashimoto M, Hikichi S, Akita M, Moro-oka Y, *Angew Chem Int Ed* 1999, **38**, 3074.
65. Alvarado Y, Busolo M, López-Linares F, *J Mol Catal A: Chem* 1999, **142**, 163.
66. Lin M-Y, Maddirala SJ, Liu R-S, *Org Lett* 2005, **7**, 1745.
67. Odedra A, Wu C-J, Pratap TB, Huang C-W, Ran Y-F, Liu R-S, *J Am Chem Soc* 2005, **127**, 3406.
68. Shen H-C, Tang J-M, Chang H-K, Yang C-W, Liu R-S, *J Org Chem* 2005, **70**, 10113.
69. Datta S, Chang C-L, Yeh K-L, Liu R-S, *J Am Chem Soc* 2003, **125**, 9294.
70. Madhushaw RJ, Lin MY, Sohel S Md A, Liu R-S, *J Am Chem Soc* 2004, **126**, 6895.
71. Datta S, Odedra A, Liu R-S, *J Am Chem Soc* 2005, **127**, 11606.
72. Fukumoto Y, Dohi T, Masaoka H, Chatani N, Murai S, *Organometallics* 2002, **21**, 3845.
73. Lo C-Y, Pal S, Odedra A, Liu R-S, *Tetrahedron Lett* 2003, **44**, 3143.
74. Jimenez-Tenorio M, Palacios MD, Puerta MC, Valerga P, *J Mol Catal A: Chem* 2007, **261**, 64.
75. Jiménez-Tenorio M, Puerta MC, Valerga P, Moreno-Dorado FJ, Guerra FM, Massanet GM, *Chem Commun* 2001, 2324.
76. Van Rooy S, Cao C, Patrick BO, Lam A, Love JA, *Inorg Chim Acta* 2006, **359**, 2918.
77. Murata M, Odajima H. Watanabe S, Masuda Y, *Bull Soc Chem Jap* 2006, **79**, 1980.
78. Slugovc C, Perner B, *Inorg Chim Acta* 2004, **357**, 3104.
79. Haskel A, Keinan E, *Tetrahedron Lett* 1999, **40**, 7861.
80. Collin J, Maria L, Santos I, *J Mol Catal A: Chem* 2000, **160**, 263.
81. Xing Y, Zhang Y, Sun Z, Ye L, Xu Y, Ge M, Zhang B, Niu S, *J Inorg Biochem* 2007, **101**, 36.
82. Joshi HK, Inscore FE, Schirlin JT, Dhawan IK, Carducci MD, Bill TG, Enemark JH, *Inorg Chim Acta* 2002, **337**, 275.
83. Inscore FE, Joshi HK, McElhaney AE, Enemark JH, *Inorg Chim Acta* 2002, **331**, 246.
84. Inscore FE, Knottenbelt SZ, Rubie ND, Joshi HK, Kirk ML, Enemark JH, *Inorg Chem* 2006, **45**, 967.

85. McElhaney AE, Inscore FE, Schirlin JT, Enemark JH, *Inorg Chim Acta* 2002, **341**, 85.
86. Joshi HK, Cooney JJA, Inscore FE, Gruhn NE, Lichtenberger DL, Enemark JH, *Proc Natl Acad Sci* 2003, **100**, 3719.
87. Uhrhammer D, Schultz FA, *Inorg Chem* 2004, **43**, 7389.
88. Kirk ML, Peariso K, *Polyhedron* 2004, **23**, 499.
89. Zhang Y, Holm RH, *Inorg Chem* 2004, **43**, 674.
90. Eagle AA, George GN, Tiekink ERT, Young CG, *J Inorg Biochem* 1999, **76**, 39.
91. Hong D, Zhang Y, Holm RH, *Inorg Chim Acta* 2005, **358**, 2303.
92. Singh UP, Sharma AK, Tyagi P, Upreti S, Singh RK, *Polyhedron* 2006, **25**, 3628.
93. Singh UP, Sharma AK, Hikichi S, Komatsuzaki H, Moro-oka Y, Akita M, *Inorg Chim Acta* 2006, **359**, 4407.
94. Cheng YH, Yeung LK, Lee HK, *J Inorg Biochem* 2003, **96**, 116.
95. Mehn MP, Fujisawa K, Hegg EL, Que Jr, L, *J Am Chem Soc* 2003, **125**, 7828.
96. Lehnert N, Fujisawa K, Solomon EI, *Inorg Chem* 2003, **42**, 469.
97. Sun Y-J, Zhang LZ, Cheng P, Lin H-K, Yan S-P, Liao DZ, Jiang Z-H, Shen P-W, *J Mol Catal A: Chem* 2004, **208**, 83.
98. Sun Y-J, Zhang LZ, Cheng P, Lin H-K, Yan S-P, Liao D-Z, Jiang Z-H, Shen P-W, *Biophys Chem* 2004, **109**, 281.
99. Desrochers PJ, Cutts RW, Rice PK, Golden ML, Graham JB, Barclay TM, Cordes AW, *Inorg Chem* 1999, **38**, 5690.
100. Kitajima N, Fujisawa K, Tanaka M, Moro-oka Y, *J Am Chem Soc* 1992, **114**, 9232.
101. Randall DW, DeBeer George S, Hedman B, Hodgson KO, Fujisawa K, Solomon EI, *J Am Chem Soc* 2000, **122**, 11620.
102. Fujisawa K, Fujita K, Takahashi T, Kitajima N, Moro-oka Y, Matsunaga Y, Miyashita Y, Okamoto K-i, *Inorg Chem Commun* 2004, **7**, 1188.
103. Matsunaga Y, Fujisawa K, Ibi N, Miyashita Y, Okamoto K-i, *Inorg Chem* 2005, **44**, 325.
104. Gorelsky SI, Basumallick L, Vura-Weis J, Sarangi R, Hodgson KO, Hedman B, Fujisawa K, Solomon EI, *Inorg Chem* 2005, **44**, 4947.
105. Chen P, Fujisawa K, Solomon EI, *J Am Chem Soc* 2000, **122**, 10177.
106. Chen P, Root DE, Campochiaro C, Fujisawa K, Solomon EI, *J Am Chem Soc* 2003, **125**, 466.

107. Sarangi R, Aboelella N, Fujisawa K, Tolman WB, Hedman B, Hodgson KO, Solomon EI, *J Am Chem Soc* 2006, **128**, 8286.
108. Chen P, Fujisawa K, Helton ME, Karlin KD, Solomon EI, *J Am Chem Soc* 2003, **125**, 6394.
109. de la Lande A, Gérard H, Moliner V, Izzet G, Reinaud O, Parisel O, *J Biol Inorg Chem* 2006, **11**, 593.
110. Hu Z, Gorge GN, Gorun SM, *Inorg Chem* 2001, **40**, 4812.
111. Aullón G, Gorun SM, Alvarez S, *Inorg Chem* 2006, **45**, 3594.
112. Cramer CJ, Tolman WB, Theopold KH, Rheingold AL, *Proc Natl Acad Sci* 2003, **100**, 3635.
113. Halcrow MA, Chia LML, Liu X, McInnes EJL, Yellowlees LJ, Mabbs FE, Davies JE, *Chem Commun* 1998, 2465.
114. Fujisawa K, Iwata Y, Kitajima N, Higashimura H, Kubota M, Miyashita Y, Yamada Y, Okamoto K, Moro-oka Y, *Chem Lett* 1999, **28**, 739.
115. Hammes BS, Carrano MW, Carrano CJ, *J Chem Soc, Dalton Trans* 2001, 1448.
116. Mairena MA, Urbano J, Carbajo J, Maraver JJ, Alvarez E, Díaz-Requejo MM, Pérez PJ , *Inorg Chem* 2007, **46**, 7428.
117. Parkin G, *Chem Rev* 2004, **104**, 699.
118. Rombach M, Vahrenkamp H, *Inorg Chem* 2001, **40**, 6144.
119. Hammes BS, Luo XM, Carrano MW, Carrano CJ, *Inorg Chim Acta* 2002, **341**, 33.
120. Lipton AS, Bergquist C, Parkin G, Ellis PD, *J Am Chem Soc* 2003, **125**, 3768.
121. Lipton AS, Ellis PD, *J Am Chem Soc* 2007, **129**, 9192.
122. Bergquist C, Bridgewater BM, Harlan CJ, Norton JR, Friesner RA, Parkin G, *J Am Chem Soc* 2000, **122**, 10581.
123. Bergquist C, Parkin G, *J Am Chem Soc* 1999, **121**, 6322.
124. Bergquist C, Fillebeen T, Morlok MM, Parkin G, *J Am Chem Soc* 2003, **125**, 6189.
125. Puerta DT, Cohen SM, *Inorg Chim Acta* 2002, **337**, 459.
126. Badura D, Vahrenkamp H, *Inorg Chem* 2002, **41**, 6020.
127. Radura D, Vahrenkamp H, *Inorg Chem* 2002, **41**, 6013.
128. Rombach M, Gelinsky M, Vahrenkamp H, *Inorg Chim Acta* 2002, **334**, 25.
129. Brand U, Rombach M, Seebacher J, Vahrenkamp H, *Inorg Chem* 2001, **40**, 6151.

130. Ruf M, Vahrenkamp H, *Chem Ber* 1996, **129**, 1025.
131. Puerta DT, Cohen SM, *Inorg Chem* 2002, **41**, 5075.
132. Bridgewater BM, Fillebeen T, Friesner RA, Parkin G, *J Chem Soc Dalton Trans* 2000, 4494.
133. Kimblin C, Hascall T, Parkin G, *Inorg Chem* 1997, **36**, 5680.
134. Seebacher J, Shu M, Vahrenkamp H, *Chem Commun* 2001, 1026.
135. Shu M, Walz R, Wu B, Seebacher J, Vahrenkamp H, *Eur J Inorg Chem* 2003, 2502.
136. Kimblin C, Bridgewater BM, Churchill DG, Parkin G, *Chem Commun* 1999, 2301.
137. Tesmer M, Shu M, Vahrenkamp H, *Inorg Chem* 2001, **40**, 4022.
138. Chiou S-J, Innocent J, Riordan CG, Lam K-C, Liable-Sands L, Rheingold AL, *Inorg Chem* 2000, **39**, 4347.
139. Chiou S-J, Riordan CG, Rheingold AL, *Proc Natl Acad Sci USA* 2003, **100**, 3695.
140. Bergquist C, Parkin G, *Inorg Chem* 1999, **38**, 422.
141. Boerzel H, Koeckert M, Bu WM, Spingler B, Lippard SJ, *Inorg Chem* 2003, **42**, 1604.
142. Bergquist C, Storrie H, Koutcher L, Bridgewater BM, Friesner RA, Parkin G, *J Am Chem Soc* 2000, **122**, 12651.
143. Bergquist C, Koutcher L, Vaught AL, Parkin G, *Inorg Chem* 2002, **41**, 625.
144. Pérez Olmo C, Bohmerle K, Vahrenkamp H, *Inorg Chim Acta* 2007, **360**, 1510.
145. Bakbak S, Bhatia VK, Incarvito CD, Rheingold AL, Rabinovich D, *Polyhedron* 2001, **20**, 3343.
146. White JL, Tanski JM, Rabinovich D, *J Chem Soc Dalton Trans* 2002, 2987.
147. Bridgewater BM, Parkin G, *Inorg Chem Commun* 2001, **4**, 126.
148. Bridgewater BM, Parkin G, *J Am Chem Soc* 2000, **122**, 7140.
149. Bakbak S, Incarvito CD, Rheingold AL, Rabinovich D, *Inorg Chem* 2002, **41**, 998.
150. Hikichi S, Ogihara T, Fujisawa K, Kitajima N, Akita M, Moro-oka Y, *Inorg Chem* 1997, **36**, 4539.
151. Fujisawa K, Kobayashi T, Fujita K, Kitajima N, Moro-oka Y, Miyashita Y, Yamada Y, Okamoto K-i, *Bull Chem Soc Jpn* 2000, 73, 1797.
152. Alvarez HM, Tran TB, Richter MA, Alyounes DM, Rabinovich D, Tanski JM, Krawiec M, *Inorg Chem* 2003, **42**, 2149.

153. Melnick JG, Parkin G, *Science* 2007, **317**, 225.
154. Jones WD, *Inorg Chem* 2005, **44**, 4475.
155. Northcutt TO, Wick DD, Vetter AJ, Jones WD, *J Am Chem Soc* 2001, **123**, 7257
156. Vetter AJ, Jones WD, *Polyhedron* 2004, **23**, 413.
157. Vetter AJ, Rieth RD, Jones WD, *Proc Natl Acad Sci* 2007, **104**, 6957.
158. Clot E, Eisenstein O, Jones WD, *Proc Natl Acad Sci* 2007, **104**, 6939.
159. Mitoraj M, Zhu H, Michalak A, Ziegler T, *Organometallics* 2007, **26**, 1627.
160. Paneque M, Pérez PJ, Pizzano A, Poveda ML, Taboada S, Trujillo M, Carmona E, *Organometallics* 1999, **18**, 4304.
161. Cao C, Fraser LR, Love JA, *J Am Chem Soc* 2005, **127**, 17614.
162. Cîrcu V, Fernandes MA, Carlton L, *Polyhedron* 2003, **22**, 3293.
163. Yeston JS, McNamara BK, Bergman RG, Bradley Moore C, *Organometallics* 2000, **19**, 3442.
164. Teuma E, Malbosc F, Pons V, Serra-Le Berre C, Jaud J, Etienne M, Kalck P, *J Chem Soc, Dalton Trans* 2001, 2225.
165. Keyes MC, Young Jr, VG, Tolman WB, *Organometallics* 1996, **15**, 4133.
166. Santa María MD, Claramunt RM, Campo JA, Cano M, Criado R, Heras JV, Ovejero P, Pinilla E, Torres MR, *J Organomet Chem* 2000, **605**, 117.
167. Boaretto R, Ferrari A, Merlin M, Sostero S, Traverso O, *J Photochem Photobiol A: Chem* 2000, **135**, 179.
168. Boaretto R, Paolucci G, Sostero S, Traverso O, *J Mol Catal A: Chem* 2003, **204–205**, 253.
169. Boutry O, Gutiérrez E, Monge Á, Nicasio MC, Pérez P, Carmona E, *J Am Chem Soc* 1992, **114**, 7288.
170. Gutiérrez-Puebla E, Monge Á, Nicasio MC, Pérez PJ, Poveda ML, Carmona E, *Chem Eur J* 1998, **4**, 2225.
171. Slugovc C, Mereiter K, Trofimenko S, Carmona E, *Angew Chem Int Ed* 2000, **39**, 2158.
172. Slugovc C, Mereiter K, Trofimenko S, Carmona E, *Chem Commun* 2000, 121.
173. Slugovc C, Mereiter K, Trofimenko S, Carmona E, *Helv Chim Acta* 2001, **84**, 2868.
174. Santos LL, Mereiter K, Paneque M, Slugovc C, Carmona E, *New J Chem* 2003, **27**, 107.
175. Carmona E, Paneque M, Poveda ML, *Dalton Trans* 2003, 4022.

176. Paneque M, Poveda ML, Santos LL, Salazar V, Carmona E, *Chem Commun* 2004, 1838.
177. Carmona E, Paneque M, Santos LL, Salazar V, *Coord Chem Rev* 2005, **249**, 1729.
178. Lara P, Paneque M, Poveda ML, Salazar V, Santos LL, Carmona E, *J Am Chem Soc* 2006, **128**, 3512.
179. Álvarez E, Paneque M, Poveda ML, Rendón N, *Angew Chem Int Ed* 2006, **45**, 474.
180. Álvarez E, Paneque M, Petronilho AG, Poveda ML, Santos LL, Carmona E, Mereiter K, *Organometallics* 2007, **26**, 1231.
181. O'Connor JM, Closson A, Gantzel P, *J Am Chem Soc* 2002, **124**, 2434.
182. Paneque M, Poveda ML, Salazar V, Carmona E, Ruiz-Valero C, *Inorg Chim Acta* 2003, **345**, 367.
183. Paneque M, Poveda ML, Carmona E, Salazar V, *Dalton Trans* 2005, 1422.
184. Pariya C, Incarvito CD, Rheingold AL, Theopold KH, *Polyhedron* 2004, **23**, 439.
185. Lam WH, Jia G, Lin Z, Lau CP, Eisenstein O, *Chem Eur J* 2003, **9**, 2775.
186. Norris CM, Templeton JL, *ACS Symp Ser* 2004, **885**, 303.
187. Reinartz S, White PS, Brookhart M, Templeton JL, *J Am Chem Soc* 2001, **123**, 6425.
188. Reinartz S, White PS, Brookhart M, Templeton JL, *J Am Chem Soc* 2001, **123**, 12724.
189. Norris CM, Reinartz S, White PS, Templeton JL, *Organometallics* 2002, **21**, 5649.
190. Norris CM, Templeton JL, *Organometallics* 2004, **23**, 3101.
191. Tsukada N, Hartwig JF, *J Am Chem Soc* 2005, **127**, 5022.
192. Jaeger DA, Mendoza A, Bragdon J, Arulsamy N, Li Z, Apkarian RP, *Langmuir* 2005, **21**, 9440.
193. Zeng Z, Twamley B, Shreeve JM, *Organometallics* 2007, **26**, 1782.
194. Uhlenbrock S, Vaartstra BA, US Patent, CODEN: USXXAM US 6127192 A, Application: US 98–141432 19980827, 2000.
195. Vives G, Rapenne G, *Tetrahedron Lett* 2006, **47**, 8741.
196. Rapenne G, *Org Biomol Chem* 2005, **3**, 1165.
197. Niu Y-H, Liu MS, Ka J-W, Jen AK-Y, *Appl Phys Lett* 2006, **88**, 093505.
198. Muller RN, Van der Elst L, Laurent S, *J Am Chem Soc* 2003, **125**, 8405.

199. Kahn O, Martinez CJ, *Science* 1998, **279**, 44.
200. Reger DL, Gardinier JR, Smith MD, Shahin AM, Long GJ, Rebbouh L, Grandjean F, *Inorg Chem* 2005, **44**, 1852.
201. Reger DL, Elgin JD, Smith MD, Grandjean F, Rebbouh L, Long GJ, *Polyhedron* 2006, **25**, 2616
202. Reger DL, Gardinier JR, Gemmill WR, Smith MD, Shahin AM, Long GJ, Rebbouh L, Grandjean F, *J Am Chem Soc* 2005, **127**, 2303.
203. Reger DL, Gardinier JR, Elgin JD, Smith MD, Hautot D, Long GJ, Grandjean F, *Inorg Chem* 2006, **45**, 8862.
204. Marder SR, Perry JW, Organic metallo-organic and polymeric materials for nonlinear optical applications, *Proc SPIE*, The International Society for Optical Engineering, Washington, DC, 1994.
205. Campo JA, Cano M, Heras JV, Lagunas MC, Pinilla E, Torres MR, *Polyhedron* 2001, **20**, 2997.
206. Rojo G, Agulló-López F, Campo JA, Cano M, Lagunas MC, Heras JV, *Synth Met* 2001, **124**, 201.
207. López-Garabito C, Campo JA, Heras JV, Cano M, Rojo G, Agulló-López F, *J Phys Chem B* 1998, **102**, 10698.
208. Campo JA, Cano M, Heras JV, López-Garabito C, Pinilla E, Torres R, Rojo G, Agulló-Lopez F, *J Mater Chem* 1999, **9**, 899.
209. Babic-Samardzija K, Hackerman N, *Anti Corros Methods Mater* 2006, **53**, 19.
210. Shipley CP, Capecchi S, Salata OV, Etchells M, Dobson PJ, Christou V, *Adv Mater* 1999, **11**, 533.
211. Niedermair F, Waich K, Kappaun S, Mayr T, Trimmel G, Mereiter K, Slugovc C, *Inorg Chim Acta* 2007, **360**, 2767.
212. Harrison BS, Foley TJ, Knefely AS, Mwaura JK, Cunningham GB, Kang T-S, Bouguettaya M, Boncella JM, Reynolds JR, Schanze KS, *Chem Mater* 2004, **16**, 2938.
213. Schanze KS, Reynolds JR, Boncella JM, Harrison BS, Foley TJ, Bouguettaya M, Kang T-S, *Synth Metals* 2003, **137**, 1013.
214. Mwaura JK, Knefely AS, Schanze KS, Boncella JM, Reynolds JR, *Polym Preprints* 2004, **45**, 907.
215. Malandrino G, Lo Nigro R, Fragalà IL, *Inorg Chim Acta* 2007, **360**, 1138.
216. Pettinari C, Marchetti F, Santini C, Pettinari R, Drozdov A, Troyanov S, Battiston GA, Gerbasi R, *Inorg Chim Acta* 2001, **315**, 88.
217. Depperman EC, Bodnar SH, Vostrikova KE, Shultz DA, Kirk ML, *J Am Chem Soc* 2001, **123**, 3133.

218. Shultz DA, Bodnar SH, *Inorg Chem* 1999, **38**, 591.
219. Shultz DA, Bodnar SH, Kumar RK, Lee H, Kampf JW, *Inorg Chem* 2001, **40**, 546.
220. Shultz DA, Sloop JC, Coote T-A, Beikmohammadi M, Kampf J, Boyle PD, *Inorg Chem* 2007, **46**, 273.
221. Omary MA, Rawashdeh-Omary MA, Diyabalanage HVK, Dias HVR, *Inorg Chem* 2003, **42**, 8612.
222. Sessoli R, Gatteschi D, *Angew Chem Int Ed* 2003, **42**, 268.
223. Christou G, Gatteschi D, Hendrickson DN, Sessoli R, *MRS Bull* 2000, **25**, 66.
224. Krusin-Elbaum L, Shibauchi T, Argyle B, Gignac L, Weller D, *Nature* 2001, **410**, 444.
225. Tejada J, Chudnovsky EM, del Barco E, Hernandez JM, Spiller TP, *Nanotechnology* 2001, **12**, 181.
226. Kim G-H, Kim T-S, *Phys Rev Lett* 2004, **92**, 137203.
227. Lescouezec R, Vaissermann J, Lloret F, Julve M, Verdaguer M, *Inorg Chem* 2002, **41**, 5943.
228. Kim J, Han S, Cho I-K, Choi KY, Heu M, Yoon S, Suh B-J, *Polyhedron* 2004, **23**, 1333.
229. Kim J, Han S, Pokhodnya K-i, Migliori JM, Miller JS, *Inorg Chem* 2005, **44**, 6983.
230. Wen H-R, Wang C-F, Song Y, Gao S, Zuo JL, You X-Z, *Inorg Chem* 2006, **45**, 8942.
231. Wang S, Zuo J-L, Zhou H-C, Choi HJ, Ke Y, Long JR, You X-Z, *Angew Chem Int Ed* 2004, **43**, 5940.
232. Tao J-Q, Wang T-W, Gu Z-G, Zuo J-L, You X-Z, *Wuji Huaxue Xuebao* 2006, **22**, 2207.
233. Wang C-F, Zuo J-L, Bartlett BM, Song Y, Long JR, You X-Z, *J Am Chem Soc* 2006, **128**, 7162.
234. Kitano T, Sohrin Y, Hata Y, Wada H, Hori T, Ueda K, *Bull Chem Soc Jpn* 2003, **76**, 1365.
235. T. Kitano, Wada H, Mukai H, Ueda K, Sohrin Y, *Anal Sci* 2001, **17**, i1113.
236. Kitano T, Sohrin Y, Hata Y, Kawakami H, Hori T, Ueda K, *J Chem Soc, Dalton Trans* 2001, 3564.
237. Shukla R, Rao GN, *Talanta* 2002, **57**, 633.
238. Caneschi S, Dei A, Gatteschi D, Poussereau S, Sorace L, *Dalton Trans* 2004, 1048.

239. Xing Y-H, Bai F-Y, Zhang S-J, Aoki K, *Jilin Huagong Xueyuan Xuebao* 2003, **20**, 6.
240. Zou W, Song J-Y, Xing Y-H, *Yingyong Huaxue* 2006, **23**, 94.
241. Santini C, Pellei M, Gioia Lobbia G, Fedeli D, Falcioni G, *J Inorg Biochem* 2003, **94**, 348.
242. Pellei M, Gioia Lobbia G, Santini C, Spagna R, Camalli M, Fedeli D, Falcioni G, *Dalton Trans* 2004, 2822.
243. Van Waasbergen LG, Fajdetic I, Fianchini M, Dias HVR, *J Inorg Biochem* 2007, **101**, 1180.
244. Santos I, Paulo A, Correia JD, *Top Curr Chem* 2005, **252**, 45.
245. Krzystek J, Zvyagin SA, Ozarowski A, Trofimenko S, Telser J, *J Magn Reson* 2006, **178**, 174.

Chapter 9

Concluding Remarks

Although the pyrazolylborate ligands have been around for 40 years, there is still more to explore including ligands that can be tridentate to polydentate, as the "third generation" scorpionates, recently reported by Reger,[1,2] which can find application in the advanced materials field.

The "classical" polypyrazolylborates, while mature ligands, still have possibility to be expanded, especially as heteroscorpionates, containing pyrazolyl arms with additional donor atoms, as for example ligands bearing an ester function on the azolyl ring, hydrogen-bond accepting donors, suitable for host–guest chemistry.[3]

The area of nonclassical scorpionates can surely grow, including ligands containing also one or two pz^x groups, in view of the rich portfolio of available diversely substituted pyrazoles.

The coordination chemistry of the nonpyrazolyl scorpionate ligands, containing a boron core, as $[Ph_2B(CH_2Z)_2]^-$ and $[PhB(CH_2Z)_3]^-$, where Z can be a variety of donor functionalities, i.e. SR,[4] PPh_2,[5,6] and NMe_2,[7] is at beginning, only a limited number of metal complexes being to date investigated.

While chelates derived from polypyrazolylborates involve six-membered rings between boron and the coordinated metal, other ligands lead to eight-membered rings. These ligands known as Tm^R,[8] and Tr^R,[9] surely will lead to interesting developments in emerging disciplines such as bioinorganic chemistry.[10,11] Imidazolyl-[12,13]

and triazolyl-borate[14] ligands, which lead to many extended structures, could be promising candidates for supramolecular engineering. Alkylborate compounds containing imidazolyl functional groups such as $[MeB(Im^{N-Me})_2X]$[15] able to act as a proton acceptors and metal-binding site have been scarcely investigated. In addition, tris-carbene borates such as those described by Smith *et al.*,[16–18] characterized by an exceptionally strong donating nature, could represent a valid alternative to Tp^x.

The neutral analogs polypyrazolylalkanes, introduced in 1970,[19] not reported here but discussed recently in two reviews in *Coordination Chemistry Reviews*,[20,21] have received great interest in the recent period; however, their coordination chemistry has far to go to be developed as that of the anionic cousins Tp^x.

The dinegative polypyrazolylberyllate ligands, $[R_nBe(pz^x)_{4-n}]^{2-}$ which should yield neutral complexes, could be a promising class of chelating ligands and together with $[R_2Al(pz)_2]^-$ and $[R_2In(pz)_2]^-$, could exhibit an interesting coordination chemistry.

References

1. Reger DL, Gardinier JR, Smith MD, Shahin AM, Long GJ, Rebbouh L, Grandjean F, *Inorg Chem* 2005, **44**, 1852.
2. Reger DL, Gardinier JR, Bakbak S, Semeniuc RF, Bunz UHF, Smith MD, *New J Chem* 2005, **29**, 1035.
3. Hammes BS, Carrano MW, Carrano CJ, *J Chem Soc Dalton Trans* 2001, 1448.
4. Ge P, Haggerty BS, Rheingold AL, Riordan CG, *J Am Chem Soc* 1994, **116**, 8406.
5. Peters JC, Feldman JD, Don Tilley T, *J Am Chem Soc* 1991, **121**, 9871.
6. Shapiro IR, Jenkins DM, Thomas JC, Day MW, Peters JC, *Chem Commun* 2001, 2152.
7. Betley TA, Peters JC, *Inorg Chem* 2002, **41**, 6541.
8. Garner M, Reglinski J, Cassidy I, Spicer MD, Kennedy AR, *J Chem Soc Chem Commun* 1996, 1975.
9. Bailey PJ, Lanfranchi M, Marchiò L, Parsons S, *Inorg Chem* 2001, **40**, 5030.

10. Rombach M, Seebacher J, Ji M, Zhang G, He G, Ibrahim MM, Benkmil B, Vahrenkamp H, *Inorg Chem* 2006, **45**, 4571.
11. Melnick JG, Parkin G, *Science* 2007, **317**, 225.
12. Janiak C, Temidzemir S, Röhr C, *Z Anorg Allg Chem* 2000, **626**, 1265.
13. Effendy, Gioia Lobbia G, Pellei M, Santini C, Skelton BW, White AH, *J Chem Soc Dalton Trans* 2001, 528.
14. Gioia Lobbia G, Pellei M, Pettinari C, Santini C, Skelton BW, Somers N, White AH, *J Chem Soc Dalton Trans* 2002, 2333.
15. Fujita K, Hikichi S, Akita M, Moro-oka Y, *J Chem Soc Dalton Trans* 2000, 1255.
16. Nieto I, Cervantes-Lee F, Smith JM, *Chem Commun* 2005, 3811.
17. Cowley RE, Bontchev RP, Duesler EN, Smith JM, *Inorg Chem* 2006, **45**, 9771.
18. Forshaw AP, Bontchev RP, Smith JM, *Inorg Chem* 2007, **46**, 3792.
19. Trofimenko S, *J Am Chem Soc* 1970, **92**, 5118.
20. Pettinari C, Pettinari R, *Coord Chem Rev* 2005, **249**, 525.
21. Pettinari C, Pettinari R, *Coord Chem Rev* 2005, **249**, 663.

APPENDIX

List of Recent Papers Not Quoted in Chapters 1–9

Title	Subject	Ref.
Why does nature use zinc — a personal view	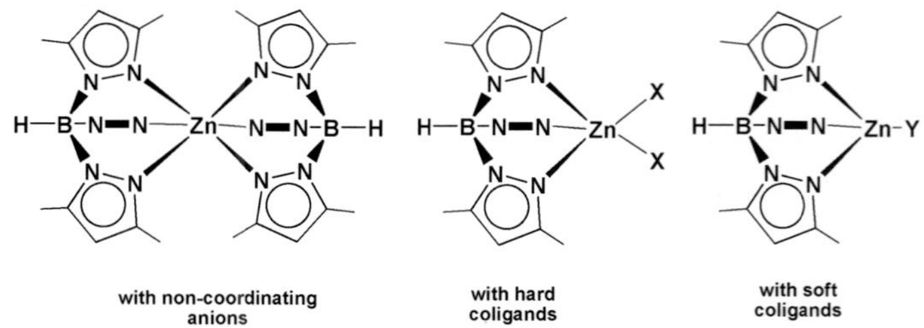	1
(*p*-methylbenzenethiolato) [tris(3-phenyl-5-methylpyrazolyl)borato] zinc(II)	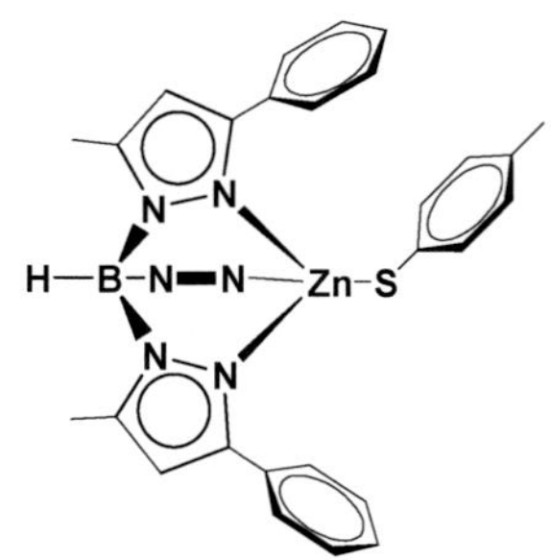	2

(*Continued*)

Title	Subject	Ref.
Synthesis, structure, and hydrothiolation activity of rhodium pyrazolylborate complexes	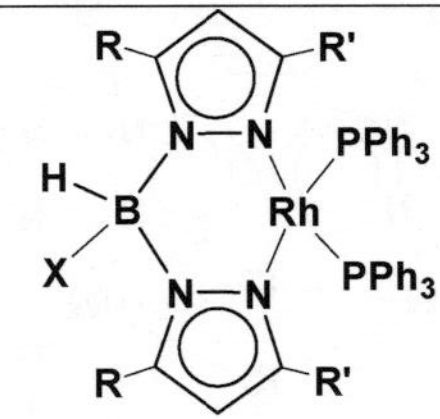 X = H or pz^x, R = H or Me, R' = H, Me or Ph	3
Labeling of the neuropeptide enkephalin with functionalized tris(pyrazolyl)borate complexes: Solid-phase synthesis and characterization of p-[Enk-OH]-$COC_6H_4Tp^{PtMe_3}$ and p-[Enk-OH]-$COC_6H_4Tp^{Me}Re(CO)_3$	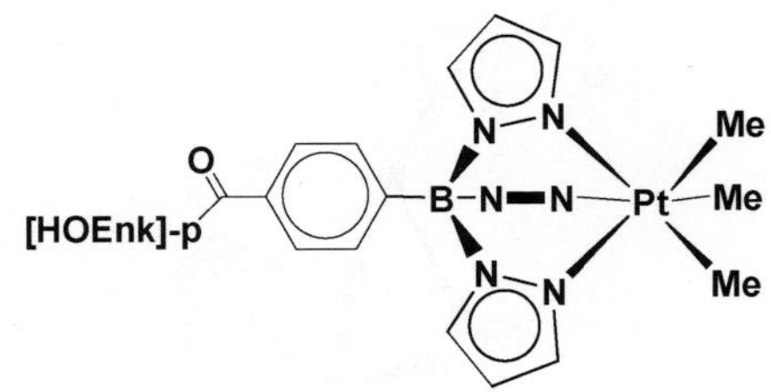	4

(*Continued*)

Title	Subject	Ref.
Catalysis design for ring-opening polymerization of cyclic esters: 1. Group 1 metal and thallium(I) trispyrazolylborate complexes with hemilabile ligands	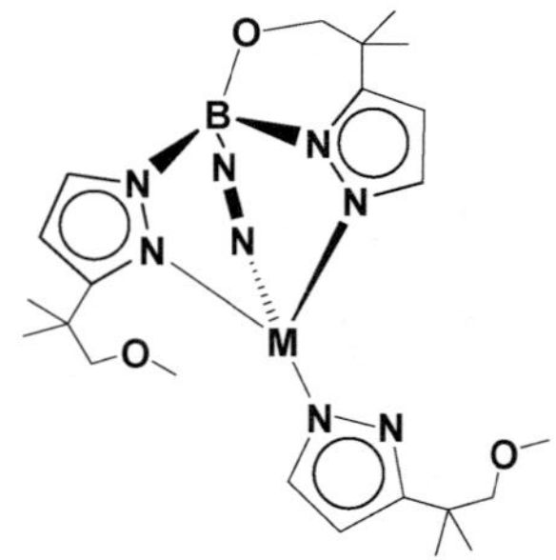	5
Metallomacrocycles from ditopic chiral scorpionate ligands	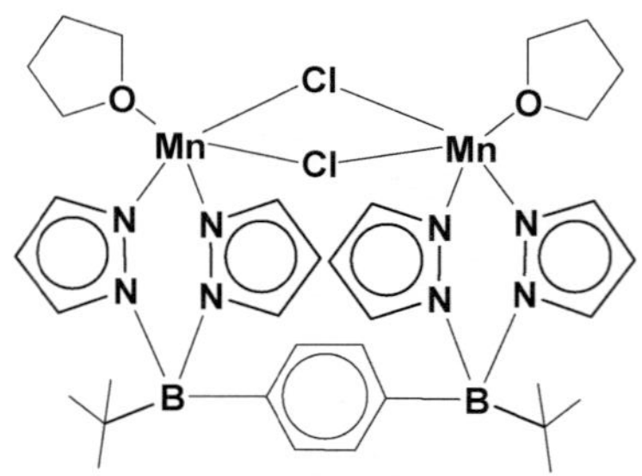	6

(*Continued*)

Title	Subject	Ref.
Small scorpionate ligands: Silver(I) organophosphane complexes of 5-CF_3-substituted scorpionate ligand combining a B-H…Ag coordination motif	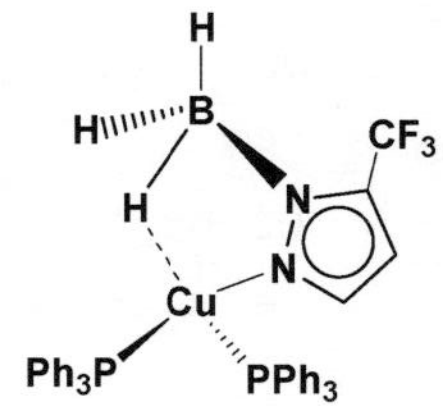	7
Copper-homoscorpionate complexes as active catalysts for atom transfer radical addition to olefins	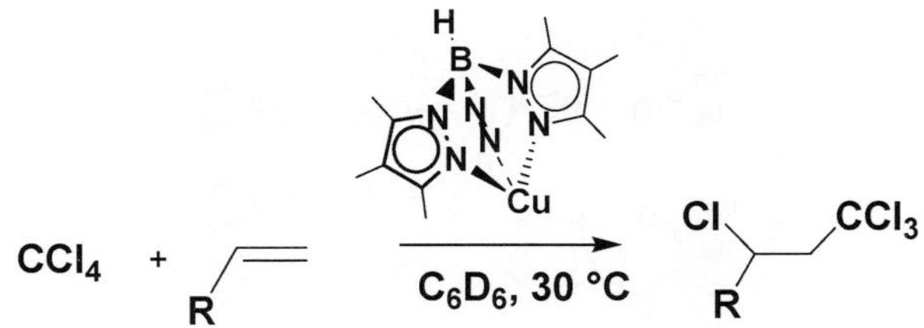	8

(*Continued*)

Title	Subject	Ref.
A novel tetrakis(pyrazolyl)borate ligand bearing triphenylphosphine oxide substituents	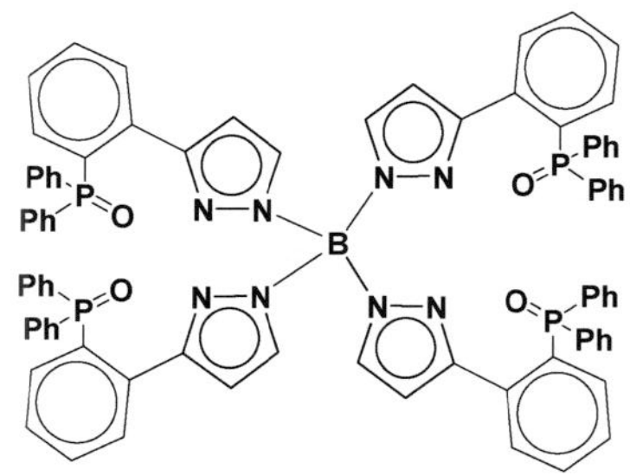	9
A theoretical and experimental study of the fluxional behaviour of molybdenum dihydrobis- and hydrotris-pyrazolylborates	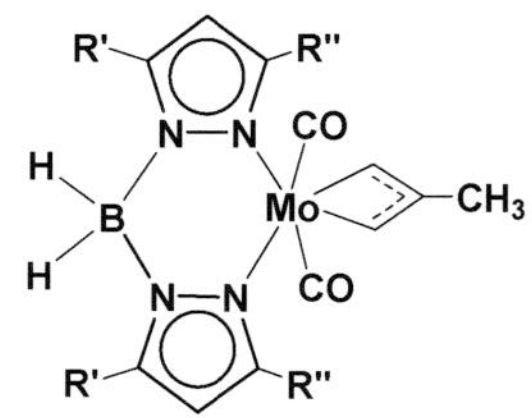	10

(*Continued*)

Title	Subject	Ref.
Trapping of anionic organic radicals by $(Tp^{Me_2})_2Ln$ (Ln = Sm, Eu)	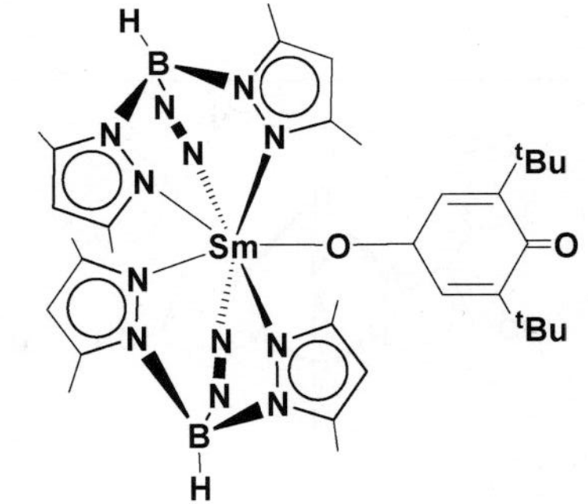	11
Preparation and reactivity with azo-species of hydride and dihydrogen complexes of osmium stabilised by tris(pyrazolyl)borate and phosphite ligands	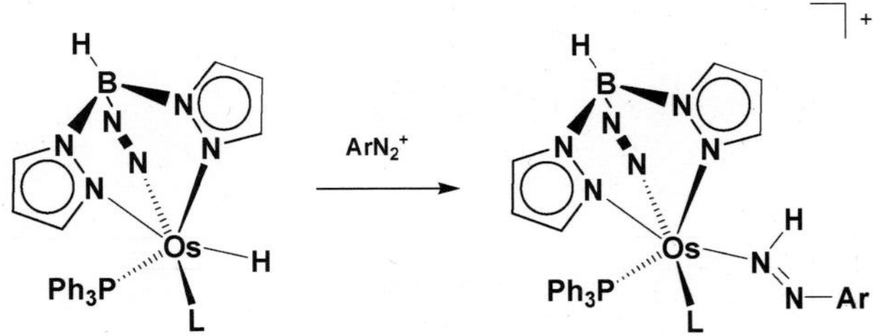	12

(*Continued*)

Title	Subject	Ref.
Heteroleptic Tm(II) complexes: One more success for Trofimenko's scorpionates	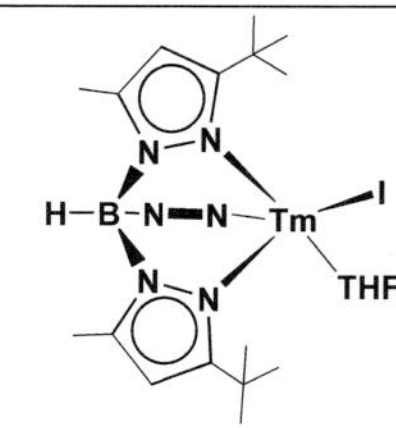	13
Sandwich compounds of cyanotrispyrazolylborates: Complexation-induced ligand isomerization	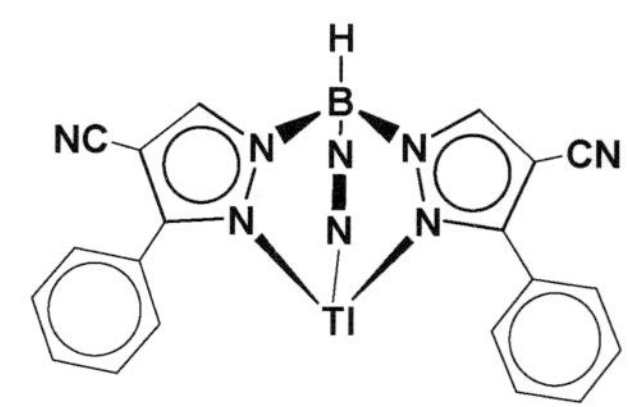	14

(*Continued*)

Title	Subject	Ref.
Manganese complexes as models for manganese-containing pseudocatalase enzymes: Synthesis, structural and catalytic activity studies	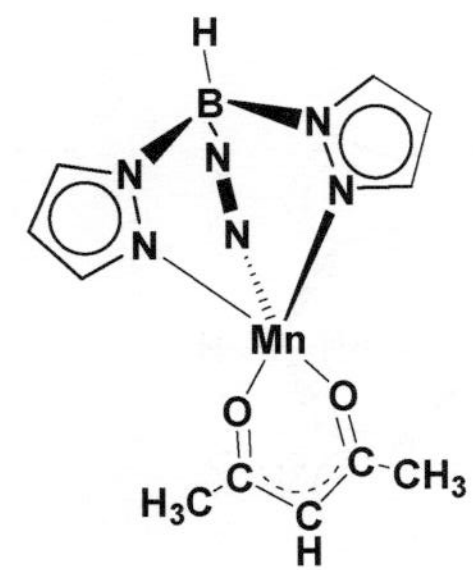 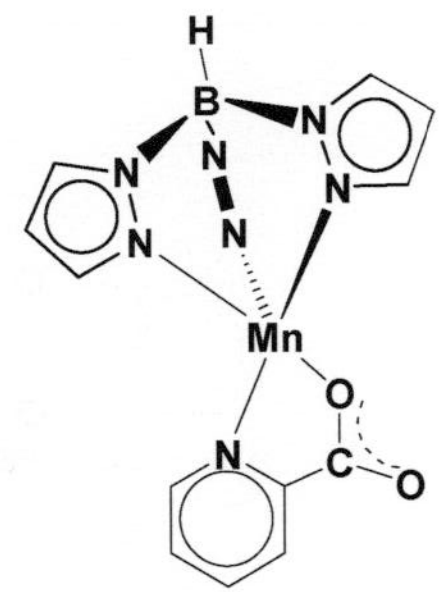	15
Toward multifunctional Mo(VI-IV) complexes: cis-Dioxomolybdenum (VI) complexes containing hydrogen-bond acceptors or donors	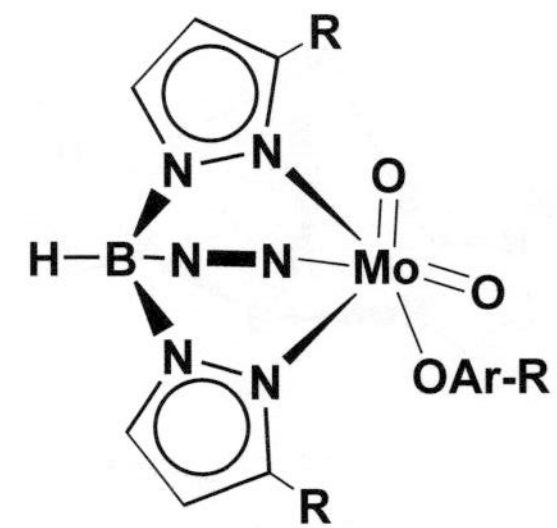	16

(*Continued*)

Title	Subject	Ref.
Nickel nitrosyl complexes in a sulfur-rich environment: The first poly(mercaptoimidazolyl)-borate derivatives	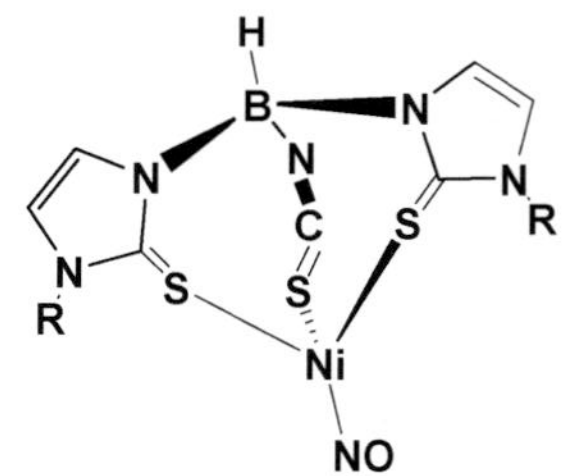	17
Mononuclear cobalt(II) carboxylate complexes: Synthesis, molecular structure and selective oxygenation study	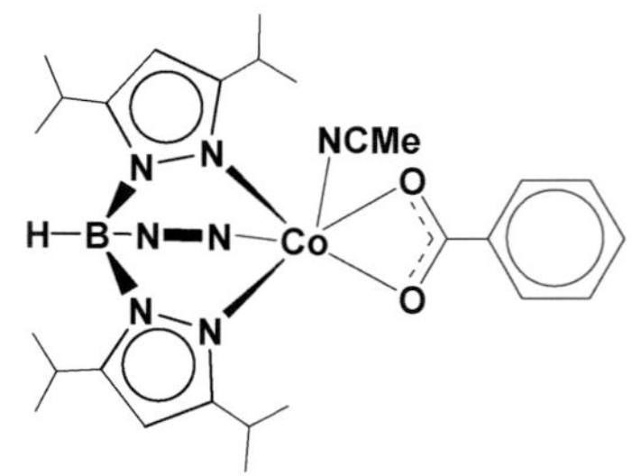	18

(*Continued*)

Title	Subject	Ref.
Silver(I)-organophosphane complexes of electron withdrawing CF_3- or NO_2-substituted scorpionate ligands	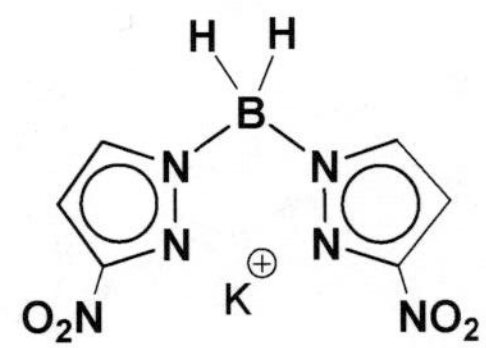	19
Characterization of polyethylene nascent powders synthesized with $TpTiCl_2(OR)$ catalysts	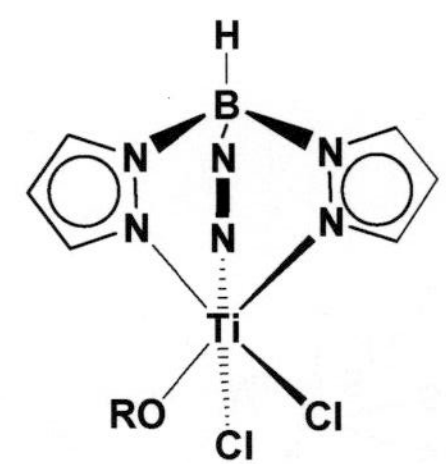	20

(Continued)

Title	Subject	Ref.
Hydroxide-promoted core conversions of molybdenum-iron-sulfur edge-bridged double cubanes: Oxygen-ligated topological P^N clusters	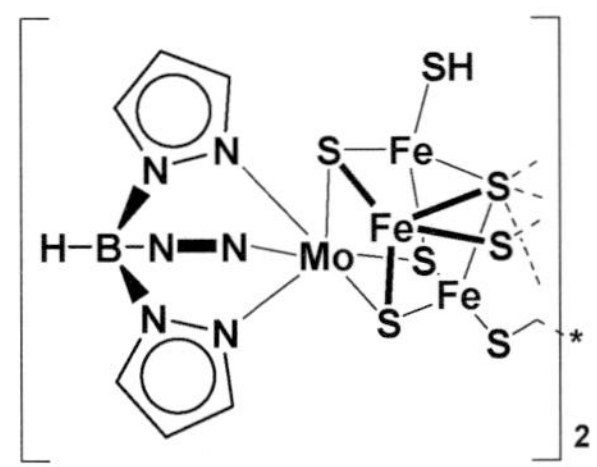	21
A homoleptic hydrotris(methimazolyl)borate complex of titanium	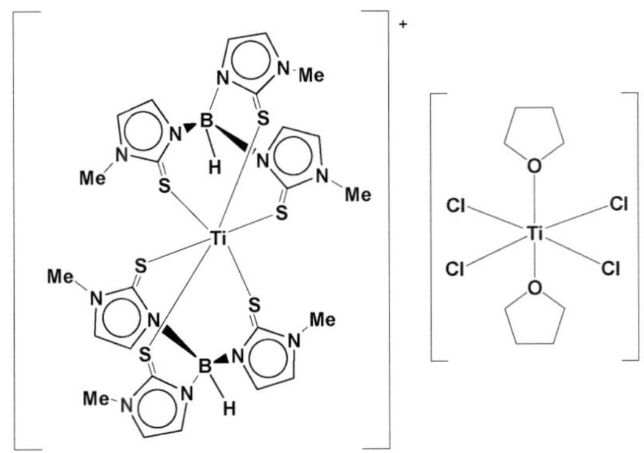	22

(Continued)

Title	Subject	Ref.
Spin-state energetics and spin-crossover behavior of pseudo-tetrahedral cobalt(III)-imido complexes. The role of the tripodal supporting ligand	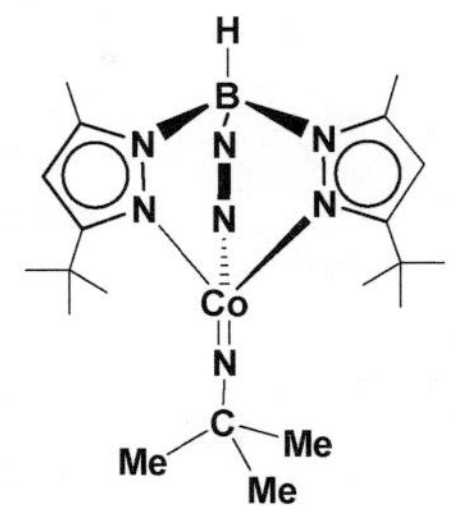	23
Insulin-like model complexes: Synthesis, spectra characterization and crystal structure of two novel oxovanadium complexes with poly(pyrazolyl)borate ligands $VO(HB(pz)_3)(H_2B(pz)_2)$ and $VO(B(pz)_4)_2$	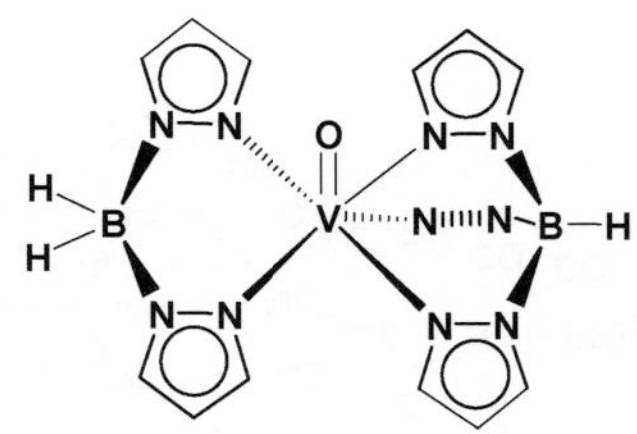	24

(*Continued*)

Title	Subject	Ref.
Cyanide-bridged linkage isomers with catecholateruthenium(II) centres bound to Mn(I) or M(alkyne) units	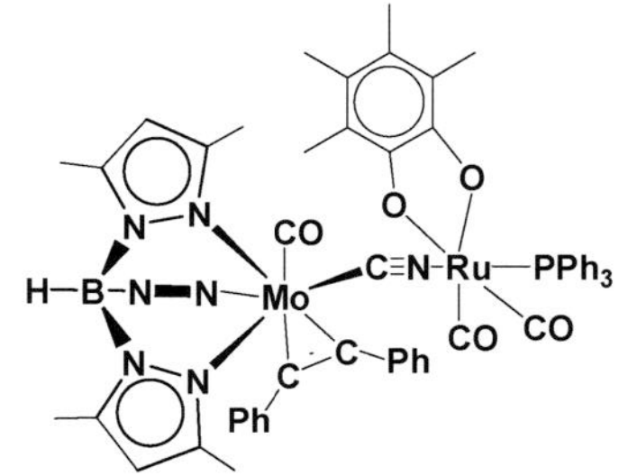	25
Synthesis, reactivity, X-ray crystal structures and electrochemical behavior of water-soluble [tris(pyrazolyl)-borato]ruthenium(II) complexes of 1,3,5-triaza-7-phosphaada-mantane (PTA)	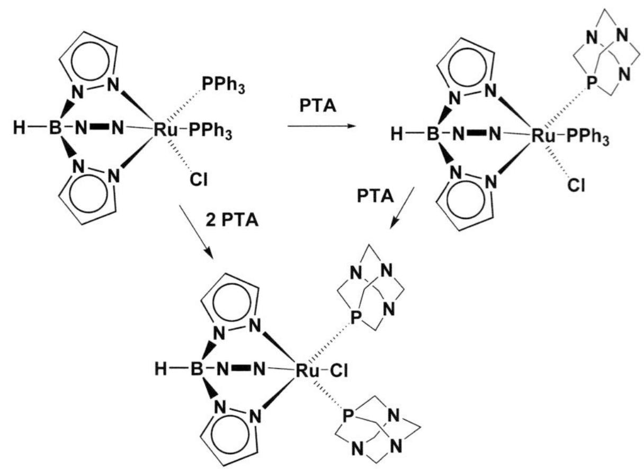	26

(Continued)

Title	Subject	Ref.
Reactivity of $TpRu(L)(NCMe)R$ (L = CO, PMe_3; R = Me, Ph) systems with isonitriles: Experimental and computational studies toward the intra- and intermolecular hydroarylation of isonitriles	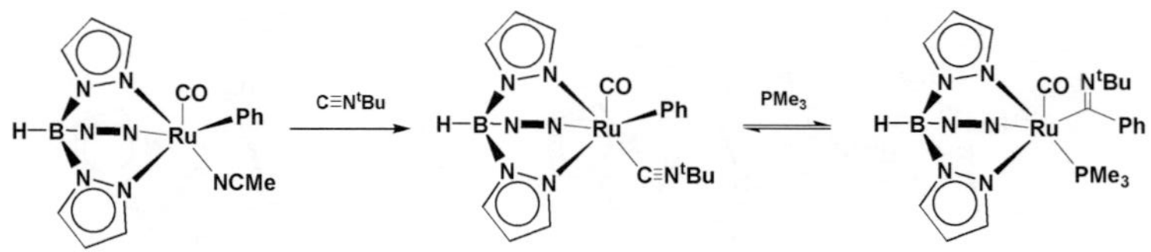	27
Synthesis, crystal structure, and reactivity of a four-coordinate iron(II) chloride complex bearing the sterically demanding tert-butyl-tris(3-isopropylpyrazolyl)borato ligand	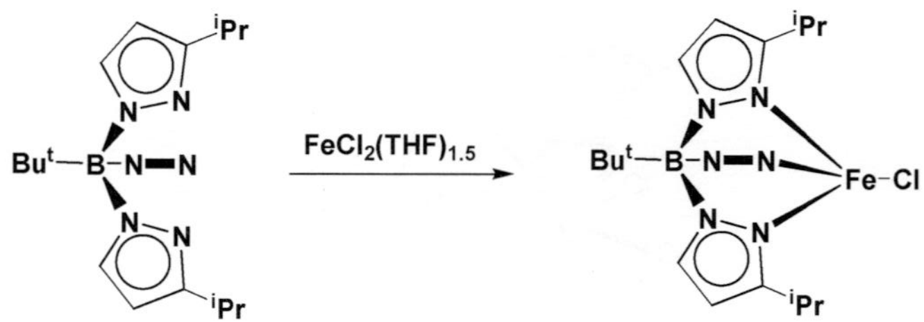	28

(*Continued*)

Title	Subject	Ref.
Ethylene polymerization using tris(pyrazolyl)borate titanium(IV) catalyst supported *in situ* on MAO-modified silica	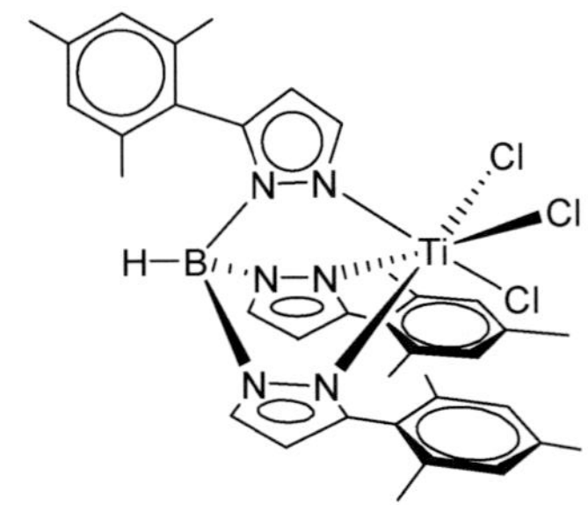	29
TpRu complexes as recoverable Lewis acid catalysts for enantioselective solvent-free cycloaddition reactions (Tp = hydrotris(pyrazolyl)borate)	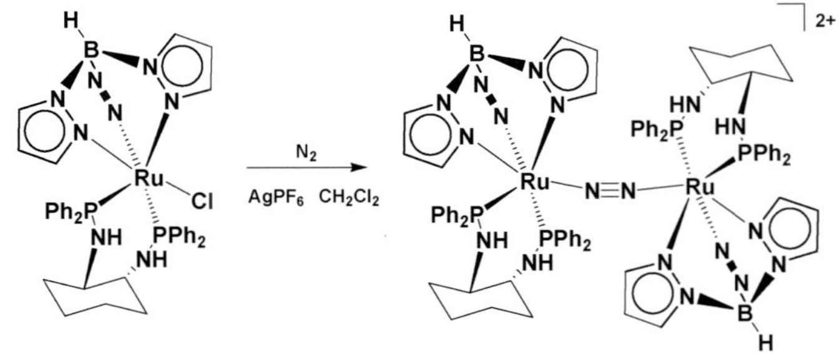	30

(Continued)

Title	Subject	Ref.
Probing the mechanism of carbon-hydrogen bond activation by photochemically generated hydridotris-(pyrazolyl)borato carbonyl rhodium complexes: New experimental and theoretical investigations	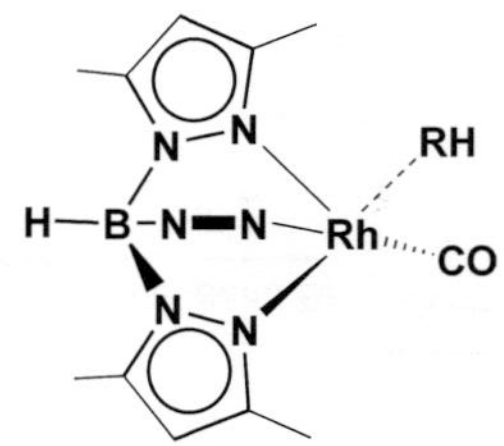	31
Redox activity of bimetallic complexes based on tris(3,5-dimethylpyrazolyl)-borato molybdenum(II) and tungsten(II) nitrosyls. Complexes derived from ethane-1,2-diol and the crystal structure of anti $[Mo(NO)\{HB(3,5\text{-}Me_2C_3HN_2)_3\}\text{-}(OCH_2CH_2O)]_2 \cdot 4CHCl_3$	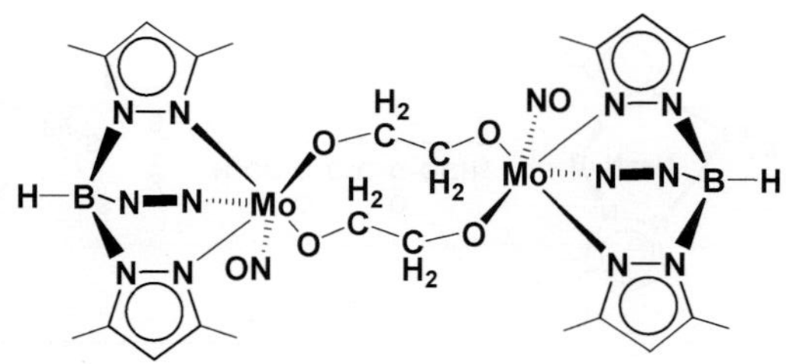	32

(*Continued*)

Title	Subject	Ref.
Synthesis, characterization, and spectroscopy of model molybdopterin complexes	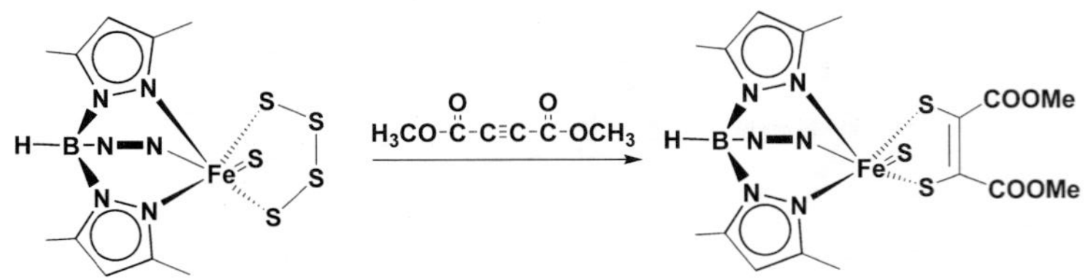	33
Dendronized scorpionate complexes of molybdenum in low and high oxidation state	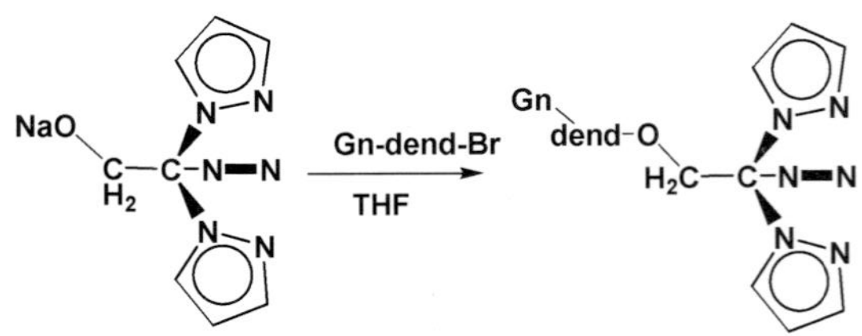	34

(*Continued*)

Title	Subject	Ref.
Applications of tripodal [S_3] and [Se_3] L_2X donor ligands to zinc, cadmium and mercury chemistry: Organometallic and bioinorganic perspectives	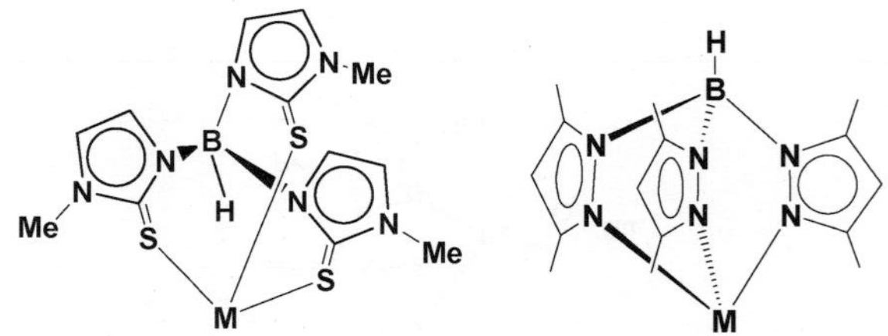	35
Activation of carbon-hydrogen bonds via 1,2-addition across M-X (X = OH or NH_2) bonds of d^6 transition metals as a potential key step in hydrocarbon fuctionalization: A computational study	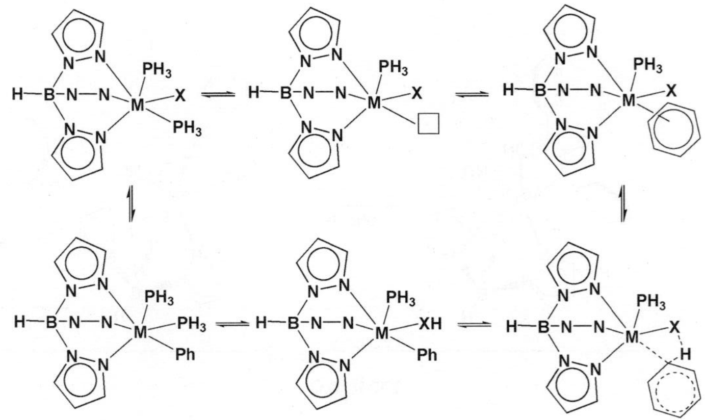	36

(*Continued*)

Title	Subject	Ref.
Syntheses of four-membered metallacyclic complexes with nitrosylruthenium and their ring-opening upon HCl addition	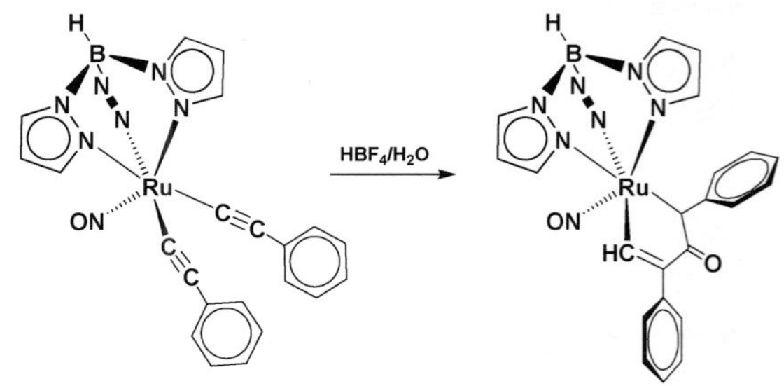	37
X-ray absorption spectroscopic studies of high-spin nonheme (alkylperoxo)iron(III) intermediates	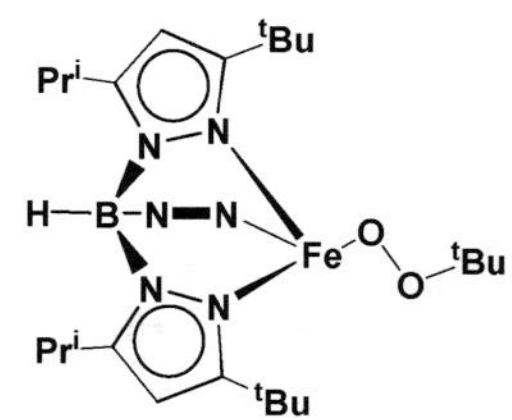	38

(*Continued*)

Title	Subject	Ref.
Modeling the inhibition of peptide deformylase by hydroxamic acids: Influence of the sulfur donor	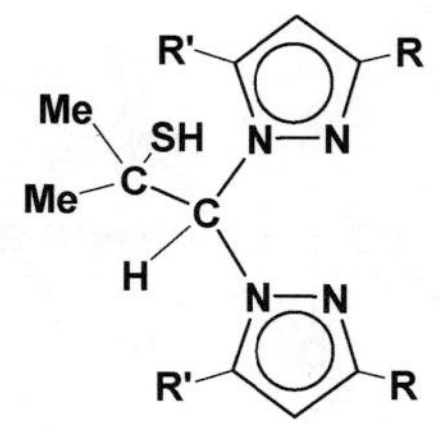 R = ^{i}Pr, R' = H; R = Ph, R' = H, R = R' =Me	39
Thermally stable gold(I) ethylene adducts: $[(HB\{3,5\text{-}(CF_3)_2Pz\}_3)Au(CH_2{=}CH_2)]$ and $[(HB\{3\text{-}(CF_3),5\text{-}(Ph)Pz\}_3)\text{-}Au(CH_2{=}CH_2)]$	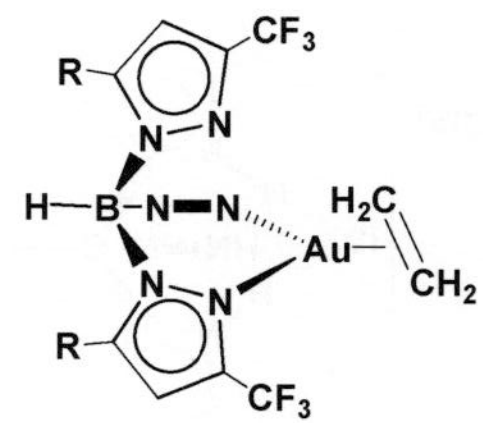 R = CF_3, Ph	40

(*Continued*)

Title	Subject	Ref.
Organometallic complexes as anion hosts (Anionic tris(pyrazolyl)borate ligands and cationic tris(pyrazole) receptors)	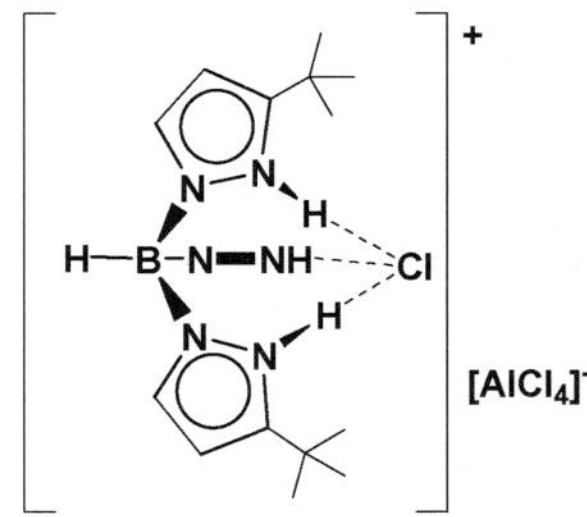	41
Hybrid density functional study of ligand coordination effects on the magnetic couplings and the dioxygen binding of the models of hemocyanin	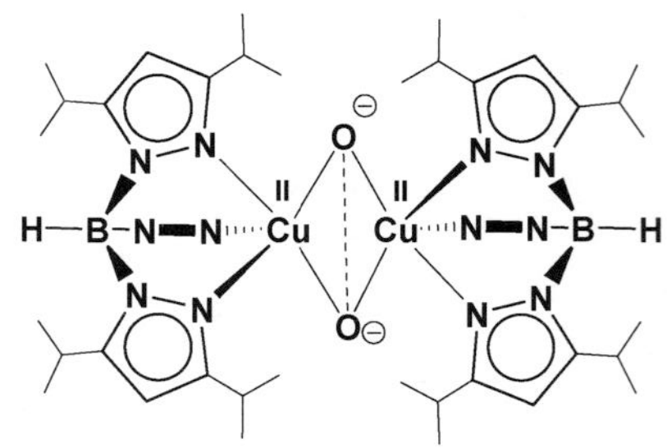	42

(Continued)

Title	Subject	Ref.
An amino acid bioconjugate of an organoplatinum tris(pyrazolyl)borate complex: Synthesis and structure of [p-(tBuO–Phe–CO)C_6H_4Tp]$PtMe_3$	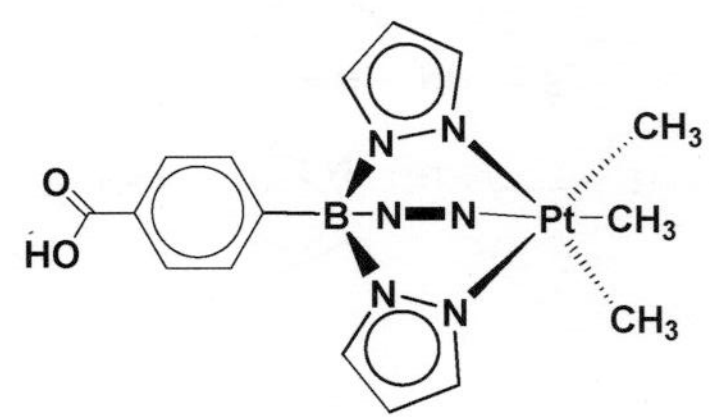	43
On the prospects of polynuclear complexes with acetylenedithiolate bridging units	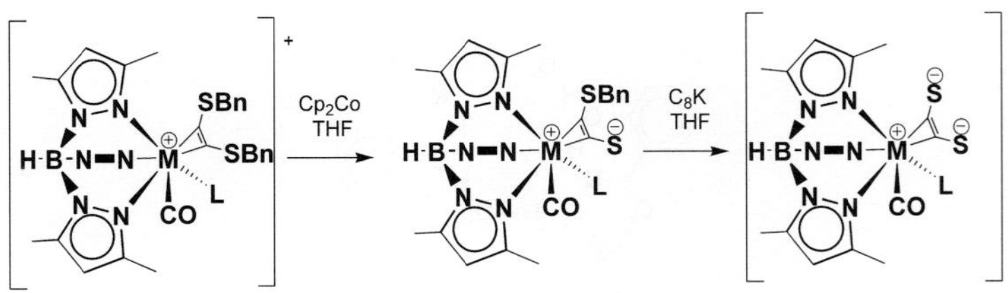	44

(*Continued*)

Title	Subject	Ref.
Enzyme inhibitor modeling with TpZn complexes of functional hydroxamates and oximates	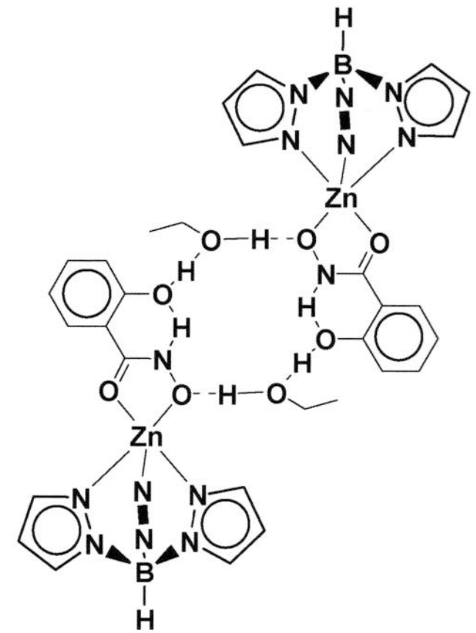	45
Nickel-cysteine binding supported by phosphine chelates	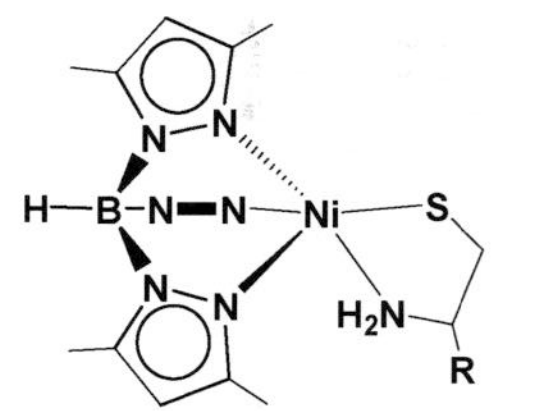 R = H, $-CO_2Et$, $-CO_2$	46

(*Continued*)

Title	Subject	Ref.
Structural, spectroscopic, and electrochemical properties of a series of high-spin thiolatonickel(II) complexes	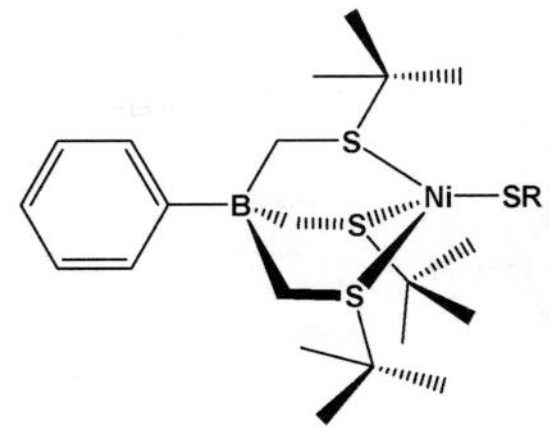	47
Experimental assessment of the relative affinities of benzene and ferrocene toward the Li^+ cation	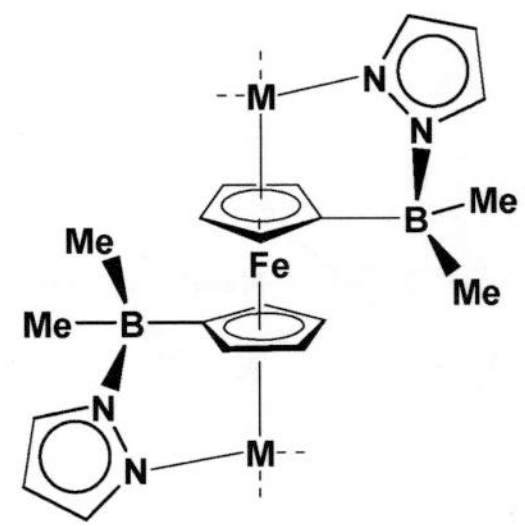	48

(*Continued*)

Title	Subject	Ref.
Carbon-hydrogen bond activation in hydridotris(pyrazolyl)borate platinum(IV) complexes: Comparison of density functionals, basis sets, and bonding patterns	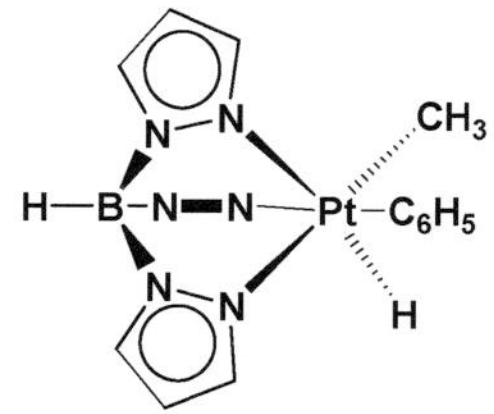	49
Combined experimental and computational study of TpRu{P(pyr)$_3$}(NCMe)Me (pyr = *N*-pyrrolyl): Inter- and intramolecular activation of C-H bonds and the impact of sterics on catalytic hydroarylation of olefins	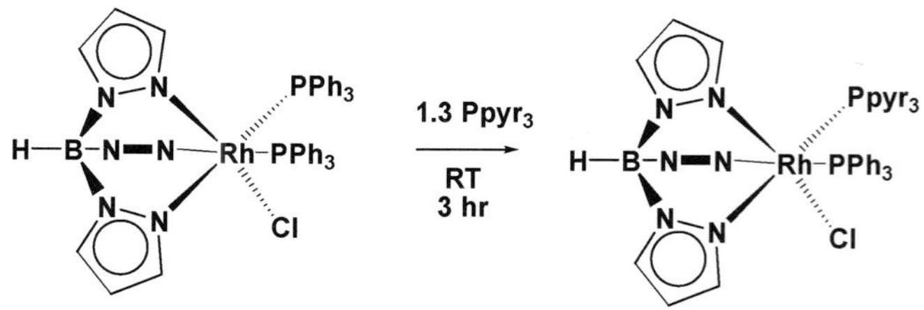	50

(*Continued*)

Title	Subject	Ref.
Reactions of a tungsten trisulfido complex of hydridotris(3,5-dimethylpyrazol-1-yl)borate (Tp*) $[Et_4N][Tp^*WS_3]$ with CuX (X = Cl, NCS, or CN): Isolation, structures, and third-order NLO properties	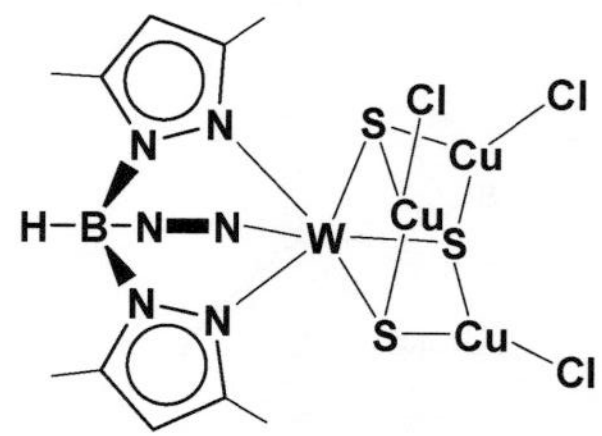	51
Organometallic tantalum tris(methimazolyl)borato complexes: $[Ta(\eta^2\text{-}RC{\equiv}CR)Cl_2\{HB(mt)_3\}]$ (R = Ph, Et; mt = methimazolyl)	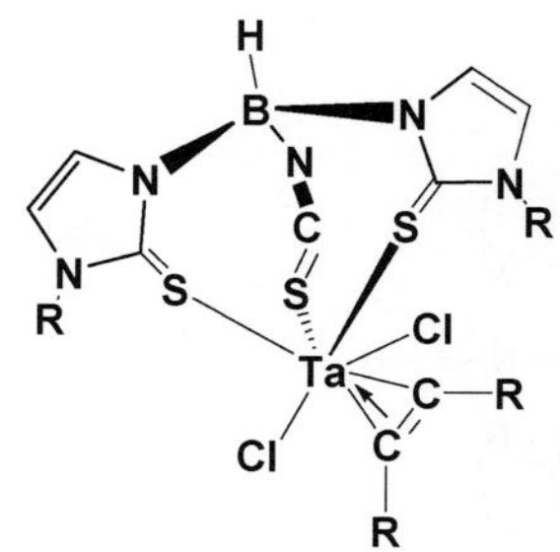	52

(*Continued*)

Title	Subject	Ref.
Understanding nonplanarity in metallabenzene complexes	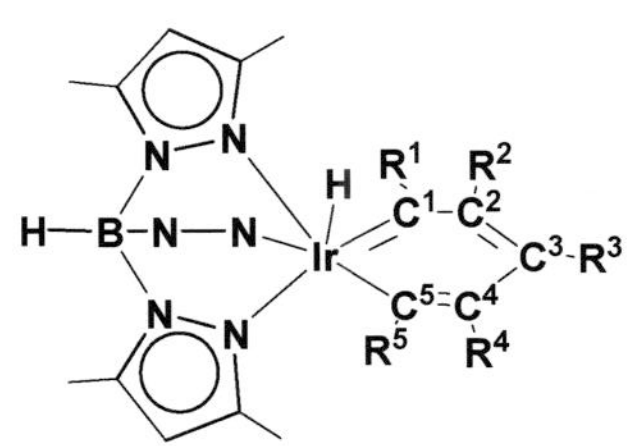 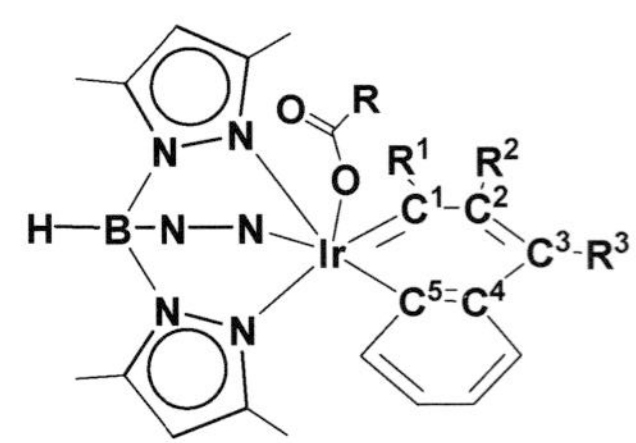	53
Investigations on the coupling of ethylene and alkynes in [IrTp^{Me_2}] compounds: Water as an effective trapping agent	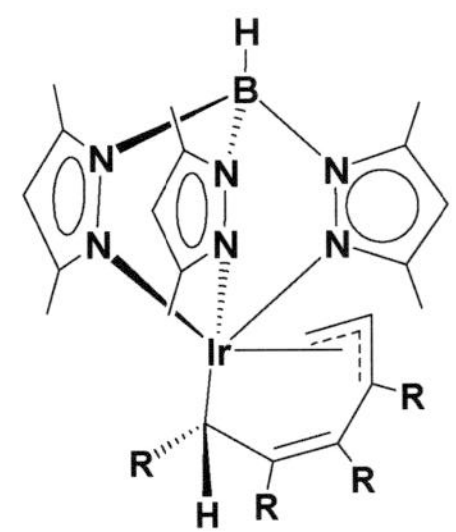 R = CO_2Me	54

(Continued)

Title	Subject	Ref.
Metallacycloheptatrienes of iridium(III): Synthesis and reactivity	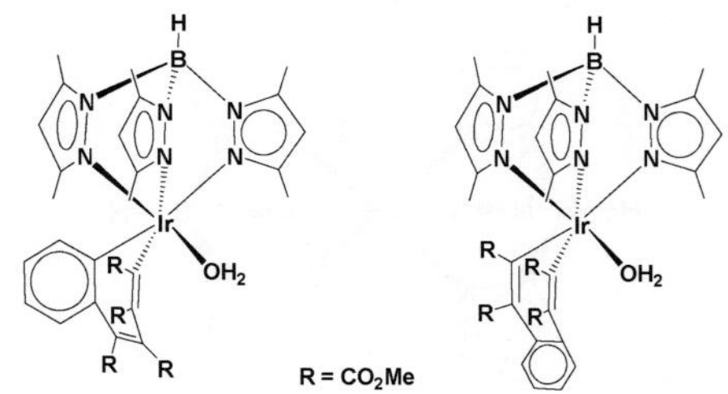	55
Rearrangement of pyridine to its 2-carbene tautomer mediated by iridium	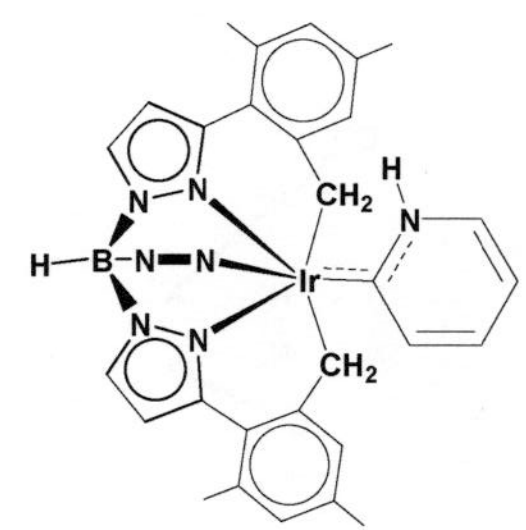	56

(*Continued*)

Title	Subject	Ref.
Transition metal migration upon attempting the Wolff rearrangement of an Ir(III) five-membered metallacycle	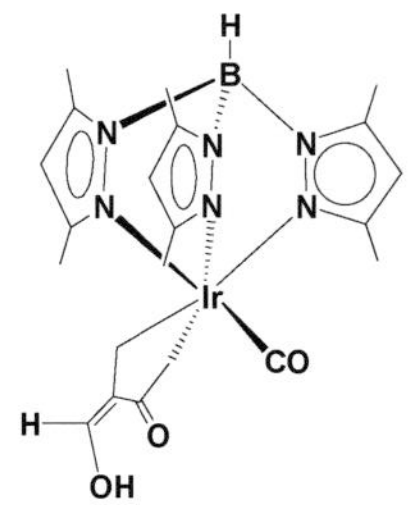	57
An MOCVD route to barium borate thin films from a barium hydro-tri(1-pyrazolyl)borate single-source precursor	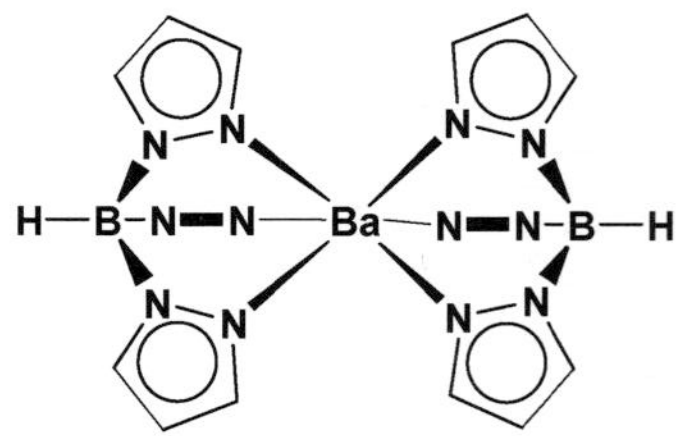	58

(*Continued*)

Title	Subject	Ref.
Combined experimental and computational studies on the nature of aromatic C–H activation by octahedral ruthenium(II) complexes: Evidence for σ-bond metathesis from Hammett studies	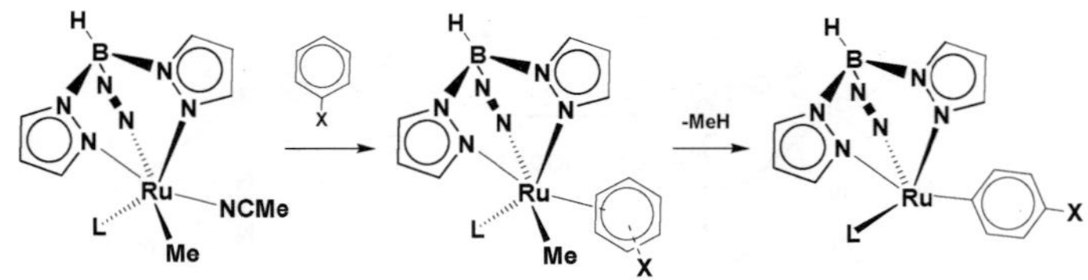	59
Mononuclear and binuclear sandwich copper(II) complexes with poly-(pyrazolyl)borate ligands: Syntheses, structures and properties	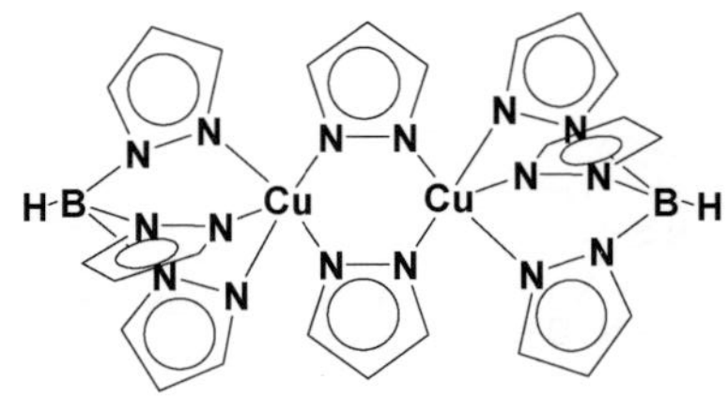	60

(*Continued*)

Title	Subject	Ref.
C–C bond-forming reactions of Ir^{III} alkenyls and nitriles or aldehydes: Generation of reactive hydride- and alkyl-alkylidene compounds and observation of a reversible 1,2-H shift in stable hydride-Ir^{III} alkylidene complexes	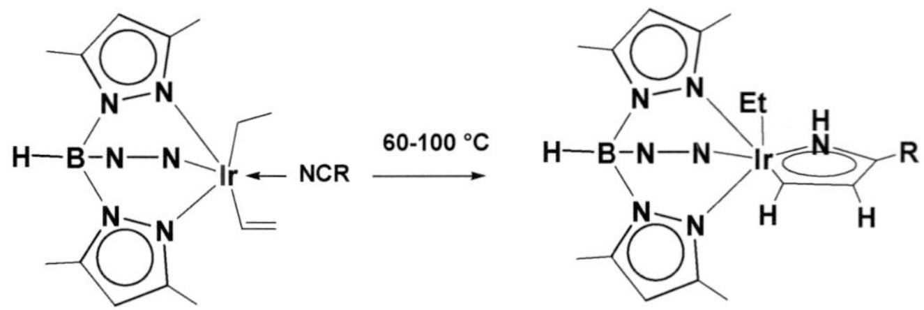	61
Formation of unusual iridabenzene and metallanaphthalene containing electron-withdrawing substituents	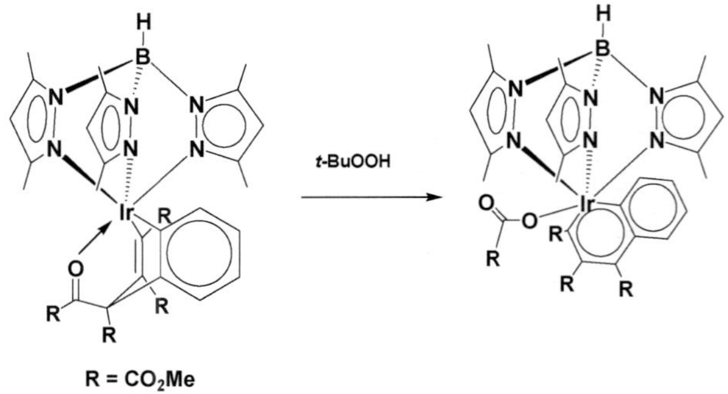	62

(*Continued*)

Title	Subject	Ref.
Coordination features of a hybrid scorpionate/phosphane ligand exemplified with iridium	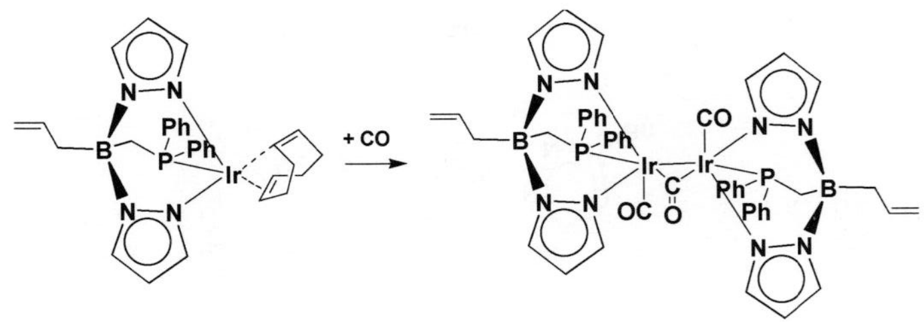	63
The synthesis of iridabenzenes by the coupling of iridacyclopentadienes and olefins	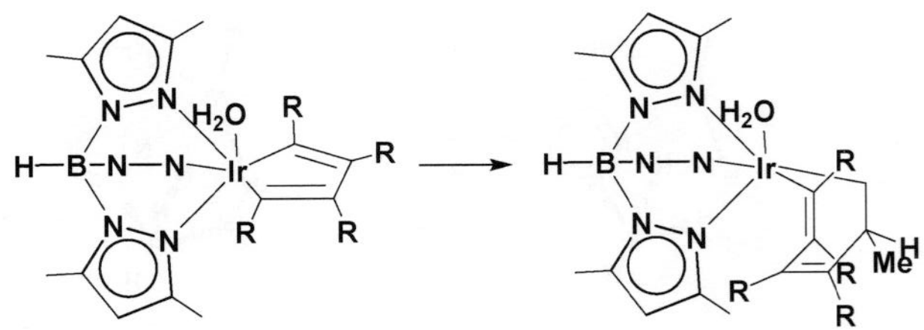	64

(*Continued*)

Title	Subject	Ref.
Facile intermolecular aryl-F bond cleavage in the presence of aryl C-H bonds: Is the η^2-arene intermediate bypassed?	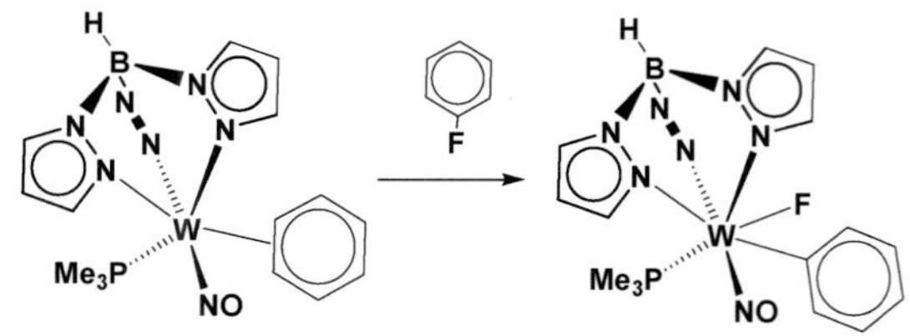	65
Sodium polypyrazolylaluminates: Synthesis, characterization, and isolation of a reaction intermediate of a trispyrazolylaluminate	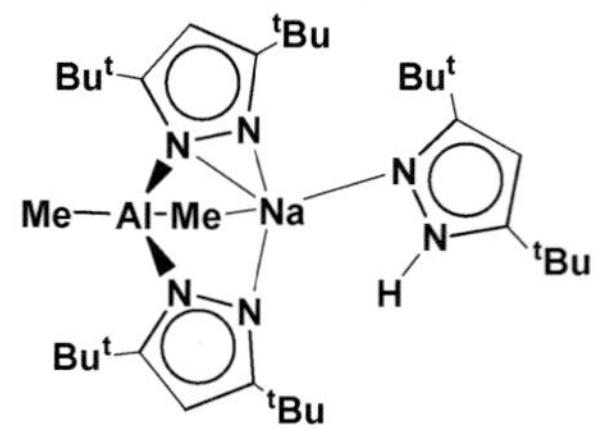	66

(*Continued*)

Title	Subject	Ref.
Ln(III) methyl and methylidene complexes stabilized by a bulky hydrotris(pyrazolyl)borate ligand	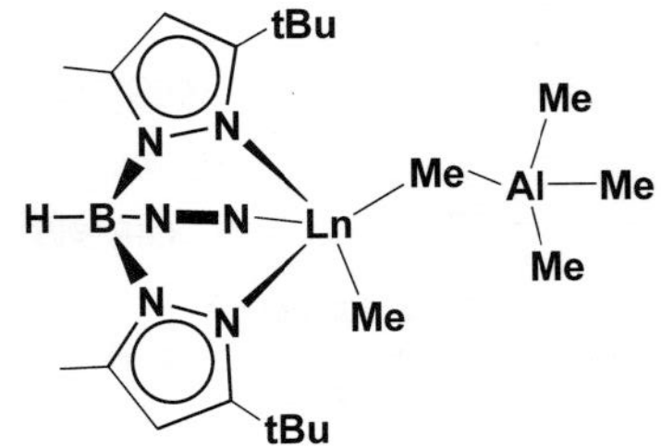	67
[Hydridotris(pyrazol-1-yl)borato-κ^3-N,N′,N″][5,10,15,20-tetrakis(4-methylphenyl)porphyrinato-κ^4-N,N′,N″,N‴]samarium(III) toluene 1.75-solvate	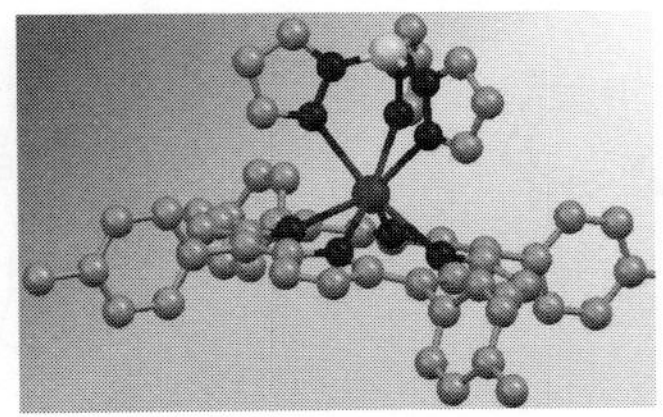	68

(*Continued*)

Title	Subject	Ref.
Tripodal chelating ligand-based sensor for selective determination of Zn(II) in biological and environmental samples	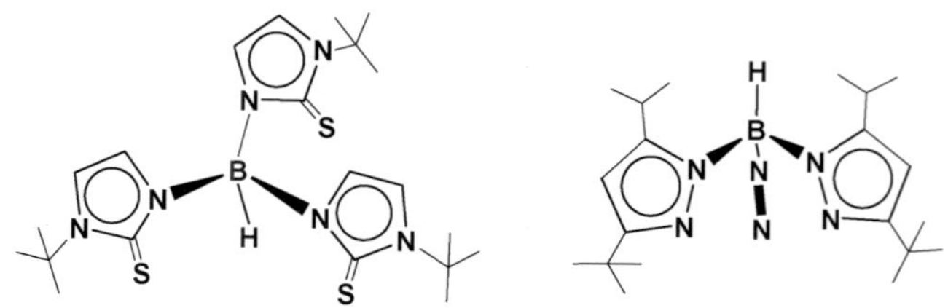	69
Platinum-catalyzed aromatic C–H silylation of arenes with 1,1,1,3,5,5,5-heptamethyltrisiloxane	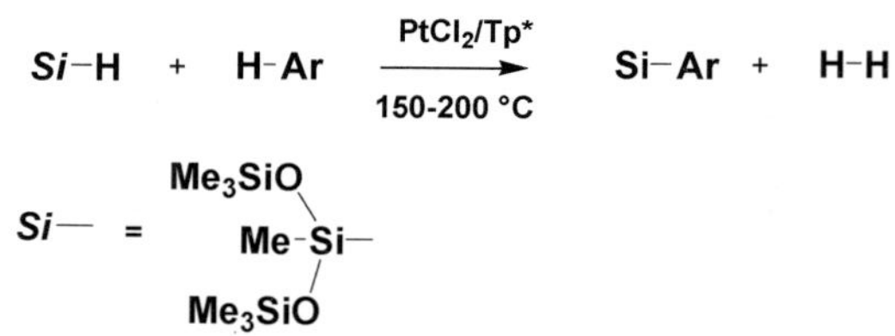	70

(*Continued*)

Title	Subject	Ref.
Synthesis and characterization of unusual heterotrinuclear Co(III)Ni(II)Co(III) hydrotris(3,5-dimethylpyrazolyl)borate complex with mixed 1,1-azide and pyrazolate bridge	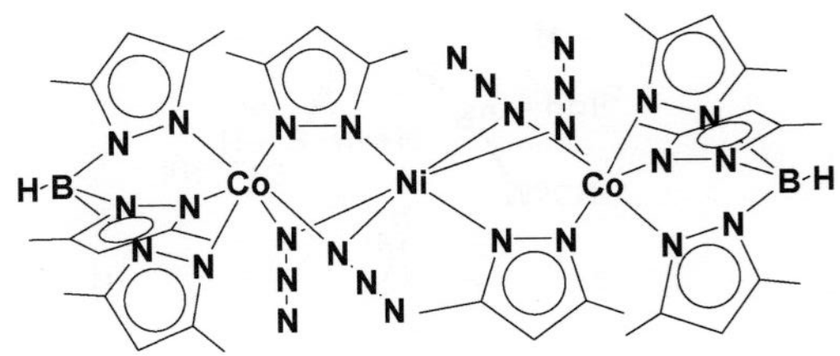	71
Synthesis and structural characterization of $[Bse^{Me}]Ni(PPh_3)(NO)$, a nickel complex with a bent nitrosyl ligand	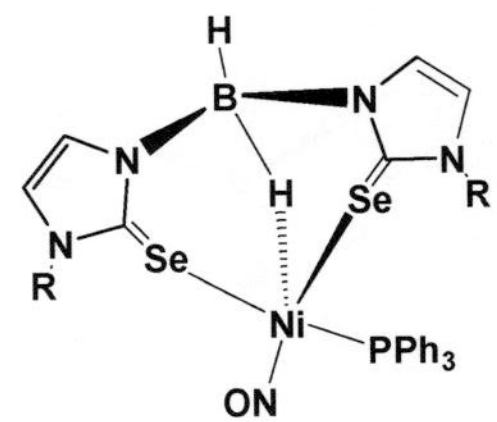	72

(*Continued*)

Title	Subject	Ref.
Comparative studies of tridentate sulfur and nitrogen-containing ligands as ionophores for construction of cadmium ion-selective membrane sensors (PVC based membranes of $KTt^{2,6\text{-xylyl}}$ and $KTp^{Ph,Me}$)	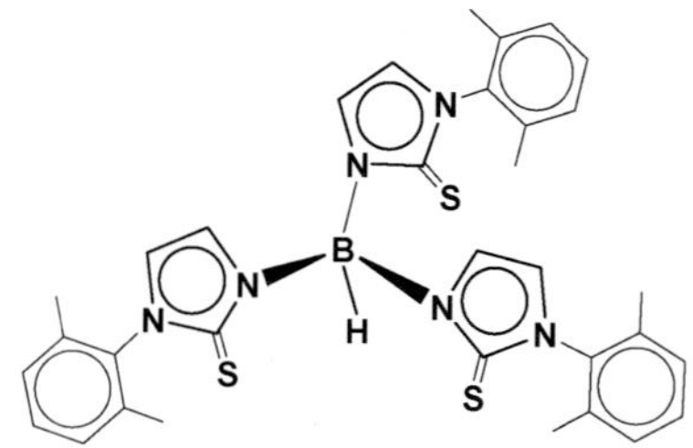	73
Interaction mechanisms of ruthenium hydride complex with water and 2,2,2-trifluoroethanol	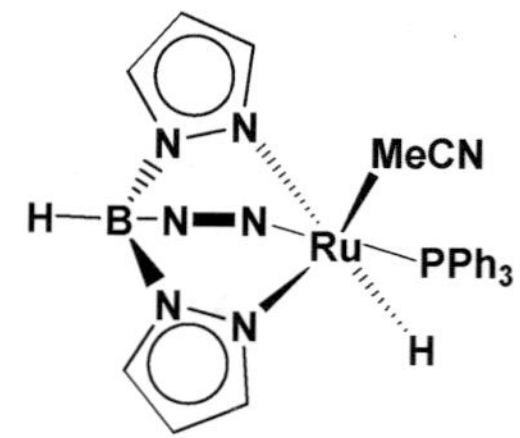	74

(Continued)

Title	Subject	Ref.
Synthesis, structure and non-thermal decomposition kinetics of scorpionates oxovanadium complexes with carboxyl acid ligand	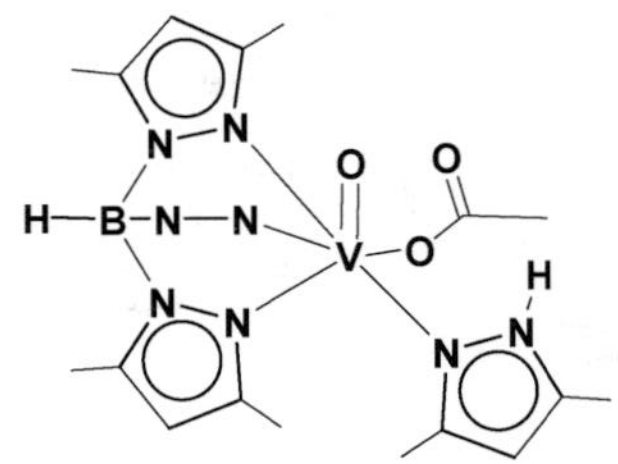	75
Preparation and structure of vanadyl coordination compound as insulin-simulated medicine	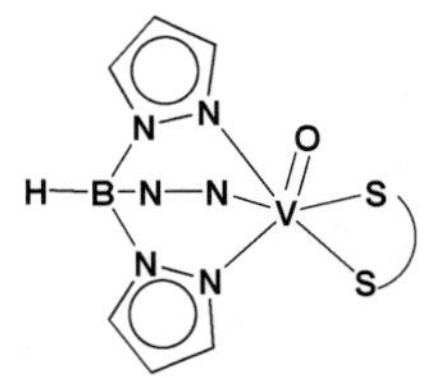	76

(*Continued*)

Title	Subject	Ref.
In situ nanostructure formation of (μ-hydroxo)bis(μ-carboxylato) diruthenium units in nafion membrane and its utilization for selective reduction of nitrosonium ion in aqueous medium	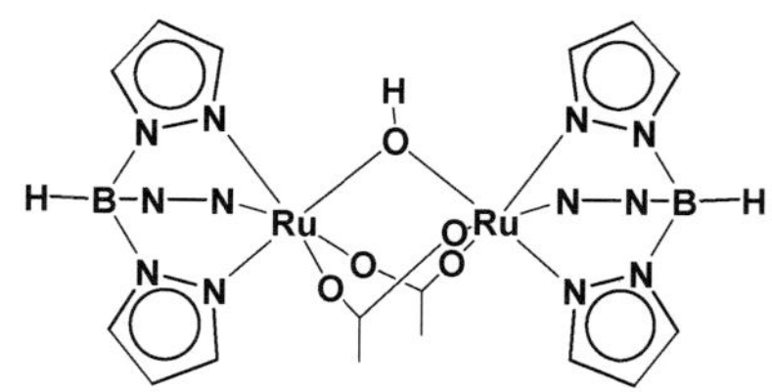	77
Preparation of star-shaped pentapyridyl ligands for the formation of giant fullerene-like molecules by coordination chemistry	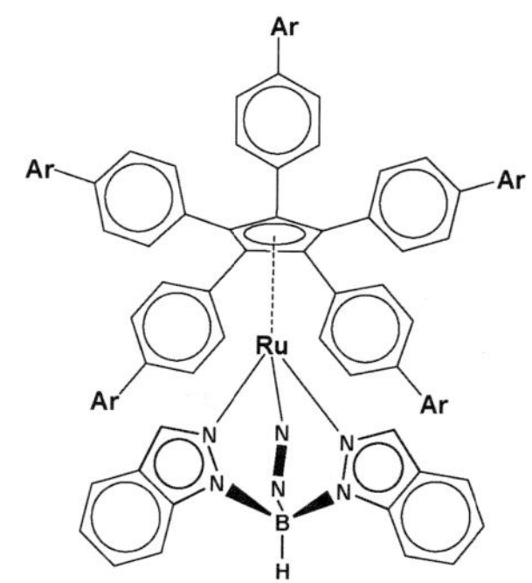	78

(*Continued*)

Title	Subject	Ref.
Alkynylselenolatoalkylidynes: $[Mo(\equiv CSeC\equiv CR)(CO)_2\{HB(pzMe_2)_3\}]$ ($R = CMe_3$, $SiMe_3$; $pzMe_2 =$ 3,5-dimethylpyrazol-1-yl)	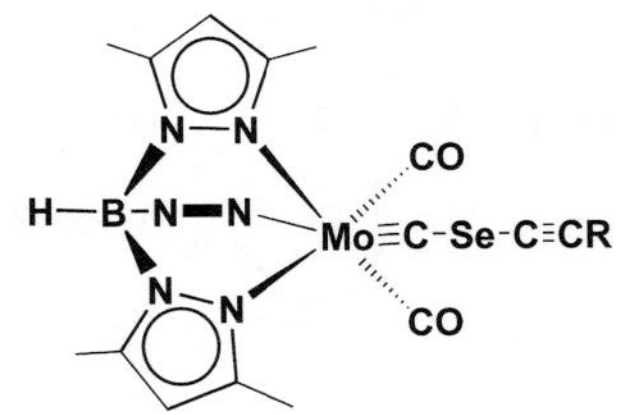	79
Unlocking the metallaboratrane cage: Reversible B–H activation in platinaboratranes	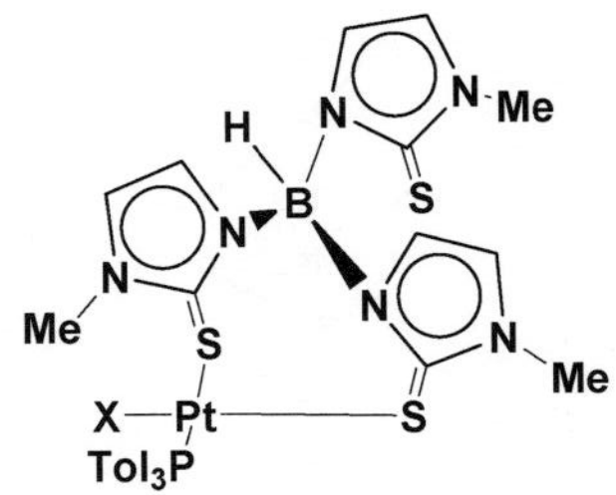	80

(*Continued*)

Title	Subject	Ref.
Syntheses, structures, and electrochemical and magnetic properties of rectangular heterobimetallic clusters based on tricyanometallic building blocks	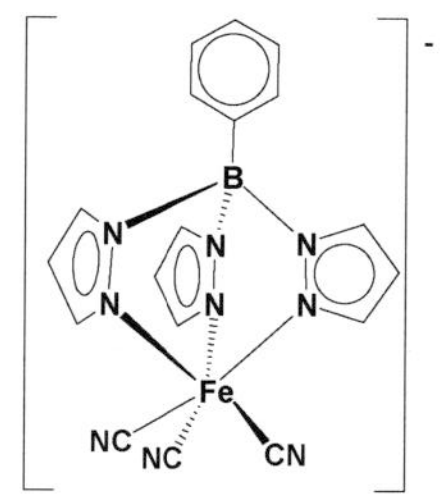	81
New scorpionate oxovanadium complexes: Syntheses and crystal structures of a series of vanadium(V) complexes with mixed ligands of poly(pyrazolyl)borates and substituted acetylacetonates	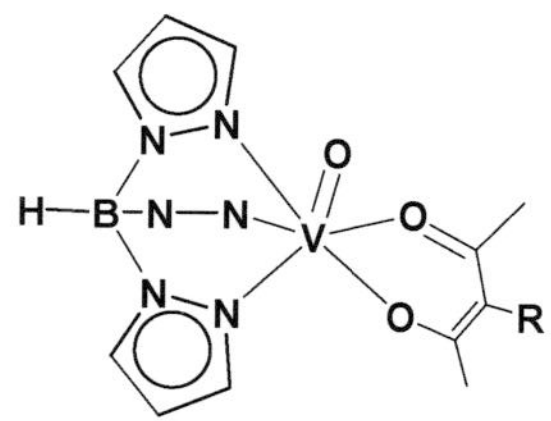 R = Me, Et	82

(Continued)

Title	Subject	Ref.
Anti-Markovnikov addition of both primary and secondary amines to terminal alkynes catalyzed by the $TpRh(C_2H_4)_2/PPh_3$ system	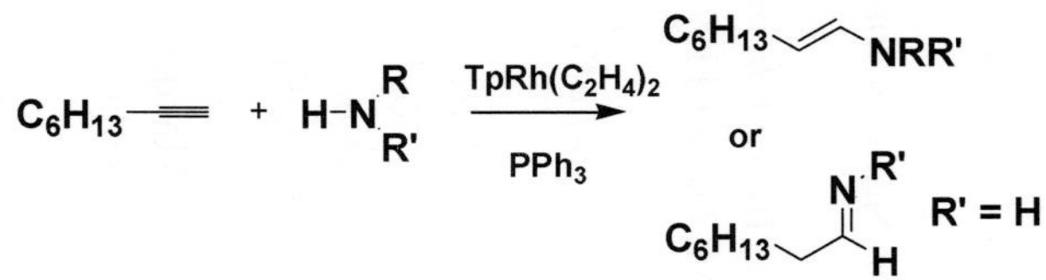	83
Mixed-ligand rhenium tricarbonyl complexes anchored on a (κ^2-H, S) trihydro(mercaptoimidazolyl)borate: A missing binding motif for soft scorpionates	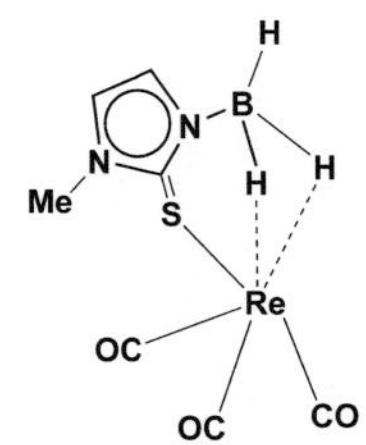	84

References

1. Vahrenkamp H, *Dalton Trans* 2007, 4751.
2. Görls H, Günther W, Notni J, Anders E, *Acta Cryst* 2007, **E63**, m2545.
3. Fraser LR, Bird J, Wu Q, Cao C, Patrick BO, Love JA, *Organometallics* 2007, **26**, 5602.
4. Kuchta MC, Gross A, Pinto A, Metzler-Nolte N, *Inorg Chem* 2007, **46**, 9400.
5. Chisholm MH, Gallucci JC, Yaman G, *Inorg Chem* 2007, **46**, 8676.
6. Zhang F, Morawitz T, Bieller S, Bolte M, Lerner HW, Wagner M, *Dalton Trans* 2007, 4594.
7. Dias HVR, Alidori S, Gioia Lobbia G, Papini G, Pellei M, Santini C, *Inorg Chem* 2007, **46**, 9708.
8. Muñoz-Molina JM, Caballero A, Díaz-Requejo MM, Trofimenko S, Belderraín TR, Pérez PJ, *Inorg Chem* 2007, **46**, 7725.
9. Kealey S, Long NJ, White AJP, Ga AD, *Dalton Trans* 2007, 4763.
10. Santa María MD, Claramunt RM, Alkorta I, Elguero J, *Dalton Trans* 2007, 3995.
11. Domingos A, Lopes I, Waerenborgh JC, Marques N, Lin GY, Zhang XW, Takats J, McDonald R, Hillier AC, Sella A, Elsegood MRJ, Day VW, *Inorg Chem* 2007, **46**, 9415.
12. Albertin G, Antoniutti S, Zanardo G, *J Organomet Chem* 2007, **692**, 3706.
13. Cheng J, Takats J, Ferguson M, McDonald RJ, *J Am Chem Soc* 2008, **130**, 1544.
14. Zhao N, Van Stipdonk MJ, Bauer C, Campana C, Eicchorn DM, *Inorg Chem* 2007, **46**, 8662.
15. Singh UP, Tyagi P, Upreti S, *Polyhedron* 2007, **26**, 3625.
16. Hill LMR, Taylor MK, Ng VWL, Young CG, *Inorg Chem* 2008, **47**, 1044.
17. Summer Maffett L, Gunter KL, Kreisel KA, Yap GPA, Rabinovich D, *Polyhedron* 2007, **26**, 4758.
18. Singh UP, Aggarwal V, Sharma AK, *Inorg Chim Acta* 2007, **360**, 3226.
19. Pellei M, Alidori S, Papini G, Gioia Lobbia G, Golden JD, Dias HVR, Santini C, *Dalton Trans* 2007, 4845.
20. Casas E, Karam A, Díaz-Barrios A, Albano C, Sánchez Y, Mendez B, *Macromol Symp* 2007, **257**, 131.
21. Hlavinka ML, Miyaji T, Staples RJ, Holm RH, *Inorg Chem* 2007, **46**, 9192.

22. Hill AF, Smith MK, *Dalton Trans* 2007, 3363.
23. Wasbotten IH, Ghosh A, *Inorg Chem* 2007, **46**, 7890.
24. Xing Y-H, Zhang B-L, Zhang Y-H, Yuan H-Q, Sun Z, Ge M-F, *Polyhedron* 2007, **26**, 3037.
25. Adams CJ, Connelly NG, Onganusorn S, *Dalton Trans* 2007, 1904.
26. Bolaño S, Bravo J, Castro J, Rodríguez-Rocha MM, Guedes da Silva MFC, Pombeiro AJL, Gonsalvi L, Peruzzini M, *Eur J Inorg Chem* 2007, 5523.
27. Lee JP, Jimenez-Halla JOC, Cundari TR, Gunnoe TB, *J Organomet Chem* 2007, **692**, 2175.
28. Graziani O, Toupet L, Tilset M, Hamon J-R, *Inorg Chim Acta* 2007, **360**, 3083.
29. Gil MP, Casagrande OL Jr, *Appl Catal A* 2007, **332**, 110.
30. Jiménez-Tenorio M, Palacios MD, Puerta MC, Valerga P, *J Mol Catal A Chem* 2007, **261**, 64.
31. Blake AJ, George MW, Hall MB, McMaster J, Portius P, Sun XZ, Towrie M, Webster CE, Wilson C, Zarić S, *Organometallics* 2008, **27**, 189.
32. Romańczyk PP, Guzik MN, Włodarczyk AJ, *Polyhedron* 2007, **26**, 1182.
33. Burgmayer SJN, Kim M, Petit R, Rothkopf A, Kim A, BelHamdounia S, Hou Y, Somogyi A, Habel-Rodriguez D, Williams A, Kirk ML, *J Inorg Biochem* 2007, **101**, 1601.
34. Sánchez-Méndez A, Ortiz AM, de Jesús E, Flores JC, Gómez-Sal P, *Dalton Trans*, 2007, 5658.
35. Parkin G, *New J Chem*, 2007, **31**, 1996.
36. Cundari TR, Grimes TV, Gunnoe TB, *J Am Chem Soc* 2007, **129**, 13172.
37. Arikawa Y, Ikeda K, Asayama T, Nishimura Y, Onishi M, *Chem Eur J* 2007, **13**, 4024.
38. Shan X, Rohde J-U, Koehntop KD, Zhou Y, Bukowski MR, Costas M, Fujisawa K, Que L Jr, *Inorg Chem* 2007, **46**, 8410.
39. Galardon E, Giorgi M, Artaud I, *Dalton Trans* 2007, 1047.
40. Dias HVR, Wu J, *Angew Chem Int Ed* 2007, **46**, 7814.
41. Pérez J, Riera L, *Chem Commun* 2008, 533.
42. Takano Y, Yamaguchi K, *Int J Quant Chem* 2007, **107**, 3103.
43. Kuchta MC, Gemel C, Metzler-Nolte N, *J Organomet Chem* 2007, **692**, 1310.
44. Seidel WW, Meel MJ, Radius U, Schaffrath M, Pape T, *Inorg Chem* 2007, **46**, 9616.
45. Tekeste T, Vahrenkamp H, *Inorg Chim Acta* 2007, **360**, 1523.

46. Desrochers PJ, Duong DS, Marshall AS, Lelievre SA, Hong B, Brown JR, Tarkka RM, Manion JM, Holman G, Merkert JW, Vicic DA, *Inorg Chem* 2007, **46**, 9221.
47. Cho J, Yap GPA, Riordan CG, *Inorg Chem* 2007, **46**, 11308.
48. Kaufmann L, Vitze H, Bolte M, Lerner H-W, Wagner M, *Organometallics* 2007, **26**, 1771.
49. Vastine BA, Webster CE, Hall MB, *J Chem Theory Comput* 2007, **3**, 2268.
50. Foley NA, Lail M, Gunnoe TB, Cundari TR, Boyle PD, Petersen JL, *Organometallics* 2007, **26**, 5507.
51. Wang J, Sun Z-R, Deng L, Wei Z-H, Zhang W-H, Zhang Y, Lang J-P, *Inorg Chem* 2007, **46**, 11381.
52. Hill AF, Smith MK, *Organometallics* 2007, **26**, 4688.
53. Zhu J, Jia G, Lin Z, *Organometallics* 2007, **26**, 1986.
54. Paneque M, Posadas CM, Poveda ML, Rendón N, Álvarez E, Mereiter K, *Chem Eur J* 2007, **13**, 5160.
55. Paneque M, Posadas CM, Poveda ML, Rendón N, Santos LL, Álvarez E, Salazar V, Mereiter K, Oñate E, *Organometallics* 2007, **26**, 3403.
56. Álvarez A, Conejero S, Lara P, López JA, Paneque M, Petronilho A, Poveda ML, Del Rio D, Serrano O, Carmona E, *J Am Chem Soc* 2007, **129**, 14130.
57. Gómez M, Paneque M, Poveda ML, Álvarez E, *J Am Chem Soc* 2007, **129**, 6092.
58. Malandrino G, Lo Nigro R, Fragalà IL, *Chem Vap Dep* 2007, **13**, 651.
59. DeYonker NJ, Foley NA, Cundari TR, Gunnoe TB, Petersen JL, *Organometallics* 2007, **26**, 6604.
60. Xing YH, Zhang XJ, Sun Z, Han J, Zhang YH, Zhang BL, Ge MF, Niu SY, *Spectrochim Acta A* 2007, **68**, 1256.
61. Alías FM, Daff PJ, Paneque M, Poveda ML, Carmona E, Pérez PJ, Salazar V, Alvarado Y, Atencio R, Sanchéz-Delgado R, *Chem Eur J* 2002, **8**, 5132.
62. Paneque M, Posadas CM, Poveda ML, Rendón N, Salazar V, Oñate E, Mereiter K, *J Am Chem Soc* 2003, **125**, 9898.
63. Camerano JA, Casado MA, Ciriano MA, Tejel C, Oro LA, *Chem Eur J* 2008, **14**, 1897.
64. Paneque M, Poveda ML, Rendón N, Álvarez E, Carmona E, *Eur J Inorg Chem* 2007, 2711.

65. Liu W, Welch K, Trindle CO, Sabat M, Myers WH, Harman WD, *Organometallics* 2007, **26**, 2589.
66. Cortes-Llamas SA, Muñoz-Hernández MÁ, *Organometallics* 2007, **26**, 6844.
67. Zimmermann M, Takats J, Kiel G, Törnroos KW, Anwander R, *Chem Commun* 2008, 612.
68. He H, He G, Ng SW, *Acta Crystallogr E*, 2007, **E63**, m2878.
69. Singh AK, Mehtab S, Singh UP, Aggarwal V, *Anal Bioanal Chem* 2007, **388**, 1867.
70. Murata M, Fukuyama N, Wada J-i, Watanabe S, Masuda Y, *Chem Lett* 2007, **36**, 910.
71. Sun Y-J, Zhao B, Cheng P, *Inorg Chem Commun* 2007, **10**, 583.
72. Landry VK, Parkin G, *Polyhedron* 2007, **26**, 4751.
73. Singh AK, Mehtab S, Singh PU, Aggarwal V, *Electroanalysis* 2007, **19**, 1213.
74. Yin C-Q, Chen Y, Feng Q-W, Bai Z-W, Wang C-F, Li Z-Y, *Huaxue Xuebao* 2007, **65**, 2320.
75. Zhang B-L, Xing Y-H, Xing Y-H, Ge M-F, Sun Z, Li Z-P, Han J, Niu S-Y, *Wuli Huaxue Xuebao* 2007, **23**, 1701.
76. Xing Y, Bai F, Li Z-P, CN 101033239 A 20070912 (Patent written in Chinese). Application: CN 1001–260 20070131. Priority: CAN 147:418657, AN 2007:1029270.
77. Kumar AS, Tanase T, Iida M, *Langmuir* 2007, **23**, 391.
78. Jarrosson T, Oms O, Bernardinelli G, Williams AF, *Chimia* 2007, **61**, 184.
79. Caldwell LM, Hill AF, Rae AD, Willis AC, *Organometallics* 2008, **27**, 341.
80. Crossley IR, Hill AF, *Dalton Trans* 2008, 201.
81. Wang C-F, Liu W, Song Y, Zhou X-H, Zuo J-L, You X-Z, *Eur J Inorg Chem* 2008, DOI: 10.1002/ejic.200701104.
82. Xing YH, Aoki K, Sun Z, Ge MF, Niu S, *J Chem Res* 2007, **2**, 100.
83. Fukumoto Y, Asai H, Masaki S, Chatani N, *J Am Chem Soc* 2007, **129**, 13792.
84. Videira M, Maria L, Paulo A, Santos IC, Santos I, Vaz PD, Calhorda MJ, *Organometallics* 2008, **27**, 1334.

Abbreviations

acacH = 2,4-pentanedione
Ad = adamantyl
Ant = 9-anthryl
apic = 2-amino-4-picoline
apy = 2-aminopyridine
Ar = aryl
Ar_F = 3,5-$(CF_3)_2C_6H_3$
ASN = 5-azonia-spiro[4.4]nonane
7-azain = 7-azaindolyl
bdt = 1,2-benzenedithiolate
$bdtCl_2$ = 3,6-dichlorobenzenedithiolate
biph = 2,2′-biphenol
bipy = 2,2′-bipyridyl
4,4’-bipy = 4,4′-bipyridyl
Bm = bis(thioimidazolyl)borate
Bn = benzyl
BNN = 1,5-borabicyclo[3.3.1]nonane
Bo = fused benzo ring
Bp = bis(pyrazol-1-yl)borate
Bp* = bis(3,5-dimethylpyrazol-1-yl)borate
Bp^x = general bis(pyrazolyl)borate
bpca = bis(2-pyridylcarbonyl)amidate

bpe = 1,2-bis(4-pyridyl)ethene
BPMU = bis(pentamethylene)urea
^{t}Bu = *tert*-butyl
Buim = 1-butylimidazole
Clsty = 4-chlorostyrene
COD = 1,5-cyclo-octadiene
COE = cyclo-octene
Cp = cyclopentadienyl
Cp* = pentamethylcyclopentadienyl
Cum = cumenyl
Cy = cyclohexyl
Cyclam = 1,4,8,11-tetraazacyclotetradecane
Cym = cymanthrene
= cyclopentadienylmanganesetricarbonyl
Cym′ = cyclopentamethylcyclopentadienyl-manganesetricarbonyl
dbabh = 2,3:5,6-dibenzo-7-aza bicyclo[2.2.1]hepta-2,5-diene
dbm = dibenzoylmethanate
DCE = 1,2-dichloroethane
DFT = density functional theory
DIEA = diisopropylethylamine
dien = diethylenetriamine
diglyme = bis(2-methoxyethyl)ether
β-dike = beta-diketonate
R,*R*-dippach = (*R*,*R*)-1,2-bis((diisopropylphosphino)amino) cyclohexane
dippae = 1,2-bis((diisopropylphosphino)-amino)ethane
dippe = 1,2-bis(diisopropylphosphino)ethane
DMAD = dimethylacetylenedicarboxylate
dme = 1,2-dimethoxyethane
DMF = dimethylformamide
dmp = 2,9-dymethylphenanthroline
dmpe = 1,2-bis(dimethylphosphino)ethane
dmpm = bis(dimethylphosphino)methane
DMPP = 3,4-dimethyl-1-phenylphosphole

DMSO = dimethylsulfoxide
dppe = 1,2-bis(diphenylphosphino)ethane
dppene = 1,2-bis(diphenylphosphino)ethene
dppf = 4,4′-bis(diphenylphosphino)ferrocene
dppm = bis(diphenylphosphino)methane
dppz = dipyridophenazine
dpq = dipyridoquinoxaline
DPSO = diphenylsulfoxide
DPVP = diphenylvinylphosphine
DTBSQ = 3,5-di-*tert*-butylsemiquinonato
EDA = ethyl diazoacetate
equiv. = equivalent or equivalents
Et = ethyl
Et_2O = diethyl ether
fc = ferrocenylene
Fc = ferrocenyl
Fl = 9-fluorenyl
Fn = fumaronitrile
FOBz = *p*-fluorobenzoate
Fu = furyl
glyme = dme = dimethoxyethane
Hdpm = dipivaloylmethane
Hex = hexyl
Him = imidazole
HIndz = indazole
HMB = hexamethylbenzene
Hmt = 2-mercaptoimidazole = methymazole
Hpz = pyrazole
Hpz* = 3,5-dimethylpyrazole
Hpz^x = general pyrazole
Htz = 1,2,4-triazole
$ImMe_4$ = tetramethylimidazol-2-yliden
Ind = indenyl
Ipc = isopinocampheyl
LDA = lithium diisopropylamide
Lut = lutidine

MAO = methylaluminoxane
mapy = 2-(methylamino)pyridine
Me = methyl
$(Me)_2$ATI = *N*-methyl-2-(methylamino)troponiminate
Meim = 1-methylimidazole
2-Meim = 2-methylimidazole
5-MeOsalen = N,N′-ethylenebis(5-methoxysalicylideneiminate)
Me_3tacn = *N*,*N*′,*N*″-trimethyl-1,4,7-triazacyclononane
MMAO = modified methylaluminoxane
MMTP = 1-methoxy-2-methyl-1-trimethylsiloxypropene
MOCVD = metal organic chemical vapor deposition
Ms = mesityl = 2,4,6-trimethylphenyl
NBD = norbornadiene
NBS = *N*-bromosuccinimide
NitOH = *p*-nitrophenol
Np = 3-neopentyl
OAc = acetate
OBz = benzoate
OTf = trifluoromethanesulfonate
OTs = tosylate
Ox = oxalate
pba = *RS*- and/or *S*-2-phenyl butyrate,
PBMA = poly(butylmethacrylate)
Pe = pentyl
Ph = phenyl
phen = 1,10-phenanthroline
4-Phpy = 4-phenylpyridine
pic = picoline
pip = piperidine
PLED = polymeric light-emitting diode
PMMA = poly(methylmethacrylate)
PPN = bis(triphenylphosphoranylidene)-ammonium
^{i}Pr = *iso*-propyl
Ptn = pterin
py = pyridine
pyl = pyrrolyl

pyr = pyrazine
pzTp = tetrakis(pyrazol-1-yl)borate
qdt = quinoxaline-2,3-dithiolate
quin = quinolin-8-olate
TBA = tribromoacetate
TCA = trichloroacetate
tdt = 3,4-toluenedithiolate
TEMPO = 2,2,6,6-tetramethyl-1-piperidinyloxy
Tevs = triethylvinylsilane
Tfa = trifluoroacetate
THF = tetrahydrofuran
tht = tetrahydrothiophene
TiBA = triisobutylaluminum
Tm = tris(thioimidazolyl)borate
TMA = trimethylaluminum
TMEDA = tetramethylethylenediamine
Tn = thienyl
Tol = *p*-tolyl
m-Tol = *m*-tolyl
o-Tol = *o*-tolyl
Tp = tris(pyrazol-1-yl)borate
Tp^* = tris(3,5-dimethylpyrazol-1-yl)borate
Tp^R = substituted tris(pyrazol-1-yl)borate
Tp^x = general tris(pyrazolyl)borate
TPA = tris(2-pyridylmethyl)amine
tpy = 2-*p*-tolylpyridyl
tren = tris(2-amino)ethylamine
Trop = tropolonate
Tu = thiourea
Van = 4-vinylanisole
Xyl = 2,6-dimethylphenyl

Index